AIP Conference Proceedings
Series Editor: Hugh C. Wolfe
Number 106

Predictability of Fluid Motions

(La Jolla Institute-1983)

Edited by
Greg Holloway
Institute of Ocean Sciences

and

Bruce J. West
La Jolla Institute
Center for Studies of Nonlinear Dynamics

American Institute of Physics
New York 1984

Copying fees: The code at the bottom of the first page of each article in this volume gives the fee for each copy of the article made beyond the free copying permitted under the 1978 US Copyright Law. (See also the statement following "Copyright" below.) This fee can be paid to the American Institute of Physics through the Copyright Clearance Center, Inc., Box 765, Schenectady, N.Y. 12301.

L.C. Catalog Card No. 83-73641
ISBN 0-88318-305-6
DOE CONF- 830240

PREFACE

The endeavor to plan rationally for future events is among the dominant characteristics of humankind. Where possible, such planning contributes to success in activities ranging from agriculture through economic, military, and recreational. As scientists strive to correct deficiencies and to understand fundamental limitations of forecasting, there occur questions of "predictability."

From one point of view, the question of predictability is philosophical; the central issue being whether the future unfolds precisely and deterministically from the present or whether there is a distribution of possible futures. Among fluid dynamicists, a prevailing view is that uncertainty at the quantum dynamical level does not significantly influence the predictability of macroscopic fluid motions. What is of significance is the uncertainty in knowledge of the present macroscopic state of a flow field. This leads to an imprecision in the statement of the initial conditions. In addition, it may not be feasible to monitor the changeable boundary conditions of an evolving flow, leading to an imprecision in the specification of the boundary conditions in time. For these reasons, there is inevitable uncertainty, through both initial and subsequent sources, which affects the accuracy of the prediction of evolving flows. Depending on characteristics of the flow, initial uncertainty can amplify sufficiently rapidly that forecasts are rendered useless beyond a given time. In studies of predictability, a fluid dynamicist is concerned with this propagation and amplification of uncertainty.

A precise and universally accepted definition of the term "predictability" is problematical. One might define "predictability in the narrow sense" as the sensitivity of dependence upon initial conditions. An unpredictable flow is one which exhibits a strong sensitivity. In this sense, predictability studies extend the analysis of flow instabilities to circumstances where the "unperturbed" flow itself may be a complicated function of space and time. It is only a small further extension to consider sensitivity of dependence upon more general spatial-temporal boundary conditions. On the other hand, "predictability in the broad sense" defies definition. Different authors appear to intend different meanings, often given only implicitly.

Weather forecasting is a natural context in which to address the predictability problem. In this context, the predictability problem is quite practical, bearing upon the possible improvement of forecasting and also upon predicting the likely range of errors in any forecast. That weather forecasting has been the driving force behind predictability studies can be seen in the articles which follow; a majority of which either directly or indirectly address the predictability of weather. Nonetheless it would be a mistake to identify predictability simply as a problem in weather forecasting. Rather, predictability is a fundamental theoretical issue in the analysis of nonlinear equations, whether they are field equations such as in fluid dynamics or low-order equations from other areas of physics.

Much of classical dynamics is divided between deterministic and statistical points of view. The epitome of the former view were particle orbits in analytical mechanics while the latter view was sometimes associated with turbulence in fluid mechanics. These two views were thought to be distinct. More recent works have established that chaotic behavior can result from systems with as few as three degrees of freedom. Thus, the point of view of a segment of the scientific community has shifted more towards the perspective that purely deterministic behavior is an illusion and only distributions, albeit very narrow in some cases, have physical significance.

Despite practical concern as well as fundamental theoretical interest, the general area of predictability studies has not become well established within the body of fluid dynamics. Although some recent textbooks have included brief discussion of predictability issues, the literature of predictability consists of isolated research articles scattered through a number of journals. To achieve a more comprehensive expression of predictability research, a symposium was convened in February 1983 on the campus of the Scripps Institution of

Oceanography of the University of California, San Diego. Two goals of the meeting were (1) to exchange views on research among participating scientists, and (2) to produce this volume of proceedings in order to make the field of study more readily available to other investigators. A list of those scientists that participated in the workshop is given below.

Articles included in this volume range over a wide variety of topics. Many of the papers concern atmospheric predictability and are based upon direct observations, numerical circulation modeling and analyses of operational forecast skill. A topic of particular concern is predictability of atmospheric mesoscale phenomena. Traditional measures of predictability such as skill indices or the kinetic energy of the difference wind field between two flow realizations are shown to be misleading in some cases. Concerns for particular kinds of flow phenomena are described including atmospheric blocking, the occurrence of persistent isolated vortices and the propagation of solitary waves. Efforts to diagnose predictability in the presence of such flows suggest concepts such as "conditional predictability."

Approaches to predictability based upon turbulence theory are described along with application of predictability concepts to turbulence phenomena. While loss of predictability is often associated with turbulence, and sometimes serves to define turbulence, similar loss of predictability can be effected by wave-wave interaction. Examples are seen in surface gravity wave and equatorial wave phenomena. For cases of quasigeostrophic dynamics, methods of functional analysis are used to establish rigorous bounds on predictability.

Onset of chaos and loss of predictability are examined in spectrally truncated models with only a few degrees of freedom. One such study considers "almost intransitivity" and the role of "eddy noise." Transitions to aperiodic motion are reported from various laboratory experiments.

A view of predictability in terms of separation of initially nearby Lagrangian particles is compared with more traditional Eulerian measures. Relationship between fluid dynamic predictability and the entropy concept from information theory is considered.

An area of special concern is to consider predictability issues in relation to developments in ocean forecasting. In part, such concerns may draw upon dynamical similarities between ocean and atmosphere circulation. However, there are differences as well. Moreover, the sparsity of oceanic data and the kinds of data (both present and as foreseen) as well as the strong role of atmospheric forcing on relatively short time scales will require careful rethinking of predictability issues in regards to ocean forecasting.

ACKNOWLEDGEMENTS

This work has been made possible by support from the U.S. National Science Foundation through the Office of the Global Atmospheric Research Programme (GARP) and by support from the U.S. Office of Naval Research. Further support has been provided by the Canadian Department of Fisheries and Oceans and by the La Jolla Institute. The meeting facility was made available by the Scripps Institution of Oceanography, University of California, San Diego. Special mention should be made of the concern and assistance of Ms. Pam Stephens at the GARP Office. Arrangements for the meeting and for the preparation of this volume were ably executed by Mrs. Grace Pitts at the Center for Studies of Nonlinear Dynamics of La Jolla Institute. During preparation of this volume, Greg Holloway was supported, in part, as a Cecil and Ida Green scholar and Bruce J. West by the independent research funds of the La Jolla Institute. Finally, but also foremost, this book could only come into existence on account of the interest, effort, patience, creative thought, and dedicated research of the scientists who gathered in La Jolla and who provided the articles which follow.

Greg Holloway
Institute of Ocean Sciences
9860 West Saanich Road
Sidney, British Columbia
Canada V8L 4B2

Bruce J. West
La Jolla Institute
Center for Studies of
Nonlinear Dynamics
Suite 2150
8950 Villa La Jolla Drive
La Jolla, California 92037

LIST OF PARTICIPANTS

Name	Affiliation	Country
R. Anthes	National Center for Atmospheric Research	USA
F. Baer	University of Maryland	USA
D. Baumhefner	National Center for Atmospheric Research	USA
A.F. Bennett	Institute of Ocean Sciences	Canada
R. Benzi	IBM Scientific Center of Rome	Italy
R.D. Blevins	GA Technologies	USA
W. Blumen	University of Colorado	USA
G. Boer	Canadian Climate Centre	Canada
J. Brindley	Leeds University	UK
P. Budgell	Center for Inland Waters	Canada
G. Carnevale	La Jolla Institute	USA
J. Curry	University of Colorado	USA
K. Denman	Institute of Ocean Sciences	Canada
R. Dole	Harvard University	USA
J.A. Domaradzki	Warsaw University	Poland
D. Durran	National Center for Atmospheric Research	USA
J. Dutton	Pennsylvania State University	USA
J. Egger	Meteorologishes Institute Universitat München	Germany (F.R.)
R. Elsberry	Naval Postgraduate School	USA
S. Flatté	La Jolla Institute	USA
U. Frisch	Observatoire de Nice	France
M. Ghil	Courant Institute for Mathematical Studies	USA
C. Gibson	University of California, San Diego	USA
A.W. Green	Naval Ocean Research and Development Activity	USA
J. Greene	GA Technologies	USA
D. Haidvogel	National Center for Atmospheric Research	USA
R.L. Haney	Naval Postgraduate School	USA
A. Hansen	Yale University	USA
J. Hart	University of Colorado	USA
J.R. Herring	National Center for Atmospheric Research	USA
R. Hide	Meteorological Office	UK
L. Ho	National Center for Atmospheric Research	USA
R.N. Hoffman	Goddard Laboratory for Atmospheric Sciences	USA
G. Holloway	Institute of Ocean Sciences	Canada
H.E. Hurlburt	Naval Ocean Research and Development Activity	USA

R. Kraichnan	Los Alamos National Laboratory	USA
B. Legras	Laboratoire de Meteorologie Dynamique	France
C. Leith	National Center for Atmospheric Research	USA
M. Lesieur	Institut de Mecanique de Grenoble	France
D. Lilly	University of Oklahoma	USA
J. Litherland	University of Washington	USA
E. Lorenz	Massachusetts Institute of Technology	USA
A. Maxworthy	University of Southern California	USA
J.C. McWilliams	National Center for Atmospheric Research	USA
C. Mechoso	University of California, Los Angeles	USA
P. Merilees	Atmosphere Environment Service	Canada
R. Miller	Tulane University	USA
C.N.K. Mooers	Naval Postgraduate School	USA
R. Moritz	University of Washington	USA
W. Munk	Scripps Institution of Oceanography	USA
P. Niiler	Scripps Institution of Oceanography	USA
W. Perrie	National Center for Atmospheric Research	USA
T.J. Phillips	NASA/Goddard Space Flight Center	USA
P. Ripa	Centro de Investigacion y de Educacion Superior de Ensenada	Mexico
P.M. Rizzoli	Massachusetts Institute of Technology	USA
J.O. Roads	Scripps Institution of Oceanography	USA
H.N. Shirer	Pennsylvania State University	USA
J. Shukla	Goddard Laboratory for Atmospheric Sciences	USA
D. Smith	Naval Postgraduate School	USA
R. Somerville	Scripps Institution of Oceanography	USA
T. Spence	Office of Naval Research	USA
P. Stephens	National Science Foundation	USA
A. Su	Naval Ocean Research and Development Activity	USA
A. Sutera	Center for Environment and Man	USA
H. Tanaka	Nagoya University	Japan
P.D. Thompson	National Center for Atmospheric Research	USA
S. Tibaldi	European Center for Medium Range Weather Forecasts	UK
J. Tribbia	National Center for Atmospheric Research	USA
G.K. Vallis	Scripps Institution of Oceanography	USA
C.W. Van Atta	University of California, San Diego	USA
E. Varley	Lehigh University	USA
R. Waltz	G A Technologies	USA

A. Warn-Varnas	Naval Ocean Research and Development Activity	USA
T.T. Warner	Pennsylvania State University	USA
B.J. West	La Jolla Institute	USA
X.P. Zhou	Institute of Atmospheric Physics	China
A.R. Robinson	Harvard University	USA

TABLE OF CONTENTS

A REVIEW OF THE PREDICTABILITY PROBLEM

Philip D. Thompson
National Center for Atmospheric Research, Boulder, CO 80307

ABSTRACT

This is a summary and review of developments in the theory of predictability — the practical motivations behind our preoccupation with this question, the formulation of problems dealing with various aspects of the question, a variety of theoretical, numerical and observational approaches to those problems, and some significant and suggestive results that have gradually come to light. It is concerned primarily with the question of meteorological predictability (from which the general question of predictability arose), but touches on some aspects of the problem that are directly applicable to the design of oceanographic observing systems and prediction.

Thank you, Mr. Chairman. Ladies and gentlemen, colleagues, distinguished guests.

The organizers of this meeting have asked me to give a review or overview — or perhaps more appropriately, an underview — of the predictability problem. By this, I understood that I was to say something about the nature of the problem, why we are pursuing it, the developments that have led to our present state of understanding or misunderstanding, and where we go from here. After a suitable show of reluctance, I agreed to undertake this chore, figuring that they were really looking for someone who knew a little about a lot of things but not much about anything. On some reflection, I see that my reluctance was not wholly unfounded: most of the protagonists in this history are not only alive and well, but are sitting here in this room, listening for any misstep I might make. The only advantage is that they can field the questions.

As nearly as I can recall, the first time I ever saw the word "predictability" in print was in an article by T.E.W. Schumann[1] in the British journal *Weather* in 1950. Briefly, it dealt with the "art" of forecasting and contained some peculiar metaphysical ideas about its uncertainties. Among others, there was some vague reference to Heisenberg's "uncertainty principle," which seemed to me totally irrelevant and certainly not having anything to do with the macroscopic behavior of fluids. So I forgot about it.

The question of meteorological predictability did not become a real one until the advent of routine numerical weather prediction in 1955. In the public's view, of course, and even in the scientific view, weather had always been regarded as unpredictable, or at least imperfectly predictable, by its very nature. The source of its unpredictability, however, was obscured by the fact that the differences between competing and, at that time, highly subjective predictions were fully as large as the errors of individual predictions. For the first time, with the introduction of numerical methods of prediction, we had an objective standard and could begin to assess the damages of various kinds of error — i.e., errors inherent in the physical models, approximations in the numerical integration of the model equations and errors in specifying the initial state. What I am suggesting is that the question of predictability arose as a natural outgrowth of numerical methods of prediction, and that it wasn't even a very sensible question before they came on the scene.

The earliest numerical predictions were based on initial data from an area covering the continental U.S., southern Canada and northern Mexico, and extending westward to about the Hawaiian Islands. Actually, there was precious little data over the Pacific Ocean. We were especially dependent on reports from Weather Ship "Papa," stationed in the Gulf of Alaska, about the best single indicator of cyclone development that might affect the West Coast in a day or two. All this came at a time when the Congress was contemplating the deactivation of Station Papa, since maintaining it was a fairly expensive proposition. This was a matter of some concern to us, simply because we had found that the absence of its reports had a deleterious effect on the quality of the forecasts over the U.S. in about two days. On a number of occasions, when Station Papa's reports had not come in by the three hour data cutoff time, we had rerun the predictions after the data had come in late. The errors in the 500 mb height analysis *without* the data from Station Papa were found to be as much as 165 meters, and the resulting errors in two day predictions were sometimes 200 meters or greater.

We also realized that this was only a particularly flagrant example of how errors in the initial conditions could contaminate the predictions. To a lesser extent, such errors would be present in any reconstruction of the initial state from a finite sample of imperfect observations. The questions that then arose were: How are errors in the initial state propagated through a prediction? Do they grow with time and, if so, on what does their rate of growth depend? How does the growth rate depend on the true initial state, or on the characteristics of the observing and analysis system?

The point of view that I adopted at that time, around 1956, was something like this:

Suppose that the prediction model were perfect, and that the model equations could be integrated without error. Then there would still be a practical limitation on the accuracy of prediction, owing to the fact that the initial analysis is subject to nonsystematic instrument error, random roundoff error in the reported observations, and sizable errors of interpolation between reporting stations. The working hypothesis, with no more basis than a vague intuitive feeling in my bones, was that the growth of error was a distinctively nonlinear phenomenon and that, as long as the idealized prediction model had the right kind of nonlinearity, it would amplify errors at about the correct rate. For this reason, I considered the simple nondivergent barotropic model and a two-level baroclinic model — the latter because it could also display the effects of strong linear instability.

Briefly, what I did was to calculate the first and second time-derivatives of the RMS wind error resulting from an ensemble of initial error fields with statistically-defined characteristic scales and amplitudes, but randomly varying phase. By extrapolating via a three term Taylor-expansion, it was found that the initial analysis error to be expected of the existing observing system would lead to a doubling of RMS wind errors in about two days, and that the predictions would be no better than a sheer guess in about a week. It was also found that doubling the density of observing stations would not only reduce the initial errors, but would also considerably reduce the rate of error-growth.

These results had a number of interesting consequences. First, although one would certainly expect that a higher-resolution network of observing stations would reduce errors of prediction and increase predictability, we had for the first time a quantitative (although crude) estimate of the increase in predictability to be expected from an increased investment in observing stations, or of the deterioration to be expected from shutting down stations. As it turned out, this was just the kind of ammunition that was needed to save Station Papa, and which was later to provide additional impetus to the planning of GARP and FGGE.

Second, a byproduct of these studies was the estimation of error-variance as a function of time and certain statistical properties of the true state — e.g., the kinetic energy spectrum, average static stability and average temperature gradient. Thus, in addition to the predicted most probable state, one could also estimate the degree of confidence that should be placed in the forecast under the conditions prevailing at that particular time. That is, the estimation of predictability contains an element of stochastic-dynamic prediction, about which more will be said later.

Finally, if and when the error of prediction approaches the limit attainable with perfect prediction models, we know that it is time to start worrying about the shortcomings of the observing system and fuller exploitation of the data that is already available. At the present time, in fact, one of the common standards against which the performance of prediction models is judged is a predictability decay-curve, which is more or less independent of the model.

The methods that I first used to estimate error-growth might be most charitably described as ham-handed, for the simple reason that I was then completely ignorant of turbulence theory. Fortunately, however, Novikov[3] of the Soviet Academy's Institute of Atmospheric Physics, picked up the basic idea and gave more general and exact expressions for the error growth in terms of spatial auto- and cross-correlation functions. These led to a slightly more pessimistic view, but not radically so. Incidentally, a good English synopsis of Novikov's work in 1960 is given in Panchev's book on stochastic processes. As far as I know, Soviet studies of predictability have not proceeded much beyond the usual exercise of examining a model's sensitivity to small variations of initial conditions: this is very possibly due to the fact that few of their people believed in two-dimensional or quasi-two-dimensional turbulence up to about ten years ago. Or possibly still don't believe it.

Well, this is where matters stood in 1960. I would next like to outline the main developments in the theory of predictability since that time. They were a little slow in coming, but is is evident that at least a few people had been chewing on the problem of error-growth. In particular, interest in this and more general questions was revived by two papers of Lorenz[4] (1963,1965). In the latter he exhibited the results of a series of numerical experiments with a 28-mode baroclinic (2-level) model that displayed at least some of the essential features of the atmosphere's behavior. Those experiments showed that the doubling time of RMSE could range from a few days to a few weeks, depending on the kinetic energy spectrum in the true initial state. Shortly thereafter, Charney *et al.*[5] (1966), published the results of a limited number of predictability experiments with a fullblown GCM. The doubling time of RMS temperature errors was found to be about five days, with a complete loss of predictability after about three weeks. These figures, which are somewhat higher than either earlier or subsequent estimates, may be more representative of the rather special initial error fields chosen for study than they are of an ensemble of random error fields. In any event, the optimism engendered by these estimates provided some impetus to the planning of GARP and FGGE.

Evidently reacting to the findings of Charney *et al.*,[5] G.D. Robinson[6] (1967) delivered himself of a few provocative remarks in his Presidential address to the Royal Meteorological Society, pointing out that nonlinear interactions between different scales of motion in a turbulent flow have the effect of transferring uncertainty about the small-scale motions to the large-scale motions. Indeed, using energy-transfer rates based on the Kolmogoroff theory of three-dimensional turbulence and Richardson's similarity law of particle dispersion, Robinson estimated that the predictability of the synoptic-scale motions would be almost completely lost after about two days. Aside from the fact that motions on scales of 100 km and greater appear to be quasi-horizontal (with lower energy-transfer rates), Robinson's estimates seemed contrary to common experience — namely, that the routine numerical forecasts showed perceptible skill after even 4 or 5 days. Although Robinson's critique was not widely accepted, it certainly succeeded in stirring up interest and promoting greater awareness of the predictability problem.

The question was put in a fresh, natural and proper perspective by one of Lorenz'[7] papers in 1969, a study of the evolutions of states following the appearance of naturally-occurring pairs of "near-analogues" — i.e., similar or "nearly" identical states, as revealed in a five-year series of daily upper-air observations. Although no close analogues were found (verifying the aperiodic character of atmospheric motions), Lorenz found that subsequent states of evolution diverged, with a doubling of RMS differences between pairs in about 2½ days. This study also revealed a "linear" regime, during which initially small errors grow roughly exponentially, followed by a "nonlinear" regime of phase-mixing when the error generally approaches an upper limit.

At this point, I would like to digress briefly to examine some of the properties of a particularly simple nonlinear system, with a view to seeing what features must be preserved in any statistical theory of predictability. In particular, we shall investigate the advantages, applicability and limitations of simple second-order closure schemes — as, for example, the quasi-normal approximation — and the necessity of allowing for correlations between amplitudes of different modes.

To see some of the issues more clearly and uncluttered by a lot of complicated physics, let us examine the behavior of the simplest nontrivial nonlinear system — that is, two-dimensional nondivergent flow. It is governed by the principle of vorticity conservation, i.e.,

$$\frac{\partial \zeta}{\partial t} + \underset{\sim}{V} \cdot \nabla \zeta = 0 \quad . \tag{1}$$

The continuity equation reduces to:

$$\nabla \cdot \underset{\sim}{V} = 0 \quad .$$

This implies that

$$\underset{\sim}{V} = \underset{\sim}{k} x \nabla \psi \qquad \zeta = \nabla^2 \psi$$

where $\underset{\sim}{k}$ is a unit vector directed normal to the plane of motion. Thus, (1) takes the form

$$\frac{\partial}{\partial t} \nabla^2 \psi + \underset{\sim}{k} \cdot \nabla \psi x \nabla \, (\nabla^2 \psi) = 0 \quad . \tag{2}$$

We suppose that the flow is contained by a fixed closed boundary B, on which ψ is necessarily some constant, say ψ=0.

For our present purposes, it is convenient to represent ψ as:

$$\psi(x,y,t) = \sum_{i=1}^{N} A_i(t) \phi_i(x,y) \tag{3}$$

in which the ϕ_i are the eigensolutions of $\nabla^2 \phi_i = -\alpha_i^2 \phi_i$ for ϕ_i=0 on B. It is easily shown that, if the eigenvalues α_p and α_q are distinct,

$$\int_A \phi_p \phi_q dA = 0 \quad .$$

That is, the eigenfunctions are orthogonal. We also require that they be normalized, so that

$$\int_A \phi_p^2 dA = 1 \quad .$$

Then, substituting from (3) into (1), multiplying by a particular eigenfunction ϕ_k and integrating over A, we obtain the evolution equations for the amplitudes A_k. They are:

$$\frac{dX_k}{dt} = \sum_{i=1}^{N} \sum_{j=1}^{N} \beta_{ijk} \, \alpha_j^2 X_i X_j \qquad (k = 1,2,3,\ldots N) \tag{4}$$

where $X_k = \alpha_k A_k$, and the nonlinear interaction coefficients β_{ijk} are given by:

$$\beta_{ijk} = \frac{1}{\alpha_i \alpha_j \alpha_k} \int_A \phi_k \underset{\sim}{k} \cdot \nabla \phi_i x \nabla \phi_j dA \quad .$$

By integrating by parts and invoking the condition $\phi_i = 0$ on B, we can readily show that
1) β_{ijk} vanishes if any two indices are equal
2) β_{ijk} is unchanged under cyclic permutation of indices, and
3) β_{ijk} changes sign under noncyclic permutation of indices.

Now, suppose that X_k is the true amplitude of ϕ_k and that X_k' is a superposed error. Then, according to (4), the evolution of the error X_k' is given by

$$\frac{dX_k'}{dt} = \sum_{i=1}^{N}\sum_{j=1}^{N} \beta_{ijk}\, \alpha_j^2 (X_i X_j' + X_j X_i' + X_i' X_j') \tag{5}$$

Thus, multiplying (5) by X_k' and summing over k, we find that the rate of change of total "error kinetic energy" is *at all times*:

$$\frac{1}{2}\frac{d}{dt}\sum_{k=1}^{N}(X_k')^2 = \sum_{k=1}^{N}\sum_{i=1}^{N}\sum_{j=1}^{N} \beta_{ijk}\, \alpha_j^2 (X_i X_j' X_k' + X_j X_i' X_k') \tag{6}$$

Owing to the properties of β_{ijk}, the contribution from the nonlinear term in (5) vanishes.

Next, let us consider the statistical behavior of errors originating in an *ensemble* of initial errors, drawn at random from a normally distributed population with zero mean value. Initially, then, when distinct X'_k's are uncorrelated in the ensemble average,

$$\frac{1}{2}\frac{d}{dt}\sum_{k=1}^{N} <(X'_k)2> = \sum_{k=1}^{N}\sum_{i=1}^{N}\sum_{j=1}^{N} \beta_{ijk}\, \alpha_j^2 \left(X_i <X_j' X_k'> + X_j <X_i' X_k'>\right)$$

$$= 0,$$

where the angle brackets denote the ensemble average. That is, the ensemble average total "error-energy" is initially neither increasing nor decreasing. Thus, whether the average error-energy grows or not depends on its second (or possibly higher-order) time derivative. The second time-derivative we obtain by differentiating (6) with respect to time and substituting for dX_k/dt and dX_k'/dt from (4) and (5). The result, which looks formidable but is perfectly straightforward, is:

$$\frac{1}{2}\frac{d^2}{dt^2}\sum_{k=1}^{N}(X_k')^2 = \sum_{k=1}^{N}\sum_{i=1}^{N}\sum_{j=1}^{N}\sum_{p=1}^{N}\sum_{q=1}^{N}$$

$$\Big\{+ \beta_{ijk}\,\beta_{pqi}\,\alpha_j^2\,\alpha_q^2\, X_p X_q X_j' X_k' \;+ \beta_{ijk}\,\beta_{pqj}\,\alpha_j^2\,\alpha_q^2\, X_p X_q X_i' X_k'$$

$$+ \beta_{ijk}\,\beta_{pqj}\,\alpha_j^2\,\alpha_q^2\,(X_i X_p X_k' X_q' + X_i X_q X_k' X_p' + X_i X_k' X_p' X_q')$$

$$+ \beta_{ijk}\,\beta_{pqk}\,\alpha_j^2\,\alpha_q^2\,(X_i X_p X_j' X_q' + X_i X_q X_j' X_p' + X_i X_j' X_p' X_q')$$

$$+ \beta_{ijk}\,\beta_{pqi}\,\alpha_j^2\,\alpha_q^2\,(X_j X_p X_k' X_q' + X_j X_q X_k' X_p' + X_j X_k' X_p' X_q')$$

$$+ \beta_{ijk}\,\beta_{pqk}\,\alpha_j^2\,\alpha_q^2\,(X_j X_p X_i' X_q' + X_j X_q X_i' X_p' + X_j X_i' X_p' X_q')\Big\} \tag{7}$$

Let us now examine this expression, evaluated at initial time. Taking the average over the ensemble of errors, we first note that the first and fourth terms on the R.H.S drop out completely, simply because β_{ijk} vanishes if any two indices are equal, and $<X_i'X_j'>$ vanishes unless i=j. In the remaining terms, of course, factors of the form $<X_m'X_n'>$ vanish unless m=n. This, in effect, performs the summation over one index.

To simplify matters still further, let us suppose that distinct X_k's are uncorrelated in an *ensemble* of *true* initial states. In that case, factors of the form $\overline{X_m X_n}$ vanish unless m=n. (We use the overbar to denote the average over the ensemble of true states, to distinguish it from the average over the ensemble of errors.) Thus, (7) reduces to:

$$\frac{1}{2}\frac{d^2}{dt^2}\sum_{k=1}^{N}<(X_k')^2> = \sum_{k=1}^{N}\sum_{i=1}^{N}\sum_{j=1}^{N}\beta_{ijk}^2\Big[\alpha_j^2\,(\alpha_i^2-\alpha_k^2)\,\overline{X}_i^2<(X_k')^2>$$

$$+\,\alpha_j^2\,(\alpha_j^2-\alpha_i^2)\,\overline{X}_i^2<(X_j')^2>$$

$$+\,\alpha_j^2\,(\alpha_k^2-\alpha_j^2)\,\overline{X}_j^2<(X_k')^2>$$

$$+\,\alpha_j^2\,(\alpha_j^2-\alpha_i^2)\overline{X}_j^2<(X_i')^2>\Big]\,.$$

In the event that $<(X_k')^2>$ is the same for all X_k', so that errors consist of "white" noise, we may symmetrize and factor to obtain

$$\frac{1}{2}\frac{d^2}{dt^2}\sum_{k=1}^{N}<(X_k')^2> = <(X_k')^2>\sum_{k=1}^{N}\sum_{i=1}^{N}\sum_{j=1}^{N}\beta_{ijk}^2\Big[(\alpha_j^2-\alpha_k^2)^2\overline{X}_i^2 \tag{8}$$

$$+\,(\alpha_k^2-\alpha_i^2)^2\overline{X}_j^2+(\alpha_i^2-\alpha_j^2)^2\overline{X}_k^2\Big]\,.$$

The R.H.S. of (8), we note, is positive definite, indicating that $d\sum_{k=1}^{N}<(X_k')^2>/dt$ is initially increasing. But $d\sum_{k=1}^{N}<(X_k')^2>/dt$ is initially zero, implying that the rate of change of average error-energy becomes positive — i.e., the variance of error increases. Referring back to our earlier results, this further implies that, although distinct X'_k's are uncorrelated initially, they must become correlated at later times through their dynamical interactions. Accordingly, (8) is not generally valid much later than initial time.

For some limited purposes, such as that of estimating the doubling time of the RMS of small initial errors, a "short time" may be long enough, in which case we would be tempted to regard (8) as valid over a relatively short time. I.e., we regard (8) as

$$\frac{1}{2}\frac{d^2E}{dt^2}=\frac{kE}{N}$$

where k, the triple summation on the R.H.S. of (8), is taken to be a known function of time, and E is just the average error-energy. For short times, the solution is

$$E = E_0\cosh(\sqrt{2k/N}\,t)\,. \tag{9}$$

This indicates that the error initially increases exponentially.

As a test of the validity of (9), we have carried out a modest numerical experiment for this occasion. The prediction model was the simplest imaginable — a 3-mode truncation of (4). We started by specifying a hypothetically correct initial state and generated a hypothetically correct prediction. We then picked ten combinations of initial amplitude-errors at random from a normally distributed population, added each of them to the true initial state, and generated ten incorrect predictions from the contaminated initial conditions. Finally, at each time stage, we calculated the error of each incorrect prediction and formed the ensemble mean-square error.

The mean-square error is plotted against time in Fig. 1. It is shown by the solid curve. For comparison, the appropriate solution (9) is shown by the dashed curve. We see that the numerical result, which assumes nothing about the correlation between errors, agrees remarkably well with the theoretical result until the MS error has approximately quadrupled (i.e., when the RMS error has about doubled). This clearly corresponds to the "linear" regime, pointed out in Lorenz's paper of 1969, during which the autocorrelation of modes dominates over the cross-correlation between different modes.

The importance of the buildup of correlations between distinct modes and the failure of simple second-order closures in the latter, nonlinear stages of error growth can be shown most simply from considerations of energy conservation. Multiplying (4) by X_k and summing over k, we see that, owing to the symmetry properties of β_{ijk},

$$\frac{d}{dt}\sum_{k=1}^{N} \langle X_k^2 \rangle = 0$$

or, in other words,

$$\sum_{k=1}^{N} \langle X_k^2 \rangle = \sum_{k=1}^{N} \langle X_{k0}^2 \rangle \qquad (10)$$

where the subscript "zero" denotes conditions at initial time. Similarly, if the true initial conditions X_{k0} are contaminated by initial errors X_{k0}',

$$\sum_{k=1}^{N} \langle (X_k + X_k')^2 = \sum_{k=1}^{N} \langle (X_{k0} + X_{k0}')^2 \rangle . \qquad (11)$$

Thus, expanding (11) and subtracting (10), we obtain

$$\sum_{k=1}^{N} \langle (X_k')^2 \rangle + 2\sum_{k=1}^{N} \langle X_k\, X_k' \rangle = \sum_{k=1}^{N} \langle (X_{k0}')^2 \rangle + 2\sum_{k=1}^{N} \langle X_{k0} X_{k0}' \rangle .$$

Initially, of course, X_k' is uncorrelated with X_k, so that the second term on the R.H.S. vanishes, and

$$\sum_{k=1}^{N} \langle (X_k')^2 \rangle = \sum_{k=1}^{N} \langle (X_{k0}')^2 \rangle - 2\sum_{k=1}^{N} \langle X_k X_k' \rangle .$$

Inspecting this equation, we see that errors can grow large only if the errors become negatively correlated with the true state.

There is a limit, however, on how correlated the errors and the true state can be. According to the Schwarz inequality,

$$-\sum_{k=1}^{N} \langle X_k X_k' \rangle < \left(\sum_{k=1}^{N} \langle X_k^2 \rangle \right)^{1/2} \left(\sum_{k=1}^{N} \langle (X_k')^2 \right)^{1/2} .$$

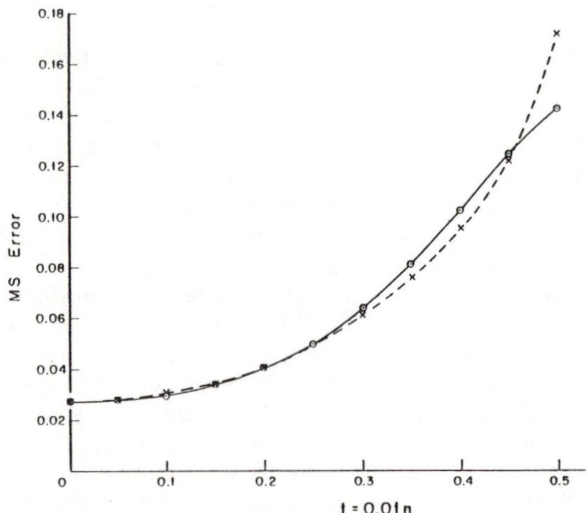

Fig. 1. Mean square error growth from a three-mode prediction model (solid curve) is compared with theoretical result (dashed curve), c.f. (9).

This has the consequence that

$$\sum_{k=1}^{N} <(X_k')^2> \; < \sum_{k=1}^{N} <(X_{k0}')^2 + \left(\sum_{k=1}^{N} <X_k^2>\right)^{1/2} \left(\sum_{k=1}^{N} <(X_k')^2\right)^{1/2} .$$

Thus, letting

$$\delta = \sum_{k=1}^{N} <(X_k')^2> \qquad \sigma = \sum_{k=1}^{N} <X_k^2>$$

we infer that

$$\delta \leqslant \frac{\sigma}{2} + \sqrt{\frac{\sigma^2}{4} + \delta_0^2} . \tag{12}$$

σ is just the square root of twice the true kinetic energy. If the initial error is small, the upper bound on the MS error is about twice the true initial kinetic energy, or "natural variance." It is also the error of "sheer guessing."

Contrary to Eq. (9), which applies in the early stages of exponential error-growth and predicts that error will grow indefinitely, (12) states that there is a definite upper bound that is approached as errors become more and more negatively correlated with the true state. In this regime, the system is highly nonlinear and cannot be described by simple second-order closure schemes of the kind outlined here. What is clearly needed is a good turbulence theory. Lorenz, in a paper that appeared later in 1969 has, in fact, proposed a modification that gives good results even in the nonlinear regime.

There is nothing in the arguments I have just given that isn't common knowledge. I have brought it up mainly as a pedagogical exercise, to put some later work in perspective. So let me now pick up the thread. That these considerations were known, if not well-known, is evidenced in Lorenz[8], (1969) a study of "error-energy" transfer between modes in a low-order spectral model. The theory is based on a generalization of the quasi-normal approximation, designed to give the correct behavior in the "nonlinear" regime. The theoretical calculations showed clearly the "back-cascade" of error-energy from small to large scales, at a rate smaller than in three-dimensional flows, with an error-doubling time of a few days and a "predictability-time" of about two weeks.

Lorenz' main results were soon confirmed and extended through the application of more general theories of two-dimensional turbulence — notably, the "eddy-damped" Markovian model of Leith[9] (1971) and the "test-field" model of Kraichnan[10] (Leith and Kraichnan, 1972). Both of these papers report the same qualitative behavior as Lorenz[7], in that a backward-cascading similarity spectrum of error-energy develops and gradually erodes the small-scale edge of the true energy spectrum. For the present level of initial errors, the doubling-time is about 2 days[9]. In Leith[9], it was also suggested that the dispersive nature of Rossby waves would inhibit the back-cascade of error-energy.

A discussion of developments in predictability theory in the late 60's and early 70's would be incomplete without some mention of "stochastic-dynamic" prediction. Epstein[11], recognizing that imperfect predictions should ideally be stated in terms of probability distributions, proposed a low-order closure scheme for predicting at least the first two moments of an ensemble of predictions, starting with the probability distribution for an ensemble of initial errors — namely, the mean and the standard error. The basic idea was pursued by Fleming[12] and Pitcher[13], who introduced some improvements in the calculation of error-energy transfer rates that were suggested by Leith's current work in two-dimensional turbulence theory. Evidently this effort got bogged down in arithmetic: Leith later proposed a simpler Monte Carlo method, but even that required computing something like ten separate predictions to generate a representative sample. One should note, however, that these were all attempts to predict explicitly the predictability to be expected at particular times and places.

In the meantime, the development of a statistical theory of error-growth and predictability was paralleled by a number of numerical experiments, some designed to test the theoretical results based on very simple models, and others to extend the results to more complicated and realistic models. Among the former are the numerical studies of Lilly[14] (1972) and Herring et al.[15] (1973) which indeed substantiate the theoretical results. In the second category, Smagorinsky[16] (1969) essentially repeated the numerical experiments of Charney et al., but using the GFDL formulation of a GCM. These lead to a doubling of RMS error in about 2½ days, which is more nearly consonant with independent estimates. Daley[17] (1981) reports comparable results from experiments with an 8-level model.

As might be expected, more recent and current work is concerned with the extension of the theory of predictability of incrementally more general models, and with exploration and testing by numerical experiment. Following the suggestions of Leith[9], Basdevant et al.[18] (1981), and Holloway[20] (1983) have investigated the effect of the dispersive Rossby waves and find that they do, in fact, inhibit the growth of error in nondivergent barotropic flows. Vallis[19] (1983), however, finds that the so-called "β-effect" in baroclinic (2-level) flows acts to decrease predictability, owing to release of energy in medium scales of motion and its rapid transfer to larger scales.

I am now treading dangerously close to the edge of work in progress, and it would be better to ask the people who are doing the work to speak for themselves. They are all here, so there's no need to misquote them.

Looking at our present position, one might feel fairly safe in saying that we now understand the basic mechanisms of error-growth in unforced, thermodynamically inactive systems. Work along that direction will undoubtedly continue to parallel developments in non-equilibrium turbulence theory — that is, once the model is given, the problem is essentially one of calculating the evolution of a probability distribution for a highly nonlinear quasi-random system, whether it is done numerically or analytically.

In the realm of real atmospheric flows, however, we haven't done much more than scratch at the surface. One knows from experience that some flow structures are very predictable, in that they persist for long periods even when they are continually being jostled by the passage of transient waves and eddies. Prime examples are the well-known "blocking flows," which may remain stationary for weeks. Owing to their preferred regions of occurrence, there is a strong presumption that they are induced by topographical or thermal forcing, or both. Some think neither. In any event, I don't think anyone has claimed that the properties of these peculiar structures, apparently defying the usual canons of error growth, are really understood — either their origin or how they maintain themselves as coherent entities in the midst of chaos.

Looking at the program for the next three days, I daresay that much will be said about these and other problems connected with real predictability. So I'll stop here. Thank you.

REFERENCES

1. T.E.W. Schumann, Weather **5**, 220 (1950).
2. P.D. Thompson, Tellus **9**, 275 (1957).
3. E.A. Novikov, Izv. Acad. Sci. U.S.S.R., Geophys. Ser. **11**, (1959).
4. E.N. Lorenz, Tellus **17**, 321 (1965).
5. J.G. Charney, et al., Bull. Am. Met. Soc. **47**, 200 (1966).
6. G.D. Robinson, Quart. Journ. Roy. Met Soc. **93**, 409 (1967)
7. E.N. Lorenz, J. Atmos. Sci. **26**, 636 (1969)
8. E.N. Lorenz, Tellus **21**, 289 (1969).
9. C.E. Leith, Jour. Atmos. Sci. **28**, 145 (1971).
10. C.E. Leith and R.H. Kraichnan, J. Atmos. Sci. **29**, 1041 (1972).
11. E. Epstein, J. Appl. Meteor. **8**, 190 (1969).
12. R. Fleming, Mon. Wea. Rev. **99**, 851-872, 927-938 (1971).
13. E.J. Pitcher, J. Atmos Sci. **34**, 3 (1977).
14. D.K. Lilly, Geophys. Fluid Dyn. **4**, 1 (1972).
15. J.R. Herring, J.J. Riley, G.S. Patterson and R.H. Kraichnan, J. Atmos. Sci. **30**, 997 (1973).
16. J. Smagorinsky, Bull. Am. Met. Soc. **50**, 286 (1969).
17. R. Daley, Atmos. Ocean **19**, 77 (1981).
18. C. Basdevant, B. Legras and R. Sadourny, J. Atmos. Sci. **38**, 2305 (1981).
19. G. Vallis, J. Atmos. Sci. **40**, 10 (1983).
20. G. Holloway, J. Atmos. Sci., **40**, 314 (1983).

TOPOLOGICAL ISSUES IN HYDRODYNAMIC PREDICTABILITY

John A. Dutton and Robert Wells
The Pennsylvania State University, University Park, PA 16802

ABSTRACT

The predictability problem arises because hydrodynamic flows with similar initial conditions may evolve differently, despite the continuous dependence of solutions on initial conditions.

The central topological issues of predictability concern the dimensions of the spaces in which solutions and solution differences reside. Concentration on the attractors of the flow reveals that such spaces may be effectively finite dimensional, and that the relevant number of dimensions may not always be large.

The mathematical approach utilizes eigenvalue problems and spectral decompositions to determine the expanding and contracting components of the flow, both directly, and through the construction of the unstable manifold via an extended version of the center manifold theorem.

A common situation in hydrodynamics allows explicit estimation of the dimension of the attractors with Lyapunov exponents. The center manifold theorem demonstrates strong convergence of truncated spectral models in attractors that bifurcate from the origin.

A version of the Poincaré map is constructed along an evolving trajectory by resolving the error field into a component transverse to the trajectory and a phase error component. The resulting equations suggest an eigenvalue problem that will reveal *a priori* those regions of phase space in which large transverse or phase errors may be expected.

INTRODUCTION

Consideration of the problem of hydrodynamic predictability directs our attention from the governing equations to various derived systems, which although familiar, must be studied from a new viewpoint.

Mathematical interest in the equations of hydrodynamic motion has centered on existence, uniqueness, stability, and smoothness of solutions to well-posed boundary value problems. Recent research has focused on transitions between solutions and the topology of solution spaces. Questions in the theory of predictability are related to those concerning stability of solutions of the equations of motion, but the emphasis must be different if we are to enhance our understanding of the phenomenology of predictability and if we are to develop effective methods to utilize that understanding in improving our capability for prediction.

To illustrate the change in emphasis, let us observe that an important mathematical issue is the continuity of solutions with respect to initial conditions. This desirable relationship between the evolving solutions and the initial conditions is demonstrated

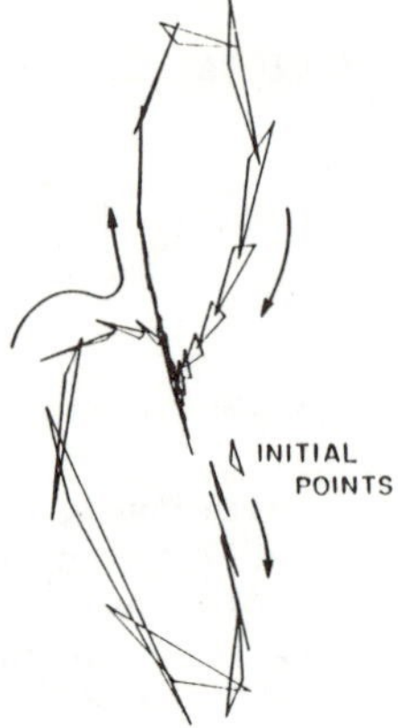

Fig. 1. Predictability of the Lorenz[7] equations for the first circuit of the attractor.

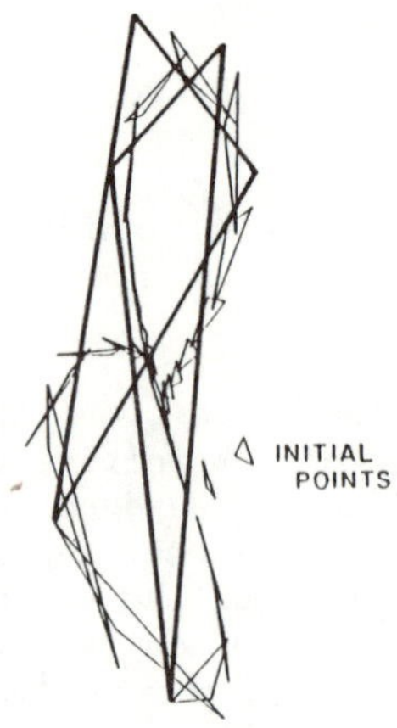

Fig. 2. Predictability after a triangle has split at the unstable stationary point at the origin.

explicitly later but the method used is aimed specifically at the continuity question and offers little practical information about growth or evolution of solutions that emanate from nearby, but distinct, initial conditions. The inequalities employed in the analysis of properties of solutions are generally too dramatic for the study of predictability problems.

The subtle aspects of predictability theory require that we turn from the functional analysis procedures that are used to study the classical mathematical problems of hydrodynamics to the study of the topological properties of the differences of solutions. As we eschew the luxury of ascertaining only whether certain norms either vanish or grow, we are faced in predictability problems with determining how such norms evolve over finite intervals, of characterizing the situations that lead to transverse convergence or divergence of bundles of solutions, and of asking whether new quantities will be more effective for the purposes of predictability theory than the classical norms.

From the topological viewpoint, then, we are concerned with defining precisely what we mean by prediction and by prediction error. We are also, as in all topological studies, concerned with the dimensions of the spaces in which our variables reside; in particular we want to ascertain whether the solutions are essentially finite dimensional, whether attractors have an even more limited number of dimensions, and whether the space of solution differences may have a manageably small number of effective dimensions. These questions lead us to new forms of eigenvalue problems.

Finally, the essential practical question is whether mathematical studies can point the way to achieving significant improvement in predictability at reasonable cost and to developing methods to assign confidence to an evolving predicted solution from information that is derived from the solution itself or from the characteristics of the neighborhoods through which it passes.

An Illustration

The well-known three-component model of convection due to Lorenz[7] can be used to illustrate and motivate the central themes of this article. Figure 1 depicts the initial behavior of solution differences as revealed by numerical integration. Perhaps surprisingly, the prediction errors first increase but then decrease as the unstable fixed point at the origin is approached after one circuit of the lower bundle of trajectories. The solution differences remain small until, as in Figure 2, one of the solutions takes a different path away from the unstable fixed point.

This example suggests some interesting questions:

1) Are large prediction errors generally created in the neighborhood of singular or special points in phase space?
2) Are large solution differences at a fixed time generally the result of phase or timing errors rather than errors in patterns or amplitudes?
3) Is it often the situation that the sequence of events, such as motion around one circuit in phase space, can be predicted with relative certainty once the trajectory has entered a specific bundle of trajectories?

Summary of Results

The methods of classical analysis of hydrodynamic problems provide a setting for beginning the study of predictability; in particular they lead to estimates of bounds for solutions and reveal that solutions depend continuously on the initial data.

We develop an eigenvalue problem and concomitant spectral models in which the eigenfunctions embody the forcing of the flow by a steady solution. The eigenvalues divide into a finite class of those with positive real parts that induces expansion away from the steady solution and an infinite class of those with negative real parts that induces contraction. The possibility that attractors are compact is examined and the interplay between the expanding and contracting components discussed. Finally, we look at Lyapunov exponents as a means of defining the dimension of the relevant attractors, and find that those dimensions are likely to be finite.

This approach leads to a study of the problem from the viewpoint of the center manifold theorem, which we extend to add further evidence for the finite dimensionality of the important part of the flow. In particular, the center manifold theorem leads to a justification for a wide range of finite truncated spectral models.

Finally, we develop concepts for examining error components transverse and parallel to a known trajectory. In principle, the derived equations allow us to map out a bundle of trajectories emanating from a neighborhood and to examine the relative importance of phase and amplitude errors.

In part of the article, the relevance to predictability is often implicit; it turns on the assumption that errors occur in directions in which solutions are expanding. If we can show that expansion occurs in only a finite number of dimensions, then we have reduced the predictability problem to manageable proportions. The attractors are the key to the topology and the dimensions of models must be large enough to describe them adequately; nevertheless, description of errors in only a limited number of dimensions may be adequate.

ANALYSIS OF HYDRODYNAMIC FLOW

The classical mathematical analysis of hydrodynamic flow produces results about existence, uniqueness, stability, and smoothness of solutions to the initial-boundary-value problem associated with the Navier-Stokes equations. This body of knowledge forms the foundation for the study of predictability. The theory as it stands today is presented and reviewed in Ladyzhenskaya[6], Shinbrot and Kaniel[12], Kaniel and Shinbrot[5], and in an illuminating summary by Serrin[11].

We consider throughout this article the nondimensional system

$$\frac{\partial v}{\partial t} + v \cdot \nabla v = - \nabla p + \nu \nabla^2 v + \mu F \qquad \nabla \cdot v = 0 \tag{1}$$

defined on a bounded domain $D \subset R^3$ and subject to the boundary conditions

$$v = 0 \quad \text{on} \quad \partial D \tag{2}$$

and the initial condition

$$v(x, 0) = v_o(x) \tag{3}$$

Here v is the vector $v = (v_1, v_2, v_3)$ with components corresponding to those of $x = (x_1, x_2, x_3)$; p combines the usual pressure-density term and the gravitational potential. The forcing $F = F(x)$ is amplified by the parameter μ; ν is the inverse Reynolds number.

Upon forming a scalar product of (1) and v and integrating, we find that the nonlinear term and the pressure gradient term both vanish owing to (2) and the solenoidal condition. Hence with the Schwarz inequality (A-2) and the Poincare inequality (A-7) we have

$$\begin{aligned} \frac{1}{2} \frac{\partial}{\partial t} ||v||^2 &= - \nu \int |\nabla v|^2 \, dx + \mu \int v \cdot F \, dx \\ &\leq - \nu \gamma_1 \, ||v||^2 + \mu \, ||v|| \, ||F|| \end{aligned} \tag{4}$$

where the integrals are taken over the domain D. From this we have

$$\frac{\partial \, ||v||}{\partial t} \leq - \nu \gamma_1 \, ||v|| + \mu \, ||F|| \tag{5}$$

and therefore

$$||v|| \leq ||v_o||\ e^{-\nu\gamma_1 t} + \frac{\mu\ ||F||}{\nu\gamma_1}(1 - e^{-\nu\gamma_1 t}) \tag{6}$$

All solutions are eventually trapped in a ball in L_2 of radius $\mu\ ||F||/\nu\gamma_1$.

As shown by Lorenz[7], if we have a representation of v in the form

$$v(x,\ t) = \sum a_n(t)\ \psi_n(x) \tag{7}$$

where the real orthonormal ψ_n are solutions of the eigenvalue problem

$$\nabla^2\ \psi_n = -\ \lambda_n\ \psi_n \qquad\qquad \nabla\cdot\psi_n = 0 \tag{8}$$

then (1) or the first line of (4) yields

$$\frac{1}{2}\frac{\partial}{\partial t}\sum_n a_n{}^2 = -\ \nu \sum_n \lambda_n\ a_n{}^2 + \mu \sum_n a_n\ \hat{F}_n \tag{9}$$

$$= -\ \nu \sum_n \lambda_n (a_n - \frac{\mu\ \hat{F}_n}{2\nu\ \lambda_n})^2 + \sum_n \frac{\mu^2 (\hat{F}_n)^2}{4\nu\ \lambda_n}$$

where the $\hat{F}_n = (\psi_n,\ F)$ are the Fourier coefficients of F.

The implications of these trapping theorems have been discussed by Lorenz[7] and Dutton[2]; note that for finite truncations the inequality implies via the fixed point theorem the existence of at least one steady solution. For the infinite-dimensional case, the existence of a (weak or generalized) steady solution $U = U(x,\ \mu,\ \nu)$ for every $(\mu,\ \nu)$ follows from the Leray-Schauder theorem (see Ladyzhenskaya[6]). For such a solution the first line of (4), together with the Poincaré inequality (A-7), implies that

$$\nu\ ||\nabla U||^2 \leq \mu\ ||U||\ ||F|| \leq \gamma_1^{-1/2}\ \mu\ ||\nabla U||\ ||F|| \tag{10}$$

or

$$\nu\ ||\nabla U|| \leq \gamma_1^{-1/2}\ \mu\ ||F|| \tag{11}$$

and thus by (A-9) we have $U \in L_4$ and we know from a theorem of Kaniel and Shinbrot[5] that $U \in C^\infty$ for $F \in C^\infty$ and hence U is smooth (infinitely differentiable) and therefore is a classical solution.

In predictability theory, we are concerned with deviations superimposed on a known solution. Let $v = U + u$ where $U = U(x,\ \mu,\ \nu)$ is a steady solution. Then

$$\frac{\partial u}{\partial t} = - U\cdot\nabla u - u\cdot\nabla U + \nu\nabla^2 u - \nabla p' - u\cdot\nabla u \tag{12}$$

$$= L_\mu u - B(u, u)$$

in which L_μ is the linear operator and $B(u, v) = u\cdot\nabla v$ is the bi-linear nonlinear operator. Now we have

$$\frac{1}{2}\frac{\partial}{\partial t} ||u||^2 = (u, L_\mu u) = - \int u\cdot(u\cdot\nabla U)\, dx - \nu \int |\nabla u|^2\, dx \tag{13}$$

and thus the Reynolds stresses acting on the shear of U provide the energy source for the disturbance.

For smooth F we know that U is smooth and hence with the Cauchy inequality ($2ab \leq a^2 + b^2$) we see that

$$\int |u\cdot(u\cdot\nabla U)|dx \leq \underset{D}{\text{Max}}|\nabla U| \sum_{i,k} \int |u_i u_k|dx \tag{14}$$

$$\leq 3\ \underset{D}{\text{Max}}|\nabla U|\ ||u||^2$$

and therefore

$$\frac{1}{2}\frac{\partial}{\partial t} ||u||^2 \leq (3\ \underset{D}{\text{Max}}|\nabla U| - \nu\gamma_1)\ ||u||^2 \tag{15}$$

or

$$||u|| \leq ||u(0)||\ \exp\left[3 \int_0^t \underset{D}{\text{Max}}|\nabla U|d\tau - \nu\gamma_1 t\right] \tag{16}$$

This result establishes stability of the steady solution when the Reynolds number

$$\frac{3\ B}{\nu\gamma_1} < 1 \tag{17}$$

where $B = \underset{D,t}{\text{Max}}|\nabla U|$, and uniqueness for any Reynolds number.

We have also established the continuity of solutions with respect to initial conditions in any interval (0, T) in which an evolving solution U has bounded gradient $|\nabla U|$, as will be certainly true for classical solutions.

At this point, we must make a strategic decision. Either we proceed to study the properties of norms of solutions in various function spaces, or we turn toward spectral models and search for appropriate finite-dimensional phase spaces. Such models allow us to eschew bounds on norms in favor of more subtle topological questions.

In making this choice we need to consider a philosophical issue. The atmosphere is a collection of a finite number of molecules and thus its dynamics is in actuality governed by a finite-dimensional system of ordinary differential equations (as pointed out by Lumley[9]). In applying the continuum hypothesis to produce the Navier-Stokes equations, we create a model that presents serious mathematical difficulties and introduces questions of existence of solutions not present in the finite-dimensional structure of the atmosphere.

Truncation of a spectral model at any finite N (perhaps equal to the number of molecules in the atmosphere) is clearly sufficient to avoid infinite-dimensional complications, but truncation at any N also introduces predictability-theoretic questions about the evolving differences of solutions to two different models.

SPECTRAL REPRESENTATIONS

The method of analysis illustrated in the previous section provides bounds for norms rather than specific information about the structure of evolving differences of solutions with different initial conditions or of those obtained from slightly different systems of equations. The exponential estimates (16) are sufficient for proving continuity, but not for providing practical information about predictability assessments.

In this section we develop systems of eigenfunctions that appear to be useful in predictability theory. The analysis shows that the properties of a particular linear operator are crucial to determining the characteristics of evolving solutions.

Using the results of the previous Section, we can in full generality turn our attention to a derived system of equations

$$\frac{\partial u}{\partial t} = L_\mu u - B(u, u) \tag{18}$$

$$\nabla \cdot u = 0 \qquad u = 0 \text{ on } \partial D$$

where

$$L_\mu u = - U \cdot \nabla u - u \cdot \nabla U + \nu \nabla^2 u - \nabla p' \tag{19}$$

and $B(u, v) = u \cdot \nabla v$. The vector $U = U(x, \mu, \nu)$ is a steady solution of the original set of equations.

Two Eigenvalue Problems

The linear part of (1) suggests the eigenvalue problem

$$L\phi = \eta\phi \qquad (L = L_\mu) \tag{20}$$

The operator L_μ has compact resolvent and hence the eigenfunctions are a complete, independent set in L_2. The adjoint operator $L^\dagger$ is defined by requiring that

$$(f,\ Lg) = (L^\dagger f,\ g) \tag{21}$$

for every pair f, g where the scalar product is

$$(f,\ g) = \int \overline{f}\ g\ dx \tag{22}$$

the overbar denoting a complex conjugate.

Because L has compact resolvent, η is in the spectrum of L if and only if $\overline{\eta}$ is in the spectrum of $L^\dagger$, and occurs with the same multiplicity. Thus we have eigenfunctions ϕ_n for L and $\phi_m{}^\dagger$ for $L^\dagger$.

If f and g are vectors and L is a matrix, then (21) gives

$$\overline{f}_i\ L_{ij}\ g_j = L_{ji}{}^T\ \overline{f}_i\ g_j = (\overline{L^T}\ f,\ g) \tag{23}$$

and so $L^\dagger = \overline{L^T}$.

If ϕ is an eigenfunction then the adjoint eigenfunction $\phi^\dagger$ satisfies

$$(L^\dagger\phi^\dagger,\ \phi) = (\phi^\dagger,\ L\phi) = (\phi^\dagger,\ \eta\phi) \tag{24}$$

and thus from

$$L^\dagger\phi_m{}^\dagger = \overline{\eta}_m\ \phi_m{}^\dagger \qquad\qquad L\phi_n = \eta_n\ \phi_n \tag{25}$$

we obtain the usual relation

$$(\phi^\dagger{}_m,\ \phi_n) = \delta_{mn} \tag{26}$$

Now we set

$$u(x,\ t) = \sum_n b_n(t)\ \phi_n(x) \tag{27}$$

in which conjugates appear in pairs to produce a real sum. Thus we obtain the spectral representation from (18) as

$$\dot{b}_n = \eta_n b_n - \sum_{\ell,m} b_\ell b_m(\phi_n{}^\dagger,\ \phi_\ell\cdot\nabla\ \phi_m) \qquad (\dot{\ }) = \partial(\ )/\partial t \tag{28}$$

These representations, although useful later, encounter the usual difficulties presented by non-orthogonal basis vectors. To circumvent them, let us observe that the energy relation associated with (18) can be written (again suppressing the parameter μ) as

$$\frac{1}{2}\frac{\partial}{\partial t}\,||u||^2 = (u,\ Lu) = \frac{1}{2}\left[(u,\ Lu) + (u,\ L^{\dagger}\ u)\right] \tag{29}$$

since (u, B(u, u)) = 0. Thus we define the self-adjoint operator

$$\mathcal{L} = \frac{1}{2}\,(L + L^{\dagger}) \tag{30}$$

and pose the problem

$$\mathcal{L}\psi = \lambda\psi \tag{31}$$

in which $\mathcal{L}$ has real eigenvalues λ_n and real eigenvectors ψ_n which form a complete orthonormal set of vectors in L_2. The eigenvalues accumlate only at $-\infty$.

Now the representation

$$u(x,\ t) = \sum_n a_n(t)\ \psi_n(x) \tag{32}$$

along with (12) shows that

$$\frac{1}{2}\frac{\partial}{\partial t}\,||u||^2 = \sum_n \lambda_n\ a_n{}^2 \tag{33}$$

and hence the growth of the solution u will depend on the distribution of positive and negative λ_n. The associated spectral model is obtained from

$$\frac{\partial u}{\partial t} = \mathcal{L}u - B(u,\ u) + \frac{1}{2}\,(L - L^{\dagger})u \tag{34}$$

as

$$\dot{a}_n = \lambda_n\ a_n - \sum_{\ell,m} a_\ell\ a_m(\psi_n,\ \psi_\ell\cdot\nabla\psi_m) + \sum_m N_{nm}\ a_m \tag{35}$$

in which the antisymmetric operator is given by

$$N_{nm} = \frac{1}{2}\,(\psi_n,\ \frac{1}{2}\,(L - L^{\dagger})\psi_m) = \frac{1}{2}\left[(\psi_n,\ L\ \psi_m) - (\psi_m,\ L\ \psi_n)\right] \tag{36}$$

Because of the antisymmetry in (n, m) of

$$D_{n\ell m} = (\psi_n,\ \psi_\ell\cdot\nabla\psi_m) \tag{37}$$

(35) yields (33) as its energy equation.

Eigenvalues

The rate of change of kinetic energy of the solution u is

$$\frac{1}{2}\frac{d}{dt}||u||^2 = (u, Lu) = (u, \mathcal{L}u) \tag{38}$$

and recalling that

$$Lu = \nu\nabla^2 u - \nabla p' - U\cdot\nabla u - u\cdot\nabla U \tag{39}$$

we find

$$L^{\dagger}u = \nu\nabla^2 u - \nabla q' + U\cdot\nabla u - \sum_{j=1}^{3} u_j \nabla U_j \tag{40}$$

which gives

$$\begin{aligned}(u, Lu) = (u, \mathcal{L}u) &= -\nu \int |\nabla u|^2 dx - \int u\cdot(u\cdot\nabla U)dx \\ &\le ||u||_4^2 ||\nabla U|| - \nu ||\nabla u||^2 \\ &\le \left(\frac{\mu ||F|| C}{\nu\gamma_1} - \nu\right) ||\nabla u||^2\end{aligned} \tag{41}$$

in which we have used (11), (A-3), and (A-9). Thus for μ small enough, $(u, \mathcal{L}u) < 0$ for every u. But then $\sum \lambda_n a_n^2 < 0$ for every set $\{a_n\}$ and hence it must be true that $\lambda_n < 0$ for every n. Eigenvalues will thus become positive as μ increases and U becomes unstable. Since the eigenvalues are distinct, have finite multiplicity, and accumulate only at $-\infty$, we can expect that the number $N_\lambda(\mu)$ of positive λ_n will be finite for all $\mu < \infty$. The import of this result is that the solutions departing from the neighborhood of the steady solution U will expand in N_λ directions and contract in all others. The implications for predictability are the main concern of this article.

An unresolved issue so far as we know is the relation between the number of eigenvalues η_n of L that have positive real part and the number λ_n of $\mathcal{L}$ that are positive. We shall see that the center manifold theorem which we consider in the next section is based on counting η_n with positive real parts; others of our results consider the number of positive λ_n.

Possible Compactness of Attractor Sets

The trapping theorem in the form (6) or (9) shows that the trajectory v(x, t) is eventually confined to a ball in L_2 specified by $||v|| \le \mu ||F||/\nu\gamma_1$. Since $u = v - U$, we know that u is eventually trapped in a suitable ball as well. All attractors of the evolving solutions u(x, t) must be inside this ball, and as illustrated by the example in the first Section, predictability questions can be phrased relative to the comparative evolution of trajectories that emanate from nearby initial points in or near the attractor.

But certainly not every point in the relevant ball B is suitable as an initial point or as a point of an evolving trajectory. The norm of u or the sum $\sum a_n^2$ may be finite but the dissipation integral $\int |\nabla v^2| dx$ or the sum $\sum \lambda_n a_n^2$ may diverge. On physical grounds these possibilities seem unrealistic. It is thus reasonable to expect that the ball can be replaced by a compact set in which both $\int |\nabla v|^2 dx$ and $\sum \lambda_n a_n^2$ converge.[1]

Curiously, the difficulties in proving compactness are caused by the possibility of large dissipation rather than large forcing by the Reynolds stresses acting on the shear of U. To see this we number the positive eigenvalues λ_n from 1 to N_λ in increasing magnitude and we number the negative eigenvalues from 0 to $-\infty$, also in increasing magnitude.

Let us suppose that $\sum \lambda_n a_n^2$ is positive; this implies that

$$\sum_{n=-\infty}^{0} |\lambda_n| a_n^2 < \sum_{n=1}^{N_\lambda} \lambda_n a_n^2 \leq \lambda_{N_\lambda} \sum_{n=1}^{N_\lambda} a_n^2 \leq \lambda_{N_\lambda} ||u||^2 \qquad (42)$$

As we shall see the existence of a finite sum on the left of (42) is sufficient to ensure compactness of attractors in the ball B.

Now let us suppose that the sum $\sum \lambda_n a_n^2$ is not positive. Then we use the smoothness of U to write

$$-(u, Lu) \leq \nu \int |\nabla u|^2 dx + 3 ||u||^2 \operatorname*{Max}_D |\nabla U| \qquad (43)$$

We denote the dissipation of the evolving component u by

$$\varepsilon = \nu \int |\nabla u|^2 dx \qquad (44)$$

and so (43) gives the inequality

$$\sum_{n=-\infty}^{0} |\lambda_n| a_n^2 \leq \varepsilon + (\lambda_{N_\lambda} + 3 \operatorname*{Max}_D |\nabla U|) ||u||^2 \qquad (45)$$

If the sums (45) were uniformly bounded (as is the sum in (42)) for trajectories in the ball B then the same bound would be valid for attractors in B. Then we would know that $|\lambda_n| a_n^2 < 1/|n|$ as $n \to -\infty$. But since $|\lambda_n| \sim Cn^{2/3}$ as $|n| \to \infty$ in a three-dimensional domain, we see that $|a_n| < (\sqrt{C}\, |n|^{5/6})^{-1}$ as $|n| \to \infty$. Hence the attractors must

---

[1] The closed unit ball in L_2 is easily seen not to be compact because the sequence of points (1, 0, 0, ...), (0, 1, 0, ...), (0, 0, 1, ...) has no limit point in the ball. But the Hilbert cube defined by points $(x_1, x_2, x_3, \ldots)$ for which $|x_n| < 1/n$ is compact. Thinness in the high-order coordinates is the essential difference.

become thin in cross-sections through their higher-order coordinates --they would be analogs of the Hilbert cube.

To prove compactness when the sum is bounded, we denote an attractor by A and choose a sequence $a^{(n)}$ of points in A. Each sequence of components $a_k^{(n)}$ is bounded for fixed k and thus has a subsequential limit $\tilde{a}_k$. We choose a convergent subsequence from $a_1^{(n)}$ and then use the corresponding elements of $a_2^{(n)}$ to choose a convergent subsequence, and so on. For the higher order components we can choose $\eta > 0$ and then choose M large enough that for every n

$$\sum_{p=-\infty}^{-M-1} (a_p^{(n)} - \tilde{a}_p)^2 \lesssim \sum_{p=-\infty}^{-M-1} \frac{2}{C\, p^{5/3}} < \eta/2 \tag{46}$$

Then we can choose Q large enough that

$$\sum_{p=-M}^{N_\lambda} (a_p^{(Q)} - \tilde{a}_p)^2 < \eta/2 \tag{47}$$

Thus we have shown that every infinite sequence in A has a limit point in A and thus A is compact. Moreover, the argument of (46) shows that the total energy in the range of indices $(-\infty, -M)$, where M is finite, is less than $\eta/4$. Certainly, we would conclude that the flow becomes effectively $N_\lambda + M$ dimensional in the attractor.

The energy equality (33) has the obvious implication that

$$\int_0^t \sum_{n=-\infty}^{N_\lambda} \lambda_n a_n^2(\tau)d\tau = \frac{1}{2} (||u(t)||^2 - ||u(0)||^2) \tag{48}$$

and hence for average values we have

$$\lim_{T\to\infty} \frac{1}{T} \int_0^T \left(\sum_{n=1}^{N_\lambda} \lambda_n a_n^2 - \sum_{n=-\infty}^{0} |\lambda_n| a_n^2 \right) d\tau = 0 \tag{49}$$

which suggests again that the excursions of a_n for large n are limited. Because we can interchange summation and integration of positive terms we write (48) as

$$\sum_{n=-\infty}^{0} |\lambda_n| \int_0^t a_n^2\, d\tau \leq \frac{1}{2} (||u(0)||^2 - ||u(t)||^2) \tag{50}$$

$$+ \lambda_{N_\lambda} \int_0^t ||u(t)||^2\, dt \leq B_0 + B_1 t$$

since $||u||$ is bounded. As before, for t large enough,

$$\int_0^t a_n^2(\tau)d\tau < \frac{B_1 t}{|n\lambda_n|} \quad , \quad n \to -\infty \quad . \tag{51}$$

Let $F_{n,k}(t) = \{0 \leq \tau \leq t \mid a_n^2 > k\}$ be the set where a_n^2 exceeds k and let μ be Lebesgue measure. Then (51) implies the inequality $k\mu(F_{n,k}(t)) < B_1 t/|n\lambda_n|$ and we find that the fraction of time that a_n^2 exceeds k is limited by $\mu(F_{n,k})/t < B_1(|n\lambda_n|k)^{-1}$. From this we can calculate that the fraction of time the total energy exceeds k in components $a_j(t)$ with $|j| > M$ is bounded by

$$\sum_{j=-\infty}^{-M} \frac{\mu(F_{j,k})}{t} < \frac{B_1}{k} \sum_{j=-\infty}^{-M} \frac{1}{|j\lambda_j|} \tag{52}$$

and can be made as small as we please by choosing M large enough.

In conclusion, we see that boundedness of the dissipation is sufficient to ensure that the attractors of the flow are compact. Even if the dissipation were not bounded, the fractional excursion time of the trajectories out of a compact subset of the attractor is limited. In a finite-dimensional model, the attractors will always be compact with finite energy because closed and bounded sets are then compact; nevertheless, we still would like the property of thinness in the higher-order components.

The advantages for predictability theory of knowing that attractors are compact merit searching for proof or confirmation. Bounding $||\nabla u||^2$ by $||u||^2$ is impossible in general. Dynamical reasoning must somehow be used to determine classes of flows for which the dissipation is bounded by relations involving energy. Observations suggest that flows reach states quite quickly in which the tails of their energy spectra are steady and contain only negligible energy. Being a property of molecular collisions in atmospheres with only finite numbers of molecules, physical dissipation, anyway, must be finite.

Qualitative Dynamics Near the Attractor

It is convenient to separate components by writing

$$a_n = \begin{matrix} A_n & \lambda_n > 0 \\ \alpha_n & \lambda_n \leq 0 \end{matrix} \tag{53}$$

and to consider the relation

$$\frac{1}{2}\frac{d}{dt}\sum_{n=-\infty}^{N_\lambda} a_n^2 = \sum_{n=1}^{N_\lambda} \lambda_n A_n^2 - \sum_{n=-\infty}^{0} |\lambda_n|\alpha_n^2 \tag{54}$$

Setting (54) equal to zero specifies a hypercone in phase space with the vertex at the origin. Points inside the cone, typified by a = A with $\alpha = 0$, move away from the origin. Those outside the cone, for example $a = \alpha$ with $A = 0$, move toward the origin. The total motion will obviously be a combination of these components, and there must be motion along the surfaces $\sum a_n^2 = \text{const}$ across the boundaries of the hypercone in order that the phase space trajectories remain confined to the attractors.

To examine this further, we use

$$Q_{nm} = - \sum_{\ell=-\infty}^{N_\lambda} a_\ell \, D_{n\ell m} \tag{55}$$

to rewrite the spectral model (35) as

$$\dot{A}_n = \lambda_n A_n + \sum_{m=1}^{N_\lambda} (Q_{nm} + N_{nm}) A_m + \sum_{m=-\infty}^{0} (Q_{nm} + N_{nm})\alpha_m \tag{56}$$

$$= \lambda_n A_n + \sum_{m=1}^{N_\lambda} P_{nm} A_m + F_n(\alpha) \qquad n = 1, 2, \ldots, N_\lambda$$

$$\dot{\alpha}_n = - |\lambda_n| \alpha_n + \sum_{m=-\infty}^{0} (Q_{nm} + N_{nm}) \alpha_m + \sum_{m=1}^{N_\lambda} (Q_{nm} + N_{nm})A_n$$

$$= - |\lambda_n| \alpha_n + \sum_{m=-\infty}^{0} P_{nm} \alpha_m + G_n(A) \qquad n = 0, -1, -2, \ldots$$

in which P_{nm} is antisymmetric and in which it is easily seen that

$$\sum_{n=1}^{N_\lambda} A_n F_n(\alpha) = - \sum_{n=-\infty}^{0} \alpha_n G_n(A) \tag{57}$$

The energy equations are

$$\begin{aligned} \frac{1}{2} \frac{d}{dt} \sum A_n^2 &= \sum \lambda_n A_n^2 + \sum A_n F_n(\alpha) \\ \frac{1}{2} \frac{d}{dt} \sum \alpha_n^2 &= - \sum |\lambda_n| \alpha_n^2 + \sum \alpha_n G_n(A) \end{aligned} \tag{58}$$

Because $\sum A_n^2$ must remain bounded for all time, $\sum A_n F_n$ cannot be uniformly positive and hence $\sum \alpha_n G_n$ cannot be uniformly negative. Over the history of the flow, there must be transfer of energy from the unstable components A_n to the stable ones α_n, where the energy is lost through the dissipation $- \sum |\lambda_n|\alpha_n^2$.

These equations can be converted to the form used in Lorenz's trapping theorem and they become

$$\frac{1}{2}\frac{d}{dt}\sum_{n=1}^{N_\lambda} A_n^2 = \sum_{n=1}^{N_\lambda} \lambda_n (A_n + \frac{F_n}{2\lambda_n})^2 - \sum_{n=1}^{N_\lambda} \frac{F_n^2}{4\lambda_n}$$

$$\frac{1}{2}\frac{d}{dt}\sum_{n=-\infty}^{0} \alpha_n^2 = - \sum_{n=-\infty}^{0} |\lambda_n| (\alpha_n - \frac{G_n}{2|\lambda_n|})^2 + \sum_{n=-\infty}^{0} \frac{G_n^2}{4|\lambda_n|} \tag{59}$$

Hence the points A move away from the origin outside a hypersphere with center coordinates $- F_n/2\lambda_n$; the points α move toward the origin outside the hypersphere with center coordinates $G_n/2|\lambda_n|$ (but we recall that both F_n and G_n are time dependent and depend on the entire trajectory a(t)). Thus the flow of the phase space swirls around these moving centers, vacating certain regions and becoming trapped in others.

The process involves the interplay between the components with positive eigenvalues and those with negative eigenvalues. We can speculate, to the extent that energy is transferred from the A components to the α components, that the A components may not be sensitive to the exact form of higher order α components. The apparent interchange of energy provided by the F_n and G_n terms in (56) and (58) may be susceptible to parameterization that would produce the correct energetics even if exact timing were lost.

To explore the topological implications of the presence of the A and α components, we turn now to an explicit characterization of the dimensionality of attractors.

Lyapunov Exponents and the Dimension of Attractors

An attractor of a hydrodynamical system is a set in phase space that is invariant and captures trajectories emanating from nearby or and possibly remote, initial points. There is considerable evidence that attractors are of smaller dimension than complete phase spaces and that they are finite-dimensional even for infinite-dimensional systems. A trivial example is a stable stationary point which is zero-dimensional; an important hydrodynamical example is the nine-coefficient model of Lorenz[8] in which solutions are attracted to a three-component geostrophic attractor.

The fact that the hydrodynamical system (18) can be represented in spectral form with A and α components corresponding to positive and negative eigenvalues, suggests that the attractor might be modeled by A components and enough of the α components to provide both motion toward the attractor and the dissipation required by the trapping theorem. We might speculate that the dimension should be of the order of $2N_\lambda$, and that the prediction errors would occur largely in the A components.

We can represent (18) with an Nth order truncated spectral model with $- N + N_\lambda + 1 \leq n \leq N_\lambda$ where $N >> N_\lambda$ of the form

$$\dot{a}_n = A_n(a) \tag{60}$$

and write the solution as a function of initial condition a_o as $a(t) = \phi_t(a_o)$. Then we can calculate the Jacobian matrix

$$J(t) = D\phi_t(a_o) = \{\partial a_\ell(t)/\partial a_m(0)\} \tag{61}$$

and solve the eigenvalue problem

$$(J - \gamma I)\,\psi = 0 \tag{62}$$

for each time t. A moving volume will expand in directions ψ_n with $|\gamma_n| > 1$ and contract in others with $|\gamma_n| < 1$, where $|\gamma_{n+1}| \leq |\gamma_n|$.

Lyapunov numbers for this flow are defined by

$$\tilde{\gamma}_n = \lim_{t\to\infty} |\gamma_n(t)|^{1/t} \tag{63}$$

and the numbers $\ln(\tilde{\gamma}_n)$ are the Lyapunov exponents. The Lyapunov dimension of the attractor is defined by

$$d_L = k - \ln(\tilde{\gamma}_1\, \tilde{\gamma}_2 \,\ldots\, \tilde{\gamma}_k)/\ln \tilde{\gamma}_{k+1} \tag{64}$$

in which k is the largest integer for which $\tilde{\gamma}_1 \ldots \tilde{\gamma}_k \geq 1$. The Lyapunov dimension of the attractor is known to correspond to the dimension obtained from other definitions in a variety of circumstances. For further exposition and examples see Farmer[3] and Farmer, Ott, and Yorke[4].

In the present case, we must calculate the Jacobian J(t) from (60) as

$$\dot{J} = \frac{\partial J}{\partial t} = \Lambda\, J \tag{65}$$

where $\Lambda = \{\partial A_\ell/\partial a_m\}$. If Λ were constant, then $J = \exp(t\Lambda)$ and the eigenvalues of J are $\exp(t\lambda)$ where the λ are the eigenvalues of Λ; both J and Λ then have the same eigenfunctions. It then follows that $\tilde{\gamma}_n$ would be $\exp(\mathrm{Re}\,\lambda_n)$.

The spectral model (28) yields (65) with the formulation

$$\Lambda_{pq} = \left[\eta_p\, \delta_{pq} + \sum_r G_{pqr}\, b_r\right] \tag{66}$$

$$G_{pqr} = -\,(\phi_p^{\dagger},\, \phi_r\cdot\nabla\phi_q + \phi_q\cdot\nabla\phi_r)$$

and we must integrate (65) along a trajectory to determine J(t) and then find the eigenvalues.

In the special case that b = 0 is a stable point, then $\mathrm{Re}\,\gamma_p < 0$ for all p and $\tilde{\gamma}_p \to \exp(\mathrm{Re}\,\eta_p) < 1$ and so $d_L = 0$ as required. If

$b = 0$ is unstable but b^o is a stable stationary point, then the eigenvalues of the linear part of (28) at b^o are just those of Λ in (66) evaluated at b^o and we again obtain the correct dimension. For more complicated cases, the stationary points are exceptional and we must integrate over trajectories moving toward the attractor to obtain the correct estimates of its dimension.

The relevance of these ideas to predictability is clear. Trajectories with nearby initial points spread apart in directions associated with positive Lyapunov exponents; they move toward the attractor in directions associated with negative exponents.

To examine this situation, let us consider a difference of solutions

$$\Delta = \phi_t(b_o + \delta) - \phi_t(b_o) = J(t)\,\delta + \ldots \qquad b_o = b(0) \tag{67}$$

which gives

$$\dot{\Delta} = \dot{J}\,\delta + \ldots = \Lambda J\delta + \ldots = \Lambda\Delta + \ldots \tag{68}$$

and as $\delta \to 0$ we have the linear version

$$\dot{\Delta} = \Lambda\,\Delta \tag{69}$$

in parallel with (65). If we let $\Delta(0)$ be the matrix of coordinates describing a small initial cube centered on b_o, then Δ will be the image of the cube, which has volume $\det \Delta$; thus Liouville's formula applied to (65) or (69) gives

$$\det J = \det \Delta / \det \Delta(0) = \exp\left[\int_0^t \operatorname{tr} \Lambda \, d\tau\right] \tag{70}$$

Now $\det J = \prod_{i=1}^{N} \gamma_i$ and so

$$\prod_i \tilde{\gamma}_i = \lim_{T\to\infty} \left|\exp\left[\frac{1}{T}\int_0^T \operatorname{tr} \Lambda \, d\tau\right]\right| \tag{71}$$

$$= \exp\left[\operatorname{Re}\left(\sum_n \eta_n + \sum_{n,r} G_{nnr}\, \overline{b}_r\right)\right]$$

where $\overline{b}_r$ is the time average along the trajectory. As μ increases so that $b = 0$ becomes unstable, the averages $\overline{b}_r$ may begin to alter the sum of the Lyanpunov exponents $\ln \tilde{\gamma}_i$. For many cases, though, $\overline{b}_r$ will remain near zero and the sum in (71) will depend on the sum $\operatorname{Re} \sum \eta_n$. But then $\prod \tilde{\gamma}_i$ becomes smaller as we increase the number of components because $\sum \eta_n$ becomes increasingly negative. It is tempting to conclude that there may be little to be gained by adding components to a model after we have enough to describe the attractor.

The implication of (71) is that we can obtain reasonable estimates without massive numerical integrations. Indeed, if $\overline{b}_r = 0$, then we can write (64) as

$$d_L = N_\lambda + 1 + k + \sum_{-k}^{N_\lambda} \mathrm{Re}(\eta_n)/|\mathrm{Re}\ \eta_{k+1}| \tag{72}$$

where k is the largest number such that the sum in the numerator is positive. The last term is clearly less than one by the definition of k and thus $d_L \leq N_\lambda + k + 2$. For many systems, then, we would expect attractors of fixed finite dimension regardless of the truncation at $N >> N_\lambda$ coefficients; but certainly we must require $N \geq d_L$ in order that we have a realistic representation of the asymptotic properties of the flow.

The task now is to compare the attractors of the original hydrodynamic system (18) and truncated models which we believe provide approximations. To do so, we turn to the center manifold theorem.

ATTRACTORS AND THE CENTER MANIFOLD THEOREM

The considerations of the previous section have shown that hydrodynamic flows can be split into components according to the sign of the real part of the eigenvalues of the linear part of their determining equation. The number of positive eigenvalues is finite for the self-adjoint operator $\mathcal{L}_\mu$ and the number of eigenvalues with positive real part is also finite for the original operator L_μ. The properties of the spectral models suggest that the flow might be dominated by a finite-dimensional space containing the components with positive real parts of eigenvalues. The center manifold theorem asserts that under certain circumstances this is exactly the case.

We are concerned now with two aspects of predictability. If we have a hydrodynamic trajectory v(x, t) confined to the basin of a minimal attractor A, then we would expect, on any reasonable definition, that the predictability of the trajectory should increase as the dimension of the attractor decreases. Thus, a particularly agreeable fact is that the dimension of an attractor which bifurcates from a stable stationary solution is bounded by one less than the number of eigenvalues of L_μ whose real parts change sign at the bifurcation. For the most important cases, this bound is finite and is often zero, one, or two.

A second aspect of the predictability problem appears when we compare the dynamics of a truncated spectral model with those of the original hydrodynamic system from which the model was derived. Clearly, with respect to a reasonable definition, the predictability of the dynamics of the hydrodynamic system in terms of the dynamics of the finite model ought to increase as the dimension of the model increases. It might be that an acceptable approximation will be obtained from a finitely truncated model only if the dimension of the truncation is absurdly large; further, it might be that the asymptotic behavior of the hydrodynamic flow as $t \to \infty$ cannot be predicted by any finite model, despite the suggestions of the Lyapunov dimension, because the attractor may be badly embedded in phase space. Nevertheless, we shall argue that when an attractor bifurcates from a stable stationary solution, finite models can indeed predict asymptotic behavior of the hydrodynamic flow. The dimension might still be absurdly large, but as we shall argue, there is again reason to hope that it may be manageably small.

These two situations are linked, as indicated above, by the center manifold theorem, which we state below. We shall modify an existing proof of the theorem in order to describe a strong sense in which the behavior of truncated models may be used to predict the behavior of the original hydrodynamic system. We consider the evolution equation (18) in a suitable Hilbert space $\mathcal{H}$ and we write its solution as

$$u(x, t) = \phi_t(u_o) \qquad t \in R \tag{73}$$

in which the local semiflow ϕ_t constitutes the hydrodynamic flow with initial condition $u(x, 0) = u_o$. As indicated in Section 3, the linear part L_μ of (18) has a pure point spectrum consisting of eigenvalues with finite multiplicities and real parts bounded above. For convenience, we assume that all eigenvalues have negative real parts for $\mu < 1$ and that finitely many have non-negative real parts for $\mu \geq 1$.

The Hilbert space $\mathcal{H}$ containing the solutions (73) is spanned by the eigenspaces of L_μ. We consider two more Hilbert spaces H and H' also spanned by these eigenspaces but we give H and H' stronger norms so that the inclusions $H \subset H' \subset \mathcal{H}$ are compact with dense images and so that both the linear operator L_μ and the quadratic operator B extend to C^∞ functions of H' into $\mathcal{H}$.

Now let $H_s \subset H$ be the space spanned by the generalized eigenspaces of $L_{\mu=1}$ corresponding to eigenvalues with negative real parts (the subscript s implies stable); let H_u be the space spanned by the generalized eigenspaces corresponding to the remaining eigenvalues (the subscript u denoting unstable). Then H_u is finite-dimensional.

The center manifold theorem (see Figure 3) ensures the existence of a submanifold M_u of $H \oplus R$ containing the origin, with tangent space at the origin equal to $H_u \oplus R$ and invariant under the solution local semiflow of the system in $H \oplus R$ given for $(\dot{\ }) = \partial(\ )/\partial t$ by

$$\dot{u} = L_\mu u - B(u, u) \tag{74}$$

$$\dot{\mu} = 0$$

Furthermore, M_u is locally attracting under the semiflow. If A_μ is a minimal attractor that bifurcates from the origin as μ exceeds one, then the set $\{(u, \mu) | u \in A_\mu\}$ must be contained in M_u, and hence dim $A \leq$ dim H_u.

We wish to approximate the manifold M_u with the corresponding center manifolds of truncated finite models obtained from (18). A

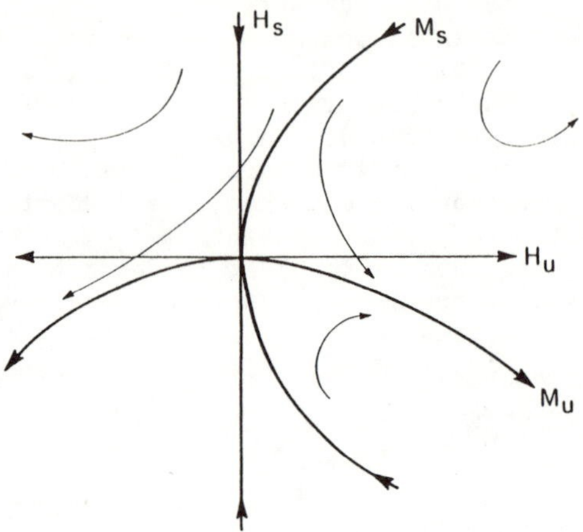

Fig. 3 Geometry of the center manifold theorem.

major defect of the available center manifold theorem in this application is that its center manifold is not unique. Thus the center manifolds of the finite models might approach a wrong center manifold of the full system -- if they approach anything at all -- and the rate of convergence is unlikely to be satisfactory. The proof of the center manifold theorem due to Lanford, presented in Marsden and McCracken[10], finds the center manifold as the unique fixed point of a certain contraction mapping and so the center manifold is unique in the right context. This proof still does not characterize the center manifold explicitly enough for us to obtain the desired convergence.

For this reason we sketch briefly a modification of Lanford's argument which produces a center manifold suitable for our purposes. Let $\gamma(x)$ be a C^∞ nonincreasing function such that

$$\begin{aligned} \gamma(x) &\equiv 1 \qquad 0 \leq x \leq \varepsilon_1 \\ \gamma(x) &\equiv 0 \qquad x \geq \varepsilon_2 \end{aligned} \tag{75}$$

in which $0 < \varepsilon_1 < \varepsilon_2$. We replace the system (74) with

$$\dot{u} = L_\mu u - \gamma(||u||^2) B(u, u) \tag{76}$$

Since the systems (74) and (76) are identical near the origin, they will have the same center manifolds near the origin, and it suffices to deal with (76).

First we note that (76) has a globally defined solution semigroup. To see this, we observe following Marsden and McCracken[10] that since the real parts of the eigenvalues of L_μ are bounded above, the semigroup $\exp(tL_\mu)$ has strong enough smoothing properties that one can show that the solution $\hat{\phi}_t(x)$ of (76) may be defined for any $\delta > 0$ on the set

$$G_\delta = \{x = (u, \mu) \mid ||u||^2 + (\mu - 1)^2 \leq \delta^2\} \tag{77}$$

for all time t such that $0 \leq t \leq t_o$, where $t_o > 0$ depends on δ. We take $\delta > 0$ large enough so that $\varepsilon_2 < \delta$. Then for $x \notin G_\delta$, the solution is given by

$$\hat{\phi}_t(x) = \hat{\phi}_t(u, \mu) = (e^{tL_\mu} u, \mu) \tag{78}$$

which is defined for all t until $t = t_1$ at which

$$\hat{\phi}_{t_1}(x) \in G_\delta \tag{79}$$

But then

$$\hat{\phi}_{t+t_1}(x) = \hat{\phi}_t(\phi_{t_1}(x)) \tag{80}$$

is defined for $0 \leq t \leq t_o$ and thus for any x, $\hat{\phi}_t(x)$ is defined for $0 \leq t \leq t_o$. Then the semigroup property shows that it is defined for any $x \in H \oplus R$ and any t in the range $0 \leq t < \infty$.

With $\hat{\phi}_t(x)$ defined for all x and t, we introduce the space

$$V = \left\{ f: H_u \oplus R \to H_s \;\middle|\; \begin{array}{ll} f \in C^k, & |D^j f(w)| \leq 1 \text{ for all} \\ 0 \leq j \leq k, & w \in H_u \oplus R \end{array} \right\} \tag{81}$$

of Lanford's proof; here $D^j f(y)$ is the jth Frechet derivative of f evaluated at y. Since

$$H = H_u \oplus H_s \tag{82}$$

we may write the semiflow $\hat{\phi}_t$ in terms of H_u and H_s components as

$$\hat{\phi}_t(y, z, \mu) = (\alpha_t(y, z, \mu), \beta_t(y, z, \mu), \mu) \tag{83}$$

with $y \in H_u$, $z \in H_s$, $\alpha_t(y, z, \mu) \in H_u$ and $\beta_t(y, z, \mu) \in H_s$. For $t \geq 0$ sufficiently small we may define a map

$$p_t: \; V \to V \tag{84}$$

by requiring that

$$\hat{\phi}_t(y, f(y, \mu), \mu) = \big(\alpha_t(y, f(y, \mu), \mu), \quad p_t\, f(\alpha_t(y, f(y,\mu), \mu), \mu)\big) \tag{85}$$

or equivalently

$$p_t\, f(\alpha_t(y, f(y,\mu), \mu), \mu) = \beta_t(y, f(y, \mu), \mu) \tag{86}$$

for all $f \varepsilon V$. For t small enough, p_t is defined and contracting. Consequently, a center manifold is now given as in Lanford's proof by

$$M_u = \{(y, f_o(y, \mu), \mu) \mid y \varepsilon H_u\} \tag{87}$$

where f_o is the unique fixed point of p_t for any $t > 0$. The crucial point here is that, in addition to supplying M_u, the semigroup property of $\hat{\phi}_t$ implies that property for p_t and hence

$$p_{t+s} = p_t\, p_s \tag{88}$$

Thus p_t must be defined for all $t \geq 0$, and it is a contraction semigroup on V. But then the unique fixed point of p_t is the same for all $t > 0$, and f_∞ is that fixed point if and only if

$$\frac{d}{dt} f(\alpha_t(y, f_\infty(y, \mu), \mu) \Big|_{t=0} = \frac{d}{dt} \beta_t(y, f_\infty(y, \mu), \mu) \Big|_{t=0} \tag{89}$$

for all $y \varepsilon H_u$. Necessity is trivial; sufficiency requires the semigroup property.

Let us write the right side of the differential equation (74) in terms of H_u and H_s components for $u = (y, z) \varepsilon H$ as

$$L_\mu u - \gamma(||u||^2)\, B(u, u) = (A_\mu y + J(y, z), B_\mu z + K(y, z)) \tag{90}$$

with the first component in H_u and the second in H_s. Here A_μ and B_μ are the linear parts with the property that A_μ: $H_u \rightarrow H_u$ and B_μ: $H_s \rightarrow H_s$ because H_u and H_s are invariant subspaces with respect to L_μ. Now (89) becomes a first-order partial differential equation on the finite-dimensional linear space H_u

$$D_y\, f_\infty(y, \mu)\cdot(A_\mu y + J(y, f_\infty(y, \mu)) = B_\mu f_\infty(y, \mu) + K(y, f_\infty(y, \mu)) \tag{91}$$

with the singular boundary conditions

$$f_\infty(0, \mu) = 0 \qquad D\, f_\infty(0, \mu) = 0 \tag{92}$$

Unfortunately, the dependent variable f_∞ is infinite-dimensional; however our observations above ensure that the characteristic problem (91) - (92) possesses a unique globally defined solution and that the graph of this solution is the center manifold resulting from Lanford's approach.

It is this fact we wish to use to relate the asymptotic behavior of the hydrodynamic system to that of certain of its truncated finite models. To obtain such models, we begin by observing that for bifurcating attractors, the restriction of the hydrodynamic system to the center manifold carries all the asymptotic information. However, this restriction is a finite-dimensional autonomous dynamical system; that is, as far as the asymptotic information regarding the bifurcating attractor is concerned, we find a perfect finite-dimensional model when we find the solution f_∞ of (91) - (92). The difficulty of course is finding the solution f_∞, and to circumvent it we turn to truncated models. We construct these by choosing

$$F_1 \subset F_2 \subset F_3 \subset \ldots \subset H_s \tag{93}$$

a sequence of finite-dimensional invariant subspaces of L_μ with dense union in H_s. Let

$$P_n : H_s \to F_n \tag{94}$$

be orthogonal projection on F_n. Then with I the identity on H_u,

$$Q_n = I \oplus P_n : H_u \oplus H_s = H \to H_u \oplus F_n \tag{95}$$

is also orthogonal projection. Now the differential equation on $H_u \oplus F_n$

$$\dot{u} = Q_n(L_\mu u - \gamma(||u||^2)\, B(u, u)) \tag{96}$$

defines a truncated finite model. The eigenvalues and eigenspaces of the linear part of (96) form a subset of those of the linear part of (18), with the unstable ones corresponding exactly. We may write the system (96) in parallel to (90) in terms of the H_u and F_n components

$$\begin{aligned} \dot{y} &= A_\mu\, y + J(y, z) \\ \dot{z} &= B_{\mu,n}\, z + K_n(y, z) \end{aligned} \tag{97}$$

with $y \varepsilon H_u$, $z \varepsilon F_n$ and

$$B_{\mu,n} = P_n B_\mu \quad , \qquad K_n(y, z) = P_n K(y, z) \tag{98}$$

As with (76), we find that the unique globally defined center manifold for the system (97) is the graph in $H_u \times (1-\varepsilon, 1+\varepsilon) \times F_n$ of the function f_n: $H_u \times (1-\varepsilon, 1+\varepsilon) \to F_n$ uniquely and globally solving the first-order partial differential equation equivalent to (91)

$$D_y f_n(y, \mu)\cdot(A_\mu y + J(y, f_n(y, \mu))) \qquad (99)$$

$$= B_{\mu,n} f_n(y, \mu) + K_n(y, f_n(y, \mu))$$

with boundary conditions

$$f_n(0, \mu) = 0 \quad , \quad D_y f_n(0, \mu) = 0 \quad . \qquad (100)$$

The final step is to consider the sense in which the functions f_n approximate the function f_∞. First we note that if in the definition of V we use $\mathcal{H}$ in place of H, then the fixed point of the contraction semiflow will still be f_∞ since f_∞ is a fixed point and the fixed point is unique. Second, we note that all of our functions f_n are in the original V, and so the derivatives $D^j f_n$, n = 1, 2, 3, ..., constitute a strongly equicontinuous set for each j = 0, 1, 2, ..., k-1. Since these derivatives have values in weakly compact sets, every subsequence must contain a weakly uniformly convergent subsequence by a suitable variant of the Arzela-Ascoli theorem. If the functions f_n are regarded as mappings into H', then we know that each subsequence must contain a subsequence S which converges strongly together with all its derivatives up to the k-1 derivative to a limit function g_S

$$\lim_{n \in S} D^j f_n = D^j g_S \quad , \quad j = 0, 1, \ldots, k-1 \quad . \qquad (101)$$

We take $k \geq 3$. Since L_μ and B are C^∞ functions from H' to $\mathcal{H}$, it follows from (91), (99), and (101) that g_S satisfies (91). As noted above, this fact implies that $g_S = f_\infty$. Thus every convergent subsequence has the same limit and every sequence has a convergent subsequence. We conclude that

$$\lim_{n\to\infty} D^j f_n = D^j f_\infty \quad , \quad J = 0, 1, \ldots, k-1 \qquad (102)$$

in the topology of strong uniform convergence from H_u to H'.

Having obtained this convergence result, we turn to its consequences. The most obvious is based on the fact that the semiflow on the center manifold for each finite system above may be transferred by means of f_n or f_∞ to a flow on H_u. This flow will be determined by a C^{k-1} vector field X_n or X_∞ on the right of the governing differential equation, and then (102) together with the choice of H' implies that

$$\lim_{n\to\infty} X_n = X_\infty \qquad (103)$$

in the uniform C^{k-2} topology.

Of course, as we have already mentioned, any attractor that bifurcates from the origin as μ crosses 1 will be contained in the flow restricted to M_u. But the flow restricted to M_u is dynamically identical with the flow determined by X_∞ on H_u, and that is a flow on a Euclidean space of dimension $\dim(H_u)$. Thus there is no intrinsic topological obstruction to embedding the attractor and its flow, as an attractor, in a Euclidean flow of dimension $\dim(H_u)$. Such obstructions exist for example to embedding a periodic solution in a one-dimensional flow or the Lorenz attractor in a two-dimensional flow. All the truncated finite models above have dimensions greater than $\dim(H_u)$ and so there is no topological obstruction to their containing the attractor. For this consequence, we need only the center manifold theorem and not its modification above; however, it is worth mentioning because one often stated objection to the use of finite truncated models is that the strange attractors of fluid mechanics are likely to be too complicated to be embedded in such small-dimensional models. A second obvious consequence, this time requiring the above modification of the center manifold theorem, is that any structurally stable attractor near the origin of the center manifold will appear in the flows of all the approximating center manifolds for large n; that is, it will appear in the flow determined by X_n on H_u for all large n, and thus it will appear in all sufficiently large truncated finite models. For example, according to Williams[16], the appearance of *a* Lorenz attractor (as opposed to *the* Lorenz attractor) is structurally stable. Thus, if a Lorenz attractor appears in the fluid flow as a bifurcating attractor, then a Lorenz attractor will appear in all sufficiently large truncated finite models. At the risk of appearing polemical, we lay particular emphasis on this consequence since it replies in part to the objection that, aside from exceptional cases, any truncated finite model must fail to capture the asymptotic behavior of a fluid because the model ignores infinitely many modes which must inevitably play a role in the energetics of any truly chaotic or turbulent motion.

Of course, *the* Lorenz attractor is not a bifurcating attractor, and it appears in a truncated finite model of dimension 3 with $\dim(H_u) = 1$. Here it would be of interest to determine whether the parameters introduced to versally unfold the stationary solutions of the Lorenz system (Shirer and Wells[13,14]) may be used to carry the standard Lorenz attractor over to a bifurcating Lorenz attractor. We have replied only in part to the objection that a finite truncated model must have absurdly large dimension in order to predict successfully the original fluid flow; we only point out that there is no topological obstruction to that dimension being as small as $\dim(H_u)$. On the other hand, we promised only to suggest a reason for hoping that relatively small-dimensional truncated finite models may suffice to capture the bifurcating attractors. The reason is the fact that the topology with respect to which the convergence of (103) takes place is very strong. Heretofore truncated finite models have been used in rigorous arguments in conjunction with some kind of weak convergence in infinite-dimensional spaces, as in the existence proof of Ladyzhenskaya[6] for the limited-time solutions of the Navier-Stokes equation. And since it requires a cofinal subsequence of a weakly convergent sequence to determine the limit within any metric error tolerance, one naturally feels that even a very large number of terms in a weakly convergent sequence will not suffice to determine such delicate matters as the presence of strange attractors. With strong convergence, on the other hand, a single term of a convergent sequence will approximate the limit within any metric error tolerance one may choose. And one may reasonably hope that this abstract replacement of infinitely many terms with a single term may translate into practical applications as the replacement of a term with a large subscript with a term with a small subscript.

PREDICTABILITY RELATIVE TO A TRAJECTORY

Solutions emanating from slightly different initial conditions may pass through different regions of the relevant phase space or may pass through the same regions at different times (as illustrated in Figure 1). The second possibility would produce forecasts in which the sequence of events was predicted correctly, but in which the phase errors associated with incorrect timing might produce large mean square or energy differences at a fixed time.

Thus we seek now to resolve solution differences into an amplitude component transverse to the trajectory and a phase component parallel to the trajectory. We are attempting to develop for predictability theory an analog of the Poincaré map that is used to study the stability of periodic solutions. If our approach is successful, it would allow determination of whether the envelope of trajectories emanating from a neighborhood of initial conditions would display divergence or convergence in transverse cross-section.

This idea allows us to formulate a specific definition: A system is predictable $(\varepsilon, \delta, T_p)$ without regard to phase if for given time t there is an interval $|t - \tau| < T_p$ such that for given $\varepsilon > 0$ there is a vector γ with $||\gamma|| < \delta$ for which

$$||\phi(x_o + \gamma, t) - \phi(x_o, \tau)|| < \varepsilon \tag{104}$$

with $\phi(x_o, t)$ being the solution at time t with initial condition x_o. Clearly this definition gives more latitude than a difference taken at the same time t. In this section, we set forth a procedure for calculating the solution difference used in (104).

The geometry is shown in Figure 4. A solution $x(\tau) = \phi(x_o, \tau)$ to the equation

$$\dot{x} = V(x) \tag{105}$$

generates a trajectory C_o. Another solution $\phi(x_o + \gamma, t)$ emanates from the initial point $x_o + \gamma$ generating a trajectory C. We define

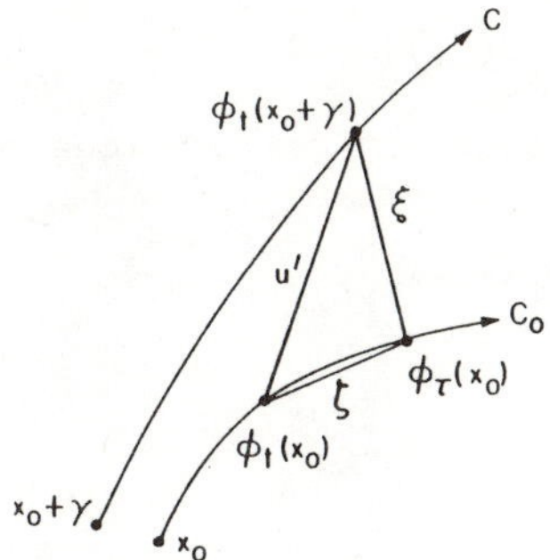

Fig. 4. Geometry and notation for solution difference u', transverse error ξ, and phase error ζ.

$$\xi = \phi(x_o + \gamma, t) - \phi(x_o, \tau) \tag{106}$$

and observe that the difference x' at time t is

$$\begin{aligned} x' &= \xi + \phi(x_o, \tau) - \phi(x_o, t) \\ &= \xi + \int_t^\tau V(\phi(x_o, t'))dt' \end{aligned} \tag{107}$$

and we shall use

$$\zeta = \int_t^\tau V(\phi(x_o, t'))dt' \tag{108}$$

We wish to arrange matters so that the difference ξ is a minimum and thus ζ is essentially phase error.

In order for our results to apply to both ordinary differential equations in R^N or partial differential equations in an appropriate space H we define the scalar product

$$\langle f, g\rangle = f \cdot g = \begin{cases} \sum_n f_n g_n & R_n \\ (f, g) & H \end{cases} \tag{109}$$

along with $||f|| = \langle f, f\rangle$.

To proceed, we must be sure that the relation between τ and t is uniquely specified. Thus given a point y representing $\phi(x_o + \gamma, t)$ near the trajectory C_o we seek $T(y) = \tau$ such that

$$||y - x(\tau)||^2 = \text{Minimum} \tag{110}$$

This condition leads upon differentiation with respect to τ, to

$$[y - x(T(y))]\cdot V(x(T(y))) = 0 \tag{111}$$

and we show that this equation has a unique solution for every y near C_o. To do so, we set

$$F(y, \tau) = [y - x(\tau)]\cdot V(x(\tau)) \tag{112}$$

and observe that

$$F(x(\tau), \tau) = 0 \tag{113}$$

and that

$$\frac{\partial F}{\partial \tau} = - \frac{\partial x(\tau)}{\partial \tau} \cdot V(x(\tau)) + \sum_k [y - x(\tau)]\cdot \frac{\partial V}{\partial x_k} \frac{\partial x_k}{\partial \tau} \tag{114}$$

Hence

$$\frac{\partial F}{\partial \tau} \Big|_{y=x(\tau)} = - |V(x(\tau))|^2 < 0 \tag{115}$$

Thus the implicit function theorem assures us of the existence of T(y) for y near x such that

$$T(x(\tau)) = \tau \qquad F(y, T(y)) = 0 \tag{116}$$

Therefore we define

$$\tau(\gamma, t) = T(\phi_o(x_o + \gamma, t)) \tag{117}$$

and we have shown in (111) that

$$\langle \xi, V(\phi(x_o, \tau)) \rangle = 0 \tag{118}$$

As in (105) we use $(\dot{\ }) = \partial(\ )/\partial t$ and we write $V_o^\tau = V(\phi(x_o, \tau))$ and $V_\gamma^t = V(\phi(x_o + \gamma, t))$. Then from (106) we have

$$\dot{\xi} = V_\gamma^t - V_o^\tau \dot{\tau} \tag{119}$$

and this leads to

$$\langle \xi, \dot{\xi} \rangle = \langle \xi, V_\gamma^t \rangle = \langle \xi, V_\gamma^t - V_o^\tau \rangle \tag{120}$$

in view of (118).

We use the transformation $\tau - t = \eta$ and the fact obtained from (118) that

$$\langle \dot{\xi}, V_o^\tau \rangle = - \langle \xi, \dot{V}_o^\tau \rangle \tag{121}$$

to convert (119) into the statement

$$- \langle \xi , \dot{V}_o^{\tau} \rangle = \langle V_o^{\tau}, V_{\gamma}^{t} - V_o^{\tau} \rangle - \langle V_o^{\tau}, V_o^{\tau} \rangle \dot{\eta} \tag{122}$$

in which we can write

$$\dot{V}_o^{\tau} = \sum_j \frac{\partial V}{\partial x_j} \frac{\partial \phi_j}{\partial \tau} \frac{d\tau}{dt} = D\, V_o^{\tau} \cdot V_o^{\tau} (1 + \dot{\eta}) \tag{123}$$

to obtain

$$\dot{\eta} = \frac{\langle V_o^{\tau}, V_{\gamma}^{t} - V_o^{\tau} \rangle + \langle \xi , D\, V_o^{\tau} \cdot V^{\tau} \rangle}{||V_o^{\tau}||^2 - \langle \xi , D\, V_o^{\tau} \cdot V_o^{\tau} \rangle} \tag{124}$$

$$= \frac{\langle V(u + \zeta), V(u+\zeta+\xi) - V(u+\zeta) \rangle + \langle \xi , D\, V(u+\zeta) \cdot V(u+\zeta) \rangle}{||V(u+\zeta)||^2 - \langle \xi , D\, V(u+\zeta) \cdot V(u+\zeta) \rangle}$$

where we now introduce u as the solution of the hydrodynamic problem (18). Upon differentiating (108) we obtain the equation governing the phase error in the form

$$\dot{\zeta} = V(u + \zeta) - V(u) + V(u + \zeta)\dot{\eta} \tag{125}$$

The companion version of (119) is

$$\dot{\xi} = V(u + \zeta + \xi) - V(u + \zeta) - V(u + \zeta)\dot{\eta} \tag{126}$$

To display explicit hydrodynamical equations we write (3.1) as

$$\dot{u} = L_{\mu} u - B(u, u) = V(u) \tag{127}$$

and then obtain an equation for $u' = \phi(u_o + \gamma, t) - \phi(u_o, t)$ in the form

$$\dot{u}' = V(u + u') - V(u) = L_{\mu} u' - B(u, u') - B(u', u) - B(u', u')$$

$$= M_{\mu}\, u' - B(u', u') \tag{128}$$

We write (124) as $\dot{\eta} = F(u, \xi, \zeta)$ and then (126) produces from similar operations

$$\dot{\xi} = M_{\mu}\, \xi - B(\zeta, \xi) - B(\xi, \zeta) - B(\xi, \xi) \tag{129}$$

$$- V(u + \zeta)\, F(u, \xi, \zeta)$$

and finally (125) gives

$$\dot{\zeta} = M_\mu \zeta - B(\zeta, \zeta) + V(u + \zeta) F(u, \xi, \zeta) \tag{130}$$

where (129) and (130) sum to (128) as expected.

The norms of these three vector functions (all solenoidal) are governed by

$$\frac{1}{2}\frac{\partial}{\partial t} ||u'||^2 = \langle u', M_\mu u' \rangle \tag{131}$$

$$\frac{1}{2}\frac{\partial}{\partial t} ||\xi||^2 = \langle \xi, M_\mu \xi \rangle - \langle \xi, B(\xi, \zeta) \rangle \tag{132}$$

and

$$\frac{1}{2}\frac{\partial}{\partial t} ||\zeta||^2 = \langle \zeta, M_\mu \zeta \rangle + \langle \zeta, V(u + \zeta) F(u, \xi, \zeta) \rangle \tag{133}$$

The last two sum to the first equation only with the addition of the required terms $\langle \dot{\xi}, \zeta \rangle$ and $\langle \xi, \dot{\zeta} \rangle$.

The issue of relative importance of amplitude and phase errors probably will be decided by consideration of the effect of the operator M_μ. Our general experience with fluids is that initial neighborhoods are strung out in intricate, connected filaments (e.g., Welander[15], whose illustrations were reproduced in ref. 1). This suggests that the largest eigenvalues of M_μ will often be associated with directions along the trajectory especially where the curvature of the trajectory is small. This is certainly true for flows leaving the neighborhood of an unstable fixed point as is intuitively obvious and is confirmed by the center manifold theorem. To the extent that this situation prevails, the action of M_μ on ξ will be small compared to its action on u' and on ζ.

Practical requirements for improving predictability can be aided by experiments based on the concepts behind (129) and (130), if not the equations themselves. It would be valuable to identify the forms of initial error that lead to amplitude and phase errors with the idea, perhaps, of arranging to devote available resources to reduction of amplitude errors.

The most illuminating information we could obtain from the decomposition (129) and (130) would be a comparison of the Lyapunov exponents for ξ and ζ with those of u'. The attractors for ξ and ζ are likely to be of different dimensions; if the dimension of the ξ attractor were small compared to that of ζ or u' it would have important implications for predictability theory and for forecast improvement.

Moreover, the exponents and the effects on ξ and ζ could be calculated along with a forecast or *a priori* throughout the relevant regions of phase space. It could then be possible to determine those regions in which either ξ or ζ forecast error would increase and to assign appropriate measures of confidence.

Finally, we have produced a differential system for determining the quantities displayed in our definition (104) of predictability, thus making it potentially useful in both theory and practice.

APPENDIX: BASIC DEFINITIONS AND INEQUALITIES

This article is concerned with analysis of vector functions u defined on a domain $D \subset R^3$ which is bounded and has a sufficiently smooth boundary ∂D. The vectors u vanish on ∂D, and we start with $u \in C^\infty(D)$, which is the class of infinitely differentiable vectors. For these we define the norm

$$||u||_p = (\int_D |u|^p \, dx)^{1/p} \tag{A-1}$$

and then upon completing the set of vectors in $C^\infty(D)$ with respect to this norm we arrive at the Hilbert space L_p. For L_2 we write $|| \; ||_2 = || \; ||$ and we have the Schwarz inequality

$$||u \cdot v|| \leq ||u|| \; ||v|| \tag{A-2}$$

Moreover with this inequality (including the form for sums), we have

$$\sum_{i,k=1,3} \int |u_i \, v_k \, \Psi_{ik}| dx \leq \sum_{i,k} [\int (u_i \, v_k)^2 \, dx \int \Psi_{ik}{}^2 \, dx]^{1/2} \tag{A-3}$$

$$\leq [\sum_{i,k} \int (u_i \, v_k)^2 \, dx]^{1/2} [\sum_{i,k} \int \Psi_{ik}{}^2 \, dx]^{1/2}$$

$$\leq [\int |u|^2 \, |v|^2 \, dx]^{1/2} \, ||\Psi||$$

$$\leq ||u||_4 \, ||v||_4 \, ||\Psi||$$

In a similar way, we arrive at a Hilbert space J by completing the space of C^∞ vectors vanishing on the boundary with respect to the norm

$$||u||_J = ||\nabla u|| = (\int_D |\nabla u|^2 \, dx)^{1/2} = [\int_D \sum_{i,k} (\frac{\partial u_i}{\partial x_k})^2 \, dx]^{1/2} \tag{A-4}$$

To obtain a relation between these two norms we consider the eigenvalue problem on D

$$\nabla^2 \psi = - \gamma \, \psi, \quad \nabla \cdot \psi = 0, \quad \psi = 0 \text{ on } \partial D \tag{A-5}$$

for which the first eigenvalue

$$\gamma_1 = \min_{||\psi|| = 1} \left[\int |\nabla\psi|^2 \, dx\right] \quad \text{(A-6)}$$

Hence for every other function than ψ_1

$$\gamma_1 ||\psi||^2 \leq ||\nabla\psi||^2 \quad \text{(A-7)}$$

where the eigenvalue γ_1 depends only on the geometry of D.

Another embedding inequality central to our concerns is

$$||u||_4{}^2 \leq C \; ||u||^{1/2} \; ||\nabla u||^{3/2} \quad \text{(A-8)}$$

in which the constant $C = 3^{-3/4}$ for $D \subset R^3$ (e.g. [11]). With (A-7) we have

$$||u||_4{}^2 \leq (C/\gamma_1 1/2) \; ||\nabla u||^2 \quad \text{(A-9)}$$

ACKNOWLEDGMENTS

We are indebted to Profs. Edward N. Lorenz and John E. Hart for their comments and suggestions following the verbal presentation of this article.

The research reported here was sponsored by the National Science Foundation with Grant ATM-7908354 to The Pennsylvania State University.

REFERENCES

1. Dutton, J.A., The Ceaseless Wind: An Introduction to the Theory of Atmospheric Motion (McGraw Hill, New York, 1976)
2. Dutton, J.A., SIAM Review 24, 1 (1982).
3. Farmer, J.D., Physica 4D 3, 366 (1982).
4. Farmer, J.D., E. Ott, and J.A. Yorke, to appear in a special issue Order in Chaos of Physica D.
5. Kaniel, S. and M. Shinbrot, Arch. Rational Mech. and Anal. 24, 302 (1967).
6. Ladyzhenskaya, O.A., The Mathematical Theory of Viscous Incompressible Flow (Gordon and Breach, London, 1963).
7. Lorenz, E.N., J. Atmos. Sci. 20, 130 (1963).
8. Lorenz, E.N., J. Atmos. Sci. 37, 1685 (1980).
9. Lumley, J.L., Stochastic Tools in Turbulence (Academic Press, New York, 1970).
10. Marsden, J.E. and M. McCracken, The Hopf Bifurcation and Its Applications (Springer Verlag, New York, 1976).
11. Serrin, J., in Nonlinear Problems, R. Langer, Ed. (Univ. of Wisconsin Press, Madison, 1962).
12. Shinbrot, M. and S. Kaniel, Arch. Rational Mech. and Anal. 21, 260 (1966).
13. Shirer, H.N., and R. Wells, J. Atmos. Sci. 39, 610 (1982).

14. Shirer, H.N. and R. Wells, Mathematical Structure of the Singularities at the Transitions Between Steady States in Hydrodynamic Systems, to appear in Lecture Notes in Physics (Springer Verlag, Berlin, 1983).
15. Welander, P., Tellus 7, 141 (1955).
16. Williams, R.F., Instit. Hautes Etudes Scientifiques, 50, 321 (1980).

COMMENTS ON STATISTICAL MEASURES OF PREDICTABILITY

Bruce J. West
Center for Studies of Nonlinear Dynamics*
La Jolla Institute, P.O. Box 1434, La Jolla, CA 92038

Katja Lindenberg
Chemistry Department, University of California
San Diego, CA 92093

ABSTRACT

Measures of predictability are determined by various moments of the conditional probability distribution for the physical observables in a flow field. A general technique is introduced whereby an *exact* stochastic differential equation for the dynamics of a flow field is constructed from the primitive equations. This is a generalized Langevin equation which in the Markov limit can be replaced by a Fokker-Planck equation. This latter equation is solved for the non-equilibrium probability density for a simple model system thereby allowing one to calculate the predictability measures exactly.

INTRODUCTION

In this paper we consider a number of general relations from nonequilibrium statistical mechanics that might be useful in considering the predictability of fluid motion. The philosophy adopted herein is based on the notion that the coarse graining of the description of a flow field results in equations of motion that are stochastic and dissipative. The fluctuations and the dissipation are manifestations of the degrees of freedom eliminated from the descriptions of the fluid flow by the coarse graining procedure.[1-4] Thus, for instance, if the flow is fully characterized by a set of N dynamical variables $\{A_\alpha(t)\}$, $\alpha = 1, 2, ..., N$, e.g., the Fourier mode amplitudes of the velocity (vorticity) field, then a possible coarse graining would be a reduction to a set of M modes ($M << N$) to describe the flow.

The evolution of the flow field can be described by a deterministic trajectory $\Gamma_t(\hat{\mathbf{a}})$ in an N-dimensional phase space with the set of axes $\hat{\mathbf{a}} = \{\hat{a}_\alpha\}$ corresponding to the full set $\{A_\alpha(t)\}$ of dynamical variables. The curve $\Gamma_t(\hat{\mathbf{a}})$ is indexed by the time t, which is a parameter in phase space. The trajectory begins at a specified point $\hat{\mathbf{a}}_0 = \{A_\alpha(t{=}0)\}$ and describes the evolution of the flow towards its final state $\{A_\alpha(t)\}$. In practical calculations only a small number M of variables ($M << N$) is used to represent the flow field. Mathematically this means that one is interested only in the projected trajectory

$$\Gamma_t(\mathbf{a}) = P\Gamma_t(\hat{\mathbf{a}}) \tag{1}$$

where $\mathbf{a} = \{a_\alpha\}$ with $\alpha = 1, 2, ..., M$ and where P denotes an appropriate integration over the variables $a_{M+1}, a_{M+2}, ..., a_N$. If we consider two trajectories with identical initial values $A_1(0), A_2(0), ..., A_M(0)$ but with different initial values $A_{M+1}(0), ..., A_N(0)$, the two projected trajectories may be different. Thus two trajectories ostensibly initiated from the same state $A_1(0), ..., A_M(0)$ in the reduced phase space can follow different phase space orbits in the reduced space. We refer to the instantaneous differences between two such trajectories as fluctuations and presume that they admit of a statistical description. The statistics enter through the specification of the initial conditions of the eliminated degrees of freedom: since these are not observable one can specify a distribution of initial conditions $A_{M+1}(0), \ldots, A_N(0)$ consistent with the "macroscopic" initial state of the flow.

---

*affiliated with the University of California, San Diego.

In order to obtain a faithful representation of the flow field dynamics one must thus consider an ensemble of realizations of orbits, each with a statistically equivalent set of fluctuations. The single path $\Gamma_t(\hat{\mathbf{a}})$ in the full phase space is therefore replaced by an ensemble of ray paths in the contracted space and this ensemble is characterized by a conditional distribution function $P(\mathbf{a}, t|\mathbf{a}_0)$. Here $P(\mathbf{a}, t|\mathbf{a}_0)d^M a$ is the probability that the set of dynamical variables $\{A_\alpha(t)\}$, $\alpha = 1, ..., M$ has values in the phase space interval $(\mathbf{a}, \mathbf{a} + d\mathbf{a})$ at time t given the initial values $\mathbf{a}_0$.

The distribution function can now be used to express the predictability of the final state of the fluid motion as described by $A_1(t), ..., A_M(t)$. A commonly employed measure of the "growth of error" is the conditional variance [5]

$$\sigma_{\alpha\beta}(t|\mathbf{a}_0) = \int [a_\alpha - <a_\alpha; t>_0]\, [a_\beta - <a_\beta; t>_0]\, P(\mathbf{a}, t|\mathbf{a}_0) d^M a\,. \tag{2}$$

where the average ray path is

$$<a_\alpha; t>_0 \equiv \int a_\alpha P(\mathbf{a}, t|\mathbf{a}_0) d^M a\,. \tag{3}$$

This measure has been used by a number of contributors to this workshop and it is not unlike the measure discused by Monin,[6] wherein he used an average over the initial distribution of data to obtain his second order quantity. A second measure, also discussed in this workshop, is Boltzmann's H-function:

$$H(t) \equiv \int P(\mathbf{a}, t|\mathbf{a}_0) \ln [P(\mathbf{a}, t|\mathbf{a}_0)/P_{ss}(\mathbf{a})]\, d^M a \tag{4}$$

interpreted in an information theoretical sense.[7] Here $P_{ss}(\mathbf{a}) \equiv \lim_{t\to\infty} P(\mathbf{a}, t|\mathbf{a}_0)$ is the steady state distribution. Note that $\lim_{t\to\infty} H(t) = 0$. Both these measures have properties to recommend them, but we do not dwell on these here. Instead we concentrate on the determination of the probability density itself.

In order to obtain the evolution of the probability in time it is necessary to specify a phase space equation. The strategy we adopt herein is the following. We begin with a set of primitive equations for the entire flow field. We then systematically eliminate all but a single degree of freedom from our description (the generalization to more than one degree of freedom is straightforward and will be done elsewhere). The dynamic equation for the macroscopic observable $A_1(t)$ is then stochastic, the fluctuations entering via the uncertainty in the initial values of the eliminated (microscopic) degrees of freedom. Under suitable statistical assumptions, the resulting stochastic differential equation can then be replaced by an equivalent phase space equation for the probability density.

FLUCTUATIONS IN FLOW FIELDS

It was pointed out in the Introduction that uncertainties in the initial state of a flow result in fluctuations in the trajectories initiated from macroscopically identical states. These uncertainties are a consequence of not including small scale structure in the flow dynamics. As we will explicitly show later, the small scale features are manifest in the macroscopic dynamics through dissipation (e.g., molecular viscosity or turbulent dissipation) in the equations of motion for the large scale flows. From the point of view of statistical mechanics, the existence of dissipation in a thermodynamically closed dynamical system implies the existence of fluctuations.[8,9] Stated more strongly, it is not physically possible to have dissipation without fluctuations and vice versa. The intimate connection between these two concepts is responsible for the invariably observed fact that the flow field achieves an aymptotic steady state. Therefore the energy from the flow lost through dissipation must in the steady state be balanced by the energy supplied to the flow by means of the fluctuations. This balance is a manifestation of a fluctuation-dissipation relation.[10]

If the physical observables at the boundaries of the system are maintained at constant values, then the fluctuation-dissipation relation allows the stabilization of the system in a non-equilibrium steady state, e.g., the formation of Rayleigh-Bernard cells in thermal gradients.[11] It is also possible for the fluid motion to achieve a meta-stable state, which although long lived, will eventually decay into an absolutely steady state, e.g., atmospheric blocking.[12] The fluctuations driving the flow field can induce a transition between such different phases of the flow field.[13-15]

Consider an abstract model system described by the equations of motion:

$$\dot{A}_\alpha(t) = - Q_\alpha(\mathbf{A}, t), \quad \alpha = - N, -N+1, ..., N-1, N \tag{5}$$

where, for example, the set $\mathbf{A} = \{A_\alpha(t)\}$ could be the Fourier amplitudes of the hydrodynamic variables representing the fluid motion and $A_{-\alpha}(t) \equiv A_\alpha^*(t)$. (We exclude $\alpha=0$ from this set.) The function Q_α can in general be decomposed as follows:

$$Q_\alpha(\mathbf{A}, t) = \lambda_\alpha A_\alpha(t) + \sum_{\beta,\gamma} d_{\alpha\beta\gamma} A_\beta(t) A_\gamma(t) \tag{6}$$

where λ_α determines the linear motion of the system and the quadratic interaction implies a hydrodynamic model system. The coupling coefficients $d_{\alpha\beta\gamma}$ determine the exact nonlinear nature of the system. Lorenz[16] established the richness of such a model description by demonstrating that in appropriate parameter regimes as few as three modes can lead to chaotic behavior, i.e., the phase space trajectory of the system is a strange attractor.

It is not our intent here to restrict the discussion to a particular low order model and to study its dynamic properties. Rather we arrange the modes in (5) in pairs of decreasing scale and eliminate from our dynamic description all but the largest scale (A_1, A_{-1}). The remaining scales of motion will ultimately appear in the dynamics as a dissipative force and also as a stochastic driving force. The procedure for doing this seems not to have been previously set down in this general hydrodynamical context.

Consider the equations of motion (5) for the gravest mode, i.e., for $\alpha = \pm 1$:

$$\dot{A}_1(t) + \lambda_1 A_1(t) = d_{111} A_1^2(t) + (d_{1,1,-1} + d_{1,-1,1}) A_1^*(t) A_1(t) \tag{7}$$

$$+ d_{1,-1,-1} A_1^{*2}(t) + g_{11}(t) A_1(t) + g_{1,-1}(t) A_1^*(t) + f_1(t)$$

and its complex conjugate. Here

$$g_{\alpha\beta}(t) = \sum_{\gamma \neq \pm 1} (d_{\alpha\beta\gamma} + d_{\alpha\gamma\beta}) A_\gamma(t) \tag{8}$$

and

$$f_\alpha(t) = \sum_{\beta \neq \pm 1} \sum_{\gamma \neq \pm 1} d_{\alpha\beta\gamma} A_\beta(t) A_\gamma(t) \, . \tag{9}$$

For this subsystem we consider the effects of molecular viscosity to be negligible. The mode $|\alpha| = 1$ can lose energy only through its coupling with the other degrees of freedom. Hence λ_1 is a purely imaginary quantity $i\omega_1$, and $\lambda_{-1} = -\lambda_1 = -i\omega_1$. Various contributions to the coupling coefficients $d_{\alpha\beta\gamma}$ have specific symmetries. For example, the Jacobian form of the nonlinear terms obtained from the potential vorticity equations indicates that $d_{\alpha\beta\gamma} = d_{\alpha\gamma\beta} =$ etc. and that $d_{\alpha\beta\gamma}$ vanishes whenever two of the indices α, β and γ have the same absolute value. This is the case we will consider here, and (7) then reduces to

$$\dot{A}_1(t) + i\omega_1 A_1(t) = f_1(t) \, . \tag{10}$$

This suggests that terms *of the forms* $g_{11}(t)A_1(t)$ and $d_{111}A_1^2(t)$ do not contribute to the hydrodynamic-like equations. However, this is not quite correct as we now show.

The equation of motion for the αth mode ($\alpha \neq \pm 1$) with coupling coefficients having repeated indices set equal to zero is

$$\dot{A}_\alpha(t) + \lambda_\alpha A_\alpha(t) = g_{\alpha 1}(t)A_1(t) + g_{\alpha,-1}(t)A_1^*(t) + f_\alpha(t) . \tag{11}$$

In general, the parameter λ_α is complex, the real part arising from molecular viscosity and the imaginary part as an effect of boundary conditions. For the purposes of the present discussion we truncate the model on a spatial scale for which molecular viscosity (and the corresponding molecular fluctuations) can be neglected. This implies that the λ_α are imaginary quantities, i.e., $\lambda_\alpha = i\omega_\alpha$, and $\lambda_{-\alpha} = -i\omega_\alpha$. These approximations do not affect the evolution of A_1 on observational time scales. The effects of molecular viscosity and fluctuations, i.e., the very fine scale structure of the system, can in principle be included in the present discussion but would obscure the main points of our presentation and would not materially affect the results of our discussion; they are therefore omitted.

The set of equations represented by (11) with $|\alpha| = 2, 3, ..., N$ contains two types of nonlinear interactions. The first type is the interaction with the $|\alpha| = 1$ mode, which we shall retain fully. The second, embodied in $f_\alpha(t)$, contains the interactions among the microscopic degrees of freedom. We shall neglect these interactions in the interests of analytic tractability, noting that the time scales of these interactions are in any case usually much shorter than $2\pi/\omega_1$. We realize that this approximation may be violated by resonant interactions among the microscopic degrees of freedom, but note that these are often artifacts of a model rather than features of a real system. With these caveats the $f_\alpha(t)$ can be set to zero and the resulting set of linear equations can be integrated explicitly. For this purpose, we define the $2(N-1) \times 1$ matrix $\mathbf{A}(t)$ by

$$\mathbf{A}^T(t) = (A_{-N}(t), A_{-N+1}(t), ..., A_{-2}(t), A_2(t), ..., A_N(t)) \tag{12}$$

where T denotes the transpose. We also define the $2(N-1) \times 2(N-1)$ matrices $\epsilon \mathbf{D}^{(\gamma)} = \{2d_{\alpha\gamma\beta}\}$ and

$$\Omega = \begin{pmatrix} \omega_{-N} & \cdots & & \vdots \\ & \ddots & & \\ & & \omega_{-2} & 0 \\ \vdots & & & \\ & 0 & \omega_2 & \\ & & & \ddots \\ & \cdots & & \omega_N \end{pmatrix} \tag{13}$$

where ϵ is an ordering parameter that shall be set equal to unity subsequently. The set (11) can then be replaced by

$$\dot{\mathbf{A}}(t) + i\,\Omega\,\mathbf{A}(t) = \epsilon\,\mathcal{D}^{(1)}(t)\mathbf{A}(t) \tag{14}$$

where

$$\mathcal{D}^{(\gamma)}(t) \equiv A_1(t)\mathbf{D}^{(\gamma)} + A_1^*(t)\mathbf{D}^{(-\gamma)} . \tag{15}$$

The solution of (14) is

$$\mathbf{A}(t) = e^{-i\Omega t}\left\{\exp\left[-\epsilon\int_0^t \hat{\mathcal{D}}^{(1)}(\tau)d\tau\right]\right\}_0 \mathbf{A}(0) \tag{16}$$

where

$$\hat{D}^{(1)}(t) \equiv e^{i\Omega t} \mathbf{D}^{(1)}(t)\, e^{-i\Omega t} \tag{17}$$

and where the subscript 0 denotes time ordering.

The components of the solution (16) can now be used to express the driving force $f_1(t)$ in (10) in terms of the unknowns $A_{\pm 1}(t)$. The other degrees of freedom enter *only* through their initial values. By assumption, $A_{\pm 1}(t)$ are the only physical observables. Therefore the initial values of the other degrees of freedom can only be specified by means of a distribution (cf. below). Thus (10) must be interpreted as a stochastic differential equation. The physical interpretation of (10) at this point is, however, difficult in this degree of generality because the stochastic term $f_1(t)$ includes the unknowns $A_{\pm 1}(t)$ in a highly nonlinear fashion. To proceed with the analysis we implement our ordering parameter ϵ and expand (16). Substitution of the perturbation solution into (10) yields

$$\dot{A}_1 + i\omega_1 A_1(t) = \epsilon C_1(t) - \epsilon^2 \int_0^t d\tau C_2(t,\tau)[A_1(\tau) - A_1^*(\tau)] + 0(\epsilon^3) \tag{18}$$

where using the symmetry of the coupling coefficients we obtain

$$C_1(t) = \sum_{\beta,\gamma} d_{1\beta\gamma}\, e^{-i(\omega_\beta + \omega_\gamma)t} A_\beta(0) A_\gamma(0) \tag{19}$$

$$C_2(t,\tau) = 4 \sum_{\beta,\gamma,\mu} d_{1\beta\gamma} d_{\gamma 1\mu}\, e^{-i\omega_\beta t} e^{-i\omega_\gamma (t-\tau)} e^{-i\omega_\mu \tau} A_\beta(0) A_\mu(0)\ . \tag{20}$$

If we neglect terms of $0(\epsilon^3)$ and higher, the resulting system is linear in $A_{\pm 1}$ and admits of an interpretation consistent with classical statistical mechanics, as we show in the next section.

To close this section we make the following observations. We have replaced the set of deterministic equations (5) with an equation for the macroscopic observable $A_1(t)$, equation (18), in which the unobservable degrees of freedom appear as initial values in the functions C_1 and C_2. Since these degrees of freedom are "unobservable", there is an attendant uncertainty in their specification. Thus one must characterize these initial values statistically and therefore the functions C_1 and C_2 are stochastic. This of course implies that (18) is a stochastic differential equation. We note that such equations have been *postulated* in flow field studies in the past,[12,17] but this is to our knowledge the first derivation of such an equation in which the fluctuations arise from the uncertainty in the initial conditions of the eliminated modes.

The fluctuations that appear in (18) are of two distinct kinds. There are the *additive* fluctuations contained in $C_1(t)$ which are independent of the state of the flow field. Such fluctuations are familiar from the traditional Langevin equation description of Brownian motion.[1,13,18] Then there are *multiplicative* or *parametric* fluctuations contained in $C_2(t)$ (as well as in higher order terms). These depend on the state $A_{\pm 1}(\tau)$ of the flow field. It is well established that such parametric fluctuations can generate instabilities.[13,19-21]

Associated with each fluctuating term is a dissipative term in such a way that a fluctuation-dissipation relation is satisfied (cf. the next section). It has been established that this is necessary to ensure that a thermodynamically closed system achieves a steady state asymptotically.[10]

Note that the above arguments can be carried through for a subspace of the flow field greater than two, that is to say when the system of interest has a phase space dimension greater than two. In this way one can construct the appropriate stochastic differential equations for a representation of the flow field involving as many degrees-of-freedom as desired.

FLUCTUATION-DISSIPATION RELATION

To specify the physical meaning and statistical properties of the initial-condition-dependent quantities $C_1(t)$ and $C_2(t, \tau)$ we must begin by choosing the distribution of initial values of the unobserved modes. We pick a distribution suggested by the fluid equilibrium studies of Onsager[22] and Kraichnan[23] for interacting point vortices:

$$P_{eq}(\mathbf{a}) = Z^{-1} \exp[- \sum_{\alpha \neq \pm 1} \Lambda_\alpha a_\alpha a_{-\alpha}] \tag{21}$$

where Λ_α is a state dependent "thermodynamic potential", Z is the partition function and the a_α are the initial values $A_\alpha(0)$.

Let us begin by analyzing the statistical properties of $C_1(t)$. Its average value $<C_1(t)>$ vanishes because (21) implies that $<a_\beta a_\gamma> = <|a_\beta|^2> \delta_{\beta, -\gamma}$ and $d_{1\beta, \pm\beta}$ vanishes. The two-time correlation function is

$$<C_1(t)C_1(\tau)> = \sum_\gamma \phi_\gamma(t-\tau) \tag{22}$$

where

$$\phi_\gamma(t-\tau) \equiv 2 \sum_\beta \frac{d_{1\beta\gamma}^2}{\Lambda_\beta \Lambda_\gamma} e^{-i(\omega_\beta + \omega_\gamma)(t-\tau)} . \tag{23}$$

is the correlation of mode γ with the other unobservable modes. Furthermore, since the number of degrees of freedom $N >> 1$, $C_1(t)$ is a sum of a large number of identically distributed mutually independent random variables. Then, since the second moment is finite (cf. (22)), the central limit theorem tells us that $C_1(t)$ is a Gaussian random variable and hence is completely specified by its first two moments. We have thus succeeded in *deriving* the statistics of the fluctuations $C_1(t)$ in (18).

Next let us consider the behavior of $C_2(t, \tau)$. Its average value does not vanish but rather is given by

$$<C_2(t-\tau)> = 2 \sum_\gamma \Lambda_\gamma \phi_\gamma(t-\tau) . \tag{24}$$

By comparing (22) and (24) we observe that the average coupling of the mode A_1 to the eliminated degrees of freedom can be expressed in terms of the correlations among the fluctuations in $C_1(t)$. Equations (22) and (24) constitute a *generalized fluctuation-dissipation relation*, which is most often seen in the context of a generalized Langevin equation.[24] To exhibit this most clearly we rewrite (18) in terms of the average of C_2 and its fluctuations:

$$\dot{A}_1(t) + i\omega_1 A_1(t) = \epsilon C_1(t) - 2\epsilon^2 \sum_\gamma \Lambda_\gamma \int_0^t d\tau \phi_\gamma(t-\tau) [A_1(\tau) - A_1^*(\tau)]$$

$$- \epsilon^2 \int_0^t d\tau [C_2(t, \tau) - <C_2(t, \tau)>] [A_1(\tau) - A_1^*(\tau)] + 0(\epsilon^3) . \tag{25}$$

In (25) we have explicitly separated out the linear dissipative term that balances the additive fluctuations $C_1(t)$. The fluctuating term containing $C_2 - <C_2>$ is multiplicative and is balanced by nonlinear dissipative terms of $0(\epsilon^3)$ and $0(\epsilon^4)$ that we have not retained in this analysis. It would not be consistent to retain these fluctuations without retaining the corresponding dissipation. Thus the lowest order self-consistent model of the fluctuating fluid flow is described by the equation

$$\dot{A}_1(t) + i\omega_1 A_1(t) = \epsilon C_1(t) - 2\epsilon^2 \sum_\gamma \Lambda_\gamma \int_0^t d\tau \phi_\gamma(t-\tau)\,[A_1(\tau) - A_1^*(\tau)]\,. \tag{26}$$

A number of observations are again in order:

1. Equation (26) has the form of a generalized Langevin equation, i.e., $C_1(t)$ is a Gaussian random force with correlations among its spectral modes given by $\phi_\gamma(t-\tau)$.

2. The system dynamics cannot be determined by a straightforward ordering of the terms in the parameter ϵ. Since averages of *products* of fluctuations must in general be balanced by *single* averages, one must clearly retain dissipative terms up to twice the maximum order (in ϵ) of the fluctuating terms retained. In (26) we have done so with $0(\epsilon)$ in the fluctuation and $0(\epsilon^2)$ in the dissipation.

3. It is especially important to exclude terms containing multiplicative fluctuations if the corresponding (nonlinear) dissipation is excluded, since it is well known that uncompensated parametric fluctuations can lead to instabilities.[3,13,18-21]

In the Introduction we pointed out that the properties of the flow field can be specified through the distribution $P(\mathbf{a}, t|\mathbf{a}_0)$ of values $\mathbf{a} \equiv (a, a^*)$ that the stochastic variable $(A_1(t), A_1^*(t))$ can assume. The next order of business is then to obtain this distribution from the stochastic differential equation (26). If one solved this equation and obtained an explicit expression for $A_1(t)$ (which is possible since (26) is linear) then the method of characteristic functions would yield the desired distribution. For our purposes here we have decided to take the Markov limit of (26), i.e., to set

$$\phi_\gamma(t-\tau) = D_\gamma \delta(t-\tau) \tag{27}$$

and to construct the corresponding Fokker-Planck equation for the distribution function (cf. the following section). The conditions under which (27) is a valid approximation to (23) have been studied by Ford, Kac and Mazur.[25]

PHASE SPACE EVOLUTION AND PREDICTABILITY

With the short correlation time assumption (27), the phase space equation for the macroscopic flow is a Fokker-Planck equation that can be written using standard techniques.[13,18] Denoting the two time point probability density $P(\mathbf{a}, t|\mathbf{a}_0)$ by P_t we have

$$\frac{\partial}{\partial t} P_t = -\frac{\partial}{\partial a}[m(\mathbf{a})P_t] - \frac{\partial}{\partial a^*}[m^*(\mathbf{a})P_t] + \frac{\nu_2}{2}\left(\frac{\partial}{\partial a} + \frac{\partial}{\partial a^*}\right)^2 P_t \tag{28}$$

where $\mathbf{a} \equiv (a, a^*)$ are the phase space axes corresponding to the values of the dynamical variables $(A_1(t), A_1^*(t))$ and where

$$m(\mathbf{a}) = -i\omega_1 a - \nu_1(a - a^*) \tag{29}$$

The constants ν_1 and ν_2 are given by

$$\nu_1 = \sum_\gamma \Lambda_\gamma D_\gamma\,, \quad \nu_2 = \sum_\gamma D_\gamma\,. \tag{30}$$

Since the dynamical equation (18) is linear and consequently $m(\mathbf{a})$ is linear, the solution of (28) is known to be a mulitvariate Gaussian:

$$P(\mathbf{a}, t|\mathbf{a}_0) = \frac{1}{2\pi[\det \mathbf{M}(t)]^{1/2}} \exp\left\{- \tfrac{1}{2}\hat{\mathbf{a}}^T(t)\, \mathbf{M}^{-1}(t)\, \hat{\mathbf{a}}(t)\right\} \tag{31}$$

where

$$\hat{\mathbf{a}}(t) \equiv \mathbf{a} - <\mathbf{a}; t> \tag{32}$$

and $<a;t> \equiv <A_1(t)>$ is the time dependent ensemble average of the macroscopic flow. The correlation matrix $\mathbf{M}$ is given in terms of the second moments $<\hat{a}^2;t> \equiv <A_1^2(t)> - <A_1(t)>^2$, etc. by

$$\mathbf{M}(t) = \begin{bmatrix} <\hat{a}^2;t> & <\hat{a}\hat{a}^*;t> \\ <\hat{a}^*\hat{a};t> & <\hat{a}^{*2};t> \end{bmatrix} . \tag{33}$$

The mean values and second moments appearing in (31)–(33) are given in terms of $\omega_2 \equiv (\omega_1^2 - \nu_1^2)^{1/2}$ by

$$<a;t> = e^{-\nu_1 t}\left[(\cos\omega_2 t - i\frac{\omega_1}{\omega_2}\sin\omega_2 t)\, a_0 + \frac{\nu_1}{\omega_2}\sin\omega_2 t\; a_0^*\right] \tag{34}$$

$$<\hat{a}^2> = \frac{\nu_2}{2\omega_1^2}\left\{e^{-2\nu_1 t}[(-\nu_1+i\omega_1)\cos 2\omega_2 t + (\omega_2 + i\frac{\nu_1\omega_1}{\omega_2})\sin 2\omega_2 t] + \nu_1 - i\omega_1\right\} \tag{35}$$

and

$$<\hat{a}\hat{a}^*;t> = -\frac{\nu_1\nu_2}{2\omega_1^2\omega_2}\left\{e^{-2\nu_1 t}[\nu_1\sin 2\omega_2 t + \omega_2\cos 2\omega_2 t] - \omega_2\right\} . \tag{36}$$

We will henceforth assume that $\omega_1 > \nu_1$ so that ω_2 is real. A similar discussion can be constructed for the case $\nu_1 > \omega_1$.

The distribution (31) can now be used to calculate the various measures of predictability mentioned in the introduction. For example, the conditional variances given by Equation (2) are precisely the second moments (35) and (36). The characteristic time scale for the loss of predictability is determined by the dissipation parameter ν_1 (for real ω_2). A more global measure of predictability is the Boltzmann H-function as defined by (4). This measure can be constructed from (31) and its steady state limit, and also decays on a time scale $(2\nu_1)^{-1}$. A more detailed analysis of these measures is not warranted at this time and will be done subsequently.

DISCUSSION AND CONCLUSIONS

In this paper we have presented the first systematic derivation of a system of macroscopic stochastic equations starting from a set of primitive equations. In the construction of these stochastic equations we have observed that the fluctuations can enter in a number of possible forms. The most familiar are state independent fluctuations, often referred to as "additive noise."[12,13] The first general study of such equations in a physical context was done by Langevin.[26] He showed that in a thermodynamically closed system the mean interaction of the system with its environment results in a net extraction of energy from the system in the form of a linear dissipation. Phenomenological models of atmospheric phenomena[12] have been *postulated* to include both additive fluctuations and a linear dissipation. We *derive* such a representation for the model system developed in the third section. The relation between the linear dissipation parameter and the mean square level of the fluctuations is known as the fluctuation—dissipation relation.[10] This relation ensures the asymptotic stability of the flow field.

The simplest form of state-dependent fluctuations contains the fluctuations as coefficients of the state variables. We demonstrate that such "multiplicative fluctuations" arise naturally in the projection of the primitive equations onto the observables of the flow field. Although not proved herein, the dissipation associated with these fluctuations in a thermodynamically closed system is generally nonlinear in the state variables[3,4] In a thermodynamically open system there is no necessary relation between the fluctuations and the dissipation, and instabilities can then result from multiplicative fluctuations.[13,18-21]

From the stochastic equations we can then proceed to write a phase space evolution equation, which in this paper we have done only in the Markov limit. The solution of this equation enables one to calculate various measures of the predictability of the flow field.

ACKNOWLEDGEMENTS

It is a pleasure to acknowledge the support of the National Science Foundation for this work (Grants No. ATM-8310672 and ATM-8310673) and the encouragement of P. Stevens of the Division of Atmospheric Sciences.

REFERENCES

1. M. Lax, Rev. Mod. Phys. **38**, 541 (1966).
2. R. Zwanzig, J. Stat. Phys. **9**, 215 (1973).
3. K. Lindenberg and V. Seshadri, Physica **109A**, 483 (1981).
4. B.J. West, Phys. Rev. **25A**, 1683 (1982).
5. E.N. Lorenz, Tellus **XXI**, 289 (1969).
6. A.S. Monin, Weather Forecasting as a Problem in Physics, MIT Press (1972), see Section 20.
7. G.F. Carnevale and G. Holloway, J. Fluid Mech. **116**, 115 (1982).
8. L.E. Reichl, A Modern Course in Statistical Physics, Univ. Texas Press (1980).
9. L. Onsager and S. Machlup, Phys. Rev. **91**, 1505 (1953); ibid. 1512 (1953).
10. H.B. Callen and T.A. Welton, Phys. Rev. **83**, 34 (1951).
11. see e.g., the lectures in Stability of Thermodynamic Systems, editors J. Casas-Vázquez and G. Lebon, Lect. Notes in Phys. **164**, Springer-Verlag, Berlin (1982).
12. J. Egger, J. Atmos. Sci. **38**, 2606 (1982).
13. K. Lindenberg, K.E. Shuler, V. Seshadri and B.J. West, in Probabilistic Analysis and Related Topics, vol. 3, pp.82, ed. A.T. Bharucha-Reid, Academic Press (1983).
14. B.J. West and K. Lindenberg, Phys. Lett. **95A**, 44 (1983).
15. K. Lindenberg and B.J. West, Physica **119A**, 485 (1983).
16. E. Lorenz, J. Atmos. Sci. **20**, 130 (1963).
17. L.D. Landau and E.M. Lifshitz, Fluid Mechanics, Pergamon, London (1959).
18. N.G. van Kampen, Stochastic Processes in Physics and Chemistry, North-Holland (1981).
19. B.J. West, K. Lindenberg and V. Seshadri, Physica **102A**, 470 (1980).
20. K. Lindenberg, V. Seshadri and B.J. West, Phys. Rev. **22A**, 2171 (1980).
21. K. Lindenberg, V. Seshadri and B.J. West, Physica **105A**, 445 (1981).
22. L. Onsager, Nuovo Cimento **6**, Supp. 279 (1949).
23. R. Kraichnan, J. Fluid Mech. **67**, 155 (1975).
24. R.F. Fox, Physics Reports **48**, 179 (1978).
25. G.W. Ford, M. Kac and P. Mazur, J. Math. Phys. **6**, 504 (1965).
26. P. Langevin, Comptes rendus **146**, 530 (1908).

OBSERVATIONAL ASPECTS OF LARGE-SCALE ATMOSPHERIC TURBULENCE AND PREDICTABILITY

G.J. Boer
Canadian Climate Centre
Atmospheric Environment Service
Downsview, Ontario

ABSTRACT

The observation that the large-scale motions of the atmosphere exhibit some of the features of two-dimensional turbulent flow has been used to justify the study of atmospheric predictability using simplfied turbulence models. Global FGGE data are used to investigate several aspects of large-scale turbulence in the atmosphere. Two rather different flow regimes are found. The high wavenumber region is dominated by the transient component, there is approximate isotropy, the spectrum has a power-law slope, the January and July spectra are remarkably similar, and there is a strong enstrophy flux down scale. This region has some of the features of an enstrophy cascading subrange. The low wavenumber region on the other hand is dominated by the stationary component, is markedly anisotropic and exhibits notable changes in the spectrum between January and July. There is a strong energy flux up scale. It is notable that there seems to be no barrier to the up scale transfer via the Rhines mechanism.

An associated diagnostic study of the error at 24, 48 and 72 hours of the Canadian Meteorological Centre forecast model is carried out in wavenumber space. Error spectra and budget terms are obtained and compared with the results expected from theory. Error behaviour at high wavenumbers and for short times is in reasonable accord with expectations. The source of error due to model imperfections is, however, a very important term at all scales. Error behaviour differs in summer and winter and the behaviour of error in the low wavenumber regime is crucial to the limit of predictability.

INTRODUCTION

The observation that the large-scale behaviour of the atmosphere displays some of the features of idealized two-dimensional turbulent flow has motivated the study of atmospheric predictability using simplified turbulence models[1-4]. This turbulence approach provides a unifying, although simplified, viewpoint from which to regard the behaviour of the atmosphere, its modelling for predictability purposes and the day to day predictability experiments performed by operational forecast systems.

In what follows, the basic ideas behind the turbulence approach are briefly noted, diagnostic calculations of global atmospheric data are presented and the results compared with simple turbulence ideas, some restrictions on the turbulence model approach to predictability are mentioned and the results of an operational forecast system are analysed.

TWO-DIMENSIONAL TURBULENCE

The conceptual model often applied to large-scale atmospheric behaviour and predictability is that of homogeneous and isotropic plane turbulence. As applied to the atmosphere, this conceptual model postulates an energy and enstrophy source located in a certain restricted wavenumber range with a sink of energy at larger scales and a sink on enstrophy at smaller scales. The source and sink regions may be connected by inertial subranges through which energy and enstrophy are transported to the sink regions.

An enstrophy cascading subrange linking the source region to the enstrophy sink at small scales is given most attention as an atmospheric mechanism in diagnostic and predictability studies[5-7].

Homogeneous and isotropic plane turbulence is characterized by statistics which are a function of the single scale parameter $k = (k_x^2 + k_y^2)^{\frac{1}{2}}$. We have

$$\tfrac{1}{2}\,\overline{V \cdot V} = \overline{u^2} = \overline{v^2} = \int E(k)\,dk \tag{1a}$$

$$\overline{\zeta^2}/2 = \int G(k)\,dk \tag{1b}$$

$$G(k) = k^2\,E(k) \tag{1c}$$

$$\frac{\partial E(k)}{\partial t} = I(k) + S_E(k) \tag{1d}$$

$$\frac{\partial G(k)}{\partial t} = J(k) + S_G(k) \tag{1e}$$

$$I(k) = -\,\partial F/\partial k \tag{1f}$$

$$J(k) = -\,\partial H/\partial k \tag{1g}$$

where the overbar represents the probability average (or the time average in data studies).

That is, the kinetic energy is equally partitioned between the velocity components which are uncorrelated and the energy and enstrophy spectra depend only on the wavenumber k. The energy and enstrophy budgets in terms of wavenumber (1d, e) state that the rate of change of energy and enstrophy for a given wavenumber depends on the non-linear interaction of wavenumber k with all other wavenumbers together with sources and sinks in wavenumber space. Since non-linear interaction terms act only to redistribute energy and enstrophy between wavenumbers they may be written as the divergence of a flux of energy or enstrophy through wavenumber space (1f, g).

An energy cascading subrange, in this simple case, is charterized by

$$E \propto \varepsilon^{2/3}\,k^{-5/3}, \quad F = -\varepsilon, \quad H = 0$$

i.e. by a $k^{-5/3}$ spectrum, a constant upscale energy flux F and a vanishing enstrophy flux H.

An enstrophy cascading subrange is characterized by

$$E \,\alpha\, \eta^{2/3} k^{-3}, \quad F = 0, \quad H = \eta$$

i.e. by a k^{-3} spectrum, a constant down scale enstrophy flux and a vanishing energy flux.

In the case of turbulence on a rotating plane, the "wave-domain" cutoff of upscale energy flux described by Rhines[8] is an additional feature of the turbulent flow.

While the atmosphere cannot be expected to exhibit precisely these features of simple turbulent flows, it is useful to analyse atmospheric behaviour from this point of view, both to gain an appreciation of actual atmospheric behaviour and to understand the consequences of abstracting atmospheric behaviour in this way for predictability studies.

As applied to atmospheric predictability, it is presumed that some error is introduced into the turbulent flow, usually at high wavenumbers. This is thought of as corresponding to atmospheric observational and analysis error. As the turbulent flow evolves, non-linear interactions redistribute this error between wavenumbers. There is also an error production term which results from the interaction of the error with the flow. The result is that error, initially confined to small scales, grows and is transported to larger scales until it eventually contaminates all scales.

TWO-DIMENSIONAL TURBULENCE ON THE SPHERE

The ideas of simple two-dimensional turbulent flow on the plane may be transferred to the sphere without major modification. As shown in Boer[9], the analogy is direct when the Fourier representation of plane turbulence goes over to the representation in terms of spherical harmonics. The expressions (1) hold also for the sphere when the wavenumber k is replaced by the spherical harmonic index n.

It is also shown that a necessary condition for homogeneity and isotropy on the sphere is that the two-dimensional spectral densities E_n^m, where m is the Fourier wavenumber and n is the order of the Legendre polynomial, are independent of m, i.e. $E(n) = E_n^m$ is a function of n only.

ANALYSIS OF GLOBAL DATA

The data used for the calculations are the FGGE III-a global data for January and July 1979. Some of the results of this section are also presented in Boer and Shepherd[7].

Figure (1) displays the vertically integrated energy spectrum for January in terms of the scale parameter n. The spectrum is further decomposed into stationary and transient parts. The general features of these spectra from global data resemble those of

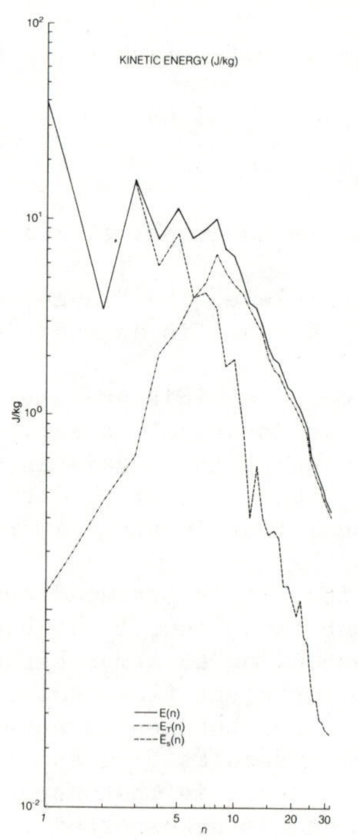

Fig. 1. Integrated spectrum of kinetic energy and its stationary and transient components for January.

Fig. 2. Integrated kinetic energy for January and July.

previous calculations using hemispheric data. The decomposition into stationary and transient parts, however, illustrates that two rather different regimes exist in wavenumber space. The high wavenumber region of the spectrum (n>10 say) is dominated by the transient component and exhibits a spectral power law behaviour with spectral slope in the neighbourhood of -3. The low wavenumber region, on the other hand, is dominated by the stationary component and does not display a clear power law behaviour.

The spectra for both January and July are shown in Figure (2). Rather remarkably, the high wavenumber, transient dominated, power law region has virtually the same energy distribution in the two months although the low wavenumber region, dominated by the time averaged component of the flow, shows marked changes.

Figure (3) displays the two-dimensional energy spectrum. If the values of E_n^m are independent of m as is required for homogeneous and isotropic turbulence on the sphere, the isolines on the diagram are horizontal. This is the case to a reasonable extent at higher wavenumbers but not at low wavenumbers.

Figure (4) displays the (resolved part) of the non-linear interaction term I(n). In the time averaged case $\partial E(n)/\partial t$ is small and

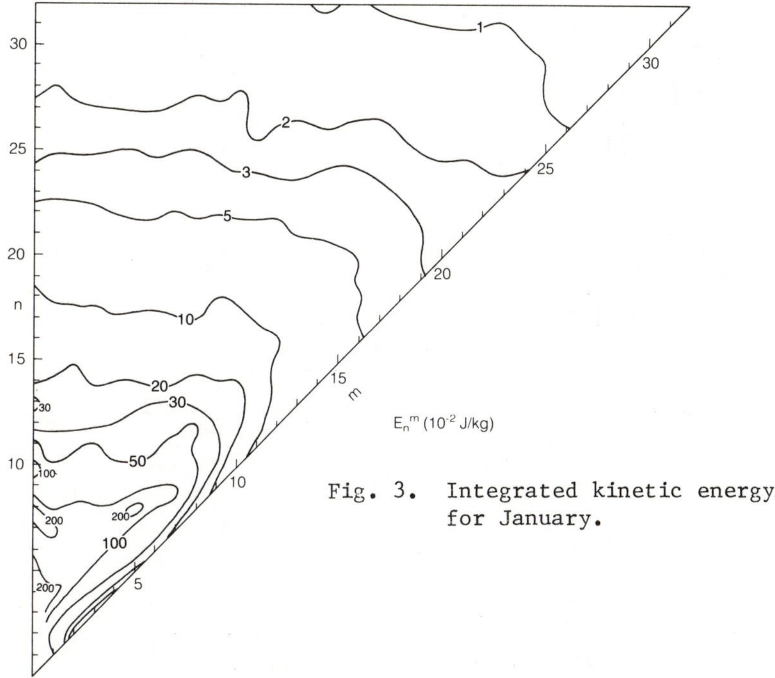

Fig. 3. Integrated kinetic energy for January.

according to equation (1e) the source/sink term S(n) is equal, but opposite in sign, to I(n). A rather broad band of wavenumbers exist for which I(n) is negative, indicating that energy is being supplied to other scales by non-linear interactions. This energy is supplied by the source term which has a corresponding distribution in wavenumber space.

Lower wavenumbers gain this energy via the non-linear interaction term and give it up to an energy sink operating at these scales.

The corresponding term for enstrophy interaction between scales is shown in Figure (5). Here the source region enstrophy is transported to smaller scales where its sink region is located.

The associated enstrophy flux through wavenumber space is shown in Figure (6). There is a strong down gradient flux at high wavenumbers. This flux, together with the features of the spectrum previously noted show that, to some extent, the features of an enstrophy cascading subrange are seen in the data in the high wavenumber region. Of course the correspondence is not complete.

The flux of energy through wavenumber space is shown in Figure (7). Here the upscale or negative energy flux is seen which is so much a feature of atmospheric flow.

The fluxes of energy and enstrophy have been decomposed into terms which depend on the product of time averaged flow components (stationary term), products of averaged and transient components (mixed term) and the product of transient components only. The flux of enstrophy at high wavenumbers is dominated by the transient

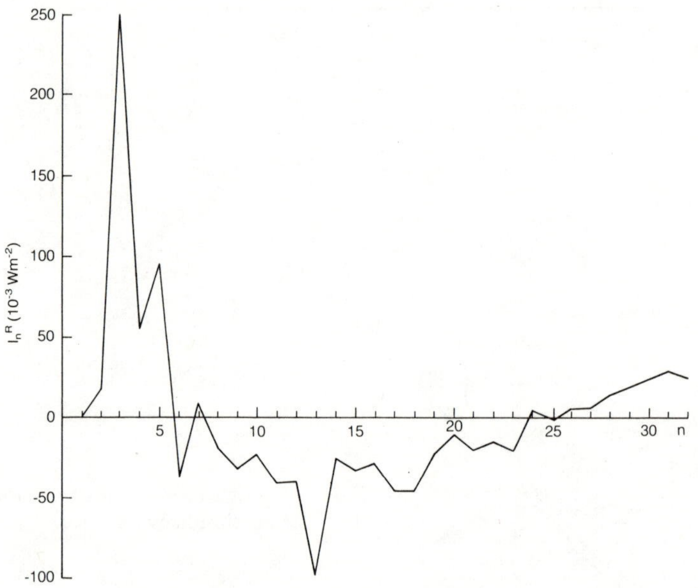

Fig. 4. The resolved part of the energy non-linear interaction term for January.

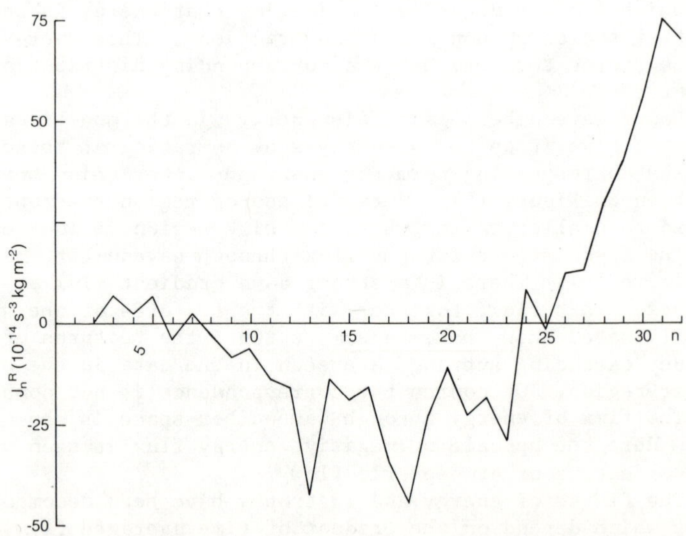

Fig. 5. The resolved part of the enstrophy non-linear interaction term for January.

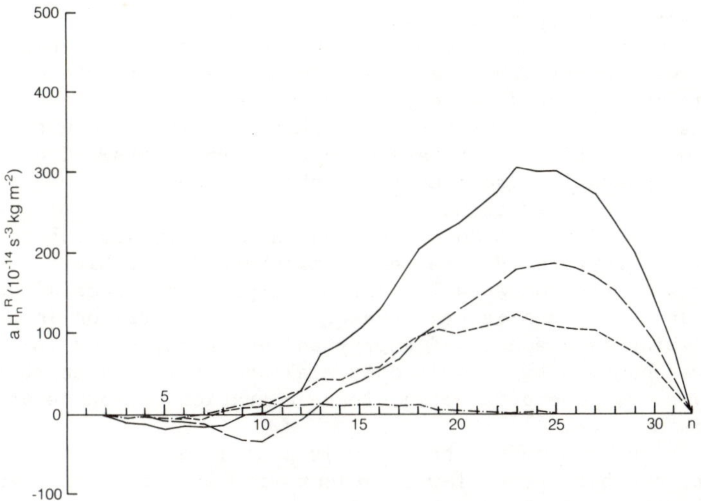

Fig. 6. The resolved part of the enstrophy flux for January. The solid line is the total flux which is made up of the transient term (long dashes), the mixed term (short dashes) and the stationary term (dot-dashes).

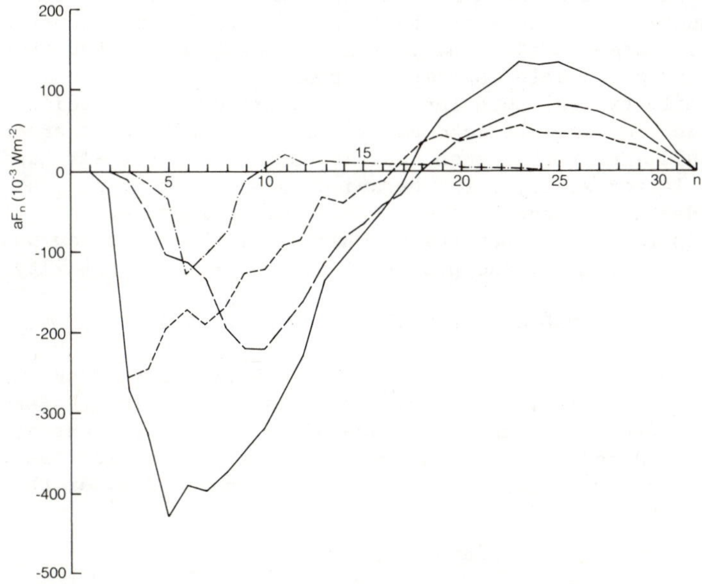

Fig. 7. As for Figure 6 but for energy flux.

component although the mixed term is not trivial. The stationary flow term does not contribute appreciably.

The upscale flux of energy is dominated by the transient term at middle wavenumbers while at low wavenumbers the remaining upscale flux is accomplished largely by the mixed mean-transient component. This is quite different behaviour than that which occurs in turbulence models which don't have mean flow components. The Rhines cutoff is apparently circumvented in the atmosphere by the transfer of energy by the mixed terms.

What implications do these results have for atmospheric predictability? Apparently, the real atmosphere does exhibit some of the features of the conceptual model of simple turbulence discussed above, at least in a general sense. There is a region in wavenumber space which is a source of energy and enstrophy which is flanked at high wavenumbers by an enstrophy sink and at low wavenumbers by an energy sink. There is not a clear separation of source and sink regions however.

At high wavenumbers the flow is dominated by the transient component, exhibits power law behaviour with spectral slope in the neighbourhood of -3, and has a downscale cascade of enstrophy. In an approximate sense therefore this region of the flow exhibits some of the features of an enstrophy cascading subrange.

At low wavenumbers, however, the flow is dominated by the stationary component and is not homogeneous and isotropic. While there is an up scale energy flux and consequent sink in this wavenumber region, it does not qualify as a simple turbulent regime. The Rhines cutoff mechanism is apparently circumvented. These low wavenumber components of the flow are vitally important in understanding atmospheric behaviour and predictability but certainly are not characterizable as simple turbulence.

While typical turbulence model approaches to predictability may provide good insight into early time and small scale error growth, it would appear that such approaches cannot correctly handle later time, larger scale, error behaviour since they do not, at present, include the consequences of the low wavenumber region. It is just in this low wavenumber region that we must look for an understanding of the theoretical and practical limits of predictability.

ANALYSIS OF OPERATIONAL FORECASTS

In previous sections, the ideas of error growth and up scale transfer in turbulence models were discussed. In this section the error budget in wavenumber space of an actual forecast system is considered and the results compared to the idealized case.

The error budget equation in wavenumber space for the rotational part of the flow is

$$\frac{\partial E(n)}{\partial t} = I(n) + P(n) + S(n) \tag{2}$$

where, in this case, E(n) is the energy associated with the error, I(n) is the non-linear transfer of error between scales, P(n) is the non-linear error production term and S(n) is the error source sink term. S(n) includes errors due to the imprecise specification of the forecast model. This source of error is not a feature of "perfect model" approaches to the study of predictability.

The forecast error analysis is carried out for January 1982 and July 1981. The forecast system is that of the Canadian Meteorological Centre. The forecast model and operational analyses are hemispheric in extent. The forecast model is a 10-level spectral model with rhomboidal 29-wave truncation and fairly complete physical parameterizations as described by Daley[10] for an earlier version of the model. The analysis is performed at triangular 30-wave truncation. The forecast error at 24, 48 and 72 hours is analysed.

Error spectra for the three forecast periods are displayed in Figure (8). The relative error, obtained by dividing the error energy by the energy of the flow at that scale is shown in Figure (9). Note that the spectra have been vertically integrated and also that the values have not been divided by the mass of the column as was done for the spectra in the previous section .

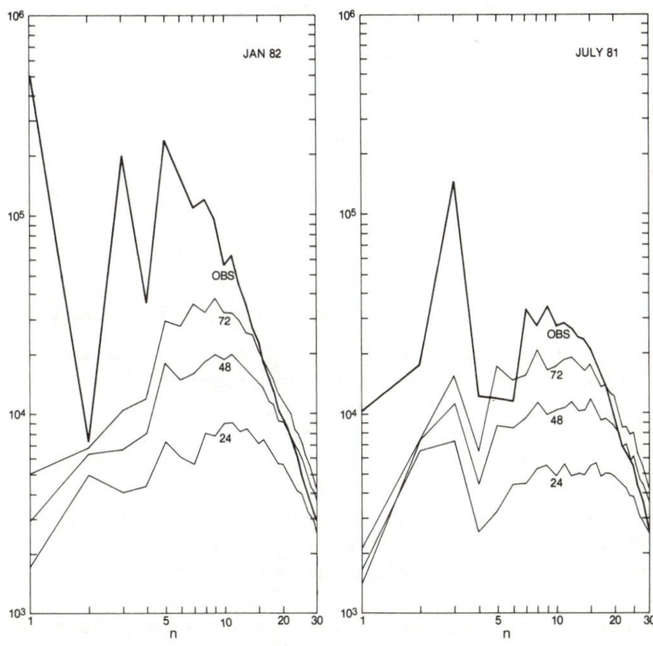

Fig. 8. Error energy spectra at 24, 48 and 72 hours together with the energy spectra of the observed flow. Units J m^{-2}.

As expected, the error at small wavenumbers grows most rapidly early in the forecast and approaches its saturation value of something near 2 in terms of relative error. In January the error isn't very effective in penetrating past n=4. For July the error behaviour is rather different. There is fairly rapid error growth at lower wavenumbers which is particularly noticeable in the relative error diagram.

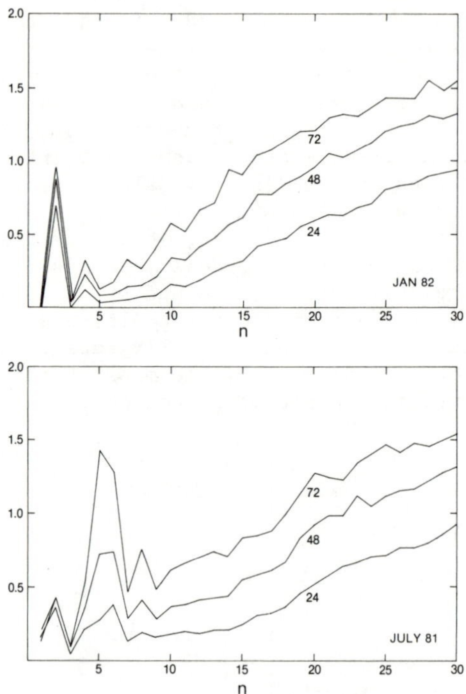

Fig. 9. Relative error spectra at 24, 48 and 72 hours.

If the predictability remaining at various scales is indicated by the area in Figure (8) between the observed flow and the error at 72 hours, it appears that the lower wavenumbers are not yet badly contaminated in January while this is not the case for July. Apparently, perhaps not surprisingly, the error behaviour and implied predictability is quite different in the two months.

Terms in the error budget equation (2) are displayed in Figure (10) for January 1981. The rate of change term $\partial E(n)/\partial t$ is calculated using simple finite differences while the source/sink term S(n) is obtained as a residual.

At 24 hours, the maximum rate of error growth is at middle wavenumbers, although the maximum relative error growth rate is at high wavenumbers. The non-linear interaction term I and the non-linear production term P are quite small since the errors are still quite small at this time. The source term S, associated with model errors, is the dominant term.

At 48 hours and subsequently at 72 hours, the maximum error growth rates increase and shift to lower wavenumbers. The high wavenumber growth rates decrease since error in these scales is reaching saturation. The importance of the non-linear interaction and production terms grows as the error grows. The error source

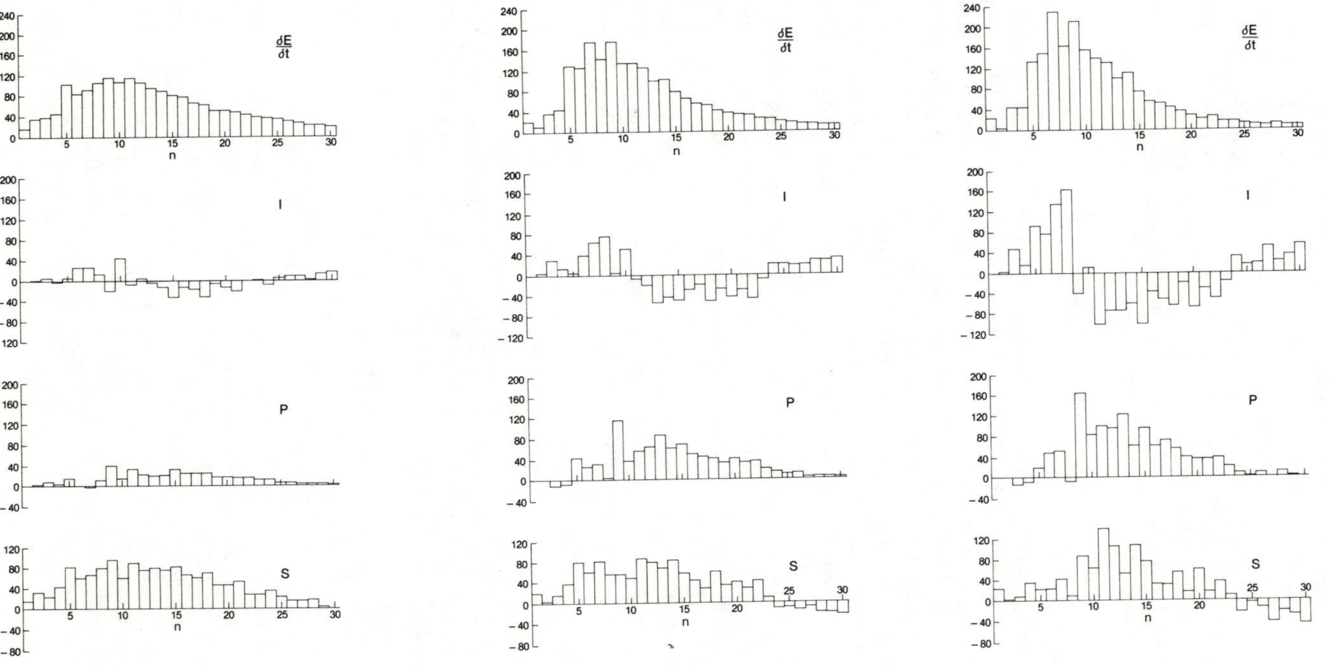

Fig. 10. Components of the error budget at 24, 48 and 72 hours respectively. Units 10^{-3} Jm^{-2} sec^{-1}.

term is still important however. Presumably at yet later times the error production term will dominate the error budget.

The kind of error behaviour seen here is roughly in accord with expectations. Clearly however, the error source term is very important in this operational case. This term introduces error more or less uniformly with wavenumber at early times in the forecast.

The nature of the forecast error is different at different seasons and the behaviour at low wavenumbers and later times will be of considerable interest and importance.

CONCLUDING REMARKS

The concepts of simple two-dimensional turbulence, which have played an important role in understanding and modelling atmospheric predictability are used as a background against which to analyse the scale dependent behaviour of atmospheric flows. These ideas are, not surprisingly, only partially applicable to the atmosphere, in particular to the high wavenumber region. The low wavenumber region does not lend itself to such a characterization and it is this low wavenumber region which is vitally important for understanding predictability at longer times.

The analysis of error in an operational forecast system shows that, for early times, the error behaviour with scale is generally in accord with expectations. Model error is clearly a very important feature of the error growth. The behaviour of the error is quite different in different seasons. Once again it is the behaviour of the error at low wavenumbers and hence at longer predictability times which will be of interest and importance in future studies.

ACKNOWLEDGMENTS

Lynda Smith has been a great help with the diagrams and the manuscript. Louis Lefaivre captured the forecast data and Christian Thibault aided in the forecast error calculations.

REFERENCES

1. E.N. Lorenz, Tellus, 21, 289-307 (1969).
2. C.E. Leith, J. Atmos. Sci. 28, 145-161 (1971).
3. C. Basdevant, *et al*., J. Atmos. Sci. 38, 2305-2326 (1981).
4. G.K. Vallis, J. Atmos. Sci. 40, 10-27 (1983).
5. F. Baer, J. Atmos. Sci. 29, 649-664 (1972).
6. T.-C. Chen and A. Wiin-Nielsen, Tellus, 30, 313-322 (1978).
7. G.J. Boer and T.G. Shepherd, J. Atmos. Sci. 40, 164-183 (1983).
8. P. Rhines, J. Fluid Mech. 69, 417-443 (1975).
9. G.J. Boer, J. Atmos. Sci. 40, 154-163 (1983).
10. R. Daley *et al*., Atmos. 14, 98-134 (1976).

PREDICTABILITY: LAGRANGIAN AND EULERIAN VIEWS

Dale B. Haidvogel
National Center for Atmospheric Research
Boulder, Colorado 80307 U.S.A.

Greg Holloway
Institute of Ocean Sciences
Sidney, British Columbia V8L4B2
Canada

INTRODUCTION

There are a variety of error measures relevant to the study and interpretation of predictability. From an Eulerian point of view, it has been common to examine the rate at which an error (difference) field initially characterized by small scales propagates to larger scales of motion and increases in amplitude relative to the climatological energy spectrum [Lorenz[1]; Kraichnan[2]; Leith[3]]. Alternatively, in the Lagrangian frame, one may examine the rate at which particles initially close together increase in separation [Morel and Larcheveque[4]].

Recently, it has been suggested that these Eulerian and Lagrangian error measures are in fact substantially equivalent [Lesieur and Chollet[5]]. Here, we examine this hypothesis, primarily by direct numerical simulation, to determine to what extent these points of view are qualitatively and/or quantitatively similar.

EQUATIONS, TECHNIQUES AND STRATEGY

Consider a homogeneous, two-dimensional, non-divergent fluid on a β-plane contained within a domain of dimensional size $(2\pi L)$ and characterized by an RMS fluid speed of (U). Using L and U as dimensional scales, we set $[x, y, t, \psi] \sim [L, L, \frac{L}{U}, LU]$. In the resulting non-dimensional system, the domain is of length (2π), the RMS fluid speed is $0(1)$, and the vorticity equation for the fluid may be written

$$\frac{D}{Dt}q = \left\{\frac{\partial}{\partial t} + J(\psi, \ )\right\} q = D \tag{1}$$

where the potential vorticity

$$q = \nabla^2\psi + \beta y,$$

and the dissipative term

$$D = -\nu\ \nabla^6\psi$$

is assumed to be of biharmonic form, $\nabla^4(\nabla^2\psi)$.

Equation (1) is characterized by two non-dimensional parameters:

$$\beta = \beta_{dim}L^2/U$$

and

$$\nu = \nu_{dim}/L^3U$$

0094-243X/84/1060067-11 $3.00

These represent, respectively, the non-dimensional strength of the β and lateral dissipative effects. Specific values for these parameters are discussed below.

For statistically homogeneous flows, it is convenient to set

$$\psi(x,y,t) = \sum_{k=-M}^{M-1} \sum_{l=-M}^{M-1} \hat{\psi}_{kl}(t) e^{2\pi i(kx+ly)} \quad , \tag{2}$$

that is, to represent the streamfunction field as a truncated double Fourier series. Under expansion (2), the vorticity equation (1) may be advanced in time using the spectral method [Orszag[6]]. Here a leapfrog time step is used (with occasional leapfrog-trapezoidal steps to avoid time-splitting), and the non-linear terms are evaluated so as to conserve total energy and enstrophy in the absence of dissipation and time-stepping error. In addition, the non-linear products are alias free.

The results and discussion which follow pertain to the specific parameter values $\beta = 2$ and $\nu = 4\text{x}10^{-7}$. The latter implies only weak dissipative effects, as will be seen below. The non-dimensional value of β might arise as follows:

	L	U	L/U
ocean	100 km	10 cm/sec	0 (10) days
atmosphere	1000 km	10 m/sec	0 (1) day.

Note that the resulting dimensional scale times would then be 0(10) and 0(1) days in the ocean and atmosphere, respectively. Results comparable to those described below have been obtained for a wide range of β values. The Eulerian and Lagrangian integrations proceed as follows:

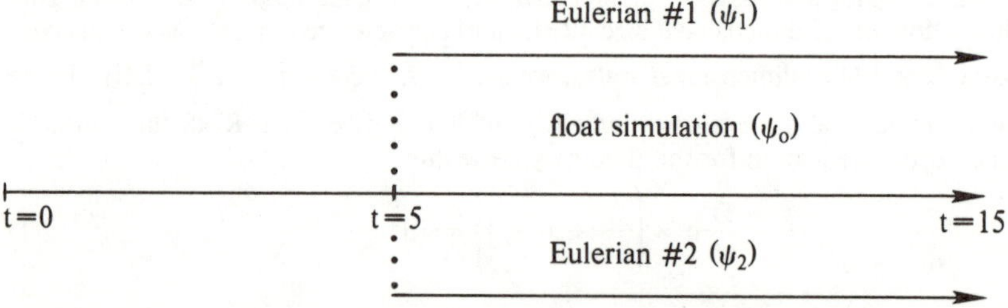

With cutoff wavenumber M=64, equation (1) is initialized with an isotropic random-phase spectrum $E(K) = K/(20+K^4)$ where the total wavenumber $K = (k^2+l^2)^{1/2}$. This form for E(K) is chosen to have a plausible power law behavior for $K \rightarrow \infty$, and to have a peak near the transfer arrest wavenumber K_β [Rhines[7]]. Here, initially, $K_\beta = 1 \sim 2$.

With $\Delta t = 0.005$, the vorticity equation is advanced in time to $t = 5$ (recall non-dimensional units!) to allow an initial adjustment of the random-phase field to a near-equilibrium enstrophy cascading state. At $t = 5$, two different things happen. First, with no perturbation to the flow field, $99(=N_p)$ Lagrangian particles are released into the fluid in clusters of three, and the integration is continued for 10 additional time units $(5<t\leqslant10)$. As the integration proceeds to compute the evolution of $\psi(x,y,t)$, the trajectories of the floats are determined via integration of the path equation (see below). Second, two independent perturbed initial streamfunction fields are generated by randomizing the Fourier phase components of $\psi_{kl}(t=5)$ for $40\leqslant K\leqslant64$. The two perturbed initial conditions are then independently advanced in time $(5<t\leqslant10)$ to determine the evolution of the two Eulerian streamfunction fields (ψ_1 and ψ_2) and the difference or error field ($\psi_d = \psi_1 - \psi_2$).

Of particular interest are the zonal and meridional length scales which characterize the evolving error field, and the separation statistics for the Lagrangian particles. Here, these quantities are estimated in the Eulerian frame as

$$\left(\overline{k^2}\right)^{-1} = \left\{ \frac{\sum_{k=-M}^{M-1} \sum_{l=-M}^{M-1} k^2 E_d(k,l)}{\sum_{k=-M}^{M-1} \sum_{l=-M}^{M-1} E_d(k,l)} \right\}^{-1} \tag{3}$$

and

$$\left(\overline{l^2}\right)^{-1} = \left\{ \frac{\sum_{k=-M}^{M-1} \sum_{l=-M}^{M-1} l^2 E_d(k,l)}{\sum_{k=-M}^{M-1} \sum_{l=-M}^{M-1} E_d(k,l)} \right\}^{-1}$$

(that is, the inverse of the mean square wavenumbers of the error energy spectrum), and in the Lagrangian frame as

$$\overline{d_x^2} = \frac{1}{N_p} \sum_{\text{clusters}} \sum_{\substack{i \neq j \\ 1 \leqslant i,j \leqslant 3}} (x_i - x_j)^2$$

$$\overline{d_y^2} = \frac{1}{N_p} \sum_{\text{clusters}} \sum_{\substack{i \neq j \\ 1 \leqslant i,j \leqslant 3}} (y_i - y_j)^2 \tag{4}$$

(that is, the mean square separation of particle pairs).

PARTICLE FOLLOWING

The streamfunction field, given in terms of the expansion (2) and advanced in time according to equation (1), is defined on a discrete (fixed) set of points in space ($0 \leqslant x,y \leqslant 2\pi$) and time ($0 \leqslant t \leqslant 15$). Kinematically, the Lagrangian particles follow the local velocity field; however, they are not constrained to occupy any specific set of grid points, but may in general lie at any location within the Eulerian grid. Since the task of recovering the particle trajectories involves knowing the fluid velocity at the instantaneous positions of the particles, it is clear that some procedure for obtaining interpolated values of velocity at arbitrary positions within the domain is necessary.

Two specific interpolation schemes have been tested and shown to give accurate (and nearly identical) results. The first scheme involves directly summing the discrete Fourier spectrum for the velocity components, i.e.,

$$[u,v]_{kl} = [-il\hat{\Psi}_{kl}, ik\hat{\Psi}_{kl}] \quad -K \leqslant (k,l) \leqslant K-1 \tag{5}$$

at the arbitrary float locations. Since the floats are not co-located with grid points, the Fourier summation cannot take advantage of the Fast Fourier algorithm. It is therefore highly accurate, but inefficient.

Much less costly is a mixed Fourier/gridpoint interpolation, rather reminiscent of the pseudospectral technique used to time integrate (1). The scheme involves two sequential steps. First, the *exact* discrete Fourier spectrum (5) for velocity (or any other desired quantity) is summed by FFT to yield a set of velocity values on the fixed, equally spaced Eulerian grid. Second, a linear combination of local grid point values is used to interpolate to the float position. (In our simulations, the use of the neighboring 4 x 4 grid point values has proven satisfactory.) The weighting factors used in the linear combination are chosen so that the interpolating function is exact at the grid points.

It is relevant to note, however, that no matter how accurate an interpolation is used, the resulting Lagrangian trajectories are not such as to conserve potential vorticity following the particles — that is, to satisfy the Lagrangian analogue of equation (1). Although the Eulerian model is constrained to conserve *total* energy and enstrophy (for $\Delta t \rightarrow o$ and $D \equiv o$), an individual Lagrangian particle suffers a loss of memory of its q value as it navigates the derived trajectory. The cause of this memory loss does not involve interpolation (although this too may contribute error), but a more subtle non-conservative effect involving the precise way in which the non-linear terms are treated in the Eulerian model itself. Haidvogel[8] has discussed this question in more detail. Here, the parameters of the model are such that the non-conservation is slight along the derived particle trajectories.

RESULTS AND DISCUSSION

At t=5 (t=0 relative to the initiation of the Eulerian and Larangian predictability experiments), the energy and enstrophy spectra have attained a near-equilibrium shape (Fig. 1). Both spectra are peaked at low wavenumber (K=1 and 2, respectively) consistent with the assumed β value, and exhibit steep power law behavior $[E(K) \simeq K^{-\psi}]$ out to the dissipative cutoff at $K \simeq 35$. Also shown in Figure 1 is the initial error energy spectrum, $E_d(K)$, resulting from random-phase scrambling over the range $40 \leqslant K \leqslant 64$.

As the three independent streamfunction fields (ψ_o, ψ_1, and ψ_2) are advanced in time ($5 \leqslant t \leqslant 15$) according to (1) they undergo a slow dissipative loss of total energy and enstrophy as a result of the scale-selective lateral dissipation D. Figure 2, for example, shows the time history of energy and enstrophy for the ψ_o field for $5 \leqslant t \leqslant 15$. Because of the steepness of E(K), only a tiny fraction ($\leqslant 0.5\%$) of the total energy is lost over the course of the simulation; however, a substantial fraction of the initial enstrophy (a much larger portion of which is resident at high wavenumbers) is lost by t=15. In the future, the addition of some forcing mechanism to equation (1) might be desirable to avoid excessive spin-down of the fluid.

In the Eulerian frame, the difference or error field ($\psi_d = \psi_1 - \psi_2$) evolves in an expected manner. After a brief interval ($5 \leqslant t \leqslant 7$) during which the total error energy and enstrophy decrease, both quantities begin to increase exponentially with error energy growing (approximately) as $e^{0.5t}$ — see Figure 3. Simultaneously, the error field propagates to larger spatial scale (lower wavenumbers). A sequence of error energy spectra, normalized by the mean of the energy spectra of ψ_1 and of ψ_2 clearly demonstrate the monotonic spreading of the error energy spectrum (Fig. 4).

In the Lagrangian simulation, the triads of particles are released at random locations, and with random orientations, into the ψ_o (x,y) field (Fig. 5a). As time proceeds, the particle clusters are advected and dispersed by the evolving circulation (Figs. 5b-d). Given sufficient time, each particle will reach one of the edges of the $(2\pi)^2$ doubly periodic domain. At that time, the particle in question may be assumed to periodically re-enter the domain (as has been assumed in constructing Figure 5), or it may continue to be tracked as it disperses outwards from its initial launch site. [The $\psi_o(x,y)$ field can be extended periodically to provide flow information at locations outside the primary $(2\pi)x(2\pi)$ domain.] Plotted in this way, cumulative particle trajectory diagrams (e.g., Fig. 6) give a strong pictorial impression of the nature of the particle dispersion process. Here, note in particular the highly anisotropic nature of the particle spreading. (Keep in mind, however, that Fig-

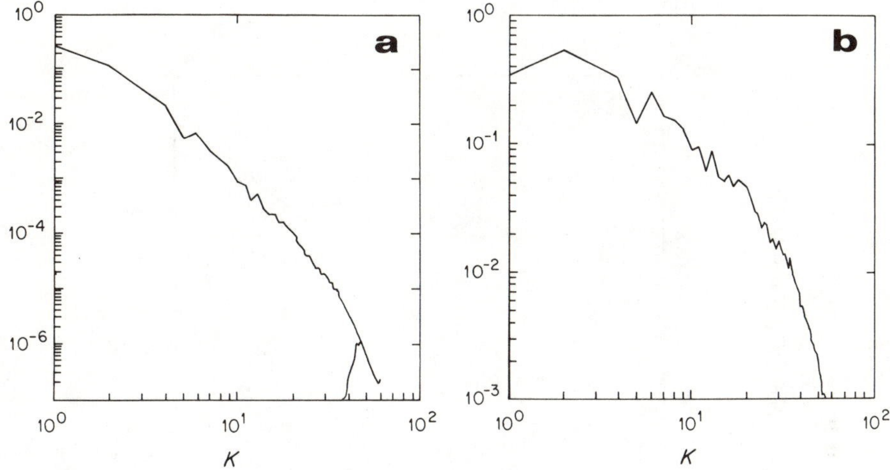

Fig. 1. The energy (a) and relative enstrophy (b) spectra at the initiation of the predictability runs. Here, K is the total circular wavenumber: $K = (k^2 + l^2)^{1/2}$. Also shown in (a) is the initial error energy spectrum ($40 \leqslant K \leqslant 60$).

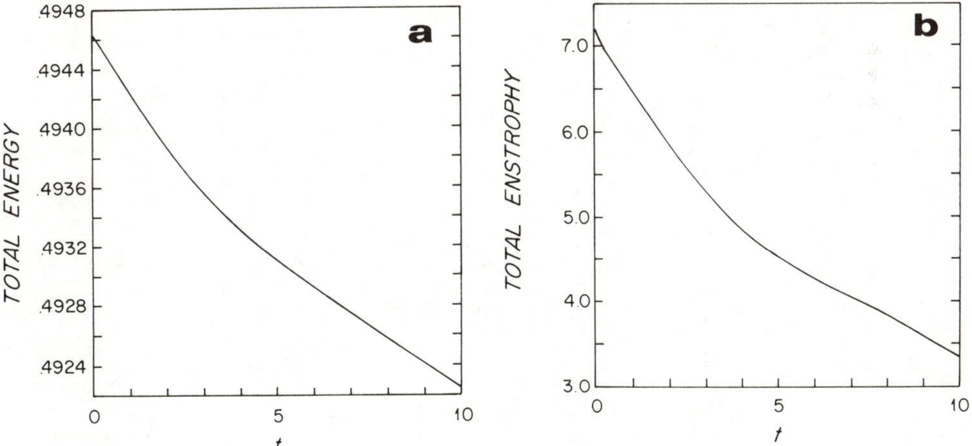

Fig. 2. Time history of total energy (a) and enstrophy (b) associated with the slowly decaying streamfunction field $\Psi_0(x,y,t)$. Here, t is the elapsed time from the beginning of the predictability simulations.

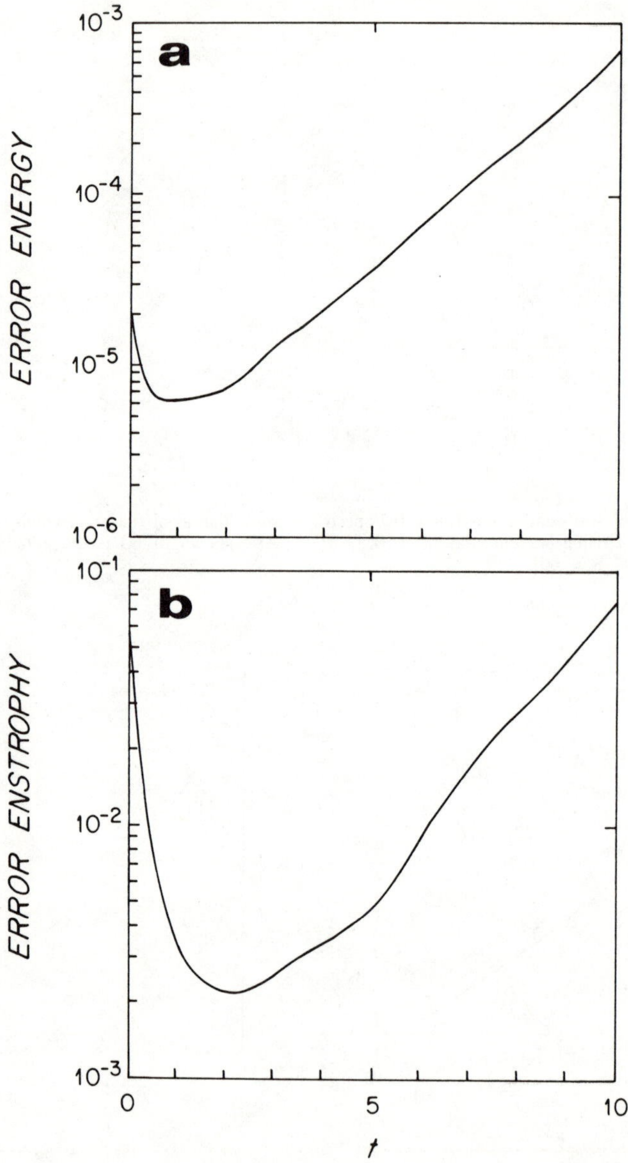

Fig. 3. Total error energy (a) and error enstrophy (b) as a function of time, t, from the beginning of the Eulerian predictability calculations.

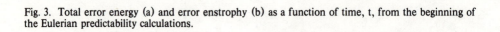

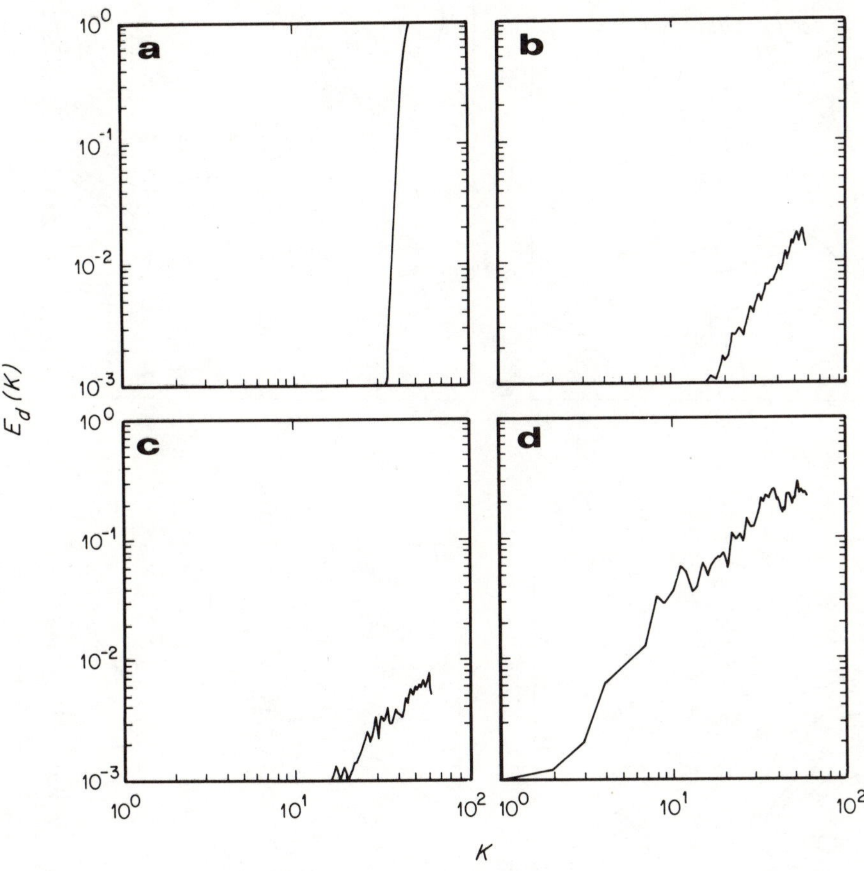

Fig. 4. Error energy spectra: (a) t = 0, at the initiation of the Eulerian predictability experiment: (b) t = 1, (c) t = 2, and (d) t = 10.

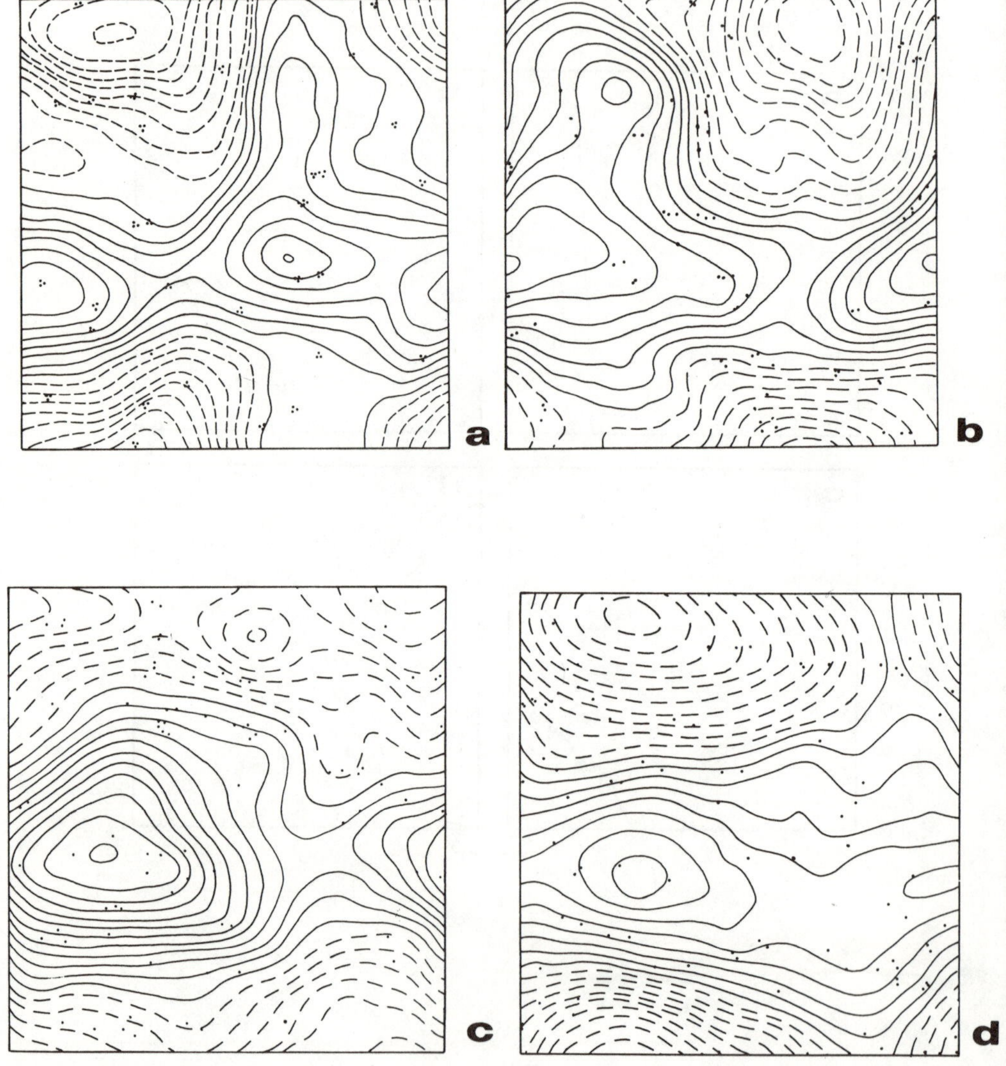

Fig. 5. Time evolution of the 33 particle triads within the (2π) x (2π) domain associated with the $\Psi_0(x,y,t)$ field: (a) t = 0, at the initiation of Lagrangian dispersal experiment; (b) t = 2; (c) t = 5; and (d) t = 10.

ure 6 summarizes the separation of *individual particles* from their initial locations; the rate at which neighboring *pairs of particles* separate — that is, the statistics $\overline{d_x^2}$ and $\overline{d_y^2}$ will be discussed presently.) The predominant zonal trajectories of the particles are at first suprising; however, despite the weak value of (βy), the relative vorticity $\nabla^2\Psi_o$ is even weaker, and has a decreasing RMS value over the course of the Lagrangian experiment (e.g., Figure 2b). The resulting potential vorticity contours are dominated by the planetary vorticity contribution, indicative of only weakly non-linear flow. Lagrangian particles which nearly conserve q, as these do, must be strongly guided by zonality of the contours.

Using expressions (3) and (4) as approximate measures of squared error length scale in the Eulerian and Lagrangian frames, respectively, the rate of error propagation to larger spatial scales can be examined. Figure 7 shows this rate of growth by plotting the Eulerian and Lagrangian quantities $[(\overline{k^2})^{-1}, (\overline{l^2})^{-1}; \overline{d_x^2}, \overline{d_y^2}]$ as a function of time for $5 \leqslant t \leqslant 15$. Initial exponential growth of the error scale characterizes both the Eulerian and Lagrangian results; the approximate growth law is exp (1.6t) in both frames. After a short period of rapid exponential growth, however, the rate of increase in $(\overline{k^2})^{-1}$ and $(\overline{l^2})^{-1}$ slows dramatically; after about four time units, the Eulerian error scales in the zonal and meridional directions have reached values (0.022, 0.013) which change very little through the end of the Eulerian simulation. In contrast, the Lagrangian particle pairs continue their exponential separation throughout the experiment, despite a noticeable reduction in the rate of growth at four time units. (The slowing of particle separation growth appears to occur as the average pair separation becomes comparable to the average eddy diameter — here, 0(1) in non-dimensional units.) By $t=15$, $\overline{d_x^2}$ and $\overline{d_y^2}$ have attained the values (5.5, 0.6) and are continuing to increase [exp (0.4t)].

It is important to point out that, whereas the Lagrangian particles are free to separate indefinitely into neighboring periodic cells such that $\overline{d_x^2}$ and $\overline{d_y^2}$ may increase without limit, the Eulerian (mean square wavenumbers)$^{-1}$ cannot increase indefinitely; rather, $(\overline{k^2})^{-1}$ and $(\overline{l^2})^{-1}$ are bounded by the value 1. In practice, this limit, signalling all error confined to the

Fig. 6. Time history ("spaghetti diagram") of all 99 particle trajectories ($0 \leqslant t \leqslant 10$). The particles have been allowed to exit the initial (2π) x (2π) periodic interval. The resulting particle domain spans the intervals $-12.3 \leqslant x \leqslant 13.6$ and $-1.8 \leqslant y \leqslant 7.5$, indicative of the enhanced zonal dispersal of the particles.

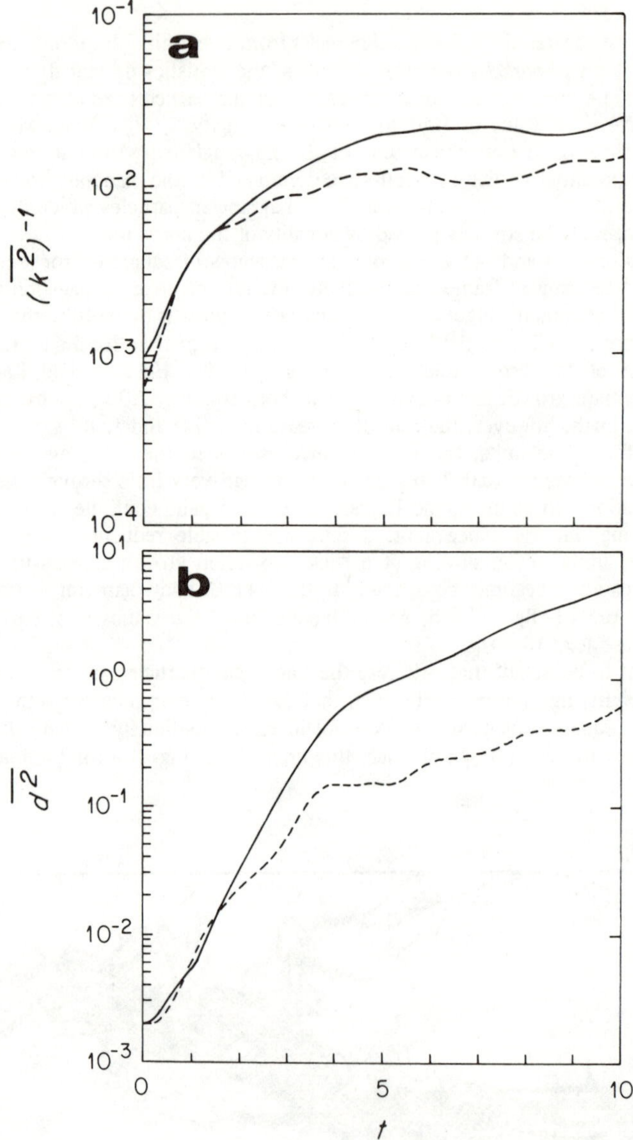

Fig. 7. Squared error length scale [as defined by (3) and (4) for the Eulerian and Lagrangian frames, respectively] as a function of time. In both (a) and (b), the solid curve corresponds to the x length scale, and the dashed line to the y length scale.

gravest mode, will not be reached so that a smaller asymptotic value for the Eulerian error scale is to be expected. Despite this limitation, these Lagrangian and Eulerian measures can be meaningfully compared until the Eulerian error field propagates to nearly box scale, or a substantial fraction thereof. This has not yet occurred here. (Simulations carried out in a spherical geometry or a closed basin would avoid this problem.)

Another respect in which the evolving Eulerian and Lagrangian error measures quantitatively differ is in the degree to which the zonal error scales exceed the meridional. The developing anisotropy is clear in both frames (Fig. 7). (Interestingly, during the brief interval of the Eulerian predictability experiment during which total error energy and enstrophy decrease, the Eulerian error field is also becoming *less* anisotropic. This trend is short-lived, however.) While it is true that $(\overline{k^2})^{-1} > (\overline{l^2})^{-1}$ and $\overline{d_x^2} > \overline{d_y^2}$, nonetheless the Lagrangian particle pairs are distinctly more anisotropic, with nearly an order of magnitude difference between the mean square zonal and meridional separations. (Recall also the cumulative particle trajectory diagram, Figure 6.)

These comparative comments have also been shown to apply in other simulations with different parameters, initial conditions, etc. Therefore, although additional realizations are needed for enhanced quantitative reliability, nonetheless the results here do strongly suggest certain systematic differences between the rate of error scale growth in the Eulerian and Lagrangian frames. To summarize, the distinction is most evident in the particle pair separation statistics which grow more rapidly, and develop substantially more zonal anisotropy than the Eulerian error scale measures examined here. Of course, other measures of Eulerian error scale are possible. Conceiving of the error field as a "front" which propagates towards lower wavenumber may be a helpful analogy; however, Figure 4 suggests that applying this analogy to determine a specific K characteristic of the error front may be difficult.

Our results support the qualitative, but not the quantitative, comparison of error growth in the Eulerian and Lagrangian frames.

ACKNOWLEDGMENT

One of the authors (DBH) acknowledges the support of the National Science Foundation, through Grant No. OCE81-09486 to the Woods Hole Oceanographic Institution and through the National Center for Atmospheric Research (NCAR). The numerical simulations described were carried out at NCAR.

REFERENCES

1. E.N. Lorenz, Tellus, **21**, 289 (1969).
2. R.H. Kraichmnan, Phys. Fluids, **13**, 569 (1970).
3. C.E. Leith, J. Atmos. Sci., **28**, 145 (1971).
4. P. Morel and M. Larcheveque, J. Atmos. Sci., **31**, 2189 (1974).
5. M. Lesieur and J.-P. Chollet, 3rd Symposium on Turbulent Shear Flows, Univ. of California at Davis (1981).
6. S.A. Orszag, Stud. Appl. Math., **50**, 293 (1971).
7. P.B. Rhines, J. Fluid Mech., **69**, 417 (1975).
8. D.B. Haidvogel, in preparation.

ON THE DYNAMICS OF ROTATING FLUIDS AND PLANETARY ATMOSPHERES: A SUMMARY OF SOME RECENT WORK

Raymond Hide
Geophysical Fluid Dynamics Labratory,
Meteorological Office (21),
Bracknell, Berkshire RG12 2SZ, England, UK

ABSTRACT

An outline is given of some recent and current research in the Geophysical Fluid Dynamics Laboratory of the UK Meteorological Office on the dynamics of rotating fluids and planetary atmospheres. Many of these investigations bear on the problem of creating a theoretical framework for atmospheric predictability studies.

INTRODUCTION

The Geophysical Fluid Dynamics Laboratory of the UK Meteorological Office carries out research on basic hydrodynamical processes in rapidly rotating fluids. Such processes underlie a wide variety of phenomena in the atmospheres and hydrospheres of the Earth and other planets. Laboratory studies play an important role in this research, along with mathematical and numerical work carried out in direct combination with laboratory investigations. Many of the major problems of dynamical meteorology and oceanography, including that of establishing a reliable theoretical framework for studies of atmospheric predictability, require for their satisfactory solution a combined attack involving the analysis and interpretation of observations in terms of basic hydrodynamical processes, and the investigation and exploitation of related systems, such as numerical models, laboratory analogues and the atmospheres of other planets. The predictability of rotating fluid systems is a central theme of our activities. These include laboratory and numerical studies of thermally and mechanically produced motions in rotating fluids under a wide variety of axisymmetric or non-axisymmetric boundary conditions, the investigation of angular momentum fluctuations of the Earth's atmosphere and associated changes in the length of the day and polar motion, and the interpretation of super-rotation of planetary atmospheres and of long-lived eddies in the atmospheres of Jupiter and Saturn.

ATMOSPHERIC ANGULAR MOMENTUM FLUCTUATIONS, LENGTH OF DAY CHANGES AND POLAR MOTION

Possibly the most striking large-scale dynamical features of the Earth's atmosphere are its average "super-rotation" relative to the solid Earth and the concentration of much of the motion in jet streams. Studies of the complex processes that produce and maintain jet streams are central to any attempt to predict large-scale atmospheric motions.

Variations in the distribution of mass within the atmosphere and changes in the pattern of winds, particularly the strength and location of the major mid-latitude jet streams, produce fluctuations in all three components of the angular momentum of the atmosphere on timescales upwards of a few days. Hide et al.[1] showed that variations in the axial component of atmospheric angular momentum during the Special Observing Periods in 1979 of the First GARP Global Experiment (FGGE, where GARP is the Global Atmospheric Research Program) are well correlated with changes in length-of-day. This would be expected if the total angular momentum of the atmosphere and 'solid' Earth were conserved on short timescales (allowing for lunar and solar effects), but not if angular momentum transfer between the Earth's liquid core and solid mantle, which is accepted to be substantial and even dominant on timescales upwards of several years, were significant on timescales of weeks or months. Fluctuations in the equatorial components of atmospheric angular momentum should contribute to the observed wobble of the instantaneous pole of the Earth's rotation with respect to the Earth's crust, but this has not been shown conclusively by previous studies.

In more recent work[2] we have re-examined some aspects of the underlying theory of non-rigid body rotational dynamics and angular momentum exchange between the atmosphere and solid Earth. Since only viscous or topographic coupling between the atmosphere and solid Earth can transfer angular momentum, no atmospheric flow that everywhere satisfied inviscid equations (including, but not solely, geostrophic flow) could affect the rotation of a spherical solid Earth. Currently available meteorological data are not adequate for evaluating the usual wobble excitation functions accurately, but we have shown that partial integration leads to an expression involving simpler functions — which *can* be reliably evaluated from available meteorological data. The length-of-day problem is treated in terms of a similar "axial angular momentum function"; and "effective angular momentum functions" are defined in order to allow for rotational and surface loading deformation of the Earth. Daily values of these atmospheric angular momentum functions were calculated from the 'initialized analysis global database' of the European Centre for Medium-Range Weather Forecasts (ECMWF) for the period 1 January 1981 - 30 April 1982, along with the corresponding astronomically observed changes in length-of-day and polar motion, published by the Bureau International de l'Heure (BIH). Changes in length-of-day during this period can evidently be accounted for almost entirely by angular momentum exchange between the atmosphere and solid Earth, and the existence of a persistent fluctuation in this exchange, with a timescale of about 7 weeks, is confirmed. The successful elucidation of this 7-week fluctuation in the atmospheric angular momentum will constitute a major advance in our understanding and ability to predict the future behavior of large-scale features of the general circulation of the atmosphere. We have demonstrated that meteorological phenomena provide an important contribution to the excitation of polar motion. Our work offers a theoretical basis for future routine determinations of atmospheric angular momentum fluctuations for the purposes of meteorological and geophysical research, including the assessment of the extent to which movements in the solid Earth associated with very large earthquakes contribute to the excitation of the Chandlerian wobble.

DIFFERENTIAL ROTATION PRODUCED BY POTENTIAL VORTICITY MIXING IN A RAPIDLY-ROTATING FLUID

Differential rotation in a partially or wholly fluid astronomical body such as a planet or star is associated with energetic processes involving the transformations between gravitational potential energy, kinetic energy and thermal energy. In the absence of the internal or external energy sources required to drive these processes, the body would rotate rigidly at a constant rate Ω_0 (say) about its fixed axis of maximum moment of inertia through its center of mass. Relative to that frame of reference, all components of the Eulerian flow velocity $\underset{\sim}{u}(R,\theta,\lambda,t) = (w,-v,u)$ would vanish, where (R,θ,λ) are spherical polar co-ordinates of a general point, R being distance from the centre of mass, θ co-latitude and λ east-longitude. Relative to any other frame which rotates steadily with constant angular speed ω with respect to this basic frame about the polar axis, including an inertial frame, for which $\omega = -\Omega_0$, we have $(w,v,u) = (0,0,-\omega R\sin\theta)$.

A major objective in the construction of theoretical models of hydrodynamical motions in planetary and stellar atmospheres and interiors is the determination from first principles of the magnitude and distribution of the mean differential rotation, as specified by

$$\overline{\Omega}(R,\theta) \equiv [\overline{u}](R,\theta)/R\sin\theta$$

$$= (2\pi T)^{-1} \int_0^T \int_0^{2\pi} (R\sin\theta)^{-1} u(R,\theta,\lambda,t)\, d\lambda dt$$

where the length of time T over which the average is taken is long in comparison with typical timescales associated with $\underset{\sim}{u}(R,\theta,\lambda,t)$ but is otherwise arbitrary. (We are here following a conventional notation of using an overbar to denote time average and square bracket to denote longitudinal average).

The dependence of $[\overline{u}]$ on R and θ would of course emerge from a full solution of the governing equations of hydrodynamics, thermodynamics, and (in the case of electrically-conducting fluids) electrodynamics, under appropriate boundary conditions. But these equations are highly intractable and have only been solved in simplified cases. Possibly the most advanced work in this connection is that done by dynamical meteorologists in their numerical studies of the general circulation of the Earth's atmosphere, in which are reproduced $[\overline{u}](R,\theta)$ and other principal features of atmospheric flow. The Earth's atmosphere is the only natural system for which observations are sufficient to enable direct determinations of $\overline{\Omega}(R,\theta)$ to be made. On average it rotates faster than the solid Earth; $[\overline{u}]$ (if measured relative to the underlying surface) is found to be positive nearly everywhere with an average value of 10 ms^{-1}, but with negative values in certain regions, including the Trade Winds at low level in the tropics. The highest values of $[\overline{u}]$ in the troposphere, about 30 ms^{-1}, are associated with mid-latitude jet streams.

In the cases of the atmospheres of Jupiter and Saturn, observations of the motions of markings on the visible surface of dense cloud going back many decades provide limited information about the dependence of [u] at the (horizontally variable) cloud level as a function of t and θ. Both planets have strong equatorial jet streams at their visible surfaces, where speeds are attained as high as about 100 ms^{-1} relative to the deep interior for Jupiter and 400 ms^{-1} for Saturn, the speeds of rotation of these interiors having been determined from radioastronomical observations. The jet streams are positive (ie westerly) in direction, and this implies that they must be produced by non-axisymmetric processes involving the action of local east-west pressure gradients. Comparable information on the dependence of [u] on t and θ for the solar atmosphere can be obtained from observations of sunspot motions and from spectroscopic data. The visible surface of the sun rotates most quickly at the equator and [u] exhibits a general decrease with distance from the equator that is more gradual than the corresponding latitudinal variation of zonal flow at the visible surfaces of Jupiter and Saturn. Some theories of the origin of planetary and stellar magentic fields invoke differential rotation in their electrically-conducting fluid interiors as the main amplification process, but there are no direct observations of [u] in these regions.

Departures from axial symmetry in the pattern of relative motion of a rapidly-rotating fluid are to be expected even when the boundary conditions are axisymmetric. But the correct quantitative representation of the effects of non-axisymmetric features on the magnitude and form of the differential rotation is by no means straightforward and presents serious technical difficulties. Some of these can be overcome by the introduction of a "mixing hypothesis", which leads to considerable theoretical simplifications without sacrificing essentials. We have investigated differential rotation in a rotating spherical shell of incompressible fluid by assuming that non-axisymmetric motions act in such a way as to smooth out latitudinal gradients in potential vorticity[3]. The latitudinal profiles of Ω thus obtained depend *inter alia* on the thickness of the shell, exhibiting strong jets near the equator when the shell is thin and at mid-latitudes when the shell is thick. Our model was developed as an improvement on one proposed much earlier by Rossby, who considered the effects of horizontal mixing of radial filaments of fluid on the profiles on the assumption that mixing eliminates gradients of the vertical component of absolute vorticity poleward of a certain arbitrary latitude. In keeping with the constraints of the Proudman-Taylor theorem, we considered the behavior of axial filaments of fluid, supposing that each filament retains its coherence and, owing to the weakness of frictional effects, undergoes little change in its potential vorticity over timescales of typical displacements perpendicular to the rotation axis. These displacements are associated with local pressure gradients which, in a rapidly rotating fluid, act at right-angles to the displacements. It is remarkable that such a simple model can reproduce many of the observed features of the differential rotation of the Earth, Jupiter, Saturn and the sun. (Whether or not internal dynamical processes such as those studied in our paper can account for the enormous value of the super-rotation of the atmosphere of Venus, at more than ten times the speed of the underlying planet, is a matter for further investigation. Some workers have argued that such high values cannot be explained without invoking the action of external couples and have developed a model based on the action of the Sun's gravitational field on non-axisymmetric density variations associated with thermal tides.)

LABORATORY AND NUMERICAL STUDIES OF THERMALLY-PRODUCED MOTIONS IN ROTATING FLUIDS

Many features of the large-scale atmospheric circulation can be reproduced in a liquid filling a cylindrical annulus rotating about a vertical axis, when the inner and outer walls of the annulus are maintained at different temperatures. Laboratory studies over a wide range of impressed conditions have revealed the existence of several possible flow regimes: axisymmetric flow at comparatively low rotation rates (or high temperature differences), regular non-axisymmeric flow at intermediate rotation rates, and irregular non-axisymmetric flow at high rotation rates. Baroclinic waves associated with meandering jet streams are characteristic features of non-axisymmetric flows. Regular flows may be in the form of either steady or vacillating waves (in which periodic changes of amplitude or shape occur). Being spatially and temporally periodic, these wave flows are "forecastable" in the meteorological sense: but different regular flows may be observed in different experiments under the same impressed conditions. The irregular flows are aperiodic and only poorly "forecastable".[4]

Studies based on the joint use of laboratory systems and their counterparts in numerical models make it possible, amongst other things, to "verify" the basic dynamical structure of numerical models of rotating baroclinic flow in a way that is virtually impossible for atmospheric numerical models — in which important small scale processes are represented by comparatively crude and uncertain parametrizations. A high resolution numerical model based on the Navier-Stokes equations for incompressible flow is currently being used in work of this type. The numerical model reproduces most of the flow phenomena seen in the laboratory systems: axisymmetric flow, steady waves, intransitivity, wavenumber transitions, hysteresis, amplitude vacillation and irregular flow. (A convincing numerical simulation of shape vacillation has not yet been produced, however). Several detailed quantitative comparisons between laboratory measurements of steady waves and corresponding numerical simulations have been carried out, with encouraging results.[5]

An important element of our current program is the investigation of the nature of steady and vacillating wave flows. Laboratory experiments have demonstrated that vacillation occurs adjacent to transitions either to a lower wavenumber flow (amplitude vacillation) or to irregular flow (shape vacillation). Amplitude vacillation[6] is a doubly periodic flow whose spectral characteristics can be interpreted in terms of an amplitude and frequency modulated wave. Accounting for the precise conditions under which steady or vacillating waves can occur is still an unsolved problem. A hypothesis to be tested (using data from laboratory experiments and numerical integrations) is that steady waves can arise only when initial wave developments are strong enough to bring about large changes in the mean flow structure.

The ability of numerical models to cover combinations of parameters not readily attainable in the laboratory is being exploited in a study of the axisymmetric flow at very low rotation rates. Of particular interest here are the magnitudes of the mean azimuthal flow and the total heat transport, which compare well with predictions based on straightforward scaling and boundary layer theory. Experiments at higher rotation rates have been carried using two small annular convection chambers, one with internal heating and one with wall heating. Effects of varying the end-wall boundary conditions have also been investigated. These experiments appear to bear out some new ideas concerning the occurrence of non-axisymmetric flows which do not directly invoke baroclinic instability theory as a starting point[7].

Recent work on annulus flows under a variety of impressed temperature fields, obtained by heating (or cooling) the fluid internally and cooling (heating) the side-walls, are mentioned below, in the section on sloping convection in the laboratory and in the atmospheres of Jupiter and Saturn.

ANALYTICAL STUDIES OF LINEAR AND NONLINEAR WAVES IN ROTATING FLUIDS

Analytical studies play an important role in the formulation and interpretation of the above mentioned laboratory experiments and numerical simulations. Linear theories assist in the interpretation of the observed transitions from axisymmetric to regular wave flow, and from regular to irregular wave flow. They also provide possible explanations (not necessarily limited to small amplitude cases) of the existence of steady waves. Non-linear analyses bear more detailed comparison with the experimental and simulated flows, and also guide the formulation of numerical models.

Linear studies. The theory of baroclinic instability reveals many flows that are unstable, and certain flows which (because of dynamical constraints) are stable although they possess available potential energy. Some of the stable flows are similar in many respect to the mean flows found in the steady wave regime. One possible theoretical model of steady waves therefore consists of a neutral wave on a stable mean flow. Another is based on the *finite amplitude* steady waves and associated mean flows (governed nevertheless by *linear* equations) which arise as exact solutions of the quasi-geostrophic potential vorticity equation. These solutions are of general interest as analytical illustrations of the celebrated "non-acceleration theorem" of wave/mean interaction theory, and they also account for many of the gross features of real and simulated steady waves[8].

Both theoretical models are consistent with the distinctive mean flow structure of the steady wave regime; and more detailed diagnostic studies will be needed to determine which is the more appropriate model.

Nonlinear studies. During the 1970's Pedlosky and Drazin made considerable progress with the mathematically demanding problem of establishing analytical descriptions of weakly nonlinear baroclinic waves interacting with a mean flow. An important later development was the discovery (made by members of this laboratory in collaboration with others[9,10]) of soliton-type solutions for the propagation of baroclinic wave packets. In subsequent studies the conditions under which such solutions (and various kinds of less ordered behavior) occur in the weakly nonlinear models were delineated. In view of the implications for predictability of fluid motion in rotating systems, it is clearly important to determine how far the weakly nonlinear models are applicable to real fluid systems. Our numerical model results imply that the state of marginal stability adopted in the existing theoretical treatments is not the most appropriate, and further analytical investigation is suggested.

The weakly nonlinear models are specializations of quasi-geostrophic formulations which are widely used in meteorological theory. Another matter for consideration is the applicability of quasi-geostrophic models themselves to the real laboratory flows: diagnostic studies using numerical data from wave simulations are at present in progress to investigate this question. The use of quasi-geostrophic equations in our theoretical work is but one example of the application of approximate forms of the Navier-Stokes equations in geophysical fluid dynamics. In meteorological modelling, for instance, approximate formulations (such as the hydrostatic set) are invariably used. In spite of this, no systematic theory of approximation is yet available, and consequently several important issues are uncertain. Thus is is not clear to what extent the various properties of the original equations should be reproduced by the approximate forms. A theoretical case study based on the quasi-geostrophic equations has recently been completed[11]; it indicates that accuracy can be improved by retaining conservation properties in approximate formulations.

SLOPING CONVECTION IN THE LABORATORY AND IN THE ATMOSPHERES OF JUPITER AND SATURN

It has now been accepted that long-lived prominent markings seen on the visible surface of dense clouds on Jupiter and Saturn, such as the Jovian Great Red Spot and three White Ovals, are manifestations of atmospheric motions, so that their explanation must be given in terms of basic processes in fluid dynamics. The very existence of such features has important implications for theories of atmospheric predictability. There have been several incomplete suggestions as to the nature of the Great Red Spot. According to one idea it is the upper end of a Taylor column produced by the interaction between atmospheric motions and deep-seated topography (which might be "hydrogen-helium ice-floes"). In the so-called "soliton" or "modon" theories the stability of the spot is accounted for on the basis of a balance between dispersion due to the latitudinal variation of the vertical component of the Coriolis parameter ("beta-effect") and horizontal advection, with the soliton drawing its energy directly from the kinetic energy of the background zonal shear and the modon from the coalescence of smaller eddies. The hypothesis that the Great Red Spot is analogous to a terrestrial hurricane invokes small scale moist convection as the basic energy source, with friction playing a key role in organizing the flow but with the beta-effect playing only a modifying role[12].

According to a recent proposal[13,14] long-lived anticylonic eddies in the atmospheres of Jupiter and Saturn, including the Jovian Great Red Spot and White Ovals, might be manifestations of fully developed "sloping" or "slantwise" convection characteristic of quasi-steady thermally driven flows in a rapidly-rotating fluid of low viscosity subject to internal heating. Baroclinic eddies of this type derive their kinetic energy directly from the potential energy due to gravity acting on the variable density field maintained by differential heating and cooling. They were first discovered in laboratory experiments by Hide and Mason and they are now also being studied with the aid of numerical models. On Jupiter and Saturn they would transport heat from the lower middle parts towards the upper outer parts of the atmospheric zones in which they occur. In such an eddy, the upper level horizontal motion is largely concentrated in a jet stream circulating around the relative quiescent core of rising fluid in an anticylonic sense, with descending motion occurring in a narrow "collar" surrounding the jet stream. Theory predicts and numerical experiments confirm that stable eddies with *cyclonic* upper level horizontal circulation, *descending* motion in the core and **ascending motion** in a collar surrounding the cyclonic jet stream, would be characteristic of fully developed slantwise convection in a rapidly rotating fluid subject to internal cooling, and it has been suggested that the cyclonic "barges" on Jupiter and Saturn might therefore be manifestations of slantwise convection transporting heat from the lower outer parts towards the upper inner parts of the atmospheric belts in which they occur[15].

The "sloping convection" hypothesis has most in common with some of the numerical general circulation studies of Jupiter's atmosphere, in which latitudinal jet streams and *transient* baroclinic eddies are produced in integrations of a numerical model similar to those used to investigate the mid-latitude circulation of the Earth's atmosphere. But it is to laboratory experiments that we owe the demonstration that long-lived baroclinic eddies can exist over a wide range of conditions and that their stability is a consequence of the action of nonlinear advective effects and not to viscosity or to the particular geometry of the boundaries, which can modify the eddies in certain details without affecting their main properties.

New laboratory and numerical experiments mentioned in this summary have been undertaken with two distinct but related objectives in mind, namely (a) the extension of knowledge and deepening of insight into sloping convection in rapidly-rotating fluids, and (b) the improvements in our understanding of the structure and dynamics of Jupiter and Saturn. In due course we shall report further experiments bearing on the isolated nature of the long-lived eddies on Jupiter and Saturn and the origin of the transient small scale eddies associated with them.

ACKNOWLEDGMENTS

This survey was presented at two recent workshops, on "Instabilities in continuous media" held in Venice, Italy, in December 1982, and on "Predictability of fluid motions" in La Jolla, California, in February 1983. It was prepared with the assistance and advice of my colleagues Mr. R.T.H. Barnes, Dr. P. Hignett, Dr. I.N. James, Dr. P.L. Read, Dr. A.A. White and Dr. C.A. Wilson, to whom I must express my indebtedness.

REFERENCES

1. R. Hide, N.T. Birch, L.V. Morrison, D. Shea and A.A. White, Nature, **286** , 114-117 (1980).
2. R.T.H. Barnes, R. Hide, A.A. White and C.A. Wilson, Proc. Roy. Soc. London, A387, 31-74 (1983).
3. R. Hide and I.N. James. Geophys. J. Royal Astronomical Soc., **74**, (in press, 1983).
4. R. Hide and P.J. Mason, Advances in Physics, **24**, 47-100 (1975).
5. P. Hignett and A.A. White, "A comparison of laboratory measurements with numerical simulations of baroclinic waves in a rotating fluid annulus", (in preparation, 1983).
6. P. Hignett, "Spectral characteristics of amplitude vacillation in a differentially-heated rotating fluid annulus", (in preparation, 1983).
7. R. Hide, P. Hignett and A.A. White, "Axisymmetric thermal convection in a rotating fluid", (in preparation, 1983).
8. A.A. White, "Finite amplitude neutral baroclinic waves and mean flows: analytical illustrations of the Charney-Drazin non-acceleration theorem," (in preparation, 1983).
9. I.M. Moroz, J. Atmos. Sci., **38**, 600-608 (1981).
10. I.M. Moroz and J. Brindley, Proc. Roy. Soc. London, **A377**, 379-404 (1981).
11. A.A. White, "Approximate forms of the equations governing nearly-geostrophic motion: Part II - A case study based on the type 1 formulation," (in preparation, 1983).
12. P.L. Read and R. Hide, Nature, **302**, 126-129 (1983).
13. R. Hide, The Observatory, **100**, 182-193 (1980).
14. R. Hide, Meteorological Magazine, **110**, 335-344 (1981).
15. P.L. Read and R. Hide, Annales Geophysicae, (in press, 1983).

BLOCKING AND VARIATIONS IN ATMOSPHERIC PREDICTABILITY

B. Legras
Laboratoire de Météorologie Dynamique, CNRS, Paris 75231

M. Ghil
Courant Institute, New York University, New York, N.Y. 10012

ABSTRACT

We consider the equivalent barotropic vorticity equation on the sphere, with simplified forcing, dissipation and topography. Twenty-five modes are retained in a spherical harmonics expansion of the stream function. Solutions are studied as a function of the nondimensional intensity of the forcing and dissipation.

Stationary solutions show the nonlinear orographic resonance of Charney and DeVore[1]. A second resonance at lower, more realistic values of the forcing appears as a result of the larger number of modes. Among the multiple equilibria associated with the second, more complex resonance are a zonal and a blocked flow pattern which present marked similarities with the synoptically defined normal and blocked Northern Hemisphere midlatitude flows.

Wave-wave interactions influence strongly the stability properties of the equilibria and time evolutions of nonequilibrium solutions. The latter show persistent sequences which occur in the phase-space vicinity of the zonal and blocked equilibria. Composite flow patterns of the persistent sequences are similar to the equilibria nearby. The mean life times of the sequences are longer than in reality due to the lack of baroclinic processes in the model. A recurrent hyper-predictable sequence with both persistent and agitated episodes, lasting a total of 500 simulated days, is also exhibited and discussed.

INTRODUCTION

We consider the general question of excitation, maintenance and life span for quasi-stationary, mid-latitude, finite-amplitude waves in the atmosphere. This theoretical question is closely connected with their practical predictability. A particularly important quasi-stationary, large-amplitude configuration is *blocking*, which in its strongest manifestations is easily observed and accounts for a large part of low-frequency atmospheric variability.

During the last few years, many studies have been devoted to the blocking problem. The theoretical explanations include two main approaches. In the first one, the generation and the maintenance of the block are seen from a point of view which is local in the phase space of the flow's waves. The block is generated by resonant amplification of a free quasi-stationary Rossby wave, superimposed on a pre-existing zonal flow which interacts with stationary orographic or thermal forcing. The resonant wave's finite amplitude is maintained by propagation away from the source of excitation and by dissipation[2] or by nonlinear interactions with other waves[3].

The second approach[1] is global in phase space: it assumes that blocking is an alternative flow regime produced by nonlinear locking of interacting waves and is irreducible to the perturbation of a zonal flow. The perturbation methods of classical atmospheric instability theory are thus not very useful in this approach, at least not yet. Studies based on the second approach rely mainly on the crude approximation of low-order models which have been used extensively in meteorology since Lorenz.[4] Low-order models proved often satisfactory in explaining simple atmospheric mechanisms, but their practical relevance is mainly founded on heuristic arguments. Near special points in parameter space, the bifurcation points, perturbation methods corroborate most of the results of low-order models for the problem at hand.[5]

In their original paper, Charney and DeVore[1] (hereafter denoted by CDV) found two stable stationary solutions of a severely truncated barotropic channel model, and associated them respectively with the zonal and blocked atmospheric regimes. They suggested also that in the real atmosphere, transitions between the two regimes are induced by superimposed synoptic perturbations. Transitions have indeed been observed by adding stochastic noise to the CVD model[6] or by considering the action of synoptic modes in a baroclinic extension of the CDV model[7].

Observations of the large-scale mid-latitude circulation do not seem to support a simple bimodal distribution of atmospheric variables which would permit associating one maximum with "normal" zonal flow and the other one with "blocked" flow.[15] Those two regimes, if they are identifiable as such, cannot be purely stationary with superimposed perturbations. They are more likely to be represented by weakly separated regions in the flow's phase space, or in parameter space, each region having its own stability and persistence characteristics.

The *purpose of the present paper* is to start drawing such a picture of the large-scale circulation, in which flows of zonal and of blocked character are not stationary, but still distinct from each other. An essential ingredient in such a picture are wave-wave interactions. Indeed, the CDV model was quasi-linear in the sense that only wave-mean flow interactions were retained, and the orography acted as an intermediary between the waves and the mean flow. This ignores the fact that the large-scale Reynolds stress due to nonlinear wave-wave interactions is responsible for a large part of energy and enstrophy fluxes in the atmosphere. This stress cannot be adequately modeled by either a simple diffusion or by stochastic noise, and it may act systematically to reinforce or destabilize the block.[8]

In the present study we shall therefore examine the effects of direct nonlinear wave-wave interactions on the regimes of large-scale atmospheric flow. We consider an extension of the $\beta-$ plane model of Charney and DeVore to spherical geometry, including a larger number of degrees of freedom. This study does not address the important problems of baroclinic stability or baroclinic forcing of the block.

In the following sections we describe the model, present the model's stationary solutions and their dependence on the problem's nondimensional parameters, and discuss the stability of these solutions and their connection with dynamic flow behavior. Nonstationary regimes with zonal and blocked characteristics are exhibited and analyzed. We study the persistence of these regimes and their predictability. A phenomenon of hyper-predictability is illustrated. Concluding remarks follow.

THE MODEL

The model is governed by the equivalent barotropic equation of vorticity conservation

$$\frac{\partial}{\partial t}(\Delta - L_R^{-2})\psi + J\{\psi, \Delta\psi + \mathbf{f}\left(1 + \frac{\mathbf{h}}{H}\right)\} = \alpha\Delta(\psi^* - \psi) \ . \qquad (1)$$

Here ψ is the stream function, Δ is the Laplacian and J the Jacobian operator, f is the Coriolis parameter, h is the topographic height and H the scale height of the atmosphere. The right-hand side induces a relaxation towards the forcing stream function ψ^* with a characteristic time α^{-1}. L_R is the external radius of deformation.

In spherical geometry, the horizontal coordinates are the longitude ϕ and the sine of the latitude $\mu = \sin\theta$. All variables are scaled by the radius of the Earth a, its angular velocity Ω and a characteristic speed U, yielding the nondimensional variables

$$L_R = \mathbf{a}\ \lambda;\ \mathbf{h} = H\mathbf{h}';\ t = t'/2\Omega;\ (\psi,\psi^*) = \mathbf{a}U(\psi',\psi_*') \ ;$$

$$\alpha = 2\Omega\alpha' \ ; \quad \mathbf{f} = 2\Omega\mu \ .$$

The nondimensional form of the equation, suppressing the primes, is thus given by:

$$\frac{\partial}{\partial t}(\Delta-\lambda^{-2})\psi + \rho\, J\{\psi,\Delta\psi\} + J\{\psi,\mu(1+\mathbf{h})\} = \alpha\Delta(\psi^*-\psi)\ , \tag{2a}$$

where the nondimensional number $\rho = U/2\Omega\mathbf{a}$ is similar to a Rossby number and measures the intensity of the forcing. It multiplies the sole nonlinear term in (2a) and, as we shall see, plays the role of a critical parameter for the behavior of the solutions.

Equation (2a) is discretized through an expansion in spherical harmonics $Y_l^m(\phi,\mu) = P_l^m(\mu)e^{im\phi}$ and truncated at $|\mathbf{m}| \leqslant \mathbf{l} \leqslant 10$. We assume equatorial symmetry as well as a sectorial periodicity (mod π) in longitude. The resulting 25 real modes are circled in Figure 1. They allow 132 triadic nonlinear interactions. Each triad conserves energy and potential enstrophy, and so does the whole truncated system in the absence of forcing and dissipation. In particular, the topography acts solely as a catalyst transferring the energy betwen different scales.

The topography (Fig. 2) is depicted as the coarsest representation of the Northern Hemisphere, with two equal continental masses separated by two equal oceans:

$$\mathbf{h} = 4\mathbf{h}_0\mu^2(1-\mu^2)\cos 2\phi\ . \tag{2b}$$

The multiplication of $\mathbf{h}$ by the nondimensional Coriolis parameter μ leads to the term $\mu\mathbf{h}$ consisting of the modes marked by a cross in Figure 1.

The mean forcing is a zonal jet, expanded in the first two zonal components marked with a double circle in Figure 1, so that the maximum speed occurs near 50°N with a value of 60 ms^{-1} for $\rho = 0.20$. The truncation above allows an exact representation of the interaction between the forcing and the orography. The forcing ψ^* models, in the absence of explicit baroclinic effects, the mean thermal wind which would be observed in the case of an idealized purely zonal circulation, with no meridional heat or mass transfers.

The right-hand side of (1) models, at the same time as the forcing by ψ^*, the dissipation across a hypothetical Ekman layer. The characteristic relaxation time α^{-1} in the mid-troposphere is of the order of 10 days.

The radius of deformation is taken equal to 1100 km throughout, except in a discussion of hyper-predictability where $L_R = \infty$. The value of 1100km is a heuristic interpolation between an internal, baroclinic radius of deformation and the external barotropic one.

Substituting the truncated expansion of the variables in Equation (2), we obtain an autonomous system of 25 ordinary differential equations for the vector $\Psi(t)$ with components $\psi_l^m(t)$. To write this system explicitly would be tedious and uninformative, so we content ourselves with giving it in compact vector-matrix form:

$$\frac{d}{dt}\Psi = \rho\Psi^T B\Psi + A\Psi + C. \tag{3}$$

The first term on the right-hand side contains the quadratic terms coming from the truncated Jacobian $J_T\{\Psi,\Delta\Psi\}$, the second term groups the linear terms from the Coriolis effect, orography and dissipation, and the last term represents the zonal forcing.

STATIONARY SOLUTIONS

We study first the nature of the model's stationary solutions and follow their behavior as the parameters α and ρ change. This will help us examine later the nonstationary flow regimes.

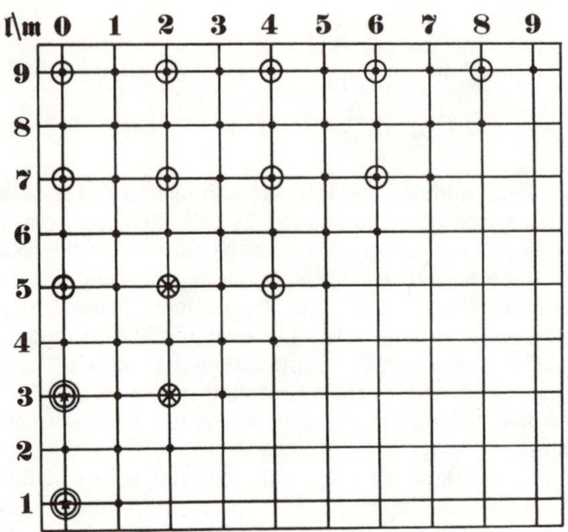

Fig. 1. Spectral truncation of the model in spherical harmonics: **m** is the zonal, and **l** the total wave number. Retained modes are circled. Topographic modes are crossed and forcing zonal modes are indicated by a double circle.

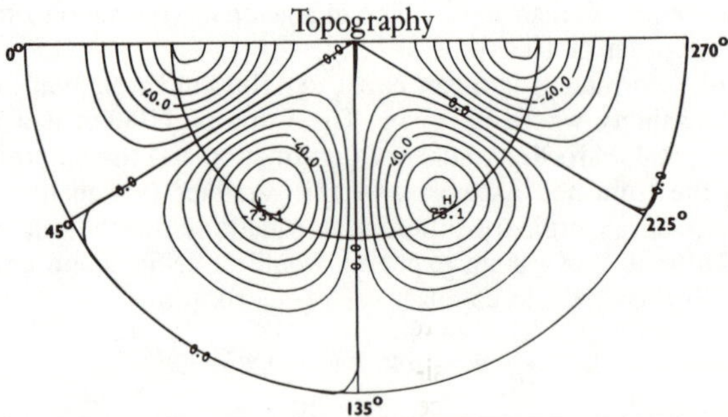

Fig. 2. Contours of the effective orography $\mu \mathbf{h}, \mu = \sin\theta$. This is a conformal conical projection of ratio 2/3. It maps a sector of the Northern Hemisphere located between 0° and 270° of longitude onto the half-disk shown. Topographic maxima and minima are indicated by H and L, respectively.

For α large, tending to infinity, the stationary solution of (3) is clearly unique and tends to ψ^*. For ρ large, tending to infinity, there still exists one solution tending to ψ^*. In the limit $\rho \rightarrow 0$, an equilibrium is reached which stays near ψ^* if α is at least of the same order as h. It is shown by Legras and Ghil[9], hereafter LGG, that these three asymptotic forms of the solutions are stable with respect to time-dependent perturbations.

Between these limits lies the interesting domain of the parameter space. Unlike in the CDV model, there is no way to solve analytically the steady-state form of Equation (3), i.e., the system of 25 nonlinear algebraic equations. Numerical solutions of the problem were found by the pseudo-arclength continuation method[10], which is a parameter-dependent refinement of a global Newton method. It enables one to explore completely a one-parameter branch of stationary solutions, eliminating the difficulties encountered at regular turning points. A simple exposition of the method and technical details on its application to the present problem are given in Appendix C of LGG.

In order to describe the distribution of stationary solutions as a function of the parameters, we have plotted in Figure 3 their total energy **F** as a function of ρ for different values of α. For $\alpha^{-1} = 1.1$ days (Curve A in Figure 3a) we are in the asymptotic domain of large α and the unique solution branch differs little from ψ^* for all ρ.

As α decreases, the relaxation no longer compensates for the destabilization of the mean flow by the orography, and waves are produced which interact with each other and feed back on the mean flow. As a result, the solution branch above is continuously distorted into a family of stationary solutions which has notably smaller energy than ψ^* and is characterized by a strong flux of energy extracted by the waves from the mean flow (curves B and C).

For $\alpha^{-1} \simeq 5.25$ days, a fold develops in this solution branch near $\rho = 0.5$, analogous to the one observed in the CDV model. This fold leads to the existence of three stationary solutions for the same value of the parameters (curve D of Figure 3a). Such a fold was also observed in Källén's[11] extension of the CDV model to the sphere. As α decreases further, we see (curve E, Figure 3b) an isolated closed branch which detaches itself from the main branch for $\alpha^{-1} \geqslant 8.5$ days and is present for $0.18 \leqslant \rho \leqslant 0.20$, down to small values of α. Next, another fold appears for $\alpha^{-1} \simeq 11$ days at $\rho \simeq 0.20$ and develops for $\alpha^{-1} \geqslant 15$ days (Curve G, Figure 3c) into a complicated structure. The first fold keeps the appearance of a unique nonlinearly distorted resonance, but the second seems to exhibit a cascade of multiple foldings (curve I, Figure 3e). The ρ–extent of the folds grows considerably as α goes to zero (cf. curves G and I in Figures 3c and 3e, respectively).

In order to compare these results with CDV, we performed a similar analysis on a quasi-linear version of our model in which all wave-wave interactions were suppressed. In this new system, the nonzonal components of a stationary solution satisfy a linear system with coefficients depending on the zonal state. This system shows nonlinear resonances for some values of the zonal flow due to the existence of small denominators. Still, this quasi-linear problem remains more complicated than that in CDV, since the system of equations for the zonal modes cannot be reduced to a single equation.

For all parameter values, the quasi-linear stationary solutions do not have components with zonal wave number $\mathbf{m} \geqslant 4$. Since the orography only affects the zonal modes with wave numbers m = 2, this means that the mean flow is barotropically stable to perturbations of order $\mathbf{m} \geqslant 4$.

Figures 3b (curve F) and 3d (curve H) show the cross-sections $E = E(\rho)$ of the quasi-linear model for $\alpha^{-1} = 10$ days and 20 days. The resonances observed in both figures for $\rho \geqslant 0.5$ are the images of the CVD orographic instability mechanism in the present model. Unlike CVD, two resonances are visible, especially in Figure 3d. It can be shown that the second resonance, at low ρ, is due to the addition of more degrees of freedom in the meridional direction: the number of resonances increases with the number of meridional modes, but most of them accumulate near $\rho = 0$.

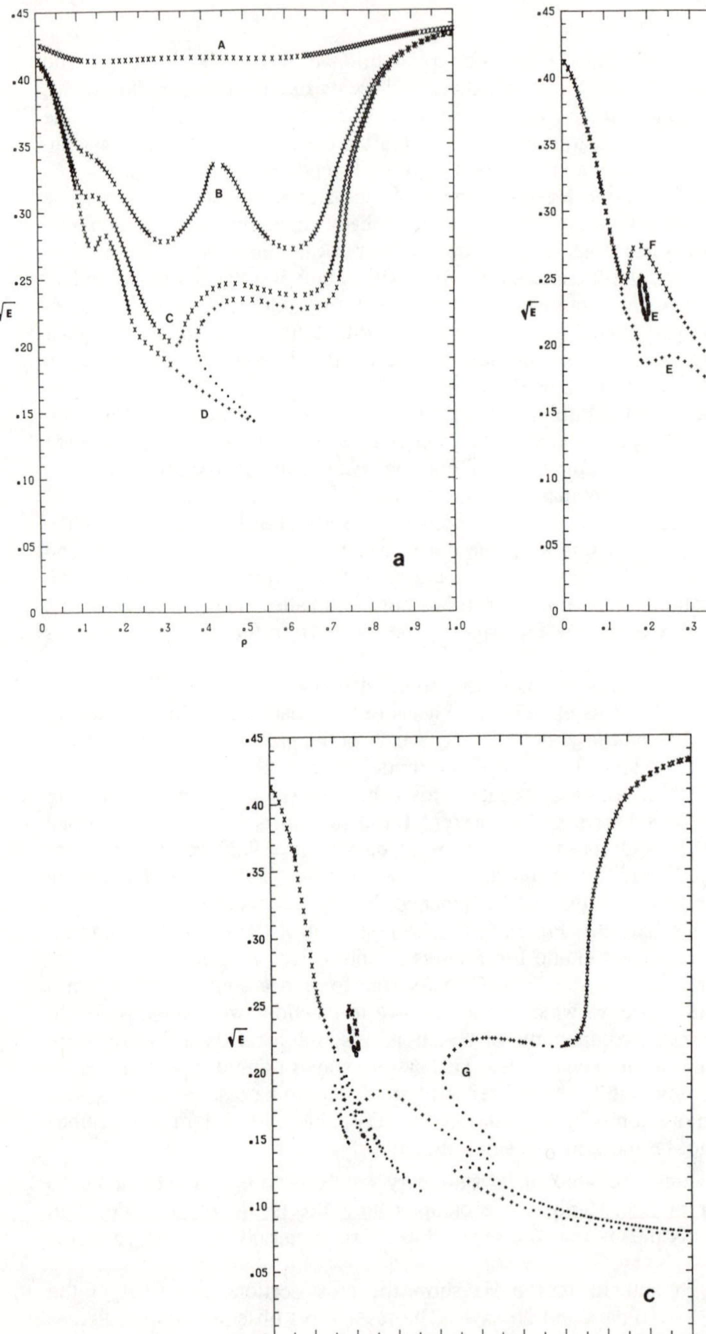
A
B
C
D
E
F
G
√E
ρ
a
b
c

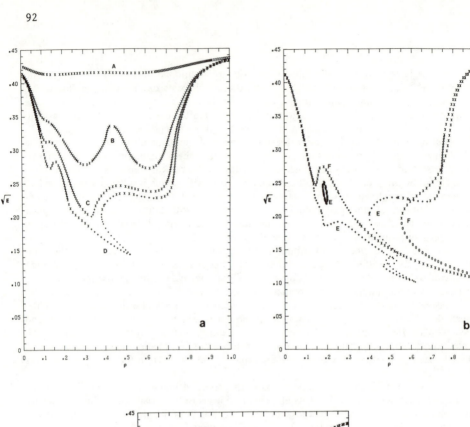

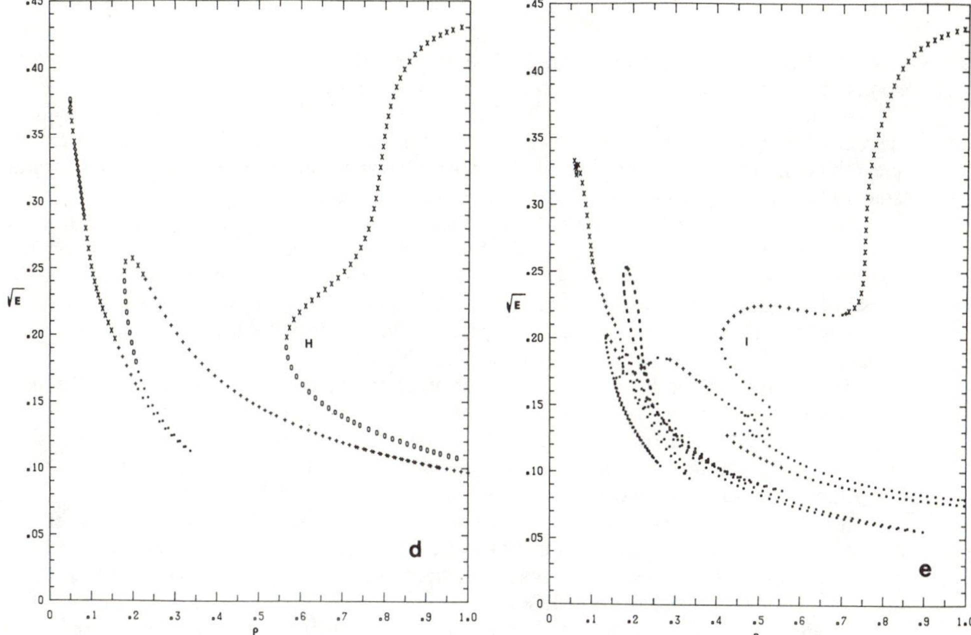

Fig. 3. Total energy E of stationary solutions as a function of the forcing parameter ρ for several fixed values of the dissipation parameter α. The stability of the solutions is denoted by the following symbols: ×, stable solution; 0, +, *, •, unstable solutions for which the eigenvalues of the linearization L_0 with positive real part are respectively: 0, one real eigenvalue; +, two complex conjugate eigenvalues; *, two real eigenvalues; •, at least three eigenvalues. Values of α^{-1} in days: (a) A, $\alpha^{-1} = 1.1$; B, $\alpha^{-1} = 3.3$; C, $\alpha^{-1} = 5.0$; D, $\alpha^{-1} = 6.7$. (b) E, $\alpha^{-1} = 10.0$; F, $\alpha^{-1} = 10.0$, quasi-linear model. (c) G, $\alpha^{-1} = 20.0$. (d) H, $\alpha^{-1} = 20.0$, quasi-linear model. (e) I, $\alpha^{-1} = 33.0$.

STABILITY AND DYNAMICS

Comparing the results of the quasi-linear model with the full models shows that the first resonance is only slightly deformed by nonlinearities. However, the occurrence of an isolated branch and of multiple foldings change completely the shape of the second resonance.

Stability of stationary solutions. The nonlinearities also change considerably the stability properties of solutions. The stability of stationary solution ψ_0 is determined by the distribution of the eigenvalues of the linear operator L_0 associated with the time-dependent perturbation problem around that stationary solution:

$$L_0(\chi) = (\Delta_T - \lambda^{-2})^{-1}\{\rho\, J_T(\psi_0,\Delta\chi) + J_T[\chi,\rho\Delta\psi_0 + \mu(1+\mathbf{h})] + \alpha\Delta_\chi\}\,. \tag{4}$$

Here $(\Delta_T - \lambda^{-2})^{-1}$ is the matrix inverse of the shifted, truncated Laplacian $(\Delta_T - \lambda^{-2})$.

The stability properties of solutions are indicated in Figure 3. Stable solutions are labeled by ×. Unstable states are labeled following the number of eigenvalues with positive real parts. Indeed, an unstable fixed point remains attracting within its stable manifold; the larger the dimension of this manifold, the greater the potential role of the fixed point in global dynamics. This simple rule is less valid when the stationary solution under discussion is imbedded in a limit cycle or an attractor with more complex stability properties.

In the quasi-linear case, for $\alpha^{-1} = 10$ days (Fig. 3b), the upper and lower branch of the CDV-like fold are stable, whereas the intermediate one is unstable. In the fully nonlinear model, only the upper part of the isolated branch and the asymptotic branches at large and small ρ are stable, all other branches are unstable.

At a given value of the parameters, the stablest solution is associated with the highest energy. This is confirmed in Figures 3c and 3d for $\alpha^{-1} = 20$ days. Here, points on the lower branches of both resonances of the quasi-linear model are also unstable. Their stability is lost to a stable limit cycle close, at given parameter values, to the corresponding fixed point. In the full model, a numerical integration starting near the lower branch at the first resonance leads to a limit cycle close to the upper branch.

The analysis of periodic solutions bifurcating from stationary ones can also be performed systematically, by various numerical means.[12] To trace out in systematic detail the next bifurcations, to aperiodic behavior, becomes rather difficult in a model of this size. Still, a large number of exploratory numerical integrations of (3) were carried out, for various parameter values and initial states (LGG).

It turns out that stable stationary solutions, as well as stable limit cycles, are observed. In a large domain of parameter space, however, the solutions remain aperiodic for an arbitrarily long time. The transition to chaos has been studied, and period-doubling sequences were found. This was described in LGG.

Inside the chaotic domain, where the dynamic behavior is very complex, some of the unstable stationary states will possess strong attractive properties: their stable manifold seems to absorb a large volume of the phase space flow. This induces recurrent, persistent sequences in the vicinity of such points, and will be illustrated subsequently.

Flow patterns. We conclude this section by studying the flow patterns exhibited by the stationary solutions along the branches of the second resonance. A striking resemblance to the pattern of atmospheric blocking is observed on some branches, together with the existence of more zonal stationary flows on other branches. By contrast, the first resonance, which is the analog of the one obtained by CDV, does not show very realistic patterns (LGG, Figures 5b-g).

We thus concentrate our attention on the second resonance area for $\alpha^{-1} = 20$ days. This value, which is twice the previously given estimate for the atmosphere, is chosen in order to obtain realistic time variations in this barotropic model.

We first present in Figure 4 an enlargement of the second resonance in Figure 3c. In this region of parameter space, the solutions' flow patterns can be classified into four families — Blocking, Zonal 1, Zonal 2 and Double Block — which correspond to the various branches as denoted in Figure 4. Representative examples of the solution families identified in Figure 4 are shown in Figure 5. Inside each family there exist amplitude and phase variations, but the general pattern of the solution remains unchanged. Transitions along branches between Blocking, the Double Block and Zonal 2 occur quite sharply in the hatched areas of Figure 4.

The Zonal 1 flow (Fig. 5a), associated with the isolated branch, has a high energy level: the maximum intensity of its zonally-averaged jet is 50 ms^{-1}. The Zonal 2 (Fig. 5b) flow is less intense, with a 35 ms^{-1} jet in zonal average. It exhibits in fact a ridge on the west side of the orography and a trough on the east side, similar to the averaged winter circulation of the northern hemisphere. It may thus be associated with the regular weather regime, and we shall do so in the sequel.

The west-coast ridge intensifies strongly in the blocking case (Fig. 5c) which shows a well developed high center on the west side of the orography. The averaged zonal wind is reduced to 18 ms^{-1} and the geopotential height difference between the trough and the ridge is about 1000 m. We show also in Figure 5d one solution of the quasi-linear model, located

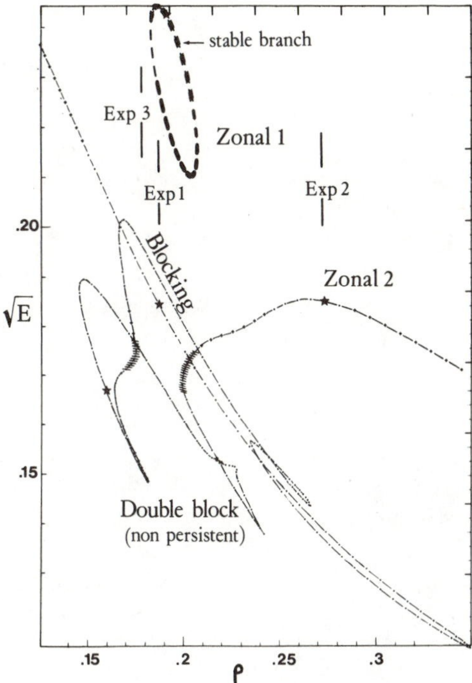

Fig. 4. Enlargement of a portion of Figure 3c. Stationary solution branches associated with the second resonance for $\alpha^{-1} = 20$ days. Same symbols as in Figure 3. Hatched segments show rapid transitions between Blocking, Double Block and Zonal 2 branches.

on the lower branch of the second resonance for the same value of the parameters as in Figure 5c. Both patterns are quite similar, although the quasi-linear solution is shifted eastward and possesses a weaker zonal wind.

The double-block type of solution is shown in Figure 5e. This family, as we shall see, is weakly attractive and does not seem to play a role in the dynamics.

The only stable solution type is Zonal 1, lying on the upper portion of the isolated branch; all others are unstable. We proceed to investigate numerically the dynamics of the time-dependent solutions. Three of these experiments, indicated in Figure 4, are presented below in detail.

PERSISTENCE AND PREDICTABILITY

Persistence. We adopt as a measure of persistence the quantity

$$P(t) = \frac{\|\Psi(t + \Delta t) - \Psi(t)\|}{\Delta t}; \tag{5}$$

here $\|\Psi_1 - \Psi_2\|$ is the Euclidean distance between the vectors Ψ_1 and Ψ_2, and Δt is a sampling time interval approximately equal to two days. $P(t)$ measures the speed on the orbit in phase space.

Figure 6a shows the variations of $P(t)$ in the first experiment, conducted for $\rho = 0.19$. We observe large variations which denote irregular motion in phase space. The minima of

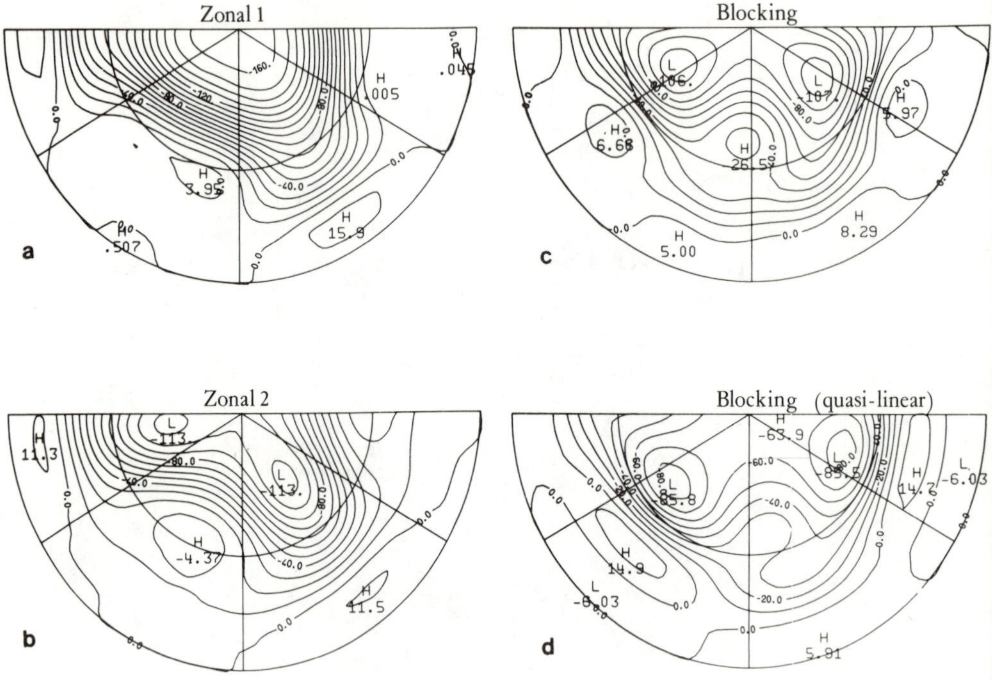

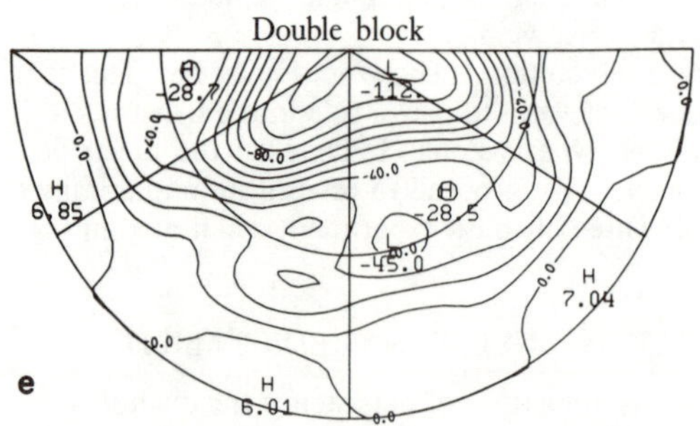

Fig. 5. Flow patterns of the stationary solutions indicated by stars in Figure 4: (1) Zonal 1; (b) Zonal 2; (c) Blocking; (d) Blocking in quasi-linear model (curve H, Figure 3d); (e) Double Block.

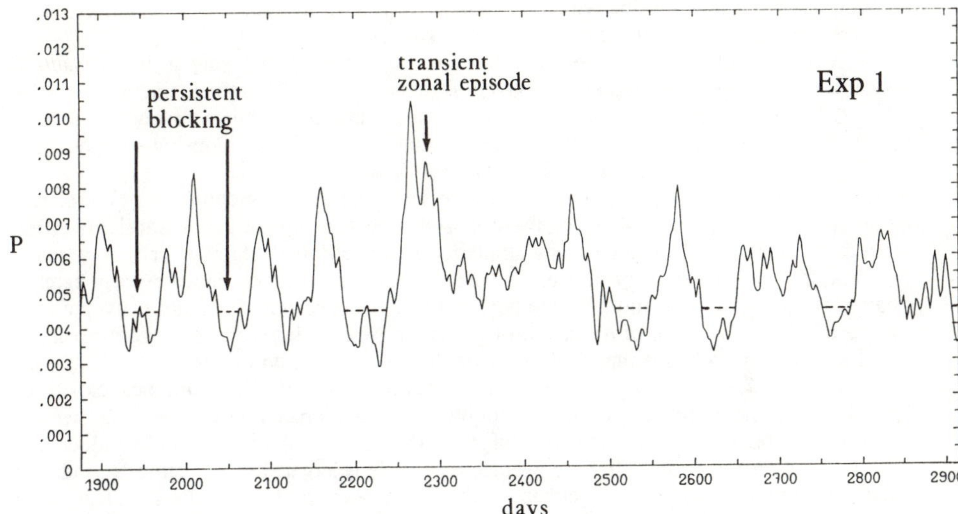

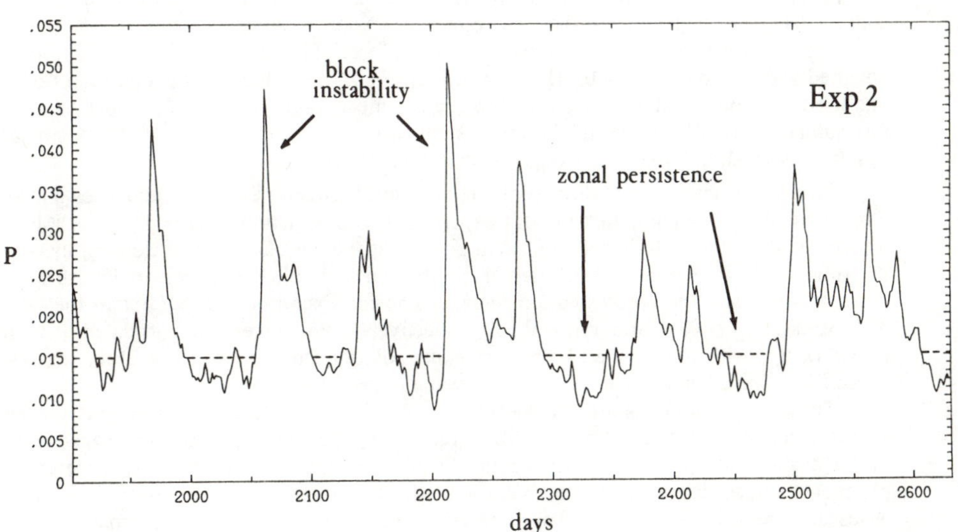

Fig. 6. Variations in time of the persistence measure P(t), Equation (5). (a) Experiment 1, $\rho = 0.19$; (b) Experiment 2, $\rho = 0.27$. Both experiments have $L_R = 1100$km. The offset value P_0 in each case is indicated by a dashed line.

the curve, where P(t) remains low for durations of up to one month, represent persistent sequences of the flow, during which the flow pattern is almost stationary. It turns out that all persistent sequences belong here to the blocking family.

A more precise definition of persistence can be given by choosing an offset value, P_0. Averaging the flow patterns during a given sequence where P(t) remains below the offset, $P(t) \leqslant P_0$, produces a *composite pattern.* Such a composite is shown in Figure 7a for an offset $P_0 = 0.0045$: it possesses strong similarities to the stationary blocking solution shown in Figure 5c. All other composites are very close to the one shown.

Two consecutive persistent sequences are separated by rather complicated episodes. It is interesting to note that most of these episodes possess almost zonal transients. These transients, however, do not bear any similarity with the Zonal 1 family for they remain at low energy. It is in fact very difficult to observe a Zonal 1 solution, since even integrations started from a vicinity of the unstable portion of the isolated branch do not converge to the stable portion. The basin of attraction of the latter appears therefore to be very small. We shall see below how this is modified by changing the deformation radius λ.

The double-block type of solution is also hardly observed. The numerical experiment just discussed had initial data close to a double-block solution, but never returned there.

Figure 6b shows the variations of P(t) in the second experiment, conducted for $\rho = 0.27$. We observe first that the average amplitude is larger than in Figure 6a, indicating that the solution wanders over a larger volume in phase space than previously. The minima of the curve, which are rather flat compared to the maxima, still define sequences of relatively persistent flow. But the observed behavior is reversed with respect to the first experiment; persistent sequences show zonal patterns and transient blocks occur during episodes of rapid motion. Unlike the first experiment, when a block tries to become established, it is immediately destabilized and the flow returns to an almost zonal pattern.

Persistent sequences are composited as before. Figure 7b shows one composite obtained with the offset $P_0 = 0.015$ given in Figure 6b; this value of P_0 is equivalent to the one used in Experiment 1, if we normalize by the mean value of P(t). The flow pattern of this composite is comparable to the zonal stationary solution in Figure 5b: strong similarities are observed, although the trough is weakened.

In order to check more directly the connection between persistent sequences and stationary solutions, we computed in both experiments the deviation $D(t) = \|\Psi(t) - \Psi_s\|$ to a prescribed stationary solution Ψ_s and then computed the correlation coefficient r_{PD} between the time series P(t) and D(t). When Ψ_s is the blocked solution shown in Figure 5c for Experiment 1, and the zonal solution in Figure 5b for Experiment 2, we obtain the rather high values $r_{PD} = 0.44$ and $r_{PD} = 0.54$, respectively. We have also correlated P(t) for Experiment 2 with te deviation from a blocking solution on the lower branch of the resonance for $\rho = 0.27$ and found the very low value $r_{PD} = 0.04$.

These correlation values corroborate the previous analysis, and show that the persistent sequences are due to the recurrent trapping of the orbit in phase space near certain unstable stationary solutions. This behavior obtains not only for the experiments and parameter values discussed above, but for a whole ρ–interval starting from the onset of chaos at $\rho \simeq 0.16$ up to $\rho \simeq 0.4$. The transition between blocked and zonal persistent regimes occurs, for $\alpha^{-1} = 20$ days, around $\rho \simeq 0.22$; it has not yet been studied in detail.

The statistics of the persistent sequences are shown in Figure 8, where the number of events per simulated year is plotted against duration of events for both Experiments 1 and 2. Equivalent offset values of $P_0^{(1)} = 0.0045$ and $P_0^{(2)} = 0.015$ are used. Zonal persistence occurs more frequently in Experiment 2 than blocking in Experiment 1. The e-folding life time of zonal persistence is also larger than for blocking: 22 days against 15 days. Both average values are larger than those suggested for atmospheric anomalies by Dole and Gordon[15], based on northern hemisphere mid-latitude winter data.

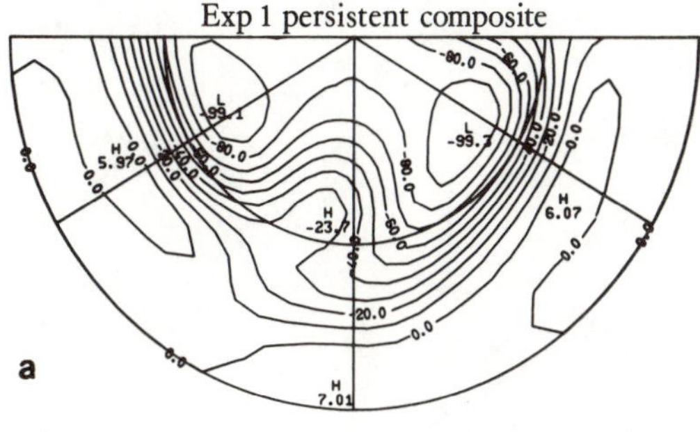

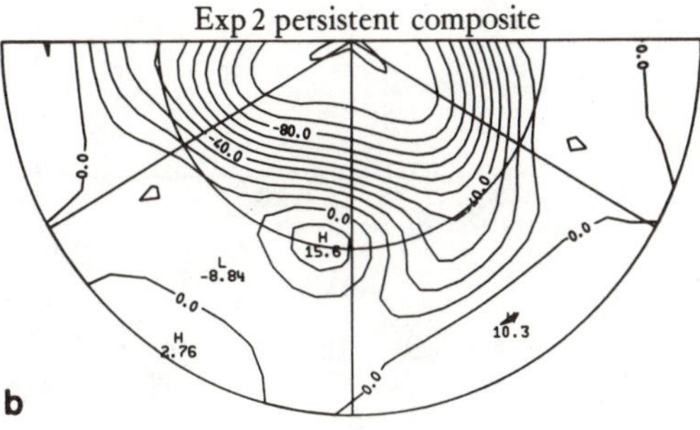

Fig. 7. Composite of a persistent sequence: (a) Experiment 1; (b) Experiment 2. Compare with Figures 5c and 5b, respectively.

The plots in Figure 8 illustrate the fact that complicated, nonlinear, deterministic dynamics produce not only red-noise-like power spectra (LGG, Figure 8), but also the type of persistence statistics often associated with random waiting times[16]. The exact values of the e-folding time for blocking and for zonal persistence in the model depend on parameter values, in particular on the ρ–distance from transition between one type of flow or the other being dominant.

Predictability. Aside from the existence of persistent sequences, we have observed in both experiments that, during the episodes of rapid variations, there exist some reproducible sequences of events which are quite easily recognized by eye when looking at the time series of stream function fields. These events occur in the same sequence, but they do not follow the same timing from one case to the other. This variation in relative duration makes such a recurrent sequence difficult to analyze by the usual objective methods. It suggests that the predictability concept should be split, at least in certain cases, into pattern predictability and phase predictability. None of these visually reproducible sequences appears however to last longer than about 20 days.

A local rate of divergence of trajectories is given by the largest real part of the eigenvalues of L_0, denoted σ_{max}. This quantity is plotted in Figure 9 for Experiment 2. The averaged e-folding divergence time for solution dispersion is about 12 days, but it varies considerably from 3 to 50 days. Comparing this with Figure 6b, we see that small values of σ_{max} are associated with the onset of persistent sequences, e.g. at 2100 days, and that large values are associated with the onset of rapidly varying sequences, e.g. at 2200 days.

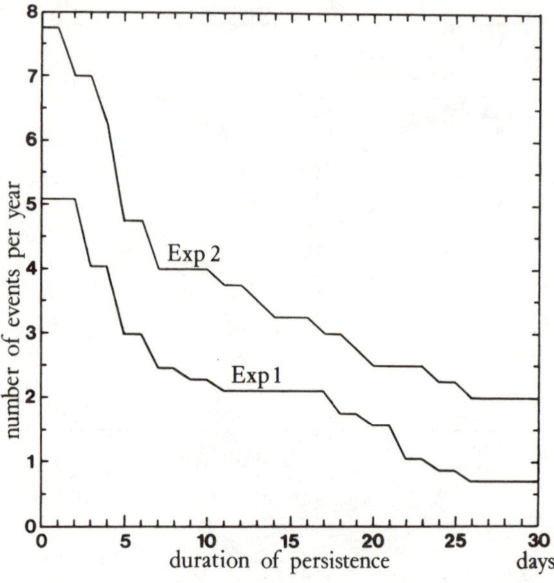

Fig. 8. Number of persistent sequences per year which exceed the number of days indicated on the abscissa, for Experiments 1 and 2 respectively. The offset value is $P_0 = 0.0045$ for Experiment 1 and $P_0 = 0.0150$ for Experiment 2.

It is interesting to note that aside from the onset periods, persistent and rapidly varying sequences do not differ significantly in their divergence rates. Similar results are obtained for Experiment 1. This suggests two kinds of competing phenomena: first, a strong instability along the unstable manifold of the fixed point, which ejects the trajectory from its vicinity: second, a contraction to the stable manifold of the fixed point, at some distance from it. This second phenomenon appears to have been studied very little. It can lead to increased predictability, as we shall see below.

Hyper-predictability. Until now, we have assumed that the Rossby radius of deformation L_R is equal to 1100 km. L_R has no effect on the distribution of stationary solutions, but influences their stability and the dynamics of time-dependent solutions. A small value of L_R favors the excitation of a large-scale stationary response to the orographic forcing, like blocking, since it reduces the phase speed of planetary waves. On the other hand, we may expect for $L_R = \infty$ to observe a more zonal circulation.

This is the case if we reproduce Experiment 1 with $L_R = \infty$. Then the basin of attraction of the stable Zonal 1 branch is large and all trajectories will eventually end on this solution. It is more interesting to perform an experiment for a third value of ρ below the interval of existence of the isolated branch, $\rho = 0.17$.

Figure 10 shows the variations of P(t) for Experiment 3. We observe, as in Experiment 1, some cases of persistent blocking, but also a long persistent sequence of Zonal 1 type close to the isolated branch.

Figure 11 shows the variations of σ_{max} for the same experiment. We observe that strong contracting properties are associated with the zonal persistent event, since σ_{max} remains negative for about 130 days. The most striking result is shown in Figure 12, where we have plotted the variations of zonal angular momentum M with time for several segments of the extended Experiment 3, when similar zonal events obtain. We observe an

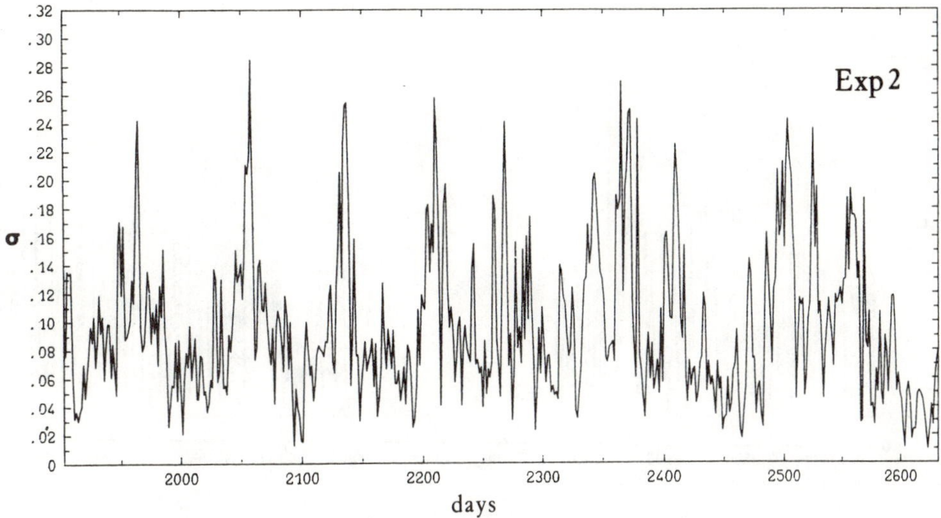

Fig. 9. Variations in time of the local rate of divergence of trajectories, σ_{max}, in Experiment 2; units of (days $^{-1}$) on the ordinate.

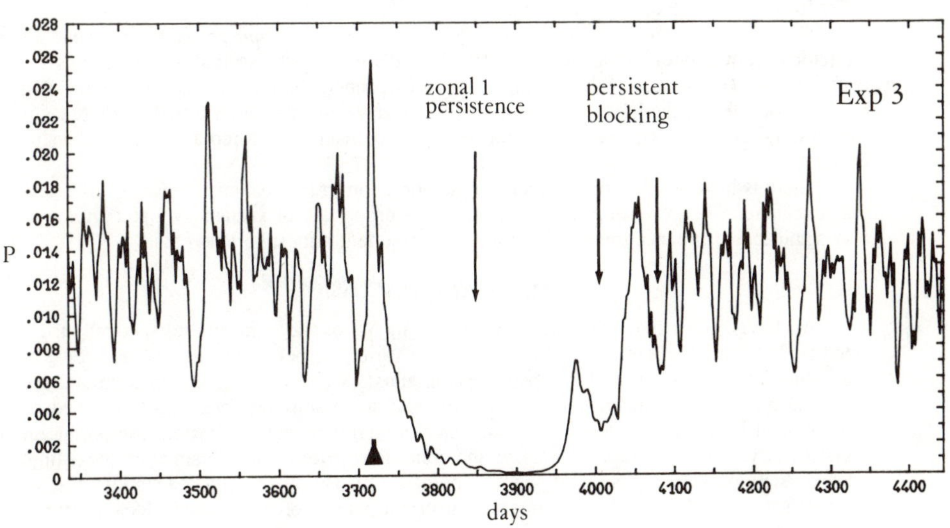

Fig. 10. Variations of P(t) in Experiment 3, $\rho = 0.17$, $L_R = \infty$. Notice the very long persistent sequence the start of which is marked by a heavy arrow point.

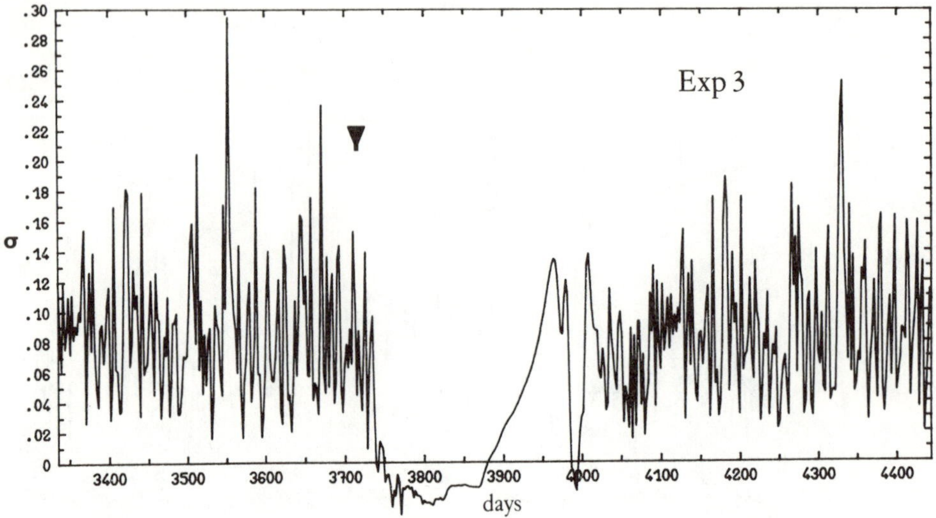

Fig. 11. Variations of σ_{max} in Experiment 3; units of (days $^{-1}$) on the ordinate. Notice σ_{max} negative near 3800 days.

almost exactly reproducible sequence, starting from the onset of the very persistent zonal event, and ending far beyond its termination. The overall length of the reproducible sequence is about 520 days, including not only zonal but also blocked episodes (cf. Fig. 10). Aside from this extraordinary event, the reproducibility properties of Experiment 3 do not differ from those of Experiments 1 and 2.

The following scenario is likely to explain this case of *hyper-predictability*. The model's attractor becomes one-dimensional close to the isolated branch, so that all trajectories are contracted to a prescribed path. At the exit of this funnel, the trajectories are so close that they do not diverge significantly for a few hundred days, even with an e-folding divergence time around 10 days. This sequence ends on a strong instability, when a second eigenvalue changes abruptly.

There is little hope to observe such a phenomenon in the real atmosphere or in a general circulation model. Still, special stabilization properties of certain synoptic situations may significantly enhance the predictability of subsequent atmospheric events.

SUMMARY AND DISCUSSION

We have shown that a nonlinear barotropic model of the atmosphere, spectrally truncated to 25 spherical harmonics, possesses a large variety of behavior patterns, according to the values of the forcing and dissipation parameters. Wave-wave interactions among the nonzonal modes were shown to destabilize the stationary solutions previously obtained by Charney and DeVore[1], and lead to the existence of additional solutions, both stationary and nonstationary. Realistic-looking blocked and zonal flow patterns are obtained as coexisting, unstable stationary solutions. The phenomenological realism of the blocked stationary solution branch might be due to the roughly barotropic nature of established blocking in the atmosphere[17].

Recurrent, persistent sequences, zonal or blocked, are observed near the corresponding unstable stationary solutions in time-dependent integrations of the model. The persistence properties of solutions depend on the intensity of the forcing and on the Rossby radius of deformation.

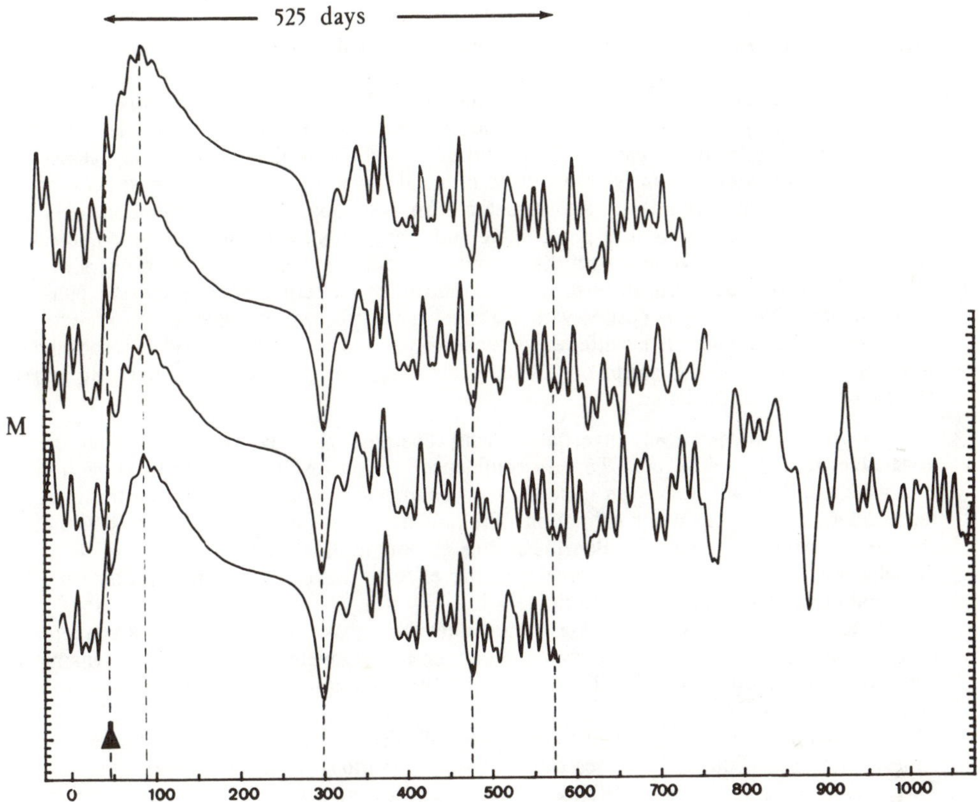

Fig. 12. Recurrent, hyper-predictable sequence in Experiment 3. Variations in time of the solution's zonal angular momentum M, over four different intervals. Each interval contains a zonally persistent sequence like the one in Figure 10, starting at the heavy arrow point. All other quantities are equally predictable for this sequence.

There is a demarcation zone in parameter space between a region where zonal flow is more persistent, and one where blocked flow is more persistent. In either case, there appears to be no preferred time scale of persistence. The mean life time of blocked and zonal episodes depends on the distance from the demarcation zone.

One can imagine that the forcing and dissipation in this barotropic model represent certain types of boundary conditions for the mid-latitude, baroclinic atmosphere. These conditions, such as equatorial sea-surface temperature and pressure anomalies, may change from one season or year to another. As a result, blocked or zonal flow will prevail at middle and high latitudes during the corresponding period, while transitions on shorter time scales between two types of flow still occur.

It is tempting to associate a hyper-predictable, mainly zonal sequence of events exhibited by the mode, for relatively high values of the forcing jet maximum, with last winter's predominantly zonal, very anomalous circulation over the North pacific and North America. Due to the model's low resolution and many simplifying assumptions, such an association might be purely fortuitous.

In particular, baroclinic processes are likely to affect the stability and persistence properties of solutions more than they affect the flow patterns themselves. It is essential to include such processes in order to study in greater detail the flow's time dependence and predictabiility.

Our model results so far indicate that the atmosphere's predictability properties are very inhomogeneously distributed in phase space. They can hardly be characterized by a few numbers, such as the linear divergence time of neighboring trajectories. For instance, predictability is low on the unstable manifold of a stationary solution, but increases considerably in contracting regions of the attractor. An operational concept of predictability should therefore be made local in phase space and time. This corresponds to the empirical observation that certain synoptic patterns are more predictable or persistent than others.

It might also be worthwhile to consider separately pattern predictability and phase predictability. The first is associated with the topological structure of the system's attractor, i.e., how its sheets connect the different regions of phase space. The second is associated with the evolution of the orbits on the attractor, and it is related to the dynamic and ergodic properties of the latter.

In order to systematically investigate such properties, truncated low-order models are a valuable, semi-empirical tool. To gain confidence in their results, one has to know how much these will be affected by an increase in the number of modes retained in a truncated representation of the equations of motion. Preliminary investigations with 100 modes (not shown) and with 250 modes (R. Benzi, personal communication) suggest that many of this model's properties persist when approaching the effective resolution of observational systems and of some general circulation models.

Even so, spherical harmonics were chosen here as a basis only for technical convenience. Center manifold theory[13] shows that under fairly general circumstances, dynamic behavior of an infinite-dimensional system near bifurcation points is actually restricted to a finite set of degrees of freedom.

One may attempt to guess the dimension of such center manifolds for large-scale atmospheric flow. Educated guesses can be based on various independent considerations. The number of parameters describing synoptic features on a weather map, the effective resolution of numerical weather prediction and general circulation models, and the study of atmospheric analogues and the divergence of subsequent weather sequences[14] all suggest an order of magnitude of 10^3–10^4 degrees of freedom.

To find special coordinates which describe, at least locally in phase space, a center manifold for the atmosphere, and which are practical to use, would seem to be an interesting pursuit. Again, the study of low-order models with arbitrary basis functions can provide some guidance in this pursuit.

ACKNOWLEDGEMENT

It is a pleasure to acknowledge discussions with C.E. Leith and R. Sadourny, and the editorial care of G. Holloway and B.J. West. This research was supported by NASA grant NSG-5130 and by the Centre National d'Etudes Spatiales.

REFERENCES

1. J.G. Charney and J.G. DeVore. J. Atmos. Sci., **36**, 1205 (1979).
2. K.K. Tung and R.S. Lindzen. Mon. Wea. Rev. **107**, 784 (1979).
3. J. Egger. J. Atmos. Sci., **35**, 1788 (1978).
4. E.N. Lorenz. J. Atmos. Sci., **20**, 130 (1963).
5. J. Pedlosky. J. Atmos. Sci., **38**, 2626 (1981).
6. J. Egger. J. Atmos. Sci., **38**, 2606 (1981).
7. B.B. Reinhold and R.T. Pierrehumbert. Mon. Wea. Rev. **110**, 1105 (1982).
8. E. Kalnay-Rivas and L.O. Merkine. J. Atmos. Sci., **38**, 2077 (1981).
9. B. Legras and M. Ghil. J. Méc. Théor. Appl., Special issue on two-dimensional turbulence, in press (1983).
10. H.B. Keller, in C. de Boor and G.H. Gollub (eds.), *Nonlinear Analysis*, Academic Press, New York, 73 (1978).
11. E. Källén. Tellus, **34**, 255 (1982).
12. M. Ghil and J. Tavantzis. SIAM J. Appl. Math., in press (1983).
13. J. Carr. *Applications of Center Manifold Theory*, Springer-Verlag, New York/Heidelberg/Berlin, 142 (1981).
14. E.N. Lorenz. J. Atmos. Sci., **26**, 636 (1969).
15. R.M. Dole and N.D. Gordon. Mon. Wea. Rev., in press (1983).
16. M. Ghil, in M. Ghil, R. Benzi and G.Parisi (eds.), *Turbulence and Predictability in Geophysical Fluid Dynamics and Climate Dynamics*, North-Holland Publ. Co., Amsterdam/New York/oxford, in press (1984).
17. R.M. Dole. *Persistent Anomalies of the Extratropical Northern Hemisphere Wintertime Circulation*, Ph.D. Thesis, MIT, Cambridge, MA 02139, 225 (1982).

PREDICTABILITY AND ENERGY TRANSFER PROPERTIES OF A QUASI-GEOSTROPHIC, UNIFORM POTENTIAL VORTICITY, LOW-ORDER DYNAMICAL SYSTEM

William Blumen
University of Colorado, Boulder, CO 80309

ABSTRACT

A low-order model of quasi-geostrophic uniform potential vorticity is considered. The depth-integrated total energy and the available potential energy on the boundaries are both conservative properties of the fluid motion. Energy exchanges between three scales of motion involving six Fourier amplitudes are analyzed. The exchange of energy between the wave components satisfies Fjørtoft's [1] theorem on scale interactions at any instant. However, the system exhibits an inherent unpredictability, and the ordering of the large, intermediate and small-scale waves changes with time. Consequently, the direction and rate of energy transfer between the wave components becomes unpredictable. It is demonstrated that these properties of the model are fundamentally different than those associated with a low-order quasi-geostrophic model involving three amplitude equations with two quadratic constraints on the motion.

INTRODUCTION

The theoretical problem of atmospheric and oceanic predictability may be posed in different ways. However, an important aspect of this problem is to establish some fundamental relationships between the characteristic energy transfer properties of fluid motions and the constraints, or conservation principles, imposed on the system. Here some transfer properties that characterize quasi-geostrophic, low-order, dynamical models constrained by two conservation principles will be examined.

In particular, the predictability and energy exhanges that characterize a model consisting of three equations for the Fourier amplitudes will be contrasted with a model consisting of six amplitude equations. Although the models considered here are far removed from real world fluid motions of the atmosphere and oceans, they do serve to isolate both physical and mathematical characteristics of quasi-geostrophic dynamics and to represent the building blocks for more complex models.

CONSERVATION PRINCIPLES

The three quadratic conservation principles of quasi-geostrophic dynamics are conservation of:

energy

$$\frac{d}{dt} \iiint |\nabla\psi|^2 \, dxdydz = 0, \tag{1}$$

potential enstrophy

$$\frac{d}{dt} \iiint |\Delta\psi|^2 \, dxdydz = 0, \tag{2}$$

available potential energy on level boundaries

$$\frac{d}{dt} \iint |\psi_z|^2 \, dxdy = 0, \tag{3}$$

where ψ represents the geostrophic streamfunction. When the boundaries are omitted, then only (1) and (2) apply. If the boundary contributions at $z = \pm\frac{1}{2}$ are retained, for uniform potential vorticity flow, then (1) and (3) apply. With $\Delta\psi = 0$, (1) may be expressed as the conservation of the depth-integrated energy

$$\frac{d}{dt} \iint \psi\psi_z \Big|_{z=-\frac{1}{2}}^{\frac{1}{2}} dxdy = 0. \tag{4}$$

The isolation of a wavenumber triad, for which the sum of the wavenumbers vanish, leads to

$$a_1^2 + a_2^2 + a_3^2 = E(0), \tag{5}$$

$$\kappa_1^2 a_1^2 + \kappa_2^2 a_2^2 + \kappa_3^2 a_3^2 = F(0), \tag{6}$$

where κ_i $(i = 1,2,3)$ denotes wavenumber, and the $a_i^2(\kappa_i, t)$ represent the energies associated with the Fourier amplitudes of ψ. These expressions represent the reduced forms of (1) and (2) when boundaries are omitted, e.g., Charney.[2] Similarly, Blumen[3,4] has shown that the isolation of a low-order system for uniform potential vorticity flow reduces (4) and (3) to

$$M_1 + M_2 + M_3 = \mathcal{E}(0), \tag{7}$$

$$\lambda_1 M_1 + \lambda_2 M_2 + \lambda_3 M_3 = \mathcal{G}(0), \tag{8}$$

where

$$M_i = (a_i^2 + b_i^2)\kappa_i \sinh \kappa_i/2 \cosh \kappa_i/2, \qquad (9)$$

$$\lambda_i = \frac{m_{Si}a_i^2 + m_{Ai}b_i^2}{a_i^2 + b_i^2}, \qquad (10)$$

$$m_{Si} = \kappa_i \tanh \kappa_i/2, \quad m_{Ai} = \kappa_i \coth \kappa_i/2. \qquad (11)$$

The quantities (m_{Si}, m_{Ai}) are characteristic vertical wavenumbers of the symmetric and antisymmetric eigenfunctions, and λ_i (κ_i, t) represents the ratio of the available potential energy on the boundary to the depth-integrated energy. This latter dynamical system consists of six coupled equations for the Fourier amplitudes (a_i, b_i) and are constrained by at least two conservation principles (7) and (8).

FJØRTOFT'S THEOREM ON SCALE-INTERACTIONS

The two conservation principles (5) and (6) place a strong constraint on the energy exchanges between the spectral components. Fjørtoft[1] has shown that "the change in energy for the component with the intermediate scale will be opposite to the changes in energy of the other two components." This constraint implies that only the intermediate scale can serve as a source or sink of energy for the other two components.

Charney[2] has provided a mechanical analogue for the interpretation of (5) and (6), which is shown in Fig. 1. This figure illustrates why three wave components must participate in any energy exchanges, and why a unidirectional energy transfer would violate the combined constraints.

Moreover, the motion described by the coupled set of three equations for the amplitudes is periodic, and consequently the motion is predictable. One way to show this is to note that (5) and (6) represent equations for a sphere with radius $[E(0)]^{\frac{1}{2}}$, and for an ellipsoid with semiaxes $[F(0)]^{\frac{1}{2}}\kappa_1^{-1}$, $[F(0)]^{\frac{1}{2}}\kappa_2^{-1}$, $[F(0)]^{\frac{1}{2}}\kappa_3^{-1}$; the motion lies along the line of intersection of the sphere and ellipsoid, shown in Fig. 2. The motion is described by closed paths and must be periodic. This problem is analogous to the application of Euler's equations to the free rotation of an asymmetrical top, for which the three moments of inertia are all different. As shown by Landau and Lifshitz[5], for example, the periodic motion may be described in terms of Jacobian elliptic functions.

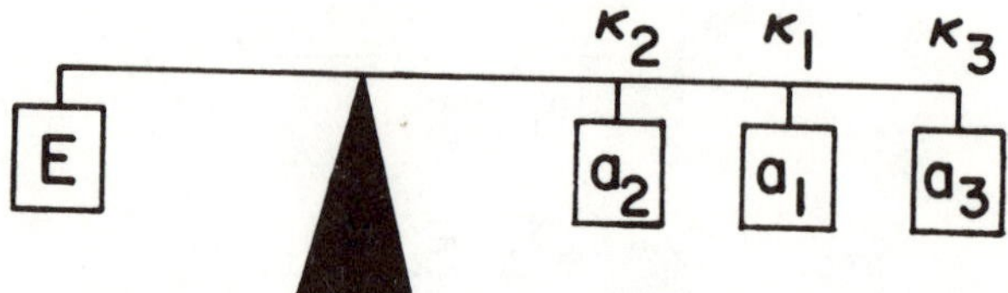

Fig. 1. Mechanical analogue of the relations defined by (5) and (6). The radius of gyration of the energy spectrum is $K^2 = F/E$ = constant. The constancy of K^2 may be preserved if the energy contained in the intermediate scale serves as either a source or a sink of wave energy. A unidirectional exchange of energy will not preserve the constancy of K^2.

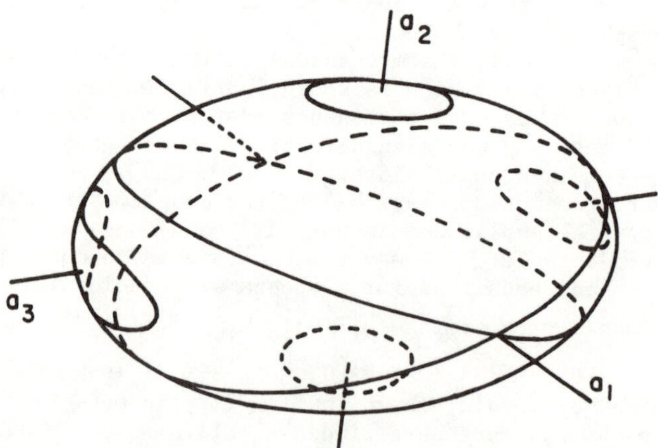

Fig. 2. Trajectories of the motion, constrained by (5) and (6), on the surface of the phase sphere. The motion is described by closed paths on the sphere, and the nature of the trajectories is determined by the initial conditions.

PREDICTABILITY OF SCALE-INTERACTIONS

At any instant, Fjørtoft's theorem applies to the motions constrained by (7) and (8). However, the λ_i are functions of the motion, so that the initial ordering, $\lambda_1 < \lambda_2 < \lambda_3$, could change with time. Figure 3 shows three of the various possible configurations. The maximum value of λ_i occurs when $a_i = 0$ and $b_i \neq 0$; the minimum occurs when $a_i \neq 0$ and $b_i = 0$. Consequently, Fjørtoft's theorem should apply for all time in a) although the λ_i may not be constant. The set-ups in b) and c) could lead to changes in the ordering of the λ_i.

For example, if $a_3 \neq 0$, but $b_3 = 0$ at a particular instant, then the ordering of the λ_i in c) reverts to $\lambda_3 < \lambda_2 < \lambda_1$. The intermediate scale is characterized by λ_2, which now serves as either the source or sink of energy for the other two wave components. Moreover, changes in the λ_i may not be predictable, because only two constraints, (7) and (8), on the six amplitude equations are evident. In principle, the motions are not constrained to periodic orbits, and may be unpredictable.

A particular example of the situation in Fig. 3c has been examined by Blumen[4]. The initial condition for the present integration is provided by an unstable normal mode solution of the complementary linear system for which (a_1, b_1) are constants. The system of six amplitude equations were integrated numerically over the range $0 \leq t \leq 4{,}096$, with $\lambda_2 < \lambda_1 < \lambda_3$ at time $t = 0$.

Blumen[4] has shown that the λ_i remain constant during the initial stages of the motion when normal mode initial conditions are used. Consequently, during this state, three more conservation principles constrain the motion together with (7) and (8). The motions are then restricted to be periodic, as in the triad interactions associated with (5) and (6). However, at some point, depending on the initial amplitudes, $\lambda_i(t) \neq$ constant. The results displayed in Fig. 4 show that the initial ordering of the λ_i can change during the evolution of the motion. A representative example is shown in Fig. 5.

The accompanying energy exchanges between the wave components is also illustrated in Fig. 5. Here the intermediate scale wave (λ_1) draws energy from both the shorter (λ_3) and longer waves (λ_2). When $\lambda_3 < \lambda_2 < \lambda_1$, the roles of the waves associated with λ_1 and λ_2 change; the intermediate wave (λ_2) now extracts energy from the wave associated with λ_1. The wave associated with λ_3 continues to serve as a source of wave energy until its energy is desicatted. The initial ordering of the λ_i is reinstated abruptly, with the result that waves associated with both λ_2 and λ_3 extract energy

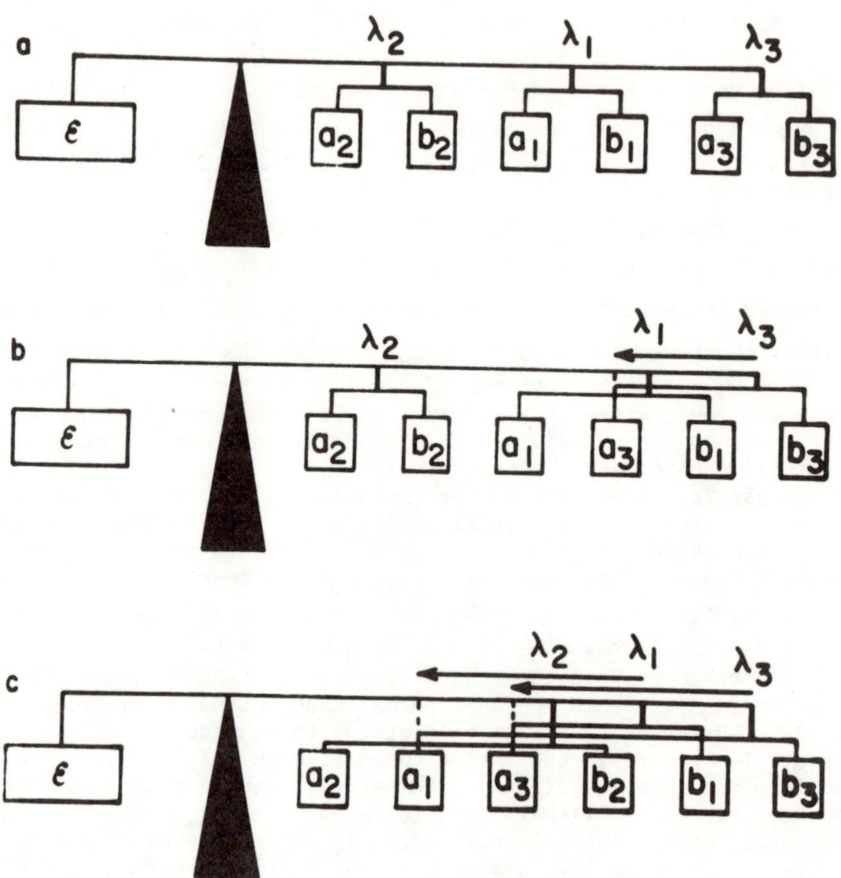

Fig. 3. Mechanical analogue of the relations defined by (7) and (8). Three of the possible arrangements that preserve the constancy of the centroid $\lambda = \mathcal{G} / \mathcal{E}$. The arrows terminate at the location of the minumum values of λ_i (i = 1 or 3) that may be attained during the motion. See text for discussion.

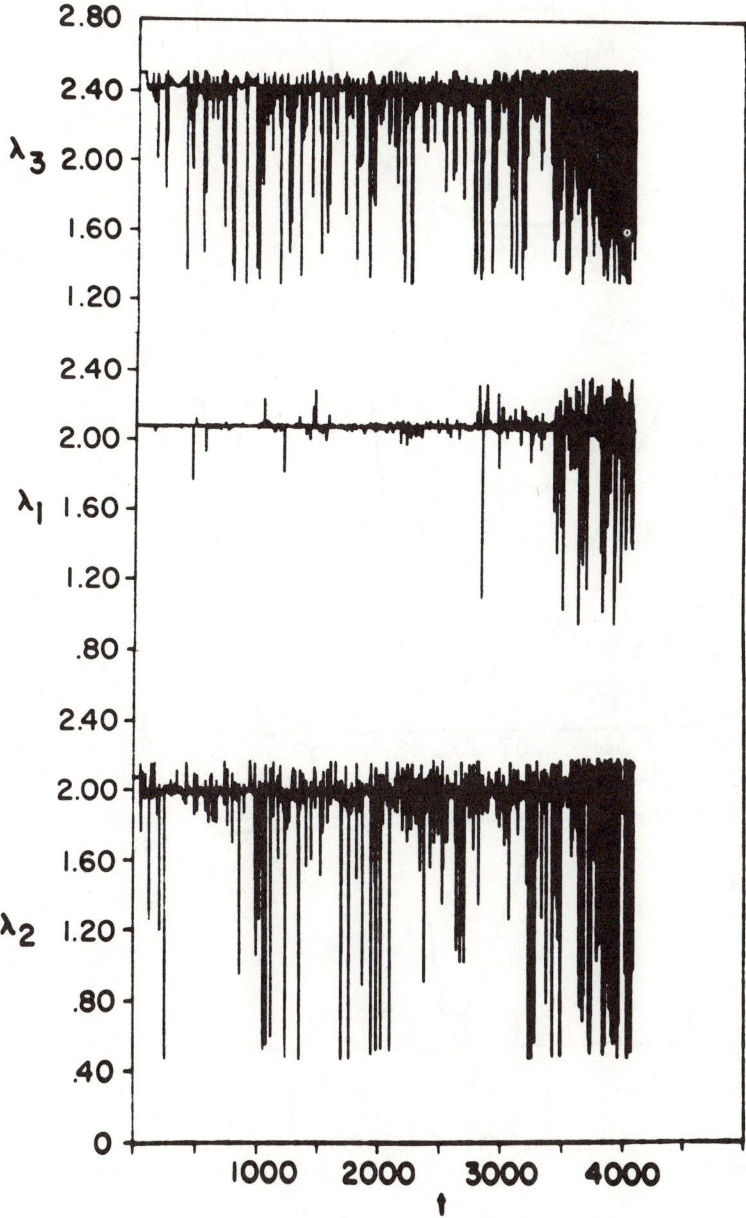

Fig. 4. λ_i (i = 1,2,3), defined by (10), as functions of time t.

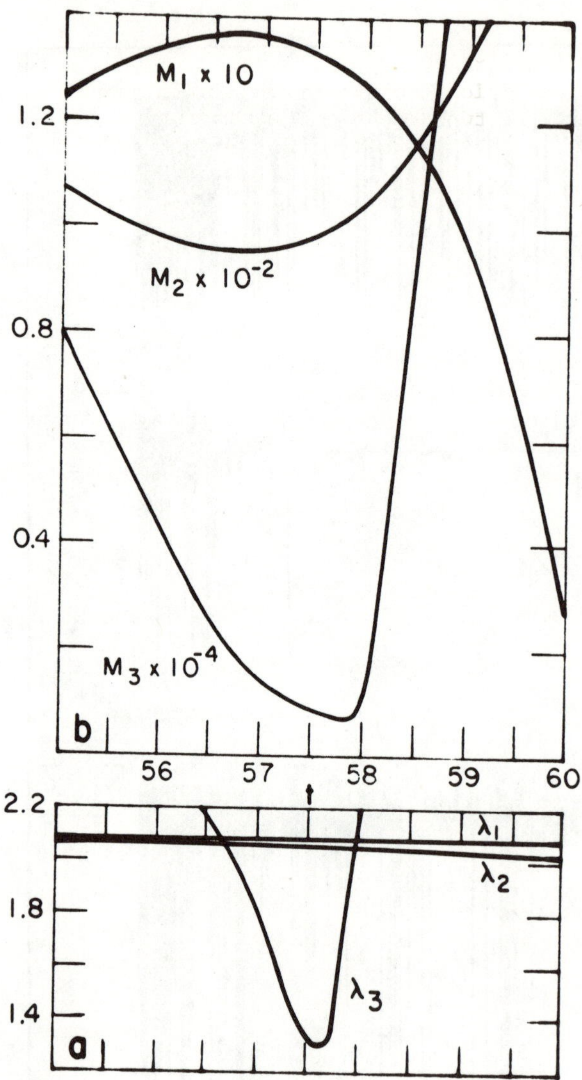

Fig. 5. (a) As in Fig. 4, except $55 \leq t \leq 60$. (b) Depth-integrated energies M_i ($i = 1,2,3$), defined by (9), as functions of time.

from the source characterized by λ_1. A similar reversal of the energy transfer occurs when λ_1 drops below λ_2 resulting in $\lambda_1 < \lambda_2 < \lambda_3$ for a short period of time.

Normal mode initial conditions constrain the motion to be periodic initially. However, the two dimensional phase plane projection of the motion, shown in Fig. 6, illustrates the tendency of the motion to become relatively unpredictable with time. When arbitrary initial conditions are specified (not shown), the initial periodicity may be by-passed.

CONCLUDING REMARKS

The theoretical predictability problem, posed by Lorenz,[6] deals with the rate of error energy transfer or uncertainty through the spectrum of motions. This rate is intimately related to the constraints that are imposed on the fluid motions. The present results serve to point-up that the rate and direction of this energy transfer through the spectrum may, in some cases, be highly variable, if not unpredictable.

ACKNOWLEDGMENTS

Financial support has been provided by the National Science Foundation under Grant No. ATM-8020138. Thanks are expressed to Kelly Kanizay for computational assistance.

REFERENCES

1. R. Fjørtoft, Tellus, 5, 225 (1953).
2. J. G. Charney, Dynamic Meteorology (Reidel, 1970), 97.
3. W. Blumen, J. Atmos. Sci., 35, 774 (1978).
4. W. Blumen, J. Atmos. Sci., 39, 2388 (1982).
5. L. D. Landau and E. M. Lifshitz, Mechanics (Pergamon, 1976), p. 165.
6. E. N. Lorenz, Tellus, 21, 289 (1969).

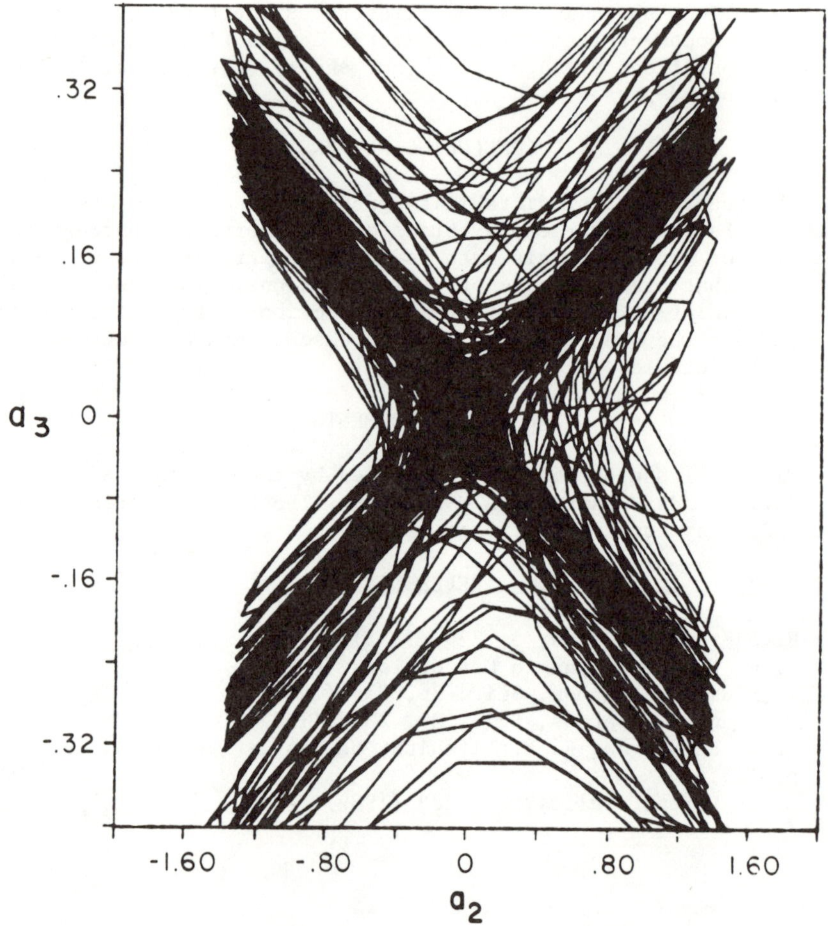

Fig. 6. Projection of the trajectory in phase space onto the a_2 - a_3 plane.

BAROCLINIC AND BAROTROPIC PREDICTABILITY IN GEOSTROPHIC TURBULENCE

Geoffrey K. Vallis†
Scripps Institution of Oceanography, La Jolla, California 92093

ABSTRACT

This paper examines the predictability properties of barotropic and baroclinic geostrophic turbulence. By geostrophic turbulence I mean rapidly rotating, stably stratified flow with many degrees of freedom, dominated to a greater or lesser extent by nonlinearity. Predictability here is concerned with the rate of divergence, measured by spectra of energy in the difference field, of two initially similar fields. In particular the paper examines the effects of a mean gradient of potential vorticity (the beta effect) and the effects of baroclinic instability on flow predictability. In barotropic (i.e., two-dimensional) turbulence, beta increases predictability by slowing nonlinear energy transfer between scales. Two layer flow is not so affected, because beta increases the range of wavenumbers over which significant energy transfer occurs due to the modification of the linear stability properties of the system. In two layer flow baroclinic and barotropic effects each account for about half of the loss of predictability.

INTRODUCTION

The study presented herein extends previous work on the predictability of barotropic flow[1,2,3,4] to include the effects of baroclinicity - in particular it is concerned with the combined effects of planetary wave propagation and baroclinic instability. For example, does the beta-effect - the slowing down of nonlinear energy transfer because of wave propagation - affect a two-layer model in the same way as a one-layer model? Does a baroclinic model have similar predictability properties to a barotropic model? Also, does the form and magnitude of the initial error affect its subsequent growth? The tool used is possibly the simplest model containing the relevant mechanisms - a two-layer quasi-geostrophic beta plane model. Section 2 describes the model used in this study. In Section 3 the equilibrium fields and spectra are discussed. Section 4 examines the predictability properties of the barotropic simulations. The two-layer simulations are described in Section 5. Section 6 summarizes and concludes.

MODEL

The equations - those of a two-layer quasigeostrophic β-plane model, with zero vertical velocity at the boundaries - may be written:

$$\frac{\partial}{\partial t}q_1 + J(\psi_1,q_1) + \beta\frac{\partial\psi_1}{\partial x} = D_1$$

$$\frac{\partial}{\partial t}q_3 + J(\psi_3,q_3) + \beta\frac{\partial\psi_3}{\partial x} = D_3 \qquad (1)$$

where

† Work supported in part by NSF, under grant ATM-82-10160.

$$q_1 = \nabla^2\psi_1 + \frac{\lambda^2}{2}(\psi_3 - \psi_1)$$

$$q_3 = \nabla^2\psi_3 + \frac{\lambda^2}{2}(\psi_1 - \psi_3) \tag{2}$$

where λ is the inverse deformation radius. Assume that the total streamfunction may be expressed as that resulting from a constant mean shear across the domain, plus an eddy component $\hat{\psi}$. Then

$$\psi_1 = -U_1 y + \hat{\psi}_1$$

$$\psi_3 = -U_3 y + \hat{\psi}_3 \tag{3}$$

Substituting (3) into (1) yields

$$\frac{\partial}{\partial t}\hat{q}_1 + J(\hat{\psi}_1,\hat{q}_1) + \beta\frac{\partial\hat{\psi}_1}{\partial x} = -\frac{\lambda^2}{2}\left(U_1\frac{\partial\hat{\psi}_3}{\partial x} - U_3\frac{\partial\hat{\psi}_1}{\partial x}\right) - U_1\frac{\partial}{\partial x}\nabla^2\hat{\psi}_1 + D_1 \tag{4}$$

$$\frac{\partial}{\partial t}\hat{q}_3 + J(\hat{\psi}_3,\hat{q}_3) + \beta\frac{\partial\hat{\psi}_3}{\partial x} = \frac{\lambda^2}{2}\left(U_1\frac{\partial\hat{\psi}_3}{\partial x} - U_3\frac{\partial\hat{\psi}_1}{\partial x}\right) - U_3\frac{\partial}{\partial x}\nabla^2\hat{\psi}_3 + D_3 \tag{5}$$

where $\hat{q}_1 = \nabla^2\hat{\psi}_1 + \frac{\lambda^2}{2}(\hat{\psi}_3 - \hat{\psi}_1)$ and similarly for $\hat{q}_3$. The carats will subsequently be dropped. The use of a constant mean shear is clearly artificial, but it is a convenient way of specifying the mean baroclinicity.

Dissipation is parameterized as follows:

$$D_1 = -\nu\nabla^6\psi_1$$

$$D_3 = -\nu\nabla^6\psi_3 - \alpha\nabla^2\psi_3$$

The $\alpha\nabla^2\psi$ term represents the effects of surface drag. The parameter α is given the meteorologically appropriate value of (1/3) days^{-1}. The term $\nu\nabla^6\psi$ is necessary to prevent the build-up of enstrophy in high wavenumbers. The value of ν is determined empirically to be such that the energy spectra appeared smooth near the cutoff wavenumber. It is found that the total energy dissipation by the high order friction is a few percent of the total dissipation. It has a larger effect than the surface drag above wavenumber 28 (20 in the single-layer experiments).

Equations (4) and (5) are integrated using an alias free spectral code with periodic boundary conditions in both directions. Truncation occurs at about wavenumber 32. The domain is square and of physical size $(2.25\ .10^7\ \text{m})^2$. A timestep of 0.75 hrs is generally used in a leapfrog scheme.

Some barotropic integrations will also be described. The governing equation now is

$$\frac{\partial}{\partial t}\nabla^2\psi + J(\psi, \nabla^2\psi) + \beta\frac{\partial\psi}{\partial x} = F + D \tag{7}$$

D is the dissipation term $\alpha\nabla^2\psi + \nu\nabla^6\psi$. ν has the same value as in the two-level experiments. α is reduced to (1/12) days^{-1}. Since baroclinic instability no longer occurs, its effects are crudely simulated using a Markovian random forcing formulation.[2] Thus, at time-step 'n'

$$F_n = R_nF_{n-1} + (1-R_n^2)^{1/2}G_nA \tag{8}$$

where $R_n = 0.98$ for a timestep of 0.75 hours.

G_n is a number with unit amplitude and a random phase, different for each wavenumber, and A is the forcing amplitude. A is zero except in the wavenumber band $|k| = 6, 7, 8$, wherein its value increases linearly with k_x (the x-wavenumber) with an average value of 10^{12} s^{-2}. The forcing differs from baroclinic instability in that G is uncorrelated with the flow and is that it provides no source of error, since the same uncorrelated G is used in each realization.

EQUILIBRIUM FIELDS AND ENERGY BUDGETS

A. Energy Transfer: Some Simple Considerations

Before presenting numerical solutions to (4) and (5) we shall show the general characteristics of energy-transfer in two layer flow can be inferred from relatively simple arguments. It is convenient to first write (5) and (6) in terms of the baroclinic (τ) and barotropic (ψ) streamfunctions defined by:

$$\psi = (\psi_1 + \psi_3)/2$$

$$\tau = (\psi_1 - \psi_3)/2$$

Adding and subtracting (5) and (6) gives

$$\frac{\partial}{\partial t}\nabla^2\psi + J(\psi,\nabla^2\psi) + J(\tau,\nabla^2\tau) + \beta\frac{\partial\psi}{\partial x} = F_\psi + D_\psi \tag{9}$$

$$\frac{\partial}{\partial t}(\nabla^2-\lambda^2)\tau + J(\psi,(\nabla^2-\lambda^2)\tau) + J(\tau,\nabla^2\psi) + \beta\frac{\partial\tau}{\partial x} = F_\tau + D_\tau \tag{10}$$

where

$$D_\psi = -\{\nu\nabla^6\psi + (\alpha\nabla^2\psi_3)/2\}$$

$$D_\tau = -\{\nu\nabla^6\tau - (\alpha\nabla^2\psi_3)/2\} \ .$$

Without loss of generality, we may set $U_1 = -U_3 = U$ giving

$$F_\psi = -U\frac{\partial}{\partial x}\nabla^2\tau$$

$$F_\tau = -U\frac{\partial}{\partial x}(\lambda^2 + \nabla^2)\psi \ .$$

Hence F_τ vanishes for $|\underline{k}| = \lambda$. The spectral barotropic and baroclinic energy budgets are obtained by first writing (9) and (10) in spectral form, and multiplying each component by $\psi^*_{\underline{k}}$ or $\tau^*_{\underline{k}}$ respectively, where * refers to the complex conjugate.

Summing over all spectral components, and adding the baroclinic and barotropic modes, then gives the integral constraint for unforced, inviscid flow

$$\frac{d}{dt}\sum_k (k^2|\psi_{\underline{k}}|^2 + (k^2+\lambda^2)|\tau_{\underline{k}}|^2) = 0 \ . \tag{11}$$

where $(\psi(x,y,t),\tau(x,y,t)) = \sum_{\underline{k}}(\psi_{\underline{k}}(t), \tau_{\underline{k}}(t))e^{i\underline{k}\cdot\underline{x}}$ and $(\psi_{\underline{k}}, \tau_{\underline{k}}) = (\psi^*_{-\underline{k}}, \tau^*_{-\underline{k}})$. The enstrophy budget is obtained by multiplying the spectral forms of (9) and (10) by $(q_1+q_2)/2$ and $(q_1-q_2)/2$ respectively. For unforced, inviscid flow the integral constraint on the enstrophy is found to be

$$\frac{d}{dt}\sum_{\underline{k}}\{k^4|\psi_{\underline{k}}|^2 + (k^2+\lambda^2)^2|\tau_{\underline{k}}|^2\} = 0 \tag{12}$$

If, and only if, beta is zero then the enstrophy for each layer (or equivalently the barotropic enstrophy $\sum_{\underline{k}} k^4|\psi_{\underline{k}}|^2$ and the baroclinic enstrophy $\sum_{\underline{k}}(k^2+\lambda^2)|\tau_{\underline{k}}|^2)$ is separately conserved. For non-zero β the quantity $(q+\beta y)^2$ is conserved in an integrated sense layer by layer. Only if β is zero is this constraint quadratic in ψ.

The advantages of writing the budgets in terms of the baroclinic and barotropic modes lie in the simplicity of the triad interactions and the form of the integral constraints. For we see immediately from (11) and (12) that a baroclinic mode is formally similar to a barotropic mode, provided we make the replacement in wavenumber:

$$k^2 \rightarrow k^2 + \lambda^2 \ .$$

A knowledge of the integral constraints can be used to deduce the general isotropic movement of energy in a two-layer model[5], in a similar fashion to their use in 2.D turbulence[7]. Thus in a purely barotropic triad energy is transferred predominantely to lower wavenumbers. In a mixed triad (ψ,τ,τ) energy will go predominantly to lower pseudo-wavenumber k' defined by $k'^2 = k^2 + \lambda^2$ in a baroclinic mode, and $k'^2 = k^2$ in a barotropic mode. Transfer of baroclinic energy may be toward higher wavenumbers, provided there is some energy conversion to a barotropic mode (which on the interval $1 \leqslant k^2 \leqslant \lambda^2$ will always have lower pseudo-wavenumber). In the oceanographic, and more general, case of two layers of different equivalent depths, two baroclinic modes may interact with a third baroclinic mode. This does not occur here, because the two layers are of equal depth.

B. The Beta Effect

On the basis essentially of scaling arguments, it has been proposed[8] that the presence of Rossby waves will slow down the transfer of energy to the gravest scales. Considering the balance of terms in the barotropic vorticity equation,

$$\frac{\partial}{\partial t}\nabla^2\psi + \beta\frac{\partial\psi}{\partial x} + J(\psi,\nabla^2\psi) = F+D$$

with approximate magnitudes

$$\beta U \sim k^2U^2 \sim F + D$$

where U is the r.m.s. velocity, linear terms dominate at scales larger than the scale K-beta, where

$$K_\beta \sim (\beta/U)^{1/2} \ .$$

In the baroclinic cases there exists also the mixed mode interaction term $\lambda^2 J(\psi,\tau)$ which scales like λ^2U^2. An *á priori* scaling argument no longer exists to enable us to eliminate nonlinear terms at the largest (horizontal) scales. However, application of the integral constraints on energy and energy does imply the increasing barotropisation of the flow at larger and larger scales, and hence the importance of the mixed mode interactions falls. Furthermore, the closure arguments[9] which show the triad correlation decay rate to be a decreasing function of beta do carry over to the baroclinic case. Certainly, then, in the barotropic case we expect enhancement of predictability in the beta case, at least in the linear regime where infinitely predictable Rossby waves dominate. In the two layer case the situation is more ambiguous — baroclinic instability of the mean flow, for example, is a mixed mode triad interaction which has very different behavior in the beta and no beta case, in that low wavenumbers are greatly stabilized by the presence of a mean gradient of planetary vorticity, and the triad interaction becomes more non-local.

C. Numerical Simulations

Numerical simulations allow one to relax the assumption of vertical homogeneity, i.e., that τ and ψ are uncorrelated - often used in semi-analytic closure theories, if only for tractability. Actually, vertical homogeneity is not a prohibitive assumption in comparisons with direct two-layer simulations, since only deep linear eddies are allowed in any case and since the general nature of the nonlinear interactions is unaltered. The advantage of numerical simulations over closure lies more in the complete absence of arbitrary phenomonological coefficients, provided sufficient resolution can be achieved.

This subsection describes the equilibrium, time-averaged fields resulting when (9) and (10) were stepped to equilibrium. The table lists the parameters used. In a barotropic simulation, it is relatively easy to isolate the effects of beta on the flow, since all forcing and dissipation mechanisms may be held constant, but it is somewhat more difficult in two layer flow since the stability properties are profoundly affected by beta. In the two layer experiments B1 and NB, the inviscid supercriticality of the shear (i.e., its value above the level required for linear instability) was set equal. It then turned out that the equilibrium, time-averaged, total energy in B1 and NB was approximately the same. In B2, the shear was reduced. All other parameters (aside from beta) were unchanged. In B1 and B2, beta is given the realistic value of $1.5 \times 10^{-11} s^{-1}$, whereas in NB it is set to zero.

TABLE

Experiment parameters and selection of results. Preditability time is the time taken for error energy to reach 90% of its final value.

Experiment	B1	B2	NB	1B	1NB
No. levels	2	2	2	1	1
Value of beta	$1.5\ 10^{-11}$	$1.5\ 10^{-11}$	0	$1.5\ 10^{-11}$	0
Shear $(u_1-u_3)MS^{-1}$	7.5	4.0	3.66	-	-
Total energy (Jm^{-2})	$7.0\ 10^5$	$1.2\ 10^5$	$7.0\ 10^5$	$4.2\ 10^5$	$4.2\ 10^5$
Supercriticality (inviscid) MS^{-1}	3.66	0.16	3.66	-	-
λ (inverse deformation radices expressed as a wave number)	10	10	10	-	-
Barotropic energy	$3.7\ 10^5$	$5.4\ 10^4$	$2.8\ 10^5$	$4.2\ 10^5$	$4.2\ 10^5$
$U_{rms}\ (ms^{-1})$	8.6	3.3	7.5	9.2	9.2
$k_\beta\ (\beta/2u)^{1/2}$ (non-dimensionalized)	3.3	5.3	-	3.2	-
Eddy turnover time (days)	.85	1.62	1.04	-	-
Error doubling time (days)	1.0	2.1	1.1	2.3	1.85
Predictability time (days)	11	19	12.5	25	20

The following energy and enstrophy spectra and transfers may be defined for the two-level model:

$$EC(\underline{k}) = (k^2 + \lambda^2)|\tau_{\underline{k}}|^2$$

$$ET(\underline{k}) = k^2|\psi_{\underline{k}}|^2 \qquad (13)$$

$$EN(\underline{k}) = (k^4|\psi_{\underline{k}}|^2 + (k^2 + \lambda^2)^2|\tau_{\underline{k}}|^2)$$

and for the barotropic model the modal energy is

$$U(\underline{k}) = k^2|\psi_{\underline{k}}|^2 \qquad (14)$$

These are respectively the baroclinic energy, the barotropic energy, the potential enstrophy, and the total energy for the one-layer model. For each of these the one dimensional isotropic spectra may be defined by summing over all wavenumbers in a band of given absolute wavenumber, i.e.,

$$EC(k) = \sum_{|\underline{k}| \leqslant |\underline{l}| < |\underline{k}|+1} EC(\underline{l})$$

and the zonal spectra by summing over all y-wavenumbers

$$EC(k_x) = \sum_{k_y} EC(\underline{k})$$

Because of symmetry, the eddy statistics for all wavenumber quadrants are identical, if the quadrants are defined by the usual x and y axes. Hence they are summed and presented simply as functions of positive wavenumbers.

The isotropic energy spectra are displayed in Figure 1. The one-layer simulations (Fig. 1c) have a spectral slope very close to k^{-4}. This has been found previously by various workers,[3] often using a much higher order viscosity. However, note that in the two-layer simulations the spectral slope is much shallower, closer to k^{-3}. These results are probably partially dependent on the friction since the simulations here are fairly viscous; it would be interesting to see comparable 1 and 2 layer simulations with a much higher resolution and less viscous model. Certainly in the two-level simulations surface-drag is relatively strong. Since drag imposes a scale independent time scale, a balance between drag and nonlinear transfer might be expected to yield a k^{-3} spectrum, since this has a time-dependent eddy turnover time where the eddy turnover time is defined by:

$$t_e = \{\frac{k^{-3}}{E(k)}\}^{1/2}$$

where k is the wavenumber, E(k) an energy spectrum. The time t_e is a constant for a k^{-3} spectra. However, integrations with a strong surface drag in the one layer case in fact *increased* the slope of the spectra. With a frictional time scale of about 3 days, the equilibrium barotropic spectra was approximately k^{-5} (Fig. 1).

An examination of the energy budgets is instructive (see Figs. 2 and 3). In B1, the case with a realistic beta, baroclinic instability of the mean shear (the only source of energy) provides a maximum input of energy at wavenumbers 5 and 6. (These are not the wavenumbers of maximum linear instability which are slightly higher. Maximum energy input occurs where $\mathrm{Im}\{k_x\psi_{\underline{k}}\tau^*_{\underline{k}}\}$ is maximized, since $FC(\underline{k}) = (\lambda^2-k^2)/k^2\ FT(\underline{k}) = \mathrm{Re}\{U(\lambda^2-k^2)ik_x\psi_{\underline{k}}\tau^*_{\underline{k}}\}$. Transfer of energy occurs, not primarily to other baroclinic wavenumbers but to the barotropic modes, where it is further transferred to the gravest modes and dissipated. That energy is transferred from baroclinic to barotropic modes can

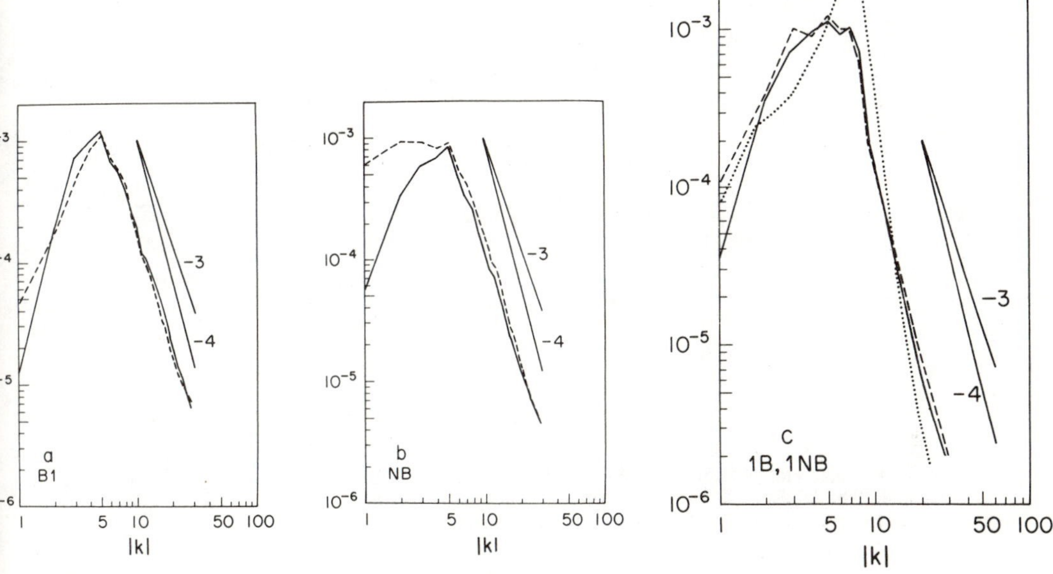

Fig. 1. Isotropic, barotropic and baroclinic energy spectra for a) B1, b) NB and c) 1B and 1NB (total energy). In a) and b) the barotropic energy is a solid line, the baroclinic energy broken. In c) 1B is solid and 1NB broken. Dotted curve in (c) is for high drag.

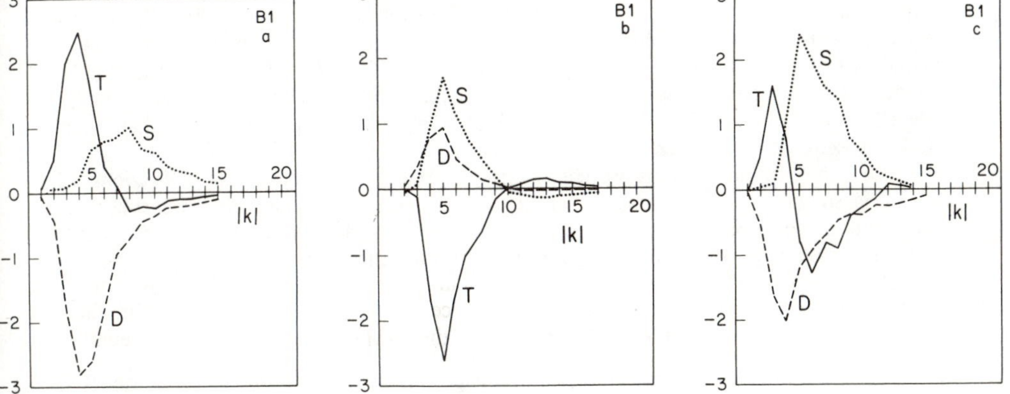

Fig. 2. Isotropic energy budgets for B1. a) is the barotropic budget, b) the baroclinic budget and c) the total. The labels S, D and T denote contributions by forcing (i.e., linear baroclinic instability), dissipation and transfer. See text for a complete explanation.

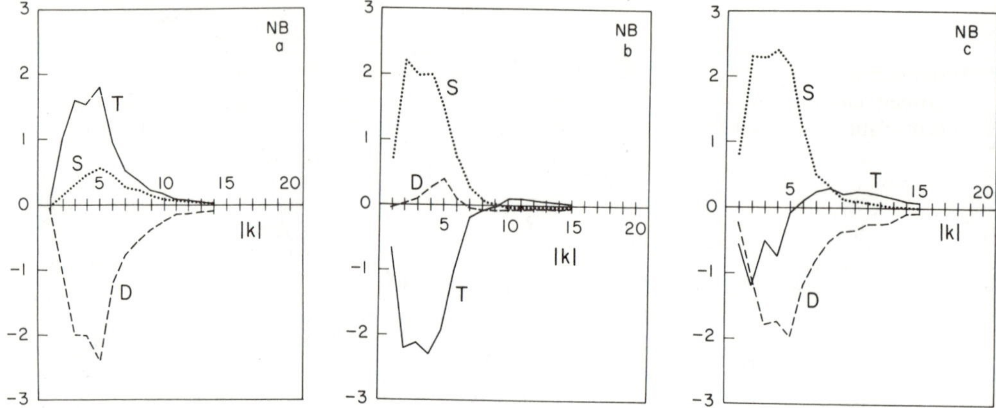

Fig. 3. Isotropic energy budgets for NB. a) is the barotropic budget, b) the baroclinic budget and c) the total. The labels S, D and T denote contributions by forcing (i.e., linear baroclinic instability), dissipation and transfer. See text for a complete explanation.

be seen by noting that the real integral of T in Figure 2b is negative, and positive in Figure 2a. In Figure 2c the area under the T curve is zero. The small amount of baroclinic energy in the largest scales appears in fact to be due to friction which here acts to *create* baroclinic energy. The formulation of surface drag used here is such that total energy must be dissipated. However, since it acts only on the lower model layer, baroclinic energy, proportional to the square of the difference of the stream function between the two layers, may be increased. The total (baroclinic plus barotropic) energy budget shows an input of energy mainly between wavenumbers 5 and 10, a transfer to larger scales where it is destroyed by friction. In the case of zero beta (NB) baroclinic linear instability provides a baroclinic energy source in much lower wavenumbers[10]. Transfer occurs almost immediately and mainly to barotropic modes (plus a little to higher baroclinic wavenumbers) where friction provides a sink. The total energy budget displays a small transfer of energy toward *higher* wavenumbers arising from the transfer on the baroclinic modes. However, the total energy budget in NB is characterized by a *smaller* amount of transfer between scales than when beta is non-zero; it has a much more local nature.

The numerical simulations are consistent with the analytic arguments of Sections 3.1 and 3.2 in showing the movement of energy to the large scales, and the predilection for zonal motion when beta is realistic. However, they also show how the position of the energy injection scale is such as to *increase* the nonlinear energy transfer when beta is nonzero.

PREDICTABILITY EXPERIMENTS

The general form of the predictability experiments is as follows. The model is integrated from random initial conditions until equilibrium is achieved. The integration is then continued for several weeks, to create the control integration. The 'forecast' is obtained by perturbing the potential vorticity at the beginning of the control experiment and then stepping the model until these fields have completely diverged from those in the control experiment. The forcing and friction are maintained throughout all integrations. The usual form of the initial perturbation is a dephasing of the spectral coefficients of the potential vorticity such that the phase difference between the control and the perturbation increases from 0 to π as $|k|$ increases from 12 to 24, except for a multiplication by a random number selected uniformly on the interval (-1, +1). Thus, the initial phase difference between control and forecast increases stochastically above wavenumber 12 until total decorrelation is achieved at wavenumber 24. The amplitude of the modes is unaltered. Most results are each from an ensemble of four experiments.

A. Error Diagnostics

For each diagnostic defined in (13) we may define the corresponding error field by substituting the error stream function ψ'_k or τ'_k and the error, or difference, forcing term. In particular, the error baroclinic and barotropic energies are

$$EC'(\underline{k}) = (k^2+\lambda^2)|\tau'_{\underline{k}}|^2$$

$$ET'(\underline{k}) = k^2|\psi'_{\underline{k}}|^2$$

where $\psi' = \psi_1 - \psi_2$ where ψ_1 and ψ_2 are the streamfunctions of the two simulations. Similarly for τ. The error energy ratio, or relative error, is defined by

$$EC'_r\left(|k|\right) = \frac{\sum_{|k|\leqslant|l|\leqslant|k|+1} EC'(\underline{l})}{\sum_{|\underline{k}|\leqslant|\underline{l}|\leqslant|\underline{k}|+1} EC(\underline{l})}$$

and

$$EC'_r(k_x) = \frac{\sum_{k_y} EC'(\underline{k})}{\sum_{k_y} EC(\underline{k})}$$

and similarly for the barotropic fields.

B. One-level Integrations

Figure 4 illustrates the isotropic error energy ratio growth for the barotropic simulations. This figure (and Figs. 9, 10 and 11) illustrate the relative error, as a function of wavenumber, at various times marked in days after the initial perturbation ($t = 0$). The random number sequence in the forcing G_n (Eq. 8) is the same for all runs. The conclusions reached from these simulations are consistent with the results of Holloway[4] and Basdevant *et al.*[5], and are presented only briefly here.

1) The effect of beta is to increase the overall predictability of the flow.

2) Predictability times are increased in the higher wavenumbers, as much as in the lower wavenumbers in spite of beta scaling out of the equations at high wavenumbers.

3) The predictability time of the gravest mode $|k| = 1$, is *reduced* by beta, to the point where it is *less* predictable than the $|k| = 2, 3, 4$ modes. In Holloway's finite equivalent depth simulations, this phenomenon seems less apparent. It is likely that this result depends heavily on the slope of the energy spectrum at very low wavenumbers, and hence on the model formation (e.g., surface drag coefficient).

4) The predictability, not shown here, of the zonal flow ($k_x = 0$) *is* increased by beta, and is greater than the predictability of the $k_x = 1, 2, 3$ modes.

5) At the injection scale (denoted with a capital I in Fig 4) the decorrelation times are large. No error energy is injected here, and so error in these scales is swept away by the energy and enstrophy cascades. This is discussed further in Section (4.3).

C. The Unattainability of Perfect Decorrelation

With the help of closure theory it may be shown that, in the presence of correlated forcing (i.e., forcing whose ensemble average correlation between 'control' and 'forecast'

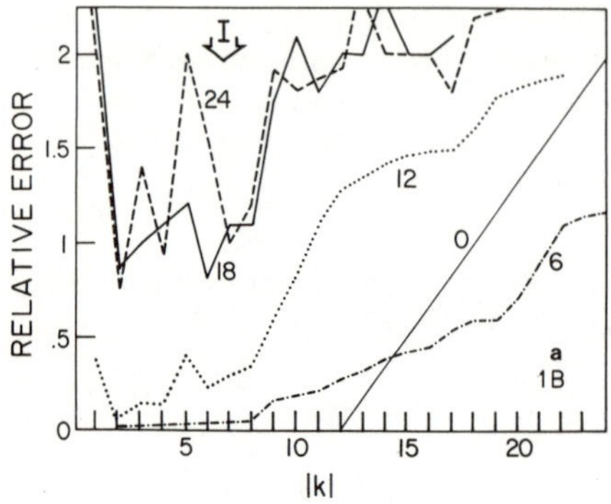

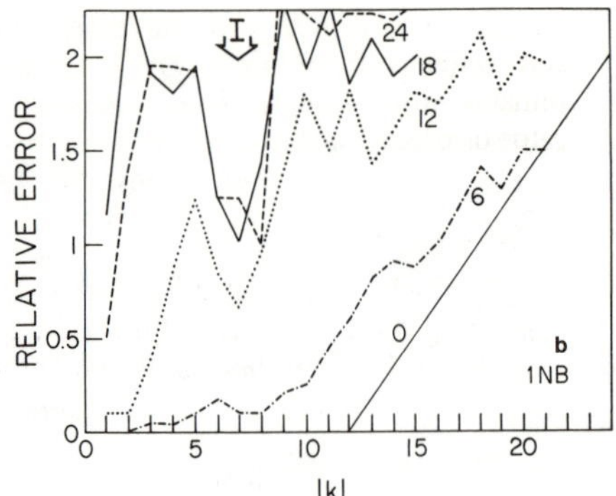

Fig. 4. Ratio of isotropic error energy to isotropic equilibrium energy at various times (marked in days) in one-layer integrations for a) 1B and b) NB. The abscissa is wavenumber. Beyond wavenumber 24, the error at time zero becomes parallel to the abscissa.

has a non-zero component) complete decorrelation between the fields may never be achieved. A Markovian quasi-normal closure for the evolution of the single field energy spectra may be written[1]

$$\frac{d}{dt}U_{\underline{k}} = \sum_{\underline{p}+\underline{q}=\underline{k}} \{\theta_{-\underline{kpq}}\ [a_{\underline{kpq}}(U_{\underline{p}}U_{\underline{q}} - U_{\underline{p}}U_{\underline{k}})]\} \tag{15}$$

$$-2\nu_{\underline{k}}U_{\underline{k}} + F_{\underline{k}}$$

The array $a_{\underline{kpq}}$ contains unchanging, geometrical interaction coefficients. $\theta_{-\underline{kpq}}$ is a triad-relaxation time, which we need not specify. The important point is that when summed over all $\underline{k}$ the first term on the right hand side of (15) vanishes, (i.e., advective terms conserve energy) informing us that energy balance in the fluid is maintained between forcing $\sum_{\underline{k}}(F_{\underline{k}})$ and dissipation $\sum_{\underline{k}}(\nu_{\underline{k}}U_{\underline{k}})$.

The corresponding evolution equation for the 'error energy' $U'_{\underline{k}}$ is

$$\frac{d}{dt}U'_{\underline{k}} = \sum\{\theta_{-\underline{kpq}}\ [a_{\underline{kpq}}(U_{\underline{p}}U'_{\underline{q}} - U'_{\underline{k}}U_{\underline{q}})$$

$$+a_{\underline{kpq}}U'_{\underline{p}}(U_{\underline{q}} - U'_{\underline{q}})\} \tag{16}$$

$$-2\nu_{\underline{k}}U'_{\underline{k}} + (F_{\underline{k}}-R_{\underline{k}})$$

where $R_{\underline{k}}$ is the correlated forcing between the two flows. If $F_{\underline{k}}$ is white delta-correlated in time then a completely correlated forcing gives $F_{\underline{k}} = R_{\underline{k}}$. In general any correlated forcing will yield $R_{\underline{k}} \neq 0$. Putting $U'_{\underline{k}} = U_{\underline{k}}$ in (16), and summing over all $\underline{k}$, yields

$$\sum_{\underline{k}}\frac{d}{dt}U_{\underline{k}} = \sum_{\underline{k}}\{-2\nu_{\underline{k}}U_{\underline{k}} + (F_{\underline{k}}-R_{\underline{k}})\}\ .$$

Unless $R_{\underline{k}} \equiv 0$, this is not consistent with (15). Thus, with any degree of correlated forcing, predictability must remain in the system at all times. The above argument does not indicate if predictability remains at all scales, or only at the forcing scale.

Note that for the above system the autocorrelation of a single flow will fall towards zero with time, because of the forcing and nonlinear effects, while predictability still remains in the system. Autocorrelation as a function of time therefore seems a poor indicator of predictability, at least in this case.

TWO-LAYER PREDICTABILITY

In considering the predictability of baroclinic flow, I shall be particularly concerned with understanding whether the restoring effects of beta carry through to the two-level case and still enhance predictability. Also, what are the effects of baroclinic instability on the flow decorrelation, and what effect does the form of the initial error have on the subsequent error growth?

Attention will be focused mainly on experiments B1 and NB.

A. Total error

It is somewhat difficult to isolate the effects of wave propagation in a two-layer model because the linear stability properties are greatly affected by beta. Experiments B1 and NB have the same supercritical shear and about the same time-averaged total energy and therefore seem comparable. If the same absolute shear had been used for the two cases, the experiment with zero beta would have been much more supercritical and would have had much more intense turbulence. Another reasonable choice would have been to compare runs with the same eddy turnover time. Since B1 has less energy in the low wavenumbers it has more energy in high wavenumbers and a smaller eddy turnover time than N.B. An increased eddy turnover time would have increased the predictability of B1 slightly, although it would still be very similar to that of NB. Examination of Figure 5 reveals that the total predictability of the flow is *not* enhanced by the beta effect, nor is there any significant difference between barotropic and baroclinic modes.

Consider the error energy budget. This may be written (where a prime denotes the difference, or error field):

$$\frac{d}{dt}\{k^2|\psi'_{\underline{k}}|^2 + (k^2+\lambda^2)\,|\tau'_{\underline{k}}|^2\}$$

$$= \mathrm{Re}\{{}_\tau F'_{\underline{k}}\tau'^{*}_{\underline{k}} + {}_\psi F'_{\underline{k}}\psi'^{*}_{\underline{k}}\} \tag{17}$$

$$+ \mathrm{Re}\{{}_\tau D'_{\underline{k}}\tau'^{*}_{\underline{k}} + {}_\psi D'_{\underline{k}}\psi'^{*}_{\underline{k}}\}$$

$$+ \mathrm{Re}\;\{{}_\tau J'_{\underline{k}}\tau'^{*}_{\underline{k}} + {}_\psi J'_{\underline{k}}\psi'^{*}_{\underline{k}}\}$$

The terms on the right hand side represent the effects of forcing by the mean shear, dissipation and transfer on the error energy, respectively. When summed over all wavenumbers, the last term on the right-hand side of (27) does *not* vanish: the Jacobian terms are able to *create* error energy, as well as distribute it across wavenumbers. Figure 6 shows the error creation by linear baroclinic instability and by non-linear interactions. The total error creation by non-linear effects is much the same when beta is zero, but the linear creation is significantly smaller for the first ten days. Recall that the effect of beta is to increase the wavenumber at which baroclinic instability is greatest, and that the energy input is proportional to the magnitude of the streamfunction. For zero beta, instability

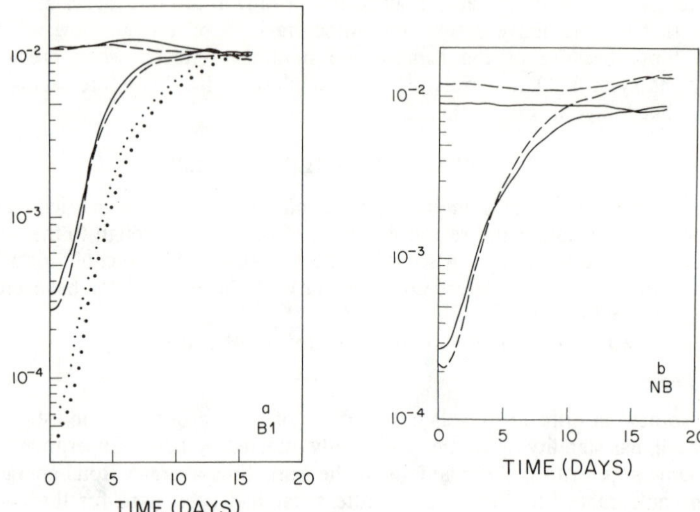

Fig. 5. Total error energy growth for a) B1 and b) NB. The barotropic energy is solid, the baroclinic broken. The dotted curves in a) are the barotropic error growths for initial errors confined to $|\underline{k}| > 24$, and $|\underline{k}| > 28$. Otherwise initial error is confined to $|\underline{k}| > 12$. The almost horizontal curves show twice the control energy for one integration from the ensemble, illustrating a typical time variation of total energy.

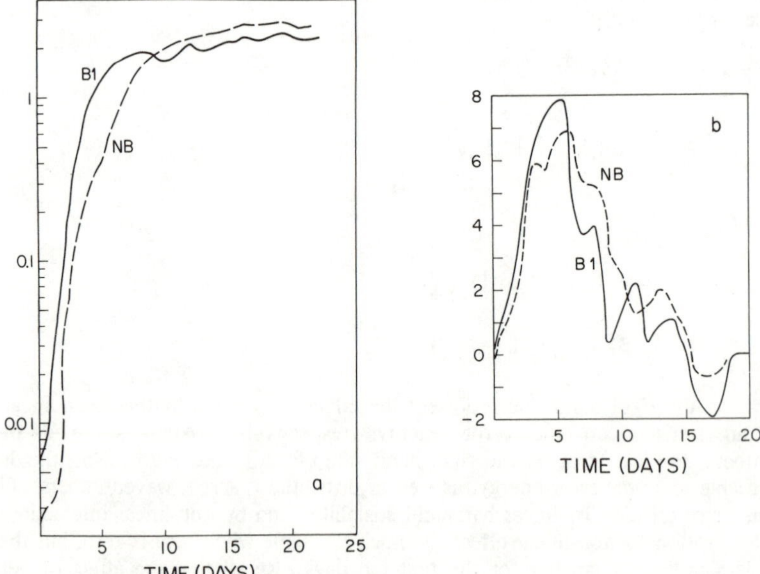

Fig. 6. Error energy creation in B1 and NB due to a) baroclinic instability of the mean flow b) non linear transfer. Units are comparable.

occurs at very low wavenumbers. For a long time the errors in these modes are very small and linear baroclinic instability is an inefficient mechanism for creating error.

Note that error creation by wave-wave interaction is no smaller when beta is non zero. Now, for zero beta, energy transfer is confined largely to low, well predicted wavenumbers. There is *less* communication between the low wavenumbers and the contaminated smaller scales, and any energy flow is toward higher wavenumbers. For the zero beta case (NB) error propagation, then, is generally against energy transfer, whereas in the realistic beta case error propagation is with energy transfer. Thus for the two layer case there is no longer any *a priori* reason why error propagation and creation will be larger when beta is zero.

The predictability times of the two-layer model are significantly shorter than those of the one-layer integrations, in spite of the barotropic energy of the two-layer models being about the same as the energy in the one-layer integrations. In the barotropic vorticity equation (7) the terms $J(\tau,\nabla^2\tau)$ and $U\frac{\partial}{\partial x}(\nabla^2\tau)$ appearing in (9) are parameterized by a forcing uncorrelated with the flow. In the two-layer simulations these provide an additional source of error, reducing the decorrelation time. In the two-layer case, balance in the error energy budget is ultimately achieved through the balance of the linear source term with the dissipation: the Jacobian terms (on average, but not necessarily instantaneously) can therefore provide no net contribution. In the one-layer case, however, where the energy source term provides no error, the dissipation is balanced by the non-linear terms which grow more or less monotonically before levelling off at a positive value. (Remember, though, they can only have a positive value because complete unpredictability is never achieved.)

The predictability time is seen to be lengthened by reducing the initial error. The lower, dotted, curves in Figure 5a are obtained with initial errors only above wavenumber 24 and above wavenumber 28 respectively. The initial error doubling rate is slightly larger when the error is confined to smaller scales - the average error doubling time in the case of the smallest initial error is 0.85 days (during the first five days of the experiment) whereas it is 1.0 days in the standard experiment. A reduced mean baroclinicity also enhances the predictability times: when the mean shear was reduced from 7.5 m/s (in B1) to 4. m/s (in B2) the predictability times rose by about a factor of two.

B. Error spectrum

The error ratio spectra are graphed in Figures 7 and 8. The figures show both the isotropic and zonal spectra of relative errors for various times (marked in days) for B1 and NB, with the 'standard' initial perturbation. The dominant feature is that of error spreading into and growing in the smaller wavenumbers. No artificial predictability at the energy injection scale is present. Predictability in the small scales (above wavenumber 10) is lost completely after about 6 days in B1, and shortly thereafter in NB. Predictability in the long waves persist much longer. Indeed the forecast of the long waves ($k = 2, 3, 4$) generally shows skill up to about 15 days. The skill-time generally increases monotonically as wavenumber decreases, for two reasons. First, error is spreading in from larger wavenumbers and unless energy transfer is completely non-local in spectral space the larger scales will be contaminated last. Secondly, the turbulent interaction rate may be expected to increase as the scale is reduced[11]; thus information will be lost in the high wavenumbers first. If beta is non-zero, interactions amongst low wavenumbers are further inhibited, and the effect will be more noticeable. Figure 9 shows the barotropic error spectra for an experiment in which all wavenumbers were dephased equally (with a random modifier). Even though the initial error ratio is the same for all wavenumbers, the forecast skill, or predictability, is lost first in the higher wavenumbers. Thus the long waves are *intrinsically* more predictable than short waves.

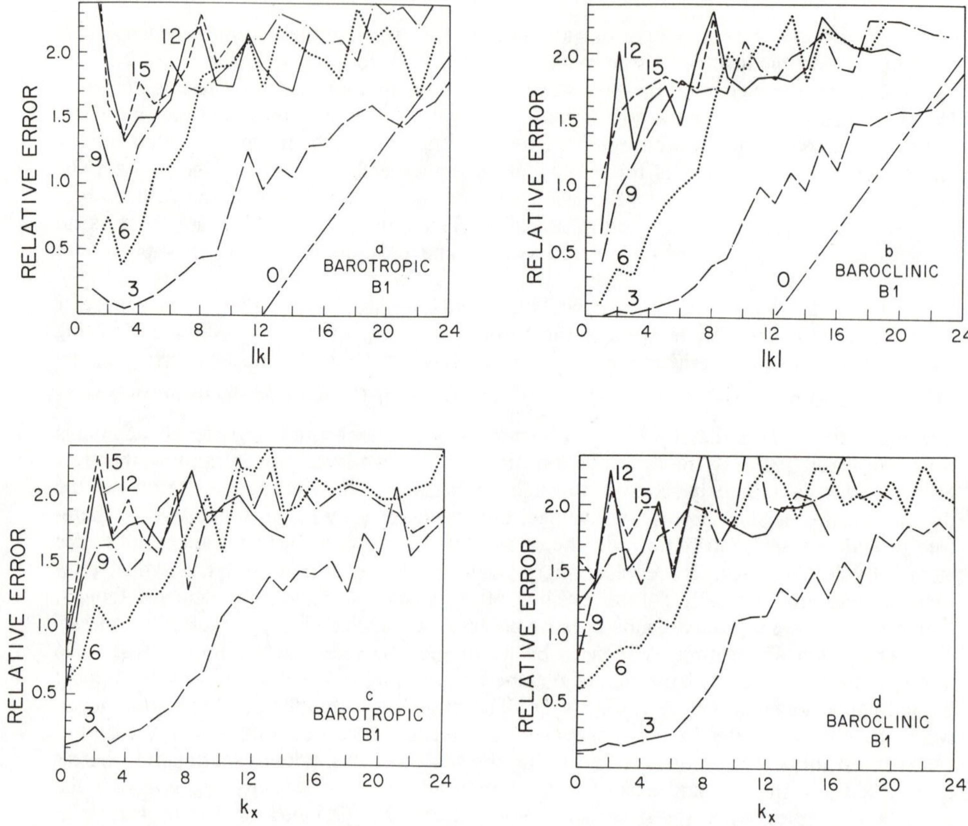

Fig. 7. Relative error growth for B1. a) Is the isotropic barotropic error, b) is the isotropic baroclinic error, c) is the zonal barotropic error and d) the zonal baroclinic error.

SUMMARY AND CONCLUSIONS

This study has been concerned with the equilibrium fields and predictability properties of two-layer flow on a beta-plane. The work extends of previous work[2,3,4] to include the effects of vertical stratification and baroclinicity.

Beta has somewhat richer effects in two-layer flow than in barotropic flow. The baroclinic instability of the long waves is inhibited by a mean gradient of potential vorticity, and energy enters the system at a higher wavenumber. When beta is zero energy enters primarily in baroclinic modes at low wavenumbers and attempts to pass to still lower wavenumbers. This can only be achieved by a conversion to barotropic energy, with a smaller amount of energy being transferred to higher baroclinic wavenumbers, an inefficient process. This contrasts with the case when beta is non-zero, where energy cascades to smaller wavenumbers. The low wavenumbers are energetically very weak when beta is non-zero.

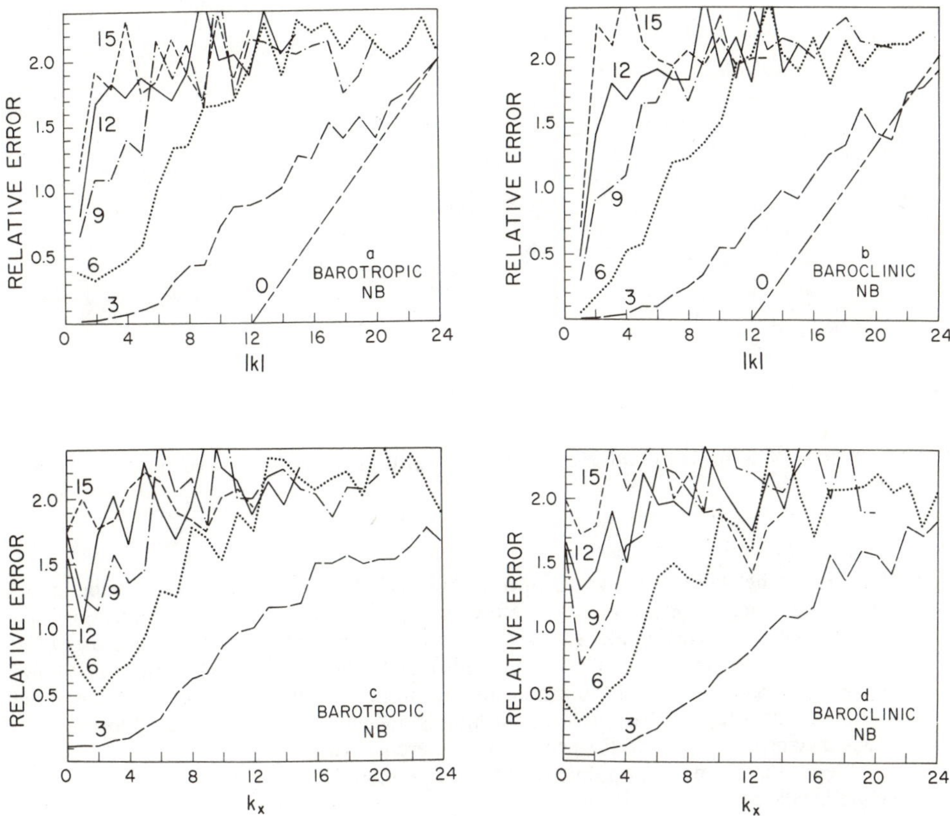

Fig. 8. Relative error growth for NB. a) Is the isotropic barotropic error, b) is the isotropic baroclinic error, c) is the zonal barotropic error and d) the zonal baroclinic error.

The predictability properties of two-layer flow are likewise rather subtly affected by beta. In contrast to the case with one-layer flow, the presence of beta does not automatically and significantly increase the decorrelation time in two-layer flow, even though we may still expect energy cascades in low wavenumbers to be slowed by beta. In one-layer simulations, energy is generally artificially injected around wavenumber seven whether beta is zero or not. The energy cascade to low wavenumbers is accompanied by an enstrophy cascade to high wavenumbers. Beta slows the energy cascade, and hence the entrophy cascade and so increases predictability in all wavenumbers. In two-layer flow energy enters at very small well-predicted wavenumbers when beta is zero; hence error is injected much less efficiently than when beta is non-zero. Secondly, and more importantly, when beta is zero energy is entering the system at low wavenumbers and cannot be transferred to still lower wavenumbers. Thus, the non-linear energy transfer is concentrated in the relatively well predicted low wavenumbers, again meaning error is created (by non-linear transfer) at least as efficiently when beta is non-zero. For in the latter case not only is there more energy transfer but it is in the same direction as error propagation.

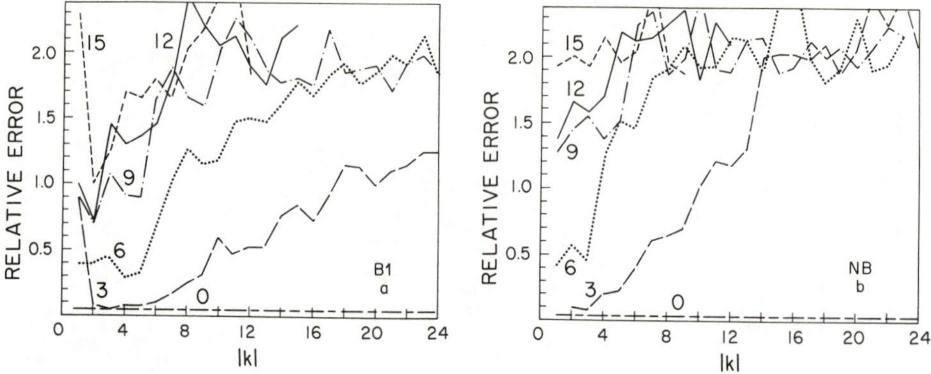

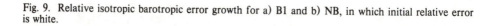

Fig. 9. Relative isotropic barotropic error growth for a) B1 and b) NB, in which initial relative error is white.

The predictability of the gravest barotropic mode (|k| = 1) is lessened by beta, at least in the simulations reported here. The effect is not important in this model since this mode has little energy. This mode varies erratically in amplitude and phase because of its extreme weakness, and so tends to be unpredictable. The predictability of the zonal flow *is* increased by beta. Although the production of strong, steady zonal currents might simply be the explanation, the phase of such currents is not obviously predictable. Closure calculations may provide a clue to the generality of these last two results.

Reducing the mean baroclinicity reduces the energy levels in the model and increases the predictability time.

The error energy ratio is generally less in lower wavenumbers, even when the initial error is distributed evenly across wavenumbers. This is the case even when beta is zero, and is due to the longer turnover, or eddy interaction time for the long waves. It is patently the case that the initial (observational) relative error in weather forecasts is larger in the smaller scales. However, the above result implies that, even with no direct forcing (e.g., by topography), the long waves are intrinsically more predictable than the short waves, a result independent of the initial error conditions.

REFERENCES

1. C.E. Leith, and R. H. Kraichnan, J. Atmos. Sci. **29**, 1041-1058 (1972).
2. D.K. Lilly, Geophys. Fluid Dynamics **4**, 1-28 (1972).
3. C.B. Basdevant, B. Legras, R. Sadourny and M. Beland, J. Atmos. Sci. **38**, 2305-2326 (1981).
4. G. Holloway, J. Atmos. Sci. **38**, (in press) (1983).
5. R.L. Salmon, Geophys. Astrophys. Fluid Dynamics **15**, 167-211 (1980).
6. G.P. Williams, J. Atmos. Sci. **35**, 1399-1426 (1978).
7. R. Fjortoft, Tellus **5**, 225-230 (1953).
8. P.B. Rhines, J. Fluid Mech. **69**, 417-443 (1975).
9. G. Holloway and M. Hendershott, J. Fluid Mech, **82**, 747-765 (1977).
10. J.S.A. Green, Quart. J. Roy. Met. Soc. **86**, 237-251 (1960).
11. E.N. Lorenz, Tellus **21**, 289-307 (1969).

ESTIMATES OF ATMOSPHERIC PREDICTABILITY AT MEDIUM RANGE

Edward N. Lorenz
Massachusetts Institute of Technology

ABSTRACT

Recent studies based upon the output of the ECMWF operational forecasting model indicates that if, after the first day of a forecast, a perfect model could be substituted for the present model, forecasts as good as those presently produced at seven days would be realized at ten days. These studies do not reveal how much improvement in one-day forecasting is possible.

We hypothesize that if all other imperfections in the forecasting procedure could be removed, the inevitable initial uncertainties in observing the small-scale features would, after D days, lead to error fields with amplitudes and spectra resembling those of the errors in present one-day forecasts. The appropriate value of D is highly dependent upon the spectrum of actual atmospheric motions. Estimates with a crude model place D at about four days, thereby implying that the present forecasting success at one week may some day be realized at nearly two weeks.

INTRODUCTION

Many studies which have addressed the problem of the predictability of the atmosphere or some other fluctuating system have been investigations of error growth. The basic question posed in these studies is the following: if at some time t_0 we could alter the state of the atmosphere by a certain amount, and subsequently allow the atmosphere to be governed again by the correct dynamics, how greatly would the state at some later time t_1 differ from the state which would have occurred at time t_1 if no alteration at t_0 had been made? The relevance of this question to predictability becomes evident when the altered state at time t_0 is identified with the observed state, taking into account the inevitable shortcomings of the observations. Since the difference between the states at time t_1 is the error which an optimal extrapolatory prediction scheme would make, the expression "error growth" is apt.

Error growth studies have many ramifications; the initial error may be systematic or random, it may be restricted geographically or distributed over the globe, it may be limited to certain scales of motion or spread over the spectrum, and it may be confined to certain atmospheric properties or allocated to all. The resulting error at a later time possesses similar possibilities.

Since it is not feasible to make deliberate alterations of the atmospheric state resembling typical errors of observation, and, in any event, if we did alter the state we could never observe the evolution of the unaltered state, most studies of error growth have been based upon numerical models of the atmosphere. The wide variety of results obtained reflects the fact that the growth rates are model-dependent. It is often taken as an article of faith that the more closely the model duplicates the readily observed features of the atmosphere, the more reliably it will reveal the growth rate.

Since it is unlikely that the very best possible prediction procedure will ever be formulated, an error-growth study should yield an upper bound to the accuracy with which prediction can be made at any range, or to the range at which prediction can meet a chosen measure of acceptability. Such an estimate will, of course, depend upon the assumed magnitude and nature of the errors of observation, and one obvious way to extend the range of acceptable prediction would appear to be to improve the observing system. As we shall see, however, there is reason to believe that the range of predictability cannot be made to approach infinity by making the observational error approach zero, so that there should be an intrinsic upper bound to predictability. A lower bound can be obtained by noting how well the best currently used prediction procedures perform. As further refinements are made in operational prediction and in error-growth studies, it may be expected that these bounds will approach one another; if they should ever be made to coincide, the technique of weather forecasting will have been perfected.

The purpose of this study is to obtain up-to-date estimates of upper and lower bounds to medium-range atmospheric predictability. For our purposes the medium range will extend from about one-half to about two weeks. We shall be concerned mainly with the extratropical troposphere, and with those atmospheric properties which would characterize a "dry" atmosphere, namely wind, pressure, and temperature. We shall consider how well these quantities may be predicted on the average, rather than in individual situations or at individual locations. We shall not undertake any new computations, and our conclusions will be drawn from the results of studies which have already been performed.

THE MIDDLE AND LATE STAGES OF ERROR GROWTH

We begin by turning to the results of a predictability study[1] which we recently performed with the output of the operational model at the European Centre for Medium Range Weather Forecasts (ECMWF). We shall describe the study briefly; for further details the reader is referred to the cited paper.

Operational forecasts j days in advance, for $j = 0, 1, ..., 10$, are prepared daily at ECMWF. (By a zero-day forecast we mean simply an analysis.) The forecasts are made with a 15-level global primitive equation model with moisture and orography. The model is a grid-point model, but, before being archived, each analyzed or predicted field is represented by a series of global spherical harmonics, truncated triangularly at wave number 40, and it is the 1722 coefficients in each of these sequencies which are stored. In our study we have used only the 500 mb height fields, analyzed on or predicted *for* each of the 100 consecutive days beginning 1 December 1981. Our data thus consists of $100 \times 11 \times 1722 = 1{,}894{,}200$ numbers.

To use these data for a predictability study, we note first that the forecasts one day in advance are reasonably good; hence the 1-day forecast for day i, prepared on day $i - 1$, may be treated as the analysis or 0-day forecast on day i, plus a reasonably small superposed error. By comparing the 2-day and 1-day forecasts for day $i + 1$, we can observe how much this error grows in one day, when both fields are governed by the operational model. Likewise we can obtain the growth during j days, for $j \leqslant 9$, by comparing the $(j + 1)$-day and j-day forecasts for day $i + j$. We also note that forecasts two or more days ahead possess some skill, so that by comparing the $(j + k)$-day and j-day forecasts for day $i + j$, with $j + k \leqslant 10$, we can observe the growth of errors of various initial magnitudes.

As a measure of the difference between two 500 mb height fields we have chosen the root-mean-square difference in height. Figure 1, which is based on a figure in Reference (1), contains the principal results. It shows the differences, in meters, between j-day and k-day forecasts for the same day, averaged over the 100 days of the study, for all pairs (j,k) with $j < k$ and $k \leqslant 10$; these are plotted against k. A heavy curve connects the points where $j = 0$, i.e., where an analysis is compared with a forecast, and it therefore summarizes the performance of the model. The indicated growth rate is the rate at which solutions of two different systems of equations — those of the model and the real atmosphere — diverge from one another. Thin curves connect points having equal values of $k - j$, and indicate the rate at which separate solutions of a single system of equations — those of the model — diverge. This is the rate which is ordinarily evaluated in predictability studies. The dashed curves are extrapolations of the thin curves; we shall presently describe the basis for extrapolating.

The latter rate is supposed to approximate the rate at which separate solutions of the real atmospheric equations diverge. If indeed it does, and if, after the first day the model could be replaced by a perfect model, the heavy curve in Figure 1 would coincide with the lowest thin curve. The actual difference between the slopes of the curves should therefore be a measure of the amount of improvement which may still be realized. In particular, 10-day forecasts should ultimately become better than present 6-day forecasts, even if the 1-day forecast is not improved at all.

To a fair approximation the separate thin curves differ only by horizontal displacements, i.e., the error growth during one day is a function of the magnitude of the error. We may therefore extrapolate the lower thin curves beyond 10 days, by displacing the higher thin curves horizontally. We conclude that 14-day forecasts should become as good as present 8-day forecasts.

This conclusion may be overly optimistic. Since the model is not perfect, it does not necessarily yield the correct growth rate, and it may give an underestimate. In that event, as the model is continually improved, and the heavy curve moves down toward the lowest thin curve, the latter curve may move up to meet it. Improvement in forecasting will then be less spectacular.

Better and better models may be anticipated in the coming years, but some improvements may be introduced immediately. First of all, the ECMWF model produces some systematic errors[2]; these may be subtracted from the forecast. Second, the model is for practical purposes a better model in the northern than in the southern hemisphere, so that we may study the performance of a better model by evaluating root-mean-square height differences for the northern hemisphere only. Introducing these "improvements", we find that the slope of the heavy curve has been reduced, but the slope of the lowest thin curve has been steepened somewhat. Our revised conclusion is that, with no further improvement at one day, 10-day forecasts should ultimately become as good as the 7-day forecasts, and 14-day forecasts should become as good as the 10-day forecasts, which can presently be made by the *improved* ECMWF model.

A familiar measure of error growth is the doubling time for small errors. The smallest error in Figure 1, about 25 m, doubles in about 3.5 days. Larger errors grow less rapidly and ultimately level off. To obtain the doubling time for truly small errors we need to extrapolate the thin curves to the left. It is obviously impossible to do this in any unique manner unless we introduce some auxiliary hypothesis. We postulated[1] that the nonlinear terms in the equation for the growth of the root-mean-square error were essentially quadratic, after which we found that small errors would double in about 2.5 days. Repeating the computation with the "improved" ECMWF model reduced the time to about 2.0 days. This doubling time is consistent with the times obtained from earlier studies[3], although somewhat shorter. It presumably approximates the doubling time for errors in other atmospheric properties which are closely coupled with height, such as wind and temperature, but it may bear little relation to the doubling time for precipitation errors.

THE EARLY STAGE FOR ERROR GROWTH

Having seen that considerable improvement in medium-range prediction is potentially realizable even without altering the one-day forecasts, it is natural to ask how much improvement is possible at one day, and how much effect any such improvement might have upon the medium range. Unlike the growth of errors represented by the lowest curve in Figure 1, which assumes an optimal prediction proceedure, the error at one day is the error produced by a currently used procedure, and results from imperfections in both the forward extrapolation and the initial analysis. Perfecting the extrapolation procedure ought to improve the one-day forecast, but improving the analysis, perhaps by establishing a superior observing system, might have a considerably greater effect. Nevertheless, we feel that

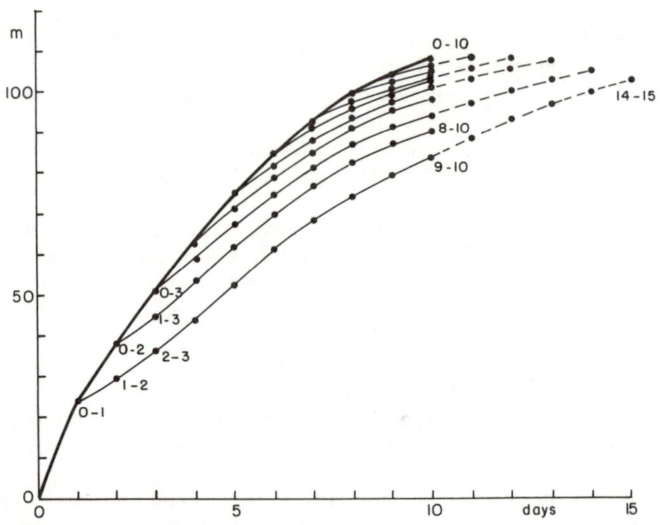

Fig. 1. Average global root-mean-square height differences, in meters, between j-day and k-day operational forecasts with the ECMWF model, for $j < k$, plotted against k. Values of (j,k) are shown beside some points. Heavy curve connects points where $j = 0$. Thin curves connect points with equal values of k- j. Dashed curves are extrapolations of thin curves; see text.

any assumption that continual reductions in the analysis error would lead to proportional reductions in the one-day prediction error is overoptimistic.

The errors represented by the points in Figure 1 are errors in the scales of motion which are resolved by the network of grid points. The *actual* error fields also include errors made by completely omitting the scales too small to be resolved. At present the actual one-day errors and very likely the analysis errors are dominated by the larger scales. However, if the observing system, and so presumably the analysis, are to be subjected to continual improvement, we can anticipate a day when this situation will no longer prevail, and the principal remaining errors will be in the unresolved or poorly resolved scales.

The two-day doubling time deduced from the ECMWF model is presumably an average over the resolved scales, with each scale weighted according to its contribution to the total error. The doubling therefore results from the self-amplification of errors in these scales. A small part of the *actual* error growth results from the influence of errors in the unresolved scales, which induce errors in the resolved scales through nonlinear interactions. Under an ideal observing system, if errors in the resolved scales are greatly reduced, the augmentation of these errors due to self-amplification will be reduced, in proportion, but the acquisition of errors from the unresolved scales will not. The error growth in the resolved scales will then be dominated by the transfer from the unresolved scales, and, until the resolved-scale errors become large, their proportional growth will be much greater.

The smaller scales themselves will amplify quite rapidly, until they approach their maximum size. At very small scales, for example, errors in the structure of a thunderstorm should amplify at least as rapidly as the thunderstorm itself, doubling in an hour or less rather than two days. It follows that any reduction in the transfer of errors from smaller to larger scales, which might be realized by reducing the initial errors in the smaller scales, will be short-lived, since small errors in the smaller scales will not remain small. In any event, developing an observing system which would resolve the mesoscale features, let alone the thunderstorms, would be a difficult and costly undertaking.

We therefore hypothesize that, when the best foreseeable observing system is put to use, the initial growth of errors in the larger scales will be dominated by the influence of the smaller scales. We further hypothesize that after D days the larger-scale errors will have grown to the point where the total error field resembles a one-day error field made by present procedures, in amplitude and spectrum. The range, beyond one day, at which predictions meeting any given measure of acceptability can be made will then be increased by D - 1 days. Our problem is to make a reasonable estimate of D.

Completed works which will lead us to a definitive estimate are hard to discover. The large global circulation models cannot be used, since they do not contain the smaller scales. Models of mesoscale or smaller-scale motions are generally too limited in an areal extent to contain the larger scales. The study[4] to which we shall turn is one in which we derived a system of second-order linear ordinary differential equations, whose dependent variables were the squared amplitudes of the wind errors in separate bands of the spectrum, each band spanning a single octave. Solutions of these equations depict the spread of errors from one scale to another.

The study is by no means ideal for our present task, partly because it was not intended primarily as an atmospheric study. The basic equation from which the ordinary differential equations were derived was the barotropic vorticity equation, which certainly does not approximate the laws governing the smaller atmospheric scales. To keep the equations manageable, the motion field was assumed to be homogeneous and isotropic. The deceleration of the error growth in each scale, which should have been brought about by nonlinear effects, was simulated by allowing the error growth to grow quasi-exponentially to a prechosen scale-dependent value, and then terminating its growth altogether. Viscosity and external forcing were omitted. Finally, a formula related to the discredited quasi-normal approximation was used to close the system.

The model was atmospheric to the extent that the prechosen spectrum of the unperturbed motion was modeled after the assumed atmospheric motion spectrum; this spectrum exerted a controlling influence on the time scale associated with each spatial scale. The principal results of the study are summarized in Figure 2, which is based on a figure in Reference (4), and shows the growth of an error field confined initially to the smallest scales. The upper curve is the assumed atmospheric spectrum, which also serves as an upper bound for the spectrum of the errors. Each curve labeled with a time (8 days, etc.) actually extends from the extreme left to the extreme right of the figure: to the left it is indistinguishable from the zero line, while to the right it is indistinguishable from the upper curve. These curves are the spectra of the errors at the indicated times.

In attempting to estimate D from the results in Figure 2, we must recall that the errors there are wind-field errors, while those in Figure 1 are height-field errors. The ratio of a wind error to a height error may be estimated geostrophically, and it is highly scale-dependent. We could scale down the right-hand portion of Figure 2 so that the curves would represent height spectra instead of wind spectra, but instead we shall circumvent the scale dependence by examining a particular scale.

In making the study[1] with the ECMWF model, we performed certain additional computations which were not described in the final write-up. These included a spectral analysis of the prediction errors. We found that the smallest scales in the archived data, with wave numbers near 40, were predicted moderately well at one day, rather poorly at two days, and not at all at three. These scales correspond to the 1250-625 km band in Figure 2. This band is indicated as being reasonably predictable at 1/2 day, slightly predictable at 1 day, and unpredictable at 1.5. Combining these results, we see that motions of this scale are already being predicted better than they can be. We are therefore faced with a contradiction, and the fault is presumably in the model which produced Figure 2.

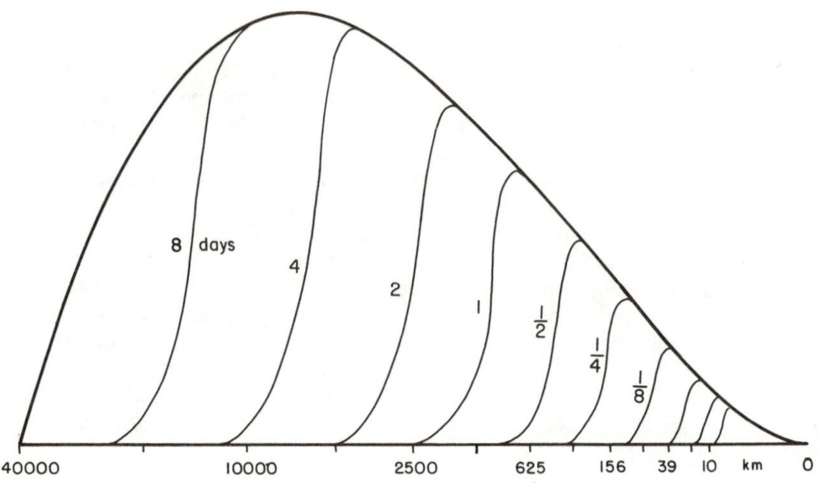

Fig. 2. Growth of errors initially confined to smallest scales, according to theoretical model. Upper heavy curve is assumed atmospheric motion spectrum; lower heavy curve is zero line. Thin curves are spectra of errors at indicated times; each thin curve coincides with lower heavy curve to the left, and upper heavy curve to the right. Areas are proportional to kinetic energy.

We have already enumerated some of the defects of the model, and it would appear possible to perform a new study with an improved model. Perhaps the model could include baroclinic effects, some inhomogeneity and anisotropy, and some forcing and damping. Perhaps the nonlinear effects could be incorporated more realistically. Perhaps a more realistic closure assumption, like those used in some subsequent works[5,6] could be introduced. However, the deduced predictability times are so highly dependent upon the assumed atmospheric spectrum that any of the above-mentioned improvements would be pointless until our estimate of the atmospheric spectrum has been made as realistic as possible.

In keeping with our intention of basing this study on the results of previously completed works, we shall turn to a study entitled "Limits of Meteorological Predictability", which we prepared in 1972 at the request of the American Meteorological Society. To the best of our knowledge the results were for internal use, and were never published. In that study we addressed the question of the effect of a possible spectral gap in the mesoscale band. We performed three sets of computations similar to those used to produce Figure 2: one with no spectral gap, one with a "weak gap", and one with a "strong gap". Although we are not even sure of the existence of a gap, let alone its structure, our "best guess" is something between the weak and strong gaps. Figure 3, which has the same format as Figure 2, has been constructed from an interpolation between the weak-gap and strong-gap computations. We see that errors initially confined to the smallest scales begin to spread up the scale just as in Figure 2, but, upon encountering the gap, they experience considerable difficulty in crossing it. Thus, it takes nearly five days for the error spectrum in Figure 3 to acquire the same form, outside of the gap, which it acquired in one day in Figure 2. Turning to motion in the 1250-625 km band, we see that it is moderately predictable at 4 days, slightly predictable at 5 days, and unpredictable at 6 days. Comparing this result with current skill in predicting these scales, we see that about 3 days can be added to the range at which they may be predicted. From this we conclude, very tentatively, that $D = 4$.

We believe that this model, crude as it may be, depicts fairly realistically the qualitative influence of a spectral gap on predictability. What may be quite unrealistic is the assumed spectrum.

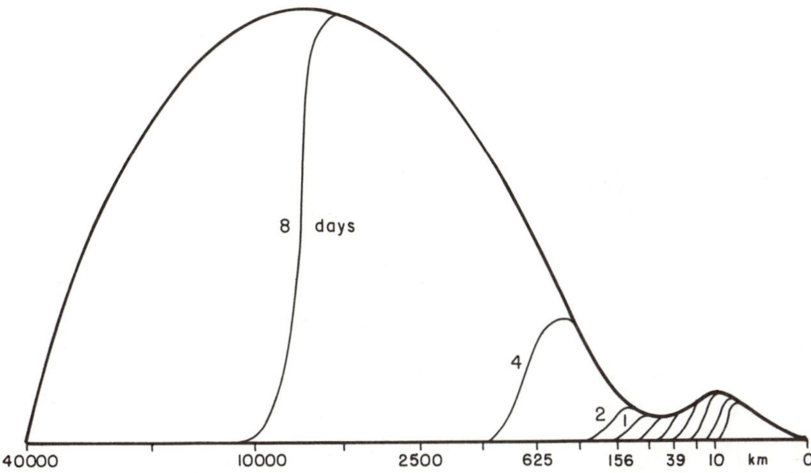

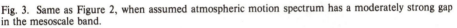

Fig. 3. Same as Figure 2, when assumed atmospheric motion spectrum has a moderately strong gap in the mesoscale band.

CONCLUDING REMARKS

We have examined some studies which together imply that major improvements in medium-range weather forecasting are possible, and,in particular, that the present forecasting success at one week may some day be realized at nearly two weeks. The studies contain estimates of the rate at which inevitable errors in the analysis will grow as the range of the forecast is extended, until they eventually render the forecast unacceptable. For the middle and later stages of error growth our estimates are on reasonably firm ground; for the early stage they are highly speculative.

The model which indicates that we might eventually forecast as well at four days as we can now forecast at one contains a rather arbitrarily chosen atmospheric spectrum, which strongly influences the numerical results. The model is also crude in other respects, and we believe that some computations with some other model, possibly a rather sophisticated mesoscale model, are in order. Nevertheless, we do not see how the final result can fail to depend upon the atmospheric spectrum which the model entails, whether it is prechosen on the basis of real observations or produced by the model itself. Since different models can produce different spectra, and since we must have confidence in the spectrum if we are to have confidence in the conclusions, it seems rather likely that the next significant refinement in our estimate of medium-range predictability will result from observations.

ACKNOWLEDGMENT

This research has been supported by the GARP Program of the Atmospheric Sciences Section, National Science Foundation, under Grant 82-14582 ATM.

REFERENCES

1. E.N. Lorenz, Tellus **34**, 505-513 (1982).
2. L. Bengtsson and A.J. Simmons, Large-scale dynamic processes in the atmosphere, B. Hoskins and R. Pearce, eds. (New York and London, Academic Press (1983).
3. J. Smagorinsky, Bull Amer. Meteor. Soc., **50**,286-311 (1969).
4. E.N. Lorenz, Tellus, **21**, 289-307 (1969).
5. C.E. Leith, J. Atmos. Sci., **28**, 148-161 (1971).
6. C.E. Leith and R.H. Kraichnan, J. Atmos. Sci., **29**, 1041-1058 (1972).

LAGGED AVERAGE FORECASTING, SOME OPERATIONAL CONSIDERATIONS

Ross N. Hoffman† and Eugenia Kalnay
Laboratory for Atmospheric Sciences
Goddard Space Flight Center, Greenbelt, Md 20771

ABSTRACT

We have previously described the lagged average forecast (LAF) method as an alternative to the Monte Carlo forecast (MCF) method[1]. The LAF differs from the MCF in the definition of the ensemble of initial states which are used to generate the ensemble of forecasts. The LAF initial states are the current analysis and the forecasts made from previous analyses verifying the current time. Thus the LAF ensemble is composed of forecasts which are made by a regular operational system of numerical weather prediction and the LAF method is therefore operationally attractive

The application of our previous ideas and results[1], to an operational model requires the resolution of what might be called the degrees of freedom problem, i.e. how to obtain a homogeneous sample large enough to calculate stable statistics. We suggest that this problem may be solved by carefully modeling the required statistics in terms of a small set of parameters and then estimating only these few parameters from the data. We also note that there may be considerable information in each initial ensemble relating to the predictability of each particular case, and that this information may be incorporated in the model of the statistics.

INTRODUCTION

In order to use the information present in past observations and simultaneously to take advantage of the benefits of stochastic dynamic prediction we have formulated and tested an ensemble average forecast method, which we have called the lagged average forecast (LAF) method[1]. Each LAF is an average of an ensemble of forecasts whose initial conditions are the current and past analyses. For example, if we label the current time as t = 0 h, then forecasts started at t = 0 h, -12 h -24 h .... -48 h all integrated to t = +72 h would be averaged to yield the 72 h LAF forecast. We will now summarize the main ideas and results of our earlier experiments.

RESULTS FROM TESTS WITH A SIMPLE MODEL

In our tests of the LAF method[1], we used a highly simplified atmospheric model[1,2]. This model is a two layer low order spectral model on a doubly periodic f-plane forced by asymmetric Newtonian heating of the lower layer. These experiments are not identical twin experiments since there is a continuous external source of forecast error. The quasigeostrophic model used for forecasting is imperfect since the observations used for initialization and verification are based on a long primitive equation 'nature' run.

In terms of forecast skill, the LAF was found to be slightly superior to the MCF, ordinary dynamical forecast (ODF) and persistence climatology forecast (PFC) (Fig. 1). Purely statistical means are sufficient to hedge the ODF towards the climate mean. the same regression techniques were also applied to the ensemble forecasts, resulting in the tempered ODF, LAF and MCF (tODF, tLAF and tMCF). The regression analysis naturally generates different weights for the individual forecasts in the LAF ensemble; the more recent the initial conditions, the greater the regression weight. The tLAF was found to be

† Universities Space Research Association Visiting Scientist.

marginally superior to all the other methods. At long range all the tempered forecasts have the same skill since they become climate mean forecasts.

Unfortunately the improvement in forecast skill obtained due to ensemble averaging is quite small. This conclusion is related to the predictability properties of the model, by which we mean the evolution of forecast error. Although the rms forecast error was found to grow roughly linearly with time until it reaches saturation (Fig. 1), we found that the typical evolution of forecast error for an individual case has a relatively short episode of rapid error growth (Fig. 2). Furthermore, the timing of this forecast breakdown varies substantially from case to case. This behavior is reminiscent of the behavior of initially small triangles on the attractor of the 3-component system of Lorenz which was shown to the workshop by Professor Dutton. In the Lorenz[3] system this behavior is due to the presence of the homoclinic orbit passing through the origin as described by Lanford[4]. These observations suggest that ensemble average forecasts can attain skill only slightly better than a statistically filtered ODF. This statement is justified by the following argument: For short forecast times, an ensemble forecast is no better than a forecast made from the ensemble average initial conditions, because nonlinear effects are absent. For long forecast times, the ensemble spreads out over the entire attractor; the ensemble forecast is therefore close to the climate mean and its superiority over a climate mean forecast is negligible. The possible advantage of an ensemble forecast is therefore confined to intermediate forecast times, specifically to those periods for each case when the forecast error is of intermediate size. As we have described, these periods are rather short.

Since the skill of the forecast model varies considerably from case to case (Fig. 2) we made *á priori* predictions of forecast skill for individual cases. Not unexpectedly we found that the spreading of the ensemble is closely correlated with the average forecast error of forecasts in the ensemble. When other factors which account for model imperfections are included in the analysis, good predictions of the time of forecast breakdown are obtained (Fig. 3).

One might not expect an ensemble forecast method with an ensemble size as small as 4 or 8, which are the sizes we have been using, to be so successful at predicting the forecast error, which is after all, a second order statistic. However if we neglect model errors, the forecast error of the discrete model is governed approximately by linear homogeneous ordinary differential equations until the errors become large. During the period when the linearization is valid we may therefore represent the evolution of any initial error in terms of the transition matrix, i.e. the matrix which obeys the linear equations and which is initially the identity matrix. A complete discussion along these lines was given by Lorenz[5]. In general the largest exponent will control the growth of error. Since an arbitrary initial error will almost always have some projection on the eigenvector corresponding to the dominant characteristic exponent we expect that even an ensemble of size 2 may be useful in predicting forecast skill.

REMARKS ON OPERATIONAL APPLICATIONS

A straightforward implementation of the LAF method in an operational setting is not possible: there are far too many parameters to be estimated from a regression analysis of a limited sample. Quite similar problems have been studied by Lorenz[6] and we refer the reader to his detailed discussion of this problem. A further difficulty here is that the sample must be further subdivided or stratified since it is not homogeneous; for example, predictability varies with season. There are two basic solutions to this problem: obtain a larger sample or reduce the number of independent parameters which must be estimated. The following discussion centers on the problems to be expected with the second approach and offers some possible solutions. We should note that the choice of the tempering weights and the coefficients relating forecast error to ensemble spread described in the previous section are both least squares problems.

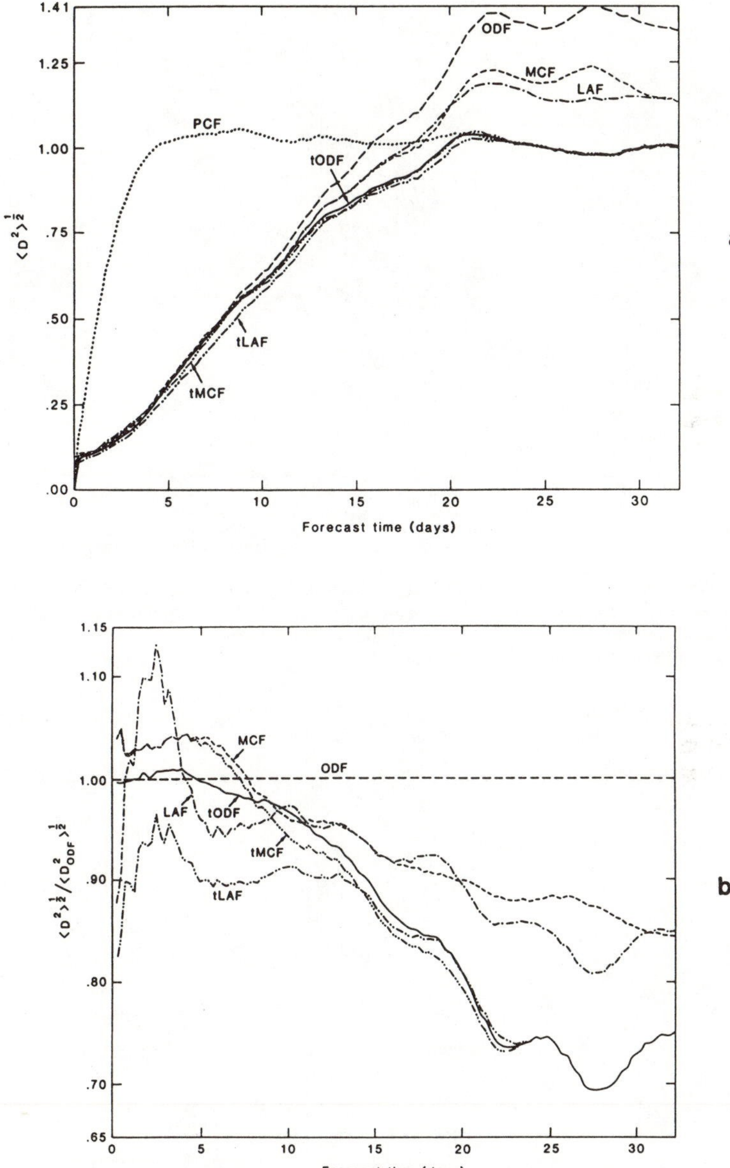

Fig. 1. Comparison of the skills of the different forecast methods. In (a) $\langle D^2\rangle^{1/2}$ and in (b) $\langle D^2\rangle^{1/2} / D^2_{ODF}\rangle^{1/2}$ are plotted as functions of forecast time, t, for the ODF (— —), MCF (– – – –), LAF (—·—·), PCF (·······), tODF (————), tMCF (—···—), and tLAF (—··—). D is a skill score which is also the distance in the model phase space from forecast to observation normalized by the expected skill of a climate mean forecast. The angle brackets indicate an average over all cases.

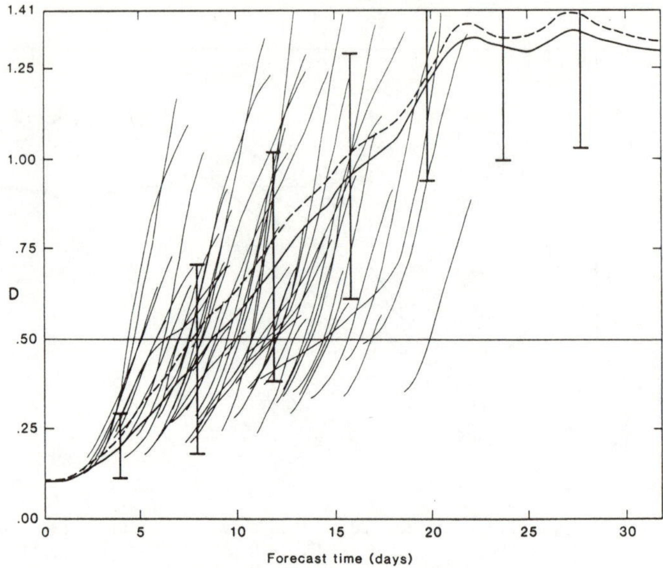

Fig. 2. Time evolution of D for the ODF experiments. Shown are selected portions of the evolution of d for each case (light ——). These have been smoothed to eliminate the roughness caused by the white noise observing error. Also shown are $<D^2>^{1/2}$ (— —) and $<D>$ (heavy ——). Error bars are calculated as $(<D^2>-<D>^2)^{1/2}$.

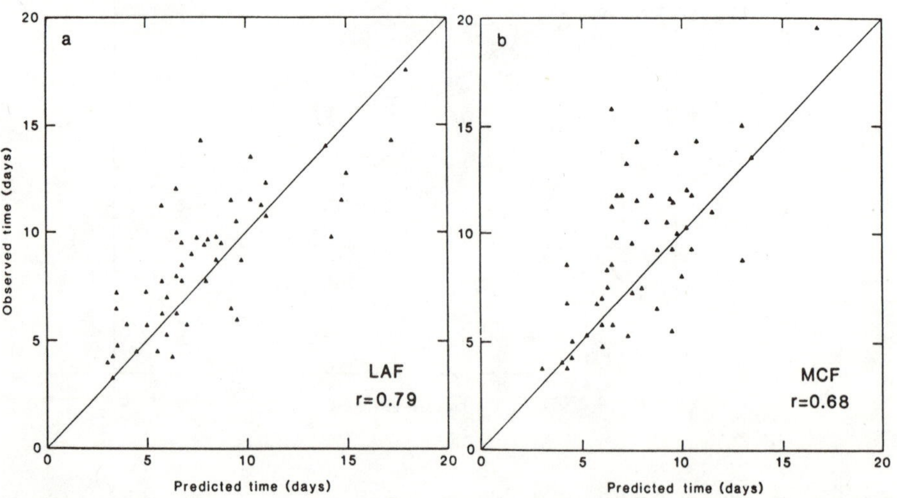

Fig. 3. Scatter plots of predicted versus observed time for the breakdown of forecast skill for the LAF (a) and the MCF (b). The time of forecast breakdown was taken to be the time when the normalized skill score reaches a value of 0.5. See Ref. (1) for details.

Let us first consider, as a simple example, the problem of tempering an ODF. Let $\underline{h}$ be the forecast deviation from climatology of the variable of interest; for example $\underline{h}$ might be a 120 h forecast anomaly of 500 mb geopotential. We consider estimates of $\underline{h}$ given by

$$\hat{h}_m = \sum_i a_{im} h_i \tag{1}$$

where $\underline{i}$ and $\underline{m}$ index the components of the representation, which might be grid point values or coefficients of spherical harmonics, normal modes or empirical functions. One interesting possibility especially for the extended range problem is to use the amplitudes of a selected set of teleconnection patterns[7]. The least squares solution, for each m, satisfies

$$<\tilde{h}_m h_n> = \sum_i a_{im} <h_i h_n> \quad n = 1,...,$$

where $\tilde{h}$ is the true value and the angle brackets indicate statistical expectation. The sum over i may be severely restricted; for example only grid points within a distance of 2000 km might be included. We must now choose between modeling the covariances and solving the normal equations at each grid point or modeling the a_{im} in terms of fewer secondary coefficients, and considering the resulting condensed least squares problem. While both of these approaches have merit, we will restrict our discussion to the approach in which the covariances are modeled. This is the approach usually taken in optimum interpolation. In a grid point model the covariances are expected to vary smoothly with respect to position so we may increase the sample used to estimate the covariances by using all the data in a given region, e.g. a 20° x 20° latitude-longitude window, under the assumption that the covariances are independent of position under the window. We would consider the estimate obtained valid at the center of the window and then move the center of the window from grid point to grid point. Another approach with the same effect would be to represent the covariances in terms of just a few position dependent parameters and to represent each of these parameters in terms of a small number of spherical harmonics, determining the coefficients of the spherical harmonics to give a best fit to the covariances observed in the sample. For example we might suppose the correlation function between a given location and any other location depends only on distance and has gaussian shape; then all the covariances are determined once the variance and correlation length are specified. Further discussion of modeling of covariances may be found in the optimum interpolation literature (see Lorenc[8] and references therein).

To further simplify the discussion we will now assume that sums like Equation (1) are truncated to the single term $i = m$: only predictions at a given grid point or of a given mode are to be used in estimating the true value. Now when several forecasts are available, which is the case when tempering a LAF, estimates of covariances between different forecasts are required. As discussed previously[1], it is nearly equivalent to estimate the differences between two forecasts or between a forecast and nature. A great deal of information about the actual initial differences is present in the initial ensemble. The remaining problem is to estimate how fast the differences grow; these growth rates are the traditional objects of predictability studies. Again, there is not a sufficient sample to get stable estimates of these growth rates for each component of the representation at each time during the forecast, and we must use some modeling assumption. Since predictability varies with the scale of motion under consideration[9,10], it might be useful to use a spectral representation where each component has an associated scale and calculate statistics for all variables associated with similar scales, i.e., by wave band. It may also be possible to fit the growth rates to simple functions of forecast time; several candidate functions are offered by other authors in this volume. In our previous work[1] we found it was adequate to assume the growth rates are constants until saturation is reached.

In practice the severe truncation assumed above would not be made, but the general problem can be reduced to the simple problem by assuming that the general correlation function is the product of a 'spatial' correlation function depending only on the location of the two grid points and an 'ensemble' correlation function depending only on the two forecasts. This is analogous to the treatment of the 'spatial' correlation function as the product of 'horizontal' and 'vertical' correlation functions which is made in optimum interpolation[8] and which may be used here.

Predictability varies with season as well as with location and scale and this would have the effect of greatly reducing our sample if we were to stratify our sample by month of the year. A temporal window technique similar to the spatial window technique described above could be used. Alternately the coefficients or covariances could be modeled as simple functions of time of year[11]. Much of the seasonal effect could be removed by using variables standardized by removing the climate mean and scaling by the climate variance, where these climate statistics vary with time of year. In the case we have been considering all that is left is to model the seasonality of the growth rate parameters.

Another sampling problem concerns the ensemble of forecasts. Intuitively, forecasts in the LAF ensemble which have no skill at $t = 0$ h, should not be included in later ensembles. Unless the covariances are modeled this would be another cause for sample stratification. However there is considerable information in the observed ensemble spread at the initial time. In fact the initial differences between pairs of forecasts is known precisely and good approximations of the differences between the various forecasts and nature may be constructed using the current analysis and estimates of its error. This information combined with the difference growth rates found using the methods described above is sufficient to calculate estimates of all the statistics needed for the tempered LAF method. When the initial differences are estimated to be large the estimated covariances will be small and poor initial forecasts will effectively be deleted from the ensemble.

For the purpose of predicting forecast skill, there is also additional information in the initial ensemble. The difference between two forecasts may be considered to be a finite difference approximation of the product of the transition matrix defined in the previous section and a vector representing the initial difference between these forecasts. The accuracy of such an estimate is expected to increase as the initial difference decreases and to increase as the difference between the current analysis and the average of the initial states decreases. Further the dimension spanned by the initial difference vectors should give some indication of the probability that the behavior associated with the largest eigenvalue has been captured by the sample.

CONCLUDING REMARKS

We feel that the LAF method is a promising technique. While our tests[1], and discussion above indicate that operational implementation of the LAF method should not be too difficult, two uncertainties remain. First, can a large enough sample be obtained for developing the required statistics? Second, is the interval between the initial states of successive operational forecasts small enough for the initial ensemble to be acceptable? With respect to this last question the interval of 24 h currently available from operational ECMWF forecasts may be too large.

The LAF may be directly applied to forecasts of time averaged quantities, e.g. 500 mb height averaged from 120 h to 240 h. There is some evidence favoring the use of the LAF method for long range forecasting of time average quantities, although the sampling problem becomes more acute as the forecast interval increases. Recent identical twin experiments[10,12] suggest that ensemble average forecasts of time averaged quantities are useful beyond the range of day to day predictability. Also, long range, ensemble forecasts can make good use of observed boundary conditions[13] filtering the synoptic scales and at very long range the ensemble of forecasts will simulate the climate ensemble associated with the particular boundary conditions[14] known at the initial time.

ACKNOWLEDGMENTS

We thank R. Livezey for sharing his insights on this problem with us. We had useful discussions with B. Legras. Ross N. Hoffman was supported by NASA contract NAS-5-27297.

REFERENCES

1. R.N. Hoffman and E. Kalnay, Tellus, **35A**, 100, in press (1983).
2. R.N. Hoffman, J. Atmos. Soc., **38**, 514 (1981)
3. E.N. Lorenz, J. Atmos. Soc., **20**, 130 (1963).
4. O.E. Lanford, Annu. Rev. Fluid Mec., **14**, 347 (1982).
5. E.N. Lorenz, Tellus, **17**, 321 (1965).
6. E.N. Lorenz, Mon. Weather Rev., **105**, 590 (1977).
7. J.M. Wallace and D.S. Gutzler, Mon. Weather Rev., **109**, 784 (1981).
8. A.C. Lorenc, Mon. Weather Rev., **109**, 701 (1981).
9. E.N. Lorenz, Tellus, **21**, 289 (1969).
10. J. Shukla, J. Atmos. Sci., **38**, 2547 (1981).
11. K. Hasselmann and T.P. Barnett, J. Atmos. Sci., **38**, 2275 (1981).
12. A.N. Seidman, Mon. Weather Rev., **109**, 1367 (1981).
13. J. Shukla, European Centre for Medium Range Weather Forecasts Seminar 1981, Problems and Prospects in Long and Medium Range Weather Forecasting 14 - 18 September. 261 (1981).
14. C.E. Leith, Nature, **276**, 352 (1978).

PREDICTABILITY OF ATMOSPHERIC LOW-FREQUENCY MOTIONS

J. Egger
University of Munich, Munich, FRG

H.D. Schilling
University of Munich, Munich, FRG

ABSTRACT

Much of the atmospheric long-term variability (periods $\geqslant$ 10 days) is contained in the largest planetary scales (zonal wave number $m \leqslant 5$). It is proposed that a considerable fraction of these low-frequency motions is induced by the interaction of planetary-scale modes with synoptic-scale modes ($m > 5$). To test this hypothesis, the synoptic-scale forcing of the planetary-scale 500-mb streamfunction is determined from data. This forcing can be fitted approximately to a Markov process of first order and depends on locality. The planetary-scale response to this forcing is determined and it is found that there is quite a good correspondence between the low-frequency variance as observed and the one computed. The predictability of such forced planetary-scale motions is discussed.

INTRODUCTION

Up to now most extended-range forecasts (predictions beyond 10 days, say) have been based on linear regression models. Dynamic models like GCMs have hardly been used partly because of economic reasons and partly because the skill of such forecasts has generally been low so far. Nevertheless, there appears to be a growing interest in extended range forecasts with dynamic models[1]. In such a forecast one does not want to predict the day-by-day fluctuations of the atmospheric motions nor is a resolution of synoptic-scale features of the flow required. Instead one aims at predicting the slow changes of the largest scales of motions since it is these scales which dominate the low-frequency part of the spectrum of atmospheric motions[2-4]. Figure 1 shows the low-frequency variance of the 500-mb streamfunction for wave modes with zonal wave number $m \leqslant 5$.

We have two maxima of the variance, one over the Pacific and another one over the Atlantic. Low values are found near 120°W and over the Himalayas. In general the variance varies strongly with locality and is rather low outside the belt 40°N to 70°N. To predict motions at these scales of space and time one needs to understand what causes these motions. There appear to be at least two mechanisms to induce low-frequency motions in the atmosphere. First, slow changes in the boundary conditions induce atmospheric responses at long time scales. A much quoted example is the impact of anomalies of the sea surface temperature on the atmosphere. These anomalies may persist for months and so will the atmosphere response. Second, one has to consider the possibility that the internal dynamics of the atmosphere will allow for such slow variations even if the boundary conditions are held fixed in time. After all, a large part of the energy of the largest modes is provided by the interaction with the shorter synoptic-scale waves. Since this interaction operates at all time scales the response may be strong at low frequencies because the planetary scale contains the quasi-stationary modes of the atmosphere.

In this paper we are going to explore the latter possibility. We propose that forcing of planetary waves by synoptic-scale waves is the source of much of the observed long-term variability of the atmosphere. To explore this mechanism we restrict our attention to the barotropic modes of the atmosphere. We assume that the barotropic vorticity equation captures the essential features of what goes on at the 500 mb surface. The equation reads

$$\frac{\partial}{\partial t}(\nabla^2-\lambda^2)\,\Psi + J(\Psi, \nabla^2\Psi + f) = -C\,\nabla^2\Psi + \nu\,\nabla^4\Psi \qquad (1.1)$$

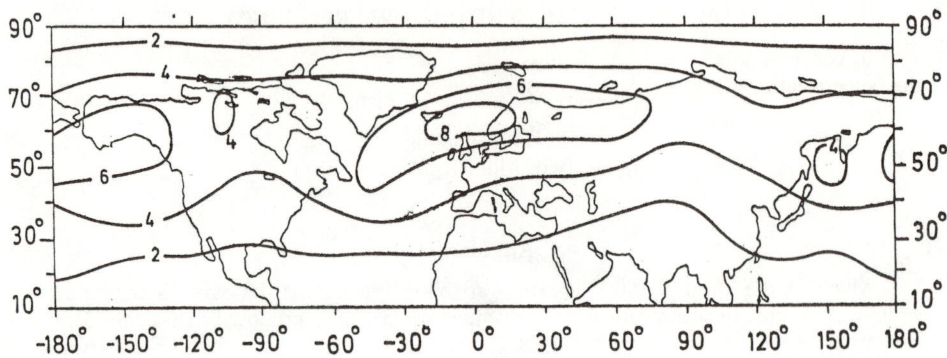

Fig. 1. Low-frequency variance of the 500 -mb planetary flow streamfunction as observed ($10^{13}m^4s^{-2}$ after Egger and Schilling[5]).

where Ψ is the streamfunction at 500 mb. The Coriolis parameter is denoted by f. The term $\lambda^2\ \partial\Psi/\partial t$ describes free-surface effects with λ^{-1} as a radius of deformation. The first term on the right-hand side crudely represents surface friction effects where ∇^2 is the Laplacian on the sphere. The second term stands for horizontal turbulent diffusion of vorticity.

It is convenient to expand the streamfunction in terms of the eigenfunctions of the Laplacian on the sphere. These eigenfunctions are the spherical harmonics

$$B_n^m = P^{|m|}_{|m|+2n-1}(\sin\Theta)e^{im\lambda} \tag{1.2}$$

(Θ latitude, λ longitude, P_l^m Legendre polynomial) so that

$$\Psi = \frac{1}{2}\sum_{m=-\tilde{M}}^{\tilde{M}}\sum_{n=1}^{\tilde{N}}\Psi_{mn}B_n^m \tag{1.3}$$

with expansion coefficients Ψ_{mn}. In (1.2) m is the zonal wave number and n the meridional wavenumber. It is seen that we admit only modes which are antisymmetric with respect to the equator.

It is somewhat arbitrary which scales are considered as planetary and which ones as synoptic. Here, the rectangular wave group with $0\leqslant m\leqslant M=5$, $1\leqslant n\leqslant N=5$ is accepted as the planetary group. It is convenient to partition the streamfunction between a planetary part and a synoptic part

$$\Psi = \Psi_p + \Psi_s \tag{1.4}$$

where

$$\Psi_p = \tfrac{1}{2}\sum_{\substack{m=-M\\ m\neq 0}}^{M}\sum_{n=1}^{N}\Psi_{mn}B_n^m + \bar{\Psi} \tag{1.5}$$

In most of what follows we shall assume

$$\bar{\Psi} = -u_o\sin\Theta\cdot a \tag{1.6}$$

with u_o constant and a the earth's radius. We ignore therefore all modes with m=0 except the "superrotational" mode. The forecast equation for Ψ_p is

$$\frac{\partial}{\partial t}(\nabla^2-\lambda^2)\Psi_p+J_p(\Psi_p,\nabla^2\Psi_p+f)+C\nabla^2\Psi_p-\nu\nabla^4\Psi_p$$
$$=-J_p(\Psi_p,\nabla^2\Psi_s)-J_p(\Psi_s,\nabla^2\Psi_p)-J_p(\Psi_s,\nabla^2\Psi_s) \tag{1.7}$$

The subscript p at the symbol J denotes the projection of a Jacobian on the planetary modes. The Jacobians on the right-hand side of (1.7) describe the impact of the synoptic scale flow on the planetary modes. In what follows we shall lump together these three terms and treat them as one synoptic forcing term J_s. We project (1.7) on the planetary basic functions (1.2) and obtain a forecast equation for the coefficients Ψ_{mn}:

$$\frac{d\Psi_{mn}}{dt}+J_{pmn}+\hat{C}\Psi_{mn}=J_{smn}/(1+\lambda^2/k^2_{mn}) \tag{1.8}$$

where J_{smn} represents the forcing of the mode (m,n) by the synoptic modes and J_{pmn} describes the interaction of the planetary modes. Frictional effects are contained in the last term on the left-hand side. The eigenvalue of the mode m,n is denoted by k^2_{mn}:

$$k^2_{mn}=(m+2n-1)(m+2n)/a^2 \tag{1.9}$$

It will be our basic strategy to solve for the planetary scale modes but to prescribe the forcing by synoptic modes according to observations. We therefore view the planetary modes as driven by the synoptic scale modes. To determine the forcing term J_{smn} from data we took two years (1972-1973) of daily geopotential height analyses as provided by the German Weather Service (DWD). The data handling and the analysis procedures are described in Egger and Schilling[5] to some detail. Here we shall just give a brief outline.

The data are available on a grid covering the Northern Hemisphere from 15°N to 85°N. The height data are expanded in the normal modes B_n^m. As a matter of fact, slightly different normal modes have been used in the actual computations which have been specially designed to fit the data domain[5]. The expansion includes twelve zonal wave numbers and five meridional wave numbers, i.e. $\tilde{M}=12$, $\tilde{N}=5$. The streamfunction Ψ is obtained from the geopotential using the linear balance equation. Finally the forcing terms J_{smn} have been evaluated so that a two-year time series of the forcing was available for each planetary mode.

Before we turn to integrations of (1.8) it is revealing to obtain some information on the statistical characteristics of these forcing terms. To that end we discuss the powerspectra of the real and the imaginary part of the forcing terms J_{smn}, respectively. It has been found that these spectra are essentially of the red noise type. We have fitted a Markov process

$$X(\tau_{j+1})=X(\tau_j)e^{-b}+((1-e^{-2b})R_{mn}(0))^{1/2}W(\tau_j) \tag{2.1}$$

to the time series of the real and imaginary part of the forcing, respectively. The variance is R_{mn} (0) and the running index j increases by one every day with $\tau_j = jDt$, $Dt = 1$ day. W (τ_j) is a white noise process with power density 1. The decay rate b is the only fitting parameter available which depends on m,n of course. The autocorrelation $R_{mn}(\tau)$ of the forcing is then approximately

$$R_{mn}(\tau)=R_{mn}(0)e^{-b\tau} \tag{2.2}$$

and the corresponding fitted powerspectrum is

$$F_{mn}(\omega) = R_{mn}(0)b/(b^2+\omega^2) \qquad (2.3)$$

Figure 2 shows the parameter b for the imaginary part of the forcing for all planetary modes. The decay rates b are of the order of one day. Therefore the forcing by synoptic scale modes has a rather short persistence. Roughly speaking the largest modes (m,n small) have the smallest values of b whereas the smallest modes have the largest values of b. To demonstrate the quality of the fitting formula we show the average overall power spectra with $1.25 \leqslant b \leqslant 1.75$ and the average at the corresponding fitted red noise spectra (Fig. 3).

The correspondence between the observed and fitted spectra is satisfactory although the fitted spectra overestimate somewhat the power of the forcing for high frequencies. Since (2.1) appears to portray the basic features of the synoptic scale forcing we can draw the conclusion that the forcing terms are essentially unpredictable beyond 2/b days, say.

The variances $R_{mn}(0)$ have been evaluated as well. They are of the order 30-100 $m^{+4}s^{-4}$. If we want to study the forcing of the planetary scale streamfunction at a certain locality we have to superimpose the contribution of all the various J_{smn} to obtain the forcing at that point. Then we can compute the power spectrum $\Pi(\lambda,\Theta,\omega)$ at any locality of the Northern Hemisphere. It is convenient to consider variances, i.e. the integral of the power over a frequency band. Following Blackmon[3] we introduce the low-frequency variance

$$G_L^2 = \frac{2}{\pi} \int_0^{\omega_{10}} \Pi(\lambda,\Theta,\omega)d\omega$$

of forcing with periods longer than ten days ($\omega_{10}=2\pi/10\text{days}^{-1}$). Figure 4 shows the low-frequency variance of the streamfunction forcing as obtained from the data. There is a fairly broad maximum of the forcing over the western part of the hemisphere. The variances are generally lower at high and low latitudes.

All the spectral techniques used so far are based on the assumption of statistical stationarity. A posteriori the assumption appears to be reasonable in view of the rather short decay times b^{-1}. Furthermore an evaluation of the forcing for other years showed good agreement with the results displayed here.

IMPACT OF FORCING AT MEDIUM RANGE

To give an impression of the impact of the forcing of planetary mode by synoptic-scale mode we have run a few numerical experiments where we compare pairs of ten-day integrations of (1.8). The forcing J_{smn} is specified in one run according to data for a randomly chosen ten-day period out of the two years of data. There is no forcing in the other run. Both integrations start from the same initial field which is obtained by running (1.8) with realistic forcing over a few weeks. Therefore the initial planetary field has an energy level which is characteristic of the synoptically forced planetary modes. Figure 5 shows the correlation coefficient of the forced and unforced runs after eight days as obtained from a sample of twenty cases. When drawing Figure 5, only the correlation of the wave modes with $m>0$ has been taken into account. We have drawn only isolines for correlations $\geqslant 0.6$ assuming that "forecasts" with correlations less than 0.6 are virtually useless. After eight days there are only a few isolated areas left where the correlation is larger than 0.6.

For example the dispersion of the forecast pairs is relatively small over most of North America. However, the forcing appears to be quite effective in separating the forecast pairs over the oceans. Note that Figure 5 does not imply that one cannot forecast planetary motions beyond eight days, since a numerical weather prediction model would predict the

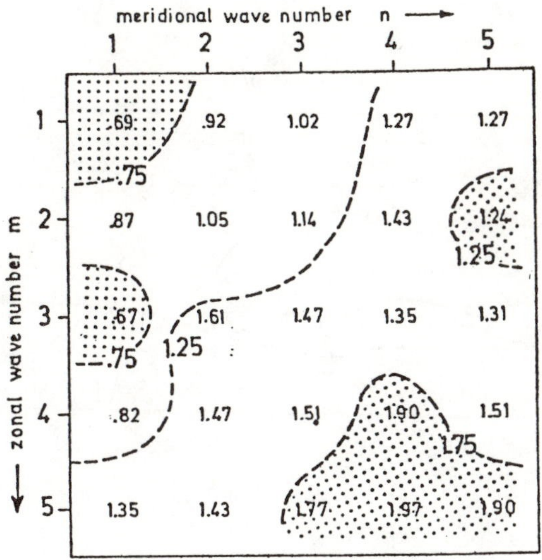

Fig. 2. Autocorrelation decay rate b (day^{-1}) of the imaginary part of the forcing for all planetary modes, after Egger and Schilling[5].

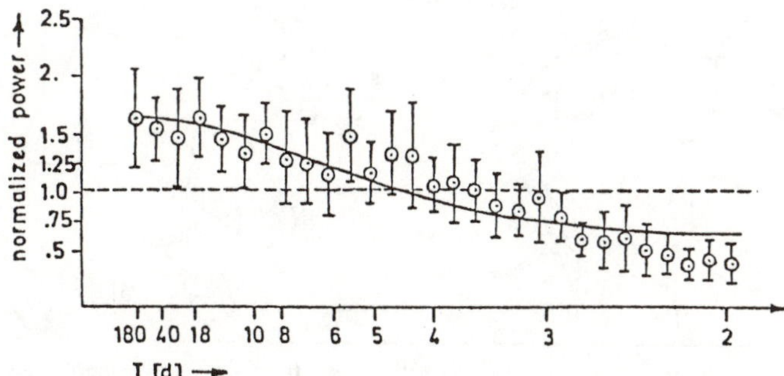

Fig. 3. Average over the power spectra of all modes with $1.25 \leqslant b \leqslant 1.75 d^{-1}$ and the mean fitting curve for these modes. The power is normalized with respect to a white spectrum for the same interval of frequencies, after Egger and Schilling[5].

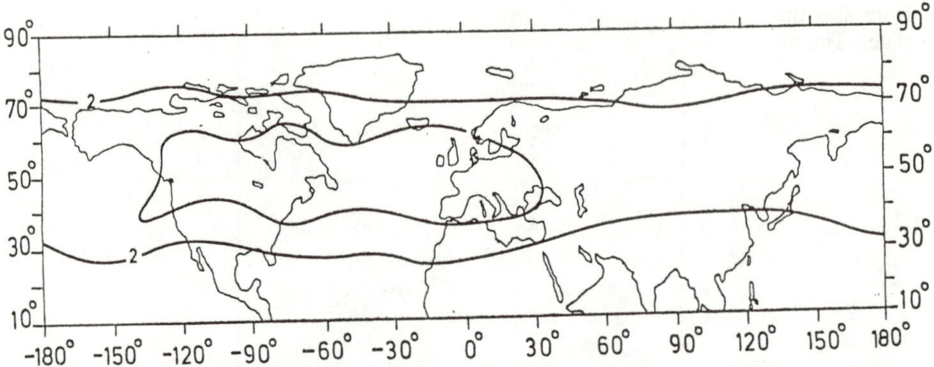

Fig. 4. Low-frequency variance of the synoptic-scale forcing in $10^2 m^4 s^{-4}$, after Egger and Schilling[5].

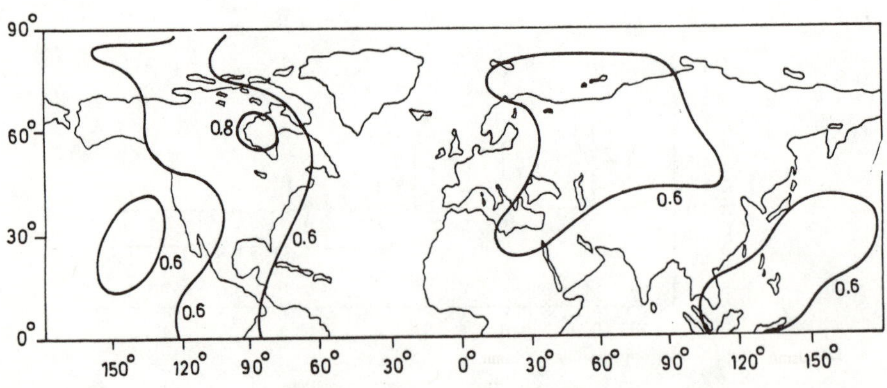

Fig. 5. Correlation coefficient of synoptic-scale forced and unforced planetary flows after eight days as obtained by integrating (1.8). Wave-modes ($m>0$) only; $C = 1/10\ d^{-1}$; $\nu = 8x10^5 m^2 s^{-1}$.

forcing as well. Nevertheless Figure 5 can be seen as an estimate of the error growth in a forecast model for the planetary modes where the impact of the synoptic scales is ignored. Note also that the dispersion of forecast pairs depends on the energy level of the initial state. The more energetic the initial state the less effective the forcing. We shall, however, not discuss this problem here since we are mainly interested in the response of planetary modes at longer time scales.

FORCED LOW-FREQUENCY VARIANCE OF PLANETARY SCALE FLOW

It remains to assess the response of the planetary modes to synoptic scale forcing for periods larger than 10 days. This could be done essentially by using the same procedures as in the foregoing section, i.e. we could integrate (1.8) numerically whereby the forcing terms would have to be updated from day to day according to observations. Of course, it would make no sense to compute correlation coefficients between forced and unforced integrations since these would be exceedingly low. Furthermore the unforced flows would vanish after a while because of damping. However, one could determine in this way the power spectra of the forced flow at any locality of the Northern Hemisphere and compare them to observed spectra.

Here we choose a simpler approach. We linearize (1.8) with respect to a zonal basic state with the wind $u_o \cos\Theta$. Then our problem is linear and we can find the response of the planetary modes to the forcing by analytical means. The linearized form of (1.8) is

$$\frac{d}{dt}\Psi_{mn} = (-i\omega_{Rmn}-\hat{C})\Psi_{mn} + J_{smn}/(1+\lambda^2/k_{mn}^2) \tag{4.1}$$

In (4.1) ω_{Rmn} is the Rossby frequency of the mode (m,n)

$$\omega_{Rmn} = \frac{m(u_o(k_{mn}^2-2/a^2)-2\Omega/a)}{a(k_{mn}^2+d^2+\lambda^2)} \tag{4.2}$$

Given J_{smn} as a function of time it is straightforward to solve (4.1). However, it proves more convenient to perform first a Fourier transform in time where

$$P_{mn}(\omega) = \int_{-\infty}^{\infty} e^{i\omega t}\Psi_{mn}(t)dt$$

$$P(\lambda,\Theta,\omega) = \int_{-\infty}^{\infty} e^{i\omega t}\Psi_p(\lambda,\Theta,t)dt \tag{4.3}$$

are the transforms of the expansion coefficients and of the planetary streamfunction, respectively.

Let $j_{smn}(\omega)$ be the transform of $J_{smn}/(1+\lambda^2/k_{mn}^2)$. Then we obtain from (4.1)

$$P_{mn} = -ij_{smn}/(-\omega+\omega_{Rmn}-i\hat{C}) \tag{4.4}$$

Switching back to physical space we have

$$P(\lambda,\Theta,\omega) = \frac{1}{2}\sum_{m=1}^{M}\sum_{n=1}^{N}(P_{mn}(\omega)B_n^m + P_{mn}^*(-\omega)B_n^{*m}) \tag{4.5}$$

where the asterisk denotes the complex conjugate. Since we know the Fourier transform j_{smn} it is now straightforward to evaluate the power spectrum $S \sim PP^*$ of the response. The powerspectra for simple cases are discussed in Egger and Schilling[5]. Here, we restrict our attention to the low-frequency variance

$$S_L^2(\lambda,\Theta) = \frac{2}{\pi}\int_0^{\omega_{10}} S d\omega \tag{4.6}$$

of the planetary flow as enforced by the synoptic-scale motion. The result is displayed in Figure 6. We have two prominent maxima of the variance, one over the Northern Pacific and Alaska, the other one over the Northern Atlantic. There are minima at 100°W and 100°E. The maxima of the variance are restricted to the belt 50°N$\leqslant\Theta\leqslant$70°N. Comparing Figure 1 and 6 we find that the computed variance is weaker than the observed one by a factor of two. The planetary response has two prominent maxima, one over the Northern Atlantic somewhat to the west of the observed maximum and another one over Alaska and the Northern Pacific in good agreement with the observed distribution. It is perhaps fair to say that the computed response is surprisingly similar to the observed low frequency variance although there are obvious differences in particular with respect to the amplitude.

The low-frequency variance displayed in Figure 6 is virtually unpredictable. After all, (2.1) suggests that the forcing by synoptic scale modes is predictable for a few days only. After that the forcing is unpredictable. Furthermore we have seen that the initial state is hardly remembered beyond ten days. Therefore, since the forcing cannot be predicted beyond a few days the response is unpredictable beyond 10 days, say. However, we can say nothing on the predictability of those low-frequency motions which are not forced by the terms on the right-hand side of (1.7). These include motions enforced by changes at the boundaries of the atmosphere as well as those induced by mountain forcing and baroclinic effects. All these make up for the difference of Figure 1 and Figure 6. It is conceivable that at least some part of these low-frequency motions is better predictable than the motions dealt with in this paper.

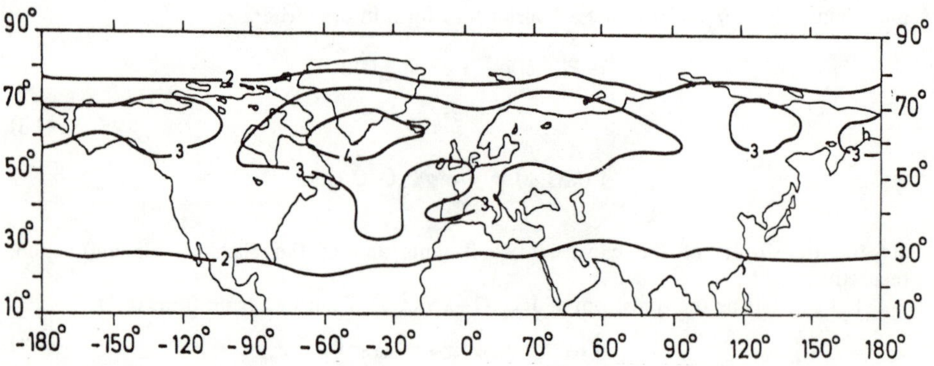

Fig. 6. Low-frequency variance of the 500 mb streamfunction Ψ_p as induced by the synoptic scale forcing (in $10^{13}m^4s^{-2}$), after Egger and Schilling[5].

Finally we want to say a few words on a shortcoming of our approach. We have assumed that the forcing by synoptic-scale waves is independent of the state of the planetary waves. This cannot be strictly true since the forcing terms on the right-hand side of (1.7) contain Ψ_s as well as Ψ_p and we have, therefore, to expect some feedback from the planetary waves on the synoptic-scale forcing. It has sometimes been tried to parameterize J_{smn} in terms of the planetary modes[6,7] in order to capture the essentials of this feedback. However, there is the basic difficulty that J_{smn} is approximately a first order Markov process. Therefore, one would have to parameterize the white noise component W in (2.1). This is impossible by definition. Therefore, if there is an influence of the planetary field Ψ_p on the forcing it must be represented by that part of the spectrum of J_{smn} which is not described by (2.1). As can be seen from Figure 3 this part is not very large and, therefore, the feedback of Ψ_p on the forcing may not be important. Nevertheless we should not rule out the possibility that there is a linkage of Ψ_p and J_{smn} which increases the predictability of Ψ_p.

REFERENCES

1. J. Shukla, J. Atmos. Sci. **38**, 2547 (1981).
2. J.S. Sawyer, Quart. J. Roy. Met. Soc. **96**, 610 (1979).
3. M. Blackmon, J. Atmos. Sci. **33**, 1607 (1976).
4. K. Fraedrich and H. Boettger, J. Atmos. Sci. **35**, 745 (1978).
5. J. Egger and H.D. Schilling, J. Atmos. Sci. **40**, (1983), in press.
6. G. Kurbatkin, Tellus, **31**, 89 (1972).
7. A. White and J. Green, Quart. J. Roy. Met. Soc. **108**, 55 (1981).

MINIMUM ENSTROPHY VORTEX

C. E. Leith
National Center for Atmospheric Research
Boulder, Colorado 80307

ABSTRACT

Classical predictability theory considers the transfer of error through the spectrum of a homogeneous turbulent fluid. Coherent flow structures embedded in chaotic turbulence pose new predictability problems. Such a structure, a minimum enstrophy vortex, is found for two-dimensional flow. It is stable against internal perturbations and may thus serve as a natural limit of self-organization by the selective decay of enstrophy.

INTRODUCTION

Limited predictability as an inherent characteristic of turbulent flows was studied over a decade ago. An early quantitative analysis was that of Lorenz[1] who used a cut-off quasi-normal model for the statistics of two-dimensional flows. Later Leith and Kraichnan[2] used the Test Field Model in both two and three dimensions and were in qualitative agreement with Lorenz for the two-dimensional case.

Such models ignore, however, the known property of turbulent flows that dissipation is distributed not homogeneously but intermittently and that localized structures often become organized and preserved in a dissipating turbulent flow. McWilliams[3] shows extreme examples of such self-organization into vortices in two-dimensional numerical simulations. Vortex structures have also been seen to emerge in two-dimensional flows in laboratory experiments[4] and are, of course, a familiar feature of atmospheric and oceanic circulations.

For a coherent structure it is useful to consider separately the external predictability of its path under the influence of other structures or a turbulent background and the internal predictability of the structure itself as determined, for example, by its mean life before destruction. The dynamics of a large collection of point vortices is believed to be chaotic and thus of limited external predictability although its internal predictability is unlimited since each vortex lasts forever. In the case of a quasistationary structure such as an atmospheric block, internal predictability becomes of great practical importance. But even for a cyclone it may be of practical interest to separate the question of its path from that of its lifetime.

This is a preliminary report on coherent structures arising from selective decay processes. It describes the simplest case of a vortex of minimum enstrophy in a two-dimensional flow. Since this minimum enstrophy vortex (MEV) is found to be stable relative to internal perturbations, it is an example of a coherent structure with extended internal predictability that may serve as a simple model of some observed vortices.

A peculiar property of two-dimensional turbulent flows is that energy tends to cascade toward and be trapped in the largest scales whereas enstrophy is cascaded toward small scales to be lost to coarse-graining or a breakdown of the two-dimensional constraint. Bretherton and Haidvogel[5], in a numerical and analytic study of two-dimensional flow over bottom topography, noted a tendency for a flow to tend, with little change in energy, toward that one with minimum enstrophy compatible with constraints. Matthaeus and Montgomery[6] have described this process in terms of a selective decay hypothesis which they also apply to magnetohydrodynamic turbulence.

The general idea is that, for a flow with several integrals of inviscid motion, some integrals will be removed by a turbulent cascade to scales sufficiently small that they will be dissipated by even a small viscosity whereas other integrals, called rugged, are less subject to decay by turbulent processes. In two-dimensional flows the decay of eddy energy is proportional to eddy enstrophy so that as enstrophy is removed energy is left trapped in the larger scales. In calculations in a periodic domain carried out by Mathaeus and Montgomery, the flow tended toward the gravest mode compatible with constraints for the domain. This example of self-organization does not, by itself, account for those interesting vortex structures that are observed to be much smaller than the domain of the flow.

Hasegawa, Kodama, and Watanabe[7] have applied the selective decay hypothesis to turbulent solutions of a slightly damped version of the Korteweg-deVries (KdV) equation. There are an infinite number of integrals for the inviscid KdV equation and no cascade processes. The addition of a small damping term, however, leads to a cascade for all but the momentum and energy integrals which remain rugged. A variational analysis for the solution that minimizes the integral of next higher order with the rugged integrals kept fixed leads to a solution or, in the case of a periodic domain, the gravest conical wave.

An analogous variational analysis for two-dimensional flows leads to an MEV. In this case the rugged integrals are the angular momentum M and the kinetic energy E, and the dissipated integral is the enstrophy G. The MEV is found to be confined to a disk with a radius R that is determined by the specification of M and E.

In the following sections I shall discuss the dynamics and kinematics of such an isolated vortex, determine by variational analysis the dependence of its shape and radius on E and M, and show that it is stable relative to any two-dimensional perturbation confined to its disk.

DYNAMICS AND KINEMATICS

The dynamics of two-dimensional flow in polar coordinates (r,θ) is governed by the vorticity equation

$$\frac{\partial q}{\partial t} + \frac{1}{r}\frac{\partial}{\partial r}(rvq) + \frac{1}{r}\frac{\partial}{\partial \theta}(uq) = 0$$

where u and v are the physical components of velocity in the θ and r directions, respectively, and the vorticity is given by

$$q = \frac{1}{r}\left[\frac{\partial}{\partial r}(ru) - \frac{\partial v}{\partial \theta}\right] .$$

The velocity is nondivergent, and thus we have

$$\frac{1}{r}\left[\frac{\partial}{\partial r}(rv) + \frac{\partial u}{\partial \theta}\right] = 0 .$$

Let an overbar indicate an average over the angle θ. Thus, for example, we have

$$\bar{q} = \frac{1}{2\pi}\int_0^{2\pi} q\, d\theta = \frac{1}{r}\frac{\partial}{\partial r}(r\bar{u})$$

$$\frac{\partial \bar{q}}{\partial t} + \frac{1}{r}\frac{\partial}{\partial r}(r\overline{vq}) = 0$$

and

$$\frac{\partial}{\partial r}\left(r\,\frac{\partial \bar{u}}{\partial t}\right) = -\,\frac{\partial}{\partial r}\,(r\overline{vq})$$

which may be integrated from r=0 to give

$$\frac{\partial \bar{u}}{\partial t} = -\,\overline{vq}.$$

Consider now the angular momentum within a disk of radius r

$$m(r) = \int_0^{2\pi} d\theta \int_0^r us^2\,ds \;=\; 2\pi \int_0^r \bar{u}s^2 ds.$$

Its rate of change is given by

$$\frac{\partial m}{\partial t} = 2\pi \int_0^r \frac{\partial \bar{u}}{\partial t}\, s^2\,ds$$

$$= -\,2\pi \int_2^r \overline{v\frac{\partial}{\partial s}(su)}\; s\,ds - 2\pi \int_0^r \overline{v\frac{\partial s}{\partial \theta}}\; s\,ds$$

$$= - -2\pi\, r^2\overline{uv} + 2\pi \int_0^r \overline{u\frac{\partial}{\partial s}(sv)}\; s\,ds$$

$$= -\,2\pi\, r^2\overline{uv}$$

where in the last step the integral vanishes owing to the nondivergence of the velocity.

We shall consider flows that are local in the sense that they are completely quiescent outside some radius R, i.e., with $q = u = v = 0$ for $r > R$. For such a local flow the total angular momentum will be $M = m(R)$ and is evidently an integral of the motion.

The other more familiar integrals of the motion are the total kinetic energy and enstrophy

$$E = \frac{1}{2} \int_0^{2\pi} d\theta \int_0^R (u^2 + v^2) r\,dr$$

$$G = \frac{1}{2} \int_0^{2\pi} d\theta \int_0^R q^2\, r\,dr.$$

We shall be primarily interested in a symmetric vortex such that $q = \bar{q}$, $u = \bar{u}$, and $v = 0$. the integrals for such a vortex become

$$M = 2\pi \int_0^R u\, r^2\,dr$$

$$E = \pi \int_0^R u^2\, r\,dr$$

$$G = \pi \int_0^R q^2\, r\,dr$$

with vorticity given by

$$q = \frac{1}{r}\frac{d}{dr}(ru) = \frac{du}{dr} + \frac{u}{r}$$

For given values of M and E we wish to find the radius R and velocity distribution u(r) of a vortex that minimizes G. If we have done so certain boundary conditions at r = R must have been satisfied.

Firstly, we must have u(R) = 0 since otherwise the discontinuity at r = R would add an infinite contribution to the enstrophy. For let us consider the ring $R - \Delta R \leqslant r \leqslant R$ within which we assume that u linearly approaches zero from its nonzero value $\hat{u}$ at $R - \Delta R$. Then to the lowest order in ΔR the vorticity and enstrophy in the ring will be

$$\hat{q} = -\frac{\hat{u}}{\Delta R} + \frac{\hat{u}}{R}$$

$$\Delta G = 2\pi \left(\frac{\hat{u}}{\Delta R}\right)^2 R\Delta R$$

which become infinite as $\Delta R \rightarrow 0$ and $\hat{u} \rightarrow u(R)$ if $u(R) \neq 0$. Clearly this is not the way to minimize enstrophy.

Secondly, we must also have q(R) = 0. This requirement is not so obvious, but if $q(R) \neq 0$ we can make a small increase δR in the radius and an associated change in the velocity u in the ring $R - \delta R \leqslant r \leqslant R + \delta R$ such that to first order δM and δE vanish but δG is negative. This contradicts the assumption that we have started with a minimum in G for fixed M and E relative to changes in u(r) and R.

Note that $u(R - \delta R)$ is given by

$$u(R - \delta R) = -q(R)\delta R$$

to first order since we have u(R) = 0. Consider the perturbed velocity $\tilde{u}$ to be a linear interpolation between $\tilde{u}(R - \delta R) = u(R - \delta R)$ and $\tilde{u}(R + \delta R) = 0$ so that the velocity continues to satisfy the outer boundary condition. In the unperturbed ring with $R - \delta R \leqslant r \leqslant R$ the mean velocity is $-(1/2)q(R)\delta R$, the mean squared velocity is $(1/12)[q(r)\delta R]^2$, and the mean squared vorticity is $q^2(R)$ to lowest order. The lowest order contributions by the ring to the integrals become then

$$\Delta M = 2\pi \left(-\frac{1}{2}q(R)\delta R\right) R^2\delta R$$

$$= -\pi q(R)R^2(\delta R)^2$$

$$\Delta E = \pi\left(\frac{1}{3}\right)\left[-\frac{1}{2}q(R)\delta R\right]^2 R\delta R$$

$$= \frac{1}{12}\pi q^2(R)R(\delta R)^3$$

$$\Delta G = \pi q^2(R)R\delta R$$

For the perturbed ring with $R - \delta R \leqslant r \leqslant R + \delta R$ the mean and mean squared velocities remain the same but the vorticity is decreased to $\tilde{q} = (1/2)q(R)$. The lowest order contributions by the perturbed ring to the integrals become then

$$\Delta\tilde{M} = 2\pi\left(-\frac{1}{2}q(R)\delta R\right)R^2(2\delta R) = 2\Delta M$$

$$\Delta\tilde{E} = \pi\left(\frac{1}{3}\right)\left[-\frac{1}{2}q(R)\delta R\right]^2 R(2\delta R) = 2\Delta E$$

$$\Delta\tilde{G} = \pi\left[\frac{1}{2}q(R)\right]^2 R(2\delta R) = \frac{1}{2}\Delta G.$$

The variations

$$\delta M = \Delta\tilde{M} - \Delta M = \Delta M = -\pi q(R)R^2(\delta R)^2$$

$$\delta E = \Delta\tilde{E} - \Delta E = \Delta E = \frac{1}{2}\pi\left[q(R)\right]^2 R(\delta R)^3$$

vanish to first order in δR, but the variation

$$\delta G = \Delta\tilde{G} - \Delta G = -\frac{1}{2}\Delta G = -\frac{1}{2}\pi\left[q(R)\right]^2 R\delta R < 0$$

does not, for $q(R) \neq 0$, as was to be shown.

VARIATIONAL ANALYSIS

We consider a variational analysis for a symmetric vortex confined to the disk $0 \leqslant r \leqslant R$ with the parameter R held fixed. We impose conditions that $u(0) = u(R) = 0$ and that $q(R) = 0$. Of the three integrals

$$M = 2\pi\int u\, r^2\, dr$$

$$E = \pi\int u^2 r\, dr$$

$$G = \pi\int q^2 r\, dr$$

we consider the first two to be rugged and seek the velocity distribution $u(r)$ that minimizes G for fixed M and E. All unlabeled integrals are over the full disk $0 \leqslant r \leqslant R$.

The first variations of the integrals for variations δu such that $\delta u(0) = \delta u(R) = 0$ are given by

$$\delta M = 2\pi\int r\delta u\, r dr$$

$$\delta E = 2\pi\int u\delta u\, r dr$$

$$\delta G = 2\pi\int q\frac{d}{dr}(r\delta u) r dr$$

$$= -2\pi\int\frac{dq}{dr}\delta u\, r dr\ .$$

The first variation equation becomes

$$\delta G + \lambda \delta E + \mu \delta M = 2\pi \int \left[-\frac{dq}{dr} + \lambda u + \mu r \right] \delta u r dr = 0$$

where λ and μ are Lagrange multipliers. It is satisfied for arbitrary δu if and only if

$$\frac{d}{dr}\left(\frac{du}{dr} + \frac{u}{r} \right) - \lambda u - \mu r = 0.$$

The vortex velocity u(r) must satisfy this inhomogeneous differential equation and thus be a combination of a particular solution and the general solution of its homogeneous part. A particular solution is given by $u = Br$ and $q = 2B$ when the constant B is such that $\mu = -\lambda B$. The homogeneous part is Bessel's equation with a solution of the form $J_1(\gamma r)$ satisfying the condition $u(0) = 0$ for $\lambda = -\gamma^2 < 0$. Thus the general solution is

$$u = AJ_1(\gamma r) + Br$$

with associated vorticity

$$q = A\gamma J_o(\gamma r) + 2B$$

It is convenient henceforth to let $z = \gamma r$ and $Z = \gamma R$.

We may next impose the outer boundary conditions. For the vorticity condition we have

$$0 = q(Z) = A\gamma J_o(Z) + 2B$$

so that

$$B = -\frac{1}{2} A\gamma J_o(Z)$$

and

$$u = A\left[J_1(Z) - \frac{1}{2} J_o(Z) z \right]$$

$$q = A\gamma \left[J_o(z) - J_o(Z) \right]$$

For the velocity condition we have

$$0 = u(Z) = A\left[J_1(Z) - \frac{1}{2} ZJ_o(Z) \right]$$

$$= \frac{1}{2} AZJ_2(Z)$$

thus $Z = \gamma R$ must be a positive root of J_2. The gravest mode provides the least enstrophy so we take

$$Z = j_{2,1} = 5.1356 \cdots$$

and also have

$$J_1(Z) = \frac{1}{2} ZJ_o(Z) = -0.33967 \cdots$$

We may now compute the integrals

$$M = 2\pi \int ur^2 \, dr$$

$$= 2\pi A\gamma^{-3} \int \left[J_1(z) - \frac{1}{2} J_o(Z) z \right] z^2 dz$$

$$= 2\pi QAR^3$$

$$E = \pi \int u^2 r \, dr$$

$$= \pi A^2 \gamma^{-2} \int \left[J_1(z) - \frac{1}{2} J_o(Z) z \right]^2 z \, dz$$

$$= 12\pi Q^2 A^2 R^2$$

where $Q = -ZJ_o(Z)/8 = -J_1(Z)/4 = 0.08492 \cdots$. The specification of M and E determine the radius R, the amplitude A, and the wavenumber γ through the relations

$$\frac{M^2}{E} = \frac{1}{3} \pi R^4$$

$$ER/M = 6QA$$

$$\gamma R = Z$$

since Z and Q are known constants.

The enstrophy integral at the minimum becomes

$$G = \pi \int q^2 r \, dr$$

$$= \pi A^2 \gamma^2 \int \left[J_o(z) - J_o(Z) \right]^2 z \, dz$$

$$= 8\pi Q^2 \gamma^2 A^2 R^2 = \frac{2}{3} \gamma^2 E$$

and reaffirms the choice of the gravest mode as giving the smallest value for γ^2 and G.

In order to establish that the vortex provides not just a stationary point but a true minimum we consider the second variations of the integrals

$$\delta^2 M = 2\pi \int r \delta^2 u \, r \, dr$$

$$\delta^2 E = 2\pi \int u \, \delta^2 u \, r \, dr + 2\pi \int (\delta u)^2 \, r \, dr$$

$$\delta^2 G = -\,2\pi \int \frac{dq}{dr}\,\delta^2 u\; r\, dr - 2\pi \int \frac{d}{dr}(\delta q)\delta u\; r\, dr$$

$$= -\,2\pi \int \frac{dq}{dr}\delta^2 u\; r\, dr + 2\pi \int (\delta q)^2\; r\, dr$$

so that we have

$$\delta^2 G + \lambda\delta^2 E + \mu\delta^2 M$$

$$= 2\pi \int \left[-\frac{dq}{dr} + \lambda u + \mu r\right]\delta^2 u\; r\, dr + 2\pi \int \left[(\delta q)^2 + \lambda(\delta u)^2\right] r\, dr$$

$$= 2\pi \int \left[(\delta q)^2 - \gamma^2(\delta u)^2\right] r\, dr\,.$$

We want this last expression to be positive to have a true minimum, and this only follows from geometrical constraints relating the relative magnitudes of δq and δu. The gravest admissible variation δu would be the one for which $(\delta q)^2$ would be relatively the smallest and the second variation most likely to be negative. But even for the variation

$$\delta u = \delta u_1\, J_1(\gamma r)$$

which is the gravest possible mode satisfying only the conditions $\delta u(0) = 0$ and $\delta M = 0$, we find

$$2\pi \int \left[(\delta q)^2 - \gamma^2(\delta u)^2\right] r dr = \pi\gamma^2 R^2(\delta u_1)^2 J_0^2(\gamma R) > 0\,.$$

Any variation δu satisfying the other conditions as well would be less grave and thus would lead to an even more positive second variation.

STABILITY ANALYSIS

In carrying out an analysis of the stability of the vortex relative to two-dimensional velocity perturbations confined to the disk $0 \leqslant r \leqslant R$, we treat G as a valid integral of the motion and note, following Arnold[8], that the functional $I = G + \lambda E + \mu M$ is an integral of the motion for which the vortex fields gives a true minimum. For two-dimensional velocity fields the integrals of motion become

$$M = \int d\theta \int u r^2 dr$$

$$E = \frac{1}{2} \int d\theta \int (u^2 + v^2) r dr$$

$$G = \frac{1}{2} \int d\theta \int q^2 r dr$$

where the vorticity is given by

$$q + \frac{1}{r}\frac{\partial}{\partial r} r u - \frac{1}{r}\frac{\partial v}{\partial \theta}\,.$$

For the first variations about the symmetric vortex we find, since $v = 0$,

$$\delta M = \int d\theta \int r\delta u \, r \, dr$$

$$\delta E = \int d\theta \int u \, \delta u \, r \, dr$$

$$\delta G = \int d\theta \int q\delta q \, r \, dr$$

$$= -\int d\theta \int \frac{dq}{dr} \, \delta u \, r \, dr$$

so that we have

$$\delta I = \int d\theta \int \left[-\frac{dq}{dr} + \lambda u + \mu r \right] \delta u \, r \, dr = 0 \, ,$$

and we see that the MEV is at a stationary point of the functional I relative to two-dimensional perturbations.

To show that it is a true minimum of I we consider next the second variation which is found to be

$$\delta^2 I = \int d\theta \int \left[(\delta q)^2 - \gamma^2 (\delta u)^2 - \gamma^2 (\delta v)^2 \right] r \, dr,$$

but as in the previous section the gravest two-dimensional mode compatible with the constraints $\delta u(0) = 0$ and $\delta M = 0$ is

$$\delta u = \delta u_1 \, J_1(\gamma r), \quad \delta v = 0 \, ,$$

and for it we find $\delta^2 I > 0$.

For $\delta M \neq 0$ there is, of course, a nearby stationary state with $\delta E \neq 0$ and $\delta G \neq 0$ given by a nearby MEV with the same radius which is accessible by admissible velocity perturbations.

SUMMARY

For a specified total kinetic energy E and angular momentum M, the minimum enstrophy vortex (MEV) is confined to a disk of radius

$$R = \left(\frac{3}{\pi} \frac{M^2}{E} \right)^{1/4}$$

and has a velocity and vorticity of the form

$$u = \left[J_1(\gamma r) - \frac{1}{2} J_o(\gamma R) \gamma r \right]$$

$$q = A\gamma \left[J_o(\gamma r) - J_o(\gamma R) \right]$$

where the wavenumber γ is such that $J_2(\gamma R) = 0$, i.e., $\gamma = Z/R$ with $Z = 5.1356 \cdots$, and the amplitude A is given by

$$A = \frac{1}{6} \frac{1}{Q} \frac{E}{M} R$$

with the constant $Q = -J_1(Z)/4 = 0.08492\cdots$

The MEV velocity and vorticity as a function of $z = \gamma r$ are shown in Figure 1 together with the shear

$$S = \frac{du}{dr} - \frac{u}{r} = q - 2\frac{u}{r}$$

which is of interest for the dynamics of the eddy enstrophy cascade process.[3]

REFERENCES

1. E.N. Lorenz, Tellus, **21**, 289 (1969).
2. C.E. Leith and R.H. Kraichnan, J. Atmos. Sci., **29**. 1041 (1972).
3. J.C. McWilliams, these Proceedings.
4. E.J. Hopfinger, F.K. Browand, and Y. Gagne, J. Fluid Mec., **125**, 505 (1982).
5. F.P. Bretherton and D.B. Haidvogel, J. Fluid Mech., **78**, 129 (1976).
6. W.H. Matthaeus and D. Montgomery, Annalys, N.Y. Acad. Sci., **357**, 203 (1980).
7. A. Hasegawa, Y. Kodama and K. Watanabe, Phys. Rev. Lett., **47**, 1525 (1981).
8. V.I. Arnold, Mathematical Methods of Classical Mechanics (Springer-Verlag, N.Y., 1980).

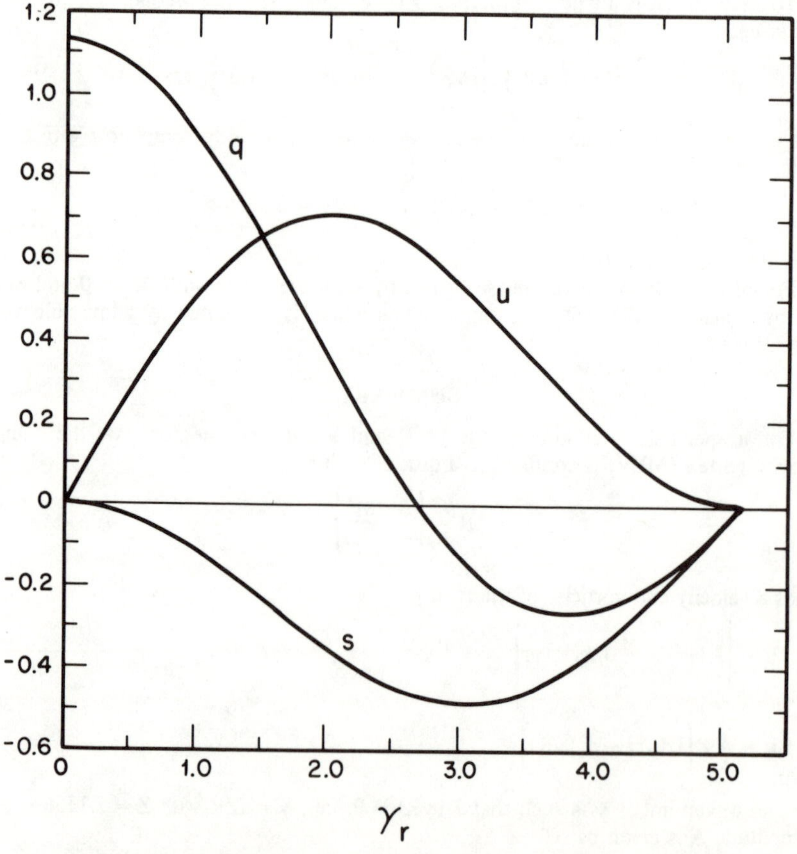

Fig. 1. Velocity, vorticity, and shear for a minimum enstrophy vortex as a function of radius.

THE RELATIONSHIP BETWEEN PRESENT LARGE-SCALE FORECAST SKILL AND NEW ESTIMATES OF PREDICTABILITY ERROR GROWTH

David P. Baumhefner
National Center for Atmospheric Research†
Boulder, Colorado 80307

ABSTRACT

Several new methods of verification are defined and tested on forecast error of current numerical models. The methods incorporate a notion of "usefulness" which is determined by placing upper and lower bounds on the error growth. The lower bound is derived from new estimates of predictability error growth produced by the NCAR Community Climate Model. The fields are spectrally decomposed to highlight forecast skill in various scales. The verification techniques are applied to forecast errors from the NMC and ECMWF operational models during 1979-82.

Results show that these techniques are highly successful in determining the relative usefulness of numerical forecasts. Present forecasts lose skill at 5 days over the Northern Hemisphere, whereas predictability estimates indicate the possibility of skillful 8-day forecasts. A significant improvement in accuracy can still be made in the largest scales.

INTRODUCTION

In principle, the objective verification of large-scale numerical weather prediction serves three major purposes. First, it should provide an accurate measure of forecast skill in agreement with subjective impressions. Secondly, it should sufficiently reduce the number of degrees of freedom to allow simple comparison. Finally, and perhaps most importantly, it should determine the relative usefulness of a forecast. The last point implies that a thorough knowledge of the potential upper and lower bounds of forecast skill is necessary for the evaluation of usefulness. The most commonly used techniques of verification manage to comply reasonably well with the first two requirements, but rarely do they provide direct information on the usefulness of the forecast. Recently, the European Centre for Medium Range Weather Forecasts (ECMWF) compared their operational forecast verifications against a defined upper limit called "useful predictability"[1,2]. This concept, which was introduced by Döös[3], uses the climatological variance of the field

† The National Center for Atmospheric Research is sponsored by the National Science Foundation.

being verified as an upper bound on forecast skill. A more liberal limit of usefulness, defined as when the forecast error approaches the value of a persistence forecast, has also been used. The lower limit of forecast skill has generally been accepted as the inherent non-linear error growth of the dynamical system starting from some initial error distribution. This has been called predictability error growth by Williamson[4] and others. A comparison of forecast scores with this lower bound was suggested by Leith[5],but has not been applied to objective verification in any widespread fashion.

The contents of this paper describe several new methods of verification that combine the essential elements of upper and lower boundedness in the form of normalized scores. The fields being verified are broken down into various spatial scales before normalization. Since these scores require a quantitative knowledge of predictability error growth throughout the spectrum of scales, new estimates of these values are obtained from predictability experiments using the NCAR Community Climate Model. Application of these verification techniques to operational forecast errors at the National Meteorological Center (NMC) and ECMWF are shown. Finally, an estimate is made, based on these techniques, of the progress that has been made in numerical weather prediction during the past years, and of the potential for improvement in the future.

VERIFICATION TECHNIQUES

The basic idea of normalizing a conventional measure of skill, such as the Root Mean Square (RMS) error, is illustrated in Fig. 1. The usual method of plotting RMS with time of forecast is shown by the heavy dashed line. To incorporate the concept of usefulness, it is necessary to compare these values with the best and/or worst possible cases of forecast skill. For example, if persistence (open dots) is a measure of the worst possible error and no RMS error is the best, then a possible score could be defined as the ratio of the forecast RMS to the persistence RMS (shown schematically as P_1). An alternate candidate, and perhaps a more realistic one, is to use the climatological value of the RMS in place of persistence (shown as C_1). We know, of course, that a "no error" forecast as a measure of the lower limit is also unrealistic from a predictability theory point of view. Therefore, it makes sense to restrict a score definition to the envelope bounded by an estimate of predictability error growth (bottom solid curve) and one of the previously defined upper bounds. In an ensemble sense, values of forecast skill are not expected to occur in the stippled area. The P_2 score, shown in Fig. 1 as a modification of the P_1 score, is one example of this type of skill measure with persistence used as the upper bound. It is defined as the ratio of the forecast-predictability difference vs. the persistence-predictability difference. Considering the strengths and weaknesses of the various upper bounds, an alternate score definition might weight the persistence error characteristics early in the forecast and approach the climatological limit of error late in the forecast. This concept of damped persistence[6] as an upper bound leads to the DP_2 score definition in Fig. 1 (solid dots). Although these skill scores have been defined in terms of the RMS of forecast error, it is possible and sometimes advantageous to transform them into the context of error variance. For example, the C_1 score becomes the forecast error variance divided by the observed climatological variance (V_1). One clear advantage of the variance scores is the additive nature of the various error sources which allows their separation into distinct components. This feature is especially valuable when the forecast error is decomposed in either space or time.

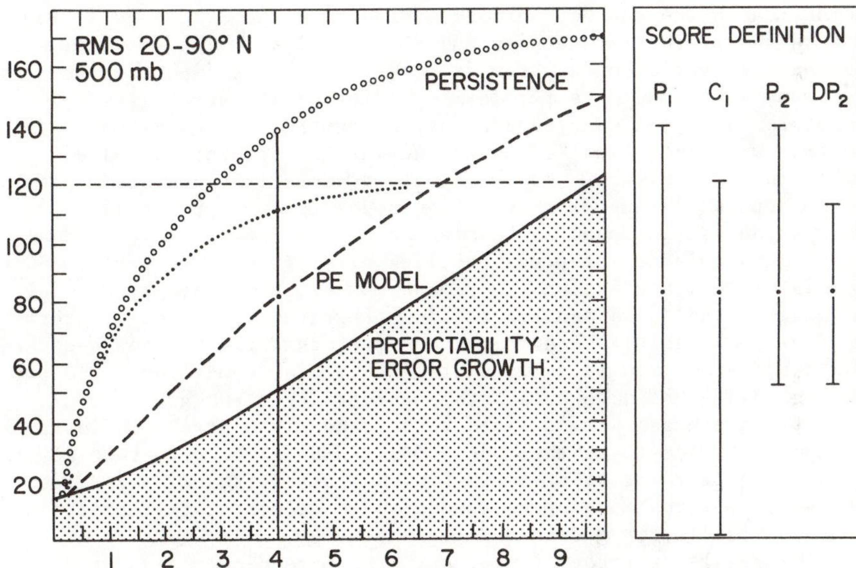

Fig. 1 Schematic diagram of several verification score definitions, for Root Mean Square (RMS) error at 500 mb. Abscissa is days, ordinate is meters. Solid curve is predictability error growth, heavy dashed curve is typical forecast error growth, open circle curve is persistence error growth. Light dashed line is climatological RMS and dotted curve is persistence damped toward climatology. Vertical score definition lines represent denominator of score ratio. Distance from dot to bottom of line represents numerator.

For the score definitions that contain an estimate of predictability error growth, it is important that these estimates are realistic and represent atmospheric behavior as closely as possible. If the error growth measure is unreliable, the application of the skill score as a measure of usefulness is seriously diminished. The predictability estimates should also be broken down into various space and time scales to permit score calculations of forecast errors at these scales.

PREDICTABILITY ESTIMATES

With the requirements of the previous section in mind, the existing predictability estimates found in the literature were inadequate from several points of view. It was therefore necessary to derive new estimates of predictability error growth from a realistic model. Recent development and experimentation with the NCAR Community Climate Model has shown that this model is capable of producing nearly the correct variance throughout the large-scale flow, both in the proper frequencies and space scales[7]. The simulation of the observed variance spectrum is perhaps the most important requirement in producing a realistic predictability error growth, therefore this model was chosen for our experiments. The model is spectral with rhomboidal 15 as the horizontal resolution and 9 levels in the vertical. The physical parameterizations are typical of sophisticated general circulation models. Further detailed description can be found in Washington[8].

The calculation of predictability error growth was performed in the classical perturbation experiment manner, discussed by Charney[9] and Smagorinsky[10]. A control run is integrated for 30 days, starting from model initial conditions. A perturbed run is then generated which starts from slightly different initial condition. The two runs are differenced to obtain the appropriate error growths. The initial differences were defined to be randomly distributed in space and the amplitude to be the order of 1 m RMS in the geopotential field. Two pairs of experiments (which started from widely different synoptic situations) were conducted to determine the uncertainty of the derived error growths.

The resulting error growth estimates were decomposed into a 2-dimensional spherical harmonic wavenumber spectrum to analyze the behavior of various scales. The decomposition was performed over the Northern Hemisphere only, therefore only symmetric modes are used. The spectrum was divided into six categories which roughly correspond to various physical characteristics of the atmosphere. The total error growth and the decompositions are illustrated in Fig. 2. Group (0) represents the zonal vortex and its entire meridional structure. The next five groups are composed of a triangular truncation of 2-dimensional wavenumbers 1-4, 5-7, etc., excluding their zonal structure. Group (1-4) contains isotropic planetary structures of the order of 20,000 km at 30°N. Groups (5-7), (8-11), (12-18) have approximately 6,000; 4,000; 2,500 km scales, respectively, and are representative of large, medium and small baroclinic structures.

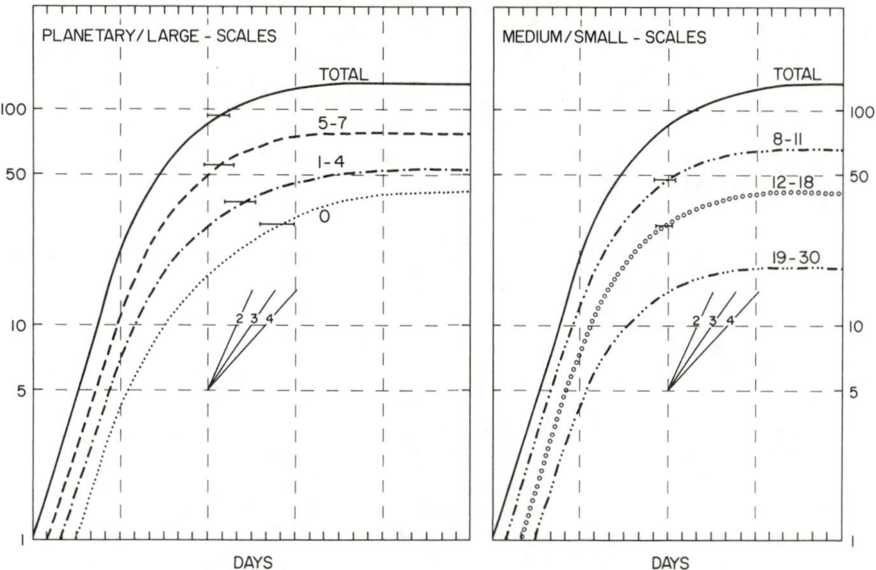

Fig. 2 Northern Hemisphere Standard Deviation in meters of 500 mb error growth from predictability experiments for total field (solid) and various scale groups. Two-dimensional wave number defines scale grouping: (0) = zonal vortex with all meridional scales; (1-4), (5-7),(8-11),(12-18),(19-30) = triangular truncation at indicated wave number, excluding zonal scales. Error growth of each scale arbitrarily positioned one day apart at origin. Slope of 2-4 day doubling times plotted in center.

The total predictability error growth and its associated wave groups from each case were adjusted, averaged, and compared to other error growth calculations to insure their representativeness when compared to the atmosphere. A smoothed error growth was established for each group and case and its asymptotic saturation level was determined. The error growth curves were then adjusted to the observed saturation level without distortion of the curve shape. This adjustment was normally less than 10% of the variance in that scale. The observed values of variance were calculated from National Meteorological Center analyses for January 78, 79, 80, and 81. The adjusted error growths were then averaged for the two cases using a common initial value of 5 m as a starting point. The group (19-30) was calculated as a residual along with the knowledge of its observed asymptotic saturation level.

Two checks were employed to insure the consistency of these values. At any given time during the error growth cycle, the sum of that there is also considerable room for improvment compared to predictability limits. It should be remembered that these results are highly dependent on the validity of the scoring techniques and predictability error growths.

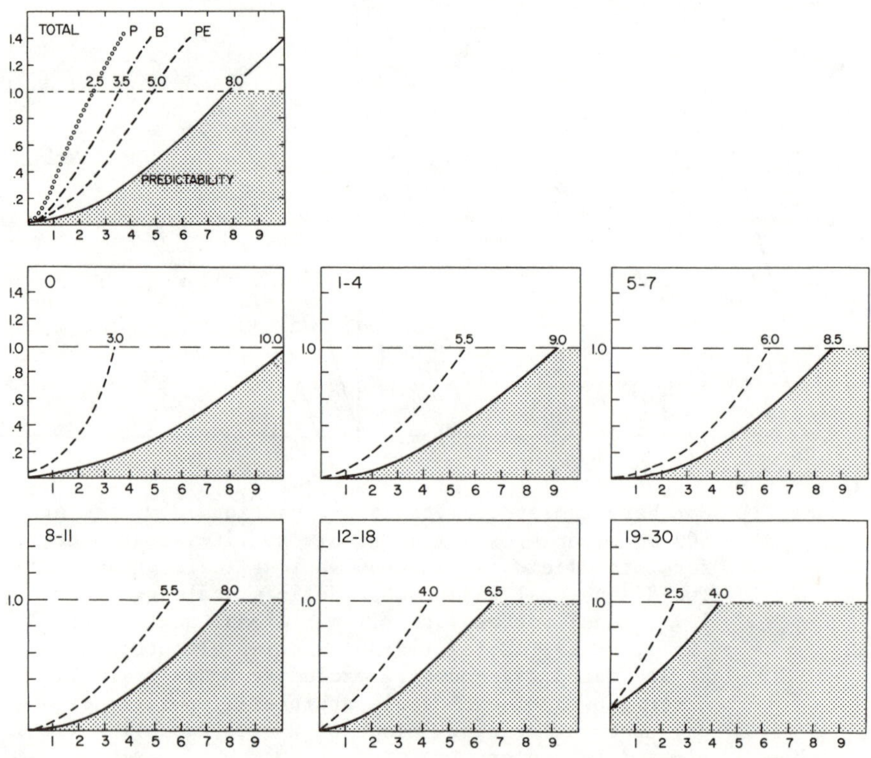

Fig. 5 Northern Hemisphere 500 mb V_1 verification score plotted as function of days for total field (top) and scale groupings (middle and bottom). Scale grouping same as Fig. 3. Predictability error = solid, current model forecast error = dashed, barotropic model forecast error = dash-dot, and persistence error = open circles. Horizontal dashed line = limit of useful skill (see text) and numbers on line indicate limit in days for each curve.

The spectral distribution of forecast and predictability error shows the most accurate scales are currently the large baroclinic structures. In fact, it may be difficult to further improve these scales in a direct fashion. The greatest potential for improvement is clearly in the largest scales, groups (0, 1-4) which account for the variances of the wave groups should equal the total error variance. In this experiment, the variance sums were accurate within 5%. The error growth rates were further compared with additional predictability experiments using a higher resolution model, but for a shorter period of integration and larger initial error distributions. All the growth rates compared favorably, with the exception of groups (0) and (1-4), which grew somewhat less rapidly in the high resolution cases.

The evolution of the adjusted predictability error growths is shown in Fig. 2. The origin of each wave group is arbitrarily offset by one day for clarity. The pattern of error growth can be divided into regimes; between 1-10m RMS all scales appear to double at less than 2 days whereas between 10-100 m RMS the doubling rate decreases with a distinct scale dependence. The largest scales clearly take a longer time to reach their asymptotic saturation limits, but at a more rapid rate than might have been expected.

The total doubling time of 1.7 days in the crucial 10-50m RMS range compares favorably with recent estimates of 1.85 days calculated by Lorenz[11] using the European Centre for Medium Range Weather Forecasts' high resolution operational model errors in the Northern Hemisphere winter. These new estimates of error doubling are considerably faster than earlier estimates of 2-5 days[12] and perhaps can be attributed to a more correct simulation of the variance spectrum.

The horizontal error bars denoted on each curve depict the range of error growth that was encountered in the two sets of experiments. For the total field, as well as the smaller scales, the range is less than two days. The largest scales exhibit the most range, which reflects the larger low-frequency variability that these scales contain. For these cases, though, the range or uncertainty of these error growths seems to be fairly small and therefore one can use the resulting error growth with some confidence.

The climatological value of RMS for each group of scales is located at the error bar crossover on the error growth plot. This has been defined in the previous section as the upper limit of "useful predictability." Squaring the climatological RMS value produces the natural variance of each group. Using the derived error growths, an initial error distribution, and the upper limit bar, it is possible to calculate the theoretical limit of forecast skill. A nearly error-free initial RMS value of 1 m in each wave group produces a limit of 11 days for the total field (Table I). The zonal vortex, group (0), has skill to 15 days and the smallest scale group (19-30) has skill extending to 7 days. A more realistic total initial value of error, based on experience with objective analysis techniques, is an RMS error of 14m which was distributed evenly throughout the spectrum. This reduces the limit of skill by 3 days for the total field (Table I) and there are significant reductions in the other scales as well. Calculations of differences between various objective analysis schemes reveal the analysis error distribution shown in Table I (right side) which has even higher values than the previous case. A further reduction in the theoretical limit of skill is seen, with the largest scales losing their skill around 8 days. This initial error distribution has the interesting characteristic of minimizing the differences between scales, such that the variations in the zonal vortex are only slightly more predictable than the medium-scale baroclinic waves.

Table I Theoretical limit of forecast skill

Scale	Small error		Optimum error		Analysis error	
T	(2.5)	11.0	(14.0)	8.0	(19.0)	7.0
0	(1)	15.0	(5.7)	10.0	(9.5)	8.5
1-4	(1)	12.0	(5.7)	9.0	(9.0)	8.0
5-7	(1)	12.0	(5.7)	8.5	(7.5)	7.5
8-11	(1)	10.0	(5.7)	8.0	(6.5)	7.5
12-18	(1)	9.0	(5.7)	6.5	(7.0)	6.0
19-30	(1)	7.0	(5.7)	4.0	(7.0)	3.5

Limit of forecast skill in days (right side of each column) for each type of initial error. Initial error in RMS meters in parentheses for each scale grouping. T = total field. Scale numbers indicate two-dimensional wave groups described in text.

All three estimates of the theoretical limit of forecast skill for the planetary scales are considerably shorter than recent estimates found in the literature[13]. The reasons for this discrepancy are not clear, but it may be partially explained by the different models, domains, and spectral techniques being used.

VERIFICATION STATISTICS

Application of the predictability error growth estimates to verification scores can be made in several ways. A direct incorporation of the information can be achieved, for example the P_2 score, or a side-by-side comparison can be presented showing both the actual forecast error and the predictability error. An example of both types is presented here.

The 500 mb geopotential forecast errors from the NMC operational model were collected for the winter months (December, January, February) of 1979, 1981, and 1982. The P_2 score was calculated over 20-90°N for each winter from the average RMS error, the average persistence values, and the new predictability error estimates. The 24, 48 and 72 hr forecasts for each winter are plotted in Fig. 3 for only the total field. The forecasts for 1979 have noticeably less skill than later years, with typical values of 40 to 50, which indicates the RMS errors are roughly half way between the upper and lower bounds of the P_2 score (see Fig. 1). Forecasts for 1981-82 are improved by nearly 20 points, clearly showing the effect of a model change at NMC.

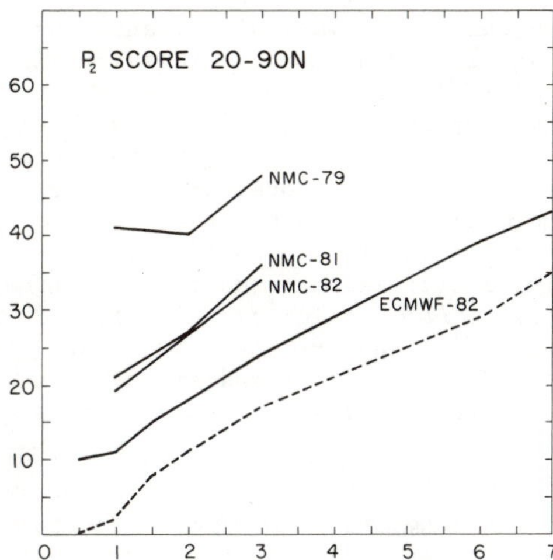

Fig. 3 Normalized P_2 verification score in percent for 500 mb calculated over 20-90°N. Horizontal axis in forecast days. Solid (dashed) lines calculated from initial error of 10 m (14 m). Top three curves for NMC forecasts from winter (DJF) 1979, 81, 82. Bottom two curves for ECMWF forecasts from winter 1982.

The forecast errors for the ECMWF model during the winter of 1982 were also converted to P_2 score values in a similar manner. The results of these calculations are shown at the bottom of Fig. 3. The ECMWF scores average 10 pts lower than the NMC forecasts for the same year. In fact, this score indicates only a 25% further improvement can be made in the 72 hr forecast, or in another sense the forecasts are 75% perfect. This is in contrast to the scores obtained in 1979 which showed a 50% level of skill. This ensemble of P_2 scores illustrates quite effectively the dramatic improvement in numerical weather prediction products in the past several years.

The P_2 scoring technique is somewhat sensitive to the value of initial error assumed to exist in the predictability error growth calculation. The P_2 scores in Fig. 3 used an initial RMS error of 10 m. The sensitivity is tested by raising this value to 14 m and recalculating the ECMWF scores for 1982 (dashed line, Fig. 3). A small uncertainty of the initial error can produce changes in the P_2 score in the order of 10%. Notice that for analysis RMS errors of 14 m the 12 hr ECMWF forecast is as accurate as can be expected from predictability theory.

A more detailed examination of the NMC model forecast error differences between 1979 and 1981 is performed using the V_1 score. In this case, a comparison is made between the forecast and predictability error growths. The January monthly average error variance is computed over the Northern Hemisphere and then decomposed in the same manner as the predictability errors in Fig. 2. These values are normalized by the climatological variance observed in each group of scales. The initial error used in these predictability error growths is defined in Table I as the "optimum error." The total score and its scale breakdowns are plotted as histograms for 24, 48, and 72 hr forecasts (Fig. 4). When the V_1 score equals unity, there is no useful skill left in the forecast. The top of the histogram represents the normalized forecast error variance and the solid middle line represents the predictability error variance. In this type of display, the stippled area denotes the true or correctable part of the error. At 72 hr, a further division of the forecast error is displayed by dividing the error variance into its time-mean or systematic contribution and its transient contribution. The dashed line shows this separation with the systematic part partitioned above the line.

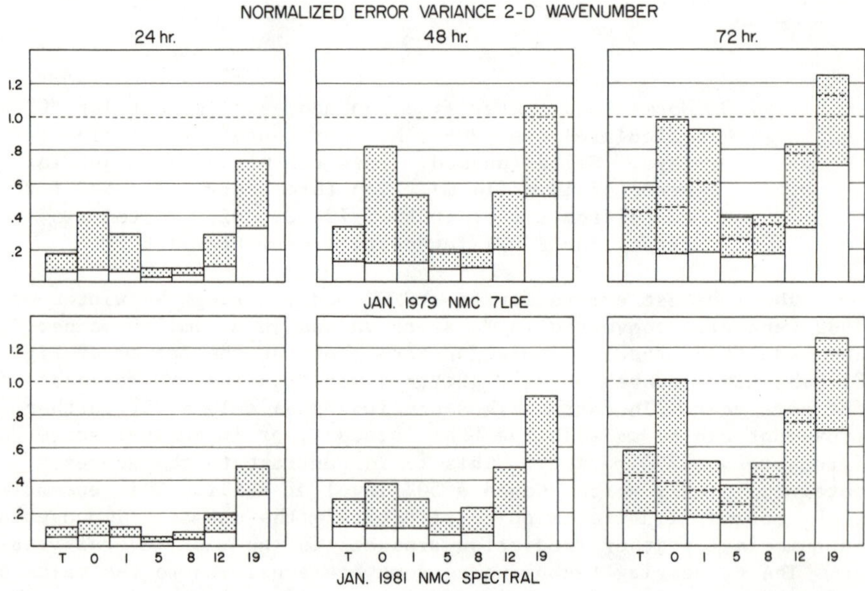

Fig. 4 Northern Hemisphere V_1 verification score plotted as histograms for 500 mb total field (T) and scale groups defined in Fig. 3. Top row from NMC 7LPE model for January 1979, bottom row from NMC spectral model for January 1981. Left column = 24 hr forecasts, middle column = 48 hr forecasts, and right column = 72 hr forecasts. Top of each histogram = normalized forecast error variance, bottom line = predictability error variance, and difference is stippled. 72 hr forecast error variance divided into stationary(top)/transient (bottom) parts by dashed line.

A distinct U-shaped signature of error is apparent from examination of these scores for both months. The largest and smallest scales have more relative forecast error than the medium scales. At 72 hr the most accurate scales are clearly the large-scale baroclinic systems (group 5-7) whereas the skill in the zonal vortex (group 0) and the smaller scales (group 19-30) is gone. This is in contrast to the relatively flat predictability error growth of the first four groups.

The systematic portion of the forecast error variance is primarily in the largest scales with values ranging from 30-50% of the total. Consequently, a large part of the U-shaped error signature is contributed by the systematic part of the error. The model's tendency to drift away from the initial conditions to its own climatology, thereby restructuring the planetary scales, is probably the largest factor in the generation of systematic error.

The differences in forecast skill from 1979 to 1981, as first seen in Fig. 3, can possibly be explained by the spectral breakdown of the error variances shown in Fig. 4. Early in the forecast period, the 1981 model is superior in all scales, with the planetary scales showing the most improvement by 20-30 pts. Late in the period, some of this advantage is lost; however, the forecasts are still better in groups (1-4, 5-7). The changes in initialization techniques and domain of integration may have played their respective roles in altering the error structure in the two years.

DISCUSSION

It is possible to speculate on the relationship of the current "state of the art" of numerical forecast skill and the ultimate limit of skill by applying some of the previous concepts. A reasonably coherent picture of this relationship can be provided by compositing the available information on forecast error characteristics of the NMC and ECMWF models and utilizing the derived predictability error growth calculations. A schematic has been constructed to illustrate this point, using the V_1 scoring technique as a basis for comparison.

The predictability error growth has been plotted (Fig. 5, solid curves) as a function of time for each group of scales starting from the "optimum error" initial conditions in Table I. The domain of calculation is again the Northern Hemisphere. The stippled area represents forecast skills that are unobtainable from a predictability point of view. Note that the initial error is a very small fraction of the total variance in all scales, except the smallest wave group (19-30). The current state of numerical forecast error composited in a subjective manner is illustrated by the dashed curve in each group of scales. The behavior of persistence and barotropic forecasts are also included in the total variance case.

The present state of forecast error shows a substantial improvement over earlier methods of prediction (Fig. 5, top). In terms of the limit of forecast skill, the present models have extended the limit of usefulness by 2-3 days. However, it appears approximately 25% of the observed variance. The smaller scale errors also grow rapidly but have a shorter predictability decay time.

As mentioned before, a major source of the large-scale error has been identified as a model drift toward its own climatology and is potentially correctable. The smallest scale errors could be reduced by a more accurate initial condition. One possible result of improvement in the largest scales might be a more accurate evolution of the baroclinic structures from a phase propagation point of view.

The new verification schemes presented here have demonstrated their ability to define the relative usefulness of numerical forecasts. They have proven to be helpful in identifying and focussing on problems of a particular model. Possible solutions for these difficulties are frequently suggested from this additional information. It is important to remember that one component of these schemes, namely the predictability error growths, only represents an estimate of the true nature of the atmosphere. There remains a strong need to refine this estimate and to establish confidence in its use.

ACKNOWLEDGMENTS

Many of the ideas presented in this paper originated in discussions with C. Leith. The verification statistics were carefully assembled by T. Bettge. Special thanks to B. Boville who produced the predictability error growth experiments from the CCM. Finally, a great many discussions with members of the Large-Scale Dynamics and Climate Sections of NCAR helped refine the concepts presented here. Partial support for this work was provided by the NOAA grant for FGGE research, No. NA83AAG00837.

REFERENCES

1. A. Hollingsworth, et al., Mon. Wea. Rev. 108, 1736-1773 (1980).
2. L. Bengtsson, Tellus 33, 19-42 (1981).
3. B. R. Döös, GARP Publication Series #6, 68 pp. (1970).
4. D. L. Williamson, J. Atmos. Sci. 30, 537-543 (1973).
5. C. E. Leith, An. Rev. Fluid Mech. 10, 107-128 (1978).
6. C. E. Leith, An. Rev. Fluid Mech. 10, 107-128 (1978).
7. R. C. Malone, et al., NCAR manuscript #0306/82-10 (1983).
8. W. Washington (editor), Climate Section NCAR Report (1982).
9. J. Charney, et al., Bull. Am. Met. Soc. 47, 200-220 (1966).
10. J. Smagorinsky, Bull. Am. Met. Soc. 50, 286-311 (1969).
11. E. N. Lorenz, Tellus 34, 505-513 (1982).
12. J. Smagorinsky, Bull. Am. Met. Soc. 50, 286-311 (1969).
13. J. Shukla, J. Atmos. Sci. 38, 2547-2572 (1981).

ASYMMETRIES IN PERSISTENCE BETWEEN POSITIVE AND NEGATIVE ANOMALIES IN PERSISTENT ANOMALY REGIONS

Randall M. Dole
Center for Earth and Planetary Physics
Harvard University
Cambridge, MA 02138

Neil D. Gordon*
Department of Meteorology and Physical Oceanography
Massachusetts Institute of Technology
Cambridge, MA 02139

ABSTRACT

We have examined the regional persistence characteristics of wintertime Northern Hemisphere 500 mb height anomalies, focusing on the behavior of strong anomalies that persist beyond the durations associated with synoptic-scale variability ("persistent anomalies"). We have placed particular emphasis on determining how the persistence characteristics of anomalies vary depending on the sign and magnitude of the anomalies.

For moderate magnitudes and durations, the numbers of positive and negative anomaly cases that occur in a given region are about the same; however, for larger magnitudes and longer durations, the number of positive cases exceeds the corresponding number of negative cases. Analyses with data that have been low-pass filtered (removing periods of less than 6 days) reveal that part (but not all) of the discrepancy between positive and negative cases results from the relatively greater likelihood that negative anomalies will experience brief interruptions by transient disturbances.

For durations beyond about 5 days, the probability that an anomaly which has lasted n days will last at least $n + 1$ days is nearly constant. This nearly constant probability of continuation resembles the behavior obtained for a linear first-order autoregressive process ("red noise"). Nevertheless, there are significant differences in persistence between the positive and negative anomalies and red noise, particularly at large magnitudes, with the positive anomalies typically more persistent than either the negative anomalies or red noise.

To examine in more detail how the persistence of an anomaly depends on its sign and magnitude, relationships between the initial value of an anomaly and its subsequent 12 h change are then studied. For a given initial anomaly value, the height changes are decomposed into two parts: a mean change and a deviation from a mean change. In each region, the anomalies tend toward only one "quasi-equilibrium" value (mean change of zero). Mean changes are considerably less rapid for large positive anomalies than large negative anomalies. The variance of the changes about the mean change (the "noise") also depends on the initial anomaly value, with the smallest noise for large positive anomalies. Both the mean changes and noise variations favor the relatively greater persistence of strong positive anomalies compared to strong negative anomalies. The observed differences in persistence between positive and negative anomalies are compared with previous synoptic descriptions of persistent phenomena (such as blocking), and some possible physical sources for the differences are discussed.

---

*Present affiliation: New Zealand Meteorological Service, Wellington, New Zealand.

Introduction

A characteristic of weather evident to even a casual observer is the tendency toward persistence. Periods marked by the unusual persistence of highly anomalous weather conditions are of particular practical importance, often attracting intense interest in the community-at-large. The problem of forecasting these persistent extreme events remains among the most fundamental and challenging in weather prediction.

Theoretical studies of predictability[1,2] also often draw a close connection between predictability and persistence. Such studies frequently model the statistics of atmospheric fluctuations by a first-order linear autoregressive ("red noise") process[3], although recent research suggests that higher-order linear models may sometimes be more appropriate[4]. Alternatively,frequency spectra are often presented, with relatively greater power at low frequencies generally being considered as indicative of relatively greater long-range predictability. These conventional statistics, while highly useful, do not provide a basis for addressing questions such as:

- What are the favored regions for the occurrence of persistent positive (or negative) anomalies? Do the locations of the favored regions vary depending on the sign, magnitude or duration of the anomalies?
- Are positive or negative anomalies typically more persistent?
- Does the probability that an anomaly will persist depend on its magnitude or duration?

In the present study, we will adopt an approach similar to, but more general than, that usually adopted in blocking studies[5,6,7,8] in order to ascertain the geographical and regional persistence characteristics of anomalies. A major objective throughout will be to provide detailed comparisons between the persistence characteristics of positive anomalies and negative anomalies; we will also compare the observed behaviors with the behaviors anticipated for certain simple statistical and dynamical models of atmospheric persistence.

Data

We will focus here on anomalies of the extratropical Northern Hemisphere wintertime 500 mb geopotential heights. The basic data set consists of twice-daily (0000 GMT and 1200 GMT) National Meteorological Center (NMC) final analyses of the Northern Hemisphere 500 mb geopotential heights for the 14 winter seasons from 1963 - 1964 through 1976 - 1977. The winter season is defined as the 90 day period from 1 December through 28 February. Prior to calculations, data were spatially interpolated by a 16-point Bessel scheme from the NMC octagonal grid to a 5 degree latitude by 5 degree longitude grid over the region from 20N to 90N. Missing or obviously incorrect analyses were replaced by linear interpolations in time. Less than one percent of the 500 mb height analyses required replacement.

Raw height anomalies are defined as the departures of the analyzed heights from the corresponding long-term seasonal trend values. The seasonal trend time series at a point is determined by a least-squares quadratic fit to the 14-winter mean time series for that point (i.e., the first value of the winter mean time series is the average of the 14 different 0000 GMT 01 December values, the second value is the average of the 1200 GMT 01 December values, etc.). The raw height anomalies h'

have been normalized by a scale factor which is inversely proportional to the sine of latitude:

$$z'_\theta = \frac{\sin 45}{\sin \theta} h' \quad (1)$$

The scaling factor is motivated by a recent study on atmospheric energy dispersion[9] showing that height field analyses provide a poor indication of the meridional component of energy propagation. This shortcoming is due to the latitudinal variation of the Coriolis parameter, which biases height field responses toward high latitudes. Quantities like streamfunction or vorticity provide better indicators of horizontal energy propagation[9]. Note that this normalization is similar to that used in obtaining a geostrophic streamfunction from height data.

Elsewhere[10], we have considered the effect of regional differences in the height variance on the geographical distributions by instead normalizing the local height anomalies by their respective standard deviations (i.e., standardizing the height anomalies). Although both the latitude-normalized anomalies (LNA's) and standard deviation-normalized anomalies (SNA's) provide some distinct information on the behavior of the height fields, for two reasons we will focus primarily on the LNA's. First, consistent with the attitude adopted in studies of blocking, we are interested in anomalies that are strong and long-lived relative to anomalies in other regions; normalizing by local standard deviations, however, tends to mask geographical variations in the intensity of the anomalies. Second, through the geostrophic relation, fields of LNA's provide a much better indication of associated wind and vorticity anomalies and therefore are more amenable to direct physical interpretation. In the geographical distribution calculations, then, the term "anomalies" refers to the latitude-normalized anomalies.

Procedure for obtaining geographical distributions

We will define a "persistent anomaly" at a point if an anomaly at that point persists beyond some threshold value for a specified duration. The method, illustrated in Fig. 1, is as follows:

1) Specify a "magnitude" criterion, M, and a duration criterion, T, where for positive anomaly cases $M \geq 0$ and for negative anomaly cases $M \leq 0$.

2) Define the occurrence of a persistent positive (negative) anomaly case at a particular grid point satisfying selection criteria (M,T) if the anomaly at that point remains equal to or greater (less) than M for at least T days.

3) Define the duration, D, for a positive (negative) case as the time from which the anomaly first becomes greater (less) than M to the time when the anomaly next becomes less (greater) than M at that point.

Note that these criteria act as lower bounds, so that all events which meet or exceed the threshold values are counted as persistent anomaly cases satisfying the specified criteria.

The numbers of persistent anomaly cases occurring over the 14 winter seasons were determined for each point for the following values of selection criteria:

$M = \pm 0$ m, ± 50 m, ..., ± 250 m
T = 5 days, 10 days, ... , 25 days,

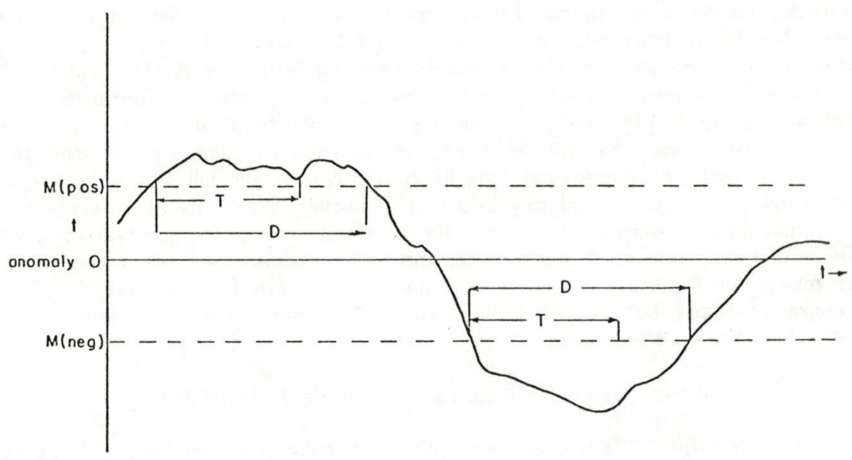

Fig. 1. Illustration of method for defining cases. A persistent positive (negative) anomaly case of duration D satisfying criteria (M, T) is defined at a point if the anomaly at that point exceeds (is less than) the threshold value M for at least T days. Examples are given for both positive anomaly cases (M > 0) and negative cases (M < 0).

where the 0 m threshold identifies runs of non-negative values for the positive cases and non-positive values for the negative cases.

In the following section, we will present results for a few values that illustrate a number of the most interesting characteristics of the geographical distributions; more detailed discussions are presented elsewhere[10,11]. For display purposes, the fields of the numbers of cases have been lightly smoothed by applying a two-dimensional nine-point spatial filter[12]. This filter effectively removes fluctuations having wavelengths of less than about 1500 km but does not otherwise affect the general character of the spatial variability.

Geographical distributions

Fig. 2a displays the geographical distribution of the number of positive cases satisfying the criteria (+150m, 5 days). These values include contributions from rather strong but short-lived events. Three major regions of maximum frequency of occurrence are evident:

1) over the central North Pacific to the south of the Aleutians (PAC);

2) over the eastern North Atlantic to the southeast of Greenland (ATL); and,

3) over the Northern Soviet Union extending northeastward to over the Arctic Ocean (NSU).

There is considerable latitudinal variability in the number of cases, despite the latitude-dependent normalization, with maxima occurring near 50N for the ATL and PAC regions and near 60N over NSU. Cases satisfying this set of criteria are rare to the south of 30N, over southern and eastern Asia and over central North America.

The corresponding distribution for negative anomaly cases satisfying (-150m, 5 days) is shown in Fig. 2b. There are several striking similarities between the positive and negative distributions. The greatest numbers of negative cases also occur over the PAC, ATL and NSU regions, although for this set of selection criteria the latter maximum is displaced somewhat westward of the corresponding positive center. There is additionally a fourth region of high negative occurrences over the extreme eastern North Pacific. For PAC and ATL, the maxima in the number of positive cases exceed the corresponding maxima for the negative cases; for NSU the maxima are comparable.

Fig. 2c shows the sum of the two previous distributions. The range in the number of cases is substantial: the four major regions have in excess of 20 cases over the 14 winter seasons, while central North America and large areas of Asia and the subtropics have fewer than four events satisfying these criteria over the period.

Fig. 3 displays similar positive, negative and sum distributions for selection criteria (+100m, 10 days) and (-100m, 10 days). These maps show a substantial reduction in the number of cases from the previous values. The three regions PAC, ATL and NSU continue to have the greatest number of cases, with the positive maxima exceeding the negative maxima for each region. There is only a weak indication, however, of the fourth region of high negative occurrences seen in the earlier analyses to the west of British Columbia. Examination of distributions with other criteria (not shown) shows that this maximum is present for all values of the magnitude criterion at 5 days but is not strongly evident for any values by 10 days. This suggests that although strong negative anomalies are frequently present in this region, they are typically relatively short-lived.

For values of the duration criterion beyond 10 days (not shown) the PAC, ATL and NSU regions continue to have the greatest number of persistent anomaly cases. There are few cases at these durations except for low values of the magnitude criterion. Aside from an overall decrease in the number of cases, the greatest qualitative

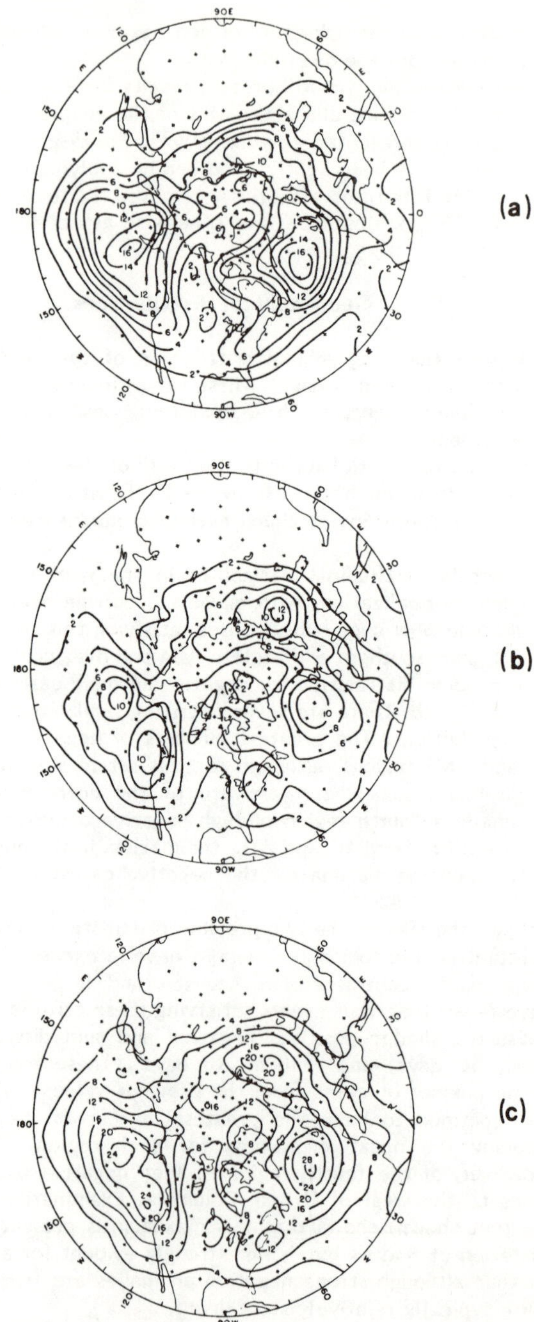

Fig. 2. The number of cases in 14 winter seasons satisfying the (a) positive anomaly criteria (+150 m, 5 days) and the (b) negative anomaly criteria (-150m, 5 days). (c) The sum of the cases in (a) and (b). Contour intervals are 2 in (a) and (b), and (4) in (c).

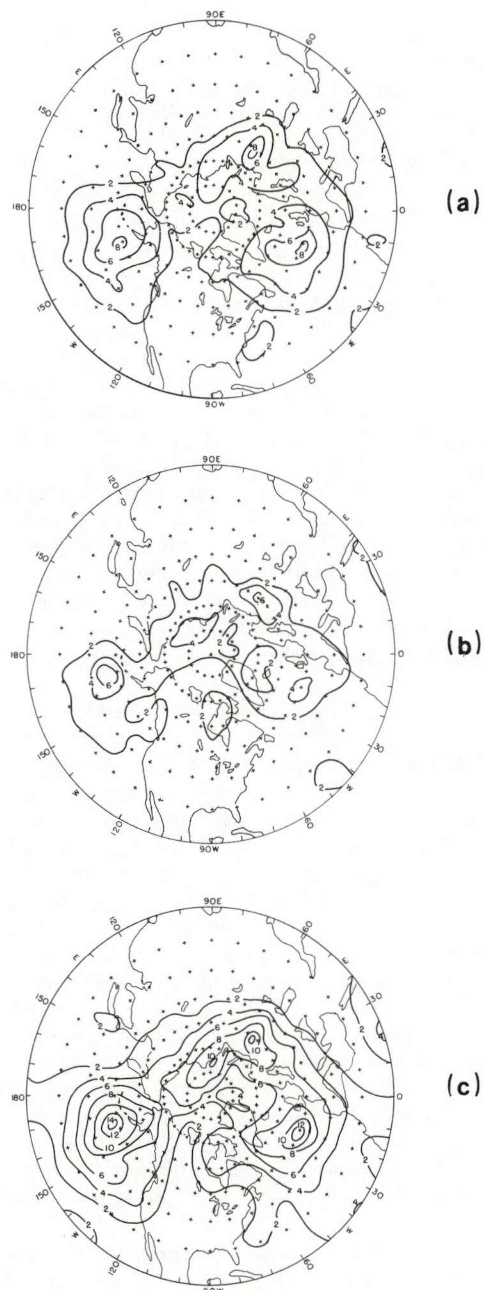

Fig. 3. The number of cases in 14 winter seasons satisfying the (a) positive anomaly criteria (+100 m, 10 days) and the (b) negative anomaly criteria (−100 m, 10 days), (c) The sum of the cases in (a) and (b). Contour intervals of 2.

change evident at low magnitudes and long durations is the relative increase in the total number of cases in NSU compared with either ATL or PAC. For relatively large magnitudes and short durations, then, the ATL and PAC maxima exceed the NSU maximum, while for relatively small magnitudes and long durations the NSU maximum is slightly larger.

Inspection of time series for individual points reveals that a number of persistent events occur which may not satisfy certain selection criteria due to brief ($\simeq$1 day) interruptions by mobile transient disturbances. In order to assess this effect, a low-pass filter that removes periods of less than about 6 days was applied to the data and the analysis described above was repeated. Fig. 4 displays the filter response function. Fig. 5 shows distributions obtained from the filtered data for selection criteria of (+100m, 10 days) and (-100m, 10 days). Comparing these distributions with the corresponding distributions for the unfiltered data (Fig. 3), we see that the three key regions (PAC, ATL and NSU) continue to have the highest number of cases. The most striking change is the relatively large increase in the number of negative cases occurring in the key regions. Whereas for the unfiltered data the number of positive cases exceeds the number of negative cases, for the low-pass filtered data the numbers of positive and negative cases are comparable. The total numbers of cases satisfying these criteria are increased by 50% - 100% over the corresponding unfiltered values, primarily as a result of the increase in the negative cases. The regional distributions, however, are otherwise almost unchanged from the unfiltered data. Low-pass distributions for other values of the selection criteria (not shown) also reveal a relatively greater increase in the number of negative cases, although at large thresholds (magnitudes greater than about 150m), there are still more positive than negative cases.

To estimate possible influences of interannual variability on the numbers and locations of persistent anomalies, geographical distributions were also calculated for anomalies that were defined with respect to the mean for each winter season (after removing the long term seasonal cycle by the method described previously). The results (not shown) are quite similar to those obtained for the earlier distributions, although the values of the maxima are slightly reduced (5% - 25%). This indicates that interannual variations in the seasonal mean provide relatively small contributions toward determining the occurrence (or non-occurrence) of persistent anomalies. Indeed, we anticipate that some of the interannual variability in the means is at least partly attributable to the occurrence of anomalous events that may themselves last a relatively small fraction of a season, as suggested by studies of sampling fluctuations in long-term means[3,13].

The gross structure of the persistent anomaly distributions closely resembles that of the daily variance of the wintertime 500 mb heights described by Blackmon[14]. The PAC, ATL and NSU regions, identified as having high numbers of major persistent anomaly cases, correspond with the major centers of large variance. This relationship is not surprising, since persistent anomalies can be expected to provide major contributions to the low-frequency variance (periods beyond 10 days) and, as Blackmon demonstrates[14], the daily height variance is dominated by low-frequency contributions.

The relatively large increase in negative events for the filtered data suggests that negative anomalies are more likely than positive anomalies to experience brief transient interruptions (predominantly from disturbances having periods of less than about 6 days). Further analyses of temporal behavior[10] indicate that this is related primarily to changes in storm activity accompanying the persistent anomalies, with relatively more frequent and vigorous eddies over the region typically associated with the negative cases.

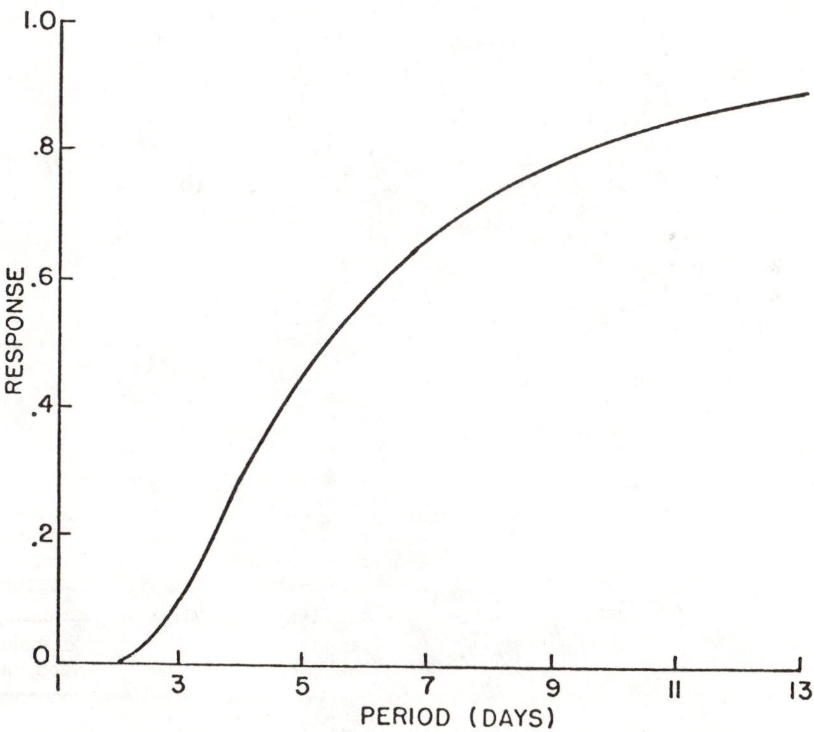

Fig. 4. Response function of the low-pass filter applied to the data.

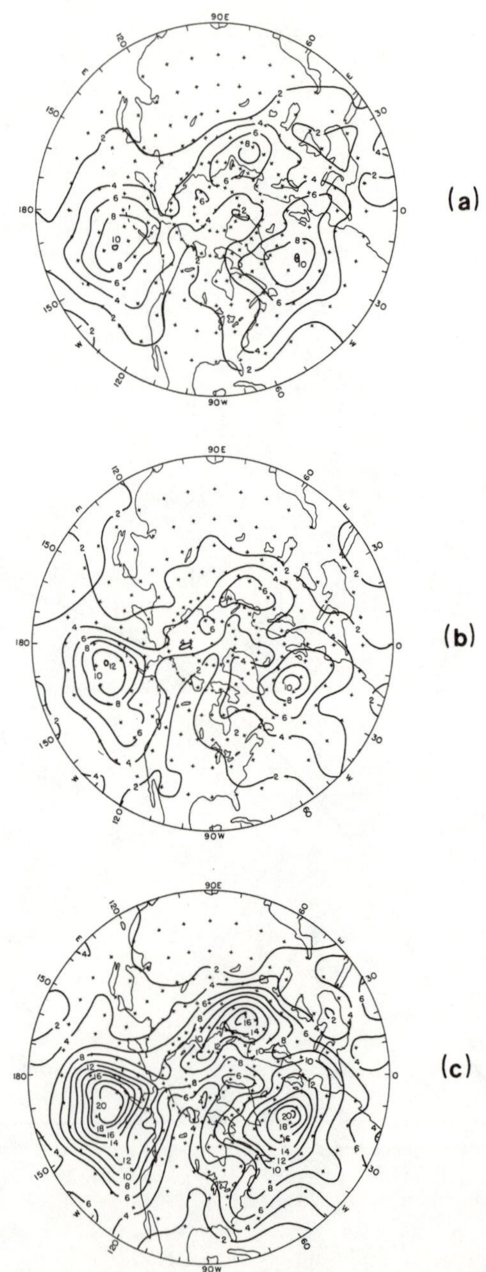

Fig. 5. As in Fig. 3, but for the low-pass filtered data.

Regional persistence characteristics

Our primary emphasis in this section will be on comparing the detailed persistence characteristics for regions characterized by relatively high and low numbers of persistent anomalies. For the former, we will focus on the eastern North Atlantic (ATL) and central North Pacific (PAC) regions, and for the latter, the central North America (AME) and eastern Asia (EAS) regions. For brevity, the results for the Northern Soviet Union (NSU) persistent anomaly region are not presented in detail, but the main results will be summarized later. Fig. 6 shows the locations of the five regions.

For ease of display, the anomalies have been normalized by their standard deviations; average values of the standard deviations for the regions are 179.3 m (PAC), 174.2 m (ATL), 162.2 m (NSU), 113.5 m (AME) and 44.8 m (EAS). Calculations are first performed at individual grid points; distributions for a region are then determined by combining the distributions obtained at each point in a 3 by 3 grid with 5 degree latitude and 10 degree longitude spacings centered on a point within the region. The center points are 50N 20W (ATL), 50N 165W (PAC), 40N 90W (AME), 30N 90E (EAS) and 60N 50E (NSU). Although the areas of the regions vary, the general statistical characteristics of the distributions at points within each region are nearly constant.

Fig. 7 presents, for each of the four regions, distributions of the numbers of runs of positive and negative anomalies that exceeded selected magnitude thresholds for at least a given duration. For comparison, estimates of the 5% and 95% confidence limits are also displayed for similar distributions generated by red noise processes appropriate for the regions; details of the method used for estimating the confidence limits are discussed elsewhere[11].

We see that after about 3-5 days the distributions for all regions form nearly straight lines, with shallowest slopes at low thresholds and largest slopes at high thresholds. On this semi-logarithmic plot, the local slope of the distribution is proportional to the probability of continuation of the run, with shallower slopes indicating a higher probability of continuation. The almost constant slopes at durations beyond a few days imply that the runs rather quickly assume a nearly constant probability of continuation. For many of the curves, a change from a steeper to a shallower slope occurs over the first few days, indicating that the probability that an anomaly will continue is initially increasing with increasing duration. This initial transitional behavior is intuitively plausible: at the start of the run, the anomaly value will generally be relatively close to the threshold, since it has recently crossed this value, and is therefore rather likely to return across the threshold. As the run continues, the expected value of the anomaly moves away from the threshold, until it eventually approaches a limiting value. This behavior is indeed observed[11].

For EAS, the distributions of positive and negative runs displayed in Fig. 7 are almost identical; for the other regions, there are systematic differences between corresponding positive and negative curves. At non-zero thresholds, the positive anomalies in these regions are typically more persistent. In general terms, the distributions for the real data follow the shape of the distributions generated by the corresponding red noise process. In particular, the red noise curves also show a tendency for relatively steeper slopes at short durations and shallower, nearly constant slopes at long durations. Nevertheless, some of the differences in behavior between positive and negative distributions and the red noise distributions appear clearly significant.

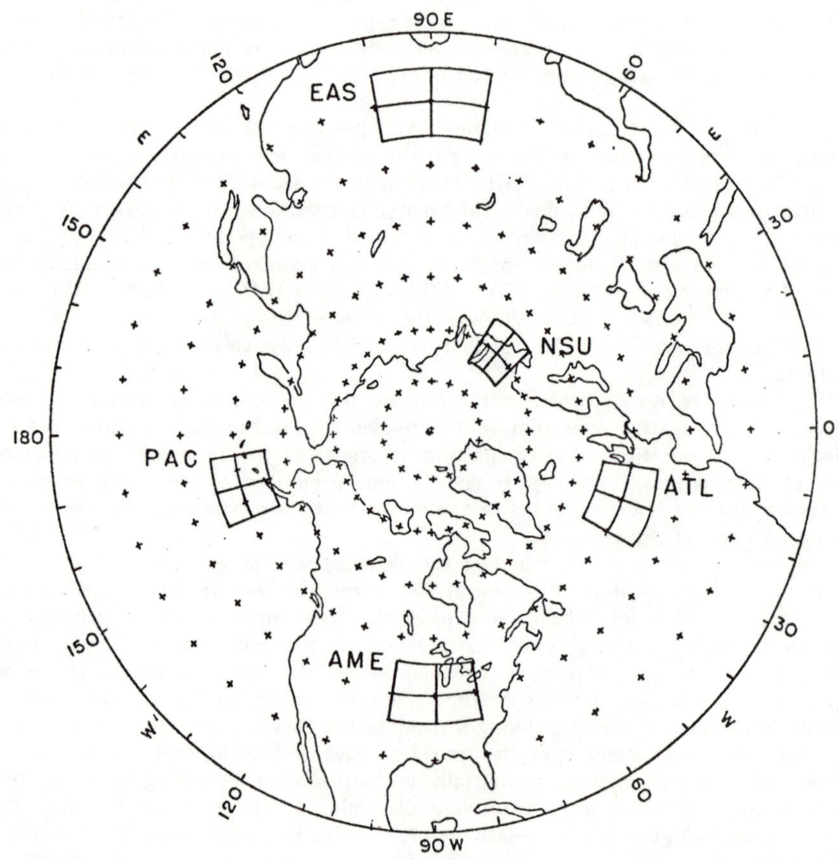

Fig. 6. The locations of the regions characterized by relatively high numbers (PAC, ATL and NSU) and low numbers (AME and EAS) of persistent anomalies. For each region, the nine grid points shown are those used in the analyses of Sections 5 and 6.

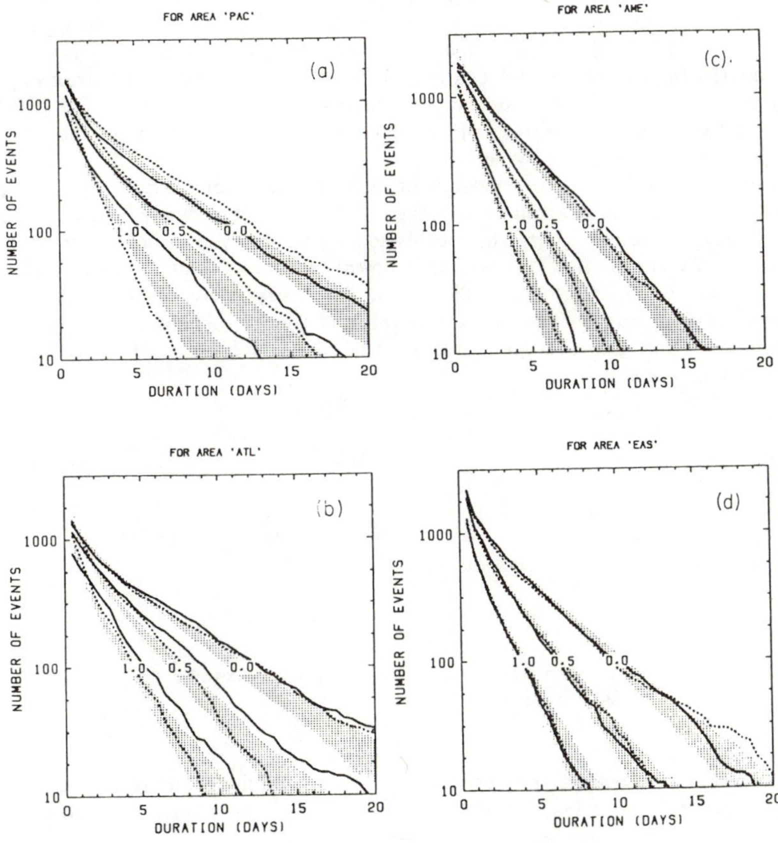

Fig. 7. Number of events in 14 winter seasons when a run of anomalies above or below a set threshhold lasts for at least a given duration, for regions (a) PAC, (b) ATL, (c) AME, and (d) EAS. The solid lines are for positive events above standardized threshholds of 0.0 (top), +0.5 (middle), and +1.0 (bottom). The dotted lines are for negative events below 0.0 (top), -0.5 (middle), and -1.0 (bottom). The shaded areas contain the corresponding estimates of the 5 to 95 percentile ranges for red-noise processes appropriate to the regions (see Appendix 1 for further details).

A question of considerable theoretical as well as practical interest is whether the anomaly z' tends to particular (possibly multiple) "preferred" values. We have examined distributions of anomaly values for evidence of multiple modes, perhaps suggesting multiple quasi-equilibria. Fig. 8 presents distributions of anomaly values for the 14 winter seasons of twice-daily data. The PAC distribution displays a marked positive skewness. The ATL distribution is more rectangular than a corresponding normal distribution, with relatively few values near the mean and in the tails and relatively many values at moderate magnitudes; indeed, a few of the distributions for individual points within the ATL region (not shown) hint at bimodality. The EAS and AME distributions, in contrast, do not differ markedly from normal distributions. The general characteristics of the regional distributions appear broadly typical of distributions at points throughout the key regions[15]. For plausible choices of the number of degrees of freedom, anomaly distributions in the PAC and ATL regions may be significantly non-Gaussian[15].

We have looked further for evidence of multiple quasi-equilibria by examining the relationship of the height anomaly tendency to the height anomaly value, searching for those anomaly values where the mean anomaly tendency vanishes. Our approach is as follows. We first define standardized anomaly intervals (bins) of width 0.1 (i.e., from -3.05 to -2.95, -2.95 to -2.85, etc.). For every entry of an anomaly value $z'(t)$ within a given bin, we determine the value of the 12 h change, $z'(t+1) - z'(t)$. We then determine the average value of the changes for each bin. Denote this average change curve by

$$\overline{\Delta z'} = f(z') \tag{2}$$

We then define a "potential" function U by

$$U(z'=x) = - \int_{x_1}^{x} f(z')\, dz' + C , \qquad x_1 < x < x_2 \tag{3}$$

where x_1 and x_2 represent, respectively, the lower and upper limits of the range of anomaly values having at least 10 observations per bin. The arbitrary constant C is chosen for graphical convenience.

By virtue of the integration, the potential function U displays less sensitivity to random sampling errors than the average change curve. Note that, since the slope of U is proportional to the average 12 h change, the maxima and minima in U correspond to anomaly values having mean tendencies equal to zero; these may be considered as quasi-equilibrium points. Furthermore, the maxima in U correspond to unstable points and the minima to stable points (in the sense that small departures will tend on average to grow or decay, respectively). Our approach is somewhat similar to that taken in a recent study employing a stochastic climate model[16], although the model examined is sufficiently simple that the potential function can be obtained directly by analytic methods.

Fig. 9 displays the average 12 h changes and the potentials as functions of the initial anomaly value. We see that for each of the four regions there is only a single minimum in the potential corresponding to a single preferred "climatic" value. We will call the value of the anomaly corresponding to this minimum z'_o. Although the precise values of the minima are rather poorly defined, examination of the change curves suggests that in some regions (particularly ATL) z'_o may be different from zero, indicating that the heights may on average be decaying toward a value other than the long-term mean.

The shapes of the potential functions also show considerable regional variability. For EAS, the potential function closely resembles a parabola, reflecting the almost exact proportionality of the 12 h change to the initial anomaly value. Potential functions in the other regions have much steeper slopes for negative than positive anomalies; correspondingly, the negative anomalies typically have larger average changes back to z'_o. The overall character of the potential function and average change curves supports our earlier results in suggesting the generally greater persistence of positive than negative anomalies in these regions.

The relationship between observed and red noise persistence

We noted in our Introduction that the statistics of atmospheric fluctuations have frequently been modeled by linear autoregressive processes, perhaps most commonly by a first-order process (often called "red noise"). Our results, however, indicate that there are systematic differences in persistence between positive and negative anomalies. These differences are generally most evident at large anomaly values. These variations in persistence cannot be accounted for by first-order (or higher-order) linear autoregressive models. Further, comparisons of the geographical distributions for unfiltered and low-pass filtered data suggest that there may also be differences in high-frequency variability between positive and negative anomalies. In this section, we will investigate in more detail the relationship between observed and red noise persistence characteristics.

The model for a discrete red noise process[17] obeys the difference equation

$$z(t+1) - \bar{z} = a\,(z(t) - \bar{z}) + e(t) \tag{4}$$

where $\bar{z}$ is the mean value of the process, a is a constant having a value between 0 and 1 and $e(t)$ is a white noise process having constant variance v^2. In the absence of the white noise, the time series decays toward a single equilibrium value $\bar{z}$ with a decay time given by $(1-a)^{-1}$.

A classical problem in time series analysis is to obtain optimal estimates of a and v from a given time series. The least squares estimate of a can be obtained by treating (4) as a linear regression problem with $(z(t+1) - \bar{z})$ as the "independent" variable[17]. An important assumption in this analysis is that the values of a and v do not depend on the value of z. We will examine whether this assumption is supported by the observations. Our approach is as follows.

For standardized anomaly intervals (bins) of width 0.1, all anomaly values initially falling within the bin, $z'(t)$, and the corresponding values 12 h later, $z'(t+1)$, are determined . For each bin having at least 10 values of $z'(t)$ and $z'(t+1)$, a least squares estimate of the parameter a is then obtained for the equation

$$(z'(t+1) - z'_o) = a\,(z'(t) - z'_o) \tag{5}$$

where z'_o is the appropriate regional "quasi-equilibrium" value determined in the previous section. An estimate of the corresponding value for v^2 is obtained from the unexplained variance in this regression calculation.

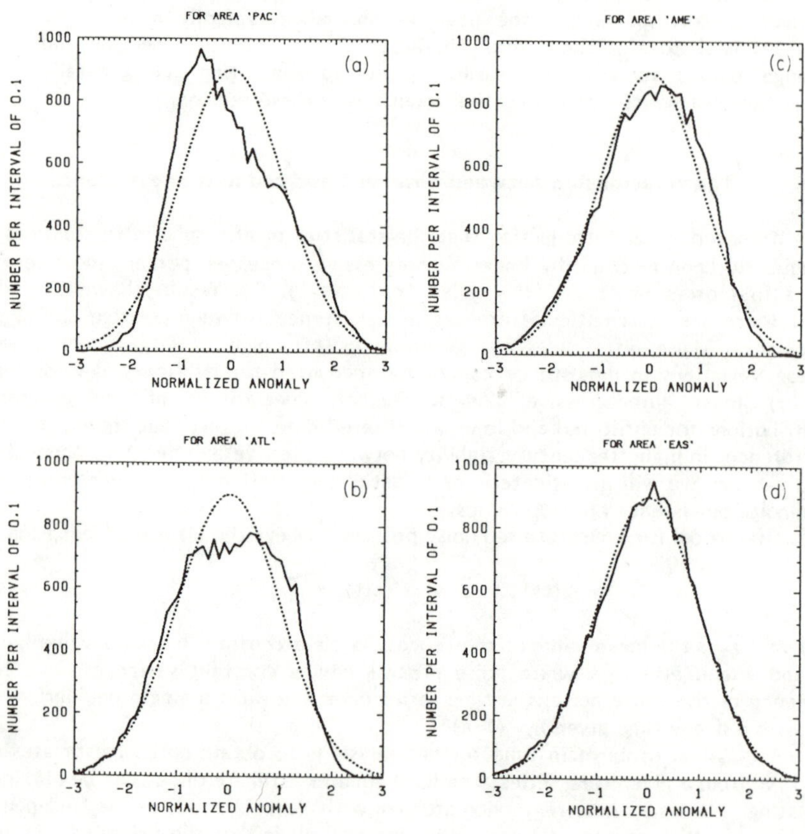

Fig. 8. Distributions of values of the standardized height anomalies (solid lines) obtained from the 14 winter seasons of twice-daily data, for regions (a) PAC, (b) ATL, (c) AME, and (d) EAS. Also shown are corresponding Gaussian distributions (dotted lines) with unit variance.

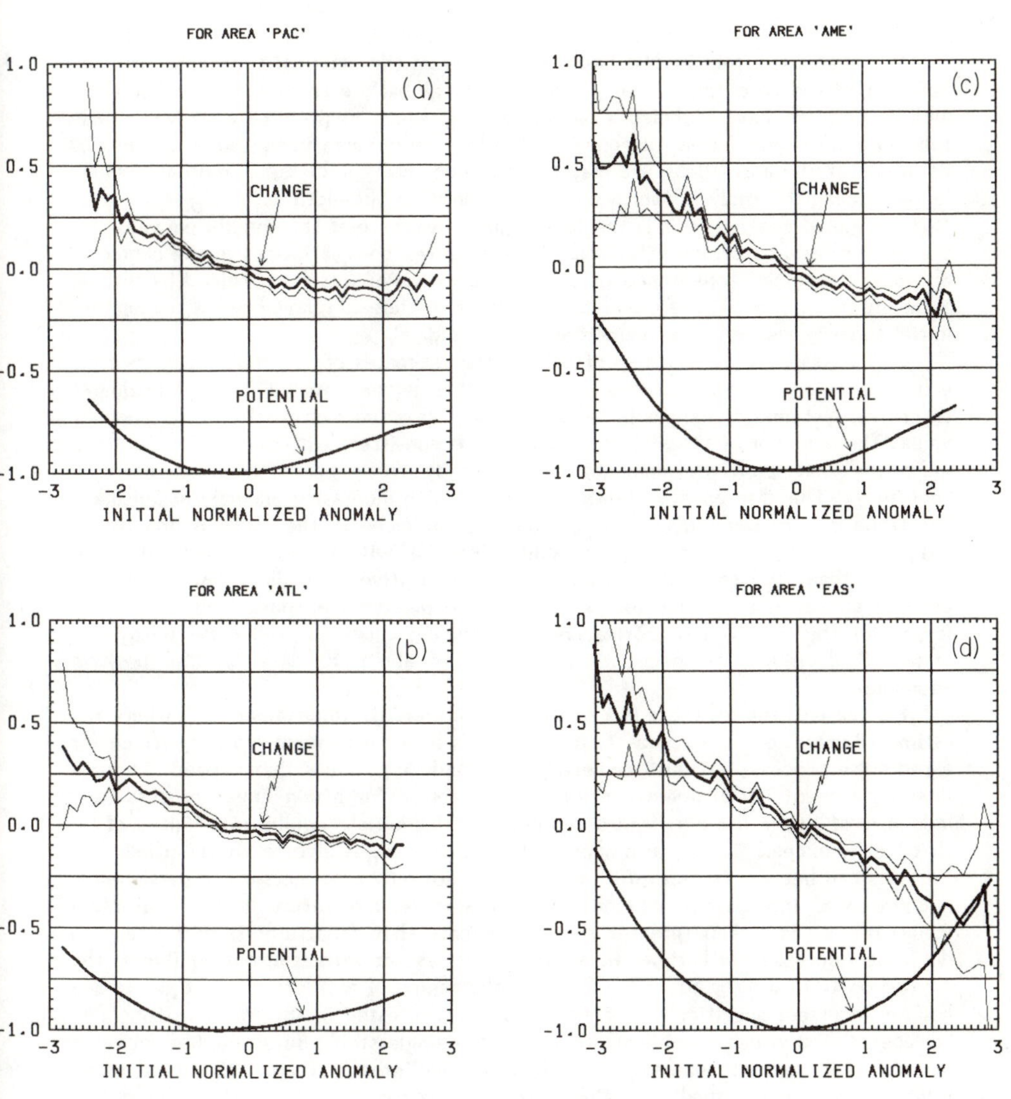

Fig. 9. The average 12 h changes in the standardized anomaly (top thick line) and the corresponding potential function U (lower thick line) defined in (3) as functions of the initial anomaly value. The thin lines above and below the change curve give the 95 % confidence interval for the average changes. The value of the arbitrary constant C in (3) is chosen such that the minimum value of U is -1. For regions (a) PAC, (b) ATL, (c) AME and (d) EAS.

Our analysis differs in two basic respects from the classical approach: first, we have attempted to obtain "local", rather than "global", estimates for a and v^2 and, second, we have included a parameter z'_o since, as previously shown, z' may tend to a value not necessarily equal to the long-term mean value (equal to zero by definition of the anomalies). We may therefore recover the classical approach results by performing the analysis over a single bin of infinite bin width with z'_o set to zero (for the global analysis, this is the least squares estimate of the equilibrium value for the time series). Later, we will discuss in some detail how variations in the choice of z'_o may affect the appearance of the results; however, results obtained by instead assuming that $z'_o = 0$ are qualitatively similar to those displayed below, except for initial anomaly values in the region between 0 and z'_o.

Fig. 10 displays, as functions of $z'(t)$, the estimates of a and v. Aside from values near z'_o, where there is considerable uncertainty in the least-squares estimates, systematic changes in a are evident as a function of the initial anomaly value. The behavior of a with respect to z' is consistent with our earlier results in indicating that positive anomalies are typically more persistent than negative anomalies. This discrepancy tends to increase with increasing anomaly magnitudes. The trend is very slight over EAS, but quite pronounced in the other regions. For ATL, the distribution of a appears almost bimodal, with values of around 0.85 for $z'(t)$ less than -0.5 and values around 0.95 for positive anomalies. Although such changes appear small, the consequences for the persistence characteristics may be large: for the ATL values, corresponding e-folding times for red noise decay are around 3 days for the negative anomalies and about 10 days for the positive anomalies.

For the PAC region, there is a small range of anomaly values near z'_o where the estimated value of a exceeds 1. Although this result may seem suprising (since for a red noise process a must lie between 0 and 1), values that exceed 1 for a limited range of initial anomalies are not prohibited for a non-linear process (and indeed must occur for a first-order process with multiple equilibria). Values of a that locally exceed 1 may also occur if the value of z'_o is incorrectly specified.

This can be seen by interpreting a in (5) not only as a measure of persistence, but also as an indicator of whether the anomalies tend to return to z'_o: absolute values of a less than (greater than) 1 indicate that departures from z'_o tend to decrease (increase) with time. In a first-order system with multiple equilibria, the transistion from $a < 1$ to $a > 1$ defines the region of attraction for z'_o. To see how an incorrect specification of z'_o may be manifested in the estimate for a, consider a first-order non-linear process with a single stable quasi-equilibrium value at $z'_o > 0$, so that anomalies initially less than (greater than) z'_o tend to be followed by larger (smaller) values. If we incorrectly assume that the quasi-equilibrium value is zero, then for initial anomaly values between 0 and z'_o, the estimate of a obtained from (5) would exceed 1 (since on average $z'(t+1) > z'(t)$ over these values). The results in the PAC region appear to reflect our uncertainty in the precise quasi-equilibrium value, and suggest that the specified value of z'_o may be slightly too high.

The estimated noise intensity v displayed in Fig. 10 also shows interesting variations. Over EAS, the trend with z' is very slight. Over the other regions, however, substantial trends are evident. In these regions, the smallest values occur for large positive anomalies; values vary by a factor of about 2-3 between maxima and minima. The anomaly values for the noise maxima show considerable regional variability; these are plausibly associated with shifts in the storm paths located adjacent to the regions. These relationships are described more fully elsewhere[10].

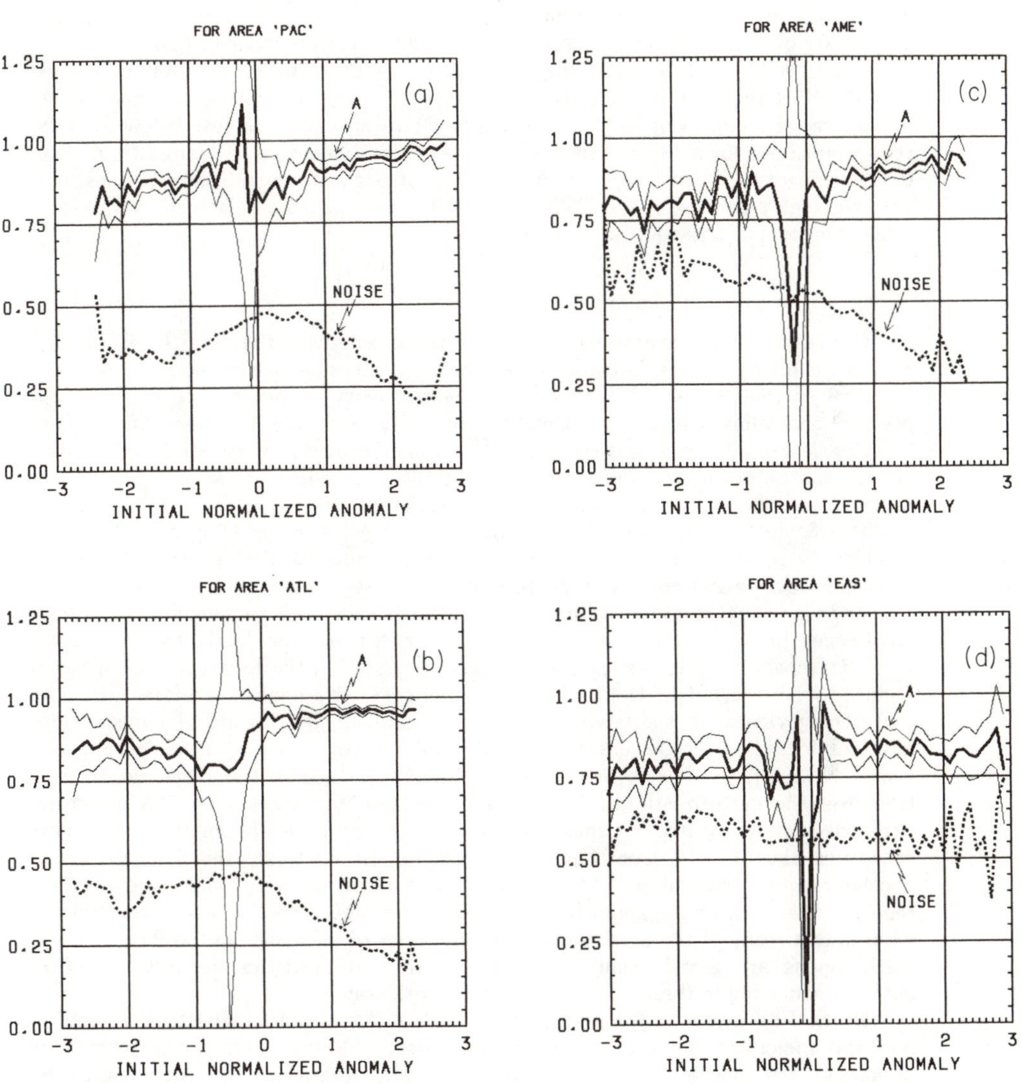

Fig. 10. The least-squares estimates of the persistence parameter a (thick solid line) in (5) as a function of the standardized anomaly value, and the 95% confidence interval for the estimates (thin solid lines). The dotted lines give the corresponding values (in standardized units) for the "noise" v, obtained as the square root of the residual variance. For regions (a) PAC, (b) ATL, (c) AME and (d) EAS.

Analyses similar to those described in this and the previous section have also been conducted for the NSU region. The results (not shown) display trends in a and v that are qualitatively similar to, although slightly less pronounced than, the trends described for the ATL and PAC regions. Principal quantitative differences suggest that the NSU region is characterized by somewhat higher overall persistence (values of a range from about 0.90 for strong negative anomalies to just below 1.00 for strong positive anomalies) and somewhat lower noise (values of v range from about 0.35 for negative anomalies to around 0.20 for strong positive anomalies). Consistent with the high persistence, the NSU potential function appears relatively flat, with a single minimum at an anomaly value near zero.

Discussion

Our results show several areas of broad agreement with the results described in previous blocking studies, but also important differences. As we might anticipate, the locations of the PAC and ATL maxima in the frequency of occurrence of persistent positive anomalies are in approximate conformity with the locations of frequent blocking described in previous studies[7,8,18]. Not described in these studies, however, are the substantial numbers of persistent negative anomalies that also occur in these regions, nor the third area of frequent persistent anomalies centered over the northern Soviet Union. Also, Rex [7] indicates that Atlantic blocking exceeds Pacific blocking by a factor of two, whereas our results suggest that persistent positive anomalies occur about equally frequently in the two regions.

Although part of these differences may be attributable to sampling variations, differences in the methods used for selecting cases are also likely to account for some discrepancies. For example, in most blocking studies the long-term mean is not removed from the height fields before defining cases. We might therefore anticipate that strong ridges or highs will show a geographical bias toward the mean ridge positions located over the eastern Atlantic and eastern Pacific; the northern Soviet Union, however, would not be particularly favored. Indeed, the mean upper-level flow over the eastern Atlantic is also characterized by a weak split flow structure, with only relatively modest anomalies required to produce the marked split flow structures that Rex [7] identifies with blocking. In contrast, the PAC persistent anomaly region, located just to the north of a more intense zonal wind maxima, requires relatively strong anomalies to produce pronounced split flows. Thus, although we find that many of the persistence characteristics of anomalies in the PAC, ATL and NSU regions are grossly similar, we anticipate that statistics derived from flow patterns occurring in these regions may vary considerably.

We find no evidence for a strongly-preferred duration for persistent anomalies, nor any indication of pronounced periodicities. Rather, for sufficiently long durations, the number of events decays nearly exponentially with increasing durations. The similarity of the observed distributions to the distributions generated by a red noise process suggests that many of the persistent anomalies may arise from fluctuations, sometimes called "climate noise"[3,13], that are generally assumed to be unpredictable on long time scales.

Nevertheless, the observed behaviors are in important respects unlike those obtained in red noise (or other linear autoregressive) models. In particular, we find that the rapidity of the decay depends on the anomaly value: generally, positive anomalies decay more slowly than negative anomalies, with the discrepancy increasing at increasing anomaly magnitudes. Also, the variance of the changes (the "noise") is typically considerably smaller for strong positive than strong negative anomalies, decreasing the probability of a rapid jump back to a more normal value with reduced persistence. The variations in both the mean height changes and the noise indicate that, once established, strong positive anomalies are relatively more apt to persist. These variations are also reminiscent of the picture presented by synopticians that strong highs (frequently associated with blocking) are relatively likely to persist for extended durations, and that they are also often accompanied by a marked local reduction in the daily variability associated with mobile disturbances.

We have suggested that variations in the intensity of the noise that depend on the anomaly value are at least partially related to variations in the storm paths adjacent to the anomalies. Another possible source for the differences between positive and negative anomalies is suggested by considering the divergence term in the vorticity tendency equation[19]:

$$\frac{\partial\zeta}{\partial t} = - \underset{\sim}{V}\cdot\nabla\,\zeta - v\,\frac{df}{dy} - (\zeta+f)\,\nabla\cdot V$$
$$- \left[\omega\,\frac{\partial\zeta}{\partial p} - \hat{k}\cdot\left(\frac{\partial \underset{\sim}{V}}{\partial p} \times \nabla\omega\right)\right] \qquad (8)$$

where ζ is the relative vorticity, f is the Coriolis parameter and ω is the "vertical motion" in pressure coordinates; all other notation is standard.

Note that the magnitudes of the relative and planetary vorticity advection terms and the terms in the brackets do not depend on the sign of the relative vorticity. For quasi-geostrophic motions, the terms within the brackets are neglected and the divergence term in (8) is linearized by replacing the absolute vorticity ($\zeta + f$) by a mean vorticity value f_o. This latter approximation is probably most reasonable at small values of the height anomalies, where we anticipate that associated vorticity perturbations will generally be weak. Very large height anomalies, however, are typically accompanied by large values of relative vorticity, with positive (negative) height anomalies associated with negative (positive) relative vorticity. We therefore anticipate from (8) that, for a given divergence, vorticity (and associated height) changes will occur more rapidly for negative than positive height anomalies, with the differences generally increasing with increasing anomaly magnitudes.

Since there are other possible sources for differences in persistence characteristics between positive and negative anomalies (e.g., asymmetries in structure) that are beyond the scope of this study, a definitive identification of the major reasons for differences cannot presently be made. The vorticity argument, however, does also provide a possible explanation for some apparent regional variations that we observed in the average change curves (Fig. 6). Although these curves appeared to disclose more differences in persistence between positive and negative anomalies in PAC and ATL than in EAS, recall that the anomalies in these figures were normalized to unit variance. A normalized deviation of 1 unit is equivalent to about 170 m - 180 m in PAC and ATL but only about 45 m in EAS; thus, the corresponding vorticity anomalies may be sufficiently small in EAS that asymmetries due to relative vorticity variations may scarcely be evident. If this is the case, then some of the differences we have observed between the regions may be due more to differences in the physical magnitudes of disturbances (which are masked by scaling by the standard deviations) than to qualitative changes in persistence characteristics. This suggests that special care must be taken in interpreting and comparing results between regions (both ours and those of others) that are derived from standardized height anomalies.

Finally, we found that the average changes always tend to reduce the local height anomalies to a single value near zero (reflecting a decay toward the long-term mean); however, fluctuations that are manifested in the variance about the mean changes (which we have termed "noise") may either decrease or increase an anomaly. If these fluctuations are generally too weak, then the anomalies may never achieve values where the differences between positive and negative anomalies (including increased positive anomaly persistence) become important; but if for strong positive anomalies the fluctuations are relatively too vigorous, then the probability will be increased of a rapid jump back to a more normal value with reduced persistence. In order to obtain the observed characteristics of persistent anomalies, then, we may need to correctly model not only the physical processes that determine the average changes, but also those that determine the variance of the changes about the average values.

Conclusions

The main points to emerge from the geographical distribution analyses are:

- There are three major regions of frequent occurrence of persistent anomalies: the North Pacific to the south of the Aleutians, the North Atlantic to the southeast of Greenland and from the northern Soviet Union northeastward to over the Arctic Ocean.
- For moderate magnitudes and durations, the numbers of positive and negative cases in each region are about the same. For very short durations (1-2 days) there are more negative than positive cases, while at longer durations and at large magnitudes, the number of positive cases exceeds the corresponding number of negative cases.
- Similar analyses performed on data that have been weakly low-pass filtered have regional distributions that are almost unchanged from the unfiltered data, although the total numbers of cases are increased by 50% - 100% over the corresponding unfiltered values, primarily as a result of an increase in the negative cases. The relatively large increase in negative events for the filtered data indicates that negative anomalies are more likely than positive anomalies to experience brief transient interruptions (from disturbances having periods of less than about 6 days). At large magnitude thresholds, however, there are still more positive than negative cases.

The regional analyses indicate that:

- In persistent anomaly regions, the probability that an event which has lasted n days will last at least $n+1$ days increases up to about $n = 5$ days and is thereafter nearly constant. The nearly constant probability of continuation is accompanied by a nearly constant average anomaly value.
- The general shapes of the persistence curves resemble those generated by a red noise process. Nevertheless, there are significant differences in behavior between the positive and negative distributions and the red noise distributions, particularly at large magnitudes and extended durations. At large magnitudes, the positive anomalies are typically considerably more persistent than either the negative anomalies or red noise.
- For each region, analyses of the relationship between initial anomaly values and their subsequent 12 h changes indicate that the height anomalies tend toward only one "quasi-equilibrium" value (mean change of zero). Mean changes are considerably less rapid for large positive anomalies than large negative anomalies. The variance of the changes about the mean change ("noise") also depends on the

anomaly value, with the smallest noise for large positive anomalies. Both the mean changes and noise variations favor the relatively greater persistence of strong positive anomalies compared to strong negative anomalies.

Our results reveal a number of typical aspects of the behavior of persistent anomalies that will require theoretical explanation. At a minimum, a comprehensive theory of persistent anomalies should be able to account for the observed geographical variations, the simple decay characteristics and the asymmetries in persistence between positive and negative anomalies.

REFERENCES

1. C.E. Leith, Ann. Rev. Fluid Mech., 10, 107 (1978).
2. R.E. Moritz and A. Sutera, Adv. Geophys., 23, 345 (1981).
3. C.E. Leith, J. Appl. Meteor., 12, 1066 (1973).
4. R.W. Katz, J. Atmos. Sci., 39, 1446 (1982).
5. R.D. Elliott and T.B. Smith, J. Meteor., 6, 67 (1949).
6. D.F. Rex, Tellus, 2, 196 (1950).
7. D.F. Rex, Tellus, 2, 275 (1950).
8. E.J. Sumner, Quart.J.R.Meteor.Soc., 80, 402 (1954).
9. B.J. Hoskins, A.J. Simmons and D.G. Andrews, Quart.J.Roy.Meteor.Soc., 103, 553 (1977).
10. R.M. Dole. Persistent anomalies of the extratropical Northern Hemisphere wintertime circulation. Ph. D. Thesis, MIT, Cambridge, MA 02139 (Available from author on request).
11. R.M. Dole and N.D. Gordon, Mon Wea. Rev., accepted for publication (1983).
12. R. Shapiro, Rev. Geophys. Space Phys., 8, 359 (1970).
13. R.A. Madden, Mon Wea. Rev., 104, 942 (1976).
14. M.L. Blackmon, J.Atmos.Sci., 33, 1607 (1976).
15. G.H. White, Mon.Wea.Rev.,108, 1446 (1980).
16. R. Benzi, G. Parisi, A. Sutera and A. Vulpiana, Tellus, 34, 10 (1982).
17. C. Chatfield, The analysis of time series: theory and practice (Chapman and Hall, London, 1975), p. 65-70.
18. W.B. White and N.E. Clar, J. Atmos. Sci., 32, 489 (1975).
19. J.R. Holton, An Introduction to Dynamic Meteorology, (second edition, Academic Press, N.Y., 1979), p. 94.

THE EMERGENCE OF ISOLATED, COHERENT VORTICES IN TURBULENT FLOW

James C. McWilliams

National Center for Atmospheric Research, Boulder, CO 80307

ABSTRACT

A preliminary report is made of some numerical calculations of two-dimensional and geostrophic turbulent flows. The primary result is that, under a broad range of circumstances, the flow structure has its vorticity concentrated in a small fraction of the spatial domain. When such vorticity concentrations occur, they tend to assume an axisymmetric shape and persist under passive advection by the large-scale flow except for relatively rare encounters with other centers of concentration. This structure can arise from random initial conditions without vorticity concentration. The concentrations subsequently evolve out of what has been traditionally characterized as isotropic, homogeneous, large-Reynolds-number turbulence with its systematic elongation of isolines of vorticity associated with the transfer of vorticity to smaller scales, eventually to dissipation scales, and the transfer of energy to larger scales. The demonstration of persistent vorticity concentrations on intermediate scales--smaller than the scale of the peak of the energy spectrum and larger than the dissipation scales--does not invalidate most of the traditional characterizations of two-dimensional turbulence, but I believe it shows them to be substantially incomplete. I also believe it is likely to alter our conceptions of the phenomena involved, as well as our understanding of the predictability of such flows.

INTRODUCTION

The dynamics of large-scale, extra-tropical, planetary fluid motions are usually approximately geostrophic and weakly non-conservative. In these circumstances most phenomena can be interpreted as hybrids of dynamical regimes: dispersive Rossby waves, geostrophic turbulence, and isolated, coherent vortices. Usually the regime boundaries can be identified by simple scale arguments. If phase speeds are greater than fluid particle speeds, then wave processes are likely to be dominant. Isolated, coherent vortices, with the correct structure to be a self-consistent solution in an otherwise quiescent fluid, are likely to persist if their particle speeds are larger than those of any disorderly flow structures in their neighborhood. If these conditions are not met, then a turbulent interpretation is usually required, at least for the transient component of the flow. Turbulence, of course, has many, not wholely consistent, definitions, but short persistence times for any particular flow configuration (on the order of a particle recirculation time) and limited predictability from imperfectly known initial conditions are often essential characteristics. The term "cascade" is often used to indicate the strong nonlinear interactions between quite disparate scales of motion which can account for these characteristics.

This paper is a preliminary report of numerical solutions which demonstrate that isolated, coherent vortices can coexist with and evolve out of actively cascading turbulence, even when there are no gross scale differences between the two components of the flow. Most attention will be directed to the simplest and most extensively studied example of geostrophic turbulence, the slow frictional decay of isotropic, homogeneous, two-dimensional turbulence, although the occurrence of isolated vortices is not inherently limited to this situation. Because this study is uncompleted, my objectives here are to illustrate the phenomena, state some hypotheses about the underlying processes, and present a tentative conceptual framework. I hope to be more definite and complete in a future report.

THE MODEL

A set of model equations which is sufficiently general to encompass all calculations to be reported here is the equivalent-barotropic, quasigeostrophic, β-plane potential vorticity balance with forcing and various types of damping:

$$\frac{\partial}{\partial t}(\zeta-\gamma^2\Psi) + J(\Psi,\zeta-\gamma^2\Psi+by) = f - \nu_0\zeta + \nu_2\nabla^2\zeta - \nu_4\nabla^4\zeta,$$
$$\zeta = \nabla^2\Psi,\ u = -\frac{\partial\Psi}{\partial y},\ v = \frac{\partial\Psi}{\partial x}. \qquad (1)$$

x, y, and t are east, north, and time coordinates. Ψ is the streamfunction, u and v are the east and north velocities, and ζ is the vorticity. γ is the inverse of the radius of deformation, b is the northward graident of the Coriolis parameter, f is the forcing function, and the ν_i are damping coefficients--referred to as Rayleigh friction, Newtonian viscosity, and hyperviscosity, respectively. J is the Jacobian operator in x and y.

We shall examine a simple, special case of (1) in particular detail; viz., decaying two-dimensional flow,

$$\gamma = b = f = \nu_0 = \nu_2 = 0,$$
$$\frac{\partial\zeta}{\partial t} + J(\Psi,\zeta) = -\nu_4\nabla^4\zeta. \qquad (2)$$

The choice of hyperviscosity is dictated by a desire to confine the effects of damping to the smallest resolved scales of the model, leaving the more energetic scales nearly inviscid.

These equations are solved as an initial-value problem in a doubly periodic domain of dimension 2π. The numerical model is a dealiased, pseudo-spectral one developed by Dr. Dale Haidvogel; it uses approximately circular truncation in the wavenumber plane and conserves area integrals of both energy and potential enstrophy when the forcing and damping terms are zero.

DECAYING TWO-DIMENSIONAL FLOW

The initial conditions are a Gaussian random realization for each Fourier component of Ψ, where at each wavenumber vector the ensemble variance is proportional to a prescribed scalar wavenumber function. This function is broad-band and peaked at intermediate wavenumbers (see Fig.2). The value selected for the hyperviscosity is 3.125 x 10^{-8}, which, because of the domain size of 2π and the initial kinetic energy of 0.5, can be interpreted as an inverse Reynolds number for the largest scales. This implies a very slow rate of energy decay, as is shown in Fig. 1a. The energy and enstrophy and centroid wavenumber are defined by

$$K \equiv \frac{1}{2} \iint dxdy\ (u^2 + v^2) = \sum_{k=1}^{N} E(k),$$
$$V \equiv \frac{1}{2} \iint dxdy\ \zeta^2 = \sum k^2 E(k), \qquad (3)$$
$$\overline{k} \equiv \sum kE(k) / \sum E(k),$$

where N is the scalar wavenumber truncation value. In this calculation it is approximately 128, which corresponds to 256 spatial grid points in each direction. E(k) is the energy spectrum. Time series

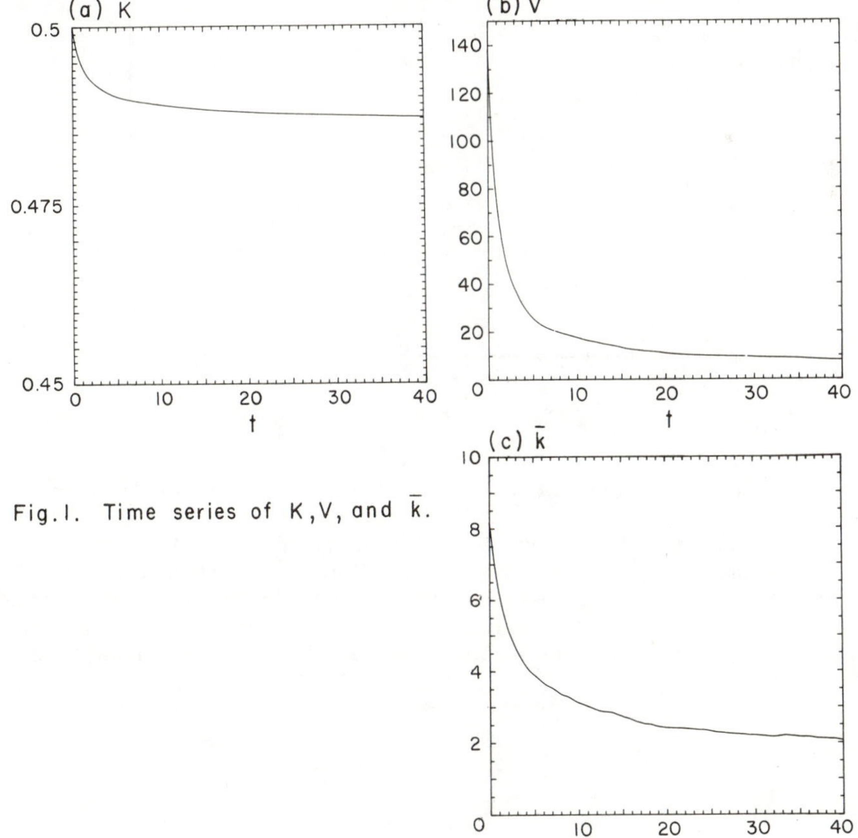

Fig.1. Time series of K, V, and $\overline{k}$.

of these quantities are plotted in Fig. 1. They show that the energy decay rate is indeed small, less than a 3% decrease over a time of 40 units, where a unit is approximately a recirculation time for eddies on the most energetic scale. The enstrophy decay rate, however, is initially quite substantial--in contrast to the energy rate, it is expected to remain finite as the hyperviscosity goes to zero--although it slows considerably after a few recirculation times. The centroid wavenumber monotonically decreases with time--energy is transferred to larger scales--although its rate also slows with time. Figure 2 shows that the energy spectra change rapidly in the first few recirculation

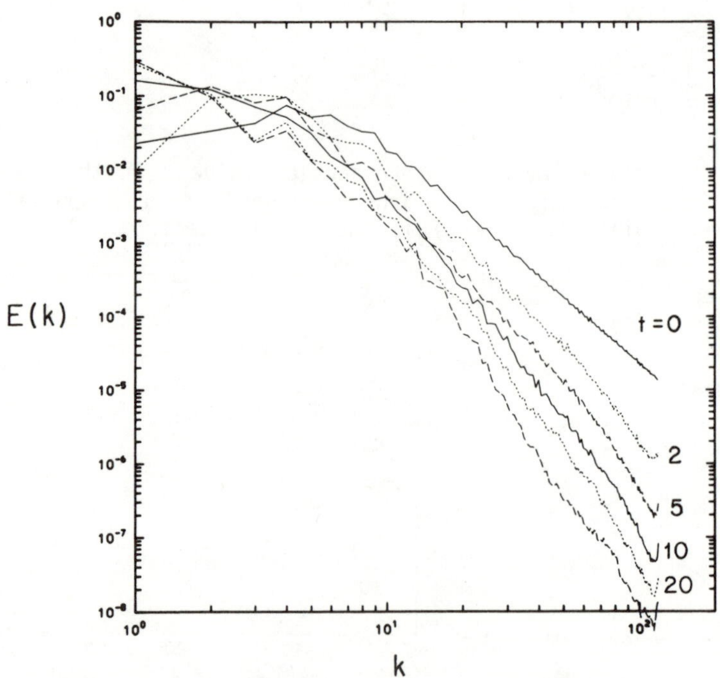

Fig. 2. E(k) at t = 0, 2, 5, 10, 20, and 40.

times, primarily by steepening at large k from the initial shape of k^{-3}. Subsequently E(k) undergoes little change in shape, except for weak fluctuations in the small k components and a slow frictional decrease in all components. All of these properties are familiar ones from previous investigations of decaying two-dimensional turbulence.

A much less familiar property is illustrated in Fig. 3.

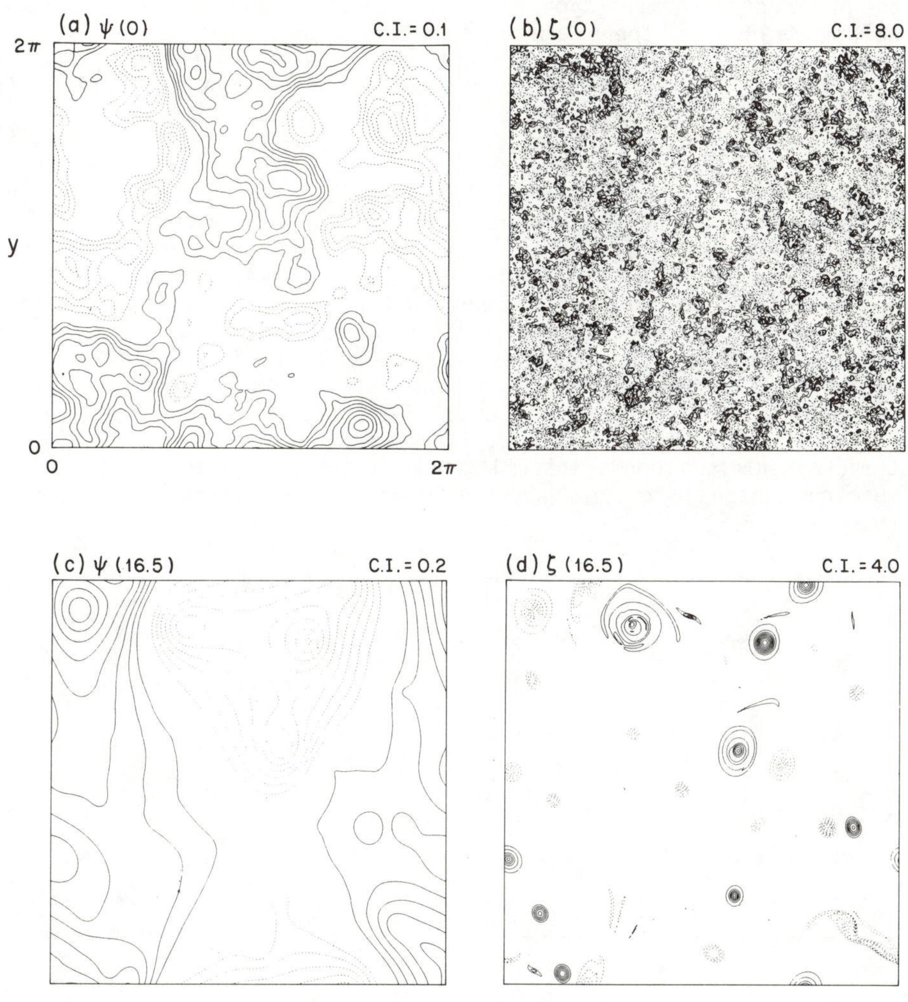

Fig. 3. Streamfunction and vorticity, initially and during the slow decay phase (t = 16.5). Contour intervals (C. I.) are as labeled. Positive contours are solid and negative dashed; the zero contour has been deleted.

While the streamfunction field exhibits a systematic evolution towards larger scales, the vorticity field evolves from an initially fairly uniform distribution in space to a collection of discrete and usually isolated vorticity extrema. These isolated vortices occur with a variety of amplitudes and sizes, although all of them are smaller than the dominant streamfunction patterns. Most of them are approximately axisymmetric, but there are obvious exceptions when two or more of them are close enough to interact strongly. The interactions can take a variety of forms: a weak vortex can be sheared out or engulfed (if of the same sign) if it too closely approaches a strong vortex; in other cases a close approach leads to strong deformations of vorticity contours followed by relaxation towards axisymmetry after separating; vortices of comparable strength but opposite sign can become attached to each other, much like a modon, and consequently undergo rapid movement through the fluid, until some other close encounter pulls them apart. The enormous qualitative difference between the distributions of the two fields is expressed by the kurtosis computed in an average over all spatial grid-point values at fixed t (Fig. 4): streamfunction remains close to the Gaussian value of 3, while vorticity, monotonically departs from it due to the decreasing of vorticity in the space between and the persistence of vorticity within the isolated vortices. The high kurtosis implies a particular kind of spatial and temporal intermittency; intermittency in general is an explanation of spectrum shapes steeper than the inertial range form k^{-3}, such as those in Fig. 2[12].

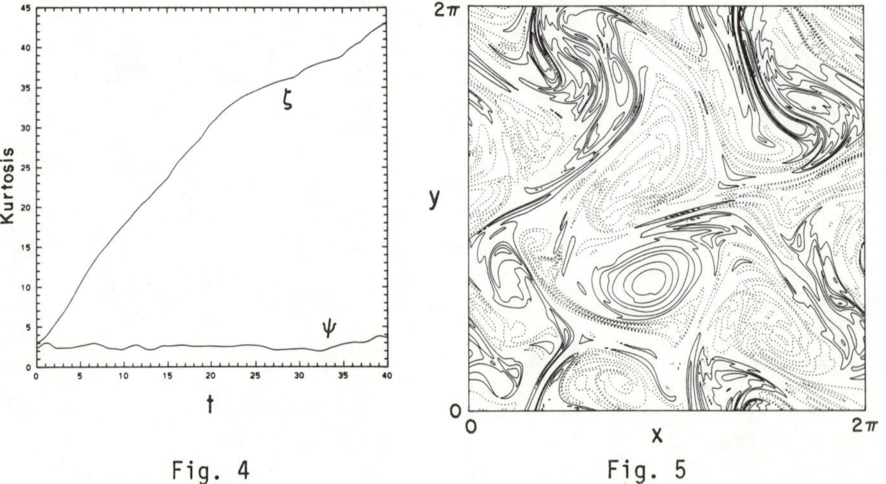

Fig. 4

Fig. 5

Fig. 4. Kurtosis for streamfunction and vorticity.

Fig. 5. Vorticity from an equilibrium solution of (1) with $\gamma = b = \nu_2 = 0.0$, $\nu_0 = 0.1$, $\nu_4 = 5.0 \times 10^{-8}$, and f a white noise forcing in time whose wavenumber spectrum is restricted to k values between 1 and 4. The plotting convention is as in Fig. 3 except that the contour interval is reduced to 1.0 due to the somewhat smaller K and V in this solution.

The properties shown in Figs. 3 and 4 are surprising, certainly to me and I suspect to others as well. The distribution in Fig. 3 can be contrasted with one more "typically turbulent", which is taken from an equilibrium solution which is sufficiently strongly forced and damped so that no vorticity concentrations or non-Gaussian kurtosis develop. The vorticity pattern in this case (Fig. 5) has the elongated contours which is often taken to be the signature of cascading, two-dimensional turbulence.

HOW DO ISOLATED VORTICES EMERGE AND PERSIST?

What follows is a largely impressionistic description of what happens to vorticity extrema in the preceding decay solution. The basis for it is a sequency of maps like those in Fig. 3, too many to be shown here.

Each of the vortices in Fig. 3d, with a magnitude of at least three contours, can be traced backwards in time, without interruption, to the initial conditions. The converse is not true. Individual vortices can cease to be traceable through interactions with other, stronger vortices during which they are either destroyed by being sheared to the point of participating in a local turbulent cascade or absorbed into the stronger vortex. A necessary condition for absorption, or merger, is that they both have the same sign of vorticity. In the initial conditions, the incipient isolated vortices are local extrema in vorticity, which occur by chance in locations sufficiently separated from other extrema of comparable or greater magnitude. As the turbulent cascade begins around them, the incipient vortices resist shearing deformations due to neighboring vorticity structures and grow in circulation (i.e., area-integrated vorticity) by mergers with weaker, like-sign vorticity extrema. Such unequal strength mergers yield an end-product whose exteme vorticity is bounded by the larger of the original extrema (usually that of the partner with greater circulation) and whose circulation is bounded by the sum of the original circulations (varying amounts of the circulation of the weaker partner can be lost to the cascade rather than absorbed). An increase in circulation without an increase in the extremum generally implies an increase in size. An example of an unequal strength merger is shown in the time sequence of vorticity maps in Fig. 6. Here only a small part of the circulation of the weaker partner is absorbed by the stronger; the rest is lost to the turbulent cascade. Examples could be adduced of other mergers which are either less or more conservative of total circulation. After undergoing shearing deformations or mergers, a surviving vortex will relax towards axisymmetry and a smooth radial profile, until it is again disturbed by encountering another vortex. Since vortices can only be lost through these encounters, their number per unit area will decrease. Vortices are lost faster than the size of the survivors grows, which is consistent with the increasing kurtosis of Fig. 4, and consequently the frequency of vortex encounters becomes rarer. Similarly, since smaller vortices are more likely to be the weaker partner--circulation appears to be the relevant measure of strength, although this must be further demonstrated--they are more likely to be destroyed in an encounter than larger vortices; hence, the average size of the survivors tends to increase. The effect, described above, of a

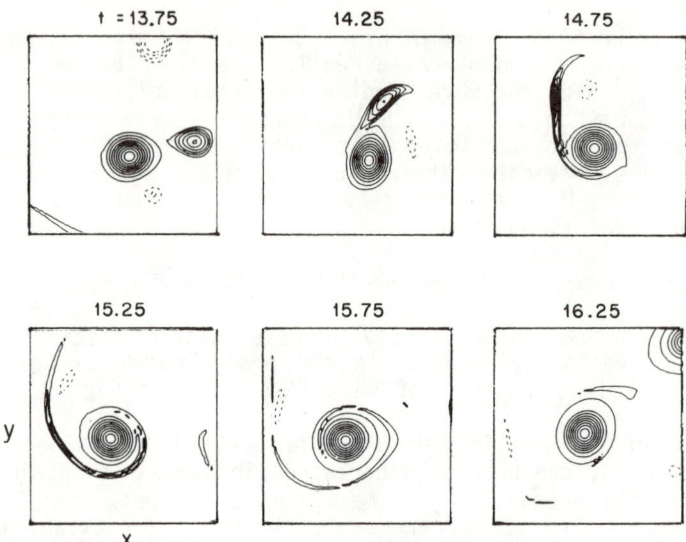

Fig. 6. Vorticity maps in a fraction of the domain, of dimension $\pi/4$, during a period of unequal strength merger. The plotting format is as in Fig. 3d.

sequence of isolated vortex interactions are illustrated by comparing vorticity patterns at different times (Figs. 3d and 7).

Many of the vortex encounters are non-destructive, however. Opposite-sign vortices can temporarily pair as a dipole, and like-sign

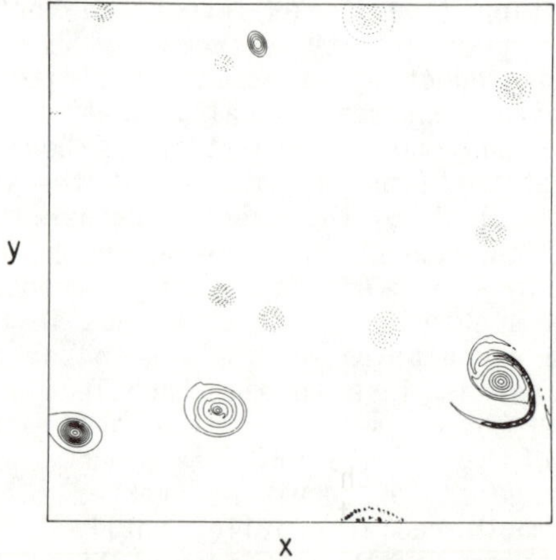

Fig. 7. ζ at t = 37, plotted as in Fig. 3d.

vortices can circle each other, often with large structural deformations, yet subsequently separate and recover their original forms. The most common condition, though, is as an isolated, axisymmetric monopole, and this is increasingly true with time. Away from encounters with other vortices, very little change in vorticity structure occurs, except for very slow weakening and broadening due to the small hyperviscosity.

Fig. 8. $\log_{10} |\zeta|$ at t = 16.5 with contours every 0.25 between -0.5 and 1.5.

Meanwhile, an active turbulent cascade occurs in the (larger) rest of the domain outside of the isolated vortices. Vorticity structures are mutually sheared by their neighbors, and vorticity gradients increase. Energy is transferred to larger scales and enstrophy to smaller scales, where it is efficiently dissipated. Thus, the vorticity which is accessible to turbulent interactions tends to disappear in time, while that which is protected in the isolated vortices persists. This yields the high kurtosis of vorticity. The isolated vortices are latent in the initial vorticity distribution, and turbulence carves them out. While the turbulent component is significantly diminished with time--after all, 94% of the original enstrophy has been lost by t = 40 (Fig. 1b)--it does not entirely disappear. This is illustrated in Fig. 8, which is simply a replotting of Fig. 3d with logarithmic contours in order to expose the weak vorticity structures between the isolated vortices. There the elongated contours have the signature of the turbulent cascade (cf., Fig. 5).

After the clear emergence of the isolated vortices from the disorderly initial vorticity distribution, which occurs around t = 4, a simplified characterization of the solution is that of finite number of point vortices, with constant circulation and no internal degrees of freedom, each of which is passively advected under the influence of the others. The flow is approximated as irrotational except at the points. Whenever some of the isolated vortices approach sufficiently closely, then a point vortex approximation may be invalid since strong deformations of the vortex structures occur (i.e., internal degrees of freedom are excited), and their (generally pairwise) interaction must be calculated in detail. When sufficient separation between vortices subsequently is restored, then the flow evolution again returns to that of point vortices, although possibly with a different number of vortices and with different amplitudes for those which underwent a close interaction. Such an idealization of the dynamics retains the properties of chaos. This is both because the trajectories of point vortices are in general unstable to small perturbations[1], hence even this type of flow is chaotic, and because pairwise isolated vortex

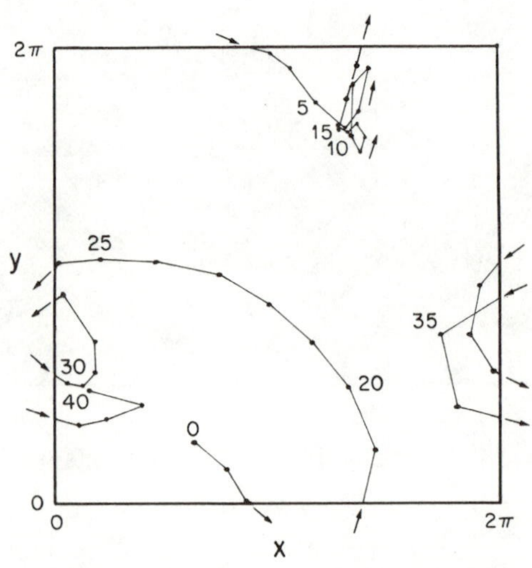

Fig. 9. The locations of a particular vorticity maximum (the strongest one in the upper right quadrant of Fig. 3d) at unit time intervals, connected by straight lines. Some times are labeled.

interactions are likely to be quite sensitive to antecedent conditions as well. On the other hand, it is intuitively plausible that the number of independent degrees of freedom involved in such a characterization, assuming one could usefully define them during vortex interactions, is substantially smaller than the total number of Fourier components, which is the basis for both the present numerical model and most closure theories of turbulence. Presumably this effectively diminished number of degrees of freedom has important implications for understanding the nature of the chaos and the limits of predictability, although I do not know what they might be. Fig. 9 gives an indication of the chaos associated with a particular isolated vortex which endures, with little change in structure, for the entire time of integration: its trajectory is certainly irregular. Occasions of sharp change in direction are associated with close interactions with other vortices, generally of the same sign. Less sharp changes in direction are associated with passive advection by more distant vortices. The period of particularly rapid translation between 18 and 25 is associated with a dipole pairing with an opposite-sign vortex.

ELEMENTS OF DYNAMICAL INTERPRETATION

There is considerable challenge in interpreting the phemomena described above. No very complete interpretation will be offered here, but several dynamical issues, probably relevant, can be raised.

An important characteristic of isolated vortices is their obvious stability. Their long lifetimes, of course, require that even energetic perturbations not induce their break-up. Many studies have been made of the stability of particular types of vortices in two-dimensional flow [2,3], but the full range of possibilities remains unexplored. I offer the hypothesis, partly on the basis of an existing study[4], that nearly axisymmetric vortices are robustly stable to perturbations whose shear, vorticity, and strain rate are small compared to the vorticity of the vortex. This condition is met for the flow in Figs. 3c,d because of the "hard-core" character of the vorticity distribution: the scale of an individual vorticity concentration is small compared to either the typical separation between concentrations or the dominant scale of the streamfunction pattern.

Another important dynamical issue is the nature of vortex interactions. Questions arise as to how many different types of interactions occur, under what conditions, and with what degree of simplicity and predictability. One important type, discussed above and illustrated in Fig. 6, is the merger of like-sign vortices. A number of studies have been made of idealized mergers[5,6], but I believe we still lack a general understanding of how close vortices must come to each other before merger is inevitable and how conservative mergers are in total circulation. Another important type is dipole pairing. For this we have the modon model, and some studies have been made of the dipole formation interaction and the interaction of two dipoles, [7,8].

I have used language in the preceding section suggesting the aptness of a decomposition of the flow into two components, cascading turbulence and isolated, hard-core vortices, both of which manifest chaotic behavior, but of different types. How might this decomposition be made explicitly? One possibility is based upon the relative magnitudes of shear and vorticity. We define the two components of shear as

$$S_1 \equiv \frac{\partial u}{\partial x} - \frac{\partial v}{\partial y} = -2\,\frac{\partial^2 \Psi}{\partial x \partial y}$$

$$S_2 \equiv \frac{\partial v}{\partial x} + \frac{\partial u}{\partial y} = \frac{\partial^2}{\partial x^2}\Psi - \frac{\partial^2}{\partial y^2}\Psi. \tag{4}$$

If an approximation is made that both vorticity and shear are slowly varying compared to the vorticity gradient in a frame of reference following a fluid parcel, then it can be shown[9] that the magnitude of the vorticity gradient will tend to grow in time at an exponential rate whenever

$$Q \equiv S_1^2 + S_2^2 - \zeta^2 \tag{5}$$

is positive. When Q is negative the time evolution is oscillatory. If we accept as an operational definition of cascading turbulence a systematic growth in vorticity gradients, then a region when $Q > 0$ is turbulent, and one where $Q < 0$ is neutral, consistent with the persistence of vorticity patterns in isolated vortices. It can easily be shown that Q must vanish in area integral, and furthermore that its positive and negative terms have identical wavenumber vector distributions (i.e., $\hat{S}_1{}^2 + \hat{S}_2{}^2 = \hat{\zeta}^2$, where the caret denotes Fourier transform). In spite of these identities, however, the domain is not equally partitioned into turbulent and neutral regions. This is because vorticity is very intermittently distributed (Figs. 3d and 4), while shear is much more uniformly distributed. This is illustrated in Fig. 10, where nonuniform contouring has been used to accommodate this difference. One can see that the broader $Q > 0$ regions tend to

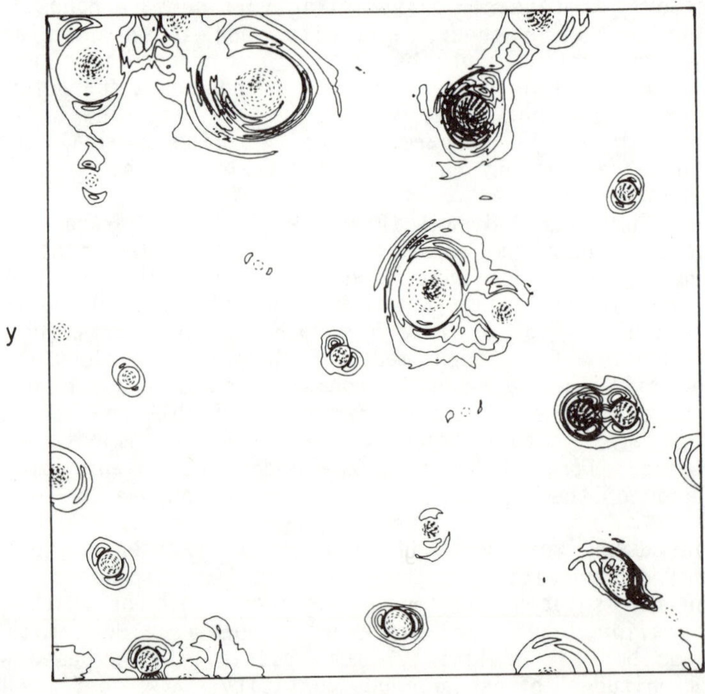

Fig. 10. Q at t = 16.5. $Q < 0$ has dashed contours at an interval of 50., and $Q > 0$ has solid contours at an interval of 10.

surround the $Q < 0$ regions--the isolated vortices (Fig. 3d)--and the strongest maxima in Q are located near strongly interacting vortices. By the preceeding arguments this would imply that the most active turbulence is in the neighborhood of the isolated vortices, but it is excluded from their cores. Note also that there is a substantial region not enclosed in either the positive or negative contours, which would suggest that much of the region is not particularly active, neither as turbulence or isolated vortices. This is because the distribution of shear is also significantly non-Gaussian (Fig. 11), although much less so than vorticity (Fig. 4).

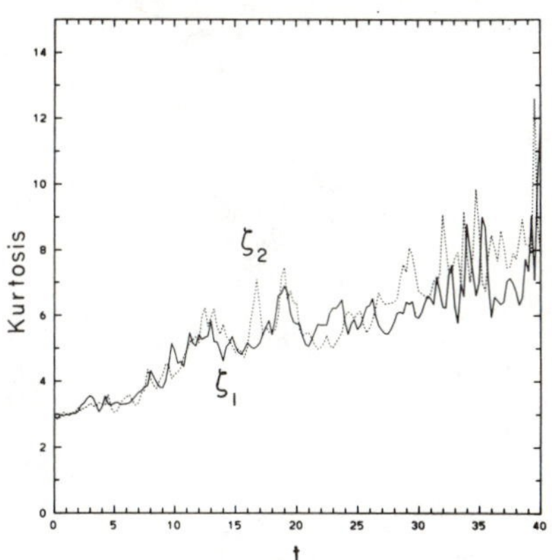

Fig. 11. Kurtosis of the two components of shear (4).

An alternative expression for the instantaneous tendency towards a turbulent cascade, as defined above, can be derived from (2), neglecting friction; viz.,

$$\frac{d}{dt}\,(1/2\ \nabla\zeta\cdot\nabla\zeta) \;=\; P,$$

$$P \;\equiv\; J(\zeta,\underset{\sim}{\nabla\Psi})\cdot\underset{\sim}{\nabla}\zeta, \tag{6}$$

where d/dt denotes the time derivative following a fluid parcel. This relation does not involve any assumptions about slowly varying structure. Its spatial distribution (Fig. 12) is complicated. Its largest magnitudes, with both signs, again occur in the vicinity of th isolated vortices. Several features can be noted by a comparison of Figs 3d, 10 and 12:

* A vortex which has undergone strong deformation and is relaxing back towards axisymmetry (e.g., the negative vortex astride the top an bottom boundaries) has predominantly $P < 0$, which reflects the contracting of vorticity contours. The structure in Q (a negative region) is not particularly informative in the sense of distinguishing this particular vortex from other, more structurally inert, vortices.

* The trailing streamers of vorticity, which represent vorticity being sheared out by a stronger neighboring vortex, are sites of active cascade, as indicated by both P and Q positive. Examples of this are the sites just above the two positive vortices in the upper-right quadrant (the upper one is just finishing the merger shown in Fig. 6) and the weak positive vortex just to the left of the strong positive vortex in the upper left quadrant.

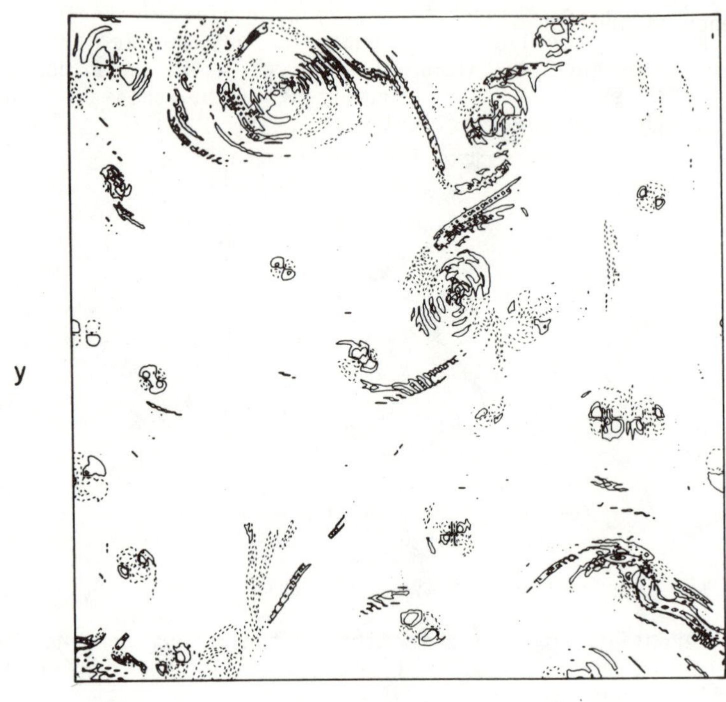

Fig. 12. $\log_{10} |P|$ at t = 16.5, contoured between 4. and 8. with an interval of 1. Positive P has solid contours, and negative P has dashed ones.

* The indications of cascade are particularly strong near the two vortices in the lower-right corner which are just beginning a merger interaction. Later they will again separate as isolated vortices, but each will have lost a fraction of its circulation to the cascade.

* Most of the currently non-interacting vortices, and even some which are interacting, have a quadrapole pattern in P. This indicates an alternation of stretching and contracting of vorticity contours as fluid recirculates. Clearly this is consistent with the "neutral" implication of $Q < 0$. This quadrapole reflects the increase, existence, decrease, or rotation of ellipticity in an isolated vortex, depending upon the relative strength and orientation of the P quadrapole and ζ ellipse.

* There are examples of elongated, weak amplitude vorticity structures, where P is strongly negative and Q is small: in the upper-right corner of the domain and above the contracting, negative vortex in the lower-left. This suggests a tendency towards formation of a weak, axisymmetric, isolated vortex; however, one does not subsequently emerge, and the interpretation of this is unclear to me.

* In both P and Q, there are substantial portions of the domain where the magnitudes are small. Together with the sparseness of the vorticity distribution, a considerable amount of spatial (and temporal) intermittency is indicated.

HOW GENERAL ARE COHERENT VORTICES IN TURBULENCE?

I don't know. However, a moderate number of solutions have been calculated for (1), and some of the parametric influences on the emergence of isolated vortices can be described:

* Their emergence does not depend upon whether a Newtonian viscosity or a hyperviscosity is used to provide dissipation.

* Their emergence does not depend upon the initial E(k).

* Their emergence is not suppressed by a finite radius of deformation, even though the transfer of energy to k smaller than γ is.

* The degree to which they occur is a strong function of Reynolds number and numerical resolution. This is indicated in Table 1, which is based upon a sequence of solutions where NG, the number of spatial grid points, is doubled and the hyperviscosity is multiplied by a factor of 16 (n.b., it has units of length to the fourth power): the final member of the sequence is the one analyzed above. One can see that the vorticity kurtosis is only slightly non-Gaussian at the coarsest resolution, is nearly proportional to NG for intermediate values, and seems to be leveling off at the highest resolution. One reason, I believe, why less has been made of isolated vortices in previous studies of decaying two-dimensional flow is that most numerical solutions have had substantially lower resolution than the one analyzed above. In retrospect, though, it is easy to see manifestations of the vortices in the earlier solutions,[10,11,12].

TABLE 1

NG	Ku(ζ) at t = 10
32	3.7
64	7.8
128	15.0
256	17.3

* In general larger vorticity kurtosis values occur in decaying or adjusting (i.e., rapidly changing E(k)) turbulence compared to equilibrium turbulence. Since only spatially homogeneous solutions to (1) have been calculated so far, this distinction has been demonstrated only for the temporal behaviour, but it is probably correct for spatial behaviour as well. If, for example, forcing were confined to a small fraction of the domain, then isolated vortices might preferentially develop elsewhere.

* In homogeneous, equilibrium solutions, the following parameter regimes favor larger vorticity kurtosis: slow rates for the non-conservative terms (i.e., small f and ν_0), scale separation between the forcing and energy-containing scales, and long correlation times in the forcing. The solution from which Fig. 5 was taken does not have these characteristics, and its kurtosis is approximately 3. Solutions with these characteristics have been calculated with kurtosis as large as 6, and no doubt larger values can be obtained with further exploration.

* Mean gradients of potential vorticity--represented by b in equation (1)--can suppress the emergence of axisymmetric, isolated vortices. Such mean gradients introduce a wave propagation mechanism and spatial anisotropy into the dynamics. For random, broad-band initial conditions, decaying flow evolves to a state of predominantly parallel flow [13] (i.e., u >> v) whenever $b \gtrsim (2K)^{1/2}$. On the other hand it has been shown that an isolated vortex, once present, can persist if $b \ll \zeta_v k_v$, where the subscript "v" denotes a characteristic value for the vortex[14]; for a flow configuration such as Figs. 3c,d, this latter can be a much less stringent condition, allowing b values larger by more than an order of magnitude.

* Baroclinic effects (resulting from the interaction of different vertical modes) on the emergence of isolated vortices are presently unknown. A more general equation than (1) is required to study them.

* There are several other circumstances, beside the ones discussed above, where isolated vortices in turbulent flow have been reported. Among these are the spatial concentration of vorticity in a numerical calculation of a two-dimensional, turbulent shear layer[15], isolated vortices in irregularly forced flow in rotating laboratory tanks[16,17], and long-lived, axisymmetric ocean eddies on scales somewhat smaller than the most energetic currents[18]. The relationship between these phenomena and the above solutions is unclear to me.

AN ALTERNATIVE MODEL FOR THE LIMITS OF PREDICTABILITY

Two-dimensional flow has been used as a model for establishing the limits of practical predictability for large-scale fluid motions in general and atmospheric weather events in particular. A common measure of predictability limits is the rate of growth of an error component in the spectrum, and near-Gaussian closure theories are used to calculate this[19,20]. For such non-Gaussian solutions as the one shown here, one must wonder whether such theories are adequate for the second-moment quantities, such as the spectrum, for which they were developed. Even if they are, the existence of vorticity concentrations with lifetimes long compared to the usual predictability limits suggests that some reconsideration of the measure of predictability may be fruitful.

REFERENCES

1. H. Aref, Annual Review of Fluid Mechanics, 15, 345 (1983).
2. V.I. Arnold, Mathematical Methods of Classical Mechanics (Springer-Verlag, N.Y., 1980).
3. P.G. Saffman, In: Transition and Turbulence, R.E. Meyers, ed. (Academic Press, N. Y., 1980), 149.
4. D.W. Moore and P.G. Saffman, In: Aircraft Wake Turbulence, J.H. Olsen, A. Goldberg, and M. Rogers, eds. (Plenum Press, N. Y., 1971), 339.
5. J.P. Christiansen, J. Comp. Phys., 13, 363 (1973).
6. J.P. Christiansen and N.J. Zabusky, J. Fluid Mech., 61, 219 (1973).
7. J.C. McWilliams and N.J. Zabusky, Geophys. Astrophys. Fluid Dynamics 19, 207 (1982).
8. J.C. McWilliams, Geophys. Astrophys. Fluid Dynamics, in press (1983).

9. J. Weiss, The dynamics of enstrophy transfer in two-dimensional hydrodynamics (La Jolla Institute, La Jolla, 1981).
10. D. Lilly, J. Fluid Mech., 45, 395 (1971).
11. B. Fornberg, J. Comput. Phys., 25, 1 (1977).
12. C. Basdevant, B. Legras, and R. Sadourny, J. Atmos. Sci., 38, 2305 (1981).
13. P. Rhines, J. Fluid Mech., 69, 417 (1975).
14. J. McWilliams and G. Flierl, J. Phys. Oceano., 9, 1155 (1979).
15. H. Aref and E. Siggia, J. Fluid Mech., 100, 23 (1980).
16. A. McEwan, Nature, 260, 126 (1976).
17. E. Hopfinger, F. Browand, and Y. Gagne, J. Fluid Mech., 125, 505 (1982).
18. J. McWilliams et al. (21 authors), In: Eddies and Marine Science, A. Robinson, ed. (Springer-Verlag, 1983).
19. C. Leith, J. Atmos. Sci., 28, 145 (1971).
20. C. Leith and R. Kraichnan, J. Atmos. Sci., 29, 1041 (1972).

COHERENT STRUCTURES IN PLANETARY FLOWS AS SYSTEMS ENDOWED WITH ENHANCED PREDICTABILITY

Paola Malanotte Rizzoli
Massachusetts Institute of Technology, Cambridge, MA 02139

ABSTRACT

The effects of nonlinearity in geophysical flows are generally modeled from the viewpoint that the motion is turbulent. In turbulence theoreis, nonlinear interactions, through the transfer of energy among the various scales of motion, cause the system to lose the memory of the initial state. This randomness postulate is violated by coherent structures, namely, solitary eddies. For them, nonlinearities play the opposite role of preserving phase correlations against the effects of dispersion. Coherent structures are therefore completely predictable systems. Model equations relevant for geophysical flows thus admit a well-defined parameter range in which the dynamics is dominated by orderly, coherent structures much more than predicted by any turbulence theory. This range is broad and the structures robust. In this nonlinear dynamical range, the permanent waves, solutions of the chosen model, are endowed with enhanced predictability, theoretically infinite in the dissipationless, unperturbed state of motion; and more generally, every eddy structure having physical space and Fourier properties similar to those of the exact, nonlinear solutions is endowed with enhanced predictability with respect to random flow evolutions obeying a turbulence dynamics.

INTRODUCTION

In recent years, a renewed interest has been focused on finding permanent form solutions to the model equations suitable for mesoscale motions in the ocean and the atmosphere. This interest has been originated by the accumulated experimental evidence for the production of coherent, long-lived structures and their robustness in the considered geophysical environment. The most important example is given by the Gulf Stream rings, whose formation, evolution and decay have been extensively documented by experiments like MODE and POLYMODE. Features of the synoptic description of the energetic eddy currents of the Gulf Stream and Ring System are emerging from this data set. In Richardson et al.[1] a remarkable "weather map" is shown for this region (their Fig. 1a), constructed from data composited over four months, roughly corresponding to a week of "atmospheric" time. This map shows nine cold-core cyclonic rings to the south and three warm-core anticyclonic rings north of the meandering current. This newly accumulated evidence suggests new ideas about ring

0094-243X/84/1060223-23 $3.00

dynamics. Until a few years ago, cold-core rings were thought to decay slowly in the Sargasso Sea, while gradually drifting to the southwest. Now the picture involves vigorous, multiple ring-stream and ring-ring interactions. After the formation, rings move generally westward (upstream) and make contact with the current again. The resulting interaction can involve advection, partial or total coalescence, permanent absorption (= ring extinction), or subsequent downstream rebirth of the ring.

During the POLYMODE Local Dynamics Experiment (LDE), a new major finding occurred, namely, the discovery of "small" mesoscale features.[2] These features, now known to exist in the upper or mid-thermocline or deep water, have radii on the order of several tens of km and extend vertically for a few hundred to almost a thousand meters. Their most remarkable aspect is their very long lifetime, as indicated by the anomalous properties of the water in their cores compared to that of the surrounding sea (see McDowell and Rossby[3] for the first of such small-scale features, observed prior to the LDE experiment). If we want to mention more "exotic" examples of long-lived coherent structures, we can even think of blocking ridge phenomena in the earth's atmosphere[4] or even go as far as Jupiter, with its Great Red Spot!

This accumulated evidence has produced the renewed interest in nonlinear, permanent-form solutions as possible models for these mesoscale, long-lived structures. Usually, in fact, geophysical fluid dynamicists resort to linearized analytical theories for infinitesimal amplitude motions or to turbulence theories, and numerical experiments, for the highly nonlinear ones. Turbulence theories bear upon the random phase approximation, and are couched in terms of flow statistics, rather than in terms of individual flow realizations. They introduce the fundamental idea of nonlinear memory loss: Fourier components "forget" the phase they were given at the instant of flow initiation. In weather prediction, this randomness postulate leads to the so-called predictability problem. In it, any forecast of how the flow evolves becomes practically useless after a finite time, the predictability time, which cannot be extended by any improvement of the initial data, short of absolute perfection.

This randomness postulate valid for turbulence theories and statistical equilibrium is however not universally obeyed. Recognition of this fact until recently rested only on the behavior of a wide class of one-dimensional, nonlinear systems. Among them, the Fermi-Pasta-Ulam (FPU)[5] model is the most famous example. The model consists of a chain of mass-point oscillators, coupled by nonlinear springs. FPU[5] wanted to investigate the approach to thermal equilibrium in the anharmonic lattice. Giving an initial condition in which energy was concentrated in one single Fourier mode, they expected the nonlinear interactions to excite higher modes until, eventually, the system would evolve towards a state of energy equipartition. Rather, they found that the system exhibited long time recurrences for which, at the recurrence time, all the energy went back to the initially excited Fourier mode. This lack of randomness was explained by Zabusky and Kruskal[6] and Zabusky.[7] They related the discretized lattice to a continuum model, the Korteweg-de Vries (KdV) equation, the solutions of which are the well-known cnoidal and solitary waves. A smooth, initial condition, like a cosine wave, left to evolve in the KdV equation, breaks up into a series of well-defined and localized pulse-like shapes, freely streaming through one another and re-emerging from the collisions unaltered in shapes, speeds and amplitudes. These nonlinear waves are the solitons, which were found to emerge from quite general initial conditions. The FPU recurrence time is the time at which all the solitons emerging from the given initial condition focus together, reproducing a state almost identical to the initial one. The soliton, being a wave of permanent form, in which nonlinearity is balanced by dispersion, is a completely predictable nonlinear structure.

In an analogous way to these one-dimensional systems, in two and three dimensions the randomness postulate is violated by the nonlinear, permanent form solutions of the same model equations used in turbulence theories. For them, nonlinearities play the opposite role of preserving phase correlations against the effects of dispersion. In a dissipationless model, these permanent waves are, in principle, completely predictable systems. (For the possible bearing of these permanent waves upon the predictability problem see, for instance, Leith.[8])

The series of studies appeared in the recent literature by different authors have mostly concentrated in finding these nonlinear solutions of the model equations for mesoscale oceanic and atmospheric motions under different approximations. For an up-to-date review on the topic, see Malanotte-Rizzoli.[9] The most important recent works have dealt with 1) the stability (or predictability) of the various types of coherent structures under superimposed perturbations in the dissipationless system; 2) the effects of dissipation upon them in the unforced model.[9,10,11,12]

Thus, in Section 2 a specific model of geophysical significance is chosen allowing for permanent form, nonlinear wave solutions in a specific parameter range. The emergence of the KdV dynamics from the chosen model is investigated in the context of the initial value problem posed by it. It is shown that, in the considered parameter range, a single free wave initial condition will evolve according to the KdV equation. The study is generalized to 2--and in principle n--interacting nonlinear waves. Their evolution is governed by coupled KdV equations.

In Section 3 specific examples are shown of predictability experiments carried out upon some of the permanent solutions allowed by the model. In these predictability experiments we follow the procedure used by Lilly[13,14,15] and we define a measure of predictability. A comparison predictability experiment is carried out upon a structure which has an isotropic Fourier spectrum similar to that of the high-amplitude permanent wave previously studied. This structure, however, is not a nonlinear solution of the model. The results show that the permanent form solutions are indeed endowed with enhanced predictability even with respect to different, but similar, initial conditions.

In Section 4, we summarize recent results.[9] Specifically, we explore the extension of the KdV dynamical range. The transitions to different types of dynamics are investigated, namely, the transition to turbulence under the forcing produced by a superimposed perturbation; and the transition to linear dynamics under the effects of dissipation. The important conclusion is that, in the category of mesoscale motions, there exists a scale range in which the dynamics is dominated by orderly, phase-coherent structures endowed with remarkable stability.[16,9] In this range, the dynamics is fully nonlinear but predictable.

Finally, in Section 5 we draw conclusions and outline some of the important questions for future research.

THE NONLINEAR (KdV) DYNAMICAL RANGE

Permanent form, nonlinear solutions to model equations describing mesoscale oceanic motions can be distinguished into two broad classes: I) analytic; II) multi-valued. Analytic solutions are those for which the potential vorticity can be expressed as an analytic functional of the stream-function (in the frame moving with the permanent wave). Multi-valued solutions are those for which this functional has a given shape in the "exterior" region, where streamlines extend from $-\infty$ to $+\infty$, which can be determined by the far field potential vorticity structure. The same functional has a different shape in the "interior" region, characterized by closed streamlines. These solutions are also called "modons." At a limiting closed streamline, the modon diameter, high derivatives of the field functions may present finite discontinuities.

Solutions relevant for geophysical flows were found I) in the category of analytic solutions by Long[17], Benney[18,19], Clarke[20], Maxworthy and Redekopp[21,22], Redekopp[23], Husuda[24], Boyd[25,26], Flierl[27], Charney and Flierl[16], and Malanotte-Rizzoli[28,29,30]; II) in the multi-valued (modon) category by Ingersoll[31,32], Stern[33], Larichev and Reznik[34], Deem and Zabusky[35], Berestov[36], and Flierl, Larichev, McWilliams and Reznik.[37] For a review, see Malanotte-Rizzoli.[9]

The permanent waves studied in this paper belong to the analytic category. Precisely, they constitute a family of nonlinear solutions (cnoidal or solitary waves) of the barotropic, quasi-geostrophic potential vorticity equation over variable relief:

$$\nabla^2\psi_t + J(\psi,\nabla^2\psi + f_o h) = 0 \tag{2.1}$$

Model (2.1) has been chosen because, although simple, its solutions range from linear waves at infinitesimal amplitude, to form-preserving nonlinear solutions at finite amplitude and dispersion; to two-dimensional turbulence over bottom relief. The form-preserving, nonlinear solutions are characterized by a small aspect ratio $\delta_1 = L_1/L_2$, if L_1 is the cross-channel wavelength or width of the wave guide; L_2 is the axial wavelength. Thus, they are asymmetric eddies in the zonal channel. They can be found both in the weak amplitude limit, $U \ll c$, if U = particle speed and c = phase speed. In this case,

$$\varepsilon = U/c = \frac{U}{(f_o \frac{|\Delta d|}{H} L_1)} \quad ,$$

the Rossby number that is the small amplitude parameter, is chosen such as $\varepsilon = \delta_1{}^2$.[28,10] They can also be found in the strong amplitude limit $U \geq c$, in which case $\delta_1{}^2$ is the small parameter of the system.[29,30] They are barotropic, but baroclinicity can be easily included, or adding a divergent term to obtain an equivalent barotropic model equation; or solving the complete quasi-geostrophic potential vorticity equation with arbitrary stratification $N^2(z)$, N^2 being the Brünt-Väisälä frequency of the system, over variable relief (Malanotte Rizzoli, unpublished results). For a complete review, see Malanotte-Rizzoli.[9]

To show how the KdV dynamics emerges from model (2.1) we choose to solve the initial value problem posed by it in the above parameter range. The basic dimensionless small parameter of the system is the aspect ratio $\delta_1{}^2$. The initial value problem can then be solved in the two limits:

a) Weak amplitude waves

ε = Rossby number = $\delta_1{}^2 \ll 1$; general topography $h(y)$

b) High amplitude waves

ε = Rossby number = 1; $U \geq c$; $\delta_1{}^2 \ll 1$

Quasi-linear topography $h(y) = \beta y + \delta_1{}^2 h^*(y)$

We show the procedure for the weak amplitude case a). This is analogous for case b) and can be used in whatever starting model. The final evolution equation, in both cases (a) and (b), is the KdV equation, which is also the model one obtains in the vast majority of cases of geophysical significance.

In the zonal channel, for weak asymmetric waves, with the scaling previously outlined, (2.1) becomes:

$$\psi_{yyt} + h_y \psi_x + \delta_1{}^2 \psi_{xxt} + \varepsilon\,(\psi,\psi_{yy}) + \varepsilon\delta_1{}^2\,(\psi,\psi_{xx}) = 0 \quad (2.2)$$

with $\delta_1{}^2 = L_1{}^2/L_2{}^2$ aspect ratio of the wave

$$\varepsilon = \frac{U}{(f_o \frac{|\Delta d|}{H}) L_1} = \frac{U}{c}$$

small amplitude parameter; H = mean depth; $|\Delta d|$ = small depth variation around the mean value.

Equation (2.2) admits permanent form solutions (steady) in the form:

$$\psi \quad \begin{matrix} A\,\mathrm{cn}^2(Bs)\phi(y) & \text{periodic} \\ A\,\mathrm{sech}^2(Bs)\phi(y) & \text{solitary} \end{matrix}$$

where $\phi(y)$ is the solution of the zero order Sturm-Liouville problem

$$\phi_{nyy} - \frac{1}{c_{on}} h_y \phi_n = 0 \qquad \phi_n(0) = \phi_n(2\pi) \tag{2.3}$$

We introduce the operator:

$$L = \partial_t(\partial_{yy}) + h_y \partial_x$$

in terms of which (2.2) becomes:

$$L\psi = -\delta_1{}^2 \partial_t(\partial_{xx})\psi - \varepsilon(\psi_x\partial_y - \psi_y\partial_x)(\partial_{yy})\psi - \varepsilon\delta_1{}^2(\psi_x\partial_y - \psi_y\partial_x)(\partial_{xx})\psi \tag{2.4}$$

Expanding:

$$\psi = \psi_1 + \varepsilon\psi_2 + \delta_1{}^2\psi_3 + \ldots$$

the zero-order problem is:

$$L(\psi_1) = 0(\varepsilon, \delta_1{}^2)$$

with $\psi_1 = A(x,t)$. $\phi(y)$ and $\phi(y)$ solution of (2.3). Then the zero-order problem becomes:

$$A_t + cA_x = \varepsilon N(A) + \delta_1{}^2 D(A) \tag{2.5}$$

where $N(A)$ and $D(A)$ are, respectively, linear and nonlinear operators acting on $A(x,t)$. We follow now a procedure originally introduced by Benney.[18] Inserting the expansion for ψ into (2.4), taking into account (2.3) and (2.5); equating equal powers of ε and $\delta_1{}^2$, the equations for ψ_2 and ψ_3 are obtained:

$$L\psi_2 = -(\psi_{1x}\partial_y - \psi_{1y}\partial_x)(\partial_{yy})\psi_1 - N(A)(\partial_{yy})\phi \tag{2.6a}$$

$$L\psi_3 = -\partial_t(\partial_{xx})\psi_1 - D(A)(\partial_{yy})\phi \tag{2.6b}$$

Choosing

$$D(A) = sA_{xxx} \qquad N(A) = 2rAA_x$$

and

$$\psi_2 = f(y)A^2 \qquad \psi_3 = g(y)A_{xx}$$

(2.6a,b) give the equations to be satisfied by $f(y)$, $g(y)$:

$$(\partial_{yy})f - \frac{1}{c}h_y f = \frac{h_{yy}}{2c^2}\phi^2 + \frac{rh_y}{c^2}\phi \tag{2.7a}$$

$$(\partial_{yy})g - \frac{1}{c}h_y g = (s\frac{h_y}{c^2} - 1)\phi \tag{2.7b}$$

The solvability condition required for boundedness of ψ_2, ψ_3, applied to (2.7a,b), gives:

$$s = \frac{c^2\int_0^{2\pi}\phi^2 dy}{\int_0^{2\pi} h_y \phi^2 dy} \quad ; \quad r = -\frac{\int_0^{2\pi} h_{yy}\phi^3 dy}{2\int_0^{2\pi} h_y \phi^2 dy}$$

and (2.5) becomes the evolution equation for $A(x,t)$:

$$A_t + cA_x = \varepsilon 2rAA_x + \delta_1{}^2 sA_{xxx} \tag{2.8}$$

namely, the KdV equation allowing for slow time evolution.

This procedure can be generalized to n-interacting solutions. For an initial state given by two of them, we put:

$$\psi = A_1(x,t)\phi_1(y) + A_2(x,t)\phi_2(y) \tag{2.10}$$

for which the initial value problem can be solved to give, with $\varepsilon = \delta_1{}^2$:

$$A_{1t} + c_1 A_{1x} = \varepsilon[s_1 A_{1xxx} + 2r_1 A_1 A_{1x} + \nu_1 A_2 A_{1x} + \lambda_1 A_1 A_{2x}]$$

(2.11)

$$A_{2t} + c_2 A_{2x} = \varepsilon[s_2 A_{2xxx} + 2r_2 A_2 A_{2x} + \nu_2 A_1 A_{2x} + \lambda_2 A_2 A_{1x}]$$

namely, two coupled KdV equations, allowing for mutual interactions of the two permanent waves (see Malanotte-Rizzoli[9] for details).

That model (2.1) indeed obeys the KdV approximate dynamics represented by (2.8) and (2.11) has been shown in a series of numerical experiments, both for the single wave initial condition and for the two interacting modes condition given by (2.10) (collision experiments). Model (2.1) obeys the KdV dynamics for asymmetric waves ($\delta_1{}^2 \ll 1$) also in the high-amplitude limit ε = Rossby number $\to 1$. The reader is referred to Malanotte-Rizzoli[11] for the single wave numerical experiments and to Malanotte-Rizzoli[9] for the collision experiments.

PREDICTABILITY EXPERIMENTS

In this section we show results of predictability experiments carried out upon the permanent form solutions of model (2.1) both in the weak and high amplitude limits.

Predictability experiments are carried out as in Lilly.[13,14,15] Specifically, call run (1) the numerical experiment in which a basic state initial streamfunction $\psi^{(1)}$ is allowed to evolve in model (2.1). $\psi^{(1)}$ is one of the permanent solutions of the model itself. Run (2) is then the experiment in which an error streamfunction $\delta\psi$ is superimposed to $\psi^{(1)}$ and the full streamfunction $\psi^{(2)} = \psi^{(1)} + \delta\psi$ is allowed to evolve. The total energy of the error $\delta\psi$ is initially several orders of magnitude smaller than the total energy of the coherent structure $\psi^{(1)}$. The error time evolution is then given observing:

$$\delta\psi = \psi^{(2)} - \psi^{(1)}$$

If $E(\kappa,\theta)$ is the two-dimensional energy spectrum of $\delta\psi$ in the polar wave-number space, we define as a measure of predictability the total energy of the error $\varepsilon(t)$, normalized with respect to the total energy of the basic state:

$$\varepsilon(t) = \int_0^{\infty}\int_0^{\pi} E(\kappa,\theta)\kappa d\kappa d\theta / E_{total}[\psi^{(1)}] \tag{3.1}$$

The time evolution of $\varepsilon(t)$ will then indicate the predictability of the chosen basic state.

We also define a predictability time as given by:

$$T^* = \frac{L}{\mu_{rms}} \tag{3.2}$$

that is, an eddy-turnover time if L is the average eddy diameter and μ_{rms} the associated root-mean square velocity which is a measure of the particle speed. Equation (3.2) in turbulence experiments is the average time in which phase-correlations are lost.

The first experiment is carried out upon a weak-amplitude solution of model (2.1). Precisely,

$$\psi^{(1)} = A \operatorname{sech}^2[B(x - \pi)]\phi_2(y) \tag{3.3}$$

if $\phi_2(y)$ is the second eigenmode over the topography $h(y) = -\sin(2y)$. For the basic state

$$|A| = 0.02 \qquad B = 0.53 \qquad \mu_{rms} = 0.02 \qquad \zeta_{rms} = 0.07$$

The evolution of $\psi^{(1)}$ in model (2.1) is shown in Fig. 1a,b, respectively at t = 0 (Fig. 1a) at after 4 basin traversals (Fig. 1b) at dimensionless time T = 160.

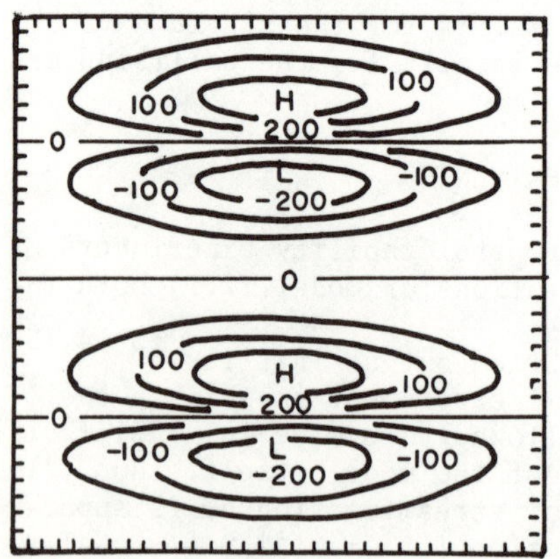

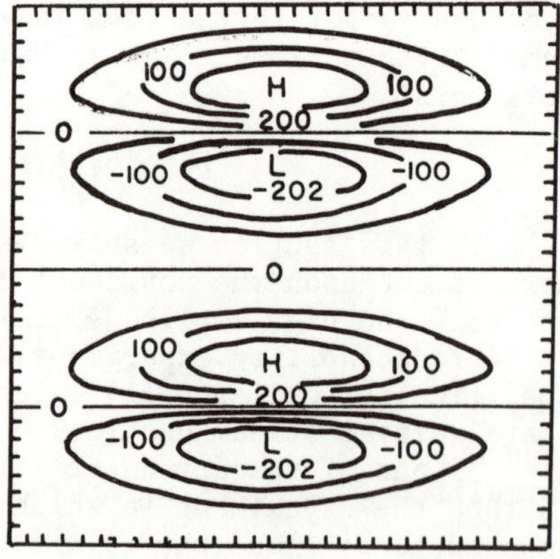

Fig. 1. Evolution of the initial condition (3.3) in model (2.1) at the initial time T = 0 (1a) and after 4 basin traversals, at dimensionless time T = 160 (1b). Contour interval 0.005. Field scaled by 10000.

Modal energies remain constant during the evolution and Fourier phases of the different modes are locked. The error $\delta\psi$ for run (2) is chosen to be a random perturbation (white noise) of $\psi^{(1)}$ at all mesh points, uncorrelated with the basic state. $\delta\psi$ is obtained through a random number generator and its scalar energy spectrum is proportional to κ^3. The error characteristics are:

$$|\delta\psi| = 10^{-1}|A| \qquad \mu_{rms} = 0.006 \qquad \zeta_{rms} = 0.01$$

$$\underset{\text{normalized}}{\varepsilon_{total}|_{t=0}} = 7.11 \times 10^{-2}$$

Fig. 2a,b shows the perturbed stream-function field $\psi^{(2)} = \psi^{(1)} + \delta\psi$, respectively at t = 0 (Fig. 2a) and after 4 basin traversals, at T = 160 (Fig. 2b).

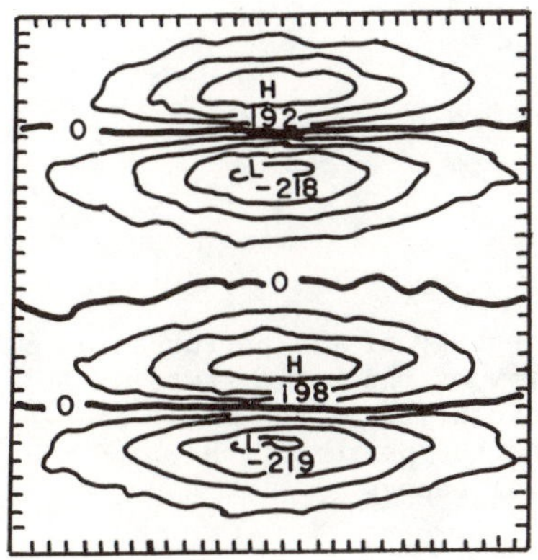

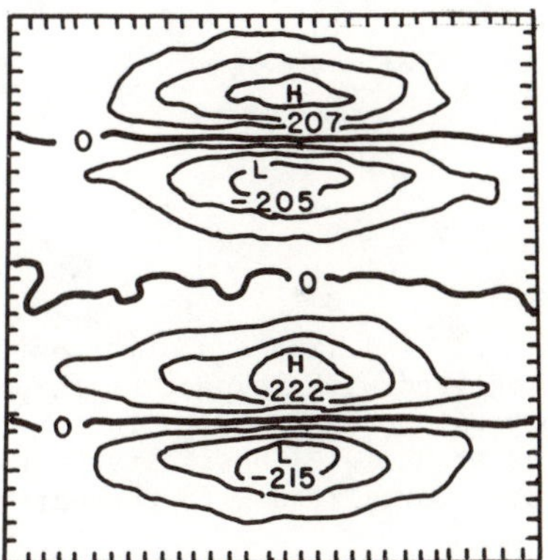

Fig. 2. Evolution of $\psi^{(2)} = \psi^{(1)} + \delta\psi = A[\text{sech}^2(B(x - \pi))\phi_2(y) + 0.1\delta\psi]$ at dimensionless time T = 0 (3a) and after 4 basin traversals, at T = 160 (3b). The field is scaled by 10000.

Figure 3a shows the time evolution for the total energy error $\varepsilon(t)$, normalized with respect to the total energy of the basic state $\psi^{(1)}$. The error energy increases until reaching a saturation value around which it then oscillates. The total growth of the perturbation energy is of about 30% of its initial value. Figure 3b gives the corresponding time evolution of the Fourier modal energies of the scalar spectrum $E(\kappa)$, respectively for the basic state (κ = 2,4) and the error (κ = 5,7,8,9). The coherent structure Fourier modes undergo short, correlated oscillations. The error modes which grow slightly are at intermediate wave numbers. The corresponding loss of energy from the basic state is also borne essentially by the intermediate Fourier modes of the basic state $\psi^{(1)}$. (Figures 1a,b; 2a,b; and 3a are taken from Malanotte-Rizzoli.[11])

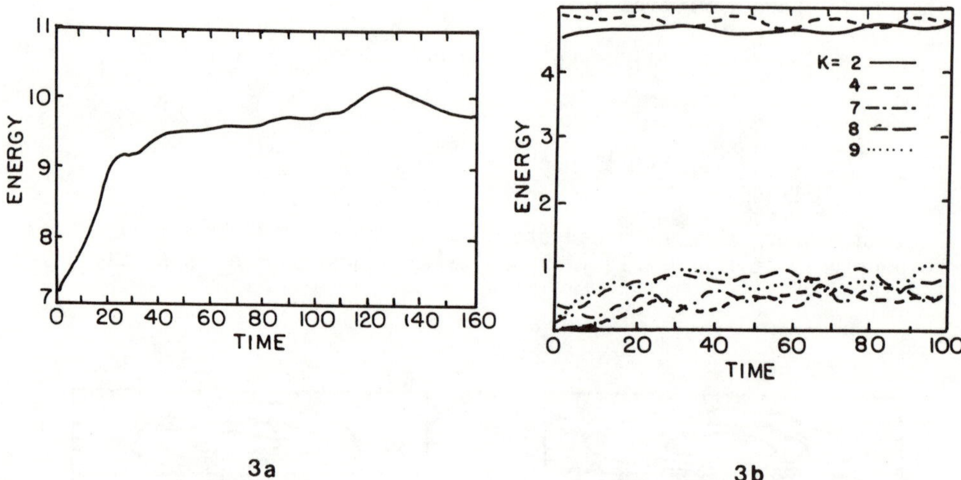

Fig. 3. (a) Time evolution of the total energy of the perturbation $\delta\psi$ normalized with respect to the total energy of the basic state $\psi^{(1)}$. Vertical scale in units 10^{-2}. (b) Time evolution of the modal energies of the $\psi^{(1)}$ basic state (K = 2,4) and the perturbation $\delta\psi$(K = 5, 7,8,9) normalized with respect to the total energy

For the high-amplitude solution, we choose:

$$\psi^{(1)} = A \operatorname{sech}^2[B(x - \pi)]\phi_1(y) \tag{3.4}$$

if $\phi_1(y)$ is the lowest eigenmode over

$$h(y) = -\sin y - \sin 2y$$

Initially $|A| = +1$; $B = 0.2$. This basic state allowed to evolve into (2.1) is known to adjust itself to the permanent form, high-amplitude solution of (2.1). When the appropriate permanent shape is reached, its properties are:

$$|A| = 1.2 \qquad \mu_{rms} = 0.61 \qquad \zeta_{rms} = 0.9$$

Fig. 4a,b show the eddy when it has reached its permanent shape (Fig. 4a), after almost one basin traversal (T = 9) and after 3 successive basin traversals (T = 37). One basin traversal is completed in T = 10 (from Malanotte-Rizzoli[11]).

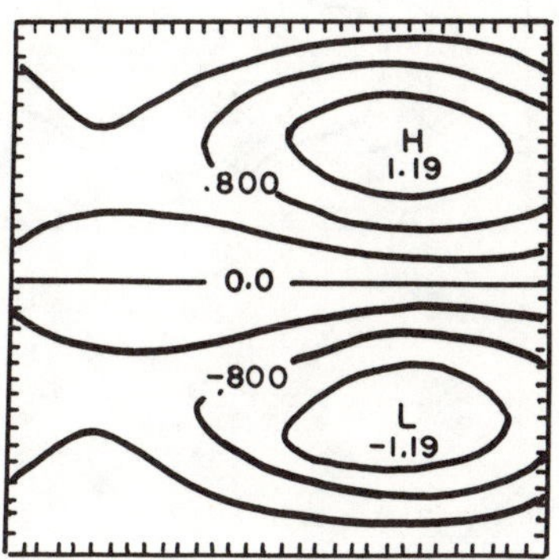

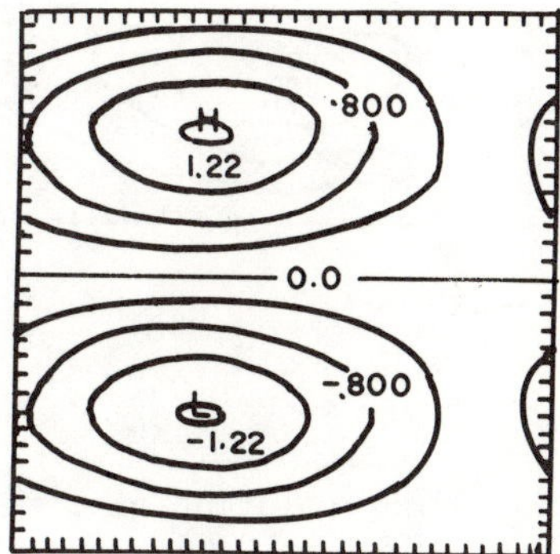

Fig. 4. Time evolution of $\psi^{(1)} = A\,\mathrm{sech}^2[B(x - \pi)]\phi_1(y)$ when it has reached its permanent shape at T = 9 (5a) and after 3 successive basin traversals at dimensionless time T = 37 (5b). Contour interval 0.3. Actual field units are shown.

$\delta\psi$ is again chosen to be a white noise error. Specifically, at time T = 0

$$|\delta\psi| = 10^{-2}|A| \qquad \mu_{rms} = 0.03 \qquad \zeta_{rms} = 0.35 \tag{3.5}$$

Run (2) is begun at T = 10, when the eddy has reached its permanent shape. Fig. 5a,b show the perturbed streamfunction $\psi^{(2)} = \psi^{(1)} + \delta\psi$ respectively at T = 0 (Fig. 5a) and T = 30 (Fig. 5b), after 3 basin traversals.

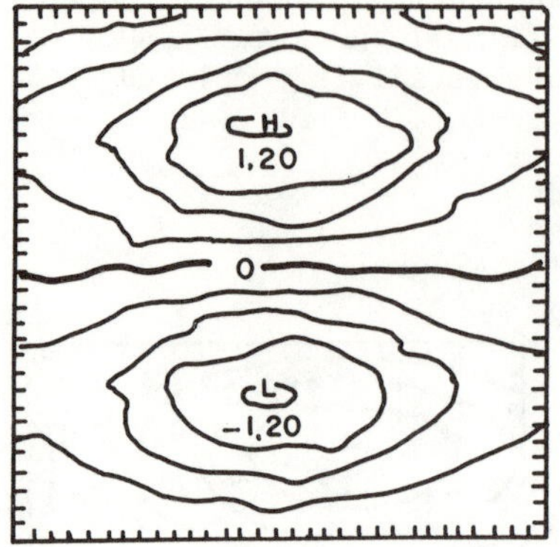

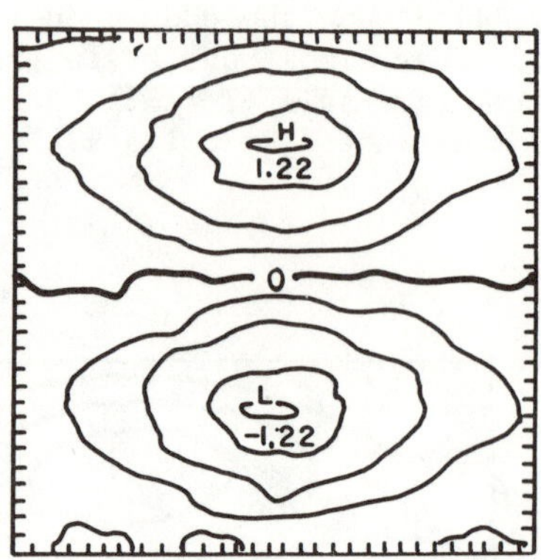

Fig. 5. Time evolution of $\psi^{(2)} = \psi^{(1)} + \delta\psi = A\ \mathrm{sech}^2[B(x - \pi)]\phi_1(y) + 10^{-2}\delta\psi$ at time T = 0 (5a) and after 3 basin traversals, at T = 30 (5b). Contour interval 0.3. Actual field units are shown.

Figures 6a,b show the time evolution of the scalar Fourier spectrum respectively for the basic state, the permanent wave (Fig. 6a) and the error $\delta\psi$ (Fig. 6b) in the perturbation experiment. Remember that, in the unperturbed run (1) the scalar Fourier spectrum of the permanent wave is constant in time. As evident from Fig. 6a,b the most energetic modes K = 1,2,3 of the permanent wave undergo time-oscillations in their energy content. Only the intermediate wave numbers of the error $\delta\psi$ grow slightly, with corresponding energy loss from the intermediate wave numbers of the basic state. The behavior is that of a completely predictable structure.

Two comparison predictability experiments were carried out. In the first, the eddy streamfunction $\psi^{(1)}$ was chosen to have a very similar shape in physical space to that of the permanent wave and, correspondingly, an essentially identical scalar Fourier spectrum. This eddy, however, is not a permanent form solution of model (2.1). The streamfunction for the basic state is now in fact given by:

$$\psi^{(1)} = A \frac{1}{1 + (x - \pi)^2/50} \cdot \phi_1(y) \qquad (3.6)$$

where $\phi_1(y)$ is again the lowest eigenmode over $h(y) = -\sin y - \sin(2y)$ and

$$|A| = 1 \qquad \mu_{rms} = 0.64 \qquad \zeta_{rms} = 0.84$$

The fact that $\psi^{(1)}$ is not a permanent form solution of (2.1) is evidenced by its time evolution. Fig. 7a,b shows the physical space pattern of (3.6) respectively at T = 0 and T = 30. As well as the physical space picture, the time evolution of its scalar energy spect-

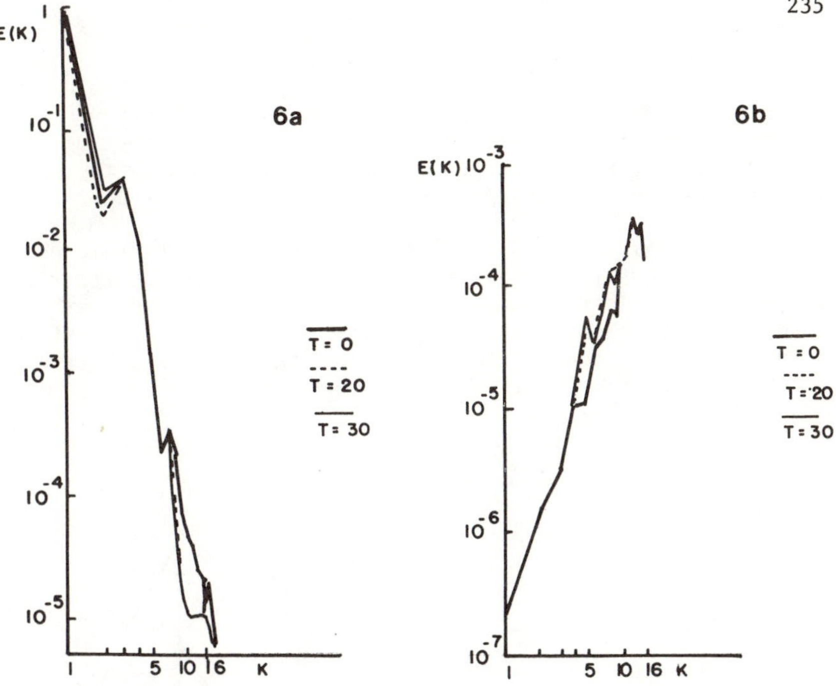

Fig. 6.(a) Time evolution of the scalar Fourier spectrum of the basic state $\psi^{(1)}$ in the perturbation experiment (2) at the labeled dimensionless times. Modal energies are normalized by the total energy. (b) Time evolution of the scalar Fourier spectrum of the perturbation $\delta\psi$ in the perturbation experiment (2) at the labeled dimensionless times. Modal energies are normalized by the total energy.

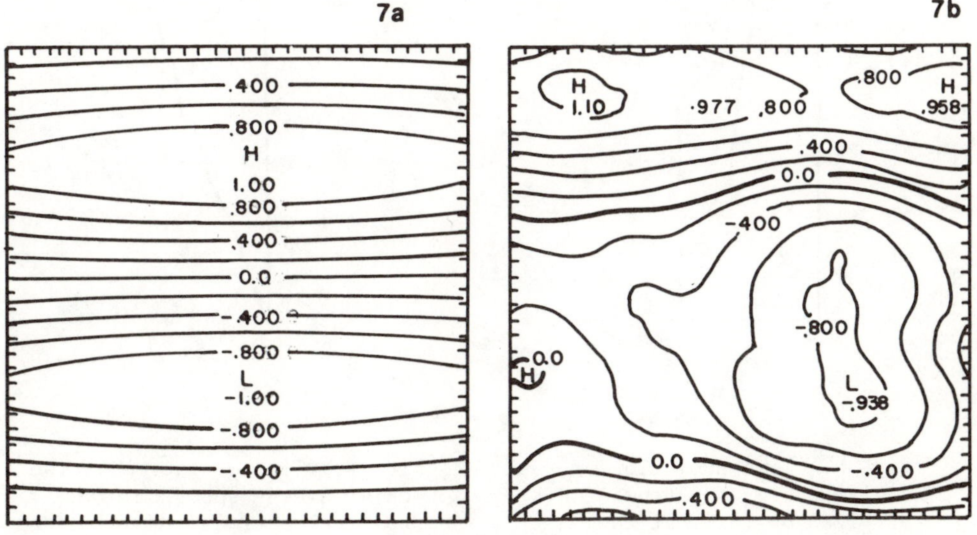

Fig. 7. (a) Physical space pattern of (3.6) at T = 0). (b) Physical space pattern of (3.6) at T = 30.

rum also shows that the eddy is not a permanent solution of the model. Figure 8 gives this time evolution for E(K): the spectrum, initially very "red," begins to flatten out with energy flowing to the high wave numbers. The major energy loss is borne by the most energetic mode K = 1.

$\psi^{(1)}$ as given by (3.6) is then perturbed by the same perturbation (3.5). Figure 9 then shows the time evolution of the error scalar energy spectrum $\varepsilon(K)$. Its behavior is that typical of turbulence predictability experiments: the error spectrum fills up at low and intermediate K's tending towards a spectral shape similar to the one of the basic state itself.

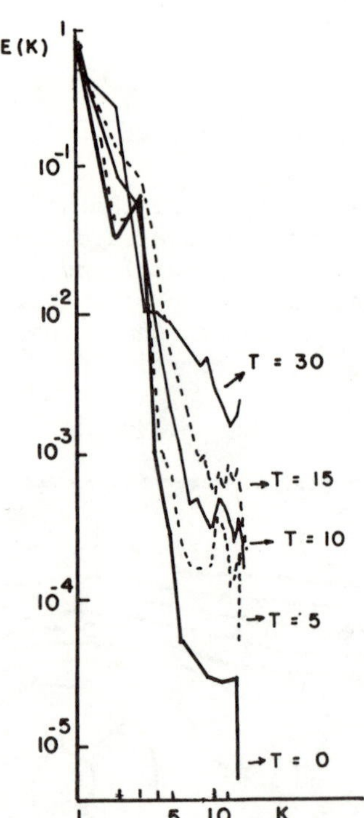

Fig. 8. Time evolution of the scalar energy spectrum of E(K) as given by (3.6) at the labeled times. Modal energies are normalized by the total energy of (3.6).

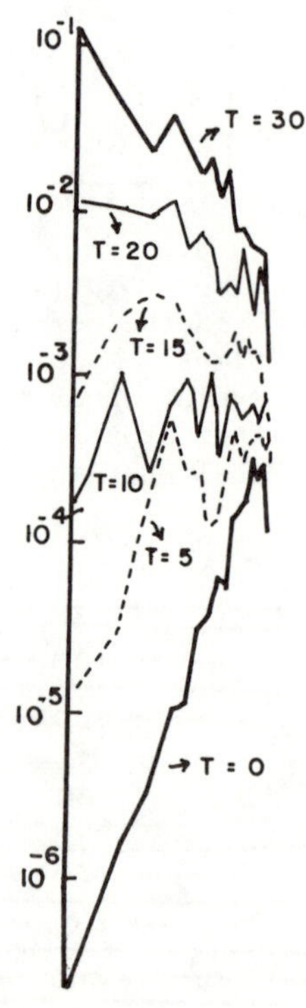

Fig. 9. Time evolution of the scalar energy spectrum of $\varepsilon(K)$ at the labeled times. Modal energies are normalized by the total energy of (3.6).

Differently from a fully turbulent field, however, the spectral evolutions of Figs. 8,9 towards turbulence seem to occur on a longer time scale. Equation (3.6) is not a permanent nonlinear solution of the model, and cannot be represented as a simple superposition of the nonlinear solutions (the cnoidal or solitary waves). The possible asymptotic evolution of (3.6) towards a superposition of different solitary waves, each of them having a different cross-channel structure, could be checked only by observing possible Fermi-Pasta-Ulam recurrences in the Fourier spectrum. This would, however, require running the numerical experiment for a much long time, of which no a priori estimate could be theoretically obtained in a simple way.

To test if the spectral evolution of (3.6) towards turbulence in the predictability experiment is actually occurring more slowly than in a fully randomized field, a second comparison experiment was carried out. In it, the basic state $\psi^{(1)}$ was chosen to be a completely random field, given by the same random number generator with the following characteristics

$$\psi^{(1)} = A \times \text{(random field)} \tag{3.7}$$

$$|A| = 0.2 \qquad \mu_{rms} = 0.6 \qquad \zeta_{rms} = 7$$

The amplitude $|A|$ was chosen requiring the field to have a μ_{rms} (average particle speed) of the same order of the μ_{rms} of the two previous fields, the solitary wave (3.4) and the eddy structure (3.6). Equation (3.7) was then perturbed by the same perturbation (3.5) and the growth in time of the total error energy $\varepsilon(t)$ was monitored.

Figure 10 shows the time evolution of the chosen predictability measure $\varepsilon(t)$ normalized with respect to the total energy of the basic state for the three predictability experiments. Curve 1 shows the error energy evolution when $\psi^{(1)}$ is the permanent wave (3.4); curve 2 when $\psi^{(1)}$ is given by the eddy shape (3.6); curve 3 when $\psi^{(1)}$ is given by the fully random field (3.7). The enhanced predictability of the permanent wave is evident: the error energy in this case is essentially constant in time. For the eddy shape (3.6), at T = 30, after 3 basin traversals, the error energy has almost reached the energy of the basic state itself, given by the straight line at vertical coordinate $\varepsilon = 1$. The fact that (3.6) is evolving slowly towards turbulence is evidenced by the comparison with curve 3. In the fully turbulent field, the error energy shoots towards the basic state energy with an exponential growth in time, reaching it at T = 15. We can therefore choose this time unit as the predictability time T* as given by (3.2) for these high-amplitude fields. The structure (3.6), having spectral properties similar to the nonlinear wave solution (3.4), is still endowed with greater predictability than a totally randomized field.

Thus, T* = 15 can be taken as a measure of the predictability time for the high-amplitude experiments and the related coherent structures. For the weak-amplitude permanent solution (3.3), taking into account the energy differences between (3.3) and (3.4), (3.6), the corresponding predictability time is T = 30. Then, the weak-amplitude solution can be seen to be completely predictable for at least 5-6 of this turbulent time scale. The high-amplitude solution

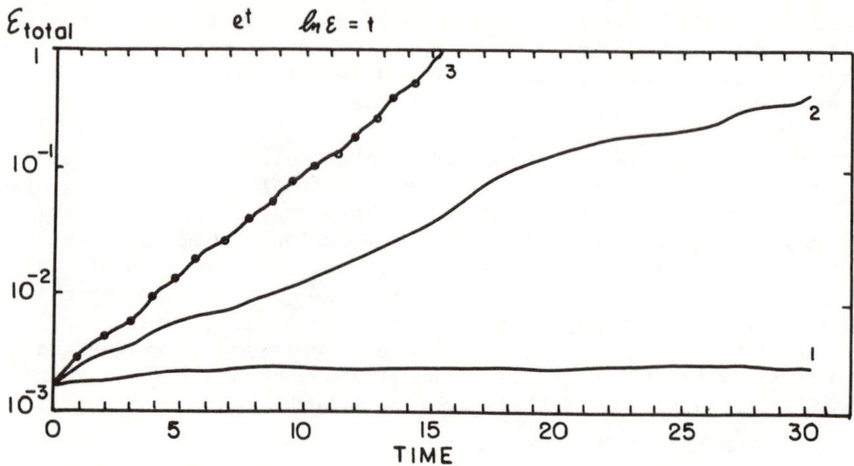

Fig. 10. Time evolution of the total normalized energy of $\delta\psi$ in the three predictability experiments: Curve 1 refers to the experiment in which $\psi^{(1)}$ is given by the permanent wave (3.4); curve 2 refers to the experiment in which $\psi^{(1)}$ is given by the eddy shape (3.6); curve 3 refers to the experiment in which $\psi^{(1)}$ is given by the random field (3.7).

is also completely predictable for at least 3-4 time units T.

The above results show that indeed nonlinear, permanent shape solutions of model equations like (2.1) are endowed with enhanced predictability with respect to flow evolution of different fields obeying the dynamics typical of two-dimensional turbulence over topography.

THE TRANSITION TO TURBULENCE OR TO LINEAR DYNAMICS

The passage from a range dominated by a KdV dynamics to a higher amplitude range dominated by turbulent behavior has been investigated through stability experiments. With them, we mean stability of the coherent waves to superimposed perturbations of variable intensity and scale contents. Also the collision experiments between two of such nonlinear waves, referred to in the previous section, can be regarded as special types of stability experiments. In them, one wave can be considered as the basic state, while the second one a special type of superimposed perturbation.

For the general stability experiments we put:

$$\psi = A(x,t)\phi(y) + \delta\psi \tag{4.1}$$

with $\phi(y)$ again given by (2.3) and $\delta\psi$ = superimposed perturbation. Then, with $\varepsilon = \delta_1{}^2$, the initial value problem (2.8) becomes:

$$A_t + cA_x = \varepsilon[2rAA_x + sA_{xxx}] + \text{Forcing induced by } \delta\psi \qquad (4.2)$$

For the collision experiments we put:

$$\psi = \delta A_1(x,t)\phi_1(y) + A_2(x,t)\phi_2(y) \qquad (4.3)$$

and the initial value problem (2.11) becomes

$$A_{1t} + c_1 A_{1x} = \varepsilon_1[s_1 A_{1xxx} + 2r_1 A_1 A_{1x} + \nu_1 A_2 A_{1x} + \lambda_1 A_1 A_{2x}]$$

$$(4.4)$$

$$A_{2t} + c_2 A_{2x} = \varepsilon_2[s_2 A_{2xxx} + 2r_2 A_2 A_{2x} + \nu_2 A_1 A_{1x} + \lambda_2 A_2 A_{1x}]$$

Exploring the extent of the KdV dynamics range means:

a) To vary gradually the perturbation $\delta\psi$ both in intensity and/or scale contents, solving numerically the <u>complete</u> model (2.1), so as to observe to which extent it obeys the <u>approximate</u> KdV dynamics (4.2).

b) To vary the relative amplitudes ε_1 and ε_2 of the two interacting solutions, allowed to evolve in the <u>complete</u> model (2.1), so as to observe to which extent it obeys the <u>approximate</u> set of coupled KdV's (4.4). In both cases, transition occurs to a different kind of dynamics, namely, the dynamics typical of quasi-geostrophic turbulence over topography. Here we summarize results of the recent literature. For the stability experiments, the permanent form basic state is given by:

$$\psi = A\,\text{sech}^2(Bx)\phi_n(y)$$

where $\phi_n(y)$ is the n-eigenmode in cross-channel direction over the given topography $h(y)$. The superimposed perturbation $\delta\psi$ is given by the random number generator. With a topography: $h(y) = -\sin(2y)$ and for the two perturbation's amplitude:

i) $|\delta\psi| = 0.1|A|$

ii) $|\delta\psi| = 0.5|A|$

Figure 4a showed the time evolution of the perturbation total energy for perturbation (i); Fig. 12 now shows the corresponding time evolution for perturbation (ii).[11] In both cases the perturbation total energy is normalized by the basic state energy. In the stable case (Fig. 4a) the perturbation energy stops growing after reaching a saturation value (units are in 10^{-2}). In the unstable case (Fig. 12), it explodes towards its theoretical limit of twice the energy of the basic state (units given by the numbers shown). The passage to a turbulence range is clear in the sudden decorrelation of the locked Fourier phases of the coherent structure.[11]

For the collision experiments, a series of them has been carried out upon the permanent solutions of Eq. (2.1). With "collision" I mean:

-evolution of two of such solutions superimposed at $T = 0$, so as to allow for their nonlinear interactions, in the total model (2.1);

-subtraction of the stronger wave (whose intensity and shape are gradually changed so as to maintain its permanent character) from the

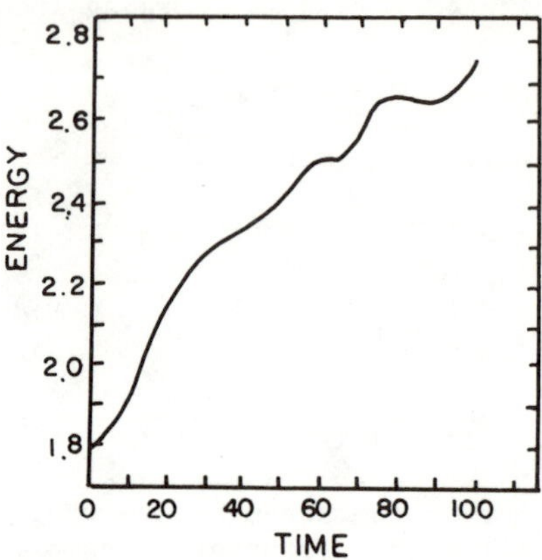

Fig. 11. Time evolution of the total energy for perturbation (ii) normalized with respect to the total energy of the basic state permanent wave, the value of which is given by the upper thin line.

total field;

-observation whether the difference wave maintains its permanent properties, and to which extent.

Table I summarizes the experiments performed:

TABLE I

Fixed wave: $\psi_2 = A_2 \text{sech}^2(B_2x)\phi_2(y)$ with $\phi_2(y)$ = second cross-channel eigenmode $|A_2| = +0.02$

colliding with $\psi_1 = \delta A_1 \text{sech}^2(B_1x)\phi_1(y)$ with $\phi_1(y)$ = first (lowest) channel eigenmode

over the topography $h(y) = -\sin(y) - \sin(2y)$, with successive amplitudes: $(\delta A_1) = -0.02; -0.05; -0.1; -0.5; -1; -2$. Transition to chaotic behavior occurs when $(\delta A_1) = -0.5$. The difference wave is gradually destroyed in physical space, with the usual sudden decorrelation of Fourier phases.[9]

All stability and collision experiments can be put into a common rationalizing frame through the theory of overlapping resonances, for which every permanent structure can be considered as composed of many nonlinear oscillators.[38,39] Then, transition to chaotic motion can be qualitatively predicted by Chirikov's criterion

$$s = \frac{\Delta w_{n,m} |\delta\psi|}{\Delta\Omega} \gtrsim 1 \tag{4.5}$$

where $\Delta w_{n,m}$ is the nonlinear frequency width induced in the basic state oscillators (n,m) by the forcing perturbation $\delta\psi$; $\Delta\Omega$ is the distance between two adjacent resonances. Through criterion (2.11) all different stability and collision experiments (viewed as stability ones) can be put into a common picture. Fig. 12[9] shows Chirikov's criterion. ε is the perturbation root mean square (RMS) vorticity amplitude ($\varepsilon = \zeta_{rms}$ of $\delta\psi$). $x = K_{sol}/K_{pert}$ is the horizontal coordinate, where K_{sol} = average wave number of the basic state (eddy diameter) and K_{pert} = average wave number of the perturbation. R indicates experiments performed by Malanotte Rizzoli; MW indicates experiments performed by McWilliams et al.[12] upon modon solutions. Even though only qualitative, the criterion separates two distinct regions in phase space. In the first, a wavelike, deterministic behavior is observed. In the second, the model obeys a turbulence dynamics. This second region is reached when the coherent structure is made to interact with other fields of appropriate intensity and scale contents. As evident from Fig. 12, the border depends on both the perturbation's intensity and its dominant length scales. Thus, perturbations having length scales larger than or as large as the basic state are much more efficient in producing its destruction. This is in agreement with the results of McWilliams et al.[12] To summarize:

Permanent structures (analytic and multivalued) are indeed rather robust. They obey a nonlinear dynamics (specifically a KdV dynamics for the solutions of model (2.1) and a turbulence dynamics in two distinct parameter ranges. The transition from order to disorder seems to occur in a rather restricted range. Upon crossing this narrow border, locked Fourier phases are suddenly and rapidly decorrelated, and the nonlinear cascade process is initiated.

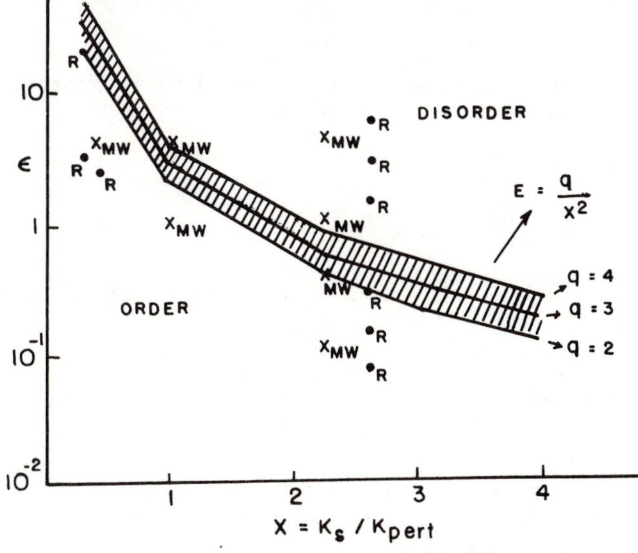

Fig. 12. Chirikov's criterion for the transition from order to disorder. Vertical coordinate $\varepsilon = \zeta_{rms}$ of $\delta\psi$. Horizontal coordinate: ratio of average diameters of basic state K_s and perturbation K_{pert}.

All the previously discussed results are valid for the dissipationless models, when the permanent basic state is forced upon by a superimposed perturbation. Let us now see in the simplest way the effects of dissipation.

Model (2.1) in dimensionful units and with a linear friction (bottom drag) is:

$$\nabla^2\psi_t + h_y\psi_x + J(\psi,\nabla^2\psi) = -K\nabla^2\psi \tag{4.6}$$

Put

$$\psi = e^{-Kt}\phi(x,y,t)$$

Equation (4.6) becomes:

$$\nabla^2\phi_t + h_y\phi_x + e^{-Kt}J(\phi,\nabla^2\phi) = 0 \tag{4.7}$$

If $\psi|_{t=0}$ is a permanent form solution of (2.1) then from (4.7) it is clear that, as $t \to 0$, at the beginning of the evolution, the solution is essentially evolving as a permanent shape wave. As $t \to \infty$, the evolution tends to the linear dynamics of topographic Rossby waves. This can also be seen passing to the system moving with the wave speed:

$$s = x - c(t)t \; ; \qquad \tau = t$$

When now $c = c(t)$ because of dissipation. Then (4.7) becomes:

$$\nabla^2\phi_s - \frac{h_y}{c + \frac{dc}{d\tau}\tau}\phi_s - \frac{e^{-K\tau}}{c + \frac{dc}{d\tau}\tau}(\phi,\nabla^2\phi) = 0 \tag{4.8}$$

For not too long a time $\frac{dc}{d\tau}\tau \ll 1$. Then (4.8) is approximately the permanent wave evolution equation if

$$\frac{1}{\gamma} = \frac{e^{-K\tau}}{c + \frac{dc}{d\tau}\tau}$$

or:

$$c = \frac{a}{\tau}(1 - e^{-K\tau}) \tag{4.9}$$

with

$$a = \frac{1}{K}c|_{\tau=0}$$

Then

$$\lim_{\tau\to 0} c = c|_{\tau=0} \; ; \quad \lim_{\tau\to\infty} c = 0$$

and c decays exponentially in time.

The physical meaning is clear. At $t \to \infty$, dissipation reduces the energy, that is, the amplitude or the speed. The permanent wave slows

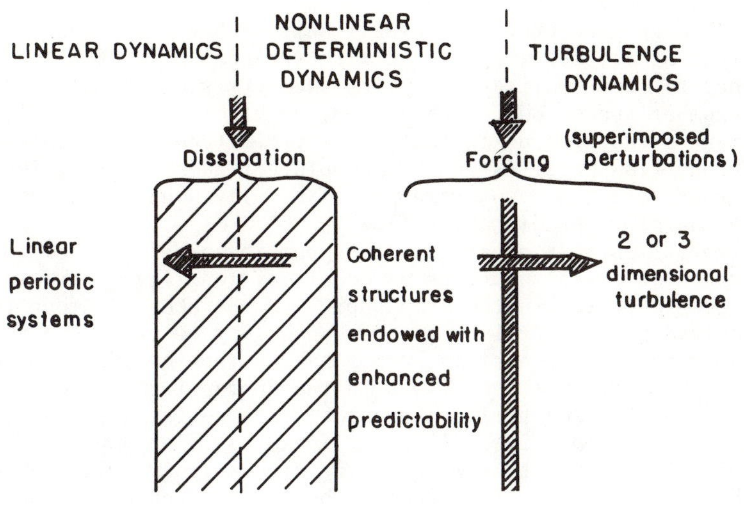

Fig. 13. Existing parameter ranges and their transitions for model equations like (2.1).

down and stops, finally entering the range of linear dynamics (infinitesimal amplitude). The asymptotic behavior is given by decay into linear topographic Rossby waves. The transition to the linear dynamics range is gradual and smooth.

These results can be summarized by Fig. 13, showing the existing dynamical ranges, and their transitions, for model (2.1) and similar model equations allowing for permanent structures as solutions.

CONCLUSIONS

We have shown that the model equations used to describe and simulate planetary motions in the ocean and the atmosphere do admit a very specific class of nonlinear, permanent form solutions in a well-defined parameter range. These model equations are the same used in turbulence theories and, in the limit of infinitesimal amplitude, they describe the linear Rossby wave dynamics. The scale range in which the dynamics is dominated by these orderly, phase coherent structures is broad. Transitions to turbulence dynamics and linear wave dynamics can be obtained through the inclusion of further effects, like forcing perturbations and dissipation of appropriate scales and intensity.

We define this parameter range of predictable nonlinear dynamics the KdV range for the ocean and the atmosphere. In it, the permanent wave solutions behave in an analogous way to the one-dimensional solitons of the KdV--and similar--equations. This means they are endowed with remarkable stability properties under superimposed perturbations and are able to collide with each other maintaining their own identities.

Through predictability experiments we have also shown that these coherent structures are endowed with enhanced predictability even with respect to flow realizations which are very similar in physical and wave-number space, but which are not solutions of the model. These coherent structures do in fact persist for time scales during which different flow patterns undergo complete randomization. They constitute therefore the counterexample of a fully nonlinear, deterministic --and therefore predictable--flow.

Very important questions still remain to be answered for the full relevance of these solutions as models of long-lived structures which have been detected in the atmosphere and the ocean. Among these, unresolved problems relate to our capability of producing them through suitable forcing mechanisms, or of having them evolve from general, random phase initial conditions in the appropriate KdV dynamical range. This raises the further problem of devising suitable criteria to recognize them amidst different features. The implication is that a seemingly turbulent flow, evolving from a general initial condition, in the KdV range, might in reality be constituted by a superposition of a number of these solutions, thus being in reality deterministic and predictable. Such a flow would exhibit long time recurrences analogous to the Fermi-Pasta-Ulam recurrences of the one-dimensional soliton models.

The answer to these problems is of paramount importance for our understanding of deterministic versus stochastic flow evolution and research along these lines is in progress.

ACKNOWLEDGEMENTS

This research was carried out with the support of the National Science Foundation Grant OCE-8118473.

REFERENCES

1. P. L. Richardson, R. E. Cheney, L. V. Worthington, J. Geophys. Res. 83, 6136 (1978).
2. B. K. Hartline, Science 205, 571 (1979).
3. S. McDowell, T. Rossby, Science 202, 1085 (1978).
4. J. McWilliams, Dyn. Atmos. Oceans 5, 43 (1980).
5. E. Fermi, J. Pasta, S. Ulam, Los Alamos Rep. LA 1940 (1955). Reprinted in Newell (1974), 143.
6. N. J. Zabusky, Phys. Rev. Letters 15, 240 (1965).
7. N. J. Zabusky, Nonlinear Partial Differential Equations (Academic Press, N.Y., 1966), p. 223.
8. C. E. Leith, Ann. Rev. Fluid Mech. 10, 107 (1977).
9. P. Malanotte Rizzoli, Advances in Geophysics 24, 147 (1982).
10. P. Malanotte Rizzoli, M. C. Hendershott, Dyn. Atmos. Oceans 4, 247 (1980).
11. P. Malanotte Rizzoli, Dyn. Atmos. Oceans 4, 261 (1980).
12. J. McWilliams, G. R. Flierl, V. D. Larichev, G. M. Reznik, J. Phys. Ocean, in press (1981).
13. D. K. Lilly, Geophys. Fluid Dyn. 3, 289 (1972).
14. D. K. Lilly, Geophys. Fluid Dyn. 4, 1 (1972).

15. D. K. Lilly, Dynamic Meteorology (D. Reidel, Boston, Mass., 1973), p. 353.
16. J. G. Charney, G. R. Flierl, Evolution in Physical Oceanography (MIT Press, Cambridge, Mass., 1981), p. 502.
17. R. R. Long, J. Atmos. Sci. 21, 197 (1964).
18. D. J. Benney, J. Math. Phys. 45, 52 (1966).
19. D. J. Benney, Studies Appl. Math. 60, 1 (1979).
20. R. A. Clarke, Geophys. Fluid Dyn. 2, 343 (1971).
21. T. Maxworthy, L. G. Redekopp, Nature 260, 509 (1976).
22. T. Maxworthy, L. G. Redekopp, Icarus 29, 261 (1976).
23. L. G. Redekopp, J. Fluid Mech. 82, 725 (1977).
24. H. Husuda, Tellus 31, 161 (1978).
25. J. P. Boyd, Review Papers of Equatorial Oceanography, FINE Workshop (NYIT University Press, 1977), p. 1.
26. J. P. Boyd, J. Phys. Oceanogr. 10, 1699 (1980).
27. G. R. Flierl, Dyn. Atmos. Oceans 3, 15 (1979).
28. P. Malanotte Rizzoli, Solitary Rossby waves over variable relief and their stability properties (Ph.D. Dissertation, Scripps Institution of Oceanography, University of California, 1978), 147 pp.
29. P. Malanotte Rizzoli, POLYMODE News 75 (1980).
30. P. Malanotte Rizzoli, "E. Fermi" Summer School of the Italian Society of Physics (North Holland Publishing Company, 1981, in press).
31. A. P. Ingersoll, Science 182, 1346 (1973).
32. A. P. Ingersoll, P. G. Cuong, J. Atmos. Sci., in press (1981).
33. M. E. Stern, J. Mar. Res. 33, 1 (1975).
34. V. Larichev, G. Reznik, POLYMODE News 19 (1976).
35. G. S. Deem, N. J. Zabusky, Phys. Rev. Letters 40, 13, 859 (1978).
36. A. L. Berestov, Izv. Acad. Sci. USSR, Atmos. Oceanic Phys. (English trans.) 15, 648, (1979).
37. G. R. Flierl, B. D. Larichev, J. C. McWilliams, G. M. Reznik, Dyn. Atmos. Oceans 5, 1 (1980).
38. V. E. Zakharov, Sov. Phys. JETP (English trans.) 38(1), 108, (1974).

PREDICTABILITY OF MESOSCALE METEOROLOGICAL PHENOMENA

Richard A. Anthes
National Center for Atmospheric Research[1]
Boulder, Colorado 80307

INTRODUCTION AND DEFINITION OF PREDICTABILITY

Suppose we desire to predict some characteristic of atmospheric flow for a time T in advance and that we have a quantitative measure of the error of this prediction. The characteristic is defined to have some degree of predictability according to this measure if the average absolute error (E) associated with predictions over an infinite ensemble of cases is less than the average absolute error (E_R) obtained by predictions based on a random selection of cases from the total ensemble of atmospheric flows. In quantitative terms, the predictability P may be defined as

$$P = 1 - E/\mathrm{Max}(E,E_R) \ , \tag{1}$$

so that the predictability of a characteristic ranges from 0 (no predictability) to 1.0 (perfect predictability).

The above definition of predictability is considerably more general than the traditional concept of predictability which has been applied to the behavior of large-scale numerical models. In the more conventional definition, predictability of models refers to the growth rate of small errors introduced by errors in the initial conditions (Lorenz[15]; Williamson[31]). According to this concept of predictability error growth, the predictability of the atmosphere has an intrinsic limit of roughly three weeks (Lorenz[15]). After this time, the differences between identical model forecasts starting with only slightly different initial conditions would be as large as the differences between two forecasts picked at random.

According to the more general definition of predictability given above, many atmospheric phenomena are predictable (to some extent) years or even centuries in advance. For example, we can be pretty sure that a prediction of temperatures at Duluth, Minnesota, for 1 January for the next 100 years equal to the climatological mean will have a significantly lower error than 100 temperatures picked at random over the Earth during that period. On short time scales, persistence is a simple method of prediction which can yield high predictability; thus, temperature is highly predictable over a time period of a minute or so using persistence.

While such simple methods as climatology and persistence often yield predictability, it may be useful to define an increment of predictability as that above the predictability associated with such simple methods alone. Thus, forecast errors are often compared to errors associated with climatology or persistence, and predictability (or at least the skill of the prediction) is said to vanish when the errors become equal to the errors associated with the simple methods. A modified definition of predictability that removes the effect of climatology would limit the ensemble of atmospheric conditions to the location, date, and time of the prediction.

An important part of the definition of predictability is the measure used to evaluate or score the prediction. A method of prediction may indicate predictability of a phenomenon by one measure but not by another. For example, it may be possible to predict the occurrence of a small-scale extreme event with skill. If the measure of success is the simple prediction of the event's existence, the event is predictable. However, measures of

[1]The National Center for Atmospheric Research is sponsored by the National Science Foundation.

forecast skill based on root-mean-square (RMS) errors computed over the area of interest may show little, if any, predictability as defined by (1). An example is given in section 2 of this paper.

Because of the importance of the measure of error in defining the predictability of atmospheric phenomena, the next section reviews some quantitative measures of forecast skill, and indicates the value of considering alternative measures of skill for mesoscale atmospheric features in addition to the conventional ones used to estimate skill in predicting large-scale atmospheric flow.

QUANTITATIVE MEASURES OF FORECAST SKILL[2]

Two nonexclusive types of verifications can be identified: those that measure the skill of forecasts and those that measure the degree to which model forecasts or simulations realistically simulate atmospheric behavior. Examples of the first, more conventional type of verification are S_1 scores, RMS errors, and threat scores. An example of the second type of verification is the model's kinetic energy spectrum. If, over a large number of cases, a model produced a spectrum similar to that of the atmosphere, it would be considered a realistic model in this respect. Statistical characteristics of the model atmosphere such as the kinetic energy spectrum may be viewed as the model's "climate." A model could have a good mesoscale climate but poor average skill scores.

A summary of useful quantitative measures of forecast skill is presented in Table 1, together with estimates of the current capability of regional models, where available.

a. Measures of forecast skill

1) S_1 scores

Synoptic-scale models have been verified over the years by calculation of objective indices or scores that reflect the skill in predicting the mass (pressure or height) fields and the precipitation occurrence and amount. The most common measure of skill in forecasting the pressure or height is the S_1 score (Teweles and Wobus[28]) which measures the skill in predicting the horizontal gradient of a scalar field. Because of the strong geostrophic (or gradient) relationship between the pressure gradient and large-scale flow in extratropical regions, the S_1 score for pressure is also a good measure of skill in predicting the synoptic-scale wind field. The S_1 score is defined as

$$S_1 = 100\,\frac{\sum |e_G|}{\sum |G_L|}\,, \tag{2}$$

where e_G is the error of the forecast pressure difference and G_L is the *maximum* of either the observed *or* forecast difference between two points. The summation is over all pairs of adjacent points in the verification region. Twenty years of experience with the S_1 score of sea-level pressure at the National Meteorological Center (NMC) have shown the practical range between essentially perfect and worthless forecasts to be 30-80 (Fawcett[10]). The S_1 scores for 30 h forecasts of SLP at NMC have decreased steadily from about 65 in 1955 to 52 in the early 1970s (Fawcett[10]). Recent operational model S_1 scores for 24 h forecasts of SLP (Fig. 1) indicate representative values of 40-50.

2) Categorical forecast scores

A *categorical forecast* is a yes or no forecast of an event, such as occurrence of a precipitation amount (usually 0.25 mm or more), at a given point during a specific time period. A measure of success of categorical forecasts is the percentage of correct forecasts. Records show annual averages of correct forecasts of precipitation occurrence over the contiguous United States of 81-86%, with little evidence of improvement over the period 1966-1977 (Ramage[20]).

[2]Portions of this section appear in a paper by Anthes.[1]

Table 1. Quantitative measures of evaluating regional model forecasts. Typical values refer to 24 h forecasts.

A. Skill of Forecasts

Measure	Typical Values (1980-1982)
S_1 Score	
Sea level pressure	45
850 mb height	40
700 mb height	30
500 mb height	25
300 mb height	20

RMS Errors	Vector Wind ($m\ s^{-1}$)	Height (m)	Temperature (°C)	Specific Hum ($g\ kg^{-1}$)
Surface	2.5	3 mb	4	?
850 mb	5	30	4	?
700 mb	7	35	3	?
500 mb	10	40	3	?
300 mb	15	45	4	?

Correlation coefficients, forecast vs. observed changes

Surface pressure		0.75
500 mb heights		0.75
Temperature		?
Threat Scores		
Precipitation	(0.25 mm or more)	0.35
	(2.5 cm or more)	0.20

Areas of indices or parameters related to thunderstorm occurrence

Convective instability	?
Lifted, K indices	?
Wind Shear	?

Characteristics of Features

Minimum pressure of cyclones	±4 mb
Maximum speed of jet streaks	?
Error in position of features (e.g., cyclone center)	?

Band-passed difference fields of conventional variables (temperature, pressure, moisture, winds)

Space-filtered
Time-filtered

Forecast of Occurrence or Nonoccurrence of an Event

Threat score (CSI) $\equiv N_1/(N_1+N_2+N_3)$

Accuracy $\equiv (N_1+N_4)/(N_1+N_2+N_3+N_4)$

Bias $\equiv (N_1+N_2)/(N_1+N_3)$

False Alarm Rate $\equiv N_2/(N_1+N_2)$

No Hit Rate $\equiv N_3/(N_1+N_3)$

Probability of detection $\equiv N_1/(N_1+N_3)$

	Event Observed	Event Not Observed
Forecast Event	N_1	N_2
Forecast No Event	N_3	N_4

B. Realism of Forecasts

Correlation Matrix

The spatial correlation between observed and forecast gridded scalar fields are calculated for various lags, or offsets, of the observed grid relative to the forecast grid. The higher the maximum correlation and the less the lag associated with the maximum, the better the forecast.

Structure Function

The structure function is a measure of the fraction of variance associated with scales of motion smaller than a given value. It ranges in value from zero at zero distance to twice the temporal variance of a variable at large distances.

Spectra

Kinetic energy
Temperature

Terms in Budget Equations

Kinetic energy
Vorticity
Water vapor
Temperature

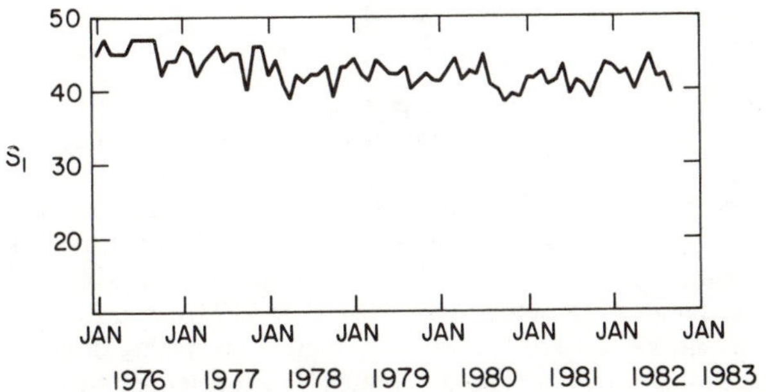

Fig. 1(a). Mean monthly S_1 scores for 24 h forecasts of sea-level pressure obtained from LFM (Newell and Deaven[17]; Deaven, 1982, personal communication).

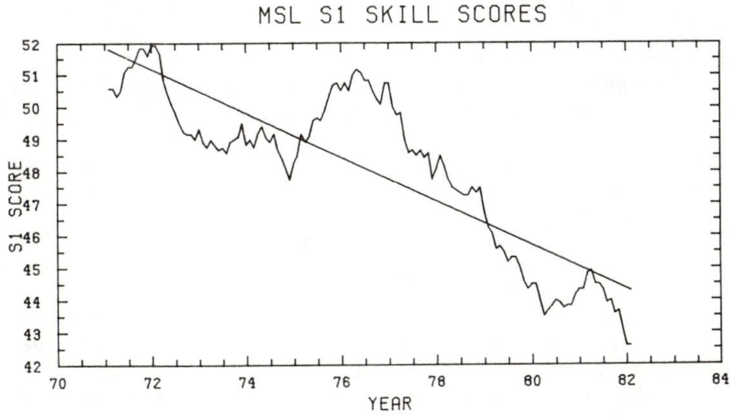

Fig. 1(b). is the same as (a) but for the Australian Numerical Meteorological Research Centre model (Gauntlett, 1982, personal communication).

Since 1965, the National Weather Service has issued *probability forecasts* of 0.25 mm or more of precipitation. The reliability of these forecasts over the entire United States has been remarkably good (Fig. 2), suggesting the value of these forecasts to those activities affected by precipitation.

3) Threat scores

In contrast to the prediction of occurrence versus nonoccurrence of measurable precipitation, the *threat score* measures the skill in predicting the area of precipitation amounts over any given threshold. The threat score TS is defined by

$$TS = \frac{CFA}{(FA + OA - CFA)} , \tag{3a}$$

where CFA is the correctly forecast area bounded by a given precipitation amount, FA is the forecast area, and OA is the observed area. A second form of the threat score, which is easily calculated from numerical models, is

$$TS = \frac{C}{F + R - C} , \tag{3b}$$

where C is the number of stations (or grid points) correctly forecast to receive a threshold amount of precipitation, F is the number of stations forecast, and R is the number of stations observing the amount.

Subjectively prepared threat scores at NMC of precipitation in excess of 2.5 cm in the period of 0-24 h have shown an annual average of around 0.20 over the United States since 1960, with no apparent trend (CAS[7]). Threat scores produced by NMC's operational regional model (the Limited area Fine-mesh Model, or LFM) for 0.25 mm of precipitation in the 12-24 h forecast period are considerably higher (averaging around 0.40) and have shown a slight increase since 1976 (Fig. 3). Figure 3 also demonstrates an annual variation of skill, with significantly lower skill in the summer.

In addition to precipitation forecasts, threat scores can be usefully applied to other parameters of significance to small-scale weather phenomena. Examples are measures of stability that are statistically related to thunderstorms, such as areas of convective instability or threshold values of various indices (e.g., lifted index, K-index). Reap and Foster[21] discuss results of a statistical technique based on numerical model output parameters to forecast the probability of thunderstorms.

4) Bias score

The *bias score* measures the tendency of a model to forecast too small or too large an area of a given amount of precipitation. In terms of an area of precipitation, it is defined as

$$B = \frac{FA}{OA} , \tag{4a}$$

while in terms of points (stations) it is

$$B = \frac{F}{R} . \tag{4b}$$

As shown in Figure 3, there was little bias in the LFM model until about 1981 when a rather strong positive bias ($B \approx 1.4$) developed. According to Hovermale (1982, personal communication), this bias occurred when an unrealistically dry boundary-layer moisture analysis was replaced by a more realistic (and more moist) analysis. Because the convective precipitation parameterization, which had been tuned to yield little bias with the dry analysis, was not altered, an excessive number of points received the threshold amount of precipitation and a positive bias developed.

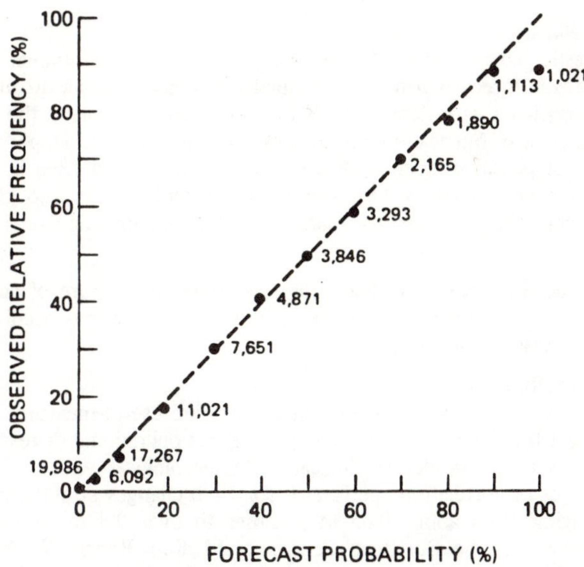

Fig. 2. Reliability of the local probability of (measurable) precipitation forecasts issued for 87 stations within the conterminous United States. The sample includes forecasts issued once daily for three projections over the period March 1978 to March 1979. The forecast probabilities were assigned the values of 0, 5, 10, 20, 30, ..., 90, 100 percent. Numbers next to the plotted points give the sample sizes (CAS[7]).

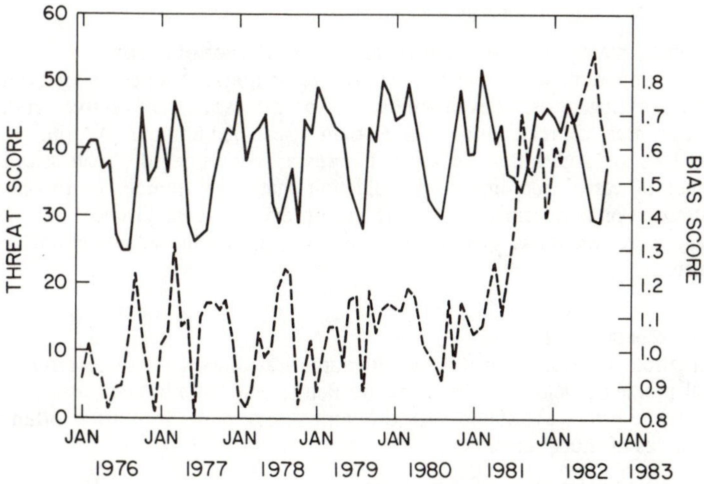

Fig. 3. Mean monthly threat score (top) and bias score (bottom) for 12-24 h LFM forecasts of 0.25 mm or more of precipitation (Newell and Deaven[17]; Deaven, 1982, personal communication).

5) Probability ellipses

If the stochastic component of mesoscale predictions is known for various phenomena, probabilistic forecasts can be developed which give users a quantitative measure of uncertainty associated with deterministic forecasts. An example is the set of probability ellipses associated with hurricane track prediction (Neumann and Hope[16]); in addition to forecasting the most probable location of a hurricane 24, 48, and 72 h in advance, the probability that the center will fall within ellipses of different sizes is also given. Similar probability forecasts could be generated for many mesoscale phenomena over land.

6) RMS errors

In addition to the S_1, threat, and bias scores, a common measure of accuracy is the RMS error. This is the most common measure of forecast skill that has been used in predictability studies of large-scale motions.

7) Correlation coefficients

Correlations between forecast and observed changes are useful measures of prediction skill, but they have not been reported extensively for either operational or research regional models. An exception is the model of the Japan Meteorological Agency. Nitta *et al.*[18] report monthly averages of correlation coefficients for 24 h changes of 500 mb heights and surface pressure ranging from about 0.65 in summer to over 0.8 in winter. For comparison, the global model of the European Centre for Medium Range Weather Forecasts, which has a horizontal resolution of approximately 200 km, reports recent (January-March 1982) correlations over 0.95 for 24 h forecast and observed changes of 500 mb heights and temperatures.

8) Characteristics of phenomena

Characteristics of significant phenomena, such as the minimum pressure of cyclones, maximum wind speed, or temperature and moisture gradients, can be tabulated over a number of forecasts and plotted against the corresponding observed values. Errors in the position of features, such as cyclones, are also of interest. This statistic is routinely reported for tropical cyclone forecasts. Hollingsworth *et al.*[12] report a threat score measuring the skill in forecasting cyclone positions.

9) Prediction matrix of occurrence versus nonoccurrence of events

It is often important to predict the occurrence or nonoccurrence of an event such as thunderstorms, cyclogenesis, development of a mesoscale convective complex, or occurrence of an area of precipitation greater than a specified amount. Within present limits, the exact location and amplitude are not necessarily considered. Scoring can be done over a number of cases according to a prediction matrix of number of forecasts of the event's occurrence or nonoccurrence versus the observed events (Table 1-A). From this table, a threat score, accuracy, false alarm rate, bias, probability of detection, and misses can be calculated.

10) Scale separation of errors

Most methods of evaluating the skill in numerical models calculate errors associated with the total predicted field. As discussed by Bettge and Baumhefner[5], separation of the total error into the errors associated with different scales in the forecast is often useful in identifying sources of model error.

b. Statistical measures of realism of simulations

Because of the increasingly random component of mesoscale predictions as the spatial scale decreases, skill scores may indicate a poor forecast and yet the forecast model may be quite realistic for understanding the evolution of a phenomenon and may even have practical utility. An extreme, hypothetical example illustrates this paradox. Suppose a small-scale model were developed that predicted the structure, intensity, and track of typhoons perfectly, except for small position errors resulting from errors in the speed of the storm. Figure 4 shows hypothetical forecast and observed sea-level pressure patterns, with the predicted storm lagging behind the observed storm (assumed to be moving toward the northeast) by 100 km.

For a two-day forecast in which the observed storm is assumed to have moved at a constant speed of 30 km h^{-1}, the error shown in Figure 4 represents an error in the model storm speed of 2.1 km h^{-1} or a time lag of 3.6 h. Although most people would agree that the above forecast would have great utility and represents significant skill, many conventional measures of skill would indicate a worthless forecast. Figure 5 shows the rapid increase of error associated with the S_1 score, RMS error of pressure and vector wind speed, and threat score for increasing position errors. In the example above, the S_1 score of SLP is 93 and the RMS pressure and wind errors are 9 mb and 12.5 m s^{-1}, respectively.

1) Correlation matrix

As the previous example demonstrates, a regional model might forecast the correct intensity and shape of a field but displace the field by some small distance. A correlation matrix scoring method (Tarbell *et al.*[27]) is a measure of skill in predicting the pattern of a scalar field such as rainfall. An observed analysis is computed on the model grid and spatial correlations between observed and predicted variables are computed for various north-south and east-west lags, or offsets of the observed grid with respect to the forecast grid. Grid points in data-void regions are not included. The result is a matrix that contains information about the skill of the model in predicting patterns. A matrix containing a few large positive correlation coefficients and a large number of small or negative correlation coefficients indicates considerable variance in the predicted and observed fields and that the model is predicting the observed structure, though not necessarily in the correct location. In the above example of the tropical storm forecast, the proper shift would yield a maximum correlation of 1.0, indicating a perfect prediction of pattern. In contrast, a smooth forecast with little structure will show a smaller maximum positive correlation and less variation for various lags. Figure 6 compares two experimental rainfall forecasts, verifying at 1200 GMT 25 January 1978. The forecast with more structure (middle panel) has a maximum correlation coefficient of 0.87 when the analysis grid is displaced five grid points to the east and four grid points to the north (Table 2).

2) Structure function

The structure of a regional model forecast or simulation can be compared quantitatively with that of the atmosphere by the structure function (Gandin[11]) defined as

$$b(\underset{\sim}{r}_1,\underset{\sim}{r}_2) = m(\underset{\sim}{r}_1,\underset{\sim}{r}_1) + m(\underset{\sim}{r}_2,\underset{\sim}{r}_2) - 2m(\underset{\sim}{r}_1,\underset{\sim}{r}_2). \qquad (5)$$

Here the m are correlation functions for the deviations of the meteorological variables from their time mean values and the $\underset{\sim}{r}$ are position vectors of the observation pairs 1 and 2. Thus, $m(\underset{\sim}{r}_1,\underset{\sim}{r}_1)$ is an autocorrelation function (the variance at station 1) while $m(\underset{\sim}{r}_1,\underset{\sim}{r}_2)$ is the covariance. We note from (5) that, as the station separation approaches zero, b approaches zero (for perfect observations). If the covariance vanishes at infinite station separation, b approaches twice the station variance. Thus, the ratio of the structure function to twice the mean station variance is a measure of the fraction of variance associated with scales smaller than the station spacing. When applied to observational data sets, as done by Barnes and Lilly,[4] all pairs of observations are considered and the correlation and structure functions are often grouped and displayed as a function of distance separation. Figure 7 shows examples of the correlation and structure function for thunderstorm conditions in Oklahoma (Barnes and Lilly[4]). A comparison of the results for temperature and mixing ratio indicates that there is much more variance associated with small scales for

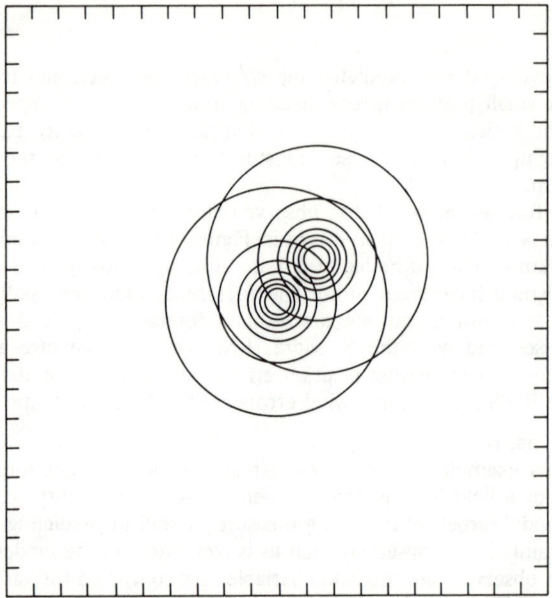

Fig. 4. Hypothetical forecast and observed sea-level pressure (mb, contour interval 15 mb) associated with a tropical cyclone. The hypothetical model is assumed perfect except for a slow bias in speed. The distance between the observed and forecast storms is 100 km and the minimum pressure is 900 km. The distance between tick marks is 50 km.

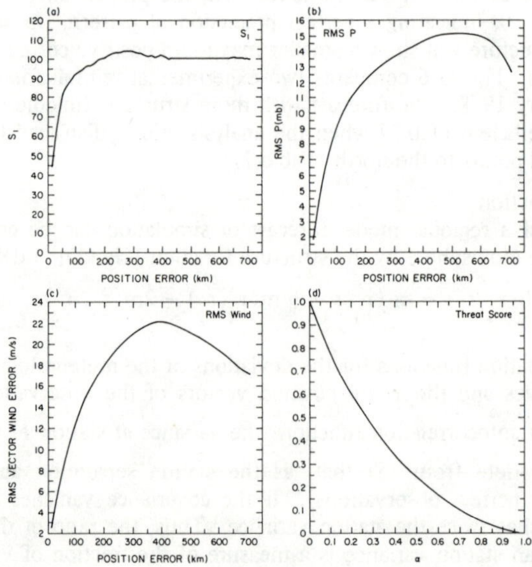

Fig. 5. Measures of forecast skill as a function of position error of hypothetical tropical cyclone model described in text. (a) S_1 score of sea level pressure, (b) root mean square error of sea level pressure, (c) root mean square error of vector wind, (d) threat score of rainfall of given amount as a function of α, the ratio of position error to the diameter of the rainfall area.

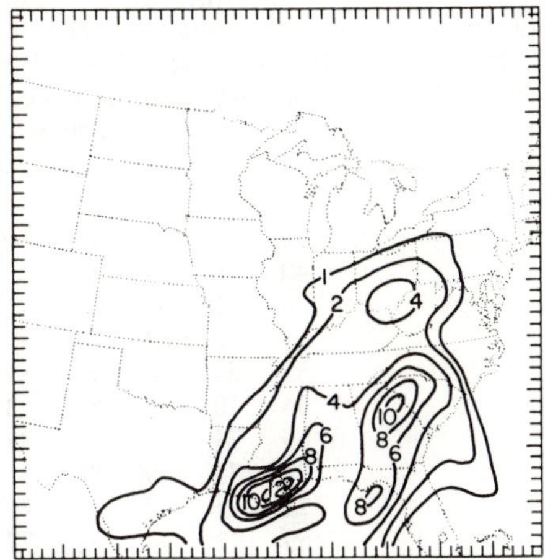

Fig. 6. 24 h precipitation (cm) over the period 1200 GMT 24 January to 1200 GMT 25 January 1978. (Left) observed (middle) forecast with latent heat included (right) forecast with latent heat neglected. (From one of 32 cases of experimental forecasts summarized by Anthes and Keyser[2]).

Table 2. Correlation matrix of observed and forecast precipitation amounts. Values are spatial correlation coefficients, in hundredths, for various north-south and east-west offsets of the observed with respect to the forecast grid. The central value of each matrix represents no shift. The observed rainfall in this example is shown in the left panel of Figure 6. Matrix A corresponds to the forecast shown in the middle panel of Figure 6, while matrix B corresponds to the forecast shown in the right panel of Figure 6.

A	36	42	47	53	60	66	72	77	81	85	*87*
	40	46	52	58	65	71	77	81	84	*87*	*87*
	46	51	58	64	71	77	81	84	85	86	85
	51	58	64	71	77	81	84	85	84	83	81
	57	64	70	77	82	84	85	84	82	80	77
	64	70	76	81	83	**84**	84	82	80	77	73
	69	74	79	82	83	83	82	80	77	74	69
	71	75	79	81	81	81	79	77	74	71	66
	72	75	77	78	79	79	77	75	72	68	62
	69	72	74	76	77	77	76	73	69	64	58
	67	70	72	74	75	75	73	70	65	60	53
B	71	76	80	83	*84*	83	82	79	75	70	64
	75	79	82	83	83	82	80	77	73	68	62
	78	81	82	82	82	81	79	75	71	66	59
	80	81	81	81	80	79	76	72	68	63	57
	81	81	81	80	79	77	74	70	65	60	54
	81	81	80	79	78	**75**	72	67	62	57	51
	80	80	79	78	76	74	70	65	60	54	48
	79	79	79	77	75	72	67	63	57	51	45
	78	78	78	76	74	70	65	60	55	48	42
	77	77	77	75	72	68	64	58	52	46	39
	76	76	75	73	70	66	61	55	49	42	35

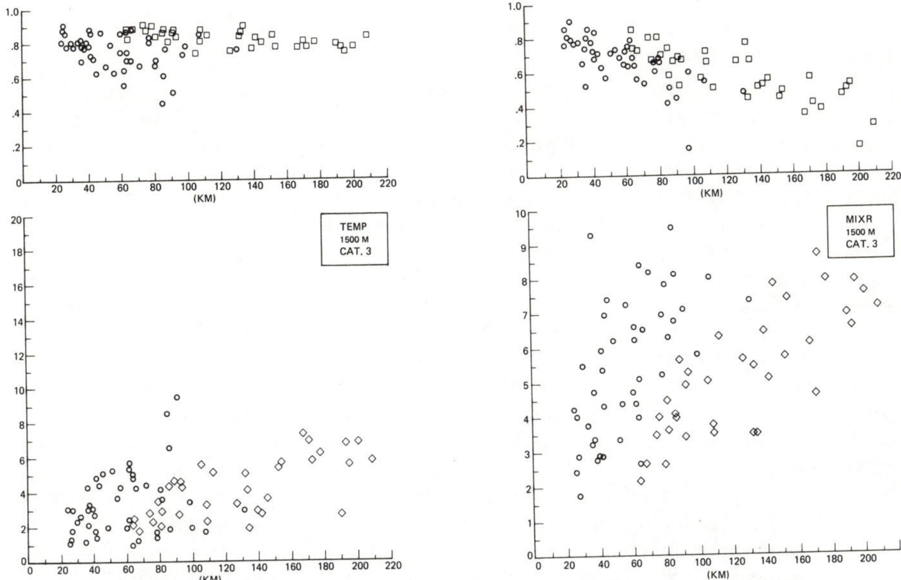

Fig. 7. Correlation function (top) and structure function (bottom) for temperature (left) and mixing ratio (right) under stormy conditions in Oklahoma. Units of structure function for temperature are K^2, and for mixing ratio $g^2\ kg^{-2}$. From Barnes and Lilly.[4]

moisture. Thus, only 22% of the total variance of temperature occurs at scales less than 200 km, while for moisture 68% of the variance is associated with these scales. Similar calculations on model data would indicate whether the model simulates scales of motion similar to those in the atmosphere.

3) Spectra

Another measure of the fidelity with which a model reproduces or simulates the atmosphere is the spectrum of various quantities such as kinetic energy, temperature, water vapor or vertical velocities.

PREDICTABILITY OF MESOSCALE PHENOMENA

During the past 25 years, major research efforts have been directed toward understanding the atmospheric general circulation and improving synoptic-scale forecasts. Technological advances, such as high-speed computers and satellites, have made possible global observations, rapid data processing, and the development of sophisticated numerical prediction models.

The progress achieved with these numerical prediction models has resulted in dramatic improvements in *synoptic*-scale *circulation* forecasts. Corresponding progress in forecasting important mesoscale phenomena, such as quantitative precipitation, has been slow and disappointing. However, recent technological and scientific advances have made it possible to envisage significant improvements in predicting mesoscale features. These advances, and a rationale for The National STORM[3] Program to achieve improvements in prediction, are presented by UCAR.[29] Portions of this section are taken from that document.

[3]STormscale Operational and Research Meteorology.

a. General considerations

Important mesoscale phenomena routinely occur in conjunction with synoptic-scale cyclogenesis. Currently, many mesoscale features are often unpredictable except by linear extrapolation based upon a close analysis of all available observations. As reviewed by UCAR[29], however, several investigations suggest that at least some mesoscale circulations are initiated by the larger-scale flow regime and then feed back upon the larger scale in an as yet unknown manner. Furthermore, there appear to be characteristic regions within an extratropical cyclone where selected mesoscale phenomena are more apt to occur. This suggests the possibility of achieving an improved mesoscale predictive capability of at least some important phenomena, either in a traditional *deterministic* sense, *or* in a *probabilistic sense* which may still be of use to the public.

There are two classes of mesoscale phenomena that we are concerned with—those that exist at the initial time of the forecast and those that will develop later within the large-scale circulation. The prediction of mesoscale phenomena that exist at the initial time of the forecast obviously requires a mesoscale observing and analysis system. Remote sensing instruments such as satellites and radars offer the potential for providing such data sets, but many problems remain before reliable systems that provide the observations of the required variables with a satisfactory accuracy and resolution will be operational.

An easier problem, from the observational point of view, is the simulation of mesoscale phenomena which develop within large-scale, routinely observed circulations. The development of mesoscale weather systems may be classified as occurring through one or both of two mechanisms: (1) forcing on the mesoscale from inhomogeneities at the earth's surface and (2) internal modifications of large-scale flow patterns that lead to smaller-scale circulations. Examples of important inhomogeneities in the terrain include variations in elevation and in ground characteristics such as albedo, moisture availability, heat capacity, and conductivity. These variations lead to differential surface heating and evaporation and produce or influence a variety of mesoscale phenomena, including land-sea breezes, mountain-valley breezes, mountain waves, heat island circulations, coastal fronts, drylines, and moist convection (Browning[6]). Reed[22] indicates that the destruction of the Hood Canal Bridge in Washington in 1979 was caused by mesoscale cyclone and orographic effects in the lee of the Cascades. In addition, variations in elevation modify the large-scale flow in a number of ways depending on the static stability, and the mean wind distribution, producing a variety of blocking effects (Schaefer[25]; Richwien[23]) and wave perturbations that often extend to great heights (Klemp and Lilly[14]).

Because the mesoscale variation in terrain parameters can be defined in appropriate detail, the development of some forced mesoscale circulation systems from large-scale initial conditions may be predictable, provided an appropriate boundary-layer model, including surface characteristics, is utilized.

The second general mechanism for the development of mesoscale structure from large-scale initial conditions is the linear and nonlinear internal processes of the atmosphere. A particularly simple example illustrates what can happen. Figure 8 shows how deformation fields associated with a synoptic-scale horizontal flow pattern can produce a mesoscale distribution of a passive scalar (Welander[30]). In more complicated atmospheric flows, shearing and stretching deformation fields may generate fronts in a day or so from a large-scale baroclinic zone (Hoskins and Bretherton[13]). Zones of convergence between two synoptic-scale air streams may generate mesoscale convective systems. Finally, large-scale flow patterns may harbor a variety of instabilities that can lead to the rapid development of mesoscale features. Examples include potential instability, trapeze instability (Orlanski[19]; Sun and Orlanski[26]), inertial instability (see Emanuel[9] for a review and application to the formation of squall lines), barotropic instability, baroclinic instability, and conditional instability of the second kind (Charney and Eliassen[8]). The development of some mesoscale

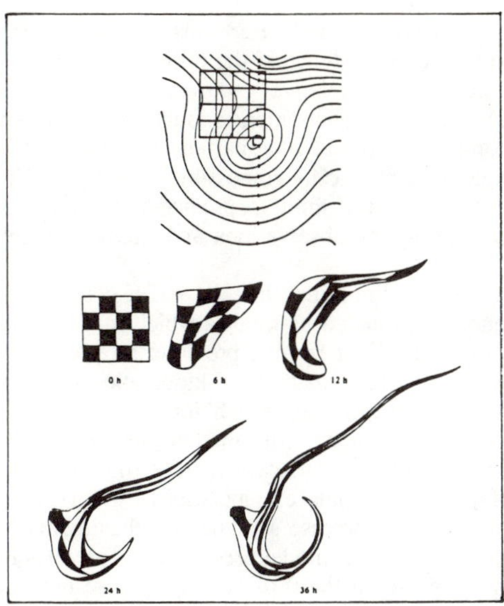

Fig. 8. Evolution of material elements in a numerical (500 mb barotropic) simulation (after Welander[30]). Despite its simple structure, the streamfield shown gives rise to a considerably more complicated material evolution. Elements are strained to the point where small-scale, irreversible processes become effective.

features from these internal physical mechanisms in a numerical model is also possible, in principle, provided that the horizontal and vertical resolution in the model is adequate and that reasonably accurate numerical techniques are employed to approximate the necessary nonlinear partial differential equations. While it is unlikely that precise forecasts of the timing and location of convective outbreaks will be possible, it is very likely that improved statements concerning the probability of such occurrences are possible with the help of existing technology.

The implications of the above discussion for the accurate simulation and prediction of many mesoscale phenomena are twofold. First, when there are no strong mesoscale circulations present initially, it may be unnecessary to observe and analyze mesoscale detail for the model's initial conditions. An accurate specification of large-scale thermodynamic and momentum fields, together with realistic physical forcing at the surface, adequate representation of diabatic effects in the free atmosphere, and the appropriate resolution, may be sufficient to predict the evolution of some mesoscale systems for hours or even a few days in advance of their development. Thus, paradoxically, mesoscale models may be more useful in the prediction of some phenomena far in advance compared to the short-range prediction of phenomena which already exist at the initial time. Second, for this latter situation in which the phenomena exist at the initial time, it may be necessary to provide detailed initial conditions that are currently beyond present operational observational capability, but appear feasible in the near future with the development and implementation of existing research technology (UCAR[29]). For these situations, a rapid data assimilation and communication system is necessary for warnings and short-range forecasts (nowcasts).

The relative role of observations and deterministic models in the resolution of the life cycle of mesoscale phenomena in existence at the initial time can be visualized by considering Figure 9, which depicts the relative accuracy of resolving the complete three-dimensional structure of a particular phenomenon (such as a squall line) with time. On this graph, 100% on the ordinate represents a perfect knowledge of the structure of the phenomena and is, of course, practically impossible to attain. The upper dashed curve represents the (unknown) theoretical limit to predictability, while the lower three curves depict hypothetical contributions of observations and models to the total predictability at the present time.

As depicted in Figure 9, at the initial time (t = 0), observations provide most of the information on the phenomenon, aided by models which may contribute some information by either providing a first-guess field from a previous forecast or by assimilating limited observations to provide a consistent analysis of kinematic and thermodynamic variables. For the early part of the forecast period, which includes the period when nowcasting is important, the observations dominate the models in providing information on the phenomena; however, this contribution decays rapidly (the observations curve essentially represents the value of persistence plus extrapolation). With time, the information provided by the models may actually increase as dynamic adjustments occur and the appropriate balance is achieved. For example, model precipitation forecasts in the 12 to 24 h range are often superior to the forecasts in the 0 to 12 h range because it takes time for realistic vertical motion and moisture fields to develop in the models.

As shown in Figure 9, models extend in time the utility of observations in describing atmospheric phenomena. A principal scientific goal of the STORM Program is to quantify Figure 9 for various mesoscale phenomena, i.e., to establish the theoretical limits of predictability of these phenomena, and to determine the observational requirements and costs for extending this current level of predictability. For example, how would the total predictability curve in Figure 9 change for various improvements in the observations or in the models?

Establishing quantitative diagrams of the type shown in Figure 9 for significant weather phenomena would have the practical result of enabling economically sound decisions concerning the establishments of new observing systems, upgrading or replacement of old ones, or implementation of new models. For example, one could provide some insight into the expected improvement of predicting flash floods on various time scales given an improved radar system, additional soundings derived from satellites, or new models with improved resolution or physics.

b. Examples of mesoscale phenomena with apparent predictability

The improvement of synoptic-scale models and parameterization of physical processes, such as surface fluxes of energy and momentum, offers the hope that improved prediction of some mesoscale phenomena is possible. This section presents examples of several mesoscale phenomena that appear to have some predictability as defined in section 1.

1) Downslope wind storms

Damaging downslope wind storms with maximum speeds exceeding hurricane force (34 ms^{-1}) are common in many parts of the world. Although associated with mesoscale waves of wavelength 50-100 km, they are generated when the large-scale (in the horizontal, but not in the vertical) flow interacts with the topography. Because the conditions favorable for their generation are reasonably well known (Klemp and Lilly[14]), they are potentially predictable well in advance of their formation given accurate large-scale forecasts. Klemp and Lilly[14] show a good correspondence between predicted and observed maximum surface wind speeds given an upstream sounding (Fig. 10).

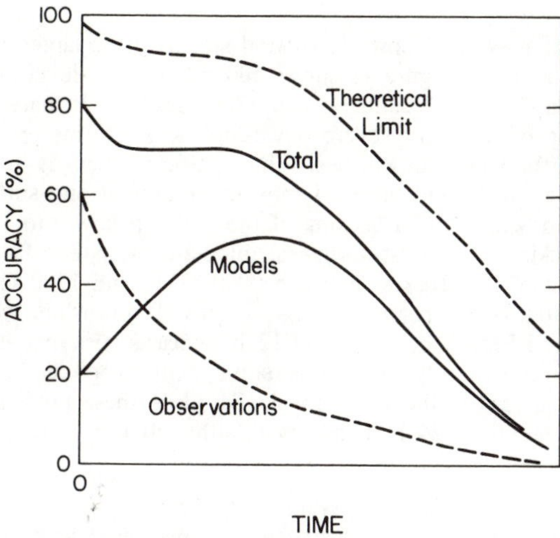

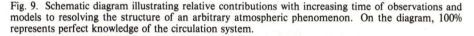

Fig. 9. Schematic diagram illustrating relative contributions with increasing time of observations and models to resolving the structure of an arbitrary atmospheric phenomenon. On the diagram, 100% represents perfect knowledge of the circulation system.

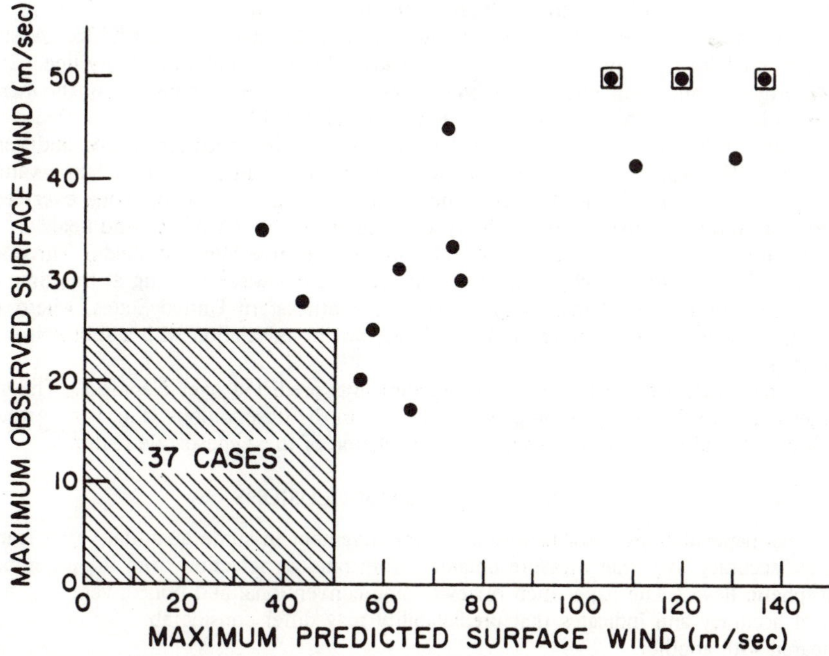

Fig. 10. Comparison of the maximum predicted surface winds with the maximum recorded by an anemometer in Boulder, Colorado, in the interval 2-5 h after the soundings were taken. A box around a data point indicates that the recorder pegged at 50 m s^{-1} (Klemp and Lilly[14]).

The predictability of mesoscale downslope wind storms using larger-scale forecasts is illustrated in Figures 11 and 12. Figure 11 shows 500 mb and sea-level pressure analyses for 0000 GMT 16 January, 0000 17 January, and 1200 GMT 17 January 1982. Between the latter two times, one of the most severe downslope wind storms on record occurred along the front range of the Colorado Rockies. The significant changes in the flow during the 24-36 h period are a rapid southeastward movement of a strong surface anticyclone from Alberta to the Ohio Valley and a backing of the 500 mb flow into a direction more perpendicular to the Rockies. The west-northwest middle-tropospheric flow and the rapid surface pressure falls east of the Rockies between 0000 GMT and 1200 GMT 17 January are among the classic synoptic-scale parameters associated with downslope wind storms.

Figure 12 shows the LFM 48 h, 24 h, and 12 h forecasts verifying at 1200 GMT 17 January. The forecasts consistently show qualitative skill in predicting the observed synoptic-scale changes that lead to the wind storm. Based on these predictions, high-wind warnings were issued more than 24 h in advance (although the extreme nature of the storm was not forecast).

2) Mesoscale features associated with severe local storms

In addition to simulating many observed features associated with fronts (Ross and Orlanski[24]), meso-α scale models have also simulated the development of low-level jets, mesoscale convective systems, upper-level jet streaks, dry lines, mountain waves, and capping inversions (lids). Anthes *et al.*[3] present results from two 24 h forecasts that illustrate the development of and relationship among the various features. Figure 13 illustrates some of the mesoscale features of a 12 h forecast verifying at 1200 GMT on 25 April 1979. Figure 13a depicts a strong frontal system in the western United States. The horizontal temperature gradient behind the cold front has intensified from a typical value of 3 K $(100\ \text{km})^{-1}$ at the initial time to over 6 K $(100\ \text{km})^{-1}$ at 12 h.

The boundary-layer winds and mixing ratio are depicted in Figure 13b. A strong west-east gradient in mixing ratio over Texas shows the development of a dry line. Shearing deformation and confluence in the boundary-layer flow were responsible for the increase in the horizontal gradient of mixing ratio by a factor of 3 in 12 h.

Figure 13b also shows a low-level southerly jet over northern Texas and central Oklahoma. The winds in this jet increased from an initial value of 4 m s^{-1} to a value of 17 m s^{-1} after 12 h. The jet formed under a strong, shallow baroclinic zone over central Texas that produced a mesoscale southerly jet in the geostrophic surface wind field.

Figure 13c shows the forecast 300 mb wind and relative humidity fields. Three areas of saturated air illustrated where the heaviest precipitation was occurring at this time (see also Figure 13a,d). The largest area is over the southeastern United States, where 12 h amounts in excess of 4 cm were predicted (Fig. 13d) in association with a subtropical low-pressure system.

A mesoscale area of moderate precipitation was predicted over Nebraska. The sharp southern edge to the precipitation area was present in the observations (Fig. 14). Anthes *et al.*[3] discuss the physical mechanisms for the formation of these mesoscale features.

SUMMARY AND CONCLUSIONS

This paper defines predictability as an improvement in some quantitative measure of forecast accuracy over that measure obtained from random selection from the ensemble of atmospheric flows. The paper then reviews some conventional and nonconventional measures of accuracy and indicates that predictability may differ considerably for one measure compared with another.

Some general factors determining the predictability of mesoscale atmospheric phenomena are discussed. The improvements in synoptic-scale prediction and

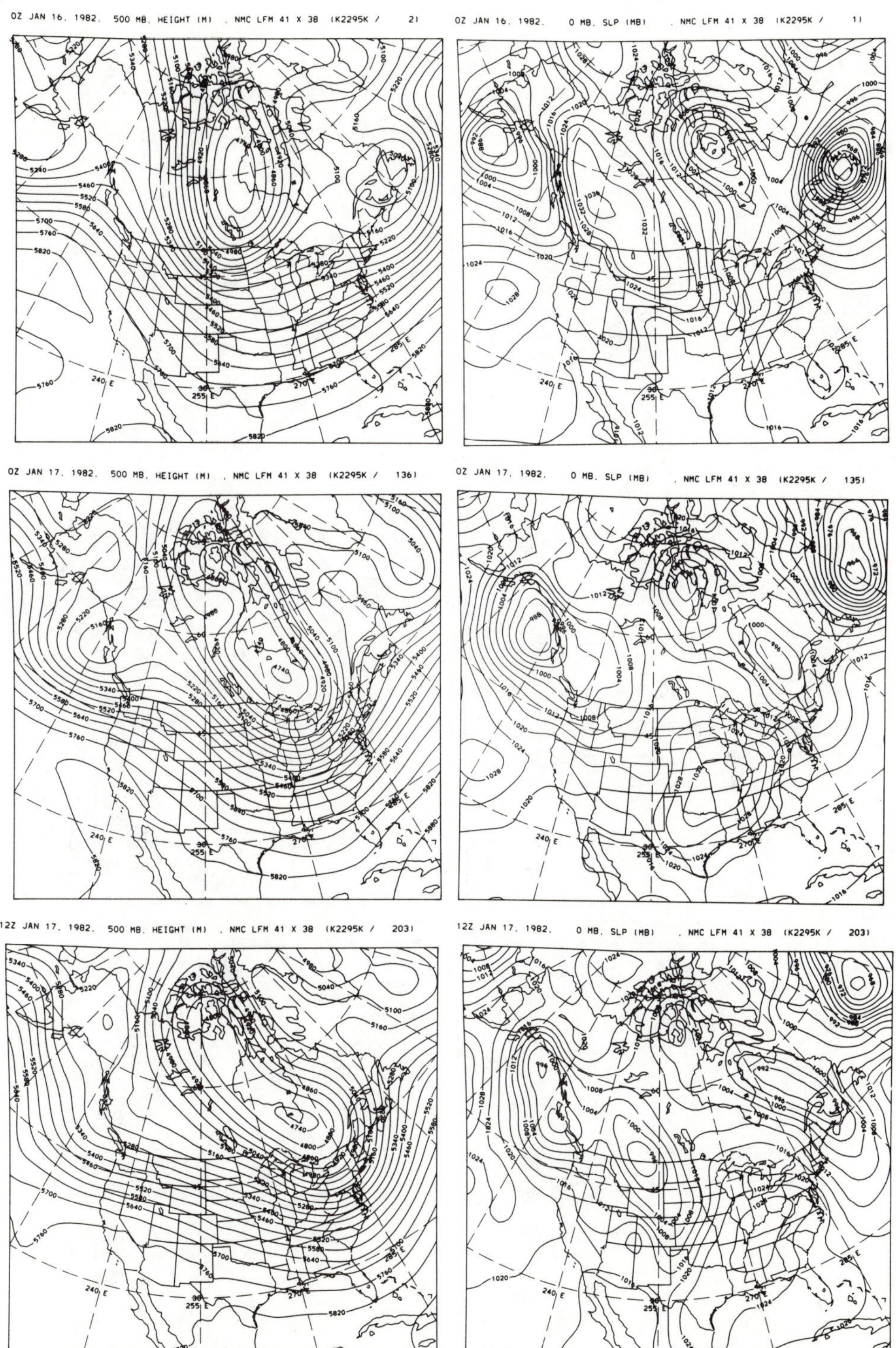

Fig. 11. Left column: NMC analysis of 500 mb heights in meters at 0000 GMT 16 January, 0000 GMT 17 January, and 1200 GMT 17 January 1982. Right column: Sea-level pressure contour interval 4 mb for same times.

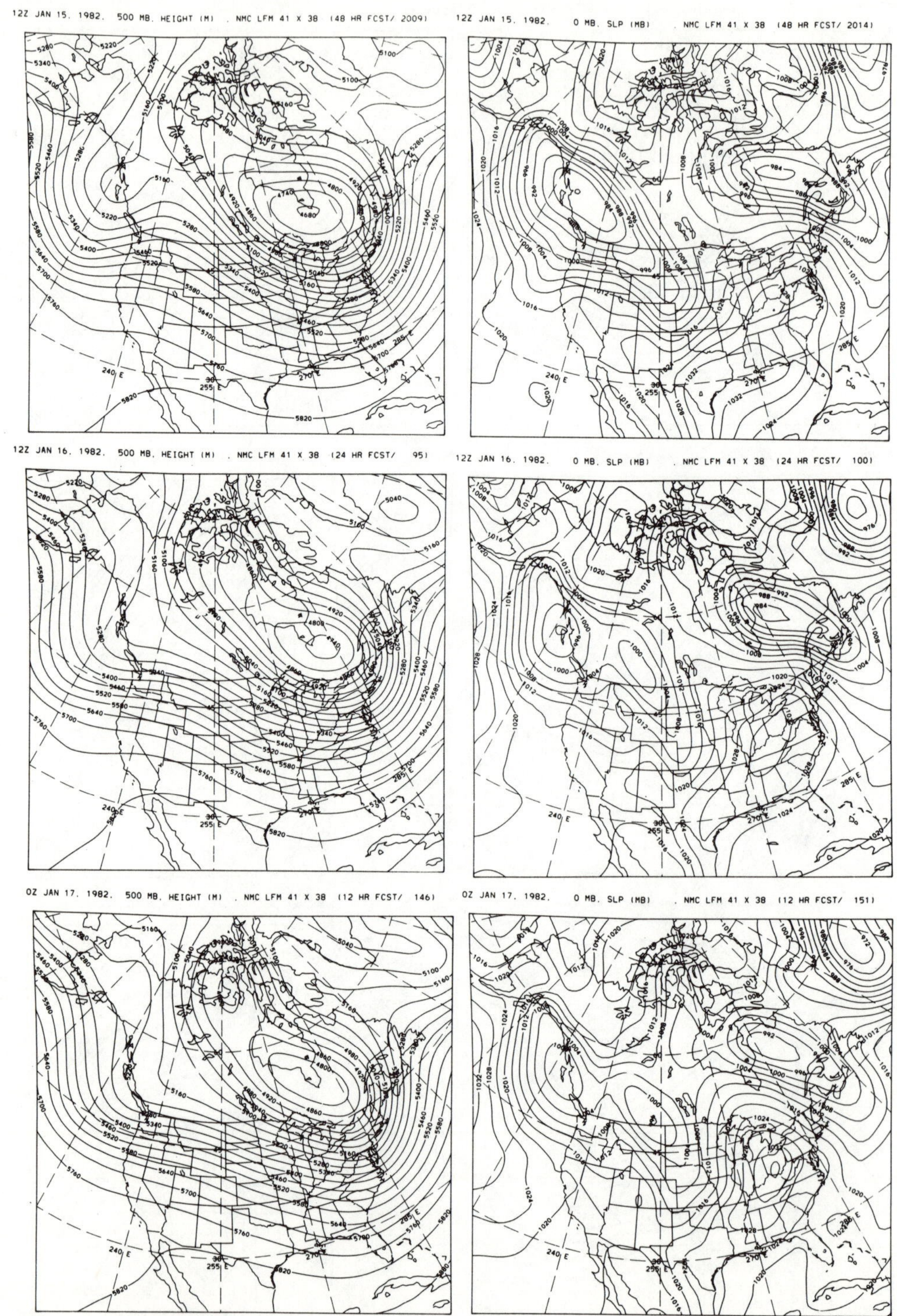

Fig. 12. Same fields as in Figure 13 except for 48 h, 24 h, and 12 h forecasts, all verifying at 1200 GMT 17 January 1982.

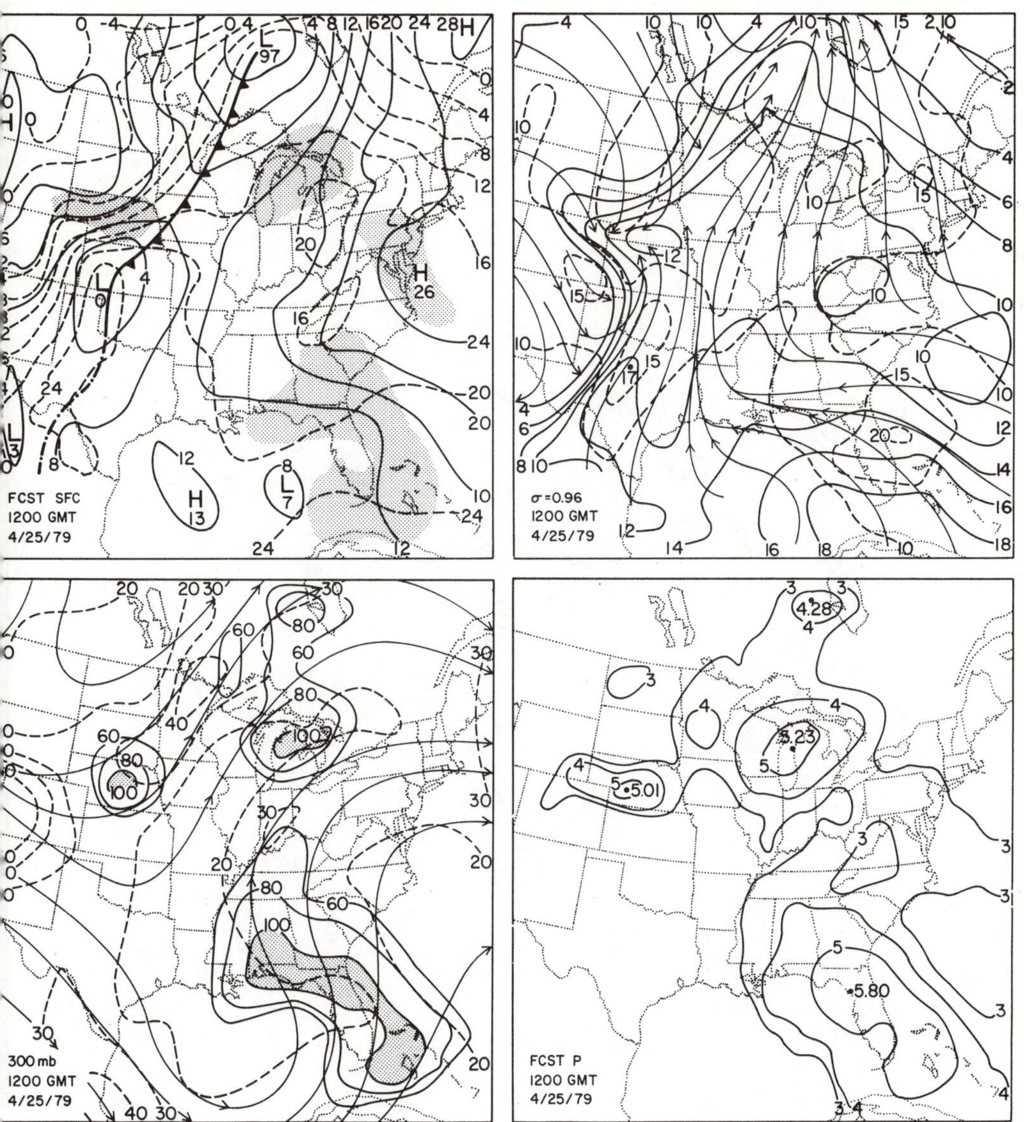

Fig. 13. 12 h forecast ending at 1200 GMT 25 April 1979 (Anthes *et al.*[3]). (a) Sea-level isobars (solid lines in mb) and isotherms (dashed lines in °C). (b) Streamlines and isotachs (dashed line in m s^{-1}) and specific humidities (solid lines in g kg^{-1}) at the $\sigma = 0.96$ level (boundary layer). (c) 300 mb streamlines and isotachs (dashed lines in m s^{-1}) and relative humidity (solid lines). (d) 12 h rainfall. Contours of $P = \ln (R + 0.01) + 4.6$ where R is expressed in cm. Values of P equal to 3, 4, 5, and 6 correspond to rainfalls of 0.19, 0.54, 1.48, and 4.05 cm, respectively.

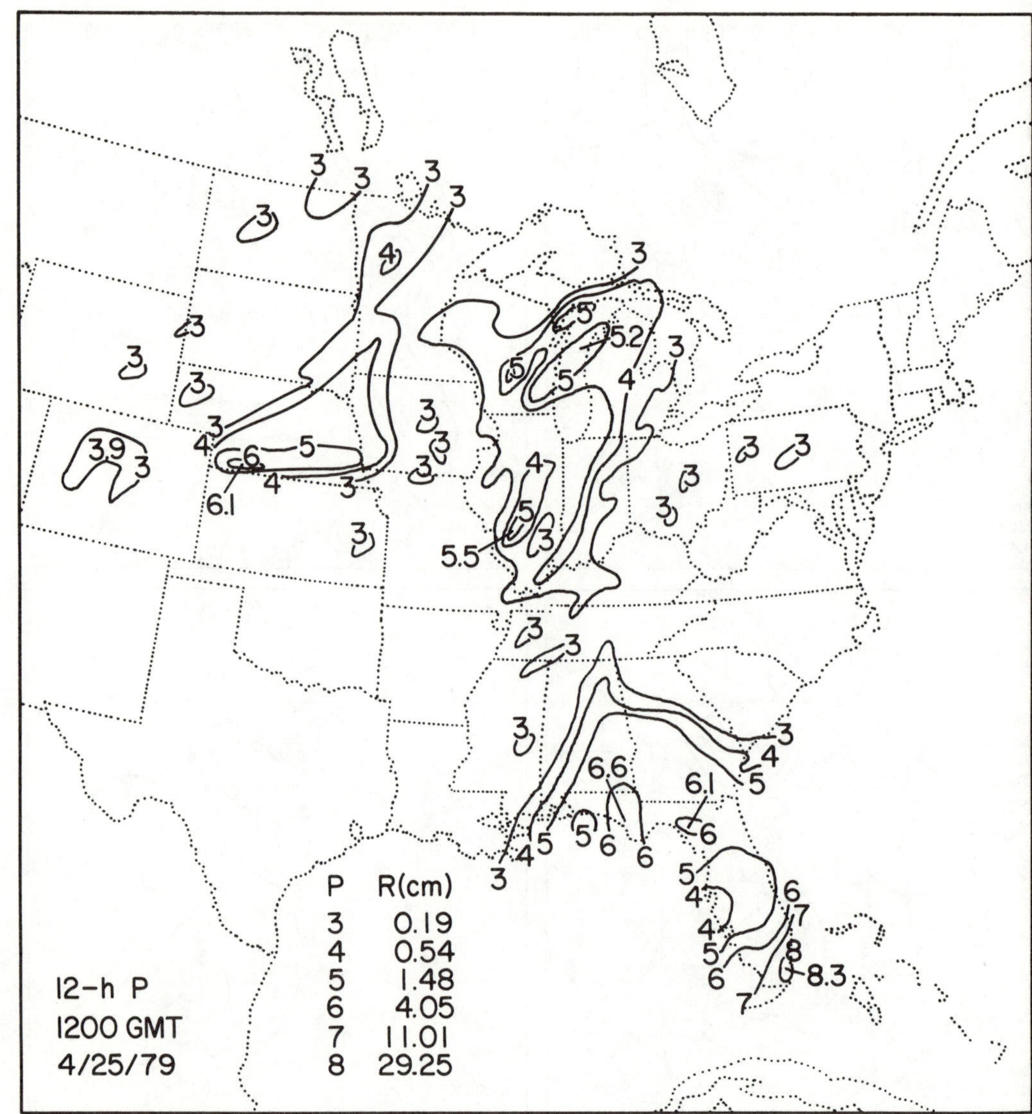

Fig. 14. Observed 12 h precipitation ending at 1200 GMT 25 April 1979 (Anthes *et al.*[3]).

parameterization of energy processes in models, together with advances in observational and computer technology, suggest that improved predictions of mesoscale phenomena of significance to people are possible. Examples illustrate the potential predictability of downslope wind storms, fronts, drylines, low-level jets, and mesoscale distributions of precipitation.

ACKNOWLEDGMENTS

I thank the following people for their generous contributions of information, figures, comments, and statistics: Stanley Barnes, David Baumhefner, Dennis Deaven, D.J. Gauntlett, Philip Haagenson, John Hovermale, Ying-Hwa Kuo, G.A. Mills, and Takashi Nitta.

Ann Modahl provided excellent editorial and typing assistance.

REFERENCES

1. Anthes, R.A.: A review of regional models of the atmosphere in middle latitudes. Mon. Wea. Rev., (to appear in 1983).
2. Anthes, R.A., and D. Keyser: Tests of a fine-mesh model over Europe and the United States. Mon. Wea. Rev., **107**, 963-984, (1979).
3. Anthes, R.A., Y.-H. Kuo, S.G. Benjamin, and Y.-F. Li: The evolution of the mesoscale environment of severe local storms: Preliminary modeling results. Mon. Wea. Rev., **110**, 1187-1213, (1982).
4. Barnes, S.L., and D.K. Lilly: Covariance analysis of severe storm environments. Preprints, Ninth Conference on Severe Local Storms, 13-21 October 1975, Norman, OK, American Meteorological Society, 301-306, (1975).
5. Bettge, T.W., and D.P. Baumhefner: A method to decompose the spatial characteristics of meteorological variables within a limited domain. Mon. Wea. Rev., **108**, 843-854, (1980).
6. Browning, K.A.: Local weather forecasting. Proc. R. Soc. Lond., A 371, 179-211, (1980).
7. CAS: Atmospheric Precipitation: Prediction and Research Problems. Report of the Panel on Precipitation Processes, National Academy of Sciences, Washington, D.C., 63 pp, (1980).
8. Charney, J.G., and A. Eliassen: On the growth of the hurricane depression. J. Atmos. Sci., **21**, 68-75, (1964).
9. Emanuel, K.: Inertial instability and mesoscale convective systems. Part 1: Linear theory of inertial instability in rotating, viscous fluids. J. Atmos. Sci., **36**, 2425-2449, (1979).
10. Fawcett, E.B.: Current capabilities in prediction at the National Weather Service's National Meteorological Center. Bull. Amer. Meteor. Soc., **58**, 143-149, (1977).
11. Gandin, L.S.: Objective Analysis of Meteorological Fields. Gidrometeorologicheskoe Izdatel'stvo, Leningrad. Translated from Russian, Israel Program for Scientific Translations, Jerusalem, 1965, 242 pp, (1963).
12. Hollingsworth, A., K. Arpe, M. Tiedthe, M. Capaldo, and H. Savijärvi: The performance of a medium-range forecast model in winter—impact of physical parameterizations. Mon. Wea. Rev., **108**, 1736-1773, (1980).
13. Hoskins, B.J., and F.P. Bretherton: Atmospheric frontogenesis models: Mathematic formulation and solution. J. Atmos. Sci., **29**, 11-37, (1972).
14. Klemp, J.B., and D.K. Lilly: The dynamics of wave-induced downslope winds. J. Atmos. Sci., **32**, 320-339, (1975).

15. Lorenz, E.N.: Three approaches to atmospheric predictability. Bull. Amer. Meteor. Soc., **50**, 345-349, (1969).
16. Neumann, C.J., and J.R. Hope: A performance analysis of the HURRAN tropical cyclone forecast system. Mon. Wea. Rev., **100**, 245-255, (1972).
17. Newell, J.E., and D.G. Deaven: The LFM-II Model — 1980. NOAA Tech. Memo. NWS NMC 66, 20 pp, (1981).
18. Nitta, T., Y. Yamagishi, and Y. Okamura: Operational performance of a regional numerical weather prediction model. J. Meteor. Soc. Japan, **57**, 308-331, (1979).
19. Orlanski, I.: The trapeze instability in an equatorial β-plane. J. Atmos. Sci., **33**, 745-763, (1976).
20. Ramage, C.S.: Have precipitation forecasts improved? Bull. Amer. Meteor. Soc., **63**, 739-743, (1982).
21. Reap, R.M., and D.S. Foster: Automated 12-36 hour probability forecasts of thunderstorm and severe local storms. J. Appl. Meteor., **18**, 1304-1315, (1979).
22. Reed, R.J.: Destructive winds caused by an orographically induced mesocyclone. Bull. Amer. Meteor. Soc., **61**, 1346-1355, (1980).
23. Richwien, B.A.: The damming effect of the southern Appalachians. Nat. Wea. Dig., **5**, 2-12, (1980).
24. Ross, R.B., and I. Orlanski: The evolution of an observed cold front. Part I: Numerical simulation. J. Atmos. Sci., **39**, 296-327, (1982).
25. Schaefer, J.T.: The life cycle of the dryline. J. Appl. Meteor., **13**, 444-458, (1974).
26. Sun, W.-Y., and I. Orlanski: Large mesoscale convection and sea breeze circulation. Part 1: Linear stability analysis. J. Atmos. Sci., **38**, 1675-1693, (1981).
27. Tarbell, T.C., T.T. Warner, and R.A. Anthes: The initialization of the divergent component of the horizontal wind in mesoscale numerical weather prediction models and its effect on initial precipitation rates. Mon. Wea. Rev., **109**, 77-95, (1981).
28. Teweles, S., and H. Wobus: Verification of prognostic charts. Bull. Amer. Meteor. Soc., **35**, 455-463, (1954).
29. UCAR: The National STORM Program—Scientific and Technological Bases and Major Objectives. Report submitted by the University Corporation for Atmospheric Research, Boulder, Colorado, to the National Oceanic and Atmospheric Administration in fulfillment of Contract NA81RAC00123. Edited by R.A. Anthes, National Center for Atmospheric Research, Boulder, Colorado, which is sponsored by the National Science Foundation, 520 pp, (1983).
30. Welander, P.: Studies on the general development of motion in a two-dimensional ideal fluid. *Tellus,* **7**, 141-156, (1955).
31. Williamson, D.L.: The effect of forecast error accumulation on four-dimensional data assimilation. J. Atmos. Sci., **30**, 537-543, (1973).

SOME PRACTICAL INSIGHTS INTO THE RELATIONSHIP BETWEEN INITIAL STATE UNCERTAINTY AND MESOSCALE PREDICTABILITY

Thomas T. Warner
Department of Meteorology
The Pennsylvania State University
University Park, PA 16802

Daniel Keyser
Louis W. Uccellini
NASA/Goddard Space Flight Center
Greenbelt, MD 20771

ABSTRACT

The application of predictability concepts to the modeling of mesoscale atmospheric processes is discussed. Special emphasis is placed on predictability implications to practical problems and on the distinction between the large-and small-scale predictability problem. We first consider some unique aspects and problems of measuring mesoscale predictive skill and then consider definitions of predictability that are appropriate for this scale. The effects on mesoscale error growth of resolved local forcing and large-scale circulations are then considered. As illustrations of practical, mesoscale predictability-related studies, we discuss some mesoscale data impact studies that have been performed using the Penn State/NCAR Mesoscale Model. Finally, error-transfer processes important to predictability studies are considered: the transfer of initial-data error among forecast variables by static and dynamic balancing procedures and by the geostrophic adjustment process during the forecast.

INTRODUCTION

Investigations of atmospheric predictability have generally focused on large-scale flows. The concepts, definitions and conclusions that are associated with the large-scale predictability questions, however, must be re-examined before it is assumed that they apply equally well to the mesoscale. Even conventional definitions of predictability may neither be appropriate nor useful when we consider the panoply of atmospheric circulation systems that prevail on the mesoscale and the extent to which they differ from synoptic/planetary-scale systems in terms of their lifetimes, triggering mechanisms and dependence on local periodic or nonperiodic forcing. A practical problem associated with evaluating prediction error on any scale is related to the ambiguous results sometimes obtained when measuring predictive skill by more than one technique. Tests of the impact of various initialization techniques on predictability have produced results which are difficult to evaluate. For example, requiring a dynamically balanced initial temperature field results in a more realistic prediction in the sense that nonmeteorological gravity-inertia wave energy will be minimized. However, standard measures of predictive skill such as sea-level pressure S1 scores can

0094-243X/84/1060271-16 $3.00

show decreased forecast accuracy compared to cases when unbalanced initial conditions are used. Further examples abound of such apparent inconsistencies and point to the need for defining operationally meaningful measures of predictability or forecast skill.

A complicating aspect of dealing with predictability is that there is not necessarily a growth in prediction error with time. While error growth in large-scale simulations occasionally approaches zero (e.g., for blocking situations), it can actually be negative for mesoscale flow with strong local forcing. Where the fluid flow responds to perfectly known local forcing by developing realistic small-scale circulation features absent from the initial state, a decrease in forecast error can result in spite of the longer term growth in uncertain energy associated with nonlinear interaction of initial errors in the large-scale, internally forced component of the motion.

The purpose of this paper is to 1) propose a definition for mesoscale predictability, 2) discuss how the predictability problem on the mesoscale differs from that on the synoptic-scale, 3) clarify the effect of resolved local forcing and large-scale circulations on mesoscale error growth, 4) show examples of the impact of initial data on mesoscale numerical forecast error, 5) illustrate how numerical models transfer initial data error among the forecast variables, and 6) show the impact of artificial balancing of model initial conditions on the initial error of a numerical forecast. These seemingly diverse subjects are all related by their bearing on the problem of atmospheric predictability, especially as applied to mesoscale motion systems. We begin by discussing why traditional concepts of predictability may not be appropriate for the mesoscale and for some applied problems. Based on these arguments, a definition of mesoscale predictability is proposed. An evaluation then follows of the impact on mesoscale predictability of mesoscale forcing and synoptic scale-mesoscale interactions. As an illustration of practical, mesoscale predictability-related studies, we discuss some data impact studies that have been performed using the Penn State Mesoscale Model. Finally, we address the important problem of the transfer of initial-data error among forecast variables by static and dynamic balancing procedures and by the geostrophic adjustment process during the forecast.

A DEFINITION OF MESOSCALE PREDICTABILITY

The following statement could represent a classical definition of atmospheric predictability:

> "A measure of atmospheric predictability is the time required by model solutions initialized with slightly different initial conditions (where the differences are typical of observational errors) to diverge to the point where the objective difference (e.g., RMS error) between the forecast and the verifying analysis exceeds the difference between climatology and the verifying analysis."

The concept of predictability thus defined, even though theoretically useful, is not especially suitable as an operational tool for quantitatively relating initial data errors to the predictive skills of the mesoscale models. Nevertheless, let us accept the idea that most definitions of predictability will employ the concept of an error growth time period during which initial state errors will cause the forecast to deteriorate to a point of zero utility. Before attempting to find an appropriate, practical definition for the mesoscale, let us consider the following points which consider the importance to the mesoscale predictability problem of 1) the fact that mesoscale predictions are often event oriented, 2) the concept of a minimum predictive skill, 3) the fact that specific data-impact questions are frequently encountered, 4) the interrelationship of synoptic-and mesoscale predictive skill, and 5) the concept of the "weak-link" in the prediction process.

a. In contrast to large-scale forecasts, where skill is measured in terms of errors in the predicted long-wave amplitude and phase, mesoscale forecasts are frequently event oriented. For example, we gauge the forecast success in terms of whether the model accurately develops a specific feature such as an intense East-Coast winter storm or a mesoscale convective complex. Furthermore, the forecasts are frequently not deemed successful unless precipitation amounts, types and spatial distrubution verify reasonably well. In other cases, we might even claim that correct prediction of the mere occurrence of a severe weather event represents a perfect forecast, while an incorrect prediction of nonoccurrence has zero skill.

b. Most numerical modelers view the concept of model predictive skill in very practical terms. On one hand, a theoretician might say the forecast has zero utility or skill when the error energy is maximized or when the objective difference between the forecast and the verifying analysis exceeds that between climatology and the verifying analysis. On the other hand, operational modelers might, for example, consider the forecast to have zero utility beyond the point in the simulation when a major observed precipitation event was not forecast by the model.

c. There is considerable interest in understanding the impact on model predictive skill of increasing the accuracy with which a single variable is specified in the initial conditions. For example, could low-level winds from SEASAT or high spatial and temporal resolution temperature and moisture soundings from VAS, increase the predictive skill of operational models? The operational predictability question is frequently even more limited in that there is interest in the impact of initial data errors on a specific forecast variable, e.g., aircraft flight-level winds, sea-surface winds or precipitation rate and duration. We even demand information about the impact of better initial data for variable A on the predictability of variable B.

d. It has been pointed out by Lorenz[1] that the ability of a model to correctly treat scale interactions has predictability implications. He notes that error growth will occur in a large-scale model because subgrid-scale events (such as thunderstorms) are unresolved, so that this error eventually appears in the large-scale solution. In this example, limited predictability of small-scale phenomena eventually influences the predictability of large-scale features. This scale interaction effect on predicatability can be even more profound for mesoscale forecasts, however. There are many mesoscale phenomena that occur only when the large-scale atmospheric characteristics produce a conducive environmental framework. Examples include squall lines, mesoscale convective complexes (MCC's), freezing rain events and coastal fronts. Thus, large-scale (global, synoptic) predictability is strongly linked to mesoscale predictability. In order to correctly predict the mesoscale event, predictability of the large-scale flow (however it may be quantified) must meet some minimum standard. Even locally forced mesoscale circulations are often greatly affected, and even generated, by the large-scale ambient flow.

e. The relationship between initial data error and model predictive skill for a particular variable will depend on whether the accuracy of the initial data is the major limiting factor in controlling forecast skill. For example, increasing the accuracy with which the initial wind field in the planetary boundary layer is specified might improve the mesoscale low-level moisture convergence and hence the rainfall forecast, but only if 1) the model contains the necessary vertical resolution at low levels, 2) the model incorporates an appropriate convective parameterization and 3) the high-quality wind data are not contaminated by unreasonably large initialization-related adjustments caused by a poorly defined initial mass field. Therefore, the link between initial state uncertainty for a particular variable and forecast skill can depend on the particular characteristics of the model physics and numerics as well as on the initial error characteristics of the other variables.

Having considered these points, we are perhaps a little better prepared to propose a definition of mesoscale predictability. Let us require that our definition have the following characteristics.

a. It should allow us to classify mesoscale phenomena in a predictability hierarchy. This requires that predictability be quantifiable.
b. It should recognize the special cases of steady and nonsteady local forcing.
c. The dependence of mesoscale predictability on the predictability of other scales should be recognized.

The following simple definition satisfies most of our requirements:

"Given a perfectly predictable large-scale atmospheric structure, a measure of mesoscale predictability is the time required for a specified error in the mesoscale structure of one or more variables to cause the prediction of a specific quantity to be sufficiently in error so that it has essentially zero utility."

This predictability measure, however, does not account for the skill embodied by a prediction prior to the time when it had zero utility. Also, the problem of the relationship between the mesoscale and the large-scale predictability is circumvented. Another definition might be:

"Given a large-scale and mesoscale atmospheric structure with specified error characteristics, a measure of mesoscale predictability is the prediction error in one variable integrated from the initial time to the time of zero forecast utility."

The former definition measures predictability in terms of a time scale while the latter measures it in terms of time-integrated error.

We can test the applicability and flexibility of the latter definition with a hypothetical example. Assume a perfect large-scale prediction and a perfect model. If we are concerned about the impact of initial moisture error on predictability of precipitation associated with an MCC event, we can use a Monte Carlo approach to relate random initial errors (say typical of observation errors) in this variable to prediction error. The measure of predictive skill would be the precipitation rate error integrated from the initial time to the time of zero utility. The time of zero utility would be determined based on the observed evolution of the MCC. If this event was the focus of the simulation, the error in the variable(s) would only be integrated for some short time after the termination of the MCC. It should also be added that we generally do not have three-dimensional verification data for mesoscale simulations, but rather have to rely on surface observations of wind, temperature, pressure and rainfall rates to define <u>mesoscale</u> prediction error.

Now that we have a more clear idea of how operationally oriented concepts of mesoscale predictability differ from conventional predictability definitions, we are in a better position to consider the material in the following sections which deals with growth in mesoscale prediction error and provides some examples of mesoscale, data-impact studies.

THE EFFECT OF RESOLVED LOCAL FORCING AND RESOLVED LARGE-SCALE CIRCULATIONS ON MESOSCALE ERROR GROWTH

Two important factors that control prediction error growth on the mesoscale are the existence of local forcing and the transfer of error from larger scales. These two factors will be considered in this section.

The existence of known, mesoscale local forcing (e.g., orography, surface heating) will generally increase the predictability of some mesoscale phenomena. Exactly how it influences error growth depends on whether the initial data contain information about the mesoscale, locally forced component of the flow. If the forecast is initialized with information about the large-scale flow only, the mesoscale forcing will normally generate realistic mesoscale atmospheric features during the course of the integration. This response to known mesoscale forcing will generate mesoscale information in the forecast where none existed in the initial state. This effect contributes to an <u>increase</u> in the mesoscale predictive skill of the simulation during the time when the large-scale flow is responding to the mesoscale forcing. After the mesoscale features have developed, the forcing will provide a constraint on the error growth. However, when the initial state contains mesoscale detail, little increase in forecast skill will be evident, but the existence of the forcing will still limit error growth. Obvious examples of locally forced mesoscale circulations are thermally driven mountain-valley circulations, coastal circulations and dynamically induced mountain waves. These will be referred to as type 1 events.

Another mechanism by which mesoscale circulations can be generated is through nonlinear interactions among large-scale waves. An example would be the development of small-scale frontal gradients with the accompanying circulations as a result of the action of large-scale deformation fields. In this same category of mesoscale circulations forced or initiated by large-scale processes would be convective events which are triggered or focused by large-scale convergence patterns or convective instability. These mesoscale phenomena, referred to as type 2 events, are similar to locally forced circulations in the sense that they can be predicted to some degree by a mesoscale model even though they are not represented in the initial conditions. Thus, there exists another mechanism by which mesoscale prediction accuracy can increase during the early stages of a forecast.

In contrast to these examples, circulations arise that resemble mesoscale turbulence in the sense that they are generated internally within the flow by dynamic instabilities. These circulations are not focused or determined by resolved local forcing or resolved large-scale features, but must exist in the initial conditions in order to have an impact on the prediction. It should be noted that these features could correspond to residual circulations that were produced originally by type 1 or type 2 events. We will call these type 3 events. Some examples are squall lines, short waves in the upper atmosphere and circulations spawned by other events such as the windstorm described by Kessler[2].

Figure 1 qualitatively illustrates the transition in mesoscale forecast error as a function of time for type 1, 2 and 3 events. This figure is based upon the assumption that 1) it is possible to draw a distinction between mesoscale and large-scale motion systems, 2) the mesoscale data and local forcing, if they exist, are known perfectly and 3) the model equations (numerics and physics) are perfect. The time scale and forecast error have not been quantified because the details of the graphs would vary greatly from one case to another. The top graph applies to type 1 and 2 events. When mesoscale information is <u>not</u> available in the initial conditions, the mesoscale predictive skill starts out at zero. As the large-scale flow either responds to mesoscale forcing or develops small-scale features by nonlinear interactions, the mesoscale predictive skill increases. When the large-scale prediction begins to deteriorate in quality because of the growth of initial error on that scale, the mesoscale forecast skill is degraded because of its close dynamic link to the large scale. Eventually, the deteriorating large-scale forecast will limit the increase in mesoscale predictive skill observed early in the forecast and cause a decrease in skill (short-dashed line). If the mesoscale solution retains some skill, even when the large solution has none, as perhaps would be the case with a sea-breeze circulation, there will be a lower limit to the skill. The latter situation would seem to only apply to type 1 events. Mesoscale predictive skill for type 2 events should approach zero when the skill of the large-scale solution does so also.

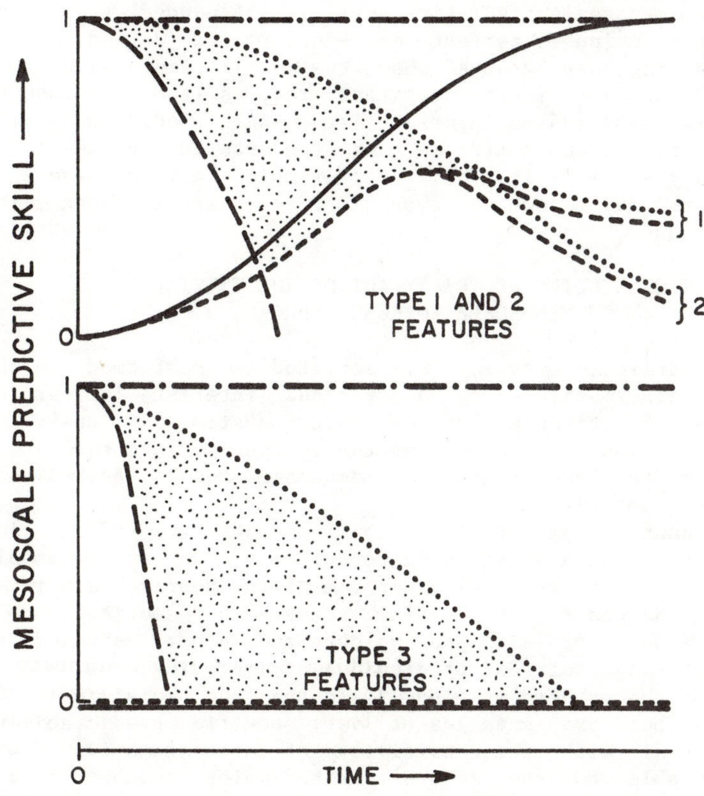

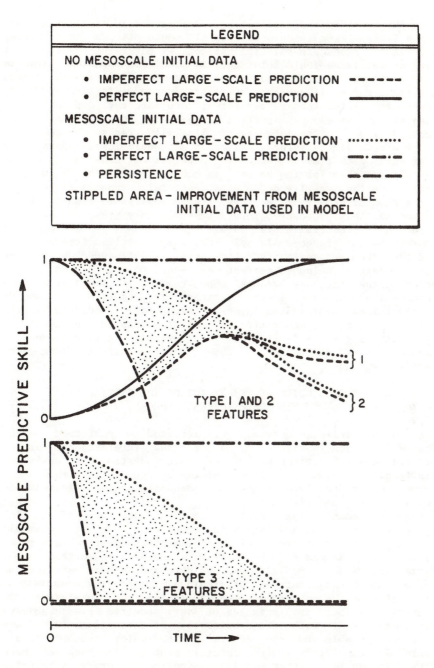

Figure 1. Qualitative estimates of relative mesoscale predictive skill as a function of time for type 1, 2 and 3 features for initial states that are defined with and without mesoscale initial data, and for perfect and imperfect large-scale predictions.

In the hypothetical case of no error growth in the large-scale solution (solid line), mesoscale skill would plateau at a higher level. If the mesoscale solution was totally determined by the local forcing or the large-scale solution, the prediction would eventually become perfect. This is easiest to visualize for type 1 events. When perfect mesoscale data are available in the initial conditions, the forecast either remains perfect or deteriorates, depending upon whether or not we assume a perfect large-scale forecast (dashed-dot and dotted line). The long-dashed line illustrates the mesoscale skill of a persistence forecast based on mesoscale initial data. The loss of skill with time is shown to be relatively slow because long time-scale synoptic forcing as well as constant local forcing will in general cause the type 1 and 2 events (especially type 1) to be of a longer period or steady. The lower figure applies to type 3 events, where we see that mesoscale forecast skill only results from the existence of mesoscale initial data. Even though these events are not forced by the large-scale solution, they do still interact with it and thus the mesoscale predictive skill is influenced by whether the large-scale solution is perfect (dash-dot) or not (dotted). The stippled areas in both graphs show the skill-time area that represents the forecast improvement that results from the use of mesoscale initial data and an imperfect large-scale prediction. The classification of mesoscale circulations or events in the described way is certainly a simplification. For example, a type 2 event or mesoscale feature may become a type 3 feature if it loses its large-scale support.

EXAMPLES OF STUDIES OF THE IMPACT OF THE INITIAL STATE ON MESOSCALE FORECAST ERROR

As noted earlier, most data impact studies performed on the mesoscale have been motivated by a practical interest in learning about the cost-effectiveness of observing systems or analysis/initialization procedures. A few of these studies that have been performed using the Penn State/NCAR mesoscale model (Anthes and Warner[3]) will be summarized.

Because quantitative precipitation forecasts are of special interest from a human and economic perspective, we are commonly concerned about the incremental improvement in hourly rain rates resulting from changes to the initial state of a numerical model. For example, Tarbell, _et al._[4] and Wolcott and Warner[5] studied the impact on short-range rainfall predictions of improving the initial specification of the divergent wind field and the moisture field, respectively. These are examples of very specific predictability problems that deal with the sensitivity of forecast skill for a particular variable to the accuracy with which another single variable is specified in the initial state. In both of these studies, it was found that by increasing the accuracy with which mesoscale structure in these variables was introduced into the model initial state, a consequent improvement resulted in the accuracy of the short-range rainfall prediction.

Results of other mesoscale predictability studies are more ambiguous because of the complex numerical and physical interactions that can occur in a realistic model forecast. For example, in an attempt to improve the amplitude of upper-and lower-level jet maxima in a forecast of a case of coupled jet streaks over the Midwest on 10 May 1973 (Uccellini and Johnson[6]), subjectively determined wind and moisture data were used to augment the model initial conditions (Burkhart and Anthes[7]). Even though the enhanced, more realistic initial conditions improved the amplitude of the upper- and lower-level jet maxima by almost 5 m s^{-1} in a 9 h forecast, the 500 mb height S1 score increased from 15.9 to 22.4 and the sea-level pressure S1 score increased from 23.3 to 38.9. Both of these S1-score trends correspond to a decreased accuracy in the predictions of these quantities.

Another case study was used to evaluate the impact on forecast skill of using the balance equation to define the temperature field. Two twelve-hour forecasts from 1200 GMT 19 November to 0000 GMT 20 November 1975 were performed, where one used observed and the other used "balanced" initial temperatures. Although the balancing procedure can have a positive impact on the forecast in terms of noise reduction, the use of derived temperatures can exert a negative impact on forecast skill. Specifically, because the balance equation is solved independently at various pressure levels in the vertical, a problem can arise that is related to the vertical consistency of the derived three-dimensional temperature field. The static stabilities of the balanced temperature profiles can be significantly different from those of the observed structure.

The two forecasts were compared in terms of root-mean-square (RMS) temperature, sea-level pressure and wind component errors as well as in terms of sea-level pressure S1 scores and gravity wave intensity. The RMS forecast errors of temperature and winds were generally smaller when unbalanced temperatures were used but the intensity of nonmeteorological gravity waves was greater. The sea-level pressure S1 score was also better when unbalanced temperatures were used, but the RMS error in this variable was greater. It is clear that an initialization procedure that has desirable properties in terms of minimizing the intensity of gravity-inertia wave noise may not be the best from the standpoint of conventional verification statistics.

ERROR ENERGY TRANSFER VIEWED AS A PROBLEM IN GEOSTROPHIC ADJUSTMENT THEORY

As noted earlier, a type of predictability study frequently encountered is the initial-data impact study, where we attempt to relate model predictive skill to the quality of the initial data for a specific variable. Some insight can be gained into this relationship, however, without performing 3D model simulations. Because we are aware of the scale-dependent dynamic relationship between the velocity field and the mass field gradients, we can apply it to help understand the probable long-term impact on prediction skill, of errors in these variables. This geostrophic adjustment theory will provide valuable insight into the dynamic relationship between the errors in these variables during the early stage of a forecast as well as an understanding of the impact of the initial errors on the error characteristics of the final predicted fields.

In order to study the effects of the geostrophic adjustment process on the redistribution among the forecast variables of initial data error, a stochastic-dynamic model has been employed. The model used in this study is a one-dimensional, shallow fluid model in u, v and h.

The stochastic-dynamic method was developed primarily with large-scale numerical models in mind (Epstein[8], Fleming[9,10]). The prognostic variables for the stochastic-dynamic equations consist of the statistical moments of the meteorological variables. Since the spectral form of the equations is used, the predicted variables are the statistical moments for the Fourier coefficients of each variable. A prediction equation exists for each moment of each coefficient of each variable. Even though these equations are truncated in a statistical sense by dropping all third and higher order moments, the calculation of the second moments still requires prediction equations for the covariances among all possible combinations of wavenumbers and variables.

The specification of the initial conditions is basically a probability statement about the dependent variables. The first and second statistical moments of the two velocity components and the fluid depth must be prescribed. Because we are concerned with the effect of the dynamic adjustment process on the transfers of data error and not about the overall adjustment toward balance, the initial first moments (i.e., the means or expected values) will be specified so that they are in geostrophic balance. The predicted second moments portray the changes in the variances as they interact during the dynamic adjustment process.

The above procedure requires the assumption that data errors be random in nature. These random errors can result from any source, whether it be instrument error, observer error, or aliasing errors related to instrument response to atmospheric "noise" on spatial scales much smaller than those resolvable by the model. The experiments have been designed so that the error-containing data are available at equally spaced grid points. The spacing of these points has been defined to correspond with the minimum resolvable scale of the stochastic dynamic model. This spacing essentially places a lower limit on the scale on which the dynamic adjustment may occur and is not unreasonable because existing synoptic-scale models have grid increments that are roughly equivalent to the rawinsonde network spacing. Also, with the advent of satellite-based measurements such as from VAS or from a LIDAR measurement system, it is not improbable that actual observations will be available on scales equivalent to the grid-point spacing of a mesoscale model.

To visualize the process of error or uncertainty exchange that occurs among variables during an operational forecast, consider one member of the ensemble of predictions that the stochastic-dynamic model represents. Let the model be initialized with values that represent the exact state of the fluid plus observation errors superimposed as perturbations. After the integration begins, the "variables" attempt to adjust to a state of dynamic compatibility. This adjustment will occur because of true imbalances in the actual fluid as well as because of errors inherent in the analysis and observation process. The scale of the true imbalance is physically determined, while the scale of the error-related imbalances is generally artificially imposed by the observation point spacing.

The mode of adjustment of the mass and velocity fields to a dynamically consistent state is, of course, scale dependent. Linear geostrophic adjustment theory provides guidance in understanding the adjustment mechanisms (Blumen[11]). Small-scale imbalances are eliminated rapidly through the adjustment of the mass field by gravity waves. Large-scale imbalances respond slowly to inertially dominated gravity-inertia waves that attempt to alter the velocity field. Thus, the velocity field dominates the adjustment process on the small scale, and the mass field dominates on the large scale. At intermediate scales, the adjustment is more mutual.

Most research on the geostrophic adjustment process has focused on the transition of a fluid that has been impulsively disturbed into an ageostrophic state by a single, isolated perturbation in the mass or momentum field. The study of an isolated disturbance simplifies the problem sufficiently to allow the development of a fairly clear understanding of the basic nature of the adjustment mechanisms involved. Blumen[12], however, has undertaken an analytical study of the adjustment that occurs when many regions of initial imbalance exist in the fluid. Blumen used a linear, nondimensional version of the shallow fluid equations employed in this study to determine the response of a fluid to random initial velocity errors with a prescribed energy density function. No initial mass perturbation existed to balance the statistically homogeneous random anomaly in the velocity component normal to the "cross-section." In spite of the fact that, in Blumen's case, the initial imbalances were randomly positioned in space, the problem is closely related to the present one in which many initial data errors of random magnitude exist. Both situations require the simultaneous adjustment across the domain of many local imbalances. The various regions of imbalance do not undergo the geostrophic adjustment process in isolation from each other, but are free to interact.

Based on the stochastic dynamic model results, it is now possible to summarize the ways in which initial data errors are transformed by the geostrophic adjustment process on the mesoscale and synoptic-scale.

1. Velocity component errors on the mesoscale maintain their original magnitude during the adjustment process. The initial error kinetic energy of the wind velocity is not redistributed or dispersed from the model domain as gravity-inertia wave energy, but rather it persists in its original location in space during the adjustment.

2. Mass field-related errors on the mesoscale are rapidly transferred to gravity-inertia wave modes. The error-energy is spread rapidly throughout the domain of integration. The ultimate fate of the gravity-inertia waves determines the fate of the uncertain energy that originated in the mass field observations.

3. Unlike the situation prevailing with the synoptic-scale, small mass and velocity errors on the mesoscale are, for all practical purposes, independent. The amount of error in one of the fields does not significantly affect the disposition of the error in the other during the adjustment phase of the prediction. The magnitude of the

mass errors determines only the amplitude of the resulting gravity-inertia waves while moderate standard errors in velocity require a very small change in the mass field in order to achieve geostrophic balance. However, transient solutions associated with gravity-inertia waves will be superimposed on the steady geostrophic mode that results from the original velocity error.

4. The lack of coupling between the initial mass and momentum errors effectively makes the concept of data error interconsistency inapplicable on the mesoscale. Even though this problem of maintaining such a state of interconsistency is germane for planetary and synoptic-scale observation networks, it need not be considered as strongly in the design of small-scale observation networks. The velocity field is shielded from the mass field observation errors by the one-sided nature of the adjustment process, and the mass field is effectively shielded from the velocity field observation errors by the very small changes in the mass field that are required for dynamic balance.

In summary, we have studied the effect of the geostrophic adjustment process on error-energy transfer during the early stages of a model simulation. A mesoscale (and synoptic scale) stochastic-dynamic, shallow fluid model calculated the error variance during a simulation, where the initial conditions assumed the existence of mesoscale observations of u, v and h with random errors superimposed. The observation point spacing was 37.5 km. As would be expected from classical geostrophic adjustment theory, the velocity errors maintained their original magnitude during the forecast whereas the mass field errors were "dispersed" as gravity-inertia wave energy. Note that the complications associated with the smoothing effect, on errors, of objective analysis algorithms and the cancellation of temperature errors when used in hydrostatic calculations of geopotential in a model, have been circumvented here.

THE IMPACT OF ARTIFICIAL BALANCING ON INITIAL ERROR

A study of the relationship between initial data error and model predictive skill must recognize that initial data errors in different variable are frequently not independent because diagnostic relationships are used to provide compatible fields of mass and velocity. For example, an investigation of the impact of initial windfield error on predictive skill would be influenced by whether the mass field variables were obtained diagnostically from the wind field. If this were the case, a transfer of initial error among the variables would therefore occur through the diagnostic relationship.

It is relatively straightforward to obtain error transfer characteristics for two simple diagnostic relationships: the geostrophic equation and the one-dimensional, linear balance equation. Use of the balance equation for synoptic-scale flows in middle latitudes usually consists of diagnosing the winds from the observed mass field, represented by the geopotential ϕ. In the tropics, and for mesoscale domains, the solution of the balance equation is often "inverted" and ϕ is solved for as a function of the observed winds. In the latter case, the initial error in the geopotential results from observation errors in the two horizontal velocity components. The difference between this simplified form of the balance equation and the geostrophic relationship is the existence of the β term corresponding to the north-south variation of the Coriolis parameter. This additional term represents another path by which error may be transmitted diagnostically.

Error-transfer relationships for the geostrophic wind equation and the linear balance equation, both obtained from the shallow fluid equations, have been used to obtain the data for Figures 2 and 3. They show the relationship between the random observation error and the error in the diagnosed variable for the geostrophic and linear balance equation initializations as a function of the scale. It is assumed that the random observation errors occur separately at each grid point. For both initialization schemes, the relationship between the errors is linear for any given scale. As the scale of the model increases, the term containing β becomes relatively more important and the two initialization methods begin to produce noticeably different error-transfer characteristics. For the mesoscale initialization, the two methods exhibit almost identical characteristics.

When the velocity field is obtained diagnostically from the mass field, either through the geostrophic relation or the balance equation, the figures show that there is a rapid increase in the velocity error for a given geopotential error as the scale decreases. For a mesoscale domain (grid increment = 37.5 km), a 12 m height error would imply a velocity error of about 40 m s^{-1}. Conversely, when the mass field is obtained from the velocity field, the mass field standard error that results from a given velocity error becomes much smaller as the space scale of the observations is decreased.

Thus, on the mesoscale, the geostrophic initialization produces very large velocity errors from geopotential errors that might be acceptably small be most criteria. Because of the nature of the geostrophic adjusment process, these large velocity errors will persist on the domain and degrade the numerical prediction. In contrast, the inverted balance equation initialization applied on the mesoscale does not create exceptionally large errors in the diagnosed mass field. The error transfer to the mass field from velocity observations with a standard error of 3 m s^{-1}, is practically negligible (σ_h = .5 m). These very small mass field errors will be dispersed as relatively harmless gravity-inertia wave energy.

As noted above, these conclusions apply to the simple <u>shallow fluid system</u> of equations. We should therefore consider how our arguments will apply to more complicated initialization problems where we have wind and temperature observations at numerous levels in the vertical that are used to define the model's initial state at a number of computation levels. In these circumstances, the geopotential error at a particular level will be a function of the integrated effect of all the temperature observation errors below. The error smoothing effect of objective analysis algorithms that define grid-point values based on randomly located observations must also be accounted for. Recall that in the data transfer study just discussed, distinct observations with independent random errors were assumed at each grid point.

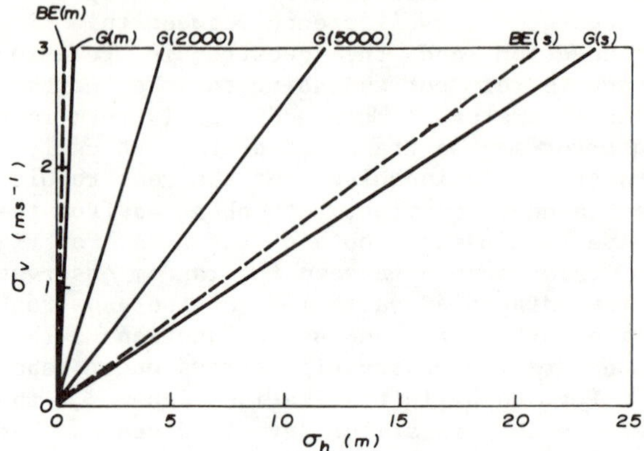

Figure 2. Relationship between mass and wind field standard errors resulting from a geostrophic initialization (G) and a linear balance equation initialization (BE) for the synoptic-scale (s) and mesoscale (m) as well as for intermediate domain sizes of 2000 km and 5000 km.

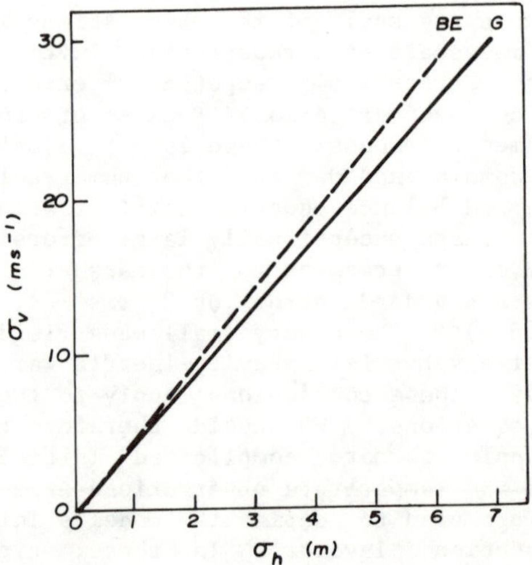

Figure 3. Relationship between mass and wind field standard errors resulting from a geostrophic initialization (G) and a linear balance equation initialization (BE) on the mesoscale. This is identical to the mesoscale results of Figure 2, except that the axis scales have been changed.

If errors in temperature observations within a radiosonde ascent were largely random, the tendency would be for them to cancel when used to hydrostatically define the geopotential. An objective analysis algorithm that incorporates information from a number of observations while computing gridpoint values will also minimize the error at the gridpoints. Both of these effects will reduce the magnitude of geopotential errors and probably increase the spatial scale of the error field as well--that is, compared to the simulated case where direct "observations" of geopotential were available at each grid point. This means that the relations described in Figures 2 and 3 almost certainly over state the error transfer that will occur when velocity is determined from the mass field when using more realistic models and data sets.

There are a number of other possible initialization schemes, many of which can be classified as "dynamic" techniques. In the dynamic initialization, the actual forecast equations are used to obtain dynamically balanced initial conditions. An example would be the method of Nitta and Hovermale[13] where a backward-forward integration that is centered on the initial time of the prediction is performed using a time differencing scheme that damps high temporal frequencies. A "first guess" is obtained utilizing a conventional initialization procedure (or using independently analyzed wind and mass data). The model then adjusts to a balanced state during the backward-forward integration cycles, while the time differencing scheme damps the unwanted higher-frequency gravity-inertia waves. Because the forecast equations are actually used as part of the initialization process, the error energy that would otherwise be transferred between the mass and wind fields during the forecast as part of the post-initialization adjustment is now transferred during the initialization. An important implication of the geostrophic adjustment process is that the procedure used to obtain the "first guess" is quite important. The initial velocity field errors in this "first guess", whether they be actual data errors or errors transferred by some diagnostic relationship, will remain after the initialization.

It is possible, with dynamic initialization techniques, to restore the original mass or velocity field at intervals during the cyclic integration. Such a procedure causes the adjustment to occur solely in the nonrestored quantity. If the velocity errors of the "first guess" are unacceptably large compared to the mass field errors, it might be tempting to restore the original mass field a sufficient number of times so that the velocity would be forced to adjust to it. On the mesoscale, such a repeated restoration of the mass field may overrule the normal adjustment laws and cause the mass field to dominate the adjustment. If the ultimate balance is geostrophic, the velocity errors would be large, even for relatively small errors in the mass field. Thus, the periodic restoration of the mass field with the intent of diminishing the velocity errors would be of no value because of the nature of the error transfer at this scale.

In summary, the goal of all initialization procedures is a set of balanced initial conditions. It is quite possible to satisfy this objective by means of techniques that may or may not amplify data errors. We have seen examples in both categories. If there is some flexibility in the selection and application of initialization procedures, the principles of error-energy transfer can provide valuable insight into the process of choosing the most accurate initialization methods.

SUMMARY

A number of topics dealing with the predictability of mesoscale atmospheric processes have been discussed. Special emphasis has been placed on predictability implications to practical problems and on the distinction between large-and small-scale predictability concepts. We first considered a number of unique aspects and problems associated with measuring mesoscale predictive skill and then proposed two possible definitions of predictability that are appropriate for the mesoscale. The effects on mesoscale error growth of local forcing and the interaction between mesoscale and synoptic-scale phenomena were discussed. As illustrations of practical, mesoscale predictability-related studies, we discussed some mesoscale data impact studies that have been performed using the Penn State/NCAR Mesoscale Model. Finally, the transfer of initial data error by the preforecast dynamic balancing and by the geostrophic adjustment process during the forecast were analyzed using stochastic-dynamic modeling techniques.

REFERENCES

1. E.N. Lorenz, Transactions of the New York Academy of Sciences, 409-432 (1963).
2. E. Kessler, Bull. Am. Meteor. Soc., 63, 1380-1386 (1982).
3. R.A. Anthes and T.T. Warner, Mon. Wea. Rev., 106, 1045-1078 (1978).
4. T.C. Tarbell, T.T. Warner and R.A. Anthes, Mon. Wea. Rev., 109, 77-95 (1981).
5. S.W. Wolcott and T.T Warner, Mon. Wea. Rev., 109, 1989-1998 (1981).
6. L.W. Uccellini and D.R. Johnson, Mon. Wea. Rev., 107, 682-703 (1979).
7. R.P. Burkhart and R.A. Anthes, Proc. Fifth Conf. on Numerical Wea. Prediction, 198-202 (1981).
8. E.S. Epstein, Tellus, 21, 739-759 (1969).
9. R.J. Fleming, Mon. Wea. Rev., 99, 851-872 (1971).
10. R.J. Fleming, Mon. Wea. Rev., 99, 927-938 (1971).
11. W. Blumen, Rev. Geophys. Space Phys., 10, 485-528 (1972).
12. W. Blumen, Tellus, 19, 174-182 (1967).
13. T. Nitta and J.B. Hovermale, Tellus, 21, 121-126 (1969).

SOME FACETS OF THE PREDICTABILITY PROBLEM FOR ATMOSPHERIC MESOSCALES

Douglas K. Lilly,
University of Oklahoma

ABSTRACT

The predictability of mesoscale weather differs from that of the larger scales in several ways, but perhaps most importantly through the greater intermittency of the small scales and the importance of relatively rare high amplitude events. For quiescent weather periods predictability is presumably roughly determinable from the structure of the velocity variance, as measured by, e.g., the energy spectrum or spatial covariance function. Recent measurements from radar and aircraft indicate that the kinetic energy spectrum follows a -5/3 power law for wave lengths between a few and a few hundred km. Two different explanations for this structure are discussed, one based on turbulence dynamics and the other assuming the dominance of internal gravity waves.

Strongly energetic turbulent events, like buoyant convective storm systems, can be expected to have a shorter predictability time than quiescent weather. For a particularly important high energy convective storm type, the rotating "supercell" storms, evidence from theory, observation, and numerical models suggests that they are unusually stable, however, and probably more predictable than ordinary turbulence concepts would suggest. This enhanced stability and predictability is attributed to the effects of helicity, acquired from the mean state and amplified by buoyancy.

INTRODUCTION

The predictability problem for meteorology has generally been defined with respect to the larger scales of motion, i.e. wave lengths of 1000 km or more. For all its complexity the atmosphere is a relatively homogeneous and mostly turbulent medium on these scales. Application of concepts of predictability based on homogeneous turbulence theory by Lorenz[1] and others has been fairly successful. At smaller scales the assumptions of homogeneity appear to be much less well founded. Although I am not aware of definitive statistics, it is apparent that intermittency in time and space is an important factor in the practical consideration of mesoscale meteorology and its prediction, and therefore should probably be entered appropriately into theoretical analysis. I don't know how to do this, but look hopefully to the Mandelbrot "fractal" concept[2,3,4].

I find it convenient to categorize mesoscale predictability into roughly four classes, on the basis of categorizations of evolution of mesoscale variables. These are: (a) nearly homogeneous turbulent flow with well-defined variance spectra; (b) frontal and jet-like near-discontinuities, arising out of larger scale processes; (c) response to small scale topographic forcing; and (d) large amplitude instabilities, such as convective storms and storm arrays. For category (a), the approach to predictability analysis is similar to that for the larger scales, modified by consideration of the mesoscale energy spectrum. For categories (b) and (c), the predictability is evidently closely controlled by that of the large scale motion field. Category (d) involves elements which may only exist in favorable large scale environments, but whose amplitude is so large that once initiated they control their own destiny to a considerable extent, and in fact may strongly impact the large scales. In another paper in these proceedings, Anthes[5] concentrates on categories (b) and (c), for which he shows evidence that the predictability of mesoscale motions is somewhat larger than would be implied by an assumption of the dominance of category (a) motions. I will first describe the observational and theoretical analysis of the mesoscale energy spectrum,

0094-243X/84/1060287-08 $3.00

and its relationship to category (a) of the predictability problem, and then discuss some aspects of category (d) events, particularly the anomalously stable and presumably predictable rotating convective storms.

THE MESOSCALE ENERGY SPECTRUM, EXPLANATION AND CONSEQUENCES

Because of the fixed scale of most meteorological instrumentation networks, the variance spectral structure at scales less than about 1000 km has not been well documented. Recently, through analysis of instrumented aircraft flights and Doppler radar, this situation has improved. Figure 1 (from Lilly and Petersen[6]) shows a composite of mid- and-upper tropospheric horizontal kinetic energy spectra obtained in various ways, for wavelengths greater than about 5 km. At the large scale end, a fairly flat spectrum between planetary wavenumbers 3 and 10 merges into a rapidly decreasing range, conforming approximately to k^{-3} , down to wavelengths of about 1000 km. The spectrum flattens somewhat, evidently becoming proportional to $k^{-5/3}$ or thereabouts, for scales less than about 200 km. The intermittency at scales less than about 10 km is so great that the average spectrum has not been well established there, but I believe that it continues at roughly the same slope into the traditional small scale 3-d inertial range at the microscales. The famous "spectral gap" in the mesoscale range is probably confined to the boundary layer, where small scale motions are forced locally.

The vertical spectra of horizontal motions are moderately well known from analysis of high resolution balloon soundings. Figure 2, from Endlich[7], summarizes spectra taken at several seasons from Cape Canaveral, Florida. It shows a spectral slope of about -2.5 for scales less than about a kilometer.

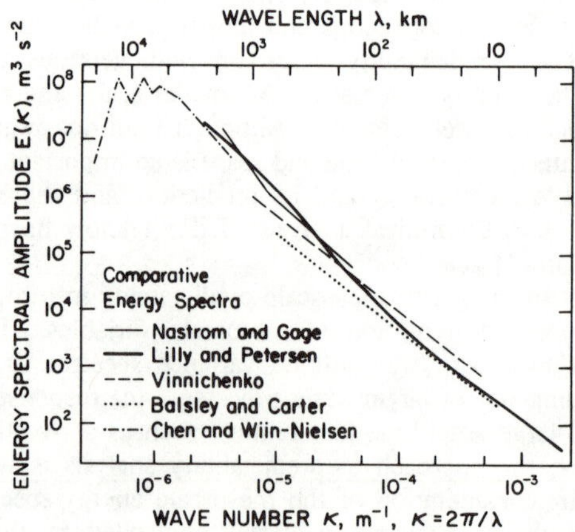

Fig. 1. Horizontal spectra of the horizontal wind in the middle and upper troposphere. The solid line curves were obtained from analysis of aircraft flight data, the dotted line curve from Doppler radar, and the others principally from sonde data. From Lilly and Petersen[6].

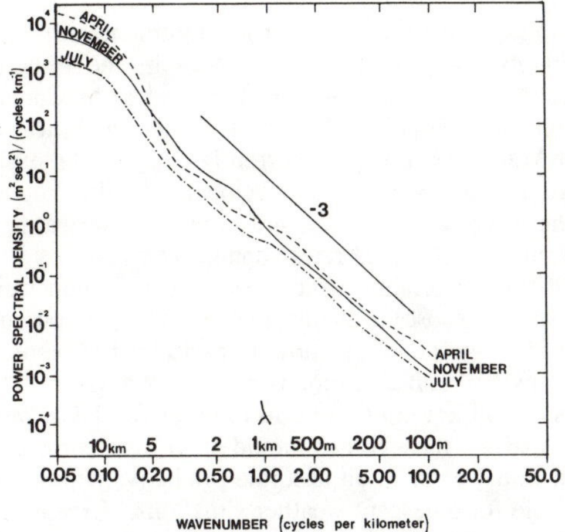

Fig. 2. Vertical spectra of the horizontal wind in the troposphere and lower stratosphere, based on high resolution balloon tracking. From Endlich[7].

Limited data for temperature spectra from Nastrom and Gage (personal communication, 1983) show spectra similar to that of kinetic energy, when weighted by the mean static stability to produce an available potential energy spectrum. Temporal vertical velocity spectra from Balsley[8] show a peak at the Brunt-Vaisala frequency, and an approximately flat spectrum function for lower frequencies.

The existence of the apparent -5/3 slope in the mesoscales is somewhat puzzling, since it clearly cannot be explained by the inertial range concepts of 3-d turbulence. Two incomplete theories have been put forth to explain it. Lilly[9] has expanded on a suggestion due to Gage[10] that the spectrum is produced by motions generated at small scales, perhaps by cloud convection. These motions become quasi-two-dimensional due to the effects of thermal stratification and then follow Kraichnan's[11] predicted $k^{-5/3}$ upscale inertial range. The transformation from three- to two-dimensional motions is really a separation of 3-d turbulent flow into a mixture of 2-d turbulence and internal gravity waves. This separation was shown by Riley, et al.[12] to occur as a part of the decay process of an initially isotropic element in a stratified fluid. After the separation it is postulated that the waves and quasi-two-dimensional (stratified) turbulence disperse according to their habits, with the wave energy eventually finding its way to the upper atmosphere, where it breaks down into new turbulence. The stratified turbulence is expected to continue to grow in scale and form the observed mesoscale spectrum. Some uncertainties remain in the theory, based on the decoupling of adjacent layers and destabilization to shearing strains.

A second hypothesis has been introduced by VanZandt[13]. He proposes that the atmospheric noise spectrum is produced by interacting internal gravity and inertio-gravity waves. He follows the concepts introduced by Garrett and Munk[14,15] in their analysis of the corresponding oceanic problem, and finds that those concepts predict consistent relationships between the vertical, horizontal, and temporal spectra of atmospheric variables. These relationships, together with the considerable success enjoyed by the Garrett-Munk analysis in ocean dynamics, constitute rather strong arguments in its favor. On the negative side, however, are doubts that the wave amplitudes in the troposphere are sufficiently strong to excite the scattering wave interactions discussed by McComas and Müller[16], which form the principal mechanistic backing of Garrett-Munk. A critical test of the Lilly-Gage vs. Van Zandt-Garrett-Munk hypotheses is currently sought. An evaluation of the relative probabilities of upward vs. downward group velocities in the troposphere would provide strong evidence. If the waves are predominantly upward-propagating, the Lilly hypothesis is more likely valid, while roughly equal rates support Van Zandt.

The resolution of this uncertainty would lead to certain conclusions regarding predictability of the rather weak mesoscale events that are the general rule. The approximate $k^{-5/3}$ spectrum observed suggests, if the flow is basically turbulent, eddy turnover times proportional to $(\text{wave length})^{2/3}$. If this time is of order 2 days for 1000 km scales, it would then be about 5 hours for 30 km scales and 30 minutes for 1 km wavelengths. Under the classical predictability analysis these times would then also correspond roughly to the e-folding growth time of initial error in an accurate predictive simulation. These time scales do not seem unreasonable for quiescent weather situations. Except for a few important fair weather phenomena, like pollution episodes, practical predictability tends to be confined to the times and places where mesoscale weather is anomalously energetic, however, when the turnover and predictability times are generally shorter than the above estimates.

Of course if the Van Zandt hypothesis is more nearly correct, the theoretical predictability of quiescent mesoscale flow is much greater than suggested above, since linear waves are nominally immortal. Only the weak resonant interactions then produce loss of predictability.

PREDICTABILITY OF BUOYANT CONVECTIVE MOTIONS, INCLUDING THOSE WITH ROTATION

A prototype mesoscale phenomenon of high energy and intermittency, probably the most important such phenomenon, is the thunderstorm. At any given time there are several thousand thunderstorms in the world, but these cover much less than 1% of the earth's surface. They are highly turbulent flows with an effective eddy turnover time of order 30 minutes. Thus it would seem unlikely that the evolution and effects of a given storm could be foreseen more than an hour or so, and then only if rather detailed observational data were obtained, such as from a network of Doppler radars. Some of the most intense storms, which are fairly frequent in regions with strong shear coinciding with strong buoyant instability, exhibit significant rotation about a vertical axis and a well-defined quasi-steady structure. These storms, called "supercells" by Browning[17,18], who first described their structure, produce most of the severe tornadoes and large hail in the U.S. midwest. They appear to be good candidates for anomalously high predictability, for reasons to be outlined further.

Numerical simulation models, supported by Doppler radar studies, have shown that the principal source of rotation in supercell type storms is the titling of horizontal vorticity, from mean vertical shear, into the vertical by convective updrafts and downdrafts. For an updraft moving with the mean flow at some level this process tends to produce rotation on the flanks of the updraft. If the updraft can find a way to move laterally to the shear vector, however, then a steady state vortex can develop coincident with the updraft position. The required condition is that the vertical wind shear relative to the updraft turns with height, i.e. that the hodograph rotates. This can be seen from the steady state equation for vertical vorticity, written in the form

$$u\frac{\partial\zeta}{\partial x} + v\frac{\partial\zeta}{\partial y} + w\frac{\partial\zeta}{\partial z} = -\frac{\partial v}{\partial z}\frac{\partial w}{\partial x} + \frac{\partial u}{\partial z}\frac{\partial w}{\partial y} + \zeta\frac{\partial w}{\partial z} \tag{1}$$

which has a solution $\zeta = w/a$, provided that $\partial v/\partial z = -u/a$, and $\partial u/\partial z = v/a$, where a is a constant with dimensions of length. Maintenance of an updraft coincident with the vortex center is accomplished by the upward suction produced under a vortex. This can be expressed through the equation for toroidal vorticity, $q_\phi = \partial v_r/\partial z - \partial w/\partial r$, written in the form appropriate for radially symmetric variables

$$\frac{dq_\phi}{dt} = \frac{\partial}{\partial z}\left(\frac{v_\phi^2}{r}\right) - \frac{\partial b}{\partial r} - \frac{v_r}{r} q_\phi \tag{2}$$

where ϕ is the azimuth angle, r the radius, and b a buoyancy variable, equal to $-g\rho'/\bar{\rho}$ in a Boussinesq system. The buoyancy term generates overturning, while the presence of a vortex overhead accelerates that overturning proportionally to the vertical gradient of centrifugal force.

The tendency toward coincidence of updraft and vortex centers in the above idealized analysis of supercell circulation suggests that this circulation may be characterized by a large value of helicity, the dot product of vorticity and velocity. Helicity is also present in a horizontal mean flow with a rotating hodograph, for which the shear has a component perpendicular to the velocity. As indicated above, rectilinear shear flow has helicity only if the velocity reference frame is off the hodograph, pointing out the fact that helicity is not independent of Galilean transforms. Despite this inconvenience, and even because of it, interpretation of convective storm structure in terms of helicity leads to useful insights.

A simple version of the conservation equation for helicity is derived from the equation of motion for Boussinesq or anelastic motion in the form

$$\partial \underset{\sim}{\mathbf{V}}/\partial t + \underset{\sim}{\mathbf{V}} \cdot \underset{\sim}{\nabla}\underset{\sim}{\mathbf{V}} + \underset{\sim}{\nabla}\pi - \underset{\sim}{\mathbf{k}} b = 0 \tag{3}$$

where b is a buoyancy variable. The vorticity equation is given by

$$(\partial/\partial t)(\underset{\sim}{\nabla} x \underset{\sim}{\mathbf{V}}) + \underset{\sim}{\mathbf{V}} \cdot \underset{\sim}{\nabla}(\underset{\sim}{\nabla} x \underset{\sim}{\mathbf{V}}) + (\underset{\sim}{\nabla} \cdot \underset{\sim}{\mathbf{V}})\underset{\sim}{\nabla} x \underset{\sim}{\mathbf{V}} - (\underset{\sim}{\nabla} x \underset{\sim}{\mathbf{V}}) \cdot \underset{\sim}{\nabla}\underset{\sim}{\mathbf{V}} - b\underset{\sim}{\mathbf{k}} = 0. \tag{4}$$

By forming the inner product of (3) with $\underset{\sim}{\nabla} x \underset{\sim}{\mathbf{V}}$ and of (4) with $\underset{\sim}{\mathbf{V}}$ and summing, we obtain

$$\partial H/\partial t + \underset{\sim}{\nabla} \cdot (\underset{\sim}{\mathbf{V}} H) + \nabla \cdot [(\underset{\sim}{\nabla} x \underset{\sim}{\mathbf{V}})(\pi - \underset{\sim}{\mathbf{V}}^2/2)] = \zeta b + \underset{\sim}{\mathbf{k}} \cdot (\underset{\sim}{\mathbf{V}} x \underset{\sim}{\nabla} b)$$

where $H = \underset{\sim}{\mathbf{V}} \cdot \underset{\sim}{\nabla} x \underset{\sim}{\mathbf{V}}$. In the absence of buoyancy, helicity is a volume-conserved quantity. The two buoyancy-related terms on the right produce equal contributions to the volume integral, since the last can be rewritten as

$$\underset{\sim}{\mathbf{k}} \cdot (\underset{\sim}{\mathbf{V}} x \underset{\sim}{\nabla} b) = \nabla \cdot (\underset{\sim}{\mathbf{k}} x \underset{\sim}{\mathbf{V}} b) + \zeta b \ .$$

The product of buoyancy and vertical vorticity is analogous to the kinetic energy generation by buoyancy flux.

Application of the helicity concept allows for some additional interpretations of the structure and development of rotating storms. Helicity is present in the mean flow in the case of a curved hodograph, but also for a rectilinear sheared flow if the velocity reference frame is off the hodograph. Thus random vertical stirring of a sheared flow should produce a response in the form of laterally moving and rotating updrafts and downdrafts. In addition, however, buoyancy should amplify the helicity if the updraft follows the vortex center, as suggested by Equation (2).

Helicity has an important effect on the evolution of turbulence. In the case of purely helical or Beltrami flow, $(\underset{\sim}{\nabla} x \underset{\sim}{\mathbf{V}}) \mathbf{x}\ \underset{\sim}{\mathbf{V}} = 0$ and the non-linear generation of vorticity consequently vanishes. André and Lesieur[19] show an integration of a set of closure equations for moments of an assumed isotropic and homogeneous field of decaying 3-dimensional turbulence. They find that a non-helical flow with an initially almost monochromatic spectrum function rapidly disperses in wave space and develops a Kolmogoroff inertial range. Figure 3 shows the resulting spectrum after 8 nominal eddy turnover times. The spectrum function is multiplied by $k^{5/3}$, so that the approximately flat region corresponds to the inertial range. Most of the kinetic energy is found near the left edge of the spectrum. Figure 4 shows the corresponding result for an initial state with high helicity. The peak on the left side shows that the kinetic energy is about 3 times larger for the same rate of dissipation than is the case for non-helical turbulence, or alternately that the dissipation rate is about $3^{3/2} \sim 5$ times smaller for the same kinetic energy in helical as compared to ordinary turbulence.

This result suggests that the helical flow developed in a rotating thunderstorm is inherently resistant to turbulent decay. Such a conclusion seems consistent with results of numerical simulation of rotating storms by Klemp and Wilhelmson[20], which are found to be relatively insensitive to both grid resolution and the formulation of sub-grid-scale stresses. By comparison with comparably energetic non-rotating thunderstorms, those showing strong rotation are typically more durable, though they sometimes split in two or recombine with other storms. Wilhelmson and Klemp[21] showed a successful simulation of a sequence of about 6 storms which developed by multiple splitting from a single storm, with the sequence remaining intact and by their model predictable for up to 6 hours.

I suggest that the exceptional predictability exhibited in the event described may be typical of rotating storms because they act like a steady-state Beltrami flow with minimal dissipation and turbulent dissipation of the energy containing scales. This suggestion is to be tested in future work by energy budget analysis of simulated (and possibly real) storms.

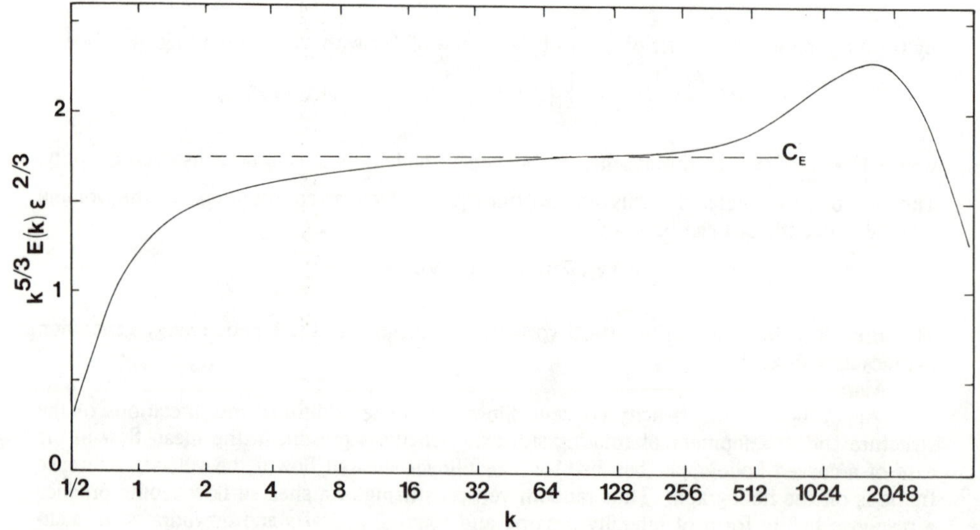

Fig. 3. Energy spectrum of decaying isotropic turbulence determined from a closure theory calculation by André and Lesieur[19]. The ordinate is weighted by $k^{5/3}$.

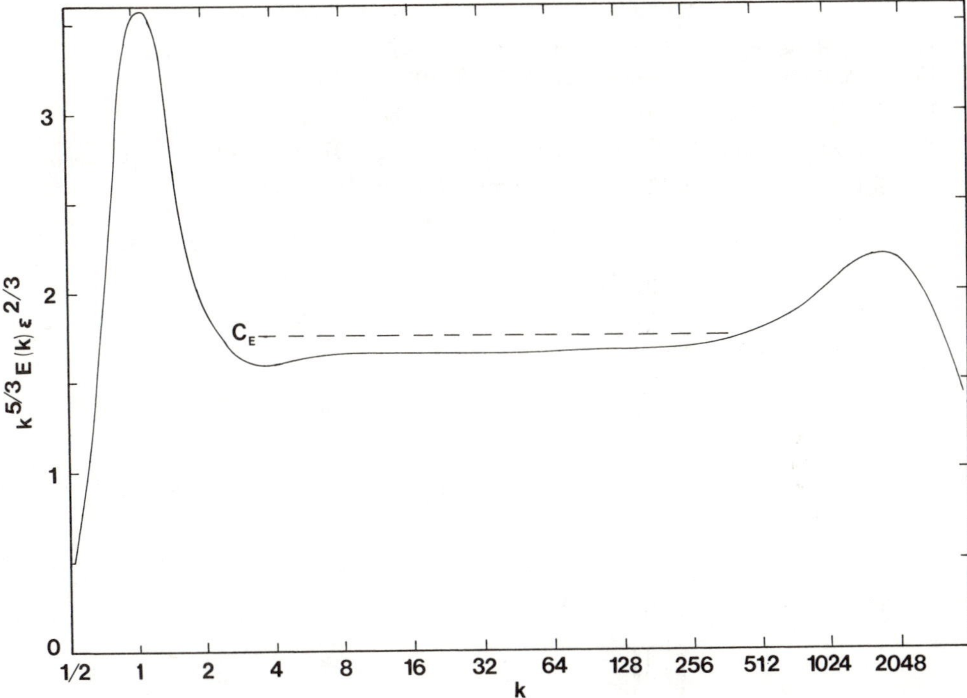

Fig. 4. Similar to Figure 3 except that the initial state contained high helicity (from André and Lesieur[16]).

REFERENCES

1. E.N. Lorenz, Tellus, **21**, 289-307 (1969).
2. B. Mandlebrot, J. Fluid Mech., **62**, 331-358 (1974).
3. B. Mandlebrot, Fractals, form, chance and dimension, (Freeman and Co., San Francisco, 1977) p. 365.
4. B. Mandlebrot, The fractal geometry of nature, (Freeman and Co., San Francisco, 1982) p. 461.
5. R.A. Anthes, these proceedings.
6. D.K. Lilly and E.L. Petersen, Tellus, **35A**, (1983).
7. R.M. Endlich, R.C. Singleton and J.W. Kaufman, J. Atmos. Sci., **26**, 1030-1041 (1969).
8. B.B. Balsley, M. Crochet, W.L. Ecklund, D.A. Carter, A.C. Riddle and R. Garello, Paper presented at the Fourth Conference on Meteorology of the Upper Atmosphere, March 22-25, 1983, Boston. American Meteorological Society.
9. D.K. Lilly, J. Atmos. Sci., **40**, (1983).
10. K.S. Gage, J. Atmos. Sci., **36**, 1950-1954 (1979).
11. R.H. Kraichnan, Phys. Fluids, **10**, 1417-1423 (1961).
12. J.J. Riley, R.W. Metcalfe and M.A. Weissman, Nonlinear Properties of Internal Waves. Ed. Bruce J. West, AIP Conf. Proceed. **79**, (1981) p. 253.
13. T.E. Van Zandt, Geophys. Res. Lett., **9**, 575-578 (1982).
14. C. Garrett and W. Munk, Geophys.Fluid Dynamics, **2**, 225-264 (1972).
15. C. Garrett and W. Munk, J. Geophys. Res., **80**, 291-297 (1975).
16. C.H. McComas and P. Müller, J. Phys. Oceanog., **11**, 970-986 (1981).

17. K.A. Browning, J. Atmos. Sci., **21**, 634-639 (1964).
18. K.A. Browning, J. Atmos. Sci., **22**, 664-668 (1965).
19. J.C. André and M. Lesieur, J. Fluid Mech., **81**, 187-207 (1977).
20. J.B. Klemp and R.B. Wilhelmson, J. Atmos. Sci., **35**, 1070-1096 (1978).
21 R.B. Wilhelmson and J.B. Klemp, J. Atmos. Sci., **38**, 1581-1600 (1981).

PREDICTABILITY AND FRONTOGENESIS

A.F. Bennett

Institute of Ocean Sciences
Sidney B.C., Canada

ABSTRACT

Rigorous but crude upper estimates are given for the rate of growth of the total energy in the difference between two solutions of the quasigeostrophic equations with different initial values. Four cases are considered. These combine non-dissipative motion and dissipative motion at unit Prandtl number, with motion between isothermal horizontal surfaces and motion between horizontal surfaces upon which the temperature is allowed to evolve. The growth rate estimates are proportional to the maximal strain rates of either of the two solutions. The strain rates may grow in time without bound, due to the action of the classical Bergeron frontogenetic mechanism. The problem of refining the growth rate estimates is discussed.

INTRODUCTION

A number of proofs of classical well-posedness (and in some cases ill-posedness) of the quasigeostrophic equations in a variety of settings have been obtained recently,[1-5] using techniques developed for the barotropic equations.[6-8] A problem is classically well-posed if it *has* a classical (that is, suitable differentiable) solution which is unique, and which depends continuously upon the data. The fundamental step in each of the above mentioned proofs of existence of a classical solution is the construction of upper bounds or *estimates* for various derivatives of the (potential) vorticity, and in some cases the (potential) temperatures on the horizontal bounding surfaces. These derivatives also bound the strain rates of the motion; the latter in turn bound the growth rate of the total energy in the difference between two solutions having different initial values or subject to different forcing. Uniqueness and continuous dependence is then readily established.

It is clear that the mathematical constructions in the proofs of well-posedness provide quantitative estimates of the *predictability* of the quasigeostropic equations, if we quantify predictability in terms of the energy in finite-amplitude perturbations of the motion. It is also clear that the fundamental step of estimating the derivatives of vorticity and temperature requires an estimate of Bergeron frontogenesis.[9] The latter, which is the only frontogenetic mechanism is quasigeostropic motion, is active where the deformation field of the motion is crowding the contours of vorticity or temperature. The estimates given below indicate that the action is particularly strong on the horizontal bounding surfaces. In fact the temperature gradient estimates "blow up" (become infinite) in a finite time, even in the presence of dissipation. Explicit examples given below suggest that the estimates may not be excessively crude. The question of refining the estimates is shown below to be reducible to a geometrical problem.

THE QUASIGEOSTROPHIC EQUATIONS

The equations are:

$$\Delta\psi + \rho^{-1}(\rho\epsilon\psi_z)_z = \omega \quad \text{for } 0<z<h \tag{1}$$

$$\psi_z = \theta^0 \quad \text{on } z = 0 \tag{2}$$

$$\psi_z = \theta^h \text{ on } z = h \tag{3}$$

$$\underset{\sim}{u} = \hat{\underset{\sim}{k}} x \underset{\sim}{\nabla} \psi \text{ for } 0 \leqslant z \leqslant h \tag{4}$$

$$\omega_t + \underset{\sim}{u}.\underset{\sim}{\nabla}\omega - K\Delta\omega = S \text{ for } 0<z<h \tag{5}$$

$$\theta_t^o + \underset{\sim}{u}.\underset{\sim}{\nabla}\theta^o - K\Delta\theta^o = H^o \text{ on } z = 0 \tag{6}$$

$$\theta_t^h + \underset{\sim}{u}.\underset{\sim}{\nabla}\theta^h - K\Delta\theta^h = H^h \text{ on } z = h \tag{7}$$

where x, y, z and t denote respectively cartesian coordinates and time, with subscripts denoting partial derivatives; ψ, ω, θ^o, θ^h and $\underset{\sim}{u} = (u,v)$ denote respectively the streamfunction, vorticity, surface temperatures and horizontal velocity corresponding to $\underset{\sim}{x} = (x,y)$. The horizontal gradient operator is denoted by $\underset{\sim}{\nabla}$; the horizontal Laplacian operator by $\Delta=\underset{\sim}{\nabla}.\underset{\sim}{\nabla}$. The vertical unit vector is denoted by $\hat{\underset{\sim}{k}}$. Body sources of vorticity, and surface heat sources are denoted respectively by S, H^o and H^h. The imposed density and inverse static stability profiles are respectively $\rho=\rho(z)$ and $\epsilon=\epsilon(z)$. The horizontal eddy diffusivities of momentum and heat are both assumed equal to K, a positive constant (i.e. unit turbulent Prandtl number, or Reynolds' analogy), while the vertical diffusivities have been assumed to vanish. Note also the neglect of the β-effect; otherwise the right hand side of (5) would have been $S-\beta v$ where β is a positive constant.

It is assumed that all fields are periodic in each of the horizontal directions x and y, with unit period in each direction. For example,

$$\psi(x \pm n, y \pm m, z, t) = \psi(x,y,z,t), \tag{8}$$

where n and m are any integers. For a detailed discussion of the reasons for adopting doubly periodic boundary conditions and the Reynolds' analogy, see [3] and [5].

Finally initial conditions are required by (5), (6) and (7):

$$\omega(\underset{\sim}{x},z,0) = \omega_I(\underset{\sim}{x},z) \tag{9}$$

$$\theta^o(\underset{\sim}{x},0) = \theta_I^o(\underset{\sim}{x}) \tag{10}$$

and

$$\theta^h(\underset{\sim}{x},0) = \theta_I^h(\underset{\sim}{x}) \tag{11}$$

where ω_I, θ_I^o and θ_I^h are doubly periodic.

ESTIMATES OF PREDICTABILITY

Let $\psi^{(1)}$ and $\psi^{(2)}$ be two solutions of (1)-(11) corresponding to the same forcing, but with different initial values. Let $\Psi = \psi^{(1)} - \psi^{(2)}$ be the difference between the two streamfunctions. The kinetic energy plus available potential energy, or total energy in Ψ is given by

$$E(t) = \tfrac{1}{2} \int \int \int_o^h \rho(|\underset{\sim}{\nabla}\Psi|^2 + \epsilon\Psi_z^2)dzdxdy \tag{12}$$

where the horizontal integration is over any unit square with a pair of opposite sides parallel to the x-axis. It may be shown that

$$\frac{d}{dt}E(t) = -K \int \int \int_o^h \rho\{(\Delta\Psi)^2 + \epsilon|\underset{\sim}{\nabla} \Psi_z|^2\}dzdxdy$$

$$+\int \int \int_o^h \rho\{(\Psi_y^2 - \Psi_x^2)\psi_{xy}^{(1)} + \Psi_y\Psi_x\left(\psi_{xx}^{(1)} - \psi_{yy}^{(1)}\right)$$

$$-\epsilon\, \Psi_z(\psi_{yz}^{(1)}\Psi_x + \psi_{xz}^{(1)}\Psi_y)\}dzdxdy \qquad (13)$$

and hence that

$$\frac{dE(t)}{dt} \leqslant G(t)E(t) \qquad (14)$$

where the growth rate estimate G is given by

$$G(t) = (5/2) \sup_{\underset{\sim}{x},\, z} \ \max_{i=1,2} \max \{|\psi_{xx}^{(i)}|, |\psi_{xy}^{(i)}|, |\psi_{yy}^{(i)}|,$$

$$\epsilon^{1/2}|\psi_{xz}^{(i)}|, \epsilon^{1/2}|\psi_{yz}^{(i)}|\} \qquad (15)$$

An immediate consequence of (14) is that

$$E(t) \leqslant E(0)e^{F(t)} \qquad (16)$$

where

$$F(t) = \int_0^t G(t')dt' \qquad (17)$$

and so if it can be shown that F(t) is finite for a given value of t, then $E(t) \to 0$ as $E(0) \to 0$. That is, solutions of (1)-(11) depend continuously upon the data. In particular if $E(0)=0$ then $E(t)=0$; that is, the solution of the initial value problem is unique. Note that the dissipation term on the right hand side of (13) has been discarded in the passage to (14); we shall return to this point subsequently but mention for now that the growth rate G is thus explicitly determined only by the strain rates $\psi_{xx}^{(i)} = v_x^{(i)}$, etc. in (15), and not by diffusion. However, it will be seen that estimates of the strain rates do depend qualitatively and quantitatively upon the diffusivity when the latter is present.

The growth rate estimate G will now be given in four cases.

Case (a) (Non-dissipative motion between isothermal surfaces)

In this case $K=0$ and $\theta_I^o \equiv \theta_I^h \equiv H^o \equiv H^h \equiv 0$. It is clear that $\theta^o \equiv \theta^h \equiv 0$ for all $t \geqslant 0$, thus ψ is determined solely by ω. In fact a consequence of (1)-(3) is that[1]

$$D_2\psi \leqslant B_1 \ln(B_2 + B_3\|\underset{\sim}{\nabla}\omega\|_{L_p}) \qquad (18)$$

where $D_2\psi$ denotes any of the second order spatial partial derivatives appearing in (15), while the L_p - norm of $\underset{\sim}{\nabla}\omega$ is defined by

$$\|\underset{\sim}{\nabla}\omega\|_{L_p} \equiv \left(\int \int \int_o^h |\underset{\sim}{\nabla}\omega|^p dzdxdy\right)^{\frac{1}{p}} \qquad (19)$$

The coefficients B_1–B_3 in (18) depend upon ρ, ϵ, h, p and $\|\omega_I\|_{L_p}$. Inequality (18) is valid only if $p > 3$. In order to estimate $\|\underset{\sim}{\nabla}\omega\|_{L_p}$ we use the characteristic representation of the solution of the hyperbolic Equation (5). Remember that K=0 in this case. In fluid dynamical parlance, we shall use the quasigeostrophic particle paths $\underset{\sim}{X}=\underset{\sim}{X}(\underset{\sim}{x},z,t,\sigma)$ where $\underset{\sim}{X} = (X,Y)$ denotes the horizontal position of a particle at time σ, given that it was at $\underset{\sim}{x}$, on level z at time t. As usual we have

$$\frac{d}{d\sigma}\underset{\sim}{X} = \underset{\sim}{u}(\underset{\sim}{X},z,\sigma) \tag{20}$$

with

$$\underset{\sim}{X}\,(\underset{\sim}{x},z,t,t\,) = \underset{\sim}{x} \tag{21}$$

Along the particle paths, the gradient of (5) becomes (assuming for simplicity that $S \equiv 0$)

$$\frac{d}{d\sigma}\,\underset{\sim}{\nabla}\omega = -(\underset{\sim}{u}_x.\underset{\sim}{\nabla}\omega,\ \underset{\sim}{u}_y.\underset{\sim}{\nabla}\omega) \tag{22}$$

and hence

$$\frac{d}{d\sigma}\|\underset{\sim}{\nabla}\omega\|_{L_p} \leqslant g(t)\|\underset{\sim}{\nabla}\omega\|_{L_p} \tag{23}$$

where

$$g(t) = \sup_{\underset{\sim}{x},z} \max\,\{|\psi_{xx}|,|\psi_{xy}|,|\psi_{yy}|\} \tag{24}$$

Combining (24), (23), (18) and (9) leads to

$$\|\underset{\sim}{\nabla}\omega\|_{L_p} \leqslant B_3^{-1}\left|(B_2 + B_3\|\underset{\sim}{\nabla}\omega_I\|_{L_p})^{e^{B_1 t}} - B_2\right| \tag{25}$$

and so

$$D_2\psi \leqslant B_1 e^{B_1 t}\ln(B_2 + B_3\|\nabla\omega_I\|_{L_p}) \tag{26}$$

The estimate (26) of the exponential growth rate of E(t) is itself growing exponentially. Such a growth rate could only be sustained briefly, since we know that

$$\begin{aligned} E(\Psi) = E(\psi^{(1)} - \psi^{(2)}) &\leqslant E(\psi^{(1)}) + E(\psi^{(2)}) \\ &\leqslant E(\psi_I^{(1)}) + E(\psi_I^{(2)}) \end{aligned} \tag{28}$$

where

$$\psi_I^{(i)} = \psi^{(i)}(\underset{\sim}{x},z,0)\ .$$

Nevertheless the estimate (26) suggests very poor predictability.

Case (b) (Non-dissipative "irrotational" motion with evolving surface temperatures)

In this case K=0, and $\omega_I \equiv \theta_I^o \equiv S \equiv H^o \equiv 0$ so $\omega \equiv \theta^o \equiv 0$, and ψ is determined solely by θ^h.

It may be shown using (1)-(3) that

$$D_2\psi \leqslant B_4\|\theta^h\|_{C^{1,\lambda}} \tag{29}$$

where B_4 depends upon ρ, ϵ and h, while the $C^{1,\lambda}$-norm or Hölder- norm of θ^h is defined by

$$\|\theta^h\|_{C^{1,\lambda}} \equiv \|\theta^h\|_{C^1} + \sup_{\underset{\sim}{x}, \underset{\sim}{x}'} \left\{ \frac{|\underset{\sim}{\nabla}\theta^h(\underset{\sim}{x}') - \underset{\sim}{\nabla}\theta^h(\underset{\sim}{x})|}{|\underset{\sim}{x}'-\underset{\sim}{x}|^{\lambda}} \right\} \tag{30}$$

$$(0<\lambda\leqslant 1)$$

where

$$\|\theta^h\|_{C^1} \equiv \sup_{\underset{\sim}{x}} \left\{ |\theta^h(\underset{\sim}{x})| + |\underset{\sim}{\nabla}\theta^h(\underset{\sim}{x})| \right\} \tag{31}$$

Once again the particle path representation may be used to derive, (assuming for simplicity that $H^h \equiv 0$

$$\frac{d}{d\sigma}\|\theta^h\|_{C^{1,\lambda}} \leqslant B_6 g(t)\|\theta^h\|_{C^{1,\lambda}} \tag{32}$$

where B_6 is a numerical factor.

Combining (32), (24), (29) and (11) leads to

$$\|\theta^h\|_{C^{1,\lambda}} \leqslant \|\theta_I^h\|_{C^{1,\lambda}} \, (1-\|\theta_I^h\|_{C^{1,\lambda}} \, B_4B_6t)^{-1} \tag{33}$$

and so

$$D^2\psi \leqslant B_4\|\theta_I^h\|_{C^{1,\lambda}}(1-\|\theta_I^h\|_{C^{1,\lambda}} \, B_4B_6t)^{-1} \tag{34}$$

In this case the growth rate estimate becomes infinite in a finite time $t = (B_4B_6 \, \|\theta_I^h\|_{C^{1,\lambda}})^{-1}$. This is due to the availability only of the linear estimate (29) instead of the stronger logarithmic estimate (18).

Case (c) (Dissipative motion between isothermal surfaces)

The assumptions are the same as in (a), except here $K>0$. We must use Hölder-norms; the horizontal strain rate is now estimated by

$$D^2\psi \leqslant B_7\|\omega\|_{C^{\lambda}} \quad (0<\lambda\leqslant 1) \tag{35}$$

where B_7 depends on ρ, ϵ and h. Since $K>0$ the vorticity Equation (5) is parabolic, but a representation of the solution may be made using the heat potentials or fundamental solutions of (5)[4]. It may be shown that

$$\|\omega\|_{C^{\lambda}} \leqslant L_1 + L_3 \tag{36}$$

and so

$$D^2\psi = B_7(L_1 + L_2) \tag{37}$$

where

$$L_1 = \|\omega_I\|_{C^o} \tag{38}$$

$$L_2 = \|\underset{\sim}{\nabla}\omega_I\|_{C^o} + 4K \, B_8\|\omega_I\|_{C^2}(K+B_7L_1)t^{1/2} \tag{39}$$

and

$$L_3 = \|\omega_I\|_{C^{1,\lambda}} + KL_1L_2t + L_2^\lambda(2L_1)^{1-\lambda}. \tag{40}$$

The coefficient B_8 depends upon K, B_7, $\|\omega_I\|_{C^o}$ and t, but in an unknown manner. It is only known that B_8 is finite if its parameters are finite. However, there is evidence that it grows algebraically in t, so the growth rate estimate in this dissipative case is algebraic rather than exponential as in the corresponding non-dissipative Case (a): compare (37) with (26).

Case (d) (Dissipative motion with evolving surface temperatures).

The assumptions are the same as in (b) except here $K>0$. As in Case (b), the growth rate estimate becomes infinite in a finite time given here by $t = (4KB_7B_8\|\theta_I^h\|_C{}^2)^{-2}$. However, since B_8 depends upon t in an unknown way this finite "blow-up" time is not explicitly known.

To conclude this section we return to the neglect of the dissipation term in the passage from (13) to (14). For the case of dissipative barotropic motion ($\psi_z \equiv 0$) subject to no-slip boundary conditions, Ladyzhenskaya[10] has retained the explicit dissipation in the estimate of the perturbation energy, and has thereby derived the following estimate for the exponent in (16)

$$F(t) \leqslant \{2\|\underset{\sim}{u}_I\|_{L_2}{}^2 + 3\left(\int_o^t \|\underset{\sim}{M}\|_{L_2} dt'\right)^2\}\nu^{-2} \ , \tag{42}$$

where $\underset{\sim}{u}_I$ is the initial velocity field, $\rho\underset{\sim}{M}$ is the force field in the momentum equations and ν is the kinematic viscosity. Note that in the absence of forcing, F is uniformly bounded in time. Ladyzhenskaya's results such as (42) are also valid for barotropic motion in doubly periodic domains, but her methods do not appear to be applicable to the baroclinic quasi-geostrophic equations. Thus it is not clear whether the qualitative differences in the growth rate estimates for barotropic and baroclinic motion are indicative of different predictability, or merely of inadequacy of the mathematical techniques in [1]-[5]. Actually, (42) is not all that informative: its right hand side is at least twice the square of the initial domain Reynolds' number.

REFINED ESTIMATES

Painstaking analysis to find the minimal values of the coefficients B_1–B_7 would refine the estimates given above. This would be of marginal significance. On the other hand qualitative refinements of the estimates would be of great significance. Consider first (18) in Case (a); this bounds the strain rates by essentially the logarithm of the L_p– norm of the vorticity gradient. The latter is not conserved by the motion, and our estimate of its evolution in time is the repeated exponential (25). Suppose we could bound the strain rates by some function of the C^o– norm (the supremum) of the vorticity itself. The latter is conserved by the motion so we would have a time-independent bound for $D_2\psi$ and hence an at most exponential growth in time of the perturbation energy E. However the following example[11], originally due to Sobolev, precludes that possibility.

Example (a)

Consider the two-dimensional domain $|\underset{\sim}{x}| \leqslant \frac{1}{2}$. The streamfunction $\psi(\underset{\sim}{x}) = xy(\ln|\ln|\underset{\sim}{x}||-\ln\ln 2)$ vanishes on the boundary, while the vorticity $\omega = \Delta\psi = xy|\underset{\sim}{x}|^{-2}\left\{4(\ln|\underset{\sim}{x}|)^{-1}-(\ln|\underset{\sim}{x}|)^{-2}\right\}$ is continuous for $|\underset{\sim}{x}| \leqslant \frac{1}{2}$ and so has a supremum. However the mixed derivative $\psi_{xy} = \ln|\ln|\underset{\sim}{x}||$ + (a bounded function) is singular at the origin, and so cannot be bounded by the supremum of ω.

Inequality (35) shows that $D_2\psi$ is bounded if ω is Hölder-continuous, while (29) shows that the sensitivity of $D_2\psi$ to the smoothness of θ^o and θ^h is an order greater. Some insight into this may be gained by inspecting the integral representation of $D_2\psi$ in terms of ω, θ^o and θ^h using the appropriate Green's function. The singularity in the volume integral of ω is $O(r^{-1}dr)$ while the singularities in the surface integrals of θ^o and θ^h are $O(r^{-2}dr)$, none of which is integrable. Standard arguments[12] show that Hölder-continuity of ω is sufficient to "smother" the singularity in the volume integral, but it is necessary to assume that $\underset{\sim}{\nabla}\theta^o$ and $\underset{\sim}{\nabla}\theta^h$ are Holder-continuous in order to "smother" the singularities in the surface integrals. Thus it is the elliptic problem (1)-(3), or in meteorological parlance the diagnostic relationship between ψ, ω, θ^o and θ^h which shows that frontogenesis should appear "first" on the horizontal bounding surfaces.

Now consider the problem of estimating the evolution of $\underset{\sim}{\nabla}\theta^h$ in Case (b). In particular we used the inequality

$$|(\underset{\sim}{u}_x.\underset{\sim}{\nabla}\theta^h,\ \underset{\sim}{u}_y.\underset{\sim}{\nabla}\theta^h)| \leqslant \sup_{\underset{\sim}{x}}(|\underset{\sim}{u}_x|,\ |\underset{\sim}{u}_y|)\sup_{\underset{\sim}{x}}|\underset{\sim}{\nabla}\theta^h|\ . \tag{43}$$

That is, we estimated the Bergeron frontogenetic term by assuming (1) that the compression axis of the deformation field was everywhere aligned with the temperature gradients and (2) that the deformation rates and temperature gradients were everywhere equal to their supreme values. This is possible if the motion is not bounded horizontally[5]

Example (b)

Let $\psi = (yz-xy)A(t)$ where $A(t)=A(0)\ (1-A(0)t)^{-1}$, which tends to infinitity as $t\rightarrow A(0)^{-1}$. This is an exact solution of this unforced non-dissipative and dissipative quasi-geostrophic equations. In particular $\underset{\sim}{u} = (x-z,-y)A(t)$, $\underset{\sim}{u}_x = (1,0)A(t)$, $\underset{\sim}{u}_y = (0,-1)A(t)$, $\theta^h = yA(t)$ and $\underset{\sim}{\nabla}\theta^h = (0,1)A(t)$, so $(\underset{\sim}{u}_x.\underset{\sim}{\nabla}\theta^h, \underset{\sim}{u}_y.\underset{\sim}{\nabla}\theta^h) = (1,0)A^2(t)$.

It is difficult to envisage such extreme frontogenesis in motion subject to any boundary conditions in the horizontal.

Clearly what is required is an estimate better than (43), not based on inequalities which hold for all functions of a certain class, but based instead on estimates which hold for velocity and temperature fields related by the diagnostic Equations (1)-(4). That is, we require a "geometrical" closure of the nonlinear equation for the temperature gradients if we are to make more refined estimates of loss of predictability due to frontogenesis.

Corrigendum Equation (23) in Reference [5] should read:

$$DE = K\int_0^t\int_0^h\int_0^1\int_0^1 \rho\left\{\omega^2 + \epsilon|\underset{\sim}{\nabla}\psi_z|^2\right\}dxdydzdt \tag{23}$$

REFERENCES

1. A.F. Bennett and P.E. Kloeden, "The simplified quasigeostrophic equations: existence and uniqueness of strong solutions," Mathematika, **27**, 287-311 (1980).
2. A.F. Bennett and P.E. Kloeden, "The periodic quasigeostrophic equations: existence and uniqueness of strong solutions," Proc. R. Soc. Edinburgh, **91A**, 185-203 (1982).
3. A.F. Bennett and P.E. Kloeden, "The quasigeostrophic equations: approximation, predictability and equilibrium spectra of solutions," Q.J.R. Meteorol. Soc., **107**, 121-136 (1981).
4. A.F. Bennett and P.E. Kloeden, "The dissipative quasigeostrophic equations," Mathematika **28**, 265-285 (1981).

5. A.F. Bennett and P.E. Kloeden, "Dissipative quasigeostrophic motion and ocean modelling," Geophys. Astrophys. Fluid Dynamics, **18**, 253-262 (1981).
6. V.I. Judovich, "A two-dimensional problem of unsteady flow of an ideal uncompressible fluid across a given domain," Matem. Sbornik, **64**(106), 562-588 (1964). (English translation: Transl. Amer. Math. Soc., **57**, 277-304 (1966).)
7. T. Kato, "On classical solutions of the two-dimensional non-stationary Euler equation," Arch. Rat. Mech. Anal., **25**, 188-200 (1967).
8. F.J. McGrath, "Nonstationary plane flow of viscous and ideal fluids," Arch. Rat. Mech. Anal., **27**, 329-348 (1968).
9. T. Bergeron, "Über die dreidimensional verknupfende Wetteranalyse I," Geophys. Publ., **5**, 1-111 (1932).
10. O.A. Ladyzhenskaya, The Mathematical Theory of Viscous Incompressible Flow, 2nd English Edition, Gordon and Breach, New York (1969).
11. V.I. Judovich, "Some bounds for the solutions of elliptic equations," Mathem. Sbornik, **59**(101) supplement, 229-244 (1962). (English translation: Transl. Amer. Math. Soc., **57**, 277-304 (1966).)
12. R. Courant and D. Hilbert, Methods of Mathematical Physics, Vol. II, Interscience, New York (1961).

PREDICTABILITY OF THE LARGE SCALES OF FREELY EVOLVING THREE AND TWO-DIMENSIONAL TURBULENCE

O. Metais,[1] J.P. Chollet[1] and M. Lesieur[2]

ABSTRACT

The predictability of freely evolving three and two-dimensional turbulence is studied within the formalism developed by Kraichnan[1] and Lorenz,[2] using the EDQNM theory. In the three-dimensional case, the time when the wavenumber characteristic of the error reaches the energy containing range is increased by more than a factor of 10 compared to the case of stationary turbulence. But the times to which the ratio r(t) of decorrelated energy over total energy equals 0.5 are only increased by a factor of 2 in the decaying case.

In two-dimensional turbulence, one verifies that the wavenumber characteristic of the error first decreases exponentially in the enstrophy cascade, the total error remaining low. Then in a second phase, it decreases as $(t-t_0)^{-1}$(t_0 being the time at which the error is initially injected), while r(t) increases rapidly. The times such that r(t) = 0.5 are, in units of large eddy turnover times at $t = t_0$, increased by a factor 1.8 when one compares the decaying case to the forced case. These results would allow the understanding of the appearance of three-dimensionality on the spanwise "coherent" vortices in the free shear layers.

INTRODUCTION

The origin of this work came from experimental evidence that the large scales of some particular flows (as free shear layers for instance) seemed to be much more "coherent" than what usual concepts of turbulence such as the return to isotropy allowed us to expect. Generally, these so-called large coherent structures have their integral scale which increases with time, and may correspond to a turbulence which evolves freely in a reference frame moving with the mean flow.

We think that a possible explanation for the coherence of these structures is an increased predictability, due to the fact that the turbulence is freely-evolving with time: indeed, the inverse cascade of error, which contaminates larger and larger scales, could possibly have some difficulties to catch up with the increasingly larger scales.

In what follows, we will study the predictability of freely-evolving turbulence, both in three- and two-dimensions. We study predictability from the statistical point of view developed by Kraichnan[1] and Lorenz:[2] one considers two random flows $u^{(1)}$ and $u^{(2)}$, both satisfying Navier-Stokes equations and having the same statistical properties. Initially, at time $t = t_0$, the two fields are

[1]Institut de Mécanique de Grenoble, BP 68, 38402 Saint-Martin d'Hères Cedex, France.

[2]Dept. of Aerospace Engineering, University of Southern California, Los Angeles, California 90089-1454. Permanent Affiliation (1).

assumed to be completely correlated in the large scales ($k < k_E$) and decorrelated in the small scales ($k > k_E$). We define the energy spectrum E(k), the correlated energy spectrum $E_w(k)$ and the decorrelated energy spectrum $E_\Delta(k)$ in the following way:

$$<u_i^{(1)}(\vec{k})u_i^{(1)}(\vec{k}')> = <u_i^{(2)}(\vec{k})u_i^{(2)}(\vec{k}')> = \frac{E(k)}{2\pi k^2}\delta(\vec{k}+\vec{k}') \quad (1)$$

$$<u_i^{(1)}(\vec{k})u_i^{(2)}(\vec{k}')> = \frac{E_w(k)}{2\pi k^2}\delta(\vec{k}+\vec{k}') \quad (2)$$

$$E_\Delta(k) = E(k) - E_w(k) \quad (3)$$

In two-dimensions, the factor ($2\pi k^2$) arising in (1) and (2) must be replaced by πk.

This problem has been studied by Leith and Kraichnan[3] in the case of stationary energy spectra, with the aid of the Test-Field-Model Theory developed by Kraichnan[4]. They could show, in the cases of a three-dimensional $k^{-5/3}$ energy cascade, a two-dimensional $k^{-5/3}$ inverse energy cascade, and a k^{-3} enstrophy cascade, that a given wave number k would finally be reached by an error initially confined to $k_E >> k$. The necessary time for the decorrelation is of the order of some turnover times $\tau(k)$.

The difference between stationary and freely evolving turbulence is that, in the latter case, the large structures of the flow increase with time: let $k_I(t)$ be a wave number characteristic of the large energy containing eddies; the evaluation of the time of predictability of these structures (such that $k_E(t)$ catches up with $k_I(t)$) must take into account the decrease of k_I.

We have then investigated the predictability of three- and two-dimensional freely-evolving turbulence, using the Eddy-Damped Quasi-Normal Markovian theory (EDQNM). This approximation appears to be a simplification of the Test-Field-Model, and is equivalent to it while considering inertial ranges.

The paper is organized as follows: in the first section, we consider the case of three-dimensional turbulence studied with the aid of the EDQNM approximation. In section II, some predictability results obtained from Large Eddy Simulations are presented. Finally in Section III, we apply the EDQNM to the predictability of decaying two-dimensional turbulence.

I. EDQNM APPLIED TO PREDICTABILITY OF THREE-DIMENSIONAL TURBULENCE

The EDQNM equations for the predictability problem were given by Leith[5], and it is not necessary to recall them in detail. They may be written

$$(\frac{\partial}{\partial t} + 2\nu k^2)E(k,t) = \tilde{f}(E,E) + F(k) \quad (1)$$

$$(\frac{\partial}{\partial t} + 2\nu k^2)E_\Delta(k,t) = \tilde{f}_\Delta(E,E_\Delta) \quad (2)$$

where $\tilde{f}$ and $\tilde{f}_\Delta$ are two functionals of E and E_Δ involving integrations over triads of wave numbers {kpq} such that {kpq} be the sides of a triangle. F(k) is an external forcing term. We have assumed that the injection of energy in the two fields $u^{(1)}$ and $u^{(2)}$ is done through correlated forces. In the decaying case, the forcing will be zero.

In order to check the validity of our numerical codes, which are described in previous papers[6,7], we have performed a predictability calculation on a forced stationary energy spectrum. The results are shown in figure 1. The results found for the inverse cascade of error in the Kolmogorov energy cascade are in excellent agreement with Leith and Kraichnan[3]. The time required for the error to reach the wavenumber $2k_I$ ($k_E = 2k_I$) is equal to

$$T_{2k_I} = 15.75\ \tau(k_I) \tag{3}$$

where $\tau(k_I) = \dfrac{1}{v_0 k_I} = \tau_0$ is the turnover time of the wavenumber k_I where the energy spectrum is maximum. k_E is the wavenumber where the error spectrum is maximum. v_0^2 is twice the kinetic energy of the stationary turbulence. One can also notice in Fig. 1, the behaviour of the spectra at low k: the energy spectrum exhibits a k^2 equipartition spectrum, while the error spectrum has a k^4 behaviour due to nonlocal interactions which will be discussed later.

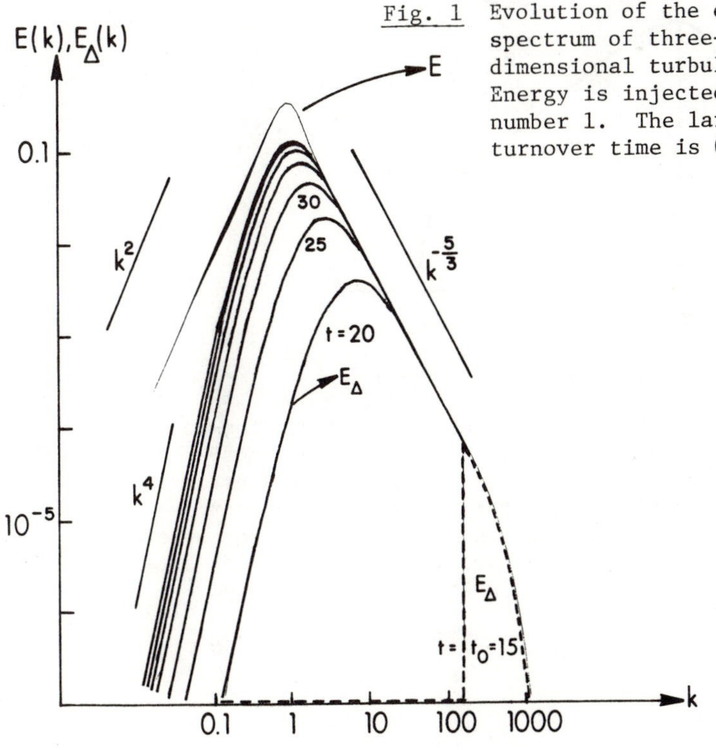

Fig. 1 Evolution of the error spectrum of three-dimensional turbulence. Energy is injected at wavenumber 1. The large eddy turnover time is 0.985.

Now, we consider the case of a three-dimensional isotropic turbulence decaying with time. It is well known[6,8] that if the characteristic exponent "s" of the energy spectrum for $k \to 0$ is lower than 4, the kinetic energy $\frac{1}{2}\langle u^2 \rangle$ and the wavenumber $k_I(t)$ (where the energy spectrum peaks) decay respectively like $t^{-\alpha_E}$ and $t^{-\alpha_\ell}$ with

$$\alpha_E = \frac{2(S+1)}{S+3}, \qquad \alpha_\ell = \frac{2}{S+3} \tag{4}$$

For S = 4, the spectrum is no more constant with time for low k, and varies like $t^\gamma k^4$. The value of the coefficient γ, determined either with the aid of the EDQNM[6] or the Test-Field Model[9] is equal to 0.16, and one then obtains:

$$\alpha_E = \frac{10}{7} - \frac{2\gamma}{7}, \qquad \alpha_\ell = \frac{2}{7} + \frac{\gamma}{7} \quad . \tag{5}$$

The latter also corresponds to the case S > 4, since a spectrum initially behaving like k^S for $k \to 0$, with S > 4, will immediately develop a k^4 range, due to the nonlocal interactions giving rise to a k^4 transfer[6].

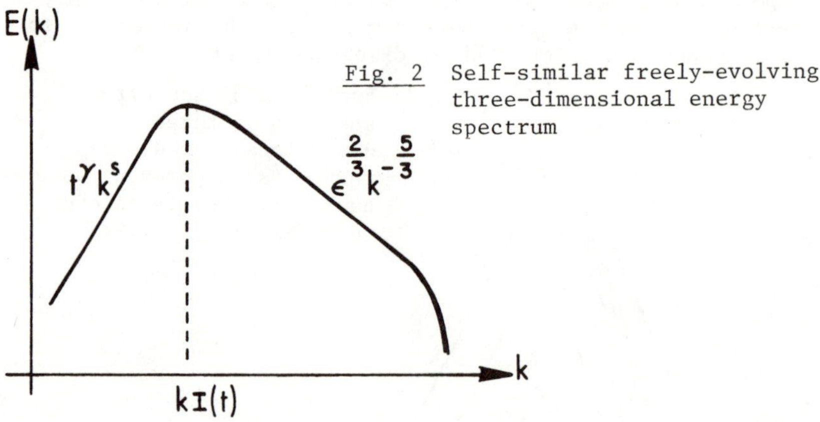

Fig. 2 Self-similar freely-evolving three-dimensional energy spectrum

We have performed a predictability calculation on a freely-evolving spectrum in both cases S = 2 (Fig. 3) and S = 4 (Fig. 4). For S = 2, the initial energy spectrum where the error is injected ($t = t_0$) is the same as the stationary energy spectrum of Fig. 1. In the case S = 4, we have taken at t = 0 an energy spectrum concentrated around $k_I(0) = 1$, and have injected the error after the energy spectrum has developed a $k^{-5/3}$ inertial range.

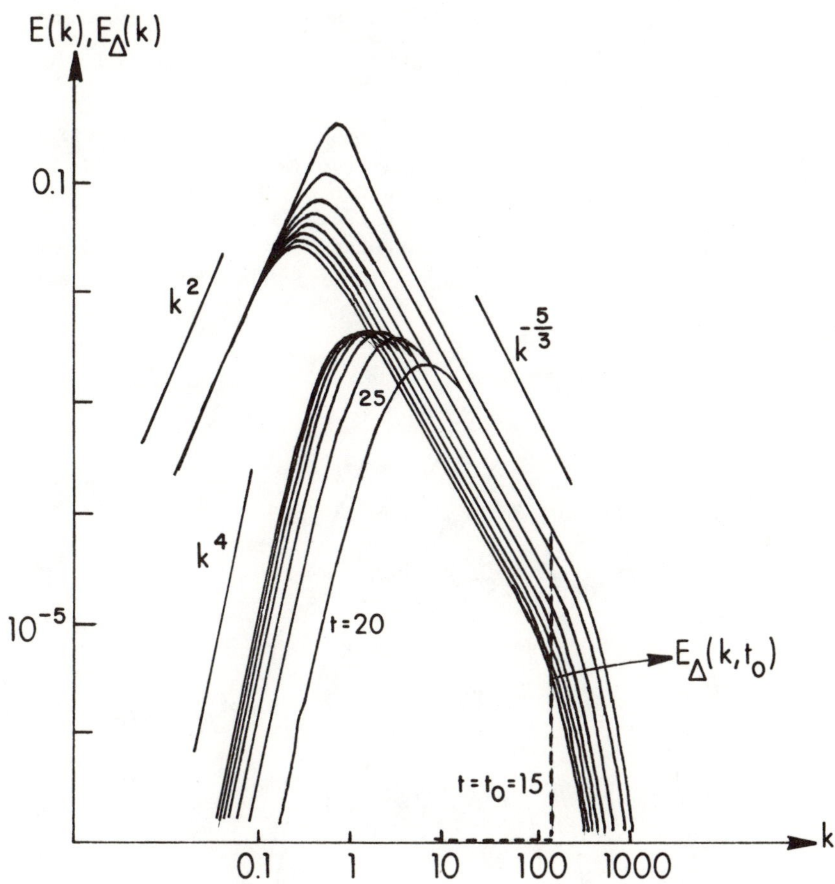

Fig. 3 Evolution of the error spectrum in a freely-evolving energy spectrum of three-dimensional turbulence, case S = 2. The large eddy turnover time τ_0 at $t = t_0$ is the same as in Fig. 1

The predictability of the "large eddies" k_I is determined by the evolution with time of the ratio k_E/k_I. We can, for instance, as already done for eq. (3), determine the predictability time as the necessary time T_{2k_I} for k_E to reach wavenumber $2k_I$. Before giving the numerical results for k_E/k_I, we are going to show how this ratio can be determined analytically as soon as only the inertial range is concerned: one considers first the error spectrum for $k \ll k_E$. An expansion of (2) with respect to the small parameter k/k_E yields, to leading order:

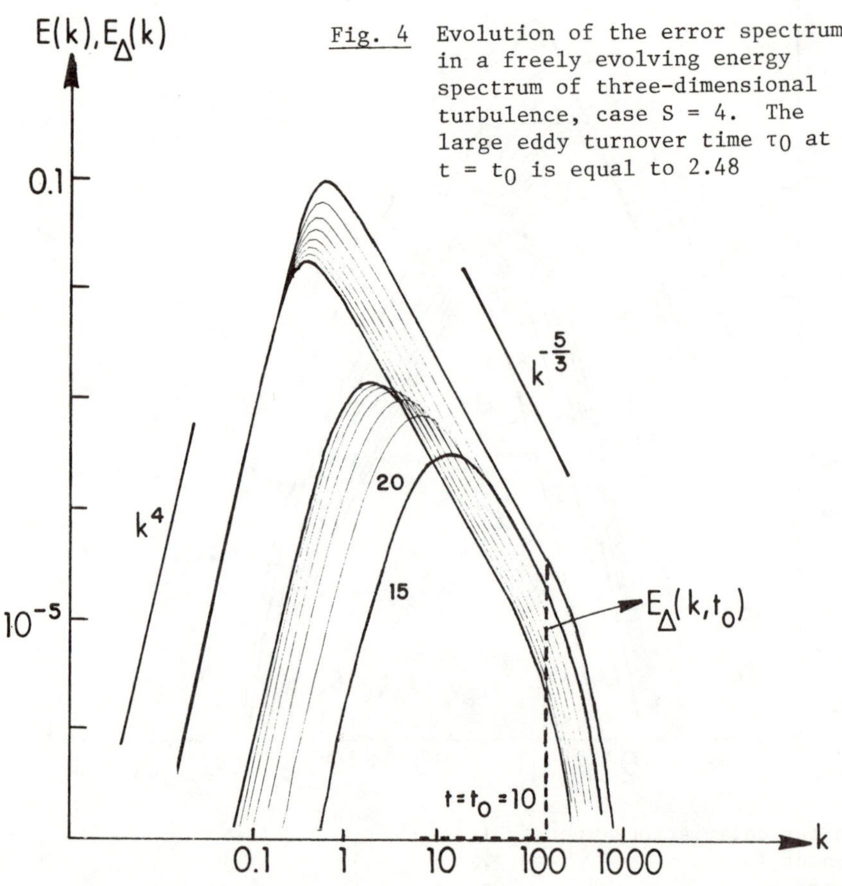

Fig. 4 Evolution of the error spectrum in a freely evolving energy spectrum of three-dimensional turbulence, case S = 4. The large eddy turnover time τ_0 at $t = t_0$ is equal to 2.48

$$\frac{\partial}{\partial t} E_\Delta(k) = \frac{14}{15}\left\{\int_{k_E}^{\infty} \theta_{opp} \frac{E^2(p)}{p^2}\, dp\right\} k^4 \tag{6}$$

where θ_{kpq} is the rate of relaxation of the triple correlations $<u_k u_p u_q>$ corresponding to the triad (kpq). These nonlocal expansions in Fourier space have been introduced by Kraichnan (see e.g. ref. 10) and developed by several authors [6,11]. Expansion leading to (6) is given in ref[3]. In the inertial range this allows us to write

$$\frac{\partial}{\partial t} E_{\Delta}(k) \sim \varepsilon k_E^{-5} k^4 \quad (7)$$

where ε is the energy flux through the cascade. From (6), it follows that the error spectrum behaves like k^4 when $k \to 0$. Assuming a schematic error spectrum represented in Fig. 5, we obtain the following system

$$\begin{cases} E_{\Delta}(k) = C(t)k^4, & k \to 0 \\ \frac{dC}{dt} \sim \varepsilon k_E^{-5} & \\ Ck_E^4 = E(k_E) & \end{cases} \quad (8)$$

ε is the dissipation rate of energy.

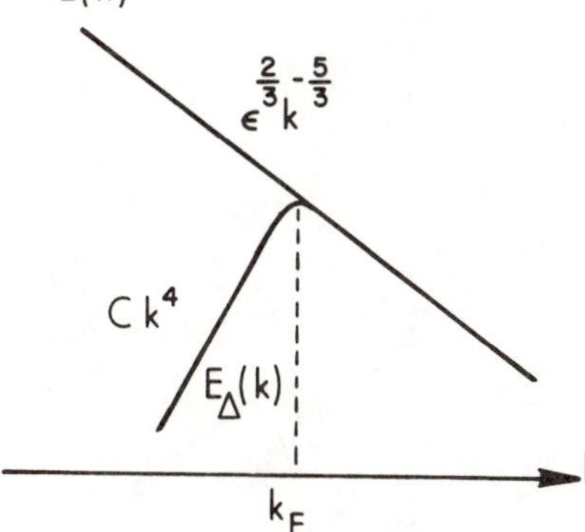

Fig. 5 Schematic error spectrum in the three-dimensional Kolmogorov energy cascade

In the stationary case, ε is constant. (8) gives

$$k_E(t) = 125 k_I \left(\frac{t-t_0}{\tau_0}\right)^{-3/2} \quad (9)$$

where the value 125 of the constant has been determined in comparison with our calculations when k_E lies in the inertial range. The exponent (-3/2) arising in (9), corresponds to an equivalent Richardson law for the diffusion of the error: by analogy with the problem of dispersion of pairs of Lagrangian tracers, one can introduce a scale k_E^{-1} characteristic of the diffusion of the error. The diffusion coefficient

$$\sigma_{\Delta} = \frac{1}{2} \frac{d}{dt} (k_E)^{-2} \quad (10)$$

will be, from (9), proportional to $k_E^{-4/3}$.

In the freely evolving case, ε is a function of α_E. The general solution of (8) is

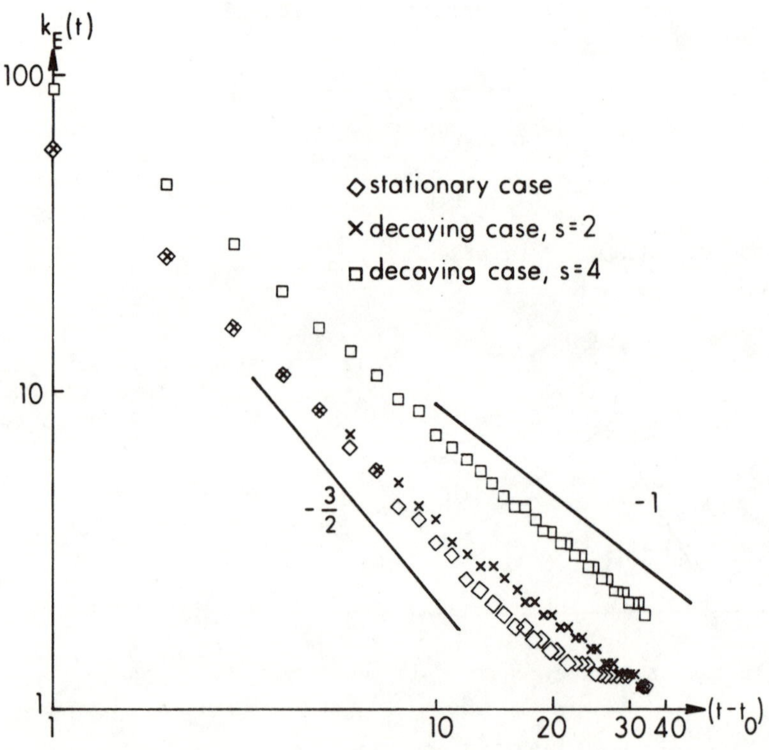

<u>Fig. 6</u> Evolution with time of the wavenumber $k_E(t)$ characteristic of the front of decorrelation, for the three numerical runs (stationary three-dimensional turbulence; decaying case S = 2; decaying case S = 4)

$$k_E(t) \sim k_I(t)\left(\frac{t}{t_0}\right)^{\left(\frac{15}{17}-\frac{21}{34}\alpha_E\right)}\left[\left(\frac{t}{t_0}\right)^{-\frac{7}{17}\alpha_E+\frac{10}{17}}-1\right]^{-3/2} . \qquad (11)$$

In the case S = 2, we have

$$k_E(t) \sim k_I(t)\left(\frac{t}{t_0}\right)^{12/85}\left[\left(\frac{t}{t_0}\right)^{8/85}-1\right]^{-3/2} \qquad (12)$$

and in the case S = 4

$$k_E(t) \sim k_I(t)\left(\frac{t}{t_0}\right)^{3\gamma/17}\left[\left(\frac{t}{t_0}\right)^{2\gamma/17}-1\right]^{-3/2}. \qquad (13)$$

Now, we turn back to the evolution of $k_E(t)$, as given by the numerical calculation: Fig. 6 shows $k_E(t)$ on a log-log plot in function of $(t-t_0)$, for the three runs: the case of stationary turbulence corresponds to an evolution of $k_E(t)$ close to the $(t-t_0)^{-3/2}$ law (though the numerical result is closer to $(t-t_0)^{-1.4}$). In the case of the decaying turbulence with S = 2, the initial $(t-t_0)^{-3/2}$ behaviour goes towards a $(t-t_0)^{-1}$ law at the end of the run. The latter exponent is not really significant (not "universal") since it corresponds more to a transient regime between the generalized Richardson law at low $(t-t_0)$ to the ultimate stage of evolution ($(t-t_0) \to \infty$) where $k_E \to k_I$ and the error will have contaminated the whole energy spectrum.

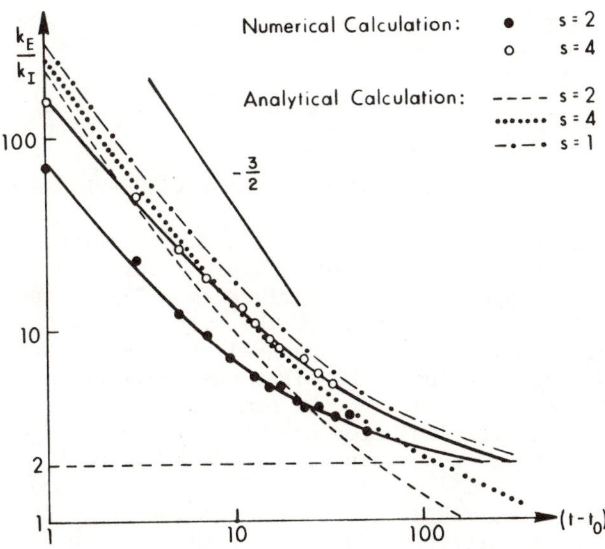

Fig. 7 Evolution with time of k_E/k_I in the decaying case. The results of the numerical calculation (extrapolated with a continuous curve) are compared with the predictions of (11), in the cases S = 2, S = 4 and S = 1

In Fig. 7, we compare, in the decaying case, the numerical results with the analytical predictions corresponding to (11). The latter are arbitrarily translated vertically on the log-log plot. It is interesting to note that the decrease of k_E/k_I, as given by the numerical results, is strongly slowed down when k_E approaches the energy containing range. Then, the inverse cascade of error goes slower than predicted by the analytical inertial range results. We have extrapolated the tendency of the numerical calculation in order to estimate the predictability times: one then finds.

$$\begin{aligned} T_{2k_I} &= 251 = 254\ \tau_0,\ S = 2 \\ T_{2k_I} &= 630 = 254\ \tau_0,\ S = 4 \end{aligned} \tag{14}$$

(we recall that τ_0 is the large eddy turnover time at $t = t_0$). (14) shows then, by comparison with (3), that the predictability time of the flow is increased by more than a factor of 10 when the turbulence decays. It must be noticed, however, that at such a time, the kinetic energy of the turbulence would be reduced by nearly a factor of 20. Finally, predictability times defined by the condition $r(t) = \frac{1}{2}$, where r is the ratio of the decorrelated energy over the kinetic energy, are substantially lower: calculation show

$$T_{r=\frac{1}{2}} \simeq T_{2k_I} = 16\ \tau_0 \text{ in the stationary case}$$

$$T_{r=\frac{1}{2}} \simeq 25\ \tau_0 \text{ in the decaying case for } S = 2$$

$$T_{r=\frac{1}{2}} \simeq 27\ \tau_0 \text{ in the decaying case for } S = 4.$$

II. LARGE EDDY SIMULATIONS RESULTS

We have performed direct numerical calculations on the basis of large Eddy Simulations of decaying Navier-Stokes equations. The model used is a spectral model derived from Siggia and Patterson[12] with a resolution of 32^3 points. The subgridscale modeling is based on a varying eddy-viscosity in spectral space[13,7]. We have performed two types of calculations: the first one consists of considering two fields $u^{(1)}$ and $u^{(2)}$ with the same cutoff k_C, and studying the evolution of an error injected near k_C (Fig. 8a). The second one consists of taking two different cutoffs $k_C^{(1)}$ and $k_C^{(2)}$ ($k_C^{(1)} < k_C^{(2)}$) and taking $u^{(1)} = u^{(2)}$ initially for $k < k_C^{(1)}$ (Fig. 8b). None of these calculations really corresponds to the predictability problem as formulated in Section I, but the results show a similar inverse cascade of error, with a k^4 error spectrum developing at low k. The same calculations show a tendency for the kinetic energy (properly extrapolated) to decay like $t^{-1.4}$, and for $k_E(t)$ to decrease like t^{-1}.

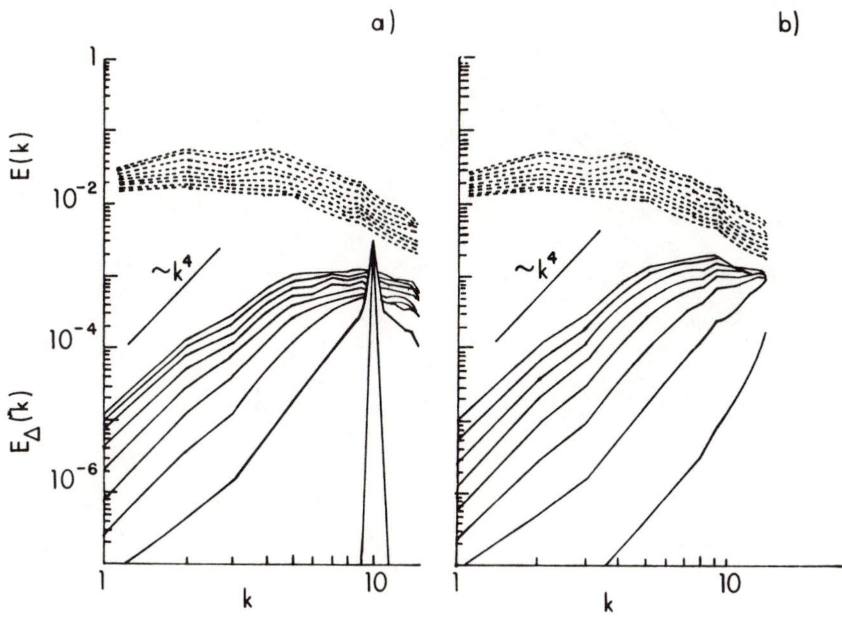

Fig. 8 Evolution of the decorrelation between two fields $u^{(1)}$ and $u^{(2)}$ in a three-dimensional decaying large-eddy-simulation. In run a), $k_c^{(1)} = k_c^{(2)} = 15$, and the error is injected at $k_c = 10$. In run b), $k_c^{(1)} = 14$ and $k_c^{(2)} = 15$. The error spectrum is shown every 0.4 sec. The initial large eddy turnover time is 0.18 sec.

III. EDQNM APPLIED TO TWO-DIMENSIONAL TURBULENCE

We have applied the EDQNM to the study of the predictability of two-dimensional turbulence, both in the stationary forcing and freely-evolving cases. In both cases, the energy spectrum at which the error is initially injected at time $t = t_0 = 70$ is a freely-evolved energy spectrum confined at $t = 0$ to $k_I = 1$ and of kinetic energy $\frac{1}{2}$. The results are presented in Fig. 9. They show a k^3 behaviour at low k both for the energy spectrum and the error spectrum, due to nonlocal interactions of the same kind as those

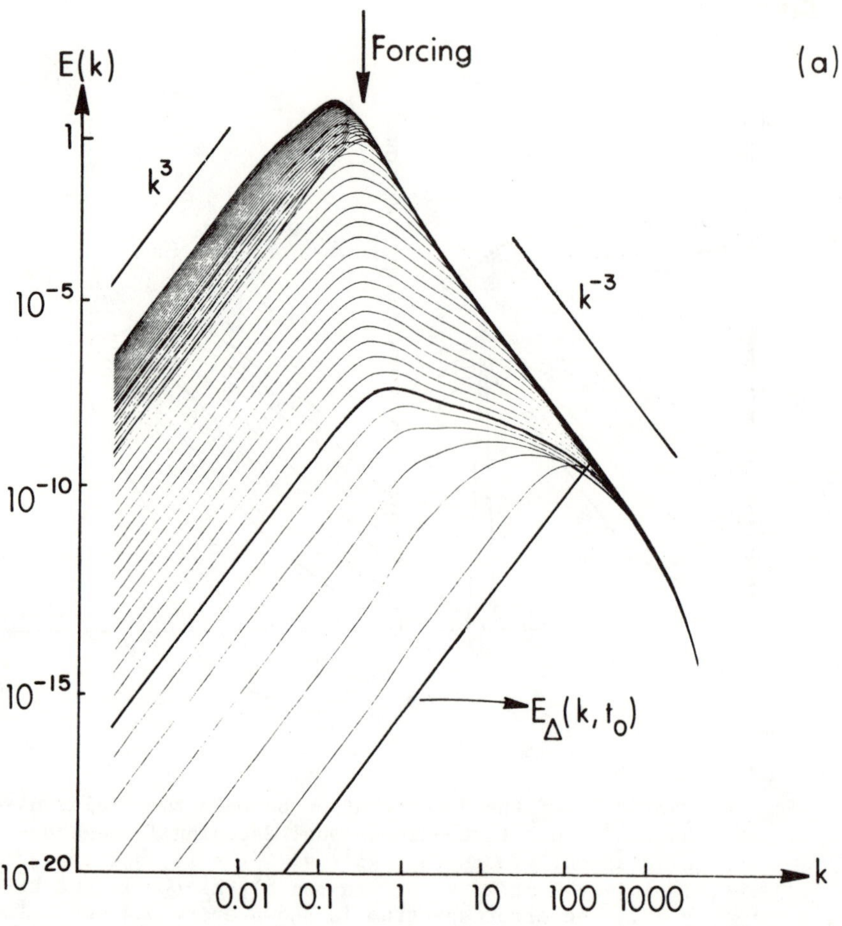

Fig. 9 EDQNM calculation showing the evolution with time of the energy and error spectra in two-dimensional turbulence. a) corresponds to the case of a stationary forcing. b) is the freely evolving case. In case a), the times of evolution are not long enough to develop a significant $k^{-5/3}$ energy cascade

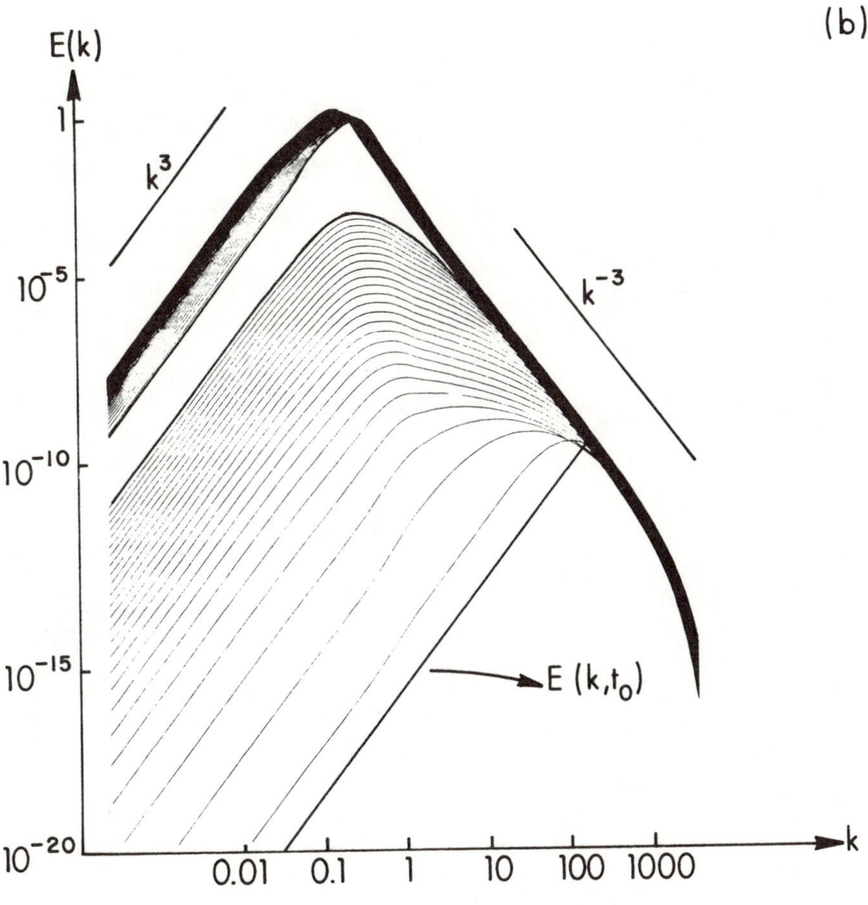

Fig. 9 (CONT.)

leading to (6). Also, one can notice that k_E rapidly reaches the energy containing range without a strong increase of the maximum of the error spectrum. Then, k_E decreases much more gently while the error increases. This drastic change of the behaviour of k_E is confirmed in Fig. 10: for $t-t_0 < 16$, k_E seems to follow an exponential law (as already shown[3] in the stationary case). Then for $t-t_0 > 16$, it follows approximately a $(t-t_0)^{-1}$ law in the decaying case. We remember that $k_I(t)$ follows a t^{-1} law[14], in such a way that k_E/k_I must follow a $t/t-t_0$ law. We have checked numerically that

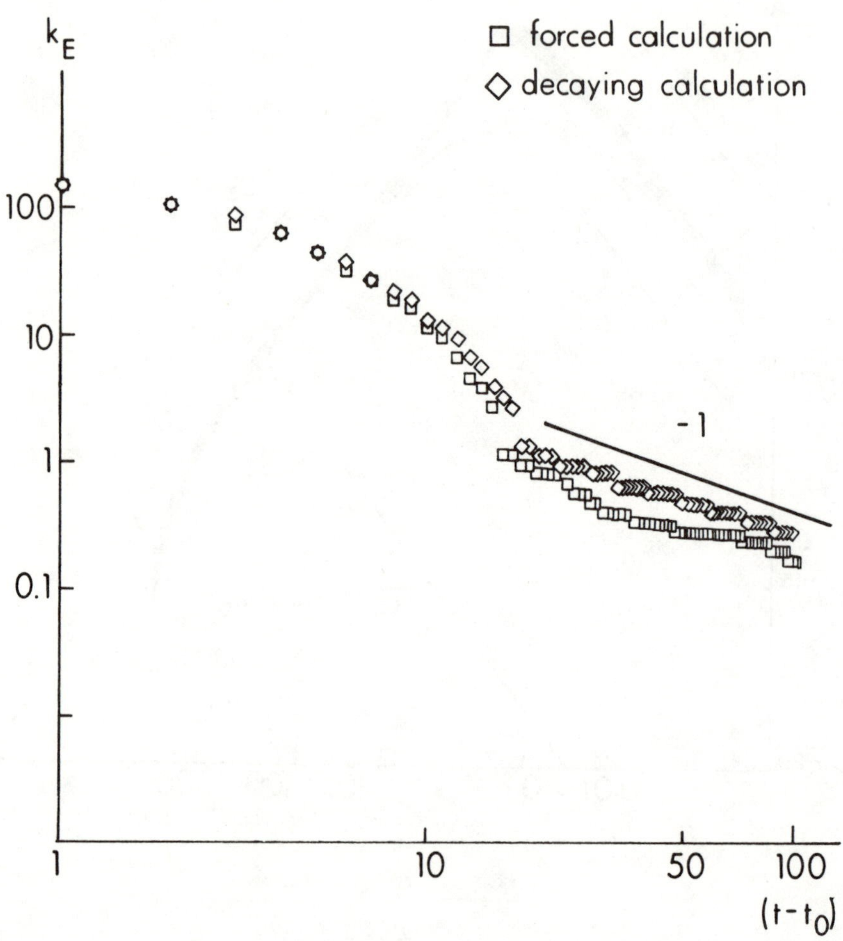

Fig. 10 Evolution with time on a log-log plot of $k_E(t)$ in the calculation of Fig. 9. The time t_0 is equal to 70, and the large eddy turnover time at $t = t_0$ is $\tau_0 = 3.36$

$$\frac{k_E}{k_I} = \frac{t}{t-t_0} \tag{15}$$

in the range $16 < t-t_0 < 100$. The predictability time T_{2k_I} is equal to t_0. Then

$$T_{2k_I} = 70 = 21\ \tau_0. \tag{16}$$

(In the case of injection, this time is of the order $9 \cdot \tau_0$). The striking fact, as shown by the calculation, is that at that time $t_0 + T_{2k_I}$, the relative error r (shown in Fig. 11) is less than 10^{-3}.

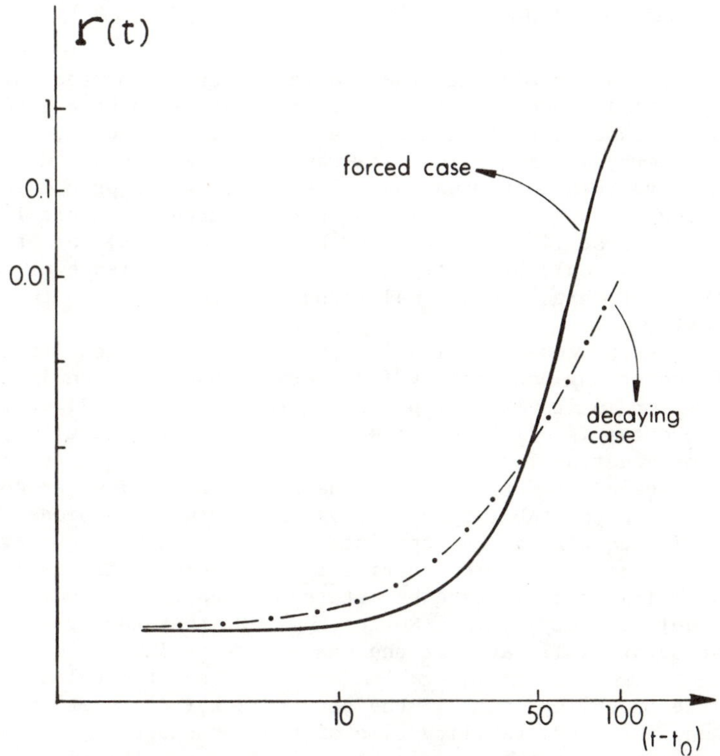

Fig. 11 Evolution with time on a log-log plot of r(t) in the calculation of Fig. 9

Then, it increases very rapidly, and should be of order 1/2 at

$$t-t_0 = 178 = 53\ \tau_0$$

Notice that at the end of the run ($t-t_0 = 100$), the forced calculation has already a relative error of order 1/2 ($T_{r=1/2} \simeq 30\ \tau_0$).

CONCLUSION

The first question which arises when interpreting these results is that the quantitative evaluation of the predictability of the flow (within the particular statistical formalism considered) depends strongly upon the various parameters which can be chosen to measure it. The essential parameters we have considered here are k_E/k_I and $r(t)$, and the corresponding predictability times are such that $k_E/k_I = 2$ and $r(t) = \frac{1}{2}$ respectively. According to the first criterion, the predictability times in the decaying case, compared with the case of stationary injection, are increased by more than a factor of 10 for three-dimensional turbulence. According to the second criterion, these times are increased by nearly a factor of 2 for three- and two-dimensional turbulence.

Now, if we want to look at the spatial coherence of the large scales in physical space, the first criterion seems to be a good measure in the case of three-dimensional turbulence: indeed, a large part of the energy is in the inertial-range scales. When the latter scales will have been contaminated by the error, which proceeds locally in Fourier space, $r(t)$ will be rather large. But the large scales $\sim k_I^{-1}$ will still remain predictable (or coherent) for extremely long times. This could be an explanation for the coherence of large scale features of three-dimensional turbulent flows such as turbulent boundary layers.

In the case of two-dimensional turbulence, things go differently because of the strong nonlocality (in Fourier space) of the inverse cascade of error in its initial phase: k_E reaches k_I rapidly, but the error spectrum is still very low. Then, k_E decreases slowly and the error spectrum level increases. In that case, the coherence of the large scales is determined by the second criterion: we could then expect the predictability of a given large two-dimensional scale, created at time t_0 with a characteristic time τ_0, to be of order of $53\ \tau_0$. If the same structure (at time t_0) had been artificially maintained in its initial state by external forces, its predictability time would be only $30\ \tau_0$. Notice however, that the freely-evolving structure will have, at the end of its evolution ($t = t_0+53\tau_0$) a turnover time of order of $3.5\ \tau_0$: indeed, the calculations show that the "doubling" time of the structure $k_I(t_0)$ is of order of $30\ \tau_0$. Then, the predictability time of the structure, in units of its turnover time τ at the end of the evolution, will be of order of $15\ \tau$. It is, then, not evident that these results would decisively modify the conclusions of predictability forced calculations in what concerns the unpredictability of the atmosphere and oceans.

What could be a possible implication of these calculations concern the large coherent structures in laboratory experiments of free-shear layers: if the three-dimensionality of the large scales could be considered as an "error" for the large two-dimensional scales, then a given large scale could not be affected by the error (≡ would not be destabilized by three-dimensional turbulence) during more than 50 turnover times. A precise comparison of these results with the free shear layer experiments will be done further in another paper.

ACKNOWLEDGEMENTS

We are grateful to F.K. Browand who convinced of us of the experimental evidence of the persistence downstream of the large two-dimensional scales in free shear layers: this evidence motivated our study. We also want to acknowledge stimulating discussions with J.R. Herring. The statistical spectral calculations were supported by the CNRS (ATP "Recherches Atmosphèriques"). M. Ménéguzzi's experience and help on the Large-Eddy-Simulations were very profitable to us. The latter simulations were performed at the NCAR during a one-year stay of one of the authors as a Research Associate. The NCAR is sponsored by the National Science Foundation.

REFERENCES

1. R.H. Kraichnan, Phys. Fluids. 13, 569 (1970).
2. E.N. Lorenz, Tellus. 21, 289 (1969).
3. C.E. Leith and R.H. Kraichnan, J. Atm. Sci. 29, 1041 (1972).
4. R.H. Kraichnan, J. Fluid. Mech. 47, 513 (1971).
5. C.E. Leith, J. Atm. Sci. 28, 145 (1971).
6. M. Lesieur and D. Schertzer, J. de Mécanique. 17, 609 (1978).
7. J.P. Chollet and M. Lesieur, J. Atm. Sci. 38, 2747 (1981).
8. G. Comte-Bellot and S. Corrsin, J. Fluid. Mech. 24, 656 (1966).
9. M. Larchevêque, J.P. Chollet, J.R. Herring, M. Lesieur, G.R. Newman and D. Schertzer, in "Turbulent Shear Flows 2" (Springer-Verlag, 1980, p. 50.
10. R.H. Kraichnan, J. Fluid. Mech. 47, 525 (1971).
11. C. Basdevant, M. Lesieur and R. Sadourny, J. Atm. Sci. 35, 1028 (1978).
12. E.D. Siggia and G.S. Patterson, J. Fluid. Mech. 86, 567 (1978).
13. R.H. Kraichnan, J. Atm. Sci. 33, 1521 (1976).
14. G.K. Batchelor, Phys. Fluids. Suppl. 12, part II, 233 (1969).

THE PREDICTABILITY OF QUASIGEOSTROPHIC FLOWS

Jackson R. Herring
National Center for Atmospheric Research†
Boulder, Colorado 80307

ABSTRACT

We examine the homogeneous quasigeostrophic system utilizing the tools of two-point closure and compare the predictability results to equivalent two-dimensional systems. As a preliminary, we discuss briefly the expected accuracy of the closure by comparisons with numerical simulations for two- and three-dimensional turbulence. Results for the error growth of baroclinic and barotrophic energy are presented. It is found that at both large and small wave numbers the error energy tends to be somewhat more barotrophic than the energy itself. As compared to two-dimensional turbulence, the quasigeostrophic system has a smaller error growth rate. The latter is here measured by the rate at which the error crossover wave number tends to overtake the peak wave number of the energy spectrum.

INTRODUCTION

Outline of equations of motion and closure formalism

Quasigeostrophic turbulence as described by Charney[1] shares many features with two-dimensional turbulence. Among these we recall the inverse cascade of energy, the forward cascade of enstrophy, and intrinsic unpredictability of both flows. The two-dimensional system is, in fact, a special (but unstable) case of the quasigeostropic equations in which initial vertical fluctuations are simply suppressed. It is of interest to ask how the predictability of the flow is modified by the introduction of the vertical structure natural to the homogeneous quasigeostrophic system. In this paper, we investigate this issue by examining the vertically homogeneous quasigeostrophic system without boundaries. Such a problem represents the simplest generalization of the Leith-Kraichnan study.[2] However, it by no means includes all the physically interesting generalizations needed to understand the implications of the quasigeostrophic equations in practice. In particular, we exclude — by invoking the homogeneous restriction — vertical boundary effects and the important tendency for a bounded, dissipative system to adjust the zero potential vorticity[3,4]. Further, the effects of baroclinic instability for the vertical homogeneous system are not as clearly related to the Eady problem[5] as the simple (two) layered system studied, for example, by Hoyer and Sadourney[6]. Nevertheless, the homogeneous system is able to rather cleanly resolve issues of the necessary vertical degrees of freedom and at the same time remain relatively simple so as to yield insights into dynamical issues.

Our tools for the present study are the moment closure techniques of the statistical theory of tubulence[7,2,8]. We record the error growth equations here for future reference. To this end, suppose $Y(\underline{k},t)$ represents the Fourier amplitude of the stream function. Then we write the equations for $Y(\underline{k},t)$ in the form:

$$dY(\underline{k})/dt = \sum C(\underline{k},\underline{p},\underline{q})Y(\underline{p})Y(\underline{q}) \tag{1}$$

Here, $C(\underline{k},\underline{p},\underline{q})$ are the usual coupling coefficients derived from the non-linearities. For homogeneous flows the wave vector set $(\underline{k},\underline{p},\underline{q})$ is constrained to form a triangle, and $C=0$ if any leg is zero. Approximate equations for the ensemble mean modal-energy, $U(\underline{k},t)=\langle Y(-\underline{k},t)Y(\underline{k},t)\rangle$, are:

† The National Center for Atmospheric Research is sponsored by the National Science Foundation.

$$1/2\ dU(\underline{k},t)/dt=\int B(\underline{k},\underline{p},\underline{q})\theta(\underline{k},\underline{p},\underline{q})U(\underline{q},t)\ \{U(\underline{p},t)-U(\underline{k},t)\}d\underline{p} \tag{2}$$

where $\underline{q}=\underline{k}-\underline{p}$, and

$$B(\underline{k},\underline{p},\underline{q})=2C(\underline{k},\underline{p},\underline{q})C(\underline{p},\underline{k},\underline{q}).$$

Here we have replaced the discrete sum (as in (1)) by an approximate integral, utilizing the well known rule,

$$\sum = \int (L/2\pi)^3 d\underline{p},$$

where L is a normalizating periodic-box containing the turbulence. In (2), $\theta(\underline{k}, \underline{p}, \underline{q})$ is a relaxation time for triple moments, whose form we discuss later. We define the correlated energy W(k,t) as:

$$W(\underline{k},t)=<(Y^1(-\underline{k},t)Y^2(\underline{k},t)> \tag{3}$$

where (Y^1,Y^2) are chosen from two equivalent (i.e., same $U(\underline{k},0)$) initial ensembles. The value of $W(\underline{k},0)$ may be specified arbitrarily. The error energy is then defined as:

$$D(\underline{k},t)=U(\underline{k},t)-W(\underline{k},t). \tag{4}$$

Equations for $W(\underline{k},t)$ in the same approximation as (2) are:

$$1/2\ dW(\underline{k},t)/dt= \int \{B(\underline{k},\underline{p},\underline{q})\theta(\underline{k},\underline{p},\underline{q})W(\underline{p},t)W(\underline{q},t) \tag{5}$$

$$- B(\underline{k},\underline{p},\underline{q})\theta(\underline{k},\underline{p},\underline{q})U(\underline{q},t)W(\underline{k},t)\}d\underline{p}$$

Equation (5) controls the development of error growth from its arbitrarily set initial value. In cases studied so far (see Kraichnan and Montgomery[9] for review), the "eddy-viscous" second term in (5) dominates the first and this leads to the eventual disappearance of correlated energy $W(\underline{k},t)$. This decrease need not be monotonic at all wave numbers; in fact, a transient initial phase of "sweeping out of error energy," $D(\underline{k},t)$ at wave number ks is observed to decrease if ks is initially largely dominated by error energy and if energy cascades into it from regions of relatively low $D(\underline{k})$.

Some selected results of closure calculations

The method described above for computing the statistics of error growth are not free of criticisms. Leith and Kraichnan[2] have noted that its failure to properly distinguish between convection and strain may lead to an overestimation of error growth, even though energy spectra (as derived from, say, an eddy damped model) are to be free from such error. For a more complete discussion of these issues, see the paper by Kraichnan in the present volume. In practice, the seriousness of such failing is not clear á priori. A further source of error is the "Markovianization" of the closure, dictated by computational ease and strict adherence to the Galilean invariance of the resulting theory. Such errors are likely to be severe at very small scales, where error growth takes place by a modulation of small scales by the large scale strain. The latter are nearly static by comparison. The modeling of large scales as white noise, as prescribed in the theory, seems particularly inappropriate for these interactions.

In order to gain some insight into the first problem, we present in Figure 1 comparison of theory (The Test Field Model[10], hereafter TFM) with direct numerical simulations for strictly two-dimensional decaying turbulence. These calculations are the same as previously published[11]; however, the predictability aspect of these calculations are new. The initial spectrum for U(k) is stated in the figure, as is the initial error spectrum,

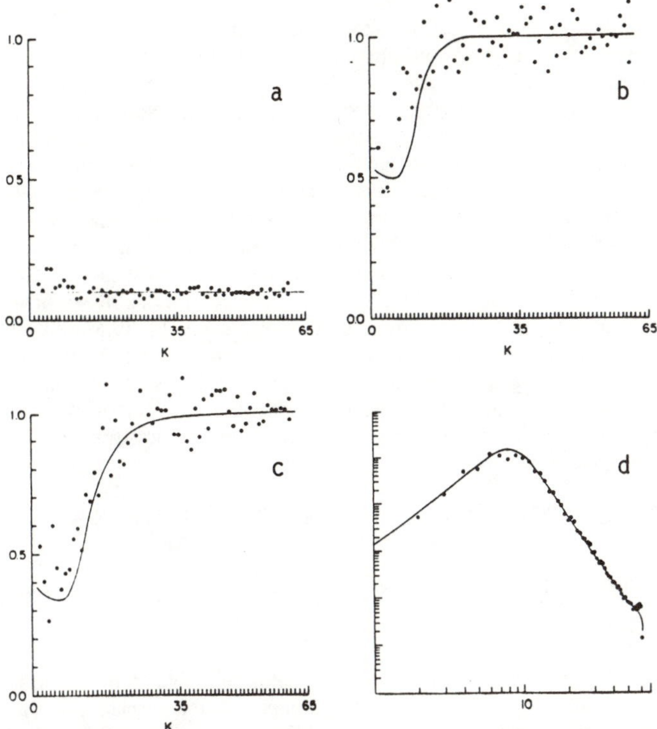

Fig. 1. Error profile (see Eq. 16) for a two-dimensional initial value problem; $E(k,0) = Ak^4 \exp(-(k/k_0)^2)$, $P(k,0)=.10$, $k_0=8$. Figure compares TFM and numerical simulation at t=0(1a), t=.4(1b), and t=.72(1c). (1d) compares E(k,0.72) for TFM and simulation. For more details of run see Reference 11, case (a).

$D(k) = .1U(k)$. The agreement here is as well as could be expected on considering the initial data's scatter for the numerical simulation. Further comparisons of theory and numerical simulations are given in Reference 11.

Let us now attempt an assessment of the Markovianization error for low Reynold's number homogeneous three-dimensional turbulence. To begin, we recall the closure's structure without Markovianization. The right hand side of (5) is then replaced by the more general (and correct?) form:

$$\int d\underline{p}d\underline{p}B(\underline{k},\underline{p},\underline{q}) \int_0^t \{W(\underline{p},t,s)W(\underline{q},t,s)G(\underline{k},t,s) \\ -W(\underline{k},t,s)G(\underline{p},t,s)U(\underline{q},t,s)\}ds \tag{6}$$

We need equations of the time-nondiagonal U's, W's, as well as Green's functions $G(\underline{k},t,t')$. The direct interaction algorithm furnishes these; we shall not record them here. The temporal structure of (2) and (5) are recovered from (6) if we assert that:

$$U(\underline{k},t,s) \simeq U(\underline{k},t,t)G(\underline{k},t,s)$$

and systematically model the turbulence force (the right hand side of (1)) as white noise in time, with an appropriate compensating amplitude factor, $(\theta(\underline{k},\underline{p},\underline{q}))^{1/2}$. Such modeling must do violence — at least to a certain extent — to the basic instability that lies at the source of error growth; the (exponential) amplification of small-scale error by large-scale, slowly evolving strain.

In order to form some practical measure of magnitude of this error, we present in Figure 2 a comparison of error growth calculation to direct numerical simulation for both the Direct Interaction Approximation, hereafter DIA, (which uses (6)) and TFM (which uses the simpler form (5)). The initial energy spectrum and error spectrum are given in the figure. We notice that the TFM under-estimates error growth, particularly at longer times. Apparently this underestimation is less significant in two dimensions (compare to Figure 1). The reasons are not clear, but may have two sources: 1) the much slower evolution of the two-dimensional spectrum, and 2) the smoother distribution of initial error in the two-dimensional comparison. If the initial error is more concentrated at small scales, the DIA results are less satisfactory. This is indicated by Figure 3. Here the lack of Galilean invariance of the DIA begins to cause serious errors. To check this point we have performed error growth calculations for the conditions of Figure 2 using an "unMarkovianized" TFM whose error (and energy) transfer is like (6), supplemented with a fluctuation dissipation relationship and with a $G(\underline{k},t,t')$ obtained from the TFM pressure scrambling algorithm. The result is the dashed line in Figure 2 which shows an improvement over the DIA comparison, as well as the Markovianized TFM. This procedure is not free from logical inconsistencies, as discussed in Reference 12.

EQUATIONS FOR QUASIGEOSTROPHIC SYSTEMS — SUMMARY OF EARLIER RESULTS

The equations of interest for the present study are:

$$(\partial/\partial t+\underline{u}\bullet\nabla)\nabla^2\psi=0 \tag{7}$$

Here, $\underline{u}(x,y,z)$ is the (strictly horizontal) velocity field, derivable from the stream function ψ by $\underline{u}=(-\partial\psi/\partial y,\partial\psi/\partial x)$, and ∇^2 is the three-dimensional Laplacian. In the coordinate frame (x,y,z) the z-axis has been stretched, $z=z(N/f)$, where N is the Brunt-Vaisala frequency and f is the Coriolis parameter. We have some time ago applied the closure formalism to (7) in the homogeneous context, and we here briefly summarize those findings before proceeding to the predictability problem. We shall use the total (kinetic $=(1/2)\langle u^2\rangle$ plus potential $=(1/2)\langle(\partial\psi/\partial z)^2\rangle$ as the basic dynamical variable. We denote its spectrum by $U(\underline{k},t)$, so that the total kinetic energy is related to U through:

$$\langle|u(\underline{k})|^2\rangle=2V(\underline{k})=\sin^2\theta U(\underline{k}). \tag{8}$$

The potential energy spectrum P(k) is:

$$2P(\underline{k})=\langle(\partial\psi/\partial z)^2\rangle(\underline{k})=\cos^2\theta U(\underline{k}). \tag{9}$$

In (8) and (9), θ is the polar angle, $\sin\theta=k_z z/(k_x^2+k_y^2+k_z^2)^{1/2}$. The closure equations (i.e., Eq. 2) for system (7) have:

$$B(\underline{k},\underline{p},\underline{q})=2(\sin\theta\,\sin\phi\,\sin\gamma)^2(p^2-q^2)(k^2-q^2)/q^2. \tag{10}$$

In (10) we specify the wave number vectors $(\underline{k},\underline{p},\underline{q})$ in polar form as:

$\underline{k}=(k_x,k_y,k_z)=(k\sin\theta,0,k\cos\theta)$,

$\underline{p}=(p_x,p_y,p_z)=(p\sin\theta_p,0,p\cos\theta_p)$, and

$\underline{q}=(q_x,q_y,q_z)=(q\sin\theta_q,0,q\cos\theta_q)$.

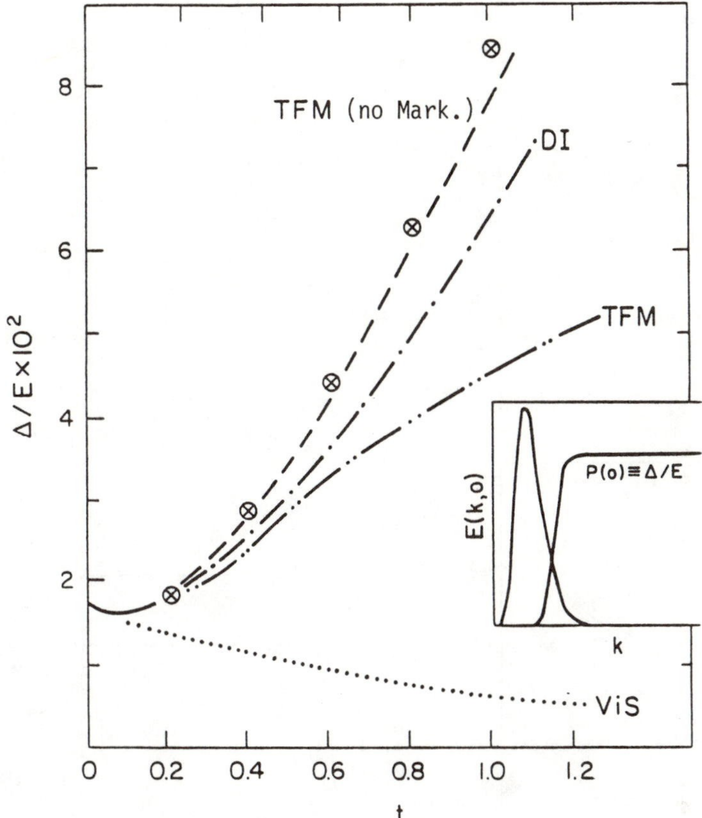

Fig. 2. Comparison of theory (DIA, TFM and non-Markovianized TFM) to numerical simulation (dashed line) for three-dimensional turbulence at low Reynold's number. The "non-Markovianized" procedure is described in the text (Eq. 6). Initial E(k,0) and O(k,0) are depicted in the inset. For more complete description of conditions of the run, see Reference 8.

Since $(\underline{k},\underline{p},\underline{q})$ form a triangle, we have:

$$\cos\theta_p = \cos\gamma \cos\theta + \cos\phi \sin\gamma \sin\theta, \text{ and} \tag{11}$$

$$\cos\theta_q = \cos\beta \cos\theta - \cos\phi \sin\beta \sin\theta. \tag{12}$$

In these equations we have for convenience assumed that $(\underline{k},\underline{p},\underline{q})$ lie in the (x-z) plane, which may be done — in the closure — with impunity if the statistical variables (such as $U(\underline{k})$) have polar symmetry.

In applying the formalism of equations (2) and (5) to the above problem, it is convenient to resolve the angular distributions of U(k) into Legendre functions:

$$U(\underline{k},t) = \sum_{l=0}^{L} P_l(\cos\theta) U_l(k,t). \tag{13}$$

In practice, we must choose L to be quite small (L = 2 or 4) if the numerics are not to get out of hand. We assume that such small values of L are reasonable in view of the near iso-

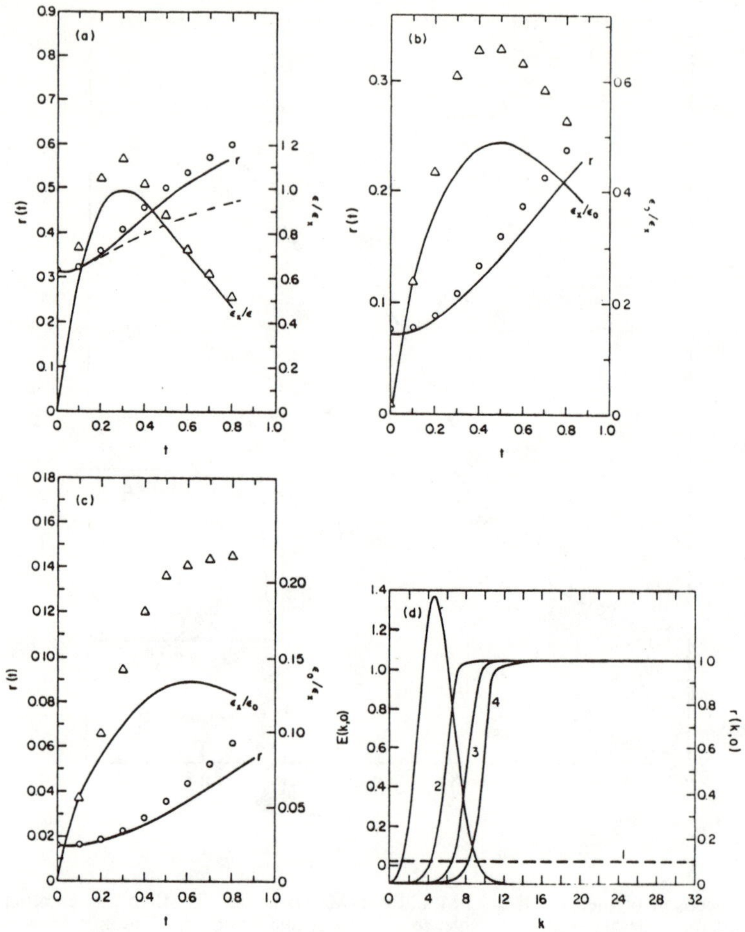

Fig. 3. Comparison of DIA total error to numerical simulation, for initial cross-over wave number, k progressively at larger k on going from (a) to (b) to (c). After Reference 8.

tropy (for which only L=0 is needed) of $U_l(k)$. There remains to introduce representation (13) into (2) and (5), and to work out equations of motion for U(k). The details are in Reference 13 and we do not repeat them here. However, to complete the story, we must specify the triple-moment eddy relaxation time $\theta(\underline{k}, \underline{p}, \underline{q})$. For this we use a simple rule:[14]

$$\theta(\underline{k},\underline{p},\underline{q}) = [\int_o^k dk'k'^2E(k')]^{1/2} + [\int_o^p dp'p'^2E(p')]^{1/2} \qquad (14)$$

$$+ [\int_o^q dqq'^2E(q'))]^{1/2}$$

Here,

$$E(k) = \int d\theta d\phi k^2 U(\underline{k}) \qquad (15)$$

is the isotropic part of the total energy. We shall use (14) and (15) for both the quasigeostrophic system and the two-dimensional comparison case that follows. We offer no justification of (14), except to point out that for isotropic turbulence, (14) is a reasonable approximation to TFM[15] and to note the plausible involvement of the rms strain in relaxation triple moments. In reality, $\theta(\underline{k}, \underline{p}, \underline{q})$ is likely to have a quite complex angular dependence.

Before presenting new numerical results, we summarize our earlier findings concerning the behavior of system (7), whose two-point moments are approximated by (2). We do this by examining an initial value problem in which the angular distribution, (13), is simply specified ($U(k,L)=0$ for $L<0$, for example). Our previous results indicated that for this case, that portion of $U(k)$ smaller than the peak value, k_E, evolved into a strongly barotropic distribution ($U_2(k)<0$), while that for which $k>k_E$ evolved into a slightly baroclinic form, followed by an isotropic inertial range. At the even higher dissipation wave numbers, the flow becomes barotrophic again. As the system evolves, the peak wave number moves into the origin, so that the spectrum becomes more isotropic as time proceeds (excluding the dissipation range). This is consistent with the entropy-increasing nature of the inviscid Markovian model for the system.[16] At small k, the energetics may be characterized by a negative eddy-viscosity, even if no minimum wave number cut-off is introduced (not true for two-dimensional turbulence). We should note that the drift of the energy-peak wave number, $k_E(t)$, to smaller values is characteristic of both two and three dimensions.

THE ERROR GROWTH PARAMETERS TO BE STUDIED

The canonical error growth problem[2] is to introduce a distribution of error, at saturated value ($D(k)=E(k)$) at large k and zero at small k. The error profile, $P(k)$, defined as

$$P(k)=(E(k)-W(k))/E(k) \tag{16}$$

develops a simple, self-similar shape, which generally moves into the origin. This drift is simples characterized by $k_{1/2}(t)$, where:

$$P(k_{1/2}(t))=(1/2)E(k_{1/2}(t)) \tag{17}$$

Generally, $-dk_{1/2}(t)/dt > -dk_E(t)/dt$, leading to an eventually unpredictable spectrum.

Here, we ask whether the presence of baroclinicity modifies the overtaking of $k_{1/2}(t)$ by $k_E(t)$, and also inquire as to the distribution of baroclinicity of the error energy (i.e., the amount at a given k by kinetic energy as compared to potential energy error). In order to make as clean a comparison as possible, we utilize the same theory and the same initial conditions for both QGT and 2DT — insofar as possible. Our theoretical approach is the EDQMN. The problem to be studied is that of self-similar decay. In two dimensions, the energy spectrum is $\sim k^3$ at small k and k^{-3} at large k. For the QGT system, the situation is more complicated. At small k, as we have noted, the QGT system is quasi two-dimensional with $U(k)\sim\sin^2\theta$). However, if the initial $E(k,0)$ is three dimensionally distributed, then (2) (for system (7)) may be shown to yield $E(k)\sim k^4\sin^2\theta$ for finite times at least. At large k, $E(k,t)\sim k^{-3}$ independent of $\sin^2\theta$. This leads us to conjecture that the near self-similar shape for QGT has $U(k)\sim k^4\sin^2\theta$, for $k\to 0$, and $E(k)\sim k^{-3}, k\to\infty$, with at transition near the peak of $E(k)$. This point has been confirmed by our earlier numerical calculation, — at least for several (large-scale) eddy turnover times. These comments leads us to set the comparison QGT initial field as:

$$E(k,0)=A_Q k^4/(k_0^7+k^7) \tag{18}$$

and to take the 2DT comparison initial data to be:

$$E(k,0)=A_2 k^3/(k_0^6+k^6). \tag{19}$$

The constants (A_Q and A_2) are set so that the initial eddy circulation times are the same for the two cases. We further take the initial error-profile (16) to be:

$$P(k)=1/(1+\exp(a(k-k_{1/2}(0)))). \tag{20}$$

For both 2DT and QGT, $a=.5$, $k_0=4$, and $k_{1/2}(0)=20$. These initial conditions are rather different from those investigated by Metais, Chollet, and Lesieur (see their paper in this volume).

In Figure 4 we show the course of $k_c(t)/k_E(t)$ for both two-dimensional and quasigeostrophic turbulence. Here, $k_E(t)$ is the peak wave number of E (k,t). Notice that the initial phase during which small scale errors are convectively swept out is much more pronounced in the quasigeostrophic system than in the two-dimensional. Some of this is undoubtedly due to the drastic readjustment of large scales. These must first become strongly barotrophic before any self-similarity develops. We may discount this initial phase by comparing rates of decrease of k_c/k_E at (different) times such that k_c/k_E are equal for the two systems (i.e., $t=2.4$ for two-dimensional compared to $t=8.0$ for the quasigeostrophic system). In this case, still we find $-d\ln(k_c/k_E)/dt=.4$ for 2DT and $\sim .16$ for QGT. We should note, however, that in both these calculations, error energy has begun to significantly penetrate the energy containing range.

Figure 5 compares the ratio of baroclinic error to total error energy to the equivalent profile of baroclinic energy. We notice a deficit of potential error energy at large scale, as compared to energy. We also show here for comparison the total energy (ordinate to the right). The sudden decrease in baroclinic energy at $k \gtrsim 150$ is associated with a hyperviscosity ($\sim(k-150)^4$) range which extends to the cutoff wave number $k=200$.

Figure 6 gives the error profile P(k) (see equation 16) (the dashed line) and its kinetic (solid line) and potential (dotted line) contributions. We notice a strongly barotropic error regime at small k and a 2/3-1/3 equipartitioning in the enstrophy inertial range. The extent of error penetration into the $k<k_E$ region is about the same as for the equivalent two-dimensional system, provided we pick for the latter that $\underline{t}$ for which k_c/k_E is the same as for the quasigeostrophic system.

Further insight concerning error growth may be obtained by considering the injection of an error sufficiently small that linearization of the error growth equation (Eqs. 4 and 5) is valid. We may then cast the error growth problem in terms of linear exponential growth on the (nearly static by comparison) existing energy profile $U(\underline{k},t)$. In this connection, the fastest growing eigenmode $X(\underline{k})$ and its potential kinetic energy distribution are of interest. Figure 7 shows this eigenmode, together with its second (Legendre) angular intensity, $U_2(k)/U_0(k)$. The basic energy profiles $U_0(k,t)$ and $U_2(k,t)$ are fixed during the eigenmode calculation. Their shapes are (18) for $U_0(k)$, with $U(\underline{k})$ strongly barotropic for $k \leqslant k_E$ and isotropic for $k \geqslant k_E$. The eigenmode profiles seem somewhat insensitive to the details of the actual angular distribution of $U(\underline{k})$. We note a very strong barotropicity for $X(\underline{k})$, $k \leqslant k_E$. For $k \geqslant k_E$, the distribution still remains significantly barotropic, even though $U(\underline{k})$ is isotropic ($U_2(k)=0$). At small k, $X_0(k)$ resembles $U_0(k)(\sim k^2)$, but for $k \geqslant k_E$, $X_0(k) \sim k^{-1}$, whereas $U(k) \sim k^{-3}$. These facts are consistent with a rapid saturation of error at large k, with relatively slow error growth for $k \leqslant k_E$.

The distribution of eigenvalues is also of some interest. This consists of an isolated real eigenvalue $\lambda=.17(k_E\sqrt{E_0})$, an apparent accumulation point at the origin, and a more uniform distribution of negative numbers. We may compare this with the "equivalent" two-dimensional problem, $\lambda=.45(k_E\sqrt{E_0})$, with otherwise a qualitatively similar distribution. These numbers are entirely consistent with the non-linear evolution calculation given earlier.

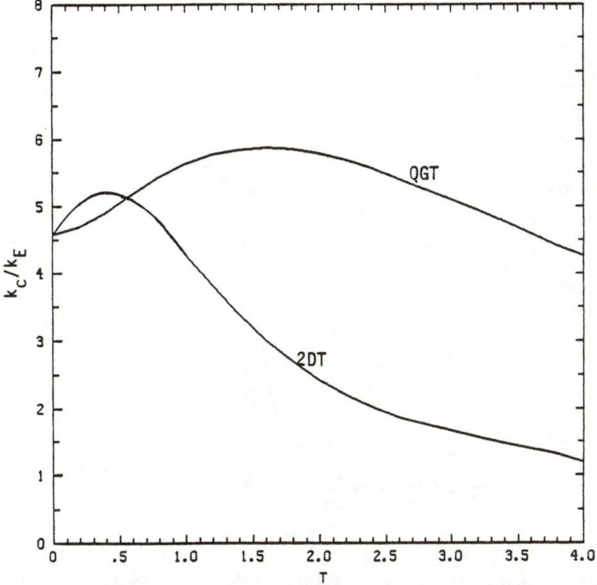

Fig. 4. Comparison of error crossover wave number to energy peak wave number for two-dimensional turbulence and equivalent quasigeostrophic turbulence. Initial conditions for runs are given in text.

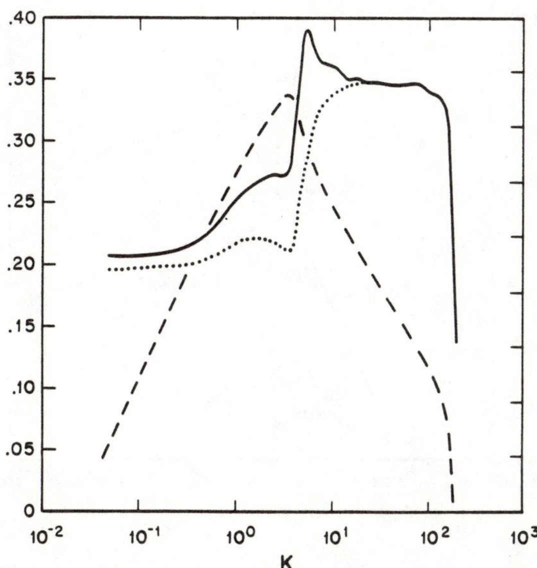

Fig. 5. Distribution of the ratio of baroclinic error energy to total error energy (dotted line) and baroclinic energy to total energy (solid line) at t=8.0 large-scale eddy circulation times. Dashed line gives the total energy profile (ordinate at right).

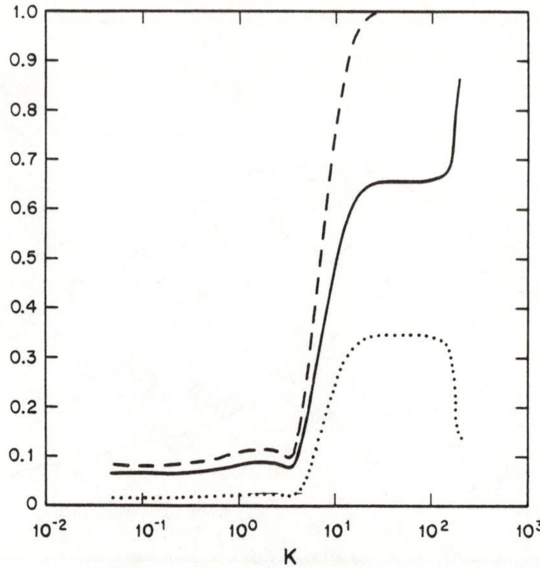

Fig. 6. Error profile (dashed line) for quasigeostrophic system at t=8.0 initial eddy circulation times. Solid and dotted lines give the kinetic and potential components.

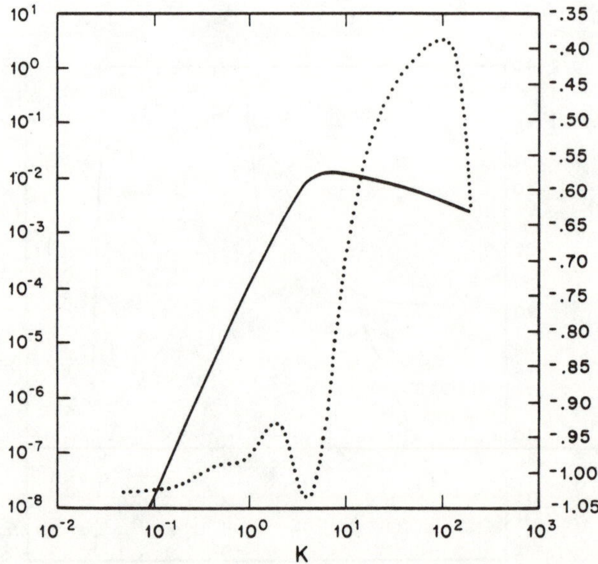

Fig. 7. Most rapidly growing eigenmode for linearized error growth problem. The energy distribution is $U_0(k)$ given by (18), $U_2(\underline{k})=-U_{0(k)}/(1+\exp(k-4)/2))$ This $U(\underline{k})$ corresponds very nearly to the evolved $U(\underline{k},t)$ at $t=20/k_E\sqrt{E_0}$. Curves shown are $X_0(k)$ (solid line) and $X_2(k)/X_0(k)$ (dotted line with scale to the right).

The comparison of the two-dimensional and quasigeostrophic error growth problems has so far been between systems with the same total energy. For the initial conditions considered (Eqs. 18 and 19), this roughly amounts to the same small scale eddy turnover time and the same enstrophy. It may be that a fairer comparison is to adjust the basic flows so that they have the same strain or velocity field $u(\underline{x})$. This would reset the basic fields so that they have the same $U(\underline{k})$ — at least insofar as differences in angular distributions would permit. Such a basis of comparison would more nearly correspond to the two systems having, as it turns out, the same enstrophy fluxes. This adjustment would increase the quasigeostrophic error growth by a factor of two, still smaller than that in strictly two dimensions.

CONCLUSION

We have examined here two rather separate issues; first, we have inquired as to the accuracy of the two-point closures as tools for understanding error growth for two and three dimensional homogeneous turbulence. The information presented on this point — although somewhat sketchy — suggests caution in trusting their results as quantitative procedures in error growth estimates. The two-dimensional comparisons indicate good accuracy for the initial phase of error growth. This is somewhat surprising considering the less satisfactory behavior in three dimensions. For the latter case, errors connected with Markovianization appear to be more of a problem. Our procedure for examining this issue — an unMarkovianized TFM — would have difficulties at higher Reynolds numbers, and for such problems we should probably employ one of the more elaborate Lagrangian history procedures — at considerably more expense.

In the second half of the paper we compared two-dimensional error growth to an equivalent problem for the quasigeostrophic system. The equivalence is by no means perfect, considering the differences in dimensionality of the two problems. Results here were restricted to rather short integration times. Surprisingly, the rate of error growth for the quasigeostrophic system is smaller than that for two-dimensional turbulence if the rate of overtaking of the peak energy wave number by the error cross-over wave number is used as a measure. Otherwise the error profiles of the two systems were closely similar. Whether this difference remains for longer times of integration or for stationarily maintained turbulence remains to be seen. The distribution of potential vs. kinetic energy error for the quasigeostrophic system was also examined. The error energy at small wave numbers appears to be more strongly barotropic than the energy itself.

We offer here a tentative suggestion as to the reason for the small quasigeostrophic error growth. We recall in this connection that much of the dynamics of energy transfer is the straining by the (two-dimensional) velocity field. The quasigeostrophic system has an additional degree of freedom which is excited by the equipartitioning property. This degree of freedom does not contribute to strain and is inefficient in effecting energy (and enstrophy) transfer — at least in the absence of boundaries. By the same argument, error growth is not enhanced. In this sense, the two-dimensional constraint represents a ducting of the turbulence along those degrees of freedom in which transfer is most efficient.

With regard to the relative smallness of error growth for the quasigeostrophic system, several additional comments should be made. First, we have here examined only the homogeneous version of this system. It may well be that the presence of boundaries would enhance error growth considerably. We should recall in this connection the role of such boundaries in baroclinic instability [5]. Secondly, it may be of interest to perform error growth calculations by injecting error into an already developed turbulent state (i.e., one for which the energy transfer is self-similarly evolving at $t=0$). This would avoid complicating effects associated with the development of the basic flow from an initial (and unrealistic) Gaussian state. Third, it would be edifying to examine error growth rates on an existing (stationary) k^{-3} range. If we assume small initial error spectra so that linearization is valid, we may make estimates based simply on assuming an efficient entrainment of error by scales less than k_c. This would give error growth characterized by the magnitude of C, where $E(k)=C(t)k^{-3}$. Our numerical calculations indicate that $C_{2DT} \simeq C_{QGT}$, leading to roughly equal error growths. This neglects possible effects of anisotropy on error growth, however. In addition, and perhaps more importantly, it assumes the validity of (14)-(15), an assumption clearly in need of closer scrutiny for the quasigeostrophic system.

REFERENCES

1. J.G. Charney, J. Atmos. Sci., **28**, 1087-1095 (1971).
2. C.E. Leith and R.H. Kraichnan, J. Atmos.Sci., **29**, 1041-1058 (1972).
3. J.C. McWilliams and J. Chow, J. Phys. Oceanography, **11**, 921-949 (1981).
4. W. Blumen, J. Atmos. Sci., **35**, 774-783 (1978).
5. J. Pedlosky, Geophysical Fluid Dynamics (Springer-Verlag, 1979), p. 624.
6. J.M. Hoyer and R. Sadoury, J. Atmos. Sci., **39**, 707-721 (1982).
7. R.H. Kraichnan, Phys. Fluids, **13**, 569-575 (1970).
8. J.R. Herring, J.J. Riley, G.S. Patterson, Jr., R.H. Kraichnan, J. Atmos. Sci., **30**, 997-1006 (1973).
9. R.H. Kraichnan and D. Montgomery, Reports on Progress in Physics, **43**, 548-619 (1980).
10. R H. Kraichnan, J. Fluid Mech., **47**, 513-524 (1971).
11. J.R. Herring, S.A. Orszag, R.H. Kraichnan and D.J. Fox, J. Fluid Mech., **66**, 417-444 (1974).
12. J.R. Herring, J. Atmos. Sci., **34**, 1732-1750 (1977).
13. J.R. Herring, J. Atmos. Sci., **37**, 970-977 (1980).
14. A. Pouquet, M. Lesieur, J.C. Andre, J. Fluid Mech., **72**, 305-319 (1975).
15. J.R. Herring, D. Schertzer, M. Lesieur, G.R. Newman, J.P. Chollet, and M. Larcheveque, J. Fluid Mech., **124,** 411-437 (1982).
16. G. Carnevale, U. Frisch, and R. Salmon, J. Phys. A., June Issue (1981).

STATISTICAL THEORY OF THE PREDICTABILITY OF EQUIVALENT BAROTROPIC MOTION ON A BETA PLANE

John Litherland
School of Oceanography, University of Washington,
Seattle, WA 98195

Greg Holloway
Institute of Ocean Sciences,
Sidney, B.C. Canada V8L 4B2

ABSTRACT

The theoretical predictability of a flow is determined by the rate of growth of error between two initially similar flows. Statistical closure theory is employed to obtain equations for the evolution of this error. Numerical evaluations of these equations reveal the effects of differential rotation (β-plane) and finite equivalent depth on error growth. Results show that both these effects result in a reduction of the rate of error growth, hence a net increase in the predictability of the flow. The anisotropy of the error field is examined and compared with that of the mean flow. Comparisons are made with the results from numerical simulation studies.

INTRODUCTION

Knowledge of the initial state of a geophysical flow is subject to uncertainty, especially for smaller scales. As a result, the predicted state of the flow will differ from the actual state at later times, even given a theoretically complete model for the flow evolution. Theoretical predictability of the flow is determined by this difference, or "error", and by the rate at which it increases.

Two approaches have commonly been taken in the study of predictability. The first employs numerical simulations in which two flows, initially differing only in the smallest scales, are allowed to evolve independently, with the error given by the difference between them[1-6]. An alternative approach uses closure theory to develop equations for calculating the evolution of the error statistics directly[7-9].

The present paper belongs to the latter group. We employ a closure model after Kraichnan's (1971) "test field model" to consider the predictability of two-dimensional, equivalent barotropic flow on a β-plane. Previous closure studies have been limited to horizontally isotropic cases thereby ignoring important effects on predictability due to differential rotation (β-plane). We treat the flow and error statistics in fully two-dimensional form, in order to investigate the anisotropy as well as other influences that result from the inclusion of β. We also relax the "rigid lid" condition by including a finite equivalent depth term as a lowest order representation of baroclinicity.

We think a good case may be made for the use of statistical closure theory as compared with direct numerical simulation. The closure equations for flow and error evolution deal with ensemble average statistics, thus eliminating the "noise" associated with the individual realizations of numerical models. The equations also show the analytical dependence on the parameters α (for finite equivalent depth) and β (for differential rotation) and hence rates of change of error growth with respect to these parameters may be calculated directly. Closure equations can be examined qualitatively to provide insight into the significant processes which govern the error evolution.

A concern when using a closure approach is whether the statistical assumptions are, in fact, justified. One method of checking this is comparison with the conceptually more straightforward numerical simulation studies. However, differing experimental conditions and theoretical assumptions have made such comparisons difficult. A goal of this study has been to coordinate our approach with the numerical study of Holloway[6], to allow a more quantitative assessment of the agreement between simulations and closure models.

CLOSURE MODEL

We deal with quasi-non-divergent, barotropic flow on the β–plane, the equation for which is

$$\partial_t(\Delta^2 + \alpha^2)\psi + J(\psi,\Delta^2\psi) + \beta\partial_x\psi = F - D \tag{1}$$

ψ is the streamfunction, $\alpha = f/\sqrt{gH_e}$ is the inverse radius of deformation, β is the latitudinal variation of the Coriolis parameter, and F and D are unspecified forcing and dissipation functions. Doubly periodic boundary conditions are imposed on a rectangular region so that ψ may be represented by a discrete Fourier series:

$$\psi(\underline{x},t) = \sum_{\underline{k}} \psi_{\underline{k}}(t)\, e^{i\underline{k}\cdot\underline{x}} \tag{2}$$

The Fourier transformed version of (1) is:

$$\left(\frac{d}{dt} + i\omega_{\underline{k}} + \nu_k\right)\psi_k = \frac{1}{2(k^2+\alpha^2)} \sum_{\underline{p}+\underline{q}=\underline{k}} (\underline{p x q})\,(q^2 - p^2)\psi_p\psi_q + f_k \tag{3}$$

where $\omega_{\underline{k}} = \dfrac{-\beta k_x}{k^2+\alpha^2}$ is the planetary wave frequency and $f_{\underline{k}}$ and $\nu_{\underline{k}}$ are forcing and dissipation functions depending on F and D. Summations are taken over wavevector triads satisfying $\underline{p} + \underline{q} = k$.

Consider two different realizations $\psi_{1,\underline{k}}$ and $\psi_{2,\underline{k}}$. Define the covariances:

$$U_{\underline{k}} = (k^2+\alpha^2) <\psi_{1,\underline{k}}\, \psi^*_{1,\underline{k}}>$$

$$W_{\underline{k}} = (k^2+\alpha^2) <\psi_{1,\underline{k}}\, \psi^*_{2,\underline{k}}> \tag{4}$$

Evolution equations may be written for these second moment quantities but will involve third moments or triple correlations. Likewise equations for the triple moments will involve fourth order correlations and so forth. Hence to close the set of equations a statistical closure assumption must be applied at some level. We use the "test field model" (TFM)[10,11] which yields the following equations for the evolution of the second moments U_k and W_k:

$$\left(\frac{d}{dt} + 2\nu_k\right)U_{\underline{k}} = 2 \sum_{\underline{p}+\underline{q}=\underline{k}} \theta_{-kpq}\, a_{kpq}\, [U_{\underline{p}}U_{\underline{q}} - U_{\underline{k}}U_{\underline{q}}]$$

$$\left(\frac{d}{dt} + 2\nu_k\right)W_{\underline{k}} = 2 \sum_{\underline{p}+\underline{q}=\underline{k}} \theta_{-kpq}\, a_{kpq}\, [W_pW_q - W_kU_q] \tag{5}$$

where a_{kpq} is a geometric coefficient given by:

$$a_{kpq} = \frac{(\underline{p\ x\ q})^2\ (p^2 - q^2)\ (k^2 - q^2)}{(k^2+\alpha^2)\ (p^2+\alpha^2)\ (q^2+\alpha^2)} \tag{6}$$

and θ_{-kpq} is a decorrelation timescale which approximates the effect of the fourth order moments in breaking down the triple correlations. Evaluation of θ_{-kpq} will be given below.

The difference or "error" energy between the two realizations may be defined by:

$$\Delta_{\underline{k}} = \frac{1}{2} (k^2 + \alpha^2) < |\psi_{1,\underline{k}} - \psi_{2,\underline{k}}|^2 > \tag{7}$$

$$= U_{\underline{k}} - W_{\underline{k}}$$

The evolution equation for the error field, Δ_k, is

$$\left(\frac{d}{dt} + 2\nu_k\right)\Delta_k = 2 \sum_{\underline{p}+\underline{q}=\underline{k}} \hat{\theta}_{-kpq}\, a_{kpq}[U_{\underline{p}}\, \Delta_{\underline{q}} + \Delta_{\underline{p}}\, U_{\underline{q}} - \Delta_{\underline{p}}\, \Delta_{\underline{q}} - \Delta_{\underline{k}}\, U_{\underline{q}}] \tag{8}$$

Evolution of decorrelation time scales θ_{-kpq} and $\hat{\theta}_{-kpq}$ are given by extension of Kraichnan[10] to include wave propagation. Let a complex quantity $\tilde{\theta}_{-kpq}$ evolve by the differential equation:

$$\frac{d\tilde{\theta}_{-kpq}}{dt} = 1 - (\mu_{kpq} + i\omega_{-kpq})\, \tilde{\theta}_{-kpq} \tag{9}$$

where

$$\mu_{kpq} = \mu_k + \mu_p + \mu_q$$

$$\omega_{-kpq} = \omega_{-k} + \omega_p + \omega_q$$

Here μ_k is an eddy damping rate estimated by:[11]

$$\mu_k = g^2 \sum_{\underline{p}+\underline{q}=\underline{k}} \theta_{-kpq}\, b_{kpq}\, U_p + \nu_k \tag{10}$$

Here g is a phenomenological coefficient of order unity.

Timescales θ_{-kpq} and $\hat{\theta}_{-kpq}$ are given by the real part of $\tilde{\theta}_{-kpq}$. For the evolution of $U_{\underline{k}}$, the flow energy, we assume quasi-stationarity for θ_{-kpq} and hence, the form:

$$\theta_{-kpq} = \frac{\mu_{kpq}}{(\mu_{kpq}^2 + \omega_{kpq}^2)} \tag{11}$$

However, the statistics of the error field evolve rapidly at first and require the time-dependent solution:

$$\hat{\theta}_{-kpq} = \text{Re}\left[\frac{\mu_{kpq} - i\omega_{kpq}}{(\mu_{kpq}^2 + \omega_{kpq}^2)}\left(1 - e^{-(\mu_{kpq} + i\omega_{kpq})t}\right)\right] \tag{12}$$

This solution approaches the quasi-stationary form. Also, while this expression could result in negative values of $\hat{\theta}_{-kpq}$, evaluations have shown that these do not occur for the parameters we employ.

In order to avoid the complication of choosing artificial forcing and dealing with its role in error growth, we allow a slowly decaying flow, setting forcing equal to zero and retaining only a high order viscosity to prevent buildup of vorticity variance at the truncation limit. We find that error growth timescales are shorter than the decay timescales of the slowly evolving flow.

EVOLUTION OF ERROR FIELD

The equations for the evolution of $U_{\underline{k}}$ and $\Delta_{\underline{k}}$ were evaluated using a roughly atmospheric energy spectrum with a peak at about k=4 and k^{-3} inertial range. The error field is initialized so that all wave numbers above the resolution limit ($k_r \approx 20$) are completely decorrelated (i.e.$\Delta_{\underline{k}} = U_{\underline{k}}$) while the lower wavenumbers were virtually error free. (Fig. 1) Several cases are considered with values of α and of β taken as 0, $.5\alpha_o$, α_o and 0, $.5\beta_o$, β_o where α_o and β_o are approximately terrestrial values.

It is evident from the plots of relative error (Δ/U) in Figure 2 that the common feature of all these cases is that the error/variance, initially restricted to the small scales, eventually invades all scales of the flow. It is interesting to note that the error in the highest wavenumbers actually decreases initially. In part, this is due to the initially error-free large scales driving the small scales coherently and thus reducing their uncertainty. Eventually, as the large scales, too, become contaminated by error, the high wavenumber error again grows.

To understand the effects of non-zero β and α on the error evolution it is helpful to look first a their effects on the evolution of the energy field alone. In the equations for both U_k and Δ_k, α and β appear only in the coefficients $\hat{\theta}_{kpq}$ and a_{kpq}, so their respective effects should be somewhat similar in both cases.

The basic processes occurring in the energy field are the non-linear transfers of energy to lower wavenumbers and enstrophy to higher wavenumbers — the classic energy and enstrophy conserving balance of two-dimensional turbulence. Plots of the energy transfer function (T_u) for cases with various values of α and β show that the presence of either parameter acts to decrease this transfer. (Fig. 3)

The closure equations are simple enough to allow explanation of these effects. Examination reveals that for $\beta = 0$, α appears only in the denominator of a_{kpq}, and that it will act to significantly decrease transfer for triads in which the magnitude of at least one component k,p, or q is of the same order as α. Likewise β appears in the denominator of θ_{-kpq} and will significantly decrease transfer for any triad in which ω_{-kpq} is of the same order as μ_{kpq}. This is approximately true if one component of the triad has a magnitude less than $k_\beta(k_\beta = \beta/\zeta_{rms})$, thus lying in the region where wave propagation tendencies inhibit non-linear transfer.[12]

From the above considerations one might conclude that α and β would affect transfer rates only for the large scales. However, the plots of transfer show significant inhibition even in high wavenumbers where $\omega_k << \mu_k$ and $k >> \alpha$. The reason is that, because the energy spectrum is peaked in the low wavenumbers, the most important energy transferring triads are generally those containing at least one low wavenumber component. Thus even for large $|k|$, $|p|$ or $|q|$ transfer will be inhibited.

In the equations for the evolution of the error variance, effects of non-zero α and β are significant. Indeed, as shown by the evolution of relative error spectra (Fig. 2) and by instantaneous error transfer spectra (Fig. 4) the error growth is significantly suppressed at the low wavenumbers. In the high wavenumbers though, the decrease in error transfer is less significant than the corresponding decrease in energy transfer. One of the conveniences of the closure model, with its analytic dependence on α and β, is that we can take partial derivatives of the transfer functions T_u and T_Δ with respect to α and β. The contour plots of these derivatives (normalized by T_u and T_Δ) show that relative change in T_Δ with respect to a change in α or β is smaller than the corresponding change in T_u at high wave numbers. (Fig. 5)

Physical interpretations of these effects are fairly easy to put forward. The presence of α implies a component of potential energy associated with baroclinic effects, particularly for the large scale motions. This effectively boosts the total amount of error-free energy initially in the low wavenumbers, and increases the time scale for contamination by a transfer of error variance from higher wave numbers. This effect is included in the

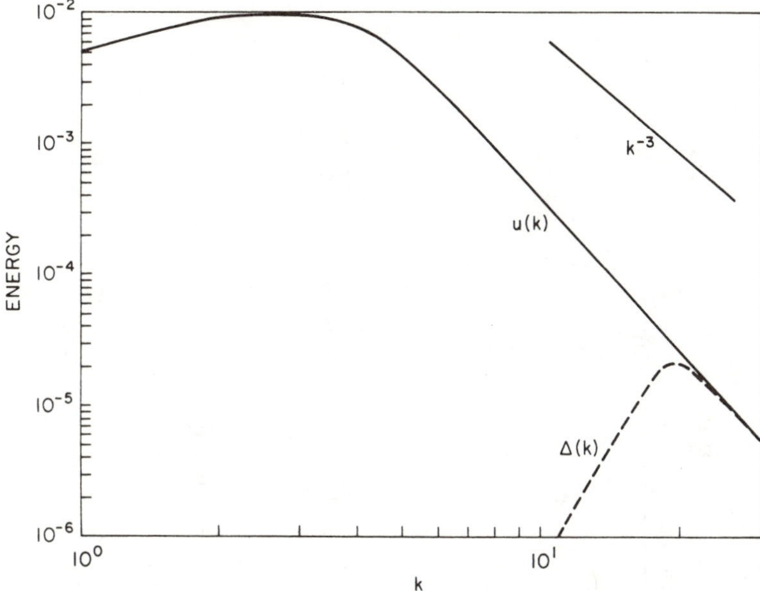

Fig. 1. Isotropic initial kinetic energy (solid line) and error kinetic energy (dotted line) of spectra used for all evolutions of the error field.

definitions of U_k and Δ_k. Also α directly slows all advective processes, including error transfer, at the larger scales. This effect is represented by the presence of α in the denominator of the geometric coefficient a_{kpq}, which we have determined does indeed slow error transfer.

Larger values of β result in higher frequencies for planetary wave propagation. These more rapid waves accelerate the decorrelation of large scale flow structures and thus slow non-linear interactions which cause error transfer. This effect is represented by the presence of β in the denominator of the decorrelation time θ_{-kpq}, which we have already examined.

The case with both α and β non-zero is interesting because α appears in a new term, the denominator of the Rossby wave frequency. Here, it acts to reduce the planetary wave frequencies, thus partially countering the effect of β. The error transfer function plot of the cases $\alpha = \alpha_o$, $\beta = 0$, $.5\beta_o, \beta_o$ (Fig. 4c) show that β does not reduce the error transfer rate by nearly as much as much as for the corresponding cases with $\alpha = 0$. Vallis (1983), in his two layer baroclinic study, also noted that β has much less effect on the baroclinic system than the barotropic.

Until now we have examined isotropic average spectra. However, the Rossby wave dispersion relation is anisotropic, and this results in significant anisotropy in the energy and error fields for cases with $\beta \neq 0$. For the energy field we find that zonal motions tend to be more energetic than meridional motions, even at high wavenumbers where wave propagation is negligible. The error field is initialized as some fraction of the energy in each mode, so that initially it has the same zonal anisotropy as the energy field. However, except at the low wavenumbers, the error quickly becomes isotropic and even slightly meridionally anisotropic. (Fig. 6) Of course, for large t,Δ_k approaches $U_{\underline{k}}$ and the error anisotropy must again approach the flow anisotropy. The isotropization of the error field may be explained by a simple hueristic argument. Consider two flow realizations whose error is

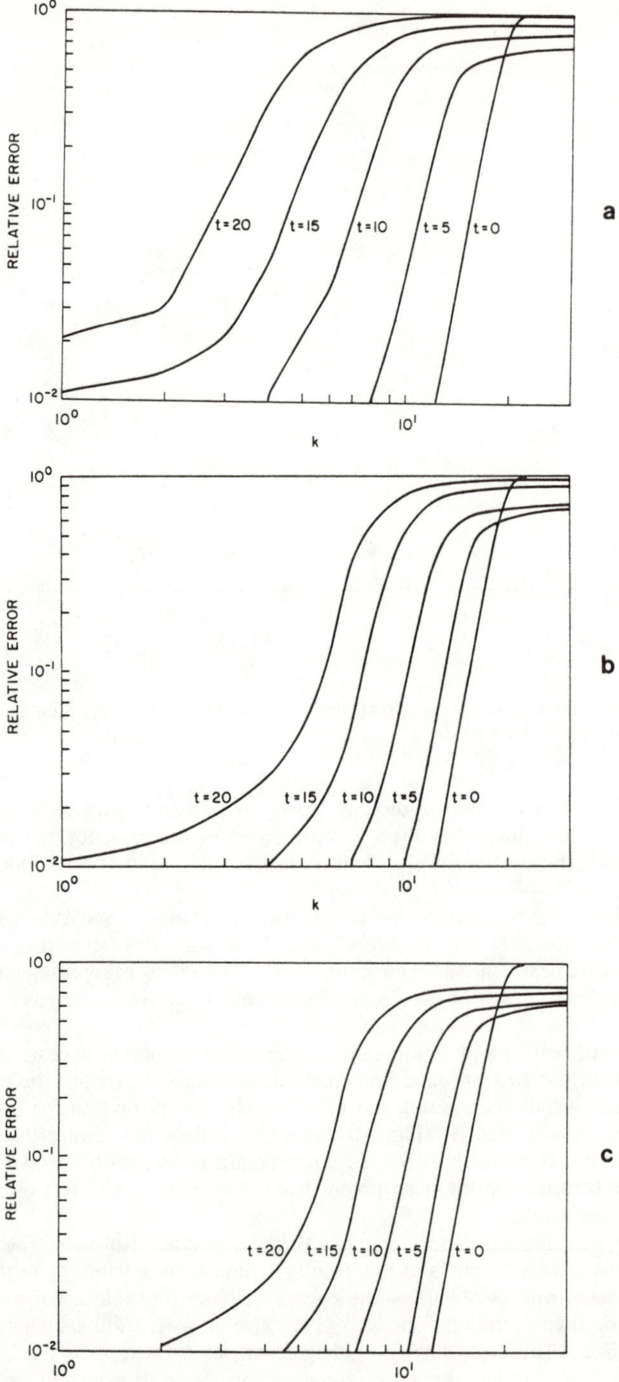

Fig. 2. Instantaneous isotropic relative error spectra $\left(e(k) = \frac{\Delta(k)}{U(k)}\right)$ at times t = 0,5,10,15 and 20 days for 3 cases a) $\alpha=0$, $\beta=0$, b) $\alpha=0$, $\beta=\beta_o$; c) $\alpha=\alpha_o$, $\beta=\beta_o$.

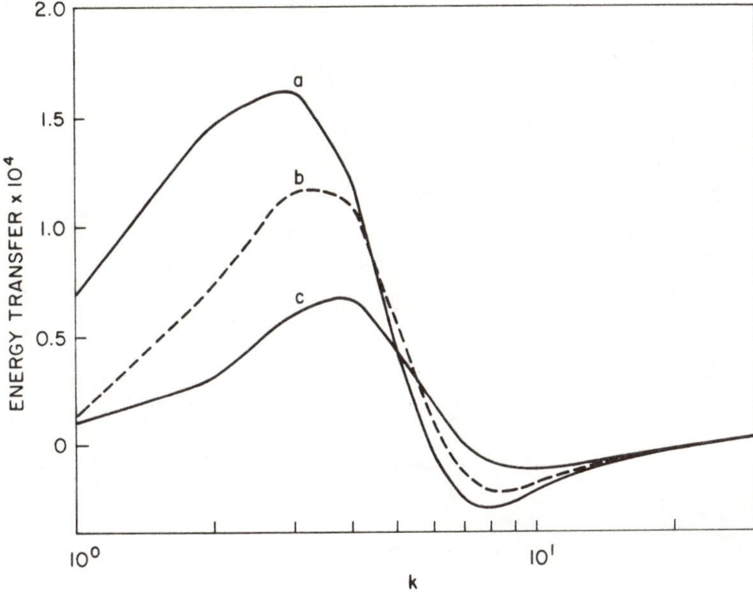

Fig. 3. Spectra of the energy transfer function at t = 5 days for the cases a) β=0, α=0; b) $\beta=\beta_o$, α=0; c) $\beta=\beta_o, \alpha=\alpha_o$. Suppression of the energy transfer by both β and α is evident throughout the entire range of wavenumbers.

initially entirely in the zonal motions as shown in Figure 7a. Suppose in the flows there exist spatial gradients of the velocities. Then the initial error in the zonal velocities results in a zonal separation of the two realizations and destroys the correlation in meridional velocities. Thus a meridional component of error is created.

A similar argument may be used to explain the slight meridional anisotropy occurring in the mid wavenumbers. Suppose we have an isotropic blob of error, represented by an isoline of error vorticity, being acted on by the zonal motions of the mid scales of the energy spectrum. (Fig. 7b) It gets stretched in such a way that it has a typical zonal scale that is larger than initially and a meridional scale that is smaller. Thus the meridional error has been moved to higher wavenumbers, increasing the meridional anisotropy there while the zonal error contributes to zonal anisotropy in the low wavenumbers. The processes for this evolution of the error and energy anisotropy are not readily apparent in the closure equations. However, if we assume that the anisotropy may be described by an angular harmonic expansion through second order:

$$U_{\underline{k}} = U(k)\,(1 - R(k)\cos 2\phi_k)$$

$$\Delta_{\underline{k}} = \Delta(k)\,(1 - R_\Delta(k)\cos 2\phi_k)$$

(where ϕ_k is the angle $\underline{k}$ makes with the $k_x$axis) then we can write an evolution equation for the anisotropy coefficient $R_\Delta(k)$[11].

$$\frac{d}{dt} R_\Delta(k) = S(k) + \int_1^{k_{max}} dp\Big(K_1(k,p)R(p) + K_2(k,p)R_\Delta(p)\Big) - \nu(k)R_\Delta(k)$$

where

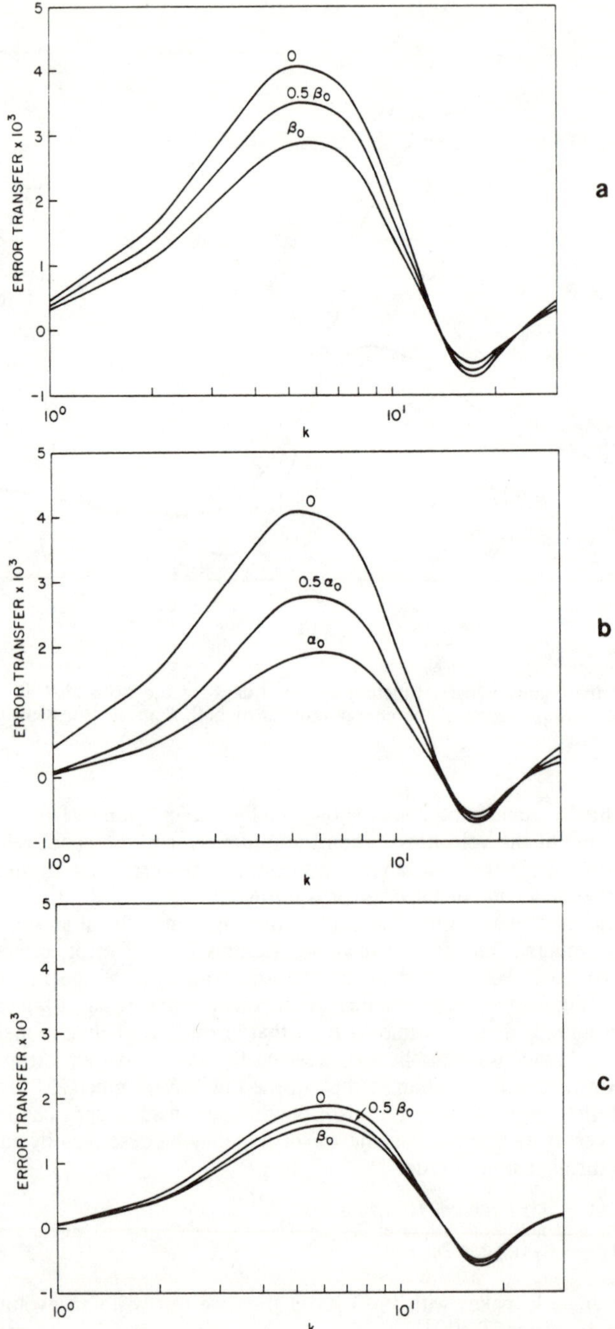

Fig. 4. Spectra of the error energy transfer function at t = 5 days. Panel a: $\alpha = 0$, $\beta = 0$, $.5\beta_o$ and β_o. Panel b: $\beta = 0$, $\alpha = 0$, $.5\alpha_o$ and α_o. Panel c: $\alpha = \alpha_o$, $\beta = 0$, $.5\beta_o$ and β_o. The effects of α and β suppress error transfer, but seem to be more important in the lower range of wavenumbers. Also when $\alpha = \alpha_o$ the effects of β are reduced.

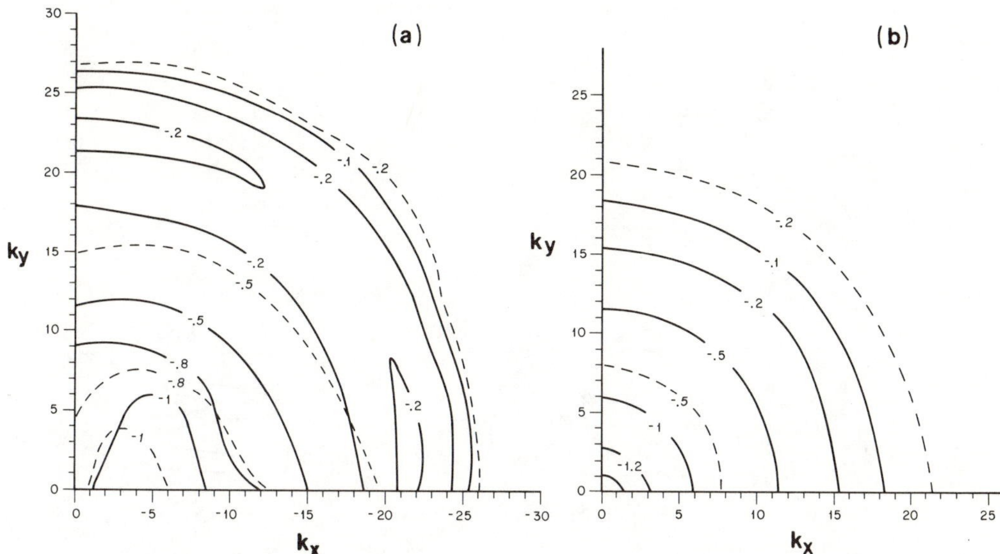

Fig. 5(a). Plot of the normalized rate of change of the energy transfer (T_u) and error energy transfer (T_Δ) with respect to β.

Fig. 5(b). As in a) but with respect to α. These plots show that α and β effect T_Δ less at higher k than they do T_u. Also note the anisotropy of the β case.

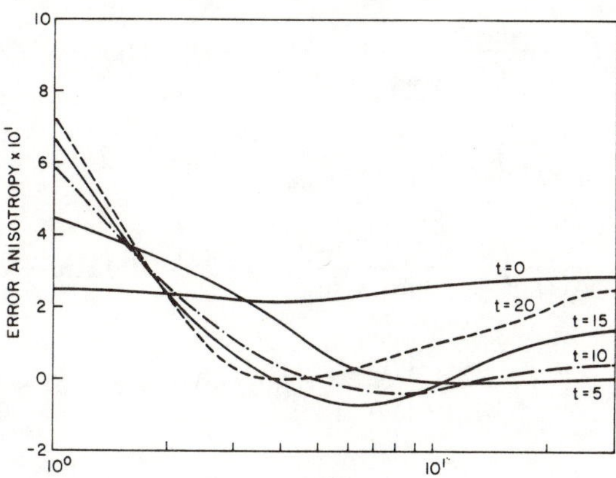

Fig. 6. Spectra of an anisotropy measure:

$$R_\Delta = \frac{((\Delta u)^2 - (\Delta v)^2)}{((\Delta u)^2 + (\Delta v)^2)}.$$

where Δu^2 and Δv^2 are the zonal and meridional error velocity variances. The error field is initialized as a fraction of the energy field so that its initial anisotropy is that of the energy field (R_u), given by the dotted line. Except in the lowest wave numbers the error field quickly becomes isotropic, then gradually approaches R_u again as $\Delta_k \rightarrow U_k$.

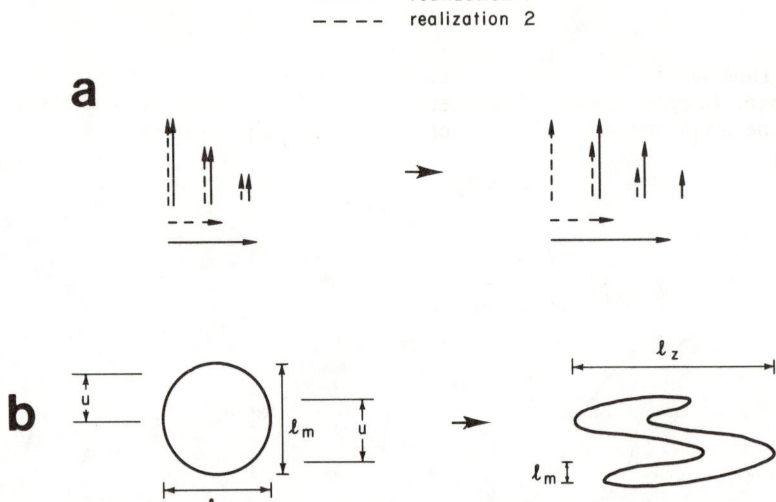

Fig. 7. Illustration of the simple heuristic arguments used to explain the behavior of the error anisotropy. (a) the generation of meridional error in a flow field which initially only has error in the zonal flow; (b) transformation of the scales of an error streamline acted on by a zonally anisotropic flow field.

$$S(k) = \frac{1}{\Delta(k)} \int\int_{\Delta} \frac{dpdq}{p \sin x} \frac{a_{kpq} I_2}{\mu_{kpq}} \{U(p)\Delta(q) + U(q)\Delta(p) - \Delta(q)\Delta(p) - \Delta(k)U(q)\}$$

$$K_1(k,p) = \frac{1}{\Delta(k)} \int_{\Delta q} \frac{dq}{p \sin x} \frac{a_{kpq} I_3}{\mu_{kpq}} \cos 2z \, \{2U(p)\Delta(q) - \Delta(k)U(p)\}$$

$$K_2(k,p) = \frac{-1}{\Delta(k)} \int_{\Delta q} \frac{dq}{p \sin x} \frac{a_{kpq} I_3}{\mu_{kpq}} \cos 2z \, \{2U(q)\Delta(p) - 2\Delta(q)\Delta(p)\}$$

$$\nu(k) = \frac{1}{\Delta(k)} \int\int_{\Delta} \frac{dpdq}{p \sin x} \left\{ \frac{I_1 a_{kpq}}{\mu_{kpq}} (U(p)\Delta(q) + \Delta(p)U(q) - \Delta(p)\Delta(q) - \Delta(k)U(q)) + \frac{I_3}{\mu_{kpq}} \Delta(k)U(q) \right\}$$

The various symbols are defined in the appendix. S(k) is a source term representing the generation of anisotropy directly at lower wave numbers. K_1 and K_2 are transfer of anisotropy coefficients multiplying the flow anisotropy R(k) and the error anisotropy $R_\Delta(k)$ respectively. The K_2 term is responsible for the isotropization for the high wavenumber error as it always acts to reduce the magnitude of R_Δ. The K_1 term represents the effect of the flow anisotropy R(k) on the error anisotropy. It causes meridional anisotropy in the middle wavenumbers. As $\Delta(k)$ approaches U(k) for all k the equation for evolution of $R_\Delta(k)$ becomes equivalent to the equation for R(k) (the flow anisotropy)[11]. Hence in this limit the error anisotropy must approach the flow anisotropy.

COMPARISON OF CLOSURE AND NUMERICAL STUDIES

Thus far we have looked at the evolution of the error field and the processes involved. In order to answer the question of what the predictability of a flow is, we may examine overall measures of the error of a realization. Several such measures may be defined, including the difference height variance:

$$E_o = (\psi_1 - \psi_2)^2 = \sum_{\underline{k}} \frac{\Delta_k}{k^2}$$

the difference energy:

$$E_1 = \frac{1}{2} |\underline{\nabla}(\psi_1 - \psi_2)|^2 = \sum_{\underline{k}} \frac{\Delta_k}{2}$$

or the difference enstrophy:

$$E_2 = (\nabla^2(\psi_1 - \psi_2))^2 = \sum_{\underline{k}} k^2 \Delta_{\underline{k}}$$

Each of these measures is weighted differently with respect to k and show different growth rates (Fig. 8a,b,c). Hence any "predictability time" is dependent upon choice of error measure.

The growth of the error measures again shows strong suppression for the cases in which α and β are not zero. Taking E_1 as a measure of the predictability we estimate an error doubling time of approximately 2.6 days when $\alpha = 0$, $\beta = 0$. This becomes approximately 6 days for the $\alpha = \alpha_o$, $\beta = \beta_o$ case so the effects of α and β are indeed significant. Error suppression by α and β is even more important for the largest scales of the flow.

Comparisons were made between the results of the closure equation studies and a numerical simulation study by Holloway[6] in which the effect of α and β were also investigated. An attempt was made to make conditions in the two studies as comparable as possible. The initial spectrum of the numerical study was slightly steeper in the inertial range, but the total energy in the two systems was nearly equal. The error evolution was similar except in the lowest wave numbers, where the error seemed to grow more quickly in the closure model. (Fig. 8) Also, the high wavenumbers seem to not recover as quickly from the initial drop in error, but this may be due to differing values for the high wavenumber viscosity. The effects of α and β in suppressing error growth are seen clearly in both approaches, emphasizing the robustness of these processes. Behavior of the error anisotropy was also found to match well.

SUMMARY

We have considered predictability for an idealized case, namely equivalent barotropic flow on a β–plane, using methods of statistical closure theory. Results should be examined not so much for specific estimates of predictability but rather for the insight into the significant processes affecting predictability. We find that inclusion of the finite equivalent depth enhances predictability, as does differential rotation, the two effects combining to double the error energy growth time scale. We find that the error energy tends to increase isotropically over most scales until near saturation. We find that the results of numerical simulation and closure study are in agreement in most aspects.

REFERENCES

1. J.G. Charney, R.G. Fleagle, V.E. Lally, H. Riehl and D.G. Wark. Bull. Am. Met. Soc., **47**, 200-220 (1966).
2. J. Smagorinsky. Bull. Am. Met. Soc. **50**, 286-311 (1969).

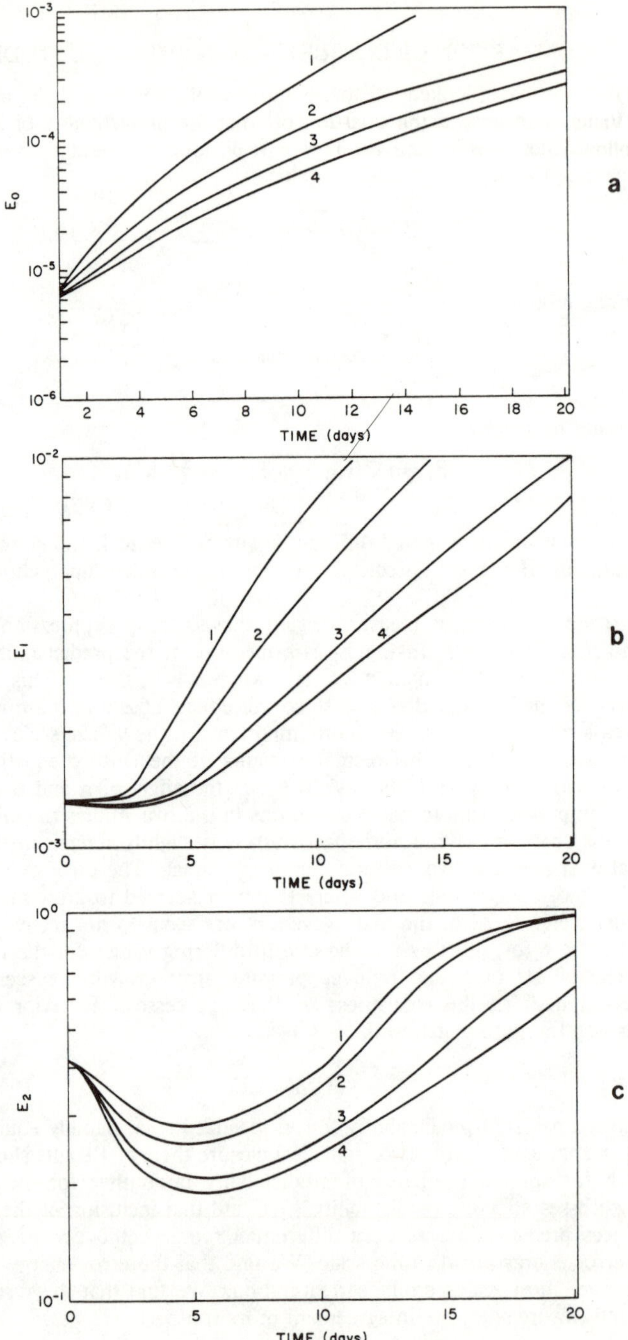

Fig. 8. Evolution of the gross error measures E_o, E_1, and E_2 in panels a, b and c respectively. In each, four cases are shown: 1) $\alpha = 0$, $\beta = 0$; 2) $\alpha = 0$, $\beta = \beta_o$; 3) $\alpha = \alpha_o$, $\beta = 0$; 4) $\alpha = \alpha_o$, $\beta = \beta_o$.

3. D.K. Lilly. Geophys. Fluid Dyn. **4**, 1-28 (1972).
4. C. Basdevant, B. Legras, R. Sadourny and M. Beland. J. Atmos. Sci., **38**, 2305-2326 (1981).
5. G.K. Vallis. J. Atmos. Sci. **40**, 10-27 (1983).
6. G. Holloway. J. Atmos. Sci., **40**, 314 (1983).
7. E.N. Lorenz. Tellus, **21**, 289-307 (1969).
8. C.E. Leith. J. Atmos. Soc. **28**, 145-161 (1971).
9. C.E. Leith and R.H. Kraichnan. J. Atmos. Sci. **29**, 1041-1058 (1972).
10. R.H. Kraichnan. J. Fluid Mech. **47**, 513-524 (1971).
11. G. Holloway and M.C. Hendershott. J. Fluid Mech. **82**, 747-765 (1977).
12. P.B. Rhines. J. Fluid Mech. **69**, 417-443 (1975).

APPENDIX: LOW-ORDER REPRESENTATION OF ANISOTROPY

For small β, the anisotropy of the energy and error fields will be small and we may adopt the representations:

$$\begin{aligned} U_{\underline{k}} &= U(k)\ (1 - R(k)\cos 2\phi_k) \\ \Delta_{\underline{k}} &= \Delta(k)\ (1 - R_\Delta(k)\cos 2\phi_k) \end{aligned} \tag{A1}$$

where ϕ_k is the acute angle made by the vector $\underline{k}$ and the $\underline{k}_x$ axis, U(k) and Δ(k) are the isotropic spectra. R(k) and $R_\Delta(k)$ are the respective anisotropy coefficients, which are small relative to unity for sufficiently small values of β. Substituting this into equation (7) yields

$$\begin{aligned} \left(\frac{d}{dt} + 2\nu_k\right)\Delta(k)\left(1 - R_\Delta(k)\cos 2\phi_k\right) = 2 \sum_{\underline{p}+\underline{q}=\underline{k}} \theta_{-kpq}\, a_{kpq} \\ \cdot\Big[\Big(U(p)\Delta(q)\left(1 - R(p)\cos 2\phi_p\right)\left(1 - R_\Delta(q)\cos 2\phi_q\right)\Big) \\ + \Big((U(q)\Delta(p)\left(1 - R(q)\cos 2\phi_q\right)\left(1 - R_\Delta(p)\cos 2\phi_p\right)\Big) \\ - \Big(\Delta(p)\Delta(q)\left(1 - R_\Delta(p)\cos 2\phi_p\right)\left(1 - R_\Delta(q)\cos 2\phi_q\right)\Big) \\ - \Big((\Delta(k)U(q)\left(1 - R_\Delta(k)\cos 2\phi_k\right)\left(1 - R(q)\cos 2\phi_q\right)\Big)\Big] \end{aligned} \tag{A2}$$

We eliminate terms of order R^2, R_Δ^2 and RR_Δ and separate the evolution of Δ(k) and $R_\Delta(k)$ by integrating over ϕ_k:

$$\int_0^{2\pi} \left[\left(\frac{d}{dt} + 2\nu_k\right)\Delta(k)\left(1 - R_\Delta(k)\cos 2\phi_k\right)\right] d\phi_k \rightarrow \left(\frac{d}{dt} + 2\nu_k\right) 2\pi\ \Delta(k)$$

and

$$\int_0^{2\pi} \left[\left(\frac{d}{dt} + 2\nu_k\right)\Delta(k)\left(1 - R_\Delta(k)\cos 2\theta_k\right)\right]\cos 2\theta_k\ k\theta_k$$

$$\rightarrow 2\pi \left[\left(\frac{dR_\Delta(k)}{dt}\right)\Delta_k(k) + R_\Delta(k)\left(\frac{d}{dt} + 2\nu\right)\Delta(k)\right]$$

thus

$$\frac{d}{dt}R_\Delta(k) = \frac{1}{\pi\Delta(k)}\left\{-R_\Delta(k)\int_o^{2\pi} d\phi_{\underline{k}} \sum_{\underline{p}+\underline{q}=\underline{k}} \theta_{-kpq}\, a_{kpq}\right.$$

$$\cdot\Big[U(p)\Delta(q)\Big(1 - R(p)\cos 2\phi_{\underline{p}} - R_\Delta(q)\cos 2\phi_{\underline{q}}\Big)$$

$$+ U(q)\Delta(p)\Big(1 - R(q)\cos 2\phi_q - R_\Delta(p)\cos 2\phi_{\underline{q}} - R_\Delta(p)\cos 2\phi_{\underline{p}}\Big)$$

$$- \Delta(p)\Delta(q)\Big(1 - R_\Delta(p)\cos 2\phi_p - R_\Delta(q)\cos 2\phi_{\underline{q}}\Big)$$

$$- \Delta(k)U(q)\Big(1 - R_\Delta(k)\cos 2\phi_{\underline{k}} - R(q)\cos 2\phi_{\underline{q}}\Big)\Big]$$

$$+ \int_o^{2\pi} d\phi_{\underline{k}}\cos 2\phi_{\underline{k}} \sum_{\underline{p}+\underline{q}=\underline{k}} \theta_{-kpq}\, a_{kpq} \qquad \text{(A3)}$$

$$\cdot\Big[U(p)\Delta(q)\Big(1 - R(p)\cos 2\phi_{\underline{p}} - R_\Delta(q)\cos 2\phi_{\underline{q}}\Big)$$

$$+ U(q)\Delta(p)\Big(1 - R(q)\cos 2\phi_q - R_\Delta(p)\cos 2\phi_{\underline{q}} - R_\Delta(p)\cos 2\phi_{\underline{p}}\Big)$$

$$- \Delta(p)\Delta(q)\Big(1 - R_\Delta(p)\cos 2\phi_p - R_\Delta(q)\cos 2\phi_{\underline{q}}\Big)$$

$$- \Delta(k)U(q)\Big(1 - R_\Delta(k)\cos 2\phi_{\underline{k}} - R(q)\cos 2\phi_{\underline{q}}\Big)\Big]$$

We approximate the summation over $\underline{p}$ by the integration

$$\int_1^{k_{max}} \int_o^{2\pi} p\, dp\, d\phi_p$$

Letting x,y, and z be the interior angles of a triangle opposite sides k,p and q we can rewrite this integral as

$$\int \int_\Delta \frac{dp\, dq}{p \sin x}$$

where the domain of integration Δ is given in Figure 9. Following Holloway and Hendershott[11] we approximate θ_{-kpq} in terms of ϕ_k by:

$$\theta_{-kpq} = \qquad \text{(A4)}$$

$$\frac{\mu(k) + \mu(p) + \mu(q)}{(\mu(k) + \mu(p) + \mu(q))^2 + \left(\frac{\beta^2}{2l^2}\right)(1 + \cos 2\phi_k \cos 2\alpha \pm \sin 2\phi_k \sin 2\alpha)}$$

where 1 is the modulus of the smallest of the wavevectors $\underline{k}$, $\underline{p}$ and $\underline{q}$ and α is the interior angle opposite 1 when $\underline{1} = \underline{k}$ and $\alpha = 0$ for $1 \neq k$. The resulting integrations over ϕ_k are of three kinds:

$$I_1(\epsilon) = \int_0^{2\pi} \frac{d\sigma}{1 + (1 + \cos\sigma \cos 2\alpha \pm \sin\sigma \sin 2\alpha)_{/\epsilon}} = \tag{A5}$$

$$\frac{2\pi\epsilon}{(\epsilon^2 + 2\epsilon)^{1/2}}$$

$$I_2(\epsilon,\alpha) = \int_0^{2\pi} \frac{\cos\sigma \, d\sigma}{1 + (1 + \cos\sigma \cos 2\alpha \pm \sin\sigma \sin 2\alpha)_{/\epsilon}} =$$

$$2\pi\epsilon \cos 2\alpha \left(1 - \frac{\epsilon + 1}{(\epsilon^2 + 2\epsilon)^{1/2}}\right)$$

$$I_3(\epsilon,\alpha) = \int_0^{2\pi} \frac{\cos^2\sigma \, d\sigma}{1 + (1 + \cos\sigma \cos 2\alpha \pm \sin\sigma \sin 2\alpha)_{/\epsilon}}$$

$$= 2\pi\epsilon \left[\frac{(2\epsilon^2 + 4\epsilon + 1)\cos 4\alpha + 1}{(\epsilon^2 + 2\epsilon)^{1/2}} - 2(\epsilon + 1)\cos 4\alpha\right]$$

where $\epsilon(k,p,q) = (\mu(k) + \mu(p) + \mu(q))^2 \; / \; (\beta^2/21^2)$ and the integral is the sum over the two sign choices. Using these expressions in (A3) yields the equation (13) for the evolution of R(k).

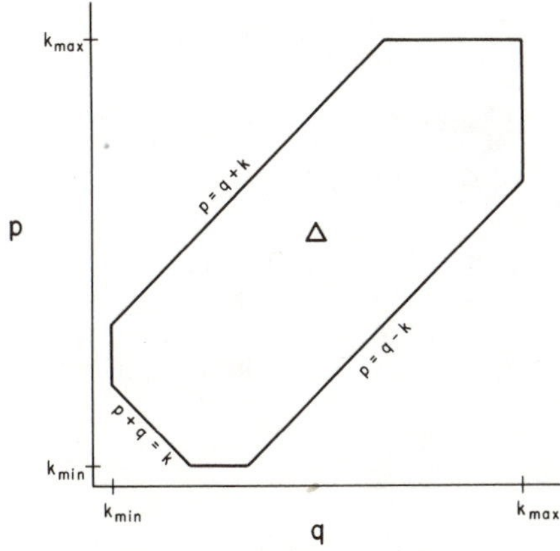

Fig. 9. Domain of integration used in obtaining the low order representation of anisotropy.

AN APPLICATION OF THE MINIMUM PREDICTABILITY TIME PRINCIPLE TO GEOSTROPHIC TURBULENCE

Roberto Benzi
Centro Scientifico IBM
Via del Giorgione 129 Roma (Italy)

Angelo Vulpiani
Dipartimento di Fisica, Universita 'La Sapienza' Roma,
and G.N.S.M.-C.N.R. Unita' di Roma.

ABSTRACT

We study the energy spectrum of the geostrophic shallow water equation using a variational principle on the predictability time. It is found that the large scale dynamics can be characterized by a non-local spectrum and by an inverse enstrophy cascade. Non-local analysis of the results does not lead to relevant changes.

INTRODUCTION

The statistical mechanics of an inviscid fluid can be studied following the standard methods and ideas of statistical mechanics of Hamiltonian systems[1,2]. If we consider a d-dimensional (d=2,3) box of size L with periodic boundary conditions we can decompose the velocity field $\underline{V}(\underline{x})$ into a Fourier series and truncating at sufficiently high values of k:

$$V_d(\underline{x}) = 1/L^{d/2} \sum_{\underline{k}} \hat{V}_d(\underline{k}) \exp(i\underline{k}\cdot\underline{x}) \quad . \tag{1}$$

One can easily see that for the set of variables $\{y_i\}$ which are the independent components of $\{\hat{V}_d(\underline{k})\}$ (there are constraints of incompressibility and of reality of $\underline{V}(\underline{x})$) a Liouville theorem holds:

$$\sum \partial/\partial y_i (dy_i/dt) = 0 \tag{2}$$

From the Liouville theorem and conservation laws:

$$E = 1/2 \sum_i y_i^2 = \text{const} \quad \text{for d=3} \tag{3}$$

$$\left.\begin{aligned} E &= 1/2 \sum_i y_i^2 = \text{const} \\ \Omega &= 1/2 \sum_i k_i^2 y_i^2 = \text{const} \end{aligned}\right\} \text{for d=2} \tag{4}$$

using the standard arguments of classical statistical mechanics (for example the maximum entropy principle or more sophisticated arguments based on central limit theory) a Gibb's distribution for $\{y_i\}$ can be obtained:

$$P(y) \sim \exp-\beta \sum_i y_i^2 \quad \text{for d=3} \tag{5}$$

$$P(y) \sim \exp-\sum_i (\alpha + \beta k_i^2) y_i^2 \quad \text{for d=2} \tag{6}$$

From Equations (5) and (6) we can compute the energy spectrum (for $k \leqslant k_{max}$):

$$E(k) \sim \begin{cases} \beta^{-1} k^2 & \text{for d=3} \\ k/(\alpha + k^2\beta) & \text{for d=2} \end{cases} \tag{7}$$

The statistical mechanics of an inviscid fluid has some connections with the behavior of fully developed turbulence. For example it is possible to show[2] from Equation (7) and the hypothesis (common to any statistical theory) of relaxation to a stationary state that there is a tendency to transfer energy up to the highest wave number in 3-dimensional case and a tendency to an inverse energy cascade in the 2-dimensional case (at least in the situation of negative temperature). It is well known that this behavior are present also in the really (viscous) turbulence.

However, it is evident that the Equation (7) is in violent disagreement with the phenomenological Kolmogorov's theory:

$$E(k) \sim \begin{cases} k^{-5/3} & \text{for } d=3 \\ k^{-3} & \text{for } d=2 \end{cases}$$

in the inertial range. Besides this disagreement there are other deep differences between the statistical mechanics of an inviscid fluid and fully developed turbulence. In the inviscid fluid case a fluctuation-dissipation theorem gives a correlation between the fluctuation and the response of the system to a perturbation. In the turbulence case the mechanism of response to a perturbation is completely different[3]. This happens because of the different nature of the two stationary states: an equilibrium state for inviscid fluid, a far from equilibrium stationary state, possibly only with an external injection of energy (and enstrophy for d=2), for turbulence.

The maximum entropy principle does not describe fully developed turbulence even qualitatively because of its substantial 'static' character. We think that this principle cannot be considered as a 'philosophical' principal but only a different mathematical formulation of the hypothesis of equiprobability of the points in the phase space. In the case where the maximum entropy principle works this probably happens because it summarizes all relevant properties of the system (as for example the Liouville theorem and ergodic hypothesis)[4]. Some authors had proposed applications of the maximum entropy principle in turbulence, for example to compute intermittency corrections[5] or to derive the -5/3 law[6]. All these approaches however have some inconsistencies and defects. We discuss these problems elsewhere.

The maximum entropy principle for its 'static' character cannot take into account the peculiar behavior of a turbulent flow which is completely different from the one of inviscid fluid. In the inviscid case the scheme of the energy exchange is: large eddies <----> small eddies; in the turbulent case the scheme is changed as: large eddies → small eddies → heat.

In a recent paper in collaboration with Vitaletti[7] we introduced a maximum randomness principle different from the maximum entropy principle. We have shown that if we state the principle for which the 'predictability time', i.e. the time in which a perturbation on the smallest scale of the motion influences the large scales, must be a minimum, we obtain the correct energy spectrum E(k). We shall give now an expression for the predictability time. If we exponentially divide the scales of the motion:

$$l_n = b^{-n} l_o \quad b > 1 \tag{8}$$

where l_o is the external length (for example the correlation length of the velocity field) the predictability time[8] is defined by:

$$T_p = \sum_{1}^{N} \tau_n \tag{9}$$

τ_n is the time in which a perturbation at scale l_n transfer 'error' at scale l_{n-1}. N is the order number of the dissipative scale η, i.e. the smallest scale involved in the turbulence:

$$\eta = l_N \quad ; l_N V_N/\nu = 1 \ . \tag{10}$$

One can believe τ_n to be the typical evolution time of the n-th scale structure[8,9] (the turn-over time at scale 1_n):

$$\tau_n = 1_n/V_n = 1/k_n V_n \quad k_n = 1_n^{-1} \tag{11}$$

where V_n is the typical velocity differences at scale 1_n:

$$V_n = \int_{k_n}^{k_n+1} E(k)dk \tag{12}$$

Note that the predictability time defined by Equation (9) is the time in which the system forgets its initial conditions. It is evident (at last on heuristic level) that T_p is correlated to Kolmogorov topological entropy h[10]:

$$T_p \cong h^{-1} \ . \tag{13}$$

Because h is proportional to the mean rate of entropy production (in the information theory meaning[10]) one has that the principle of minimum predictability time is a maximum chaotic principle. Note that our principle of minimum predictability time has a 'dynamical' nature, while the maximum entropy principle concerns only 'static' properties of a stationary states.

Let us briefly review our approach in the case of a three dimensional turbulence. We are now seeking an expression of V_n that minimizes the predictability time with the constraint of the conservation of mean energy E of the flow:

$$E = 1/2 \sum V_n^2 \tag{14}$$

(indeed we are interested in the properties of the turbulent motion in the stationary state and far from dissipative scale). By the standard technique of Lagrange's multipliers we obtain:

$$\partial(T_p + \lambda_1 E)/\partial V_n = 0 \tag{15}$$

by Equation (15) we obtain:

$$V_n = \lambda_1^{-1/3} k_n^{-1/3} \tag{16}$$

and for the spectrum E(k):

$$E(k) \approx V_n^2 k_n^{-1} \approx \lambda_1^{-2/3} k_n^{-5/3} \tag{17}$$

It is very easy to see that $\lambda_1 \sim \epsilon^{-1}$. From this result and Equation (17) we obtain the well known Kolmogorov spectrum. In the paper in Reference 7 we applied this variational principle also to turbulence in two dimensions, with intermittency, with thermal convection, in three dimensions with helicity, and in two-and three-dimensional magnetohydrodynamics. In all the cases we obtain the same results obtained by other means (phenomenological, experimental, numerical etc.) or very weak disagreements.

APPLICATION TO GEOSTROPHIC TURBULENCE

In this paper we apply this variational principle to the equation:

$$\partial q/\partial t + J(\psi, q) = \text{dissipative} + \text{forcing terms} \tag{18}$$

where $q = \Delta\psi - \psi/R^2$ and R is the Rossby deformation radius. This equation is relevant in several different atmospheric and oceanic motions[11]. It refers to the shallow water equation for the geostrophic flow of a layer with a free surface.

The inviscid form of Equation (18) conserves the two quantities:

$$E = \int_D (1/2(\text{grad}\psi)^2 + 1/2(\psi/R)^2)dxdy = \text{energy} \tag{19}$$

$$\Omega = 1/2\int_D(\Delta\psi - \psi/R^2)^2\, dxdy = \text{potential enstrophy} \tag{20}$$

where D is the domain of the flow. In terms of the discrete variables V_n and k_n, Equations (19) and (20) become:

$$E = 1/2 \sum V_n^2(1 + 1/k_n^2R^2) \tag{21}$$

$$\Omega = 1/2 \sum k_n^2V_n^2(1 + 1/k_n^2R^2)^2 \tag{22}$$

The predictability time T_p is defined as in the first section to be:

$$T_p = \sum 1/k_nV_n \ . \tag{23}$$

Following the case of two dimensional turbulence[1] we can compute V_n by applying the variational principle:

$$\partial(T_p + \lambda_1E + \lambda_2\Omega)/\partial V_n = 0 \ . \tag{24}$$

Using Equations (21), (22) and (23) in Equation (24) we obtain:

$$V_n^3 = 1/k_n[\lambda_1(1+1/k_n^2R^2) + \lambda_2k_n^2(1+1/k_n^2R^2)^2] \ . \tag{25}$$

The ratio $(\lambda_1/\lambda_2)^{1/2}$ defines a scale k_o in the system. Thus the energy spectrum $E(k_n) \approx V_n^2\, k_n^{-1}$ depends on two scales variables k_o and R^{-1}. Three different cases are therefore possible:

a) $k_o >> 1/R$;

b) $k_o << 1/R$;

c) $k_o \sim 1/R$.

Corresponding to these three cases we obtain:

$$\text{case a)}\ \ E(k) \sim \begin{cases} \lambda_2^{-2/3}k^{-3} & \text{for } k >> k_o \\ \lambda_1^{-2/3}k^{-5/3} & \text{for } 1/R << k << k_o \\ \lambda_1^{-2/3}R^{4/3}k^{-1/3} & \text{for } k << 1/R \ . \end{cases} \tag{26}$$

$$\text{case b)}\ \ E(k) \sim \begin{cases} \lambda_2^{-2/3}k^{-3} & \text{for } k >> 1/R \\ \lambda_2^{-2/3}R^{8/3}k^{-1/3} & \text{for } k_o << k << 1/R \ . \\ \lambda_2^{-2/3}R^{8/3}k^{-1/3} & \text{for } k << k_o \end{cases} \tag{27}$$

$$\text{case c)}\ \ E(k) \sim \begin{cases} \lambda_2^{-2/3}k^{-3} & \text{for } k >> k_o \\ k^{-1/3}R^{4/3}/(\lambda_1 + \lambda_2/R^2)^{2/3} & \text{for } k << k_o \end{cases} \tag{28}$$

In the cases b and c no inertial ranges with a -5/3 slope develop and an inertial range with -1/3 slope characterizes the large scale spectrum. The Lagrange's multipliers λ_1 and λ_2

refer to transfer of energy and enstrophy respectively. It follows that a direct enstrophy cascade is always acting for very large value of k while a inverse cascade of energy and enstrophy acts at large scale. In particular in case b the large scale spectrum is characterized by an inverse enstrophy cascade governed by the term ψ/R^2 in the dynamical equation. In case c both energy and enstrophy are transferred to large scales. In analogy with two dimensional turbulence[12] k_o is the scale of both energy and enstrophy input into the system. In geophysical application k_o can be interpreted as the scale of active baroclinic instability[13]. Therefore the straightforward physical interpretation of our results is that when baroclinic instability develops on a much larger scale of Rossby deformation radius on scale comparable with it the turbulent power spectrum is characterized by Equations (27) and (28) with a slope -1/3 for large scale. On the other hand for R much larger than the characteristic scale of baroclinic instability the turbulence spectrum resembles for most of the k-range the strictly two dimensional case. Note that for $R \rightarrow \infty$ we obtain from Equation (26) the Kraichnan-Batchelor two dimensional spectrum.

The slope -1/3 in the inertial range is strongly non-local[2]. A possible way to discuss in our analysis the non-local transfer of energy and enstrophy can be accomplished by using a non-local expression for τ_n^2:

$$\tau_n = \left(\sum_{J=1}^{n} k_J^2 V_J^2 \right)^{-1/2} . \tag{29}$$

In terms of (29) the predictability time T_p becomes:

$$T_p = \sum_{n} 1/\left(\sum_{J=1}^{n} k_J^2 V_J^2 \right)^{1/2} . \tag{30}$$

We use this formulation of T_p in analyzing the change in the power spectrum for case a (cases b and c can be discussed in a similar way). Equation (24) becomes:

$$-\partial T_p/\partial V_n = k_n^2 V_n \sum_{m \geqslant n} 1/\left(\sum_{J=1}^{m} k_J^2 V_J^2 \right)^{3/2} =$$

$$= \lambda_1 (1 + 1/k_n^2 R^2) V_n + \lambda_2 k_n^2 (1 + 1/k_n^2 R^2)^2 V_n \ . \tag{31}$$

For $k_n >> k_o$ we simplify expression (31) to:

$$\sum_{m \geqslant n} 1/\left(\sum_{J=1}^{m} k_J^2 V_J^2 \right)^{3/2} = \lambda_2 \tag{32}$$

The left hand side of Equation (32) can be estimated to be:

$$\sum_{m \geqslant n} 1/\left(\sum_{J=1}^{m} k_J^2 V_J^2 \right)^{3/2} = \ln(k_N/k_n) \Big/ \left(\sum_{J=1}^{q} k_J^2 V_J^2 \right)^{3/2} \tag{33}$$

where $q > n$ and k_N^{-1} is the Kolmogorov dissipative scale. Expression (33) is the discrete version of the classical average theorem for integrals:

$$\int_a^c f(x)dx = (c-a)f(\bar{x}) \quad \text{where } a < \bar{x} < c \ .$$

From Equations (33) and (32) it follows:

$$\sum_{J=1}^{q} k_J^2 V_J^2 \cong \lambda_2^{-2/3} (\ln k_N/k_n)^{2/3} \ . \tag{34}$$

Differentiating (34) respect to k_n we obtain:

$$V_n^2 \sim \lambda_2^{-2/3} k_n^{-2} (\ln k_N/k_n)^{-1/3} \tag{35}$$

with a corresponding power spectrum:

$$E(k) \sim \lambda_2^{-2/3} k^{-3} (\ln k_N/k)^{-1/3} \ . \tag{36}$$

This form of E(k) is slightly more local than the original spectrum k^{-3}. Similar computations can be done for $1/R << k_n << k_o$ and $k_n << 1/R$. After some algebraic calculations we obtain the same expression for E(k) given in Equation (26) plus logarithm corrections.

CONCLUSIONS

We have shown in this paper that the turbulent energy spectrum for a single layer shallow water system in geostrophic approximation can differ substantially from the pure two dimensional case. The presence of Rossby deformation radius effects large scale dynamics developing a rather non-local energy spectrum with a slope -1/3. When the characteristic scale of energy and enstrophy input into the system is greater or equal to R an inverse enstrophy cascade takes place. It could be interesting to understand if this inverse enstrophy cascade might be associated with the existence of large scale coherent structures and vortices as observed in atmospheric and oceanic circulations.

REFERENCES

1. R. Salmon, G. Holloway and M.C. Hendershott. J. Fluid Mech. **75**, 691 (1976).
2. H. A. Rose and P.L. Sulem. J. Physique **39**, 441 (1978).
3. R.H. Kraichnan, Adv. Math. **16**, 305 (1975).
4. O. Penrose, Rep. Prog. Phys. **42**, 1937 (1979).
5. H. Mori and H. Fujisaka, in SYSTEM FAR FROM EQUILIBRIUM, Ed. L. Garrido (Springer, Berlin 1980) p. 181.
6. T.M. Brown. J. Phys. **15A**, 2285 (1982).
7. R. Benzi, M. Vitaletti, and A. Vulpiani. J. Phys. **15A**, 883 (1982).
8. E.N. Lorenz, Tellus **21**, 289 (1969); D.K. Lilly, Geophys. Fluid Dyn. **3** , 889 (1972) and **4** 1, (1972).
9. U. Frisch, Ann. N.Y. Acad. Sci., **357**, 359 (1980).
10. M. J. Rabinovitch, Sov. Phys. Usp. **21**, 443 (1978).
11. J. Pedlosky, GEOPHYSICAL FLUID DYNAMICS (Springer, Berlin 1979).
12. R.H. Kraichnan, Phys. Fluid. **10**, 1417 (1967); G.K. Batchelor, Phys. Fluid. (suppl. 2) **12**, 233 (1969).
13. R. Salmon, Geophys. Astr. Fluid Dyn. **10**, 25 (1978).

ON THE DYNAMICS OF PREDICTABILITY

Hampton N. Shirer
Department of Meteorology
The Pennsylvania State University

ABSTRACT

The predictability of the three-component Lorenz model of shallow convection augmented by a horizontal heating term is examined by studying the characteristics of a six-component companion set of equations governing both the sum and the difference of two trajectories in phase space. In this set of equations, the sum represents the average of two trajectories, and the difference provides a measure of the error. With the six-component system, the growth of errors, and hence predictability, can be studied directly.

When the two convective states are stable, two individual trajectories will either converge to the same solution, or they will diverge, one approaching the first solution, the other approaching the second solution. Numerical integrations of the error/average equations demonstrate that there are large sets of initial conditions from which all of the trajectories approach either a converging or a diverging solution; there are four such stable (and five unstable) solutions available here. Sensitivity to initial conditions and predictability problems occur primarily when the initial conditions are chosen close to the boundaries of the four basins of attraction of the stable solutions. As the value of the forcing is increased, the initial values must be farther from these boundaries for predictability problems to be avoided. In many cases, the solution approaches the unstable conductive solution before evolving toward one of the four stable states; thus, the critical portion of the flow requiring the most accurate representation is the part near the unstable state. Thus, in some cases, the loss of predictability of a system is not linked directly to the separation of two solutions, but is linked directly to the necessity of very precisely knowing either the initial or the subsequent values of the temporal solutions.

INTRODUCTION

Loss of predictability is a fundamental characteristic of fully turbulent flows. Nonlinear terms in the equations of motion redistribute spectrally the components of the flow causing nearby initial values to grow apart (Thompson[24]; Lorenz[13]; Smagorinsky[22]; Robinson[19]; Leith[9]). In essence, these flows become unpredictable because it is not possible to obtain either initial conditions with sufficient accuracy or numerical models with sufficient resolution for acceptable forecasts to be produced. Here, then, unpredictability is the same as error growth, and many linear and nonlinear studies have been performed to determine the bounds on error growth and the practical limits of predictability of the atmosphere (Thompson[24]; Lorenz[13-17]; Smagorinsky[22]; Robinson[19]; Leith[9]; Jacobs[4,5]; Merilees[18]; Lilly[10]; Williamson[25]; Baer and Alyea[1]; Blumen[2]; Shukla[21]).

This view of predictability, however, is too limited. There are many instances when accurate knowledge of all details is not necessary for an acceptable forecast of the flow. Observations of many laboratory systems indicate that there are many situations in which a few spectral components dominate the fluid motion. For example, Krishnamurti[6-8] has shown this dominance to characterize Rayleigh-Bénard convection; a more extensive review of this and other systems is given in Swinney and Gollub[23]. Many scales of atmospheric motion, ranging from cloud streets to upper air flow also exhibit this behavior. These relatively simple flows typically are found in the early stages of the transition to turbulence, which often occurs via a small number of bifurcations from steady to temporal solutions.

In the early portions of the transition to turbulence, when the possible number of stable states is limited, the flow should exhibit a high degree of predictability. Indeed, Lorenz[12] has noted that flows dominated by either periodic or quasi-periodic components contain a high degree of intrinsic predictability; but either their attainable predictability, limited by observational inaccuracies, or their practical predictability, limited by mathematical insufficiencies, may be restricted severely. Moreover, as discussed by Dutton and Wells[3], the number of degrees of freedom available to the temporally periodic solutions of an infinite-dimensional partial differential equation is actually finite-dimensional owing to the finite dimension of the unstable manifold. A poorly designed numerical simulation of such a process might create spurious loss of predictability if the dimension of the unstable manifold of the solutions to the numerical model is too large. In this case, the numerical model has too many degrees of freedom, and predictability may be enhanced by prudent limitation of the degrees of freedom to those appropriate to the periodic or multiply periodic components that dominate the flow under consideration. Indeed, Thompson[24] and Shukla[21] have presented evidence for this idea when they found that averaged fields were predictable for longer time intervals than were unaveraged ones; also Blumen[2] noted that error growth was explained by only a small number of harmonics.

Even in these early phases of the transition to turbulence predictability problems may exist. These problems arise when there is more than one attracting element in the limit set so that the flow must choose the state toward which to evolve. If the final observed state does not depend critically on the values of the initial conditions, then the flow would be classified as predictable; but if the final state does depend sensitively on the values of the initial conditions, then the flow would be unpredictable (Jacobs[5]).

Thus, predictability problems arise when two nearby trajectories straddle initially the boundary between two basins of attraction. Predictability problems would be expected if the two basins of attraction (or their numerical approximations) were interwoven in a complicated manner, making necessary very accurate knowledge of the initial data for acceptable predictions.

It might be expected that the shapes and sizes of the basins of attraction would vary with the values of the external parameters. Truncated spectral models, because they involve only a few ordinary differential equations, provide ideal nonlinear systems for the investigation of the properties of the basins of attraction, particularly, for those flows that are dominated by a few wavenumbers. Indeed, direct study of the relative evolution of two trajectories can be accomplished by the conversion of a single spectral model into a companion set governing both the sum and the difference of the components of the two trajectories. The temporal behavior of the sum provides a measure of the average value of the two trajectories, while the temporal behavior of the difference provides a measure of the distance or error between them. Error growth or decay, traditionally viewed as predictability loss or gain, can be studied directly with this system. Thus, these error/average equations govern the dynamics of predictability.

In this article, we explore some aspects of the predictability problem at the most elementary level: when there are only one unstable and two stable steady states to the governing system. In this case, the flow is intrinsically predictable (Lorenz[12]). We convert the three convective equations of Lorenz[11] into a six-component nonlinear set of error/average equations and investigate the relationships between the initial and final states as functions of the thermal forcing. We will find that only certain subsets of initial conditions are troublesome, although the sizes of these subsets vary with the values of the forcing. Thus, in certain cases, intrinsically predictable flows may be practically unpredictable, as noted in principle by Lorenz[12]. This type of behavior surely occurs in either large grid-point or large spectral models as well, but detailed determination of the subsets of these would be impractical. This suggests that appropriately designed low-order spectral models be used to identify those situations for which accurate prediction would be relatively easy and those for which it would be extremely difficult. This information might allow effective management of resources by separating situations for which accurate knowledge of initial conditions would be needed from those for which exceptionally accurate knowledge would be needed.

THE ERROR/AVERAGE EQUATIONS

Shallow convection forced thermally in both the horizontal and the vertical can be modeled by the three-component system

$$\dot{x} = -P\,x + P\,y - H \tag{1}$$

$$\dot{y} = -x\,z + r\,x - y \tag{2}$$

$$\dot{z} = x\,y - b\,z + (H/3P)x \tag{3}$$

in which r is proportional to the vertical temperature difference, H is proportional to the horizontal temperature difference, P is the Prandtl number, x is a Fourier coefficient of the stream function and y and z are Fourier coefficients of the temperature perturbation. Additional details are given in Lorenz[11] and Shirer and Wells[20].

If we denote $\underset{\sim}{X}_1(t) = (x_1(t), y_1(t), z_1(t))$ and $\underset{\sim}{X}_2(t) = (x_2(t), y_2(t), z_2(t))$ as two trajectories having two different initial values $\underset{\sim}{X}_1(0)$ and $\underset{\sim}{X}_2(0)$, then we may define the error components e_1, e_2, e_3 by the differences

$$e_1 = \tfrac{1}{2}\,(x_1 - x_2) \tag{4}$$

$$e_2 = \tfrac{1}{2}\,(y_1 - y_2) \tag{5}$$

$$e_3 = \tfrac{1}{2}\,(z_1 - z_2) \tag{6}$$

and the average components a_1, a_2, a_3 by the sums

$$a_1 = \tfrac{1}{2}\,(x_1 + x_2) \tag{7}$$

$$a_2 = \tfrac{1}{2}\,(y_1 + y_2) \tag{8}$$

$$a_3 = \tfrac{1}{2}\,(z_1 + z_2) \tag{9}$$

Then (1)-(3) expressed in terms of the variables $\underset{\sim}{e}$ and $\underset{\sim}{a}$ are

$$\dot{e}_1 = -P\,e_1 + P\,e_2 \tag{10}$$

$$\dot{e}_2 = -e_1\,a_3 - e_3\,a_1 + r\,e_1 - e_2 \tag{11}$$

$$\dot{e}_3 = e_1\,a_2 + e_2\,a_1 - b\,e_3 + (H/3P)e_1 \tag{12}$$

$$\dot{a}_1 = -P\,a_1 + P\,a_2 - H \tag{13}$$

$$\dot{a}_2 = -a_1\,a_3 - e_1\,e_3 + r\,a_1 - a_2 \tag{14}$$

$$\dot{a}_3 = a_1 a_2 + e_1 e_2 - b a_3 + (H/3P)a_1 \tag{15}$$

From (10)-(15) we see immediately that the error and average components do not evolve independently and that (10)-(15) might be expected to contain different nonlinear behavior than (1)-(3). Thus, the dynamics of convection would be modeled by (1)-(3), but the dynamics of the predictability of the convective states by (10)-(15).

Whether or not the characteristics of the limit set of (10)-(15) are simple, it can be demonstrated easily that the error growth is bounded because, as discussed by Lorenz[11], individual trajectories must eventually become trapped in a finite region of phase space. If we set

$$a_2' = a_2 + H/3P \tag{16}$$

$$a_3' = a_3 - r - P \tag{17}$$

$$E = \tfrac{1}{2}(e_1^2 + e_2^2 + e_3^2) \tag{18}$$

$$A = \tfrac{1}{2}\left[a_1^2 + (a_2')^2 + (a_3')^2\right] \tag{19}$$

then we find that

$$\begin{aligned} \dot{E} &= e_1 \dot{e}_1 + e_2 \dot{e}_2 + e_3 \dot{e}_3 \\ &= -P e_1^2 - e_2^2 - b e_3^2 - e_1 e_2 a_3' + e_1 e_3 a_2' \end{aligned} \tag{20}$$

$$\begin{aligned} \dot{A} &= a_1 \dot{a}_1 + a_2' \dot{a}_2' + a_3' \dot{a}_3' \\ &= -P a_1^2 - a_2'^2 - b a_3'^2 - e_1 e_3 a_2' + e_1 e_2 a_3' \\ &\qquad -[H(3P+1)/(3P)]a_1 - b(r+P)a_3' \end{aligned} \tag{21}$$

and that

$$\begin{aligned} \dot{E} + \dot{A} &= -P e_1^2 - e_2^2 - b e_3^2 - P[a_1 + H(3P+1)/(6P^2)]^2 \\ &\quad - a_2'^2 - b[a_3' + 0.5(r+P)]^2 + H^2(3P+1)^2/(36P^3) \\ &\qquad + b(r+P)^2/4 \\ &= Q \end{aligned} \tag{22}$$

Thus, from (22), we see that errors eventually become bounded within the ellipsoid $Q = 0$. At worst then, error growth will reach a finite asymptotic limit that increases in magnitude as the value of either the thermal forcing H or r increases.

From (20), we also find that errors will grow initially if

$$P e_1^2 + e_2^2 + b e_3^2 < - e_1 e_2 a_3' + e_1 e_3 a_2' \tag{23}$$

and decay otherwise. Thus, we see that there is a set of initial conditions for which small errors grow initially, and so for these initial conditions the flow might be labeled unpredictable. From (23) we note that the magnitude of the initial error growth can be minimized by reducing the magnitude of e_1, which is the error component for the stream function coefficient. Thus, the attainable predictability of the convective flow may depend more sensitively on the magnitudes of the velocity components than on those of the thermal components.

When $r<1$ and $H=0$, the system (10)-(15) has a unique trivial solution: all trajectories converge to the conductive state. But when $r>1$ and $H=0$, (10)-(15) possesses nine steady states: three that we label convergent and six, divergent. In general, both trajectories may converge to one of the three possible solutions (1)-(3), in which cases the error components $\underset{\sim}{e}$ vanish, or they may diverge, one trajectory approaching one solution to (1)-(3), the other trajectory approaching a different one. The steady states of (10)-(15) are given by (Fig. 1)

$$e_2 = e_1 \tag{24}$$

$$e_3 = [2e_1\, a_1 + (4H/3P)e_1]/b \tag{25}$$

$$a_2 = a_1 + H/P \tag{26}$$

$$a_3 = [a_1^2 + e_1^2 + (4H/3P)a_1]/b \tag{27}$$

$$e_1^3 + [3a_1^2 + (8H/3P)a_1 + b(r-1)]\, e_1 = 0 \tag{28}$$

$$a_1^3 + (4H/3P)\, a_1^2 + [3e_1^2 - b(r-1)]\, a_1 + (4H/3P)\, e_1^2 + bH/P = 0 \tag{29}$$

From (28) and (29), we see that the convergent solutions are given by

$$e_1 = 0 \tag{30}$$
$$a_1^3 + (4H/3P)\, a_1^2 - b(r-1)\, a_1 + bH/P = 0$$

and the divergent ones, with nonzero components, are given by

$$e_1^2 = b(r-1) - 3a_1^2 - (8H/3P)\, a_1 \tag{31}$$
$$a_1^3 + (2H/3P)\, a_1^2 - [b(r-1)/4 + H^2/9P^2]\, a_1 + bH(r-1)/6P - bH/8P = 0$$

In the special case $H = 0$, there is one unstable convergent solution $e_1 = a_1 = 0$, two stable convergent solutions (denoted by - and + in Figure 1)

$$e_1 = 0 \tag{32}$$
$$a_1 = \pm\, [b(r-1)]^{1/2}$$

two stable divergent solutions (denoted by x and * in Figure 1)

$$e_1 = \pm [b(r-1)]^{1/2} \tag{33}$$
$$a_1 = 0$$

and four unstable divergent solutions (denoted by o in Figure 1)

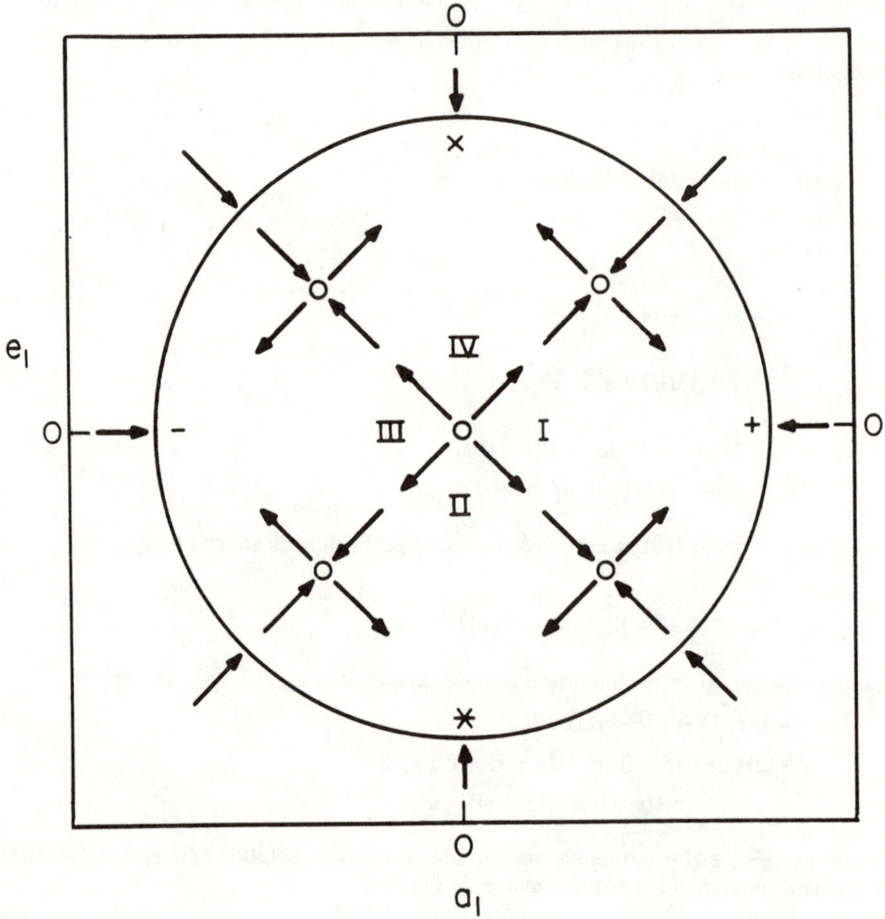

Fig. 1. The e_1 and a_1 components of the steady states (32)-(34) of (10)-(15) and the projections of their associated expanding and contracting directions. The symbols + and - denote the stable positive and negative convergent solutions (32), x and * denote the stable positive and negative divergent solutions (33) and o denotes the unstable solutions (34).

$$e_1 = \pm[b(r-1)]^{1/2}/2 \tag{34}$$
$$a_1 = \pm[b(r-1)]^{1/2}/2$$

The location of these solutions and their associated basins of attraction provide the key to the attainable predictability of convective flows represented by (1)-(3).

BOUNDARIES OF THE BASINS OF ATTRACTION AND PREDICTABILITY

In the neighborhood of each of the five unstable solutions to (10)-(15), there are certain directions along which small perturbations depart from, and other directions along which they decay toward, the solution. The expanding directions are tangent to the unstable manifold, and the contracting directions tangent to the stable manifold, of each solution. Because two eigenvalues of (10)-(15) linearized about the trivial solution are positive when $r>1$ (and $H=0$), there are two expanding directions near the trivial solution; the projections of these directions are shown in Figure 1 and lie on the lines $e_1 = a_1$ and $e_1 = -a_1$. The four unstable, divergent solutions locally have one expanding and five contracting directions because only one eigenvalue of the linearized system is positive; the projections of these directions are depicted schematically in Figure 1. From (22) we see that all directions are contracting when $e_1^2 + a_1^2 > b(r+P)^2/(4P)$; this is shown schematically in Figure 1.

From Figure 1, we can see that, apparently, the projection of the boundaries separating the four basins of attraction might be expected to be given by the lines $e_1 = a_1$ and $e_1 = -a_1$. Initial conditions chosen in one region between the lines would be expected to approach the stable solution within that region. In this case, predictability problems would be expected only for those initial conditions chosen near the trivial solution or chosen near the lines $e_1 = a_1$ and $e_1 = -a_1$.

If Figure 1 is correct at least qualitatively, then we see that the classification of systems and their predictability suggested by Lorenz[13] must be modified. In many cases, trajectories originating in regions II and IV that approach a stable divergent solution have the property that the two individual trajectories $\underset{\sim}{X}_1$ and $\underset{\sim}{X}_2$ separate and remain farther apart than they were initially. Trajectories originating in regions I and III have the property that eventually the individual trajectories $\underset{\sim}{X}_1$ and $\underset{\sim}{X}_2$ approach each other. Thus, if we were to choose initial conditions near the origin but always from regions II or IV, then we would label the convective system (1)-(3) as Type 3 because in these systems initial errors grow and remain larger than their initial values. But if we choose initial conditions always from regions I or III, than we would label the system as Type 1 because in these systems errors eventually become smaller than their initial values. Type 3 systems are viewed as ones having a finite, unextendable limit of predictability, but Type 1 systems are ones having no limit. Here, though, we can determine when individual trajectories will separate and where they each will go, and so (10)-(15) exhibits a high degree of predictability in all four regions. From this discussion, we conclude that examination of error growth alone may lead to incorrect labeling of an intrinsically predictable system as an unpredictable one. Thus, we suggest that classification of predictability should depend on whether the solutions of a system are temporally periodic or aperiodic (Lorenz[12]); this classification in turn depends on the topological structure of the attractors and their basins.

We determined the actual location of the basins of attraction of the four stable states by performing a large number of integrations of (10)-(15) using the fourth-order accurate Hamming's modified predictor-corrector method. The parameter values used were $H=0$, $b=8/3$, $P=10$, and $r=5$, 10, 15, 20. The convective states of (1)-(3) are stable for these values of r, with loss of stability occurring via Hopf bifurcation at $r=24.74$. Two sets of experiments were conducted in which the initial values of e_1 and a_1 were varied in a grid over the range $-9 \leqslant e_1 \leqslant 9$, $-9 \leqslant a_1 \leqslant 9$.

In the first set, the initial values of e_2, e_3, a_2, and a_3 were chosen to satisfy the steady equations (24)-(27). In this way, varying only the magnitudes of e_1 and a_1 leads to varying smoothly the initial values of the error and average components along a hypersurface passing through all nine steady states. Thus, trajectories can be judged to originate either close to or far from a stationary point by simply examining the values of e_1 and a_1 alone. As a result, depiction in Figures 2a-d of the location of the basins of attraction via projection onto the e_1-a_1 plane leads to consistent results. In Figure 2 we used the following convention: initial conditions from which trajectories approached the positive or negative convergent solutions (32) are denoted, respectively, by horizontal and vertical hatching; initial conditions from which trajectories diverged to the positive or negative solutions (33) are denoted, respectively, by the positively and negatively sloped hatching.

We see from Figure 2 that the basins of attraction of the four solutions divide the e_1-a_1 plane into a pattern that is symmetric about both the e_1 and a_1 axes. Initial conditions chosen nearby one of the four stable solutions converge to it; moreover, when $r=5$, 10 (Figs. 2a,b) the actual location of the projections of the basins of attraction agree quite well with those given in Figure 1. The lines $e_1=a_1$ and $e_1=-a_1$ do in fact divide the e_1-a_1 plane into four quadrants, with initial conditions originating in one region producing trajectories that approach the stable state that is within that region.

Once the initial conditions are chosen far enough from the origin, additional boundaries also exist. Predictability problems would only be expected when initial conditions are chosen near these boundaries; if an initial condition is not known with sufficient accuracy, then there might be some uncertainity as to which side of the boundary a trajectory actually originates. The initial values having the greatest problems are those near where two boundaries intersect, such as near the origin; in these portions of phase space four basins of attraction are close to one another. For initial conditions that are far from these boundaries, the evolution of the trajectories is not dependent on precise knowledge of initial conditions, and no predictability problems exist there. Convergence times do vary smoothly, however, having minimum values near the centers of the regions, and larger values near the boundaries.

As the value of r is increased the projections of the basins of attraction divide the e_1-a_1 plane into more regions. When $r=15$ (Fig. 2c), additional regions appear near the lines $e_1=a_1$ and $e_1=-a_1$, reducing the sizes of the attracting quadrants around the stable points. Here the initial conditions associated with the greatest predictability problems are those near the lines tangent to the stable manifold of the unstable divergent solutions (34) (cf Fig. 1). This apparent link between the stable manifold and troublesome initial conditions demonstrates the importance of locating the positions of the stable manifolds for the solutions to the equations of interest; for finite dimensional systems, these locations can be found using a method discussed by Dutton and Wells[3].

When $r=20$ (Fig. 2d), the lines bounding the new regions that appeared in Figure 2c have moved toward the stable solutions, creating a nearly equal subdivision of each quadrant into four separate regions. But these interior boundaries are no longer straight, implying that the basins of attraction are becoming interwoven in a complicated manner. In this case, predictability problems would become quite severe near these winding boundaries where initial conditions must be known with great accuracy. Furthermore, as the value of r is increased toward that of the Hopf bifurcation point, even more complicated topological structures might be expected for the basins of attraction. Thus, loss of predictability may be associated with the approach of the values of a parameter to that of a bifurcation point. Nevertheless, trajectories originating near a stable solution still approach it.

A second set of experiments was performed in which only knowledge of the physical structure of the average components is assumed. The results are given in Figure 3. Again the parameter values were $H=0$, $b=8/3$, $P=10$ and $r=5,10,15,20$ and the initial values of a_1 and e_1 were varied over the range $-9\leqslant e_1 \leqslant 9$, $-9\leqslant a_1 \leqslant 9$. For simplicity, the initial

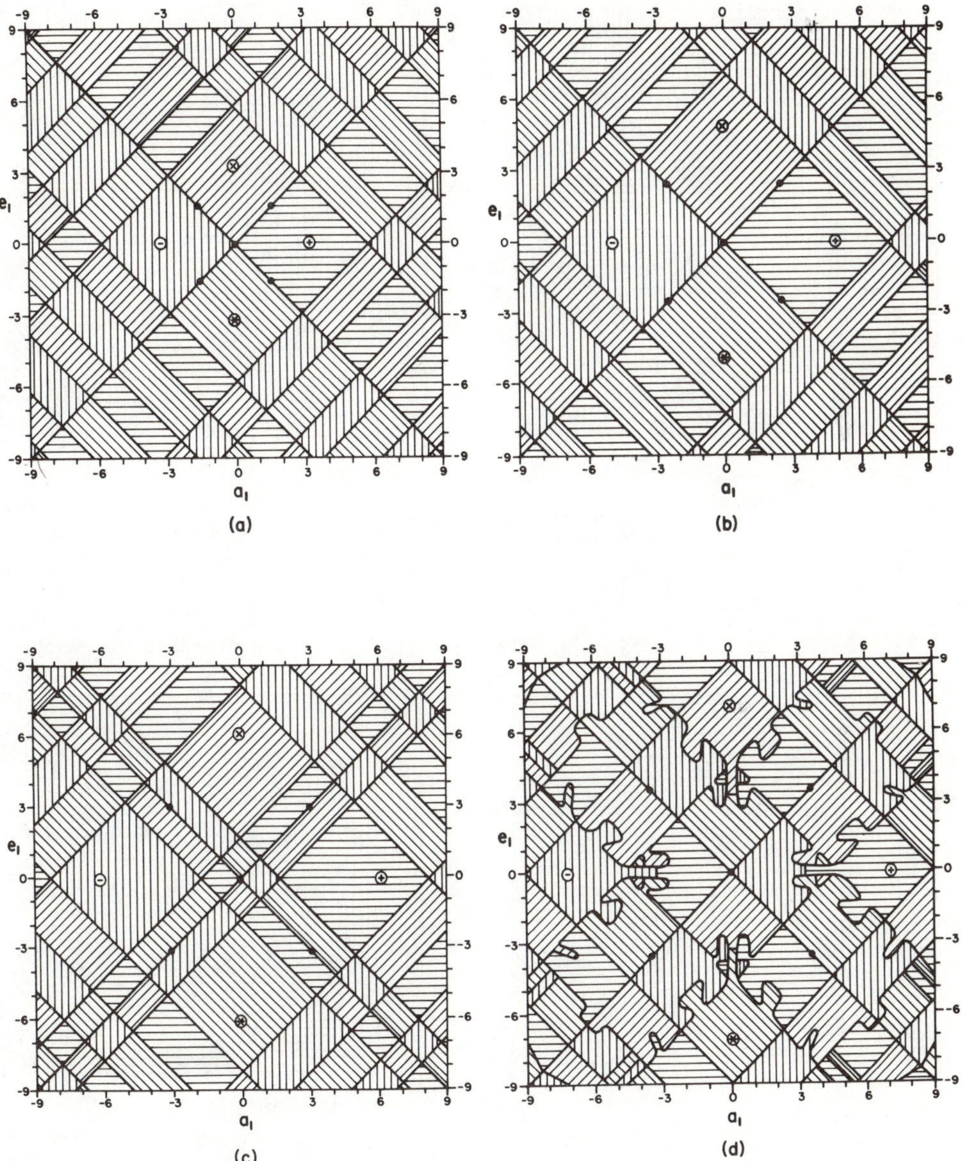

Fig. 2. The e_1 and a_1 components of the domains of attraction of (10)-(15) when $e_2(0)$, $e_3(0)$, $a_2(0)$ and $a_3(0)$ satisfy (24)-(27). The labeling of the steady states is the same as in Figure 1; initial conditions of trajectories that approach the + solution are horizontally shaded, that approach the - solution are vertically shaded, that approach the x solution are shaded with positively sloped lines, that approach the * solution are shaded with negatively sloped lines. Here $H=0$, $b=8/3$, $P=10$. In (a) $r=5$; in (b) $r=10$; in (c) $r=15$; in (d) $r=20$.

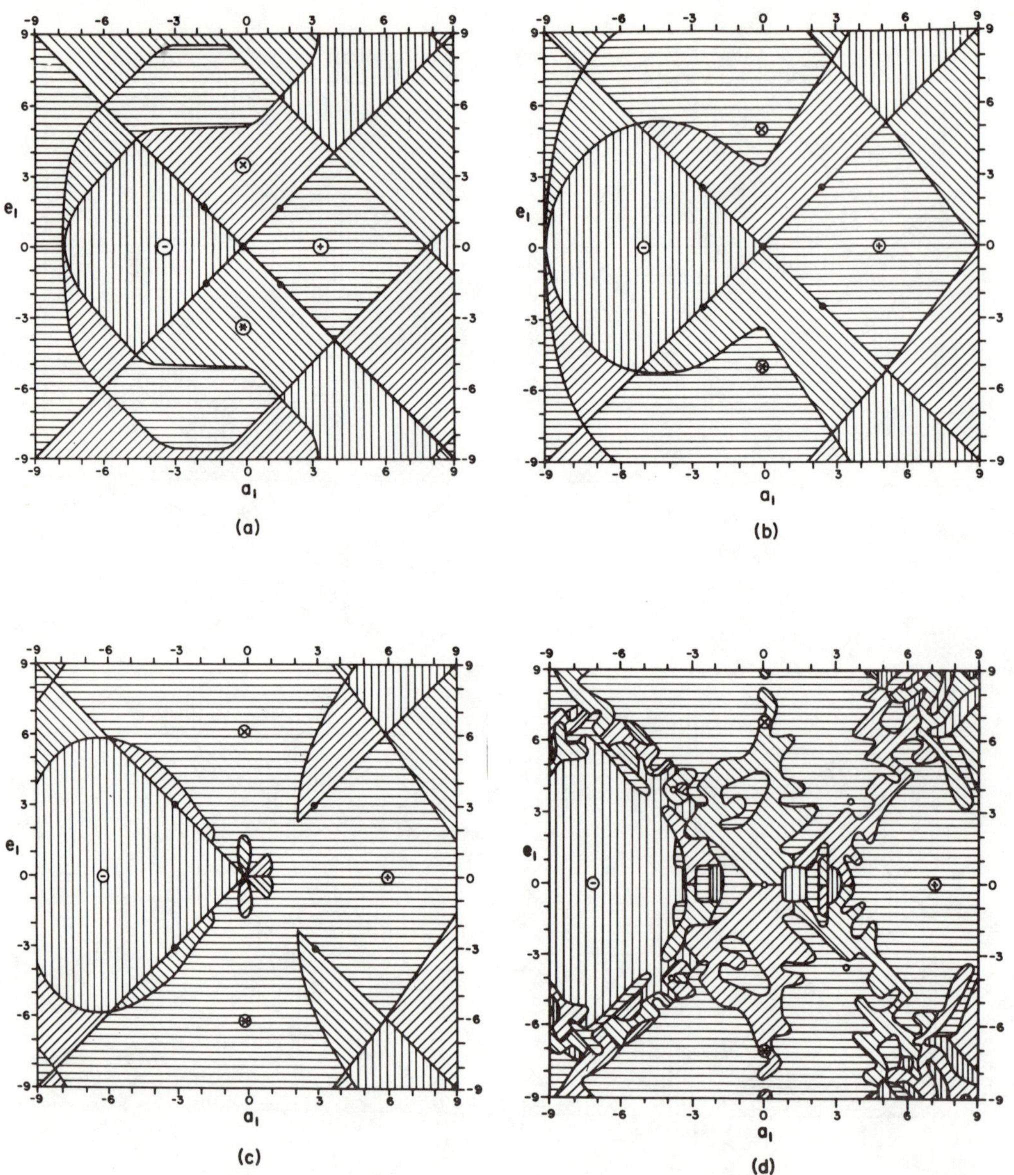

Fig. 3. The e_1 and a_1 components of the domains of attraction of (10)-(15) when the initial components are equal and the initial average components are a percentage of the convergent solution (32). The labeling of the steady states and the attracting regions is the same as in Figure 2. Here H=0, b=8/3, P=10. In (a) r=5; in (b) r=10; in (c) r=15; and in (d) r=20.

values of the error components were chosen to be equal, $e_1(0) = e_2(0) = e_3(0)$, but the initial values of the average components were chosen to be a percentage of the convergent solution (32), that is $a_2(0) = a_1(0) = c[b(r-1)]^{1/2}$ and $a_3(0) = |c|(r-1)$. This initial choice for $\underset{\sim}{e}$ corresponds to errors in knowledge of the actual values of the stream function and the temperature; the initial choice for $\underset{\sim}{a}$ corresponds to correct knowledge of the relationship between the components of the stable convective flow.

In these experiments, the initial conditions are weighted toward the convergent solutions so that there is a preference for trajectories to approach one of them. Because the initial values are projected onto the e_1-a_1 plane in Figure 3, only those values that are apparently near a convergent solution correspond to values actually near it. Initial values that are apparently near a divergent solution, however, are actually not near it; consequently, it is not surprising that such trajectories may not approach it (Fig. 3b-d).

As in Figure 2, the lines $e_1 = a_1$ and $e_1 = -a_1$ provide projections of the boundaries for the basins of attraction of the four stable solutions; in Figure 3a for $r=5$ the e_1-a_1 plane is divided into quadrants in much the same manner as depicted in Figures 1 and 2a. As the value of r is increased, the sizes and shapes of the attracting regions vary (Fig. 3b). When $r=15$ (Fig. 3c), however, some predictability problems begin to occur, particularly for initial conditions that are near the trivial solution. Small divergent regions lie within the positive convergent one so that initial values must be known very accurately for correct prediction of the flow. Again the worst initial conditions are those near the origin, far from a convective state. We conclude that predictability problems can be reduced by using knowledge of the actual structure of the steady states to facilitate the choice of initial values.

The situation changes dramatically when $r=20$ (Fig. 3d). The basins of attraction are intertwined in a manner that is more complicated than that seen in Figure 2d. Thus the relative evolution of two individual trajectories depends sensitively on the values of the initial conditions. The regions of greatest sensitivity are primarily those near the lines $e_1 = a_1$ and $e_1 = -a_1$, the tangents to the stable manifold of the unstable divergent solutions (cf Fig. 1).

Thus, predictability problems can be anticipated when initial conditions of the error/average system are chosen to be too close to a boundary separating two basins of attraction. As the values of the forcing increase, the initial conditions must be chosen farther from these boundaries for predictability problems to be reduced (cf Figs. 2d, 3d). This demonstrates the necessity of determining the boundaries of the basins of attraction of elements of the limit set of the governing partial differential system: once these boundaries are determined, the initial conditions that lead to predictability problems can be found. A method discussed by Dutton and Wells[3] for inferring the location of neighboring diverging trajectories in partial differential equations may be used also for inferring the location of these boundaries.

The results given here suggest that for predictability to be enhanced in an operational model, the boundaries of the basins of attraction must be found and initial conditions chosen to avoid these boundaries; this might be accomplished via a dynamic initialization process in which the measured values are varied toward the interior of an appropriate region. It is not practical to locate boundaries in large grid point and spectral models suitable for operational forecasting; thus, these boundaries must be located with the aid of a low-order companion model, and in many cases severely truncated spectral models would be ideal candidates.

In any case, these results illustrate that the initial conditions having the greatest problems are the ones near the unstable trivial solution, for the boundaries of all four basins of attraction pass through this state; small variations in these initial conditions will lead to large changes in the observed flow, and the flow may be incorrectly interpreted as being unpredictable over a wider range of initial conditions. Thus, it is not surprising that unpredictability has been linked to instability of a basic state (e.g. Lorenz[16]; Lilly[10]).

The complicated behavior of the basins of attraction in Figures 2 and 3 can be traced to the tendency of the evolving flow to approach the supposedly unobservable, unstable trivial solution. The behavior of the trajectory near the trivial solution plays a major role in the final evolution of the flow, and it is this portion of the flow that must be represented the most accurately.

These observations are illustrated in Figure 4, in which the projection on the $e_1 - a_1$ plane of four individual trajectories of (10)-(15) are shown. Here $r=15$, $H=0$, $b=8/3$ and $P=10$ so that the four initial conditions correspond to four points on Figure 3c. The initial conditions are chosen to be near the trivial solution, and in each case they have the same initial error, $e_1(0) = e_2(0) = e_3(0) = 0.5$, but varying initial averages. In Figure 4a we show the trajectory for $a_1(0) = a_2(0) = -1.0$ and $a_3(0) = 2.29$, which is relatively far from the negative convergent solution (32). The trajectory spirals eventually toward the convergent negative solution, and the projection of the trajectory never crosses the lines $e_1 = a_1$ or $e_1 = -a_1$, but stays in region III of Figure 1. In Figure 4b, we show the trajectory for $a_1(0) = a_2(0) = a_3(0) = 0$. Here we see that the projection of the flow evolves initially toward the positive divergent solution (33), but then returns very near the trivial solution. After the trajectory leaves the neighborhood of the trivial solution, it approaches the negative divergent solution, remaining always in region II. When $a_1(0) = a_2(0) = 1.0$, but $a_3(0) = 2.29$, which is relatively far from the positive convergent solution, then in Figure 4c we see that the projection of the trajectory evolves initially toward the positive convergent solution, but again approaches the neighborhood of the trivial solution, and then finally falls into region IV and approaches the positive divergent solution. Finally, when $a_1(0) = a_2(0) = 2.0$, but $a_3(0) = 4.58$, which is 33% of the positive convergent solution, the trajectory always remains in region I.

In all cases shown, the crucial part of the error/average trajectory was the portion following one loop of a stable state in which the trajectory passed near the trivial solution. The quadrant in which the trajectory ultimately remained depended completely on which quadrant (the projection of the) trajectory was in when it left the neighborhood of the trivial solution. Thus, the relationship between the trajectory and the expanding directions in the neighborhood of the unstable convergent solution is crucial to the accurate prediction of the flow. This provides more evidence that a key to determining the predictability of a flow lies with the knowledge of the location of the expanding directions and more generally of the location of the unstable manifold (Dutton and Wells[3]).

CONCLUSION

Once the attracting set contains temporal solutions the situation will become more complicated than that discussed here. But as long as the system is not fully turbulent, and is on an intermediate step of the hierarchy of transitions leading to turbulence, then the predictability issues are the same as those discussed here. They are that (i) predictability will be most difficult for those trajectories that either begin nearby or evolve toward boundaries of basins of attraction of solutions in the limit set, (ii) predictability will be enhanced if knowledge of the physical structure of a stable state is built into the initial conditions, and (iii) in many cases, the most crucial portion of the flow is the part near the unstable trivial solution.

ACKNOWLEDGMENTS

I am deeply indebted to Professor John A. Dutton who stimulated my interest in the nonlinear dynamics of the error/average equations.

In addition, I gratefully acknowledge the many helpful suggestions given me by Professor Robert Wells during the evolution of this study. The research reported here was partially funded by the National Science Foundation through grant ATM 79-08354.

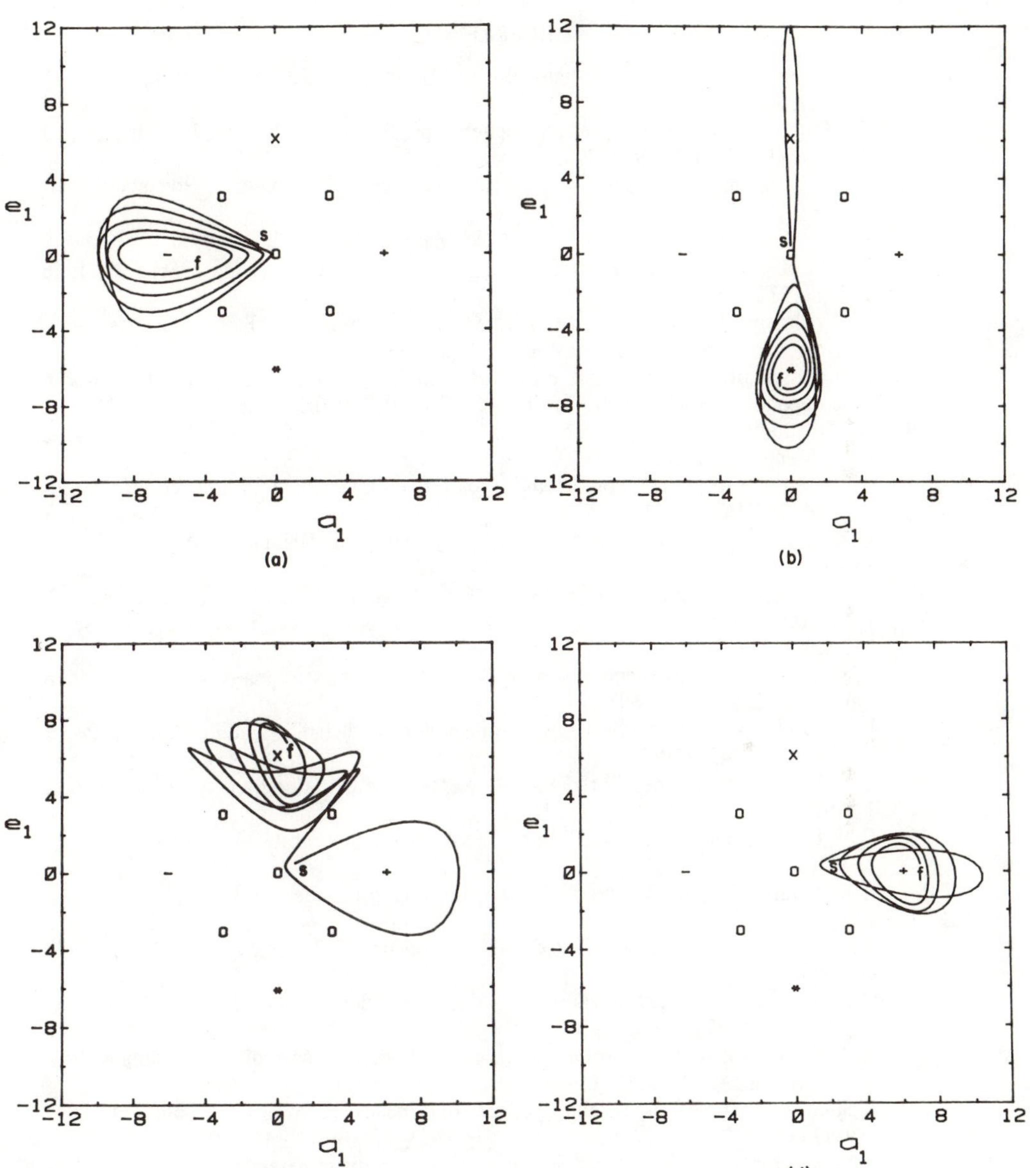

Fig. 4. Projection on the e_1—a_1 plane of a trajectory of (10)-(15) for H=0, r=15, b=8/3, P=10, and $e_1(0)=e_2(0)=e_3(0)=0.5$; the letter s denotes the beginning of a trajectory, the letter f the end of a trajectory. In (a) $a_1(0)=a_2(0)=-1.0$ and $a_3(0)=2.29$; in (b) $a_1(0)=a_2(0)=a_3(0)=0$; in (c) $a_1(0)=a_2(0)=1.0$ and $a_3(0)=2.29$; and in (d) $a_1(0)=a_2(0)=2.0$ and $a_3(0)=4.58$.

REFERENCES

1. F. Baer and F.N. Alyea, Predictability and spectral truncation. J. Atmos. Sci., **31**, 920-931, (1974).
2. W. Blumen, Experiments in atmospheric predictability: Part I. Initialization. J. Atmos. Sci., **33**, 161-169, (1976).
3. J.A. Dutton and R. Wells, Topological issues in hydrodynamic predictability, this volume, 1983.
4. S.J. Jacobs, A note on predictability. J. Atmos. Sci., **29**, 767-768, (1972).
5. S.J. Jacobs, Upper bounds for forecast errors I: Theory. J. Atmos. Sci., **29**, 1246-1251, (1972).
6. R. Krishnamurti, On the transition to turbulent convection. Part 1. The transition from two- to three-dimensional flow. J. Fluid Mech., **42**, 295-307, (1970).
7. R. Krishnamurti, On the transition to turbulent convection. Part 2. The transition to time-dependent flow. J. Fluid Mech., **42**, 309-320, (1970).
8. R. Krishnamurti, Some further studies on the transition to turbulent convection. J. Fluid Mech., **60**, 285-303, (1973).
9. C.E. Leith, Atmospheric predictability and two-dimensional turbulence. J. Atmos. Sci., **28**, 145-161, (1971).
10. D.K. Lilly, A note on barotropic instability and predictability. J. Atmos. Sci., **30**, 145-147, (1973).
11. E.N. Lorenz, Deterministic nonperiodic flow. J. Atmos. Sci., **20**, 130-141, (1963).
12. E.N. Lorenz, The predictability of hydrodynamic flow. New York Acad. Sci. Ser 2, **25**, 409-432, (1963).
13. E.N. Lorenz, The predictability of a flow which possesses many scales of motion. Tellus, **21**, 289-307, (1969).
14. E.N. Lorenz, Three approaches to atmospheric predictability. Bull. Amer. Meteor. Soc., **50**, 345-349, (1969).
15. E.N. Lorenz, Atmospheric predictability as revealed by naturally occurring analogues. J. Atmos. Sci., **26**, 636-646, (1969).
16. E.N. Lorenz, Barotropic instability of Rossby wave motion. J. Atmos. Sci., **29**, 258-264, (1972).
17. E.N. Lorenz, On the existence of extended range predictability. J. Appl. Meteo., **12**, 543-546, (1973).
18. P.E. Merilees, A note on a predictability function. J. Atmos. Sci., **29**, 991-993, (1972).
19. G.D. Robinson, The predictability of a dissipative flow. Quart. J. Roy. Meteo. Soc., **97**, 300-312, (1971).
20. H.N. Shirer and R. Wells, Improving spectral models by unfolding their singularities. J. Atmos. Sci., **39**, 610-621, (1982).
21. J. Shukla, Dynamical predictability of monthly means. J. Atmos. Sci., **38**, 2547-2572, (1981).
22. J. Smagorinsky, Problems and promises of deterministic extended range forecasting. Bull. Amer. Meteo. Soc., **50**, 286-311, (1969).
23. H.L. Swinney J.P. Gollub, Hydrodynamic Instabilities and the Transition to Turbulence (Springer-Verlag, 1981), 292 pp.
24. P.D. Thompson, Uncertainity of initial state as a factor in the predictability of large scale atmospheric flow patterns. Tellus, **9**, 275-295, (1957).
25. D.L. Williamson, The effect of forecast error accumulation on four-dimensional data assimilation. J. Atmos. Sci., **30**, 537-543, (1973).

LABORATORY EXPERIMENTS ON THE TRANSITION TO BAROCLINIC CHAOS

J.E. Hart
Department of Astro-Geophysics, University of Colorado
Boulder, Colorado 80309

ABSTRACT

Laboratory experiments on the transition to aperiodic or chaotic motion in a two layer rotating fluid system are described. Theoretical results based on the single wave quasi-geostrophic theory indicate that the transition should be abrupt, via a snap-through bifurcation. The experimental results indicate either a quasi-periodic, or a period doubling transition depending on the basic potential vorticity gradient.

INTRODUCTION

In recent years there has been much interest in relatively confined fluid systems that demonstrate some of the scenarios for transition to turbulence or chaotic motion arising from primarily topological considerations. For example, behavior similar to the period doubling cascade of Feigenbaum[1] has been observed in convection experiments of Maurer and Libchaber[2] among others. In addition, the quasi-periodic transition of Ruelle and Takens[3], where broadband noise is preceded by motion involving two irrationally related frequencies (motion on a 2 - torus), seems to occur in certain Taylor couette experiments of suitable aspect ratio (Fenstermacher et. al[4]). This paper reports on some preliminary experiments concerned with the transition sequence in a simple laboratory baroclinic flow. The work was motivated in part by the existence of an asymptotic theory for the problem, primarily developed by Pedlosky[5] and Pedlosky and Frenzen[6], which for one of our geometries predicts a bifurcation to a chaotic Lorenz[7] attractor. We wanted to see if the laboratory flow followed this or any of the other "standard" scenarios.

In terms of the predictability problem, we had wanted to find out how the dimension of the baroclinic flow changes as a function of external parameters. Perhaps, over a range of parameters, the flow is essentially a low dimensional dynamical system with possibly many slaved modes. Indeed the chaotic Lorenz attractor is of low fractal dimension and reflects a principal dynamical interaction between a single longitudinal wavenumber and the zonally symmetric component of the motion, even though these both may have fairly complicated cross-stream structure. Unfortunately the direct determination of the dimension of the experimental chaotic attractor is very difficult, especially so with a rather noisy probe system we use. Thus we concentrate here on the transition question only.

THEORETICAL BACKGROUND

The basic experimental configuration is shown in Figure 1. A cylinder rotating at basic rate Ω contains two immiscible fluids with density $\rho_2 > \rho_1$. Motions are driven by a weak differential rotation of the upper surface with $\omega \ll \Omega$. The aspect ratio of the experiment is slightly less than one. The cross-section shown below indicates parabolic top and bottom topography. This is to emphasize the important role of the basic equilibrium fluid depth distribution with no differential driving. Of course with any basic rotation Ω, the interface will deform into a parabola and if one wants to simulate an f-plane geometry with no basic radial potential vorticity gradients, the two parabolic surfaces must match the curvature of the interface so that the depth in each layer is constant with radius. A β–plane experiment can be effected by using a flat bottom, and a parabolic top with Ω chosen so that the interfacial parabola is exactly half as strong as the lid. However, for

0094-243X/84/1060369-07 $3.00

both the f-plane and the β– plane cases Ω is fixed for each lid and it is difficult to scan the control parameters for the experiment as defined below. Therefore in this paper we consider only cases with a flat top and bottom. In this instance notice that there is an asymmetry between co-rotation and counter-rotation of the upper surface that is not present in the f-plane cases.

When the Rossby number Ro= $\omega/2O$ is small the motions in the two layers are described approximately by the quasi-geostrophic potential vorticity equations:

$$\partial\xi_i/\partial t+J(\psi_i,\xi_i) = -Q\ (\xi_i + 2\ \delta_{il}) \tag{1}$$

where i=1 or 2 for the two layers, ψ_i are the geostrophic streamfunctions, and ξ_i are the potential vorticities:

$$\xi_i= \nabla^2\psi_i+(-1)^i F(\psi_1-\psi_2) \tag{2}$$

The two control parameters are the rotational Froude number F, and a frictional parameter Q that is inverse in the Rossby number:

$$F = 4\Omega^2 L^2/g(\delta\rho/\rho)H \quad Q = \sqrt{(\nu/\Omega H^2)}/Ro \tag{3}$$

The derivation of these equations and a more detailed description of the experiment can be found in Hart[8,9].

Typically, for a given F, as R_o is increased or as Q is decreased, the simple axisymmetric solution to the above equations becomes unstable to an azimuthally traveling baroclinic wave. Near the linear neutral curve F(Q), the exponentially growing disturbance of linear theory equilibrates to constant amplitude. This regime is often called the steady wave state. As Q is decreased further, the wave amplitude in the laboratory experiments can become periodic, quasi-periodic, or chaotic. The Pedlosky-Frenzen[6] weakly non-linear expansion of (1) for the f-plane case goes from steady waves, to chaotic, and then through a series of period halving bifurcations to singly periodic wave amplitude as Q is decreased. Pedlosky[10] later showed that this transition scenario was very sensitive to small amounts of β. Since the flat bottom case we are studying here is sort of a $+\beta$ lower layer and $-\beta$ upper layer configuration, one might expect a different response from that for the f-plane. A simple numerical solution of (1) is obtained by expanding ψ_1, say, as:

$$\psi_1=3r^2/8 + \sum_m A_m(t)\exp(in(\Theta-ct))J_n(l_{mr}) + \sum_m B_m(t)J_o(\gamma mr) \tag{4}$$

That is, we consider one azimuthal wave n along with M radial modes including a correction to the zonal component of the flow. Apart from the different basic height profile and more radial modes, this calculation is quite similar in form to the asymptotic one. Indeed the numerical integrations for positive Ro predict a transition from steady waves to chaos, followed by a reverse period doubling cascade as Ro is increased, just as the f-plane theory did. For the value F=20, this transition is well inside the linearly unstable region of F-Q space. Figure 2 shows that the computed chaotic solution is akin to a Lorenz attractor. This can be seen by plotting the zonally averaged interface height $\underline{h}(t)$ vs. $\underline{h}(t+\tau)=h_1$, where τ is a fixed delay time. The resulting projection of the attractor fills in on sheets. A Poincare section through the attractor yields an almost one-dimensional map typical of the chaotic Lorenz solution. It is not clear how robust this scenario is when more wavenumbers are considered. The chaotic transition takes place well away from the weakly non-linear region. On the other hand wave selection mechanics tend to select the lowest wave number as the dominant response, so there is at least the hope that a 1-wave model might be accurate. It turns out however, that the experimental flows make their transition to chaos in a very different manner, suggesting that the single wave approximation may not be appropriate.

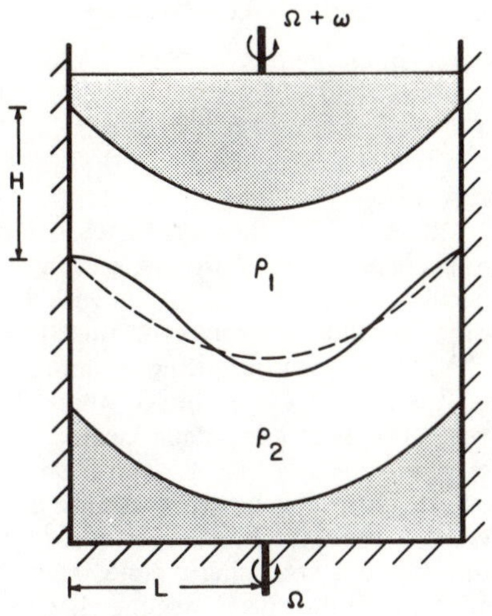

Fig. 1. Experiment cross-section.

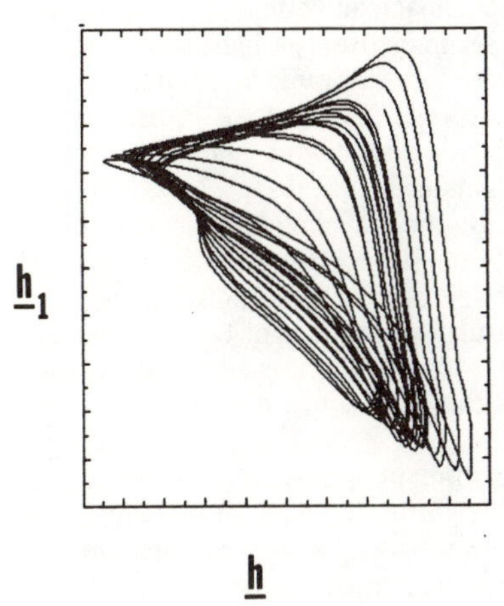

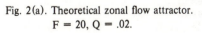

Fig. 2(a). Theoretical zonal flow attractor.
F = 20, Q = .02.

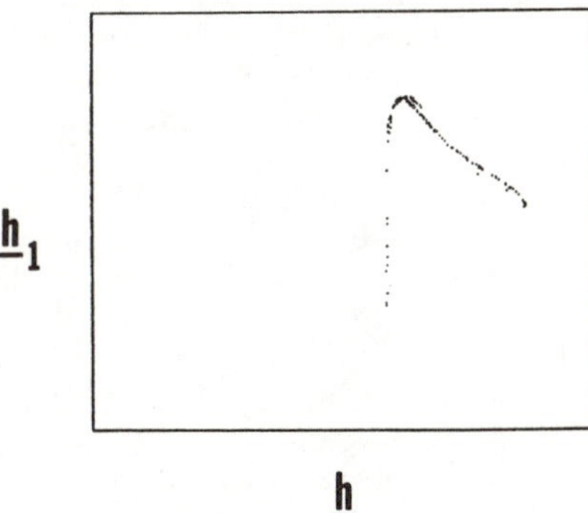

Fig. 2(b). Poincare section through attractor (2a).

EXPERIMENTAL RESULTS

The experimental flows are monitored by measuring the interface height deflection using a .001 inch platinum wire stretched through the interface. Measuring the impedance to ground through the lower slightly conducting alcohol mixture (the upper fluid is a silicone oil of viscosity chosen to match that of the alcohol) gives a signal that can be digitized and processed in various ways.

With a probe at a radius of about .7L, the signal h(t) is dominated by the traveling wave crests passing by the probe. For values of Ro just on the unstable side of the linear neutral curve, steady wave oscillations are found. At large values of Ro these become modulated. Thus as far as the traveling wave is concerned this could be construed as motion on a torus provided the modulational frequency is unrelated, as one would suspect physically, to the wave drift. This is shown in Figure 3, where the projected attractor and Poincare sections are displayed. The Poincare section clearly shows the incommensurate winding motion on the underlying toroidal attractor. The fuzziness here is due to instrumental noise. That is, the steady wave regime, theoretically 2 dots on the section, has similar breadth. Figure 4(a-c) show the wave height spectra for three values of Ro encompassing the steady wave regime (a), the vacillating regime (b) with its second incommensurate modulational frequency (f_1), and the chaotic regime characterized by dominant spectral peaks at both the traveling wave frequency and the vacillation frequency, along with a broadband component.

Looking at the wave height at r=.7L for details of the transition to chaos proves to be very difficult because the non-linear processes are masked by the very large traveling wave signal. Indeed the asymptotic theory of Pedlosky[5] shows that the azimuthal wave phase can be removed in the expansion, and has no effect on the equilibration dynamics. The equilibration problem focuses entirely on the wave amplitude, vertical phase, and the amplitude and structure of the wave induced correction to the zonal component of the motion. Looked at this another way, one could imagine an experiment (that is hard to do) with a differentially rotating lid turning at a rate $\omega/2$ and a counter-rotating bottom rotating at $-\omega/2$. For the f-plane one would then get a non-drifting wave. For our $\mp\beta-$ plane a similar adjustment to the boundaries would yield stationary waves. Thus one apparent frequency associated solely with wave drift around the cylinder would disappear. One could effect the same result by moving the probe along with the traveling disturbance in the actual experiment, but this too is technically out of the question. However, in any case it would seem that the steady wave regime does indeed represent a fixed point in the nonlinear dynamics, and so the vacillating regime with modulations in time of the wave pattern can be thought of as a limit cycle in phase space, not a 2-torus. The question is then, what is the transition scenario in terms of the wave amplitude or some other quantity that doesn't include the wave drift?

In the cylinder, it is possible to monitor a signature of the zonally averaged flow directly by measuring the interface height right at the axis of rotation. The traveling wave disturbance is exactly zero on the axis, so no difficult filtering of the signals dominated by the drifting waves need be done. If one could simultaneously measure the interface height at one point on the edge of the cylinder, where the traveling wave eigenfunctions are also zero, the difference would give the zonal baroclinic mass flow as a function of time. So physically one can think of the interface height on axis as giving a measure of the level of the zonal vertical shear from which the baroclinic waves receive their energy.

Unfortunately, the on-axis signal levels are less, and hence subject to more instrumental noise. It should perhaps be said that great care has been taken in the positioning of the probe, and in the constancy of the various mechanical drives. The basic rotation is maintained by a tachometer feed-back torque motor with a .003% measured rotation rate flutter. The differential rotation is produced by a 16000 step/revolution steeper motor

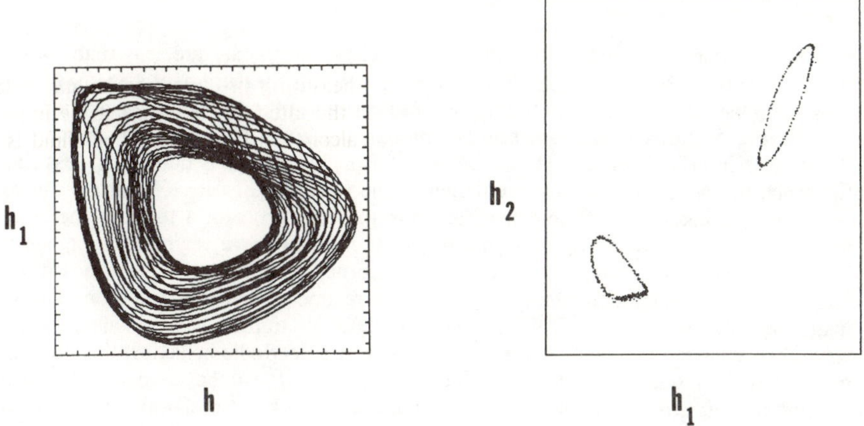

Fig. 3(a). Wave attractor for F = 19.7, R_o = .1.

Fig. 3(b). Poincare section through 3a.

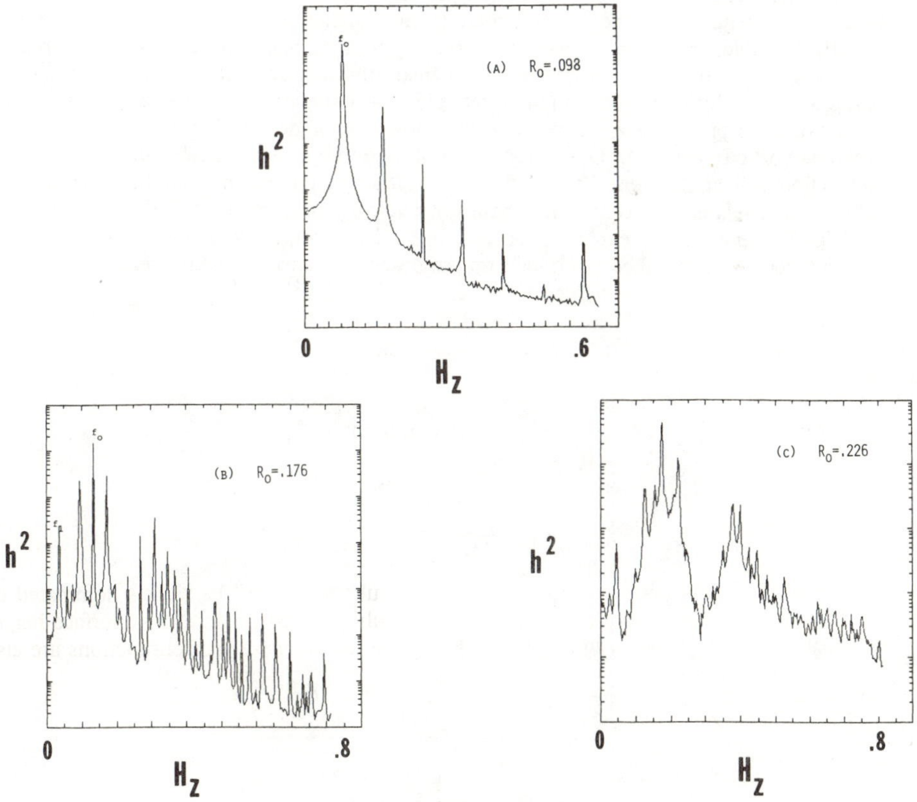

f_o
(A) R_O=.098
h^2
0
.6
H_Z

f_o
f_n
(B) R_O=.176
h^2
0
.8
H_Z

(C) R_O=.226
h^2
0
.8
H_Z

Fig. 4. Wave height spectra F=20.

driven by a computer controlled clock. Its fluctuations in rate are less than .001%. All these precautions do not cure the basic problem with this measurement system, namely that very small particles of dust on the interface can cause significant signal degradations.

Figure 5 shows a Poincare section through the zonal flow attractor characterized by the on-axis interface height. The value of the Rossby number is quite a bit less than that for which the broadband spectrum emerges in the wave height data. What one can see in a series of data taken for different parameters is the emergence of a torus in the zonal flow data. That is, the zonal signal is constant in the steady wave regime, undergoes regular limit cycle type oscillations in the vacillating regime, and prior to wave chaos develops a second incommensurate frequency. The chaotic regime is characterized by motion off this instrumentally fuzzy torus. This is clearly a different transition mechanism than the 1-wave theory gave. The flow evolves more slowly into aperiodic behavior, developing first a periodic then a quasi-periodic zonal flow oscillation. It will be necessary to determine experimentally, using multiple probes, and theoretically, the role of wave-wave interactions in this transition scenario.

All the discussion so far has centered on the co-rotating or positive Ro case. A very few experiments have been done with counter-rotation. The linear stability properties in this case are considerably different and lead to the expectation of more active wave-wave interactions. That is, in the co-rotating case wave number 1 is always the most unstable linear wave, whereas in the counter-rotating case higher wavenumbers can be the most linearly unstable. Figure 6 shows some on-axis interface height-time traces. It is reasonably clear that a transition to irregularity is made through at least a period doubling, and perhaps with a little imagination one can see a 4-periodic motion in Figure 6c. These results, though preliminary, suggest that the chosen "scenario" is very sensitive to the basic potential vorticity gradient. One is reminded of the different transition routes observed in a convection cell at different Prandtl numbers, although the mechanism here is no doubt different and related to a change in a basically inviscid characteristic of the flow. It is hoped that further multi-probe experiments, and numerical integration of the N-wave potential vorticity equations will shed more light on the observed transition phenomena.

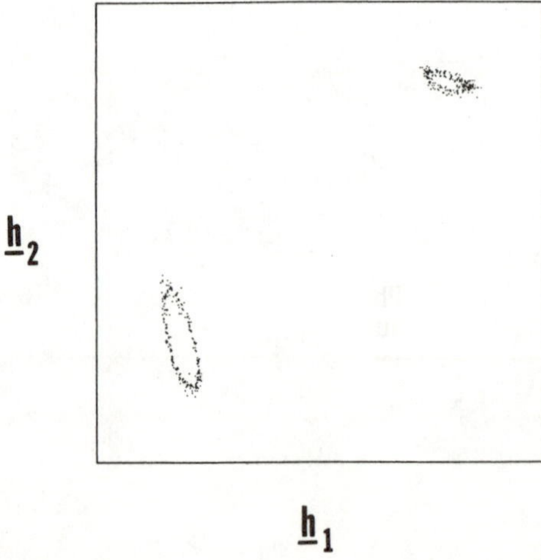

Fig. 5. Poincare section through the zonal flow attractor. $F=20$, $R_o=.18$.

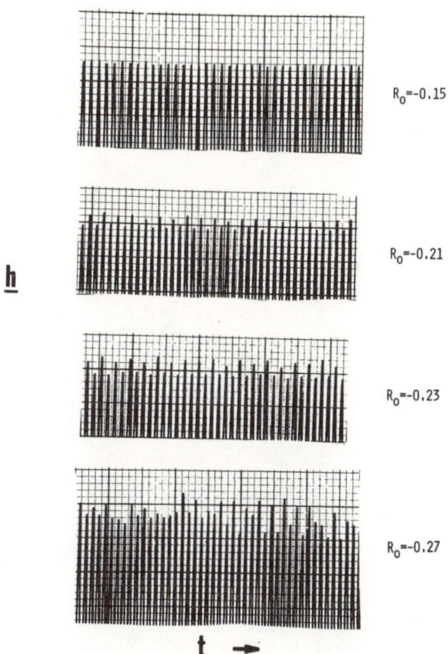

Fig. 6. Zonal height-time traces F = 37.

ACKNOWLEDGMENT

This research was sponsored by the Atmospheric Sciences section of the National Science Foundation.

REFERENCES

1. M.J. Feigenbaum, J. Stat. Phys. **19**, 25, 1978.
2. J. Maurer and A. Libchaber, J. Physique Lett. **41**, L515, 1980.
3. D. Ruelle and F. Takens, Commun. Math. Phys. **20**, 167, 1971.
4. P.R. Fenstermacher, H.L. Swinney, and J.P. Gollub, J. Fluid Mech. **94**, 103, 1979.
5. J. Pedlosky, J. Atmos. Sci. **28**, 587, 1971.
6. J. Pedlosky and C. Frenzen, J. Atmos. Sci. **37**, 1177, 1980.
7. E.N. Lorenz, J. Atmos. Sci. **20**, 130, 1963.
8. J.E. Hart, Geophys. Fluid Dyn. **3**, 181, 1972.
9. J.E. Hart, J. Atmos. Sci. **33**, 1874, 1976.
10. J. Pedlosky, J. Atmos. Sci. **38**, 717, 1981.

PREDICTABILITY IN MODELS OF BAROCLINICALLY UNSTABLE SYSTEMS

J Brindley and Irene M Moroz
School of Mathematics, University of Leeds, Leeds LS2 9JT

ABSTRACT

Geophysical fluid dynamics abounds with flows whose mean states are baroclinically unstable. The consequent flow fields are much complicated by the existence at finite amplitude of disturbances which are unstable on their mean, and a primary objective of theoretical modelling of such flows is to expose in as simple a manner as possible those physical processes whose effect is critical in determining at least the qualitative behaviour of the flow fields.

We present and review a wide range of models of baroclinic instability which trace the evolution of finite amplitude disturbances to unstable flows. Several non-dimensional parameters are shown to be influential in determining the qualitative character of the evolution, which may be highly predictable, as in the case of steady or periodic behaviour, or apparently unpredictable, according to parameter values.

The models considered are mainly low order mathematical systems, arising from some form of truncation of a model representation of the flow, and it is a matter of some importance to assess the extent to which they are capable of representing the actual behaviour of a real fluid system. The question of the evaluation of the model outputs against data from actual observations is addressed, and we review progress in and prospects for the identification and assessment of predictability in the models.

1. INTRODUCTION

Baroclinic flows are ubiquitous in geophysical fluid dynamics; they are frequently unstable, and the disturbances, often (baroclinic) wave-like, arising from instabilities of these flows, are enormously important in influencing the large scale flow patterns of the Earth's atmosphere and oceans. Full descriptions of the character of baroclinic waves, and their relevance to geophysical fluid dynamics are to be found in recent monographs by Pedlosky[1] and Gill[2], and we refer the general reader to those sources.

It is not surprising then that a great deal of effort has been applied to the development and study of models, laboratory, numerical and theoretical, of baroclinic flows, in which the fundamental physical process of conversion of available potential energy into eddy kinetic energy is preserved and exposed in as simple a form as possible; substantial reviews[3,4] describe this effort. The hope, of course, is that an understanding of the essential properties of baroclinic instabilities, and of the subsequent finite amplitude behaviour and further transitions of the baroclinic waves, will best be obtained from the examination of relatively simple baroclinic flows, unencumbered by the complicated boundary and forcing conditions endured by the real geophysical fluids. Many mathematical models exist for the description of the stability characteristics of baroclinic flows and the evolution of wave-like disturbances to such flows, and it is with some of these models that we are mainly concerned here. Fundamental to nearly all approaches is an assumption that, at least locally and temporarily, the flow is in approximate geostrophic balance, that is that pressure gradients are in balance with Coriolis and gravitational accelerations as described by a set of linear partial differential equations containing no dissipative effects. The time evolution of these geostrophically balanced states depends on non-geostrophic effects, and these may be taken into account in a self-consistent approach leading to the quasi-geostrophic equations[1]. All other physical effects, associated for example with diffusion of heat or momentum, together with any nonlinear consequences of departures from some simple basic state, are regarded as secondary to this primary geostrophic balance, and their influences are customarily treated as perturbations to an appropriate solution of the quasi-geostrophic equations.

The near balance of Coriolis and pressure terms makes it advisable to describe the evolution of vorticity fields rather than velocity fields, and a common starting point for mathematical adventure is a conservation equation, taking the form

$$\frac{DQ}{Dt} = 0 \quad , \tag{1.1}$$

where Q is called the quasi-geostrophic potential vorticity, and is defined in several slightly differing forms in the literature. Pedlosky[1] provides a good description of the derivation of this equation from the Navier-Stokes equations, and of the conditions required for its validity. We are concerned with the problem of the stability of basic steady flows satisfying (1.1), and the subsequent evolution in time and space of flow patterns as described by the bifurcational behaviour of appropriate unsteady solutions. Our approach is limited to a consideration of normal mode solutions to the perturbation equations arising from (1.1); it makes use of multiple scales techniques based on appropriate small parameters measuring disturbance amplitude and the various physical effects. The structure of the resulting amplitude evolution equations is found to depend sensitively on these parameters.

Recent work[5] on initial value problems in baroclinic fluids suggests that early growth of perturbations to a baroclinic flow may depend also on the continuous spectrum of neutral modes of the differential operator constituted by (1.1) and suitable boundary conditions. We should not expect this result to influence our conclusions insofar as they relate to predictability over longer time scales, since the message conveyed by the results of our paper is essentially pessimistic; over substantial ranges of parameter values the asymptotic solution spaces of even the simplest normal mode models contain strange attractors. Our definition of unpredictability is essentially tied to this concept; time evolution is unpredictable if strange attractors appear in the asymptotic attracting set, since the exponential divergence of trajectories in such cases will ultimately amplify to arbitrary size even the most infinitesimal differences in initial conditions. However, we should not necessarily claim that the absence of strange attractors ensures predictability!

2. MATHEMATICAL MODELS

Amongst the simplest flows satisfying (1.1) are the so-called layer models, pioneered by Phillips[6], in which variations in the vertical direction of properties of a basic horizontal flow are represented in a quantized way; the fluid is regarded as consisting of a number of superposed layers, in each of which the density and velocity is constant. The vast majority of investigations have been restricted to 2-layer models (Fig.1(a)), though we indicate in 5 that a number of features relevant to predictability depend on the more realistic representation available with a 3-layer model.

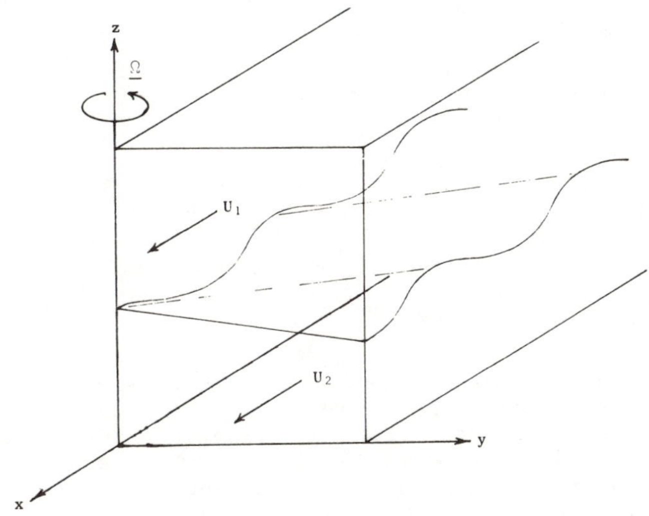

Fig.1(a) Two-layer model of baroclinic instability; in basic state, upper layer moves with constant uniform speed U_1, lower layer with different uniform speed U_2. Baroclinic waves occur on interface.

A second simple basic flow assumes vertical variations of density and velocity with linear gradients, the vertical velocity shear being in balance with a horizontal density gradient through the thermal wind equation (Fig.1(b)); this model is associated with the name of its originator, Eady[7]. Both models have obvious physical shortcomings, and more subtle mathematical difficulties associated with their highly singular nature; nevertheless their essential simplicity has made them enormously popular as vehicles for developing an understanding of the central process of baroclinic instability[4,8].

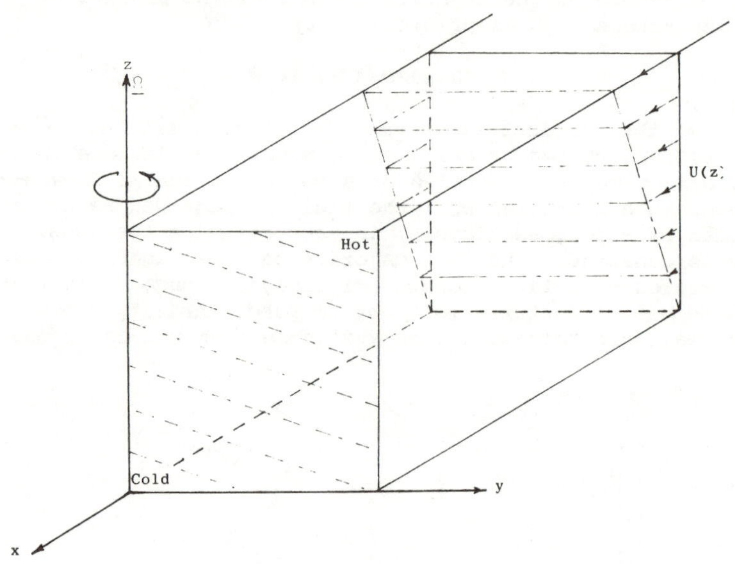

Fig.1(b) Eady model of baroclinic instability; in basic state U(z) is sheared to conform with the horizontal density gradient. Constant temperature (density) surfaces, shown by dashed lines, constant pressure surfaces assumed horizontal, i.e. z = constant.

Another model having continuously varying velocities and temperatures was introduced by Charney[9]. Though this model is in many ways more realistic than either the 2-layer model or the Eady model[10,11], it has nevertheless been somewhat neglected, largely because of the complicated nature of the eigenfunctions of the linear problem, and the occurrence of critical layers.

3. LINEAR STABILITY PROBLEM

The objective of a linear analysis of the stability of basic flows satisfying (1.1), in which the stream function is expanded in a series of appropriate spatial modes, each having associated with it a time dependence of the form exp(ict), is a <u>dispersion relation</u>

$$c = c(\underline{K}) , \ldots \qquad (3.1)$$

relating the complex phase speed c to a wave-number vector $\underline{K}$ describing the horizontal structure of the eigenfunction.

It is usual to assume a normal mode expansion, dependent for its completeness on the existence of a basic flow with no horizontal variations. Though such horizontal variations undoubtedly occur in practice, and indeed horizontal gradients of potential vorticity are known to have an important influence on the dispersiveness of baroclinic waves, nevertheless most direct attempts at including such variations in models with continuous vertical variations have been restricted to perturbation approaches in which the gradients are small; a very full discussion of the normal mode approach is given by Killworth[8]. Success in the development of a weakly nonlinear theory for baroclinic waves is somewhat dependent on the existence of simple analytic forms for the eigenfunctions of the linear problem, and a device which permits us to model the physical effects of internal potential vorticity gradients without losing the benefits of the method of separation of variables in the linear problem is the introduction of sloping boundaries. Such boundaries physically force the thickness of the layer, and hence the length of vertical vortex tubes, to vary in the horizontal, and lead mathematically to a linear problem formally identical with that arising if the vertical component of angular velocity of the reference frame is allowed to vary horizontally. We refer the reader to Pedlosky[1] for detailed discussion of the analogy between boundary slope and horizontal potential vorticity gradient, much influenced in natural geophysical systems by the "β-effect" of variation of Coriolis parameter with latitude.

The linear stability theory then provides the springboard for non-linear analysis, pivoted about the neutral stability curve or surface generated by setting $c_i = 0$ in (3.1). We focus attention on the Eady model, in which the crucial stability parameter takes the form of a Burger number, B, measuring the relative effects of de-stabilising buoyancy forces and stabilising rotational effects; a wave of horizontal wave number $\underline{K} \equiv (k, \ell)$ will be unstable for $B < B_0(k, \ell)$. For inviscid flows with no boundary slopes and a given cross-stream wave number ℓ, the variation of B_0 with k is shown schematically in Fig.2.

A cornerstone of the work of the last 20 years in geophysical fluid dynamics has been the belief that the effects of viscosity in the fluid are well modelled by the so-called Ekman suction effect, by which interior vortex tubes are stretched or compressed by flux into or out of Ekman boundary layers according as the vorticity of the interior flow is less than or greater than that of the boundary. Thus dissipative effects can be modelled in a simple way whereby the inviscid equations of motion are supplemented by modified boundary conditions embodying the appropriate normal velocity component. This assumption, although undoubtedly a good physical approximation, quite changes the character of the mathematical problem from that of the Navier-Stokes equations, in which viscous effects are modelled by high order spatial derivations; the implications for predictability may be substantial.

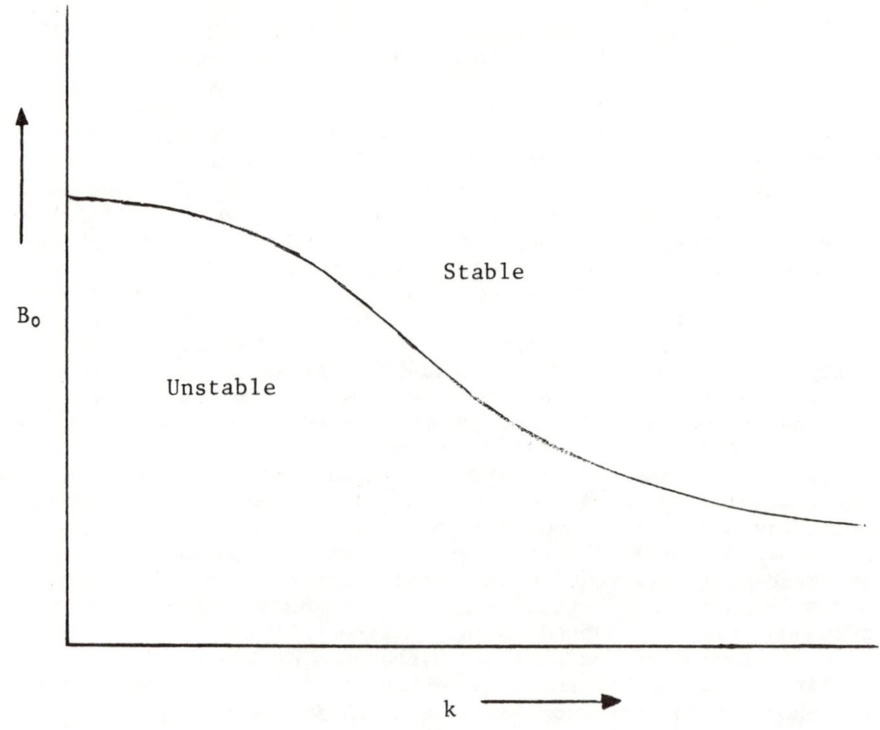

Fig.2 Schematic form of neutral stability curve in the (B,k) plane for an inviscid fluid and horizontal boundaries; note the maximum of the curve at k = 0.

When the effects of horizontal vorticity gradients and viscosity are taken into account in the artificial way described above, by modifying the boundary conditions satisfied by an inviscid flow, it is clear that surfaces of neutral stability analogous to Figure 2 now exist in a space of 5 dimensions implied by the entry of parameters S_u, S_ℓ and τ , representing respectively the upper and lower boundary slopes and dissipative effects in some suitably non-dimensional way, into the (B,k) relationship obtained by setting $c_i = 0$ in (3.1).

This extended linear problem has been discussed at length elsewhere[12]; here it suffices to present as much of its essence as is necessary to establish a notational basis for the non-linear theory summarised in §4.

Thus the quasi-geostrophic equation obtained from (3.1) for the perturbation pressure field, p , is, in non-dimensional form,

$$\left[\frac{\partial}{\partial t} + z\frac{\partial}{\partial x} + J_{(x,y)}(p,\ldots)\right]\left[p_{zz} + B(p_{xx} + p_{yy})\right] = 0 \tag{3.2}$$

where $J(p,\Phi) \equiv p_x\Phi_y - p_y\Phi_x$.

Inviscid boundary conditions on the side walls of a channel infinite in extent in the x-direction are

$$p_x = 0 \quad \text{on} \quad y = \pm\frac{1}{2} ,$$

$$\text{and} \quad \lim_{x\to\infty}\frac{1}{2x}\int_{-x}^{x}\frac{\partial^2 p}{\partial y\partial t}\,dt = 0 \quad \text{on} \quad y = \pm\frac{1}{2} , \tag{3.3}$$

and the conditions at the upper and lower sloping boundaries become respectively[13]

$$\left(\frac{\partial}{\partial t} + z\frac{\partial}{\partial x}\right)p_z - (1 - s_u,s_\ell)p_x = -J_{(x,y)}(p,p_z)$$

$$\pm Br(p_{xx} + p_{yy}) \quad \text{on} \quad z = \pm\frac{1}{2} \tag{3.4}$$

The physical origins of the parameters B, s_u, s_ℓ, r have been introduced above. Their <u>precise</u> definitions need not concern us; their role in the mathematical problem is clear from equations (3.2) - (3.,4).

Substitution of a normal mode form

$$p(x,y,z,t)$$
$$= A(P\cosh 2qz + Q\sinh 2qz)\sin\{m\pi(y+\tfrac{1}{2})\}\exp ik(x-ct), \tag{3.5}$$

where

$$4q^2 = B(k^2 + m^2\pi^2) , \quad m = 1,2,\ldots , \tag{3.6}$$

P,Q are complex constants of $O(1)$, and A is an amplitude measure, gives, on application of the boundary conditions (3.4), the characteristic equation for c.

$$4q^2c^2 + 2qc(\tanh q + \coth q)\{i\lambda - \tfrac{1}{2}(s_u - s_\ell)\}$$

$$+ \{i\lambda - \tfrac{1}{2}(s - s_\ell)\}^2 - (q\tanh q-1)(q\coth q-1)$$

$$- \tfrac{1}{4}(s_u + s_\ell)^2 + \tfrac{1}{2}(s_u + s_\ell)\{2-q(\tanh q + \coth q)\} = 0, \tag{3.7}$$

where

$$\lambda = 4\,q^{2}/k \quad . \tag{3.8}$$

The condition $c_i = 0$ then gives an equation

$$\mathcal{G}(B, s_u, s_\ell, r, k) = 0 \tag{3.9}$$

describing the neutral stability surface.

Though this surface is set in a 5-dimensional parameter space, its projection on to a 3-dimensional subspace is frequently of a form shown for example in Figure 3, obtained by setting $s_u = s_\ell = 0$. The inviscid stability curve corresponding to Figure 2 is seen as the intersection of the surface with the (B,k) plane.

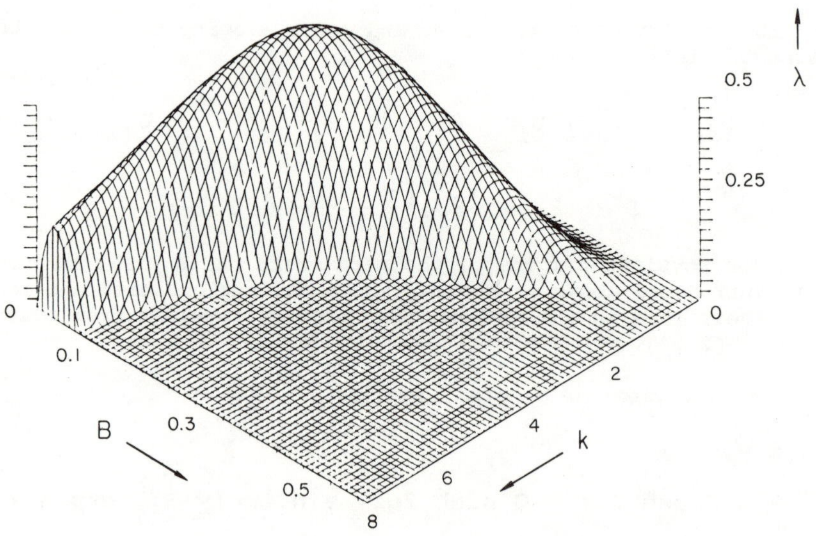

Fig.3 Form of neutral stability surface in (B,λ,k) space for $s_u = s_\ell = 0$; for $B \neq 0$ the maximum λ occurs at non-zero k and for $\lambda \neq 0$ the maximum B occurs at non-zero k.

We see from Figure 3 that for a fixed $\lambda \neq 0$, or for a fixed $B \neq 0$, the neutral stability curves in the (B,k) plane or the (λ,k) plane will have a form schematically like Figure 4, indicating the existence of a finite width waveband of excited waves for values of the "stability parameter", either B^{-1} or λ^{-1}, above some critical value B_c^{-1} or λ_c^{-1} . This form justifies our consideration of an excited "wave-packet" of modes, centred on the wavelength of the first excited mode as, say, B is decreased, rather than a single mode of form (3.5). We summarise in 4 the time and space evolution of such wave packets, as well as of the time evolution of "wave-trains" represented by (3.5).

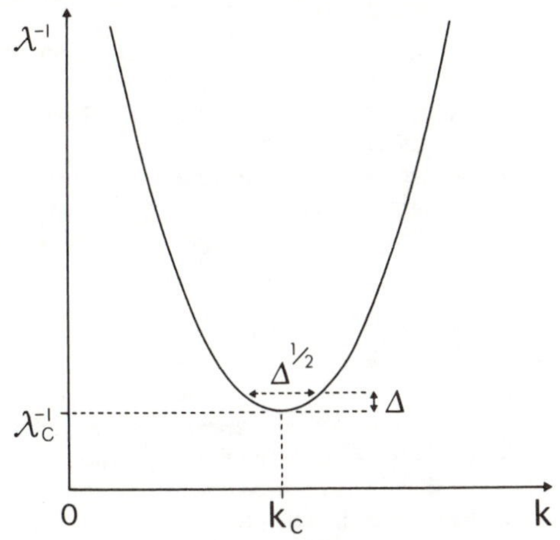

Fig.4 Schematic form of (B^{-1},k) or $\lambda^{-1},k)$ curve for fixed non-zero λ or b respectively.

4. WEAKLY NONLINEAR INSTABILITY THEORY

In the weakly nonlinear theory we assume that the value of the stability parameter, say λ , is slightly supercritical, so that

$$|\lambda - \lambda_c| = \Delta \quad , \qquad \text{where } \Delta \text{ is small} \tag{4.1}$$

The importance of the parameters r, s_u and s_ℓ to the qualitative nature of the nonlinear problem depends critically on their order of magnitude relative to the amplitude of the perturbation (3.5) which we can denote by ϵ ; in order to balance nonlinear terms with linear growth terms for small supercriticality ϵ must be related to Δ so that $\Delta = O(\epsilon^2)$.[14] We then examine the cases in which s_u, s_ℓ, r are $O(1)$, $O(\epsilon)$ and zero; in fact we restrict ourselves to the geometry in which $s_u = -s_\ell = s$, in which the mathematical problem is formally the same as that for a 2-layer model with a β-effect.

The procedure used is a straightforward multiple scales analysis in which the amplitude of the perturbation (3.5) is assumed to vary slowly in time and/or space, so that

$$A = A(X,T) \tag{4.2}$$

where X,T are appropriately scaled slow variables. The different order of magnitude combinations for r and s demand the use variously of variables

$$X_1 = \epsilon x \ , \quad X_2 = \epsilon^2 x \ , \quad T_1 = \epsilon t \ , \quad T_2 = \epsilon^2 t \ . \tag{4.3}$$

Evolution equations for $A(X,T)$ describing the slow development in time and space of the amplitude of wave packets or wave trains then arise as solvability conditions at successive orders of ϵ for the elements $p^{(i)}$ in an expansion of the form

$$p = p^{(0)} + \epsilon\, p^{(1)} + \epsilon^2\, p^{(2)} + \ldots \qquad (4.4)$$

for p .

Details of this analysis are contained more fully elsewhere[12,15]. The resulting equations are summarised in Tables I(a) and I(b) and we review below the implications for predictability of the behaviour associated with several of the more important cases; it is convenient to arrange our remarks so as to cover successively the cases in which the viscosity paramemeter takes the values, 0, ϵ, 1 .

(i) Dissipation negligible - Inviscid fluids
($r=0$; $s=0$, $O(\epsilon)$, $O(1)$)

Some of the earliest nonlinear analyses[16,17] showed that, in an inviscid fluid, the amplitude of a wave train ultimately oscillates periodically on a slow time scale in a manner dependent on the initial conditions. This ultimate behaviour is (at least theoretically) predictable, except for an uncertainty in phase, is qualitatively robust, and is true for all orders of the dispersion parameter s .

The evolution of the amplitude of a wave packet in an inviscid fluid has been shown to be describable variously by the self-induced transparency, the sine-Gordon, and the nonlinear Schrodinger equations, according to the value of s[18,19,20]. All these equations are of course completely integrable and all admit soliton solutions in an infinite domain. The future behaviour of a wave packet of given form $A(X,0)$ at $T = 0$ is then totally predictable for future time. In a finite domain the Hamiltonian property of the equation guarantees (FPU) reccurrence[21] which might be expected to have exploitable predictability properties. Note that this is certainly not the same as saying that the future behaviour of a particular disturbance $A(x,0)$, varying on the fast x-scale, is predictable. Much recent work[5] has questioned the adequacy of the normal mode approach in modelling local sources of baroclinic instability, and it appears that much of the initial energy of a finite amplitude source finds its way into the continuous spectrum of neutral modes of the linear stability problem; only after a time long compared with the disturbance e-folding time (perhaps equivalent to 10-12 days in the atmosphere) do the normal modes having maximum linear growth rate dominate in the total solution. Any relevance of these results to the atmosphere would certainly be confined to the long baroclinic-Rossby waves, which persist for many days in a recognisable form, and often exhibit oscillations in amplitude reminiscent of the behaviour described here.

(ii) Small dissipative effects
($r = O(\epsilon)$, $s = 0$, $O(\epsilon)$, $O(1)$)

This range of equations, the "middle row" of both Table 1a

TABLE 1(a). WAVE TRAIN EQUATIONS AND BEHAVIOR

$A(T_1,T_2)$ = Wave Amplitude $B(T_1,T_2)$ = Mean Flow Correction $\mathscr{C}$ - Complex Coefficient (real Coefficients Suppressed)

INCREASING DISPERSION PARAMETER S →

INCREASING DISSIPATION PARAMETER r ↓

r \ s	s = 0	s = 0(ε)	s = 0(1)
r = 0	REGULAR $A_{T_1} = A - AB$ $B_{T_1} = \|A\|^2_{T_1}$	PERIODIC (DEPENDENCE ON INITIAL CONDITIONS) $A_{T_1T_1} + \mathscr{C}A_{T_1} = A - AB$ $B_{T_1} = \|A\|^2_{T_1}$	OSCILLATIONS $A_{T_1T_1} = A - AB$ $B_{T_1} = \|A\|^2_{T_1}$
r = 0(ε)	LORENZ ATTRACTOR EQNS. STEADY, PERIODIC (SINGLY & MULTIPLY) APERIODIC SOLUTIONS HIGHLY SENSITIVE TO INITIAL CONDITIONS $A_{T_1T_1} = A - A_{T_1} - AB$ $B_{T_1} + B = \|A\|^2_{T_1} + \|A\|^2$	COMPLEX LORENZ EQNS. $A_{T_1T_1} = \mathscr{C}A - \mathscr{C}A_{T_1} - AB$ $B_{T_1} + B = A^2_{T_1} + A^2$	STEADY EQUILIBRATION $A_{T_2} = A - \|A\|^2A$
r = 0(1)	$A_{T_2} = \mathscr{C}A - \|A\|^2A$	STEADY EQUILIBRIUM (INDEPENDENT OF INITIAL CONDITIONS) $A_{T_2} = \mathscr{C}A - \|A\|^2A$	$A_{T_2} = \mathscr{C}A - \mathscr{C}\|A\|^2A$

and Table 1b is perhaps the most interesting mathematically but at the same time the most depressing for those hopeful of predictability. It has been shown[22,23] that, in the absence of dispersion ($s = 0$), and for certain sidewall boundary conditions slightly less restrictive than those above at (3.3), the evolution equations are transformable into precisely the system of three ordinary nonlinear differential equations, (4.5), examined by Lorenz[24] as a highly truncated model, first proposed by Saltzman[25], for Rayleigh-Benard convection, viz.

$$\begin{aligned} \dot{X} &= -\sigma X + \sigma X \quad , \\ \dot{Y} &= -XZ + r_a X - Y \\ \dot{Z} &= XY - bZ \quad , \end{aligned} \tag{4.5}$$

where X, $Z \propto A,V$, and Y is defined by the middle equation; γ, r_a and b are constants depending on B, r and s.

It is well known that the asymptotic solution space of these equations contains a strange attractor; the implications for predictability are substantial, in that the solution behaviour, though deterministic, is known to be arbitrarily sensitive both to initial conditions and to parameter values. Brindley and Moroz[26] showed that the use of the more physically acceptable boundary conditions (3.3) give rise to a system of equations closely related to the Lorenze system, in which one of the differential equations is replaced by a (formally infinite) set of equations (4.6) of identical form which govern the evolution of "slaving" modes

$$\begin{aligned} \dot{X} &= \sigma X + \sigma Y \quad , \\ \dot{Y} &= -X \sum_{1}^{\infty} Z_n + r_a X - Y \quad , \\ \dot{Z}_n &= \gamma_n XY - b_n Z_n \quad , \quad n = 1,2,\ldots \end{aligned} \tag{4.6}$$

The sensitivity of these equations resembles that of the Lorenz equations themselves, a result borne out by the work of Pedlosky and Frenzen[27], who integrated numerically a very similar system and found a sequence of behaviours similar to those associated with the Lorenz system. Dramatically different results are obtained when dispersion is included[23]; for small dispersion equations arise (4.7) which are formally similar to the Lorenz equations but which have complex cofficients:

$$\begin{aligned} \dot{X} &= -\sigma X + \gamma Y \quad , \\ \dot{Y} &= -XZ + (r_{aR} + i r_{aI})X - (1 - i\alpha)Y \quad , \\ \dot{Z} &= \frac{1}{2}(XY^* + X^*Y) - bZ \quad . \end{aligned} \tag{4.7}$$

Two of the dependent variables are now complex, and the system becomes fifth order[28,29]. The strange attractor familiar to the student of the Lorenz equations is now confined to a very small

TABLE 1(b). WAVE PACKET EQUATIONS AND BEHAVIOR.

$A(T_1,T_2)$ = Wave Amplitude $B(T_1,T_2)$ = Mean Flow Correction $\mathscr{C}$ - Complex Coefficient (real Coefficients Suppressed)

INCREASING DISPERSION PARAMETER S →

INCREASING DISSIPATION PARAMETER r ↓

r \ s	s = 0	s = 0(ε)	s = 0(1)
r = 0	NONLINEAR SCHRODINGER EQUATION WITH USUAL SPACE AND TIME INTERCHANGED $A_{T_1T_1} + iA_{X_2} = A - AB:$ $B_{T_1} = \|A\|^2{}_{T_1}:$	$A_{T_1T_1} + i\left(A_{T_1} + A_{X_2}\right) = A - AB:$ $B_{T_1} = \|A\|^2{}_{T_1}:$	CLASSIC N.L.S; SOLITON SOLUTIONS $A_{X_1X_1} + iA_{T_2} = A - AB:$ $B_{X_1} = \|A\|^2{}_{X_1}:$ SINE-GORDON/SELF-INDUCED TRANSPARENCY EQUATIONS SOLITON SOLUTIONS $\left(\frac{\partial}{\partial T_1} + g_1 \frac{\partial}{\partial X_1}\right)\left(\frac{\partial}{\partial T_1} + g_2 \frac{\partial}{\partial X_1}\right) A = A - AB:$ $\left(\frac{\partial}{\partial T_1} + g_2 \frac{\partial}{\partial X_1}\right) B = \left(\frac{\partial}{\partial T_1} + g_1 \frac{\partial}{\partial X_1}\right) \|A\|^2$
r=0(ε)	SPATIAL "LORENZ FAMILY" $A_{T_1T_1} = A - A_{T_1} - iA_{X_2} - AB:$ $B_{T_1} + B = \|A\|^2{}_{T_1} + \|A\|^2:$	$A_{T_1T_1} = \mathscr{C}A - \mathscr{C}A_{T_1} - iA_{X_2} - AB:$ $B_{T_1} + B = \|A\|^2{}_{T_1} + \|A\|^2:$	UNDER INVESTIGATION $iA_{T_2} + \mathscr{C}A_{X_1X_1} + \mathscr{C}A_{X_1} = A - AB:$ $B^{(1)}_{X_1} + B^{(2)} = \|A\|^2{}_{X_1} + \|A\|^2:$
r=0(1)	RELATED TO THE TIME-DEPENDENT GINZBERG LANDAU EQUATION (N.L.S. WITH COMPLEX COEFFICIENTS) PERIOD AND APERIODIC (SPATIAL AND TEMPORAL) $A_{T_2} = \mathscr{C}A_{X_1X_1} + \mathscr{C}A - \|A\|^2A:$	$A_{T_2} + \mathscr{C}A_{X_1} = \mathscr{C}A_{X_1X_1} + \mathscr{C}A - \|A\|^2A$	$A_{T_2} + \mathscr{C}A_{X_1X_1} + \mathscr{C}A - \mathscr{C}\|A\|^2A:$

region of parameter space near the limit of vanishing dispersion, and is totally lost for even very small dispersion; solution behaviour is almost always periodic at values of B and r which would give rise to a strange attractor for $s = 0$. A similar variation of the strange attractor when a higher order system is used has of course been noted by other authors[30,31] using a larger number of modes to model Rayleigh-Bernard convection. Thus the Lorenz attractor may be a spurious hazard in real baroclinic flows, where, as we have stated in the introduction, dispersive effects associated with potential vorticity gradients are always present.

The evolution of wave packets for this range of dissipative values is described by partial differential equations whose form is not well known. They are closely related, for small dispersion, to the Lorenz family described above. Earlier numerical work[32] showed that the temporal behaviour at a point in space of solutions to these partial differential equations, in the case with no dispersion, is identical to the behaviour of the solution of the Lorenz equations at parameter values shifted by an amount related to the spatial structure. It seems, however, that the spatial structure is itself ill-defined; in more recent numerical integrations (supported by analytical work) it appears to adopt the smallest scale available to the numerical model in a manner which is basis dependent. The character of the solutions depends solely on grid resolution and initial data, and the wave packet model in this case is ill-posed physically. For strong dispersion ($s = O(1)$) the same difficulties do not arise, but our knowledge of solutions to the evolution equations is still rudimentary.

(iii) Strong dissipation
($r = O(1)$, $s = 0$, $O(\epsilon)$, $O(1)$)

The wave train problem in the case $s = 0$ was examined by Pedlosky[33] and Drazin[34], and in all cases, with or without dispersion, the behaviour shows an equilibration of wave amplitude to a steady value independent of initial conditions[12]; such a state is totally predictable. The wave packet equations take the form of time dependent Ginzburg-Landau equations with complex coefficients[26]; they are known to admit steady, periodic and aperiodic behaviour[35,36,37] depending on coefficient values. The coefficients themselves are complicated functions of s, r and B, and it is clear that predictability is likely to be highly variable in a similarly complicated way.

5. DISCUSSION

We have commented on the form of the evolution equations for differing parameter magnitudes, and on the predictability of solutions in each case for the very simple mathematical models described. It is perhaps most easy to assess the importance and significance of these results by centering the discussion around a number of questions which might be asked by the sceptic; some of the most pressing are:

(1) Is it possible to use the results obtained in a predictive way for any real baroclinic flow problem? In other words, is it possible to attribute particular values to parameters and coefficients in the model equations with such certainty that accurage prediction, at least of qualitative behaviour, can be made? For example, can we predict unequivocally that amplitude vacillation will occur at one particular point in parameter space, but that aperiodic behaviour will occur at another point, for rotating heated annulus experiments (see Fig.5 after Pfeffer et al[38]). It is interesting that this is in a sense a question concerning the "predictability of predictability".

(2) How can we assess the validity of a model system by comparing its solution behaviour with observational data from experiments on laboratory baroclinic systems? Since many model equations have regions of steady state and periodic solutions, it may well be that the existence and character of any <u>aperiodic</u> solution will provide the crucial test in model evaluation. What measures of an aperiodic output constitute an unambiguous "signature".

(3) What is the relevance of either the mathematical models, or of simple laboratory experiments, to naturally occurring geophysical flows?

The answer to the first question is almost certainly negative, at least at present. However, we have been able to show recently[39,40] that, using a representation as simple and crude as a three layer model, and using only linear stability theory in the first instance, it is possible to construct a regime diagram containing all the qualitative structure obtained in a laboratory experiment. Our early results show that the spatial relationship of regions of steady waves, vacillation and aperiodic flow is correct in that traverses of parameter space for, e.g., increasing Ω and fixed ΔT (along line AB in Figure 5) give the same <u>sequence</u> of qualitative behaviour in the model as in the experiment. This suggests that the model can be useful in establishing the existence and relative location of regions in parameter space in which the predictability characteristics of the flow will be qualitatively different, and more importantly in relating this existence and location to the value of parameters measuring real physical effects.

Before we can be totally confident, however, we must address the second question. It may be that <u>many</u> simple mathematical models yield regime diagrams of the correct topological form, but do not correctly represent the chaotic behaviour in the aperiodic regions of parameter space. We need to define an unmistakable and unambiguous "signature" of the chaos, and to establish a means of reading the signature.

A popular approach to this question is to examine the time series behaviour of one or more variables of the flow, in model and experiment. Appropriate measures of aperiodic time series as a means of classifying strange attractors have been the objective of

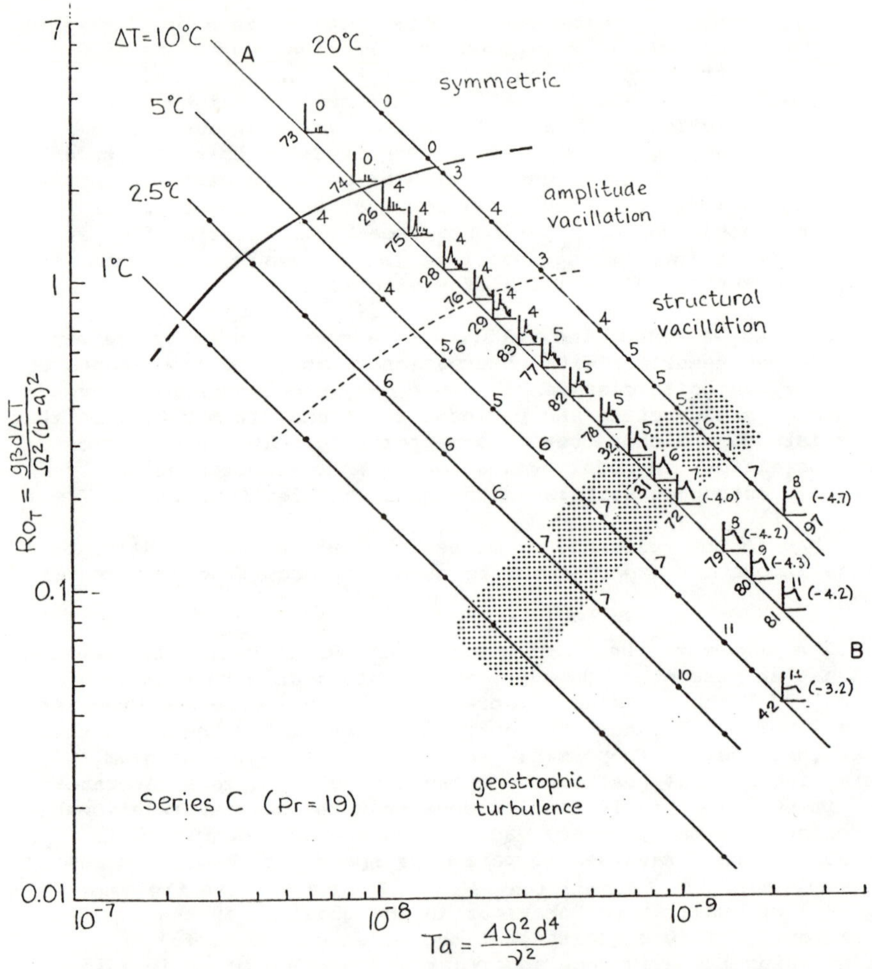

Fig.5 Regime diagram for a particular series of experiments performed by G.Buzyna et al at Florida State University. Thermal energy spectra (log E vs log k) at mid-radius and mid-depth are shown in the figure at the appropriate locations in dimensionless-parameter space. The dominant wave number is indicated in these and in most of the experiments shown. Also indicated are the slopes of the energy profiles at the short wave end of the spectrum in the geostrophic turbulence regime. The shading indicates the region of transition to geostrophic turbulence. The parameters R_{OT} and T_a against which the results are plotted are related to our parameters, at least qualitatively, so that $R_{OT} \propto B$ and $T_a \propto r^{-4}$.

a good deal of recent research in dynamical systems theory[41,42,43]. Early concepts of limit capacity and topological entropy proved too computationally costly for useful application; more simple methods[44,45] hold out more promise and we are at present actively analysing outputs of models and experiments to test these ideas.

The third question is a good deal harder to answer. We have commented earlier on the probable inadequacy of the normal mode approach in modelling small scale and rapidly varying baroclinic instability in the atmosphere or oceans, where the consequences of finite amplitude "initial" perturbations are considerable[5]. It is likely, however, that the very occurrence of local strong baroclinicity and/or large amplitude initial disturbances is itself much influenced by the quasi-steady long baroclinic-Rossby waves, which are at least plausibly modelled by the rotating annulus experiments and therefore perhaps share the predictability characteristics of the mathematical models. This gives rise to the idea of two stages in prediction, one associated with the long waves themselves, and the other with the rapid evolution of non normal-mode short waves. Indeed a coupling of these two stages involving feedback mechanisms, is sometimes suggested by atmospheric behaviour and may be an important feature of predictability in the atmosphere not well modelled in the annulus experiments or the models reviewed here.

In summary then, what can we deduce from our results? We have shown that certain low order systems of nonlinear ordinary or partial differential equations, making some pretence of modelling the essential physics of baroclinic instability, have solution behaviour with widely varying predictability characteristics, ranging between steady states with global stability at one extreme to infinitely sensitive strange attractor states at the other. We are well on the way to a complete classification of behaviour according to parameter values, and within the limits of weakly non-linear approximations, for the model equations themselves. But how is this result relevant to real flows? An acceptance that such low order and relatively crude models can suffice for an understanding of the gross and robust features of baroclinic flow is supported by a vast body of experimental and theoretical work. However, much as is apparently the case in the rather simpler but similar problems of Taylor-Couette flow[46], or Rayleigh-Benard convection[47], it seems that detailed features of transitional and time dependent behaviour are highly sensitive to model assumptions. Thus the theoretical description, and indeed even the observational description, of time-dependent behaviour, including vacillation, transition between relatively stable states, and aperiodic behaviour, is far from completion; this time dependent behaviour lies of course at the very heart of the predictability problem.

It seems then, if there is to be a prosperous future for low order models of baroclinic flows (in contrast to large-effort numerical approaches, attempting to model the full fluid dynamic and thermodynamic problem, whose results are themselves extremely difficult to evaluate) they must be shown to be able to produce the correct details of time-dependent flow, and in particular to be able to differentiate between intrinsic and externally forced time-dependent behaviour. Quantitative argument with real flows is not a reasonable expectation, but correct qualitative behaviour is the sine qua non of a good model, and this must be the objective of the theoretician. The physical understanding which will accompany success in this venture makes worthwhile the substantial programme of delicate experiment, coupled with a suitably imaginative, and even speculative, theoretical modelling which is demanded.

REFERENCES

1. J.Pedlosky, 1979a. Geophysical Fluid Dynamics.. Springer, Berlin and New York.
2. A.E.Gill, 1982. Atmosphere-Ocean Dynamics. Academic Press, New York.
3. R.Hide and P.J.Mason, 1975. Adv. Phys. 24, 57-100
4. J.E.Hart, 1979. Ann.Rev.Fluid Mechs., 11, 147-172.
5. B.F.Farrell, 1982. J.Atmos.Sci., 39, 1663-1686.
6. N.A.Phillips, 1954. Tellus 6, 273-286
7. E.T.Eady, 1949. Tellus, 1, 35-52.
8. P.D.Killworth, 1980. Dyn.Atmos.Oceans, 4, 143-184.
9. J.G.Charney, 1947. J.Meteor. 4, 135-162.
10. J.Pedlosky, 1979b. J.Atmos.Sci. 36, 1908-1924.
    J.Pedlosky, 1981. J.Atmos.Sci. 38, 717-731.
11. P.A.Card and A.Barcilon, 1982. J.Atmos.Sci. 39, 2128-2135,
12. I.Moroz and J.Brindley, 1983a. To appear in Studies in App. Maths.
13. P.J.Mason, 1975. Phil. Trans. Roy. Soc. Lond. A278, 397-445.
14. M.A.Weissman, 1979. Phil.Trans.Roy.Soc. Lond. A290, 639-681.
15. I.M.Moroz, 1981. Ph.D.Thesis, University of Leeds
16. J. Pedlosky, 1970. J.Atmos.Sci., 27, 15-30.
17. P.G.Drazin, 1970. Q.J.R. Met.Soc. 96, 667-676.
18. J.D.Gibbon, I.N.James and I.M.Moroz, 1979. Proc.Roy.Soc. Lond. A367, 219-237.
19. I.M.Moroz and J.Brindley, 1981. Proc.Roy.Soc. Lond. A377, 379-404.
20. I.M.Moroz and J. Brindley, 1983b. to appear in "Non-linear Waves", ed. L.Debnath, Camb.Univ.Press.
21. E. Fermi, J. Pasta and S. Ulam, 1955. Los Alamos Report LA1940, reprod. in "Non-linear Wave Motion", ed. A.C.Newell, Amer.Math.Soc. 1974.
22. J.D.Gibbon and M.J.McGuinness, 1980. Phys.Lett. 77A, 295-299.
23. J.D.Gibbon and M.J.McGuinness, 1982. Physica 5D, 108-122.
24. E.N.Lorenz, 1963. J.Atmos.Sci. 20, 130-141.
25. B.Saltzman, 1962. J.Atmos.Sci. 19, 329-
26. J.Brindley and I.M.Moroz, 1981. Phys.Lett. 83A, 259-262.
27. J.Pedlosky and C.Frenzen, 1980. J.Atmos.Sci. 37, 117-1196.
28. J.Brindley and I.M.Moroz, 1980. Phys.Lett. 77A, 441-444.
29. A.C.Fowler, J.D.Gibbon and M.J.McGuinness, 1982. Physica 4D, 139-163.
30. J.B.McLaughlin and P.C.Martin, 1975, Phys.Rev. 12A, 186-203.
31. O.Manley and Y.Treve, 1981. Phys.Lett. 82A, 88-90.
32. M.E.Alexander, J.Brindley and I.M.Moroz, 1982, Phys.Lett. 87A, 240-244.

33. J.Pedlosky, 1971. J.Atmos.Sci. 28, 587-597.
34. P.G.Drazin, 1972. J.Fluid Mech, 55, 577-588.
35. L.M.Hocking and K.Stewartson, 1972. Proc.Roy.Soc. Lond. A326, 289-313.
36. H.T.Moon, P.Huerre and L.G.Redekopp, 1982. Phys.Rev.Lett. 49, 458-60.
37. Y.Kuramoto and S.Koga, 1982. Phys.Lett. 92A, 1-4.
38. R.L.Pfeffer, G.Buzyna and R. Kung, 1980. J.Atmos.Sci. 2577-99.
39. I.M.Moroz and J.Brindley, 1982. Phys.Lett. 91A, 226-230.
40. I.M.Moroz and J.Brindley, 1983c, in preparation.
41. D. Ruelle, 1978. Ann.N.Y. Acad. Sci. 316, 408.
42. O.E.Lanford, 1981. in "Hydrodynamic Instablities and the Transition to Turbulence" ed. H.L.Swinney and J.P.Gollub, Springer, Berlin and New York.
43. F. Takens, 1981. in "Dynamical Systems and Turbulence, Warwick 1980". Lecture Notes in Mathematics, vol.898, ed. A.Dold and B.Eckman, Springer, Berlin and New Yorks.
44. J.Guckenheimer, 1982. Nature, 298, 358-361.
45. H.Froehling, J.P.Crutchfield, D.Farmer, N.H.Packard and R.Shaw, 1981. Physica 3D, 605-617.
46. R.Di Prima and H.L.Swinney, 1981. in "Hydrodynamic Instabilities and the Transition to Turbulence", ed. H.L.Swinney and J.P.Gollub, Springer, Berlin and New York.
47. F.H.Busse, 1981. in "Hydrodynamic Instabilities and the Transition to Turbulence", ed. H.L.Swinney and J.P.Gollub, Springer, Berlin and New York.

ON THE RELATIONSHIP BETWEEN THE SYSTEMATIC ERROR OF THE ECMWF FORECAST MODEL AND OROGRAPHIC FORCING

by

S. Tibaldi
ECMWF, Shinfield Park, Reading, Berkshire, UK.

ABSTRACT

The work by Wallace, Tibaldi and Simmons[1] on the relationship between orographic forcing and ECMWF model systematic error is briefly reviewed and some recent numerical experimentation results corroborating their conclusions are presented.

Some hypotheses are also put forward on the mechanisms responsible for such results on the basis of model diagnostics.

INTRODUCTION

Different numerical models used for weather prediction beyond the short range (2 to 3 days) and for climate simulations (usually referred to as General Circulation Models, GCM's) are subject to a variety of errors.

A natural way to investigate the deficiencies shown by those models that are routinely used to forecast the weather on a given time range (a week to ten days, say) is to compare ensemble mean of forecast fields to correspondent ensemble mean of observed (analyzed) fields, (e.g. 500 mb geopotential height). The difference field between such observed and forecasted values is usually referred to as the Systematic Error (SE). Conversely, models used to simulate climate are often evaluated on the basis of how well, during suitable long integrations, they reproduce some time-meaned fields, compared to observed atmospheric (time-mean) quantities. Their mean error can therefore be called climatic error.

It is not difficult to see that, if one extends in time the range of the integrations of a forecasting model to periods of the same order of magnitude of a season (2 to 3 months), the SE tends toward the climatic error. This is why very often the existence of SEs is also referred to as the problem of the model's climate drift. Wallace, Tibaldi and Simmons[1], hereafter referred to as WTS, have put forward the hypothesis that a considerable proportion of models' SEs (and, in particular, SEs of the ECMWF operational model) can be explained in terms of lack of orographic forcing. Section 2 will review this work, supplementing their previous conclusions with some more recent results, while Section 3 will attempt to shed some light onto the mechanisms through which this orographic influence takes place.

OROGRAPHIC FORCING AND SE OF THE ECMWF OPERATIONAL MODEL

Fig. 1 (from WTS) shows Northern Hemispheric maps of ensemble mean error (SE) for the 500 mb geopotential height for day 1, 4, 7 and 10. The sequence, therefore, describes the progressive climate drift of the ECMWF operational model ; the ensemble mean was constructed using 100 daily analyses and corresponding day 1, 4, 7 and 10 forecasts verifying on those days. The period spans from 1 December 1980 to 10 March 1981 and is, therefore, characteristic of winter conditions only.

The main characteristics of the evolution of the SE that can be recognised in Fig. 1 are :

a) The error evolves from a "small scale" configuration to a "large scale" pattern.

b) The smaller scale day 1 error bears a more than passing resemblance with the underlying orographic features, with important (but not unique) maxima in direct correspondence to the Rockies, the Himalaya and the Euro-Carpatian mountains.

c) The larger scale day 10 error pattern is distinctly correlated with the planetary scale wave pattern, in that negative errors always occur over ridges and positive error over troughs ; this suggests an inadequate representation, in the model's atmosphere, of those forcings that are mostly responsible for the maintenance of the ultra-long planetary waves (e.g. orography).

d) There is more area covered by negative error than there is by positive error, indicating a progressive unrealistic tropospheric cooling of the model's atmosphere.

e) Negative error centres occupy, on average, more northerly positions, while positive error centres lie on more southerly positions, indicating (in the geostrophic assumption) that an error in the mean Westerlies also develops, in that the model's atmosphere ends up being, on average, too westerly after 10 days. This might also be related to the fact that observed low frequency planetary waves have a north-south structure that puts trough maxima preferably to the south and ridge maxima preferably to the north.

Fig. 2 shows the 10 day geopotential height SE at 1000 and 300 mb. This, in turn, suggests that

f) The SEs are mostly equivalent barotropic in character, in that their large-scale features change little with height.

WTS work proceeds to investigate further the hypothesis that those forcings responsible for the maintenance of the ultra-long planetary waves are possible sources of SE. This is accomplished by the use of a simple barotropic model based on the numerical solution

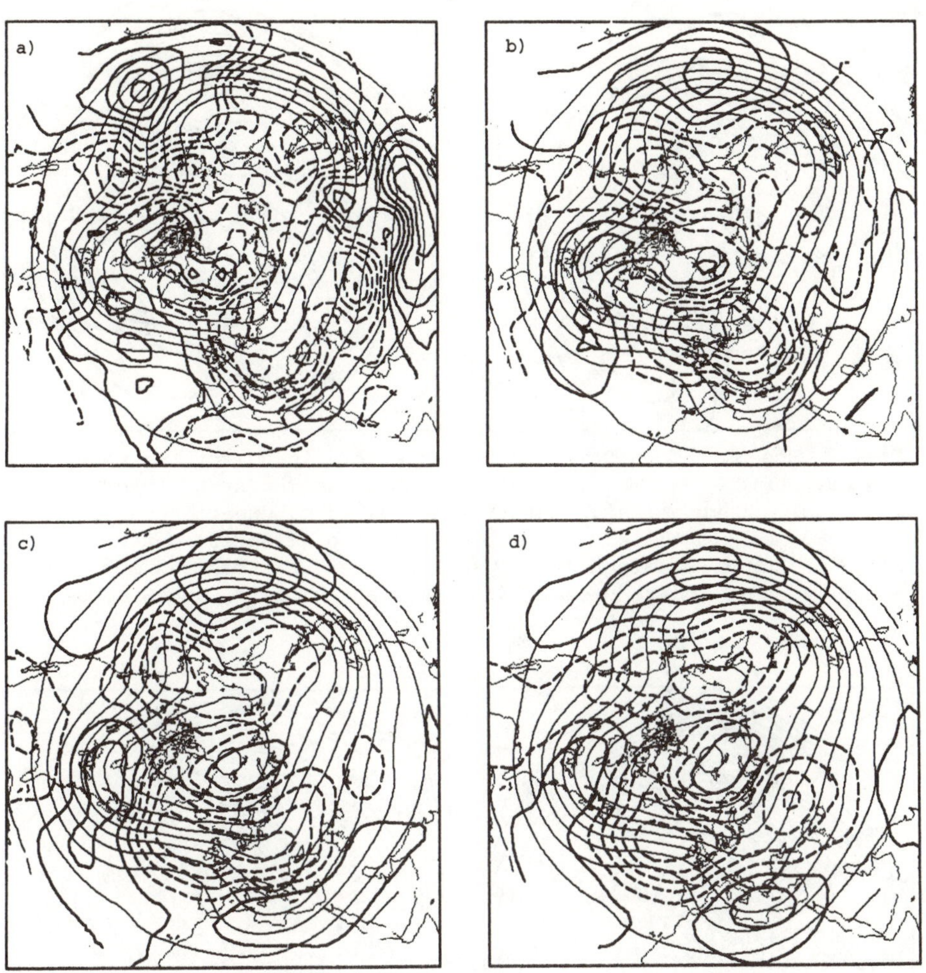

Fig. 1 Ensemble mean forecast error fields for ECMWF operational forecasts of 500 mb height for the 100 day period (1 December 1980-10 March 1981 inclusive). (a) Day 1 forecasts, contour intervals 5m; (b) Day 4 forecasts, contour interval 16 m; (c) Day 7 forecasts, contour interval 30 m; (d) Day 10 forecasts, contour interval 30 m. Background field (lighter contours) is the mean 500 mb height field based on ECMWF operational analyses for the same period, contour interval 80 m. Negative contours are dashed. (From Wallace, Tibaldi and Simmons[1]).

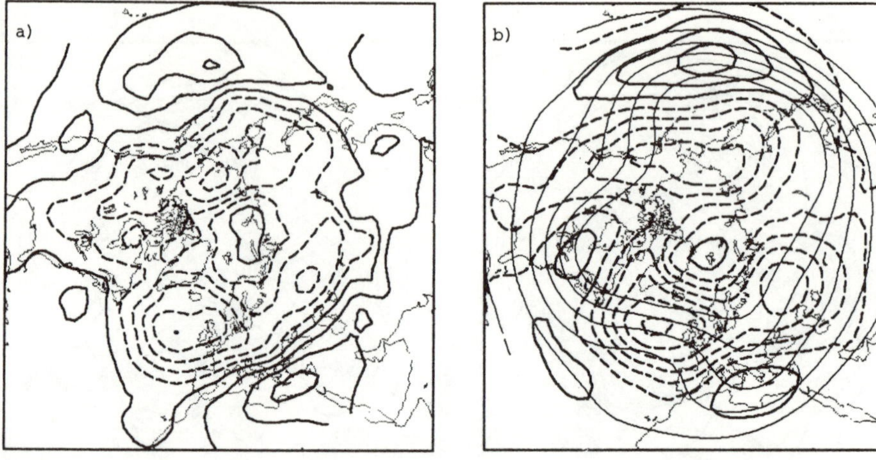

Fig. 2 Ensemble mean Day 10 forecast error fields for the same 100 day period as Fig. 1. (a) 1000 mb height, contour interval 20 m; (b) 300 mb height, contour interval 40 m, superimposed on the mean 300 mb height field for the same period (lighter contours) based on ECMWF operational analyses, contour interval 160 m. Negative contours are dashed. (From Wallace, Tibaldi and Simmons).

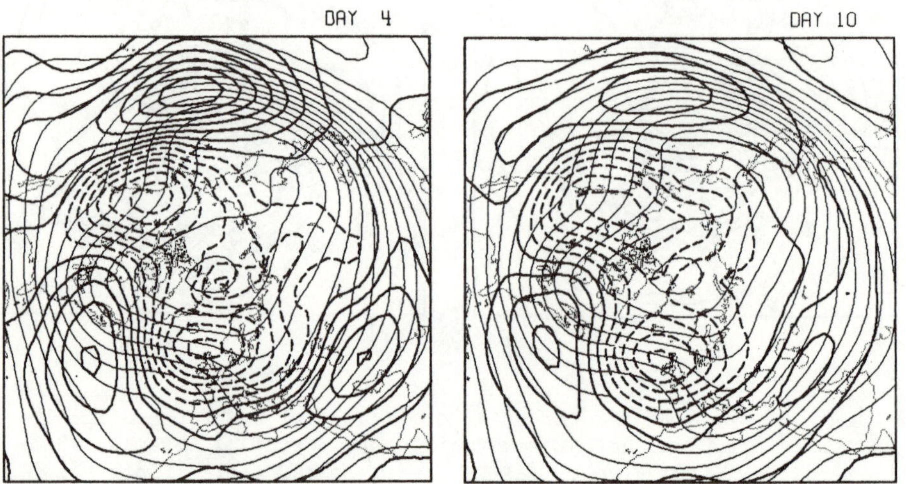

Fig. 3 Perturbation height fields (heavy solid, dashed) for the Northern Hemisphere at Day 4 (left, contour interval 20 m) and Day 10 (right, contour interval 40 m). These represent the response of the barotropic model to the constant additional forcing proportional to the ECMWF operational model day-1 error growth. (From Wallace, Tibaldi and Simmons).

of the vertically integrated full vorticity equation. The model is applied, for reasons of convenience, to the 300 mb level[2]. Firstly a forcing is computed that is sufficient, on its own, to maintain the observed 300 mb mean flow for the same 100 day period. Then an additional forcing is computed, proportional to the 24 hour rate of growth of the day 1 SE, and superimposed to the mean flow forcing. The barotropic model is then integrated in time for 10 days to evaluate, in this simple framework, the response of a simplified barotropic atmosphere to the erroneous forcing responsible for the day 1 SE. The comparison between Fig. 3 and Fig. 1c and 1d shows that, with the notable exception of the east-European SE negative centre, the "forcing" hypothesis is very well confirmed. Further experiments with the same barotropic model (Fig. 4) also suggest that erroneous tropically-located forcings (e.g. originated by an ill-treatment of condensation processes) could be held responsible for a considerable part of the other (main) west-European SE negative centre. This would exclude, therefore, the misrepresentation of orography as the <u>unique</u> or even main cause for this particular feature of the SE.

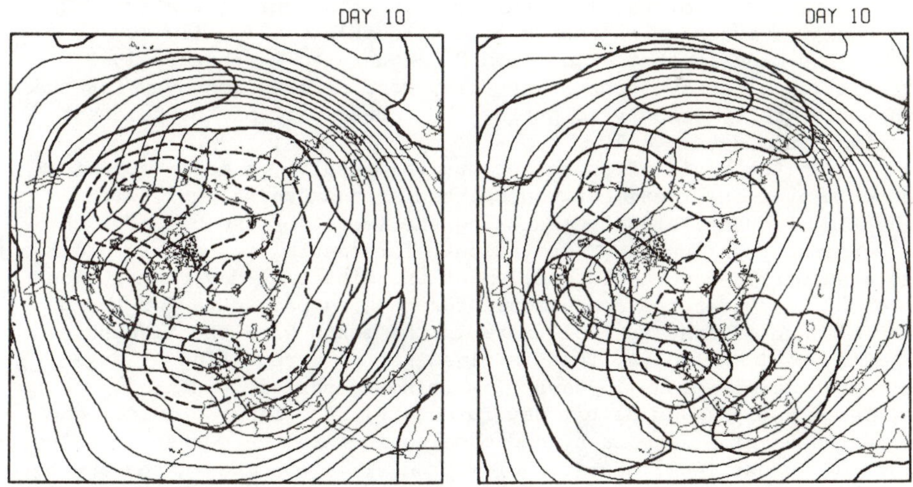

Fig. 4 The Day 10 response to forcing located north of 20°N (left) and between 20°N and 20°S (right). Other details are as for Fig.3. (From Wallace, Tibaldi and Simmons).

The simple, but effective, way proposed by WTS to test the "orographic cause" hypothesis is to try model runs with a somewhat enhanced large-scale orographic forcing. A suitable rationale to provide guidance on how to enhance the large scale orography is the recognition of the absence, in GCM's, of the integrated effects on the large-scale flow of the sub-grid scale orographic features ; of those features, that is, that have direct or indirect influence on the large scale flow, but are not being suitably represented by a grid-point average terrain height because of resolution limitations. Notable examples of such features are "blade"-type mountains (Alps, Northern Rockies), very efficient in blocking the lower-tropospheric large-scale flow, but covering, little area. Deep valleys that, mostly in winter, develop local mesoscale circulations that participate little to the large-scale flow are another example. The large-scale circulation, in turn, sees an "envelope"-type mountain rather tan an "average". An example of the product of such a mountain enhancement procedure is shown in Fig. 5 where an "envelope"-type orography, obtained adding a proportion (in this case twice) of the sub-grid scale terrain height variance to the grid-square mean value, is compared to a "mean"-type one.

WTS reported very encouraging results from the use of such orography in the ECMWF GCM used for operational medium-range forecasting. Marked improvements could be noticed not only in the mean circulation of the last thirty days of a typical winter 50-day integration but also in the SE of an ensemble of 21 daily 10 day forecasts made in the period from 20 January to 9 February 1982. Prediction objective scores were also considerably improved for the ensemble, more markedly so during the latter part of the forecast period, from day 4 to 10.

On the basis of those encouraging results, a major numerical experimentation experiment was started around August 1982, with the aim of quantitatively assessing the impact of the use of such an orography on the ECMWF operational forecasting system. In addition to a summer 50-day run to complement the winter 50-day run, it was decided to repeat the operational forecasting procedure for the whole month of January 1981, notably the month with the highest ever SE for the ECMWF model. This involved not only repeating 41 days of data assimilation (10 days of warming-up period, plus 31 January days) and 31 ten day forecasts, but doing so twice, once with the envelope orography and once again with the current (at the 10th of August 1982) operational system. This was made necessary by the large number of improvements that the operational system had undergone from January 1981 to August 1982. Furthermore, an interesting by-product of this experiment was the opportunity to quantitatively assess, in a clean "twin" experiment, the overall improvement of the ECMWF forecasting system over the 19-month period January 1981-August 1982. This will be the subject of a separate paper.

Fig. 6 shows the mean 500 and 1000 mb NH geopotential height maps for the last 30 days of the summer 50-day run. The overall impression one derives from the comparison between both the envelope

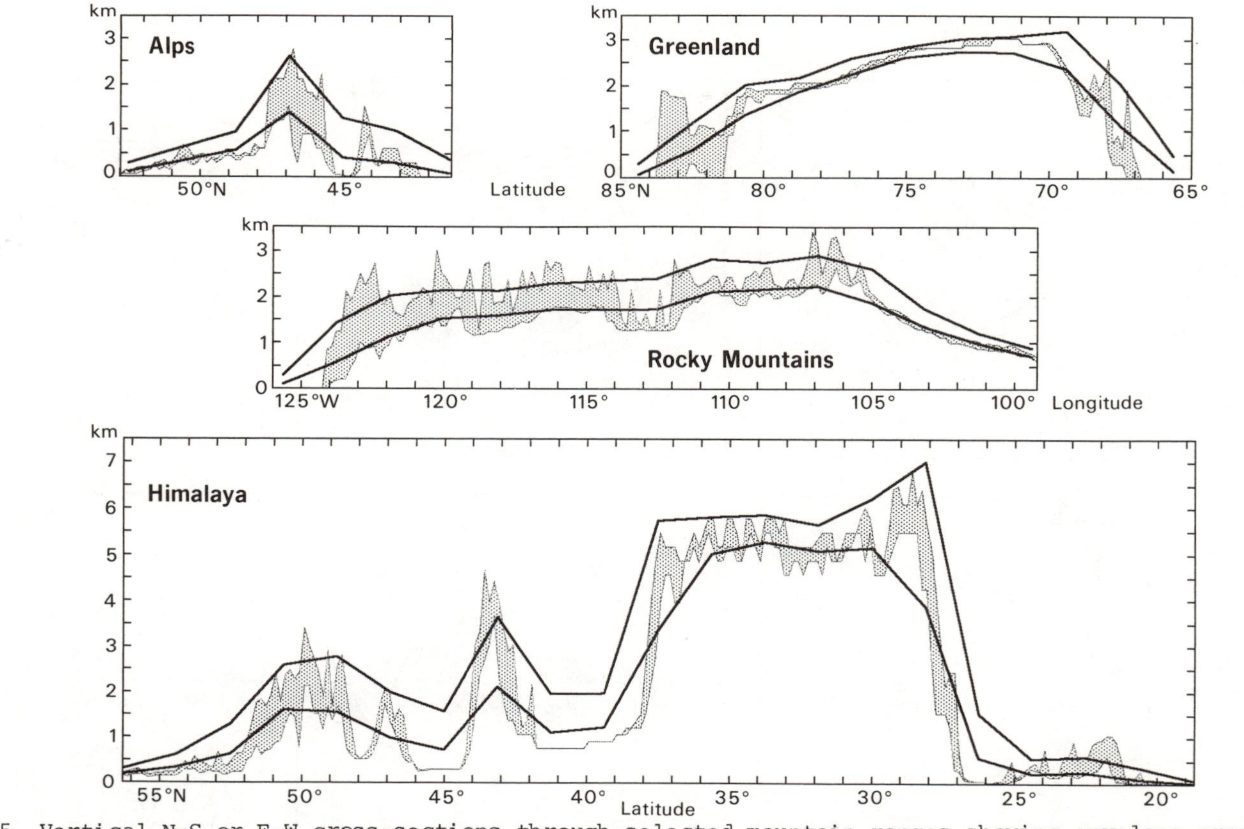

Fig. 5 Vertical N-S or E-W cross-sections through selected mountain ranges showing envelope orography (upper curve) and average orography (lower curve). Shading indicates the range of terrain heights between the maximum and minimum value in the U.S. Navy high resolution grid. (a) N-S along 11.25°E, (b) N-S along 33.75°W, (c) E-W along 41.25°N and (d) N-S along 85-25°E. (Wallace, Tibaldi and Simmons).

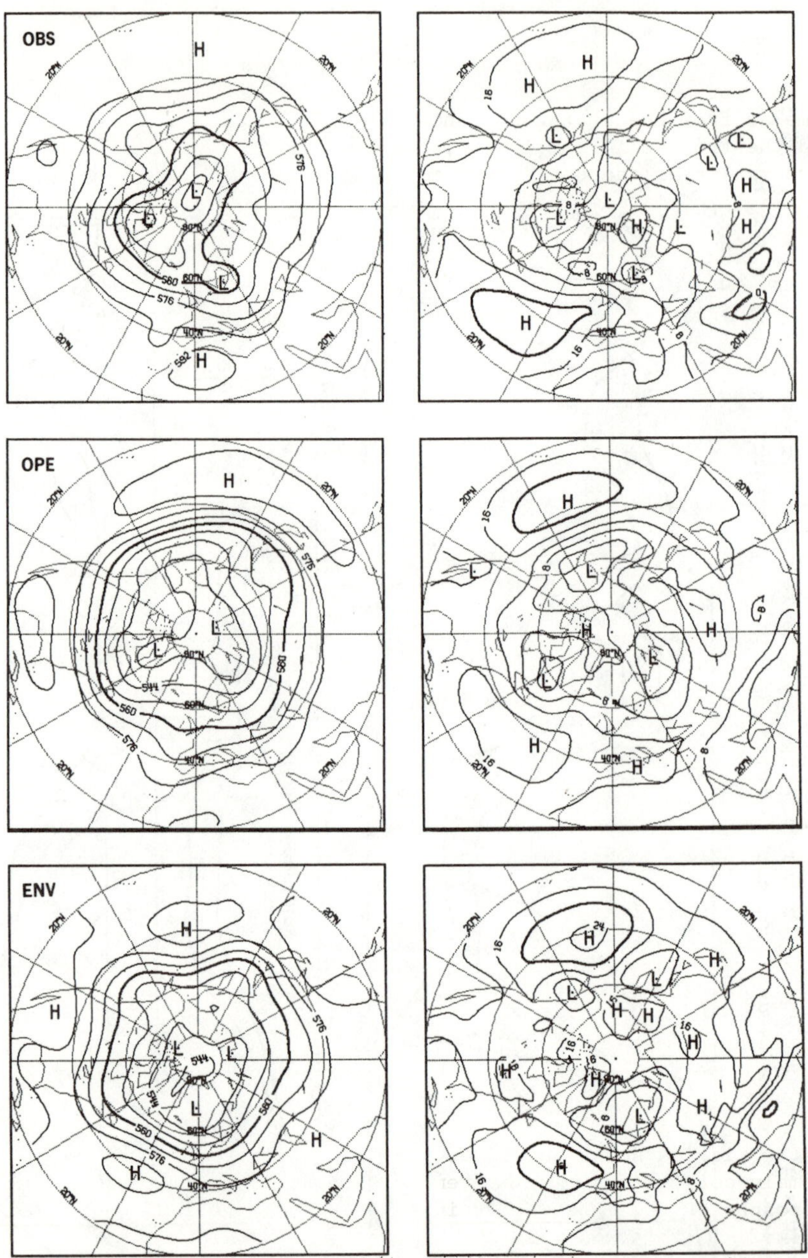

Fig. 6 Averages for Days 21-50 of extended integrations from initial conditions for 11 June, 1979, with the ECMWF operational model with the operational orography (OPE, middle), and with the envelope orography (ENV, bottom) together with FGGE observed data for the same period (OBS, top). Left: 500 mb height, contour interval 80 m, 5200 and 5600 m contours thickened. Right: 1000 mb height, contour interval 40 m, 0 and 200 m contours thickened.

and the operational orography integrations and the observed mean fields for the same period is, *'mutatis mutandis'*, essentially the same as from the winter run shown in WTS. The planetary scale waves, and the troughs in particular, are more pronounced and, therefore, more realistic and their axes are longitudinally better positioned than in the control run. It is also (unfortunately) plain to see that other deficiencies of the integration, e.g. the too intense and too narrowly confined in latitude jet stream, are left virtually unaltered.

The result of this summer 50-day integration is, however, important because it contributes to dissipate legitimate fears that the improvement brought about by the enhancement in orography in the model's winter climatology was coming from a "compensating error" (or "tuning") effect, with mountains providing enough planetary wave forcing to compensate for insufficient forcings of an altogether different nature, for example land-sea thermal contrast. Witnessing a comparable improvement in a period of the year for which the land-sea thermal contrast forcing changes sign is a comforting argument (albeit not conclusive) against such a possibility.

Fig. 7 shows the 500 mb height SE maps for days 4, 7 and 10 of the 31 ten-day forecast ensemble. There is, at first glance, very little need for comment, because it is plain to see that not only the absolute values of the positive and (even more so) negative centres have been greatly reduced, but the overall bias (tropospheric cooling) and north - south structure (enhanced westerlies) have been almost wiped out.

When the enthusiasm for such a remarkable improvement cools down, however, one starts noticing that, concerning the two major European negative centres the eastern one has been hardly changed at all at day 10. Its onset, however, has been substantially postponed, of a period of a few days. It is perhaps worth recalling at this point that this negative centre is the one that was not reproduced by the barotropic model's simple experiments (see Fig. 3). This would possibly indicate a different nature of the primary cause of this particular feature. However, the partitioning between the part of the SE that is removed by the use of the envelope orography and that part that is, so to speak, left over, is strongly reminiscent of the two different responses of the barotropic model to the extra-tropical and tropical part of the additional forcing derived form the day 1 systematic error (Fig. 4).

Fig. 8a shows the impact of the envelope orography on the anomaly correlation of geopotential height averaged over the NH troposphere. The improvement in "predictability" (unsophisticatedly defined, in this case, as the time it takes for the anomaly correlation to go below 60 %) is around 6 hours.

The remainder of Fig. 8 (b to f) contains a spectral breakdown of the standard deviation of geopotential height averaged over the same atmospheric volume. This shows that the largest contribution to the improvement comes form the zonal part and from the ultra-long waves. It is worth remembering at this point that ultra-long waves explain most of the variance of NH tropospheric

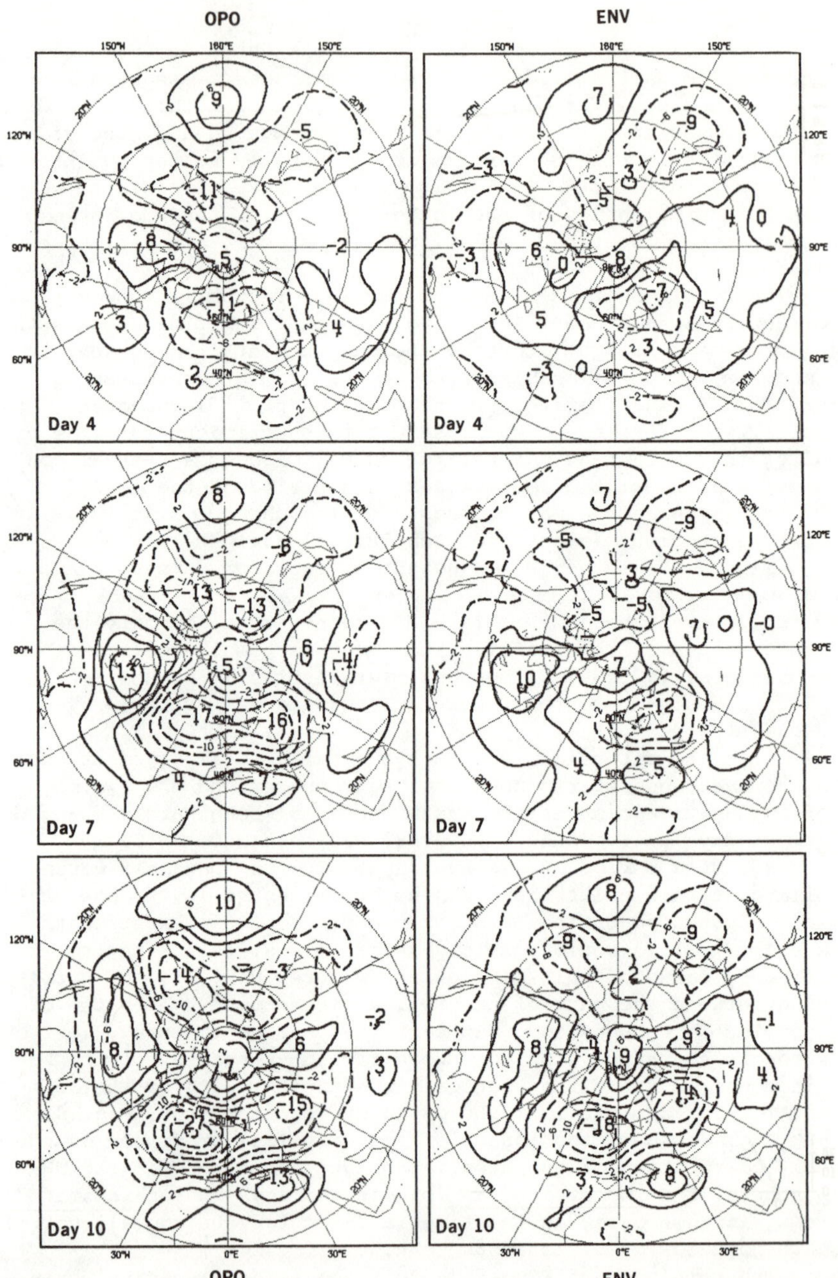

Fig. 7 Ensemble 500 mb geopotential height forecast error fields for the 31 cases of January 1981 with the operational average-type orography (OPO) and the envelope orography (ENV). From top to bottom, Day 4, Day 7 and Day 10.

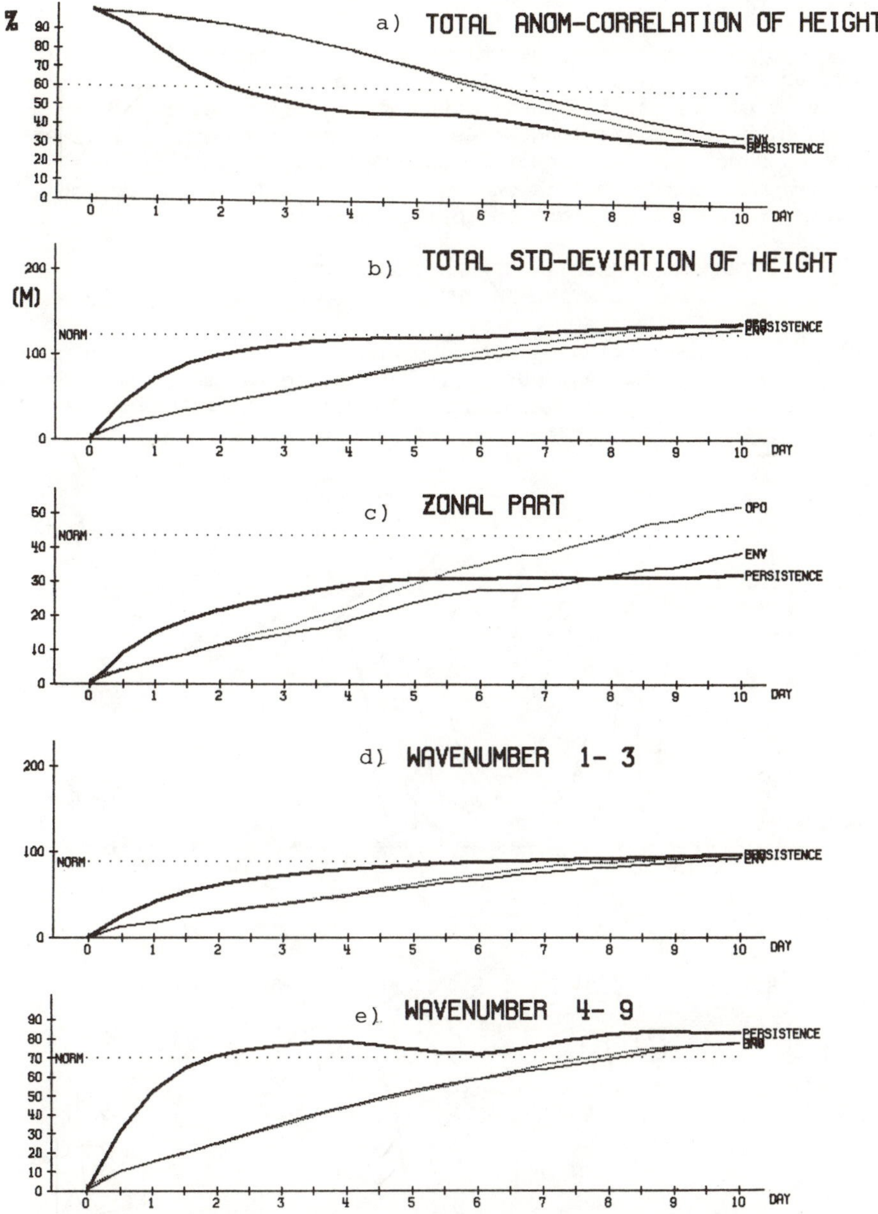

Fig. 8 (a) Vertically averaged (200-1000 mb) pattern correlation between geopotential height anomalies in forecast charts and the corresponding verification charts as a function of forecast interval averaged over the ensemble of 31 forecasts of January 1981 and computed for the area extending from 20° to 82.5°N. Heavier curves denote the envelope forecasts and dashed curves the corresponding operational forecasts. (b) Same as (a) but for the standard deviation of geopotential height (c), (d) and (e) show a spectral breakdown of the total standard deviation shown in (b).

geopotential height (see the values corresponding to the norm, in Figs. 8b to f). While it is intuitively easy to understand a strong influence of the earth's (and consequently, the model's) orography on the planetary ultra-long waves, it is at first (and in a linear frame of mind) more difficult to understand such a strong impact on wavenumber zero, that is on the zonal mean. The next section will try to shed some light on to this problem.

THE RELATIONSHIP BETWEEN IMPROVEMENTS IN THE ZONAL FLOW AND IN THE ULTRA-LONG WAVES

Fig. 9 shows height-latitude cross-sections of the mean zonal wind error in the two ensembles. It is easy to see that the presence of the envelope orography approximavely halves the tropospheric mean wind error. This effect is confirmed by the standard deviation

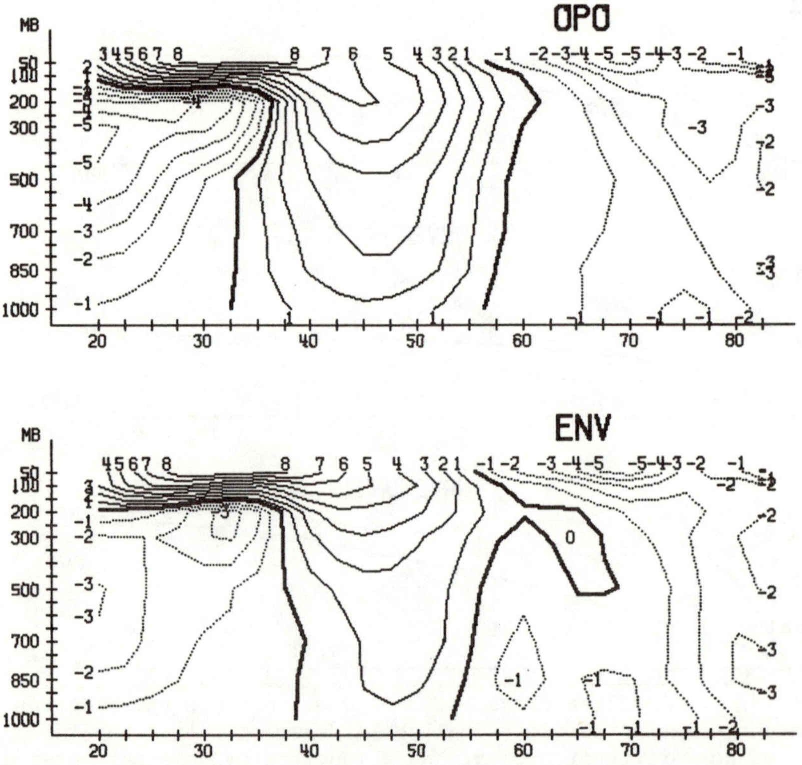

Fig. 9 Latitude-height cross-sections of Northern Hemisphere zonally averaged westerly component of wind error. Ensemble average over the 31 January cases and time average over the second five days of forecast validity (days 5 to 10). Units of mean error are msec^{-1} OPO: operational orography; ENV: envelope orography.

of wind error and by its zonal component alone shown in Fig. 10a and b. Even larger effects can be seen in plots, similar to Figs. 8, 9 and 10, but for the stationary part of the fields only (an example is given in Fig. 10c). Since the enhancement of orography has no direct bearing on mean zonal quantities, in a completely linear atmosphere it would be virtually impossible to bring about a change in the mean zonal transport properties by a modification of the lower boundary condition. It is well known however,[3, 4]

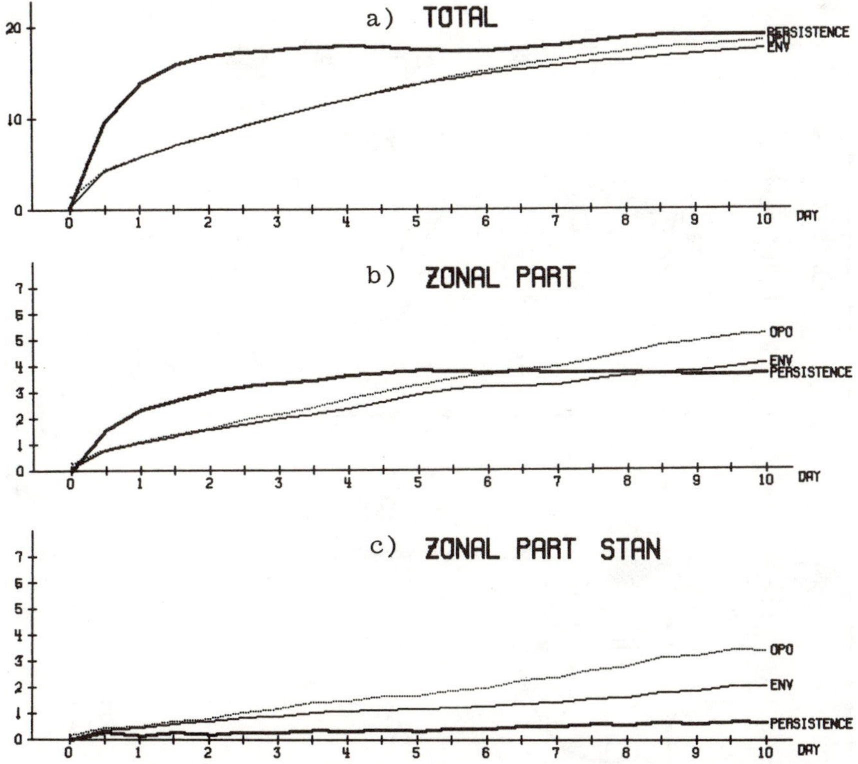

Fig. 10 Mean tropospheric (1000-200mb) standard deviation from observation of wind vector as a function of forecast period. Area mean over the NH, 20° to 82.5°. Units are $msec^{-1}$. (a) total field; (b) zonal part only; (c) zonal part, standing waves only.

that the mean zonal flow is maintained by a mainly barotropic, non-linear feed back of Kinetic Energy from the eddies. It is, therefore, a substantial modification of this conversion process that we expect to see in the model's diagnostics. Fig. 11 shows the barotropic energy conversion between the zonal flow and all the eddies averaged over the second half (Day 5 to Day 10) of the 31 10-day forecasts in January 1981. One should remember that here

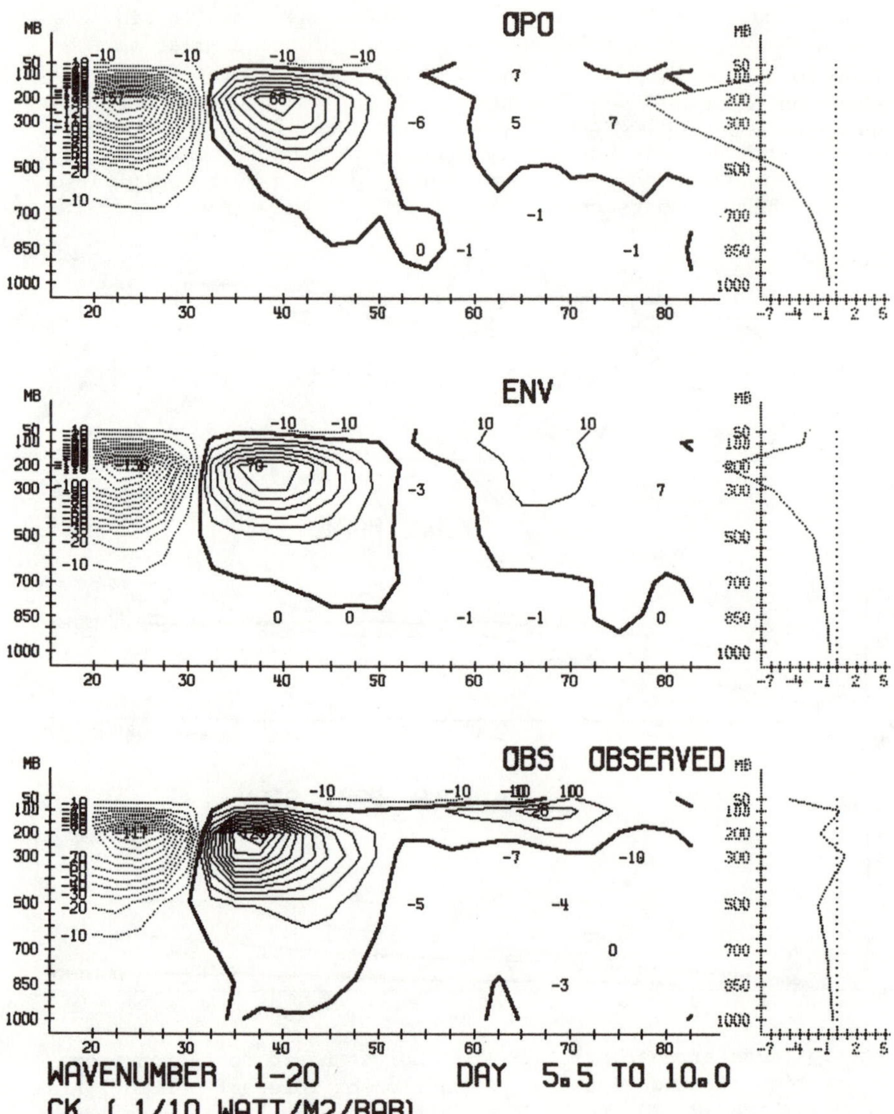

Fig. 11 Latitude-height cross-sections of barotropic conversion of kinetic energy from the zonal flow to the eddies averaged over the 31 cases and over the second five days of the forecast period. All wavenumbers from 1 to 20 are included. On the right the integrals over latitude of the same quantity are shown as a function of height only. Units are 10^{-1} Wm^{-2} bar^{-1} throughout. OPO: operational orography; ENV: envelope orography; OBS: analysed values.

positive values imply a transfer of Kinetic Energy from the zonal flow to the eddies. The northern hemispheric conversion integrals shown on the right as a function of pressure indicate that maximum negative conversions occur in the analyzed troposphere (bottom diagram) at 500 and 200 mb, with almost zero or slightly positive minima at 300 and 100 mb. The control model atmosphere (top diagram) shows a large negative maximum at 200 mb, more than 10 times larger than observed. Since the net integrated balance results from the difference between the two large maxima of opposite sign around 200 mb at about 25°N and 37°N, one expects a high degree of sensitivity in such a diagnosed quantity ; however, the total model error is of high proportion indeed. The envelope ensemble, though, reduces the error to about 40 % of the total spurious negative (from the waves to the zonal flow) conversion. This is, therefore, coherent with a reinforcement of planetary waves amplitude at the expense of a spuriously too westerly mean zonal flow.

This is not to say that local maxima of zonal KE have been reduced. Fig. 12 in fact shows that both the zonal KE and the eddy KE were, in the control run, displaying too weak upper-tropospheric maxima (358 and 288 KJ m^{-2} bar^{-1} instead of the observed 472 and 338 for the eddies and 898 instead of the observed 1089 for the zonal flow). All these discrepancies are improved upon by the use of the envelope orography, although a spectral breakdown of the eddy KE shows that too much KE is being accumulated in WN 1 to 3 and too little in WN 4 to 9, suggesting a possible overshooting effect in the "envelope" ensemble.

An analysis of the spectral distribution of KE at various tropospheric levels for the NH mid-latitude belt averaged during the second five days of the ensemble of forecasts (see Fig. 13) shows that the enhanced orographic forcing restores the correct spectral "signature" amongst the ultra-long waves (WN 1 to 4), and even more so for stationary waves only (see Fig. 14). The effect appears to be bigger at lower levels (but note the different scales of the y-axes). An "overshooting" effect of the envelope orography is evident both in the 300 mb KE spectra (Figs. 13d and 14d) and even more so in the WN band 4 to 12 for the total spectrum (Fig. 13d), again supporting the idea of an excessive correction of this lack of forcing.

SOME THOUGHTS ABOUT PREDICTABILITY

For the purpose of this discussion, it will be useful to distinguish clearly between theoretical (ultimate) predictability of the atmosphere and practical forecasting skill of a given forecasting model, as it can be measured by some particular objective skill score. Which skill score one has in mind is not important at this stage, but practical examples being used at ECMWF are : the time it takes for the correlation coefficient of geopotential height (or of temperature) anomaly (field minus its own climatological value), averaged over a given forecasting volume, to reach the 60 % value (see Fig. 8a) ; the time it takes for the standard deviation

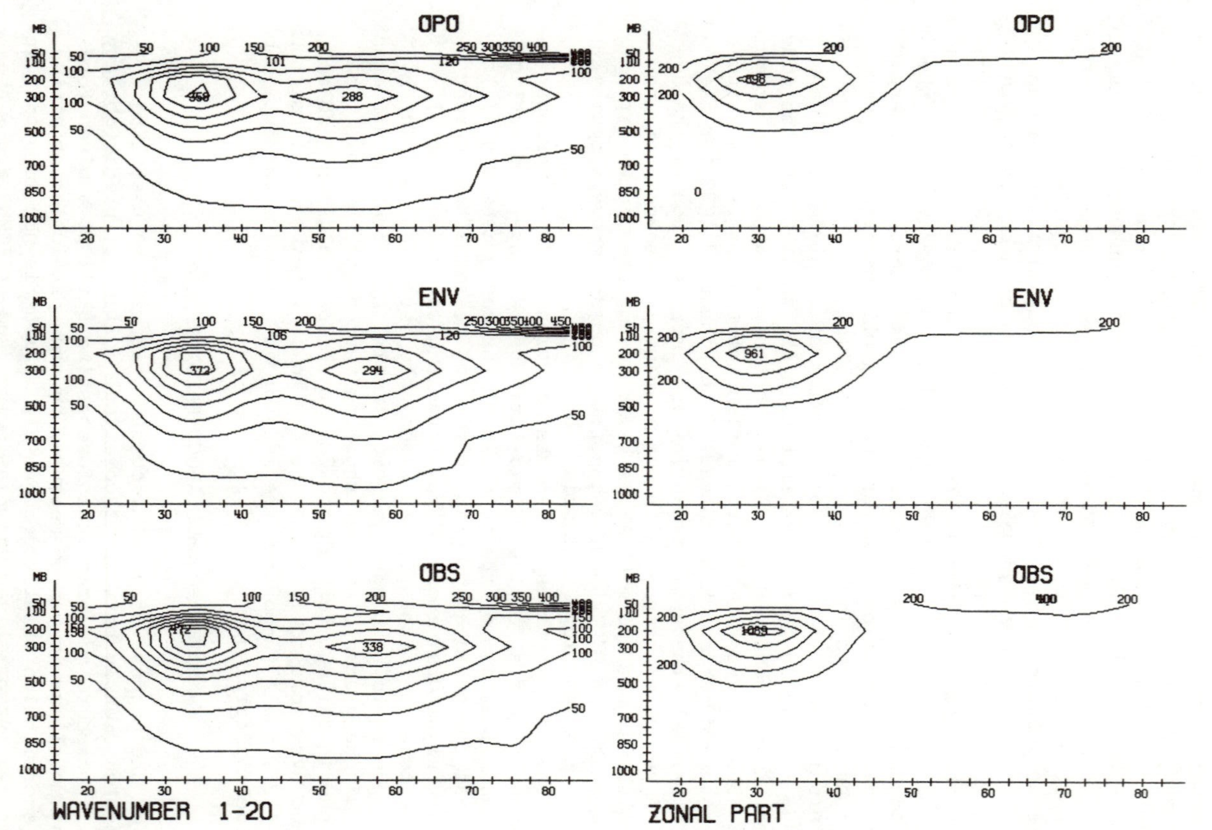

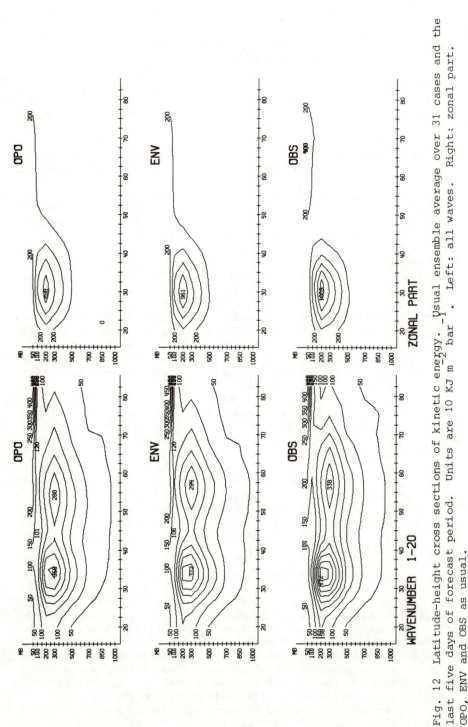

Fig. 12 Latitude-height cross sections of kinetic energy. Usual ensemble average over 31 cases and the last five days of forecast period. Units are 10 KJ m^{-2} bar^{-1}. Left: all waves. Right: zonal part. OPO, ENV and OBS as usual.

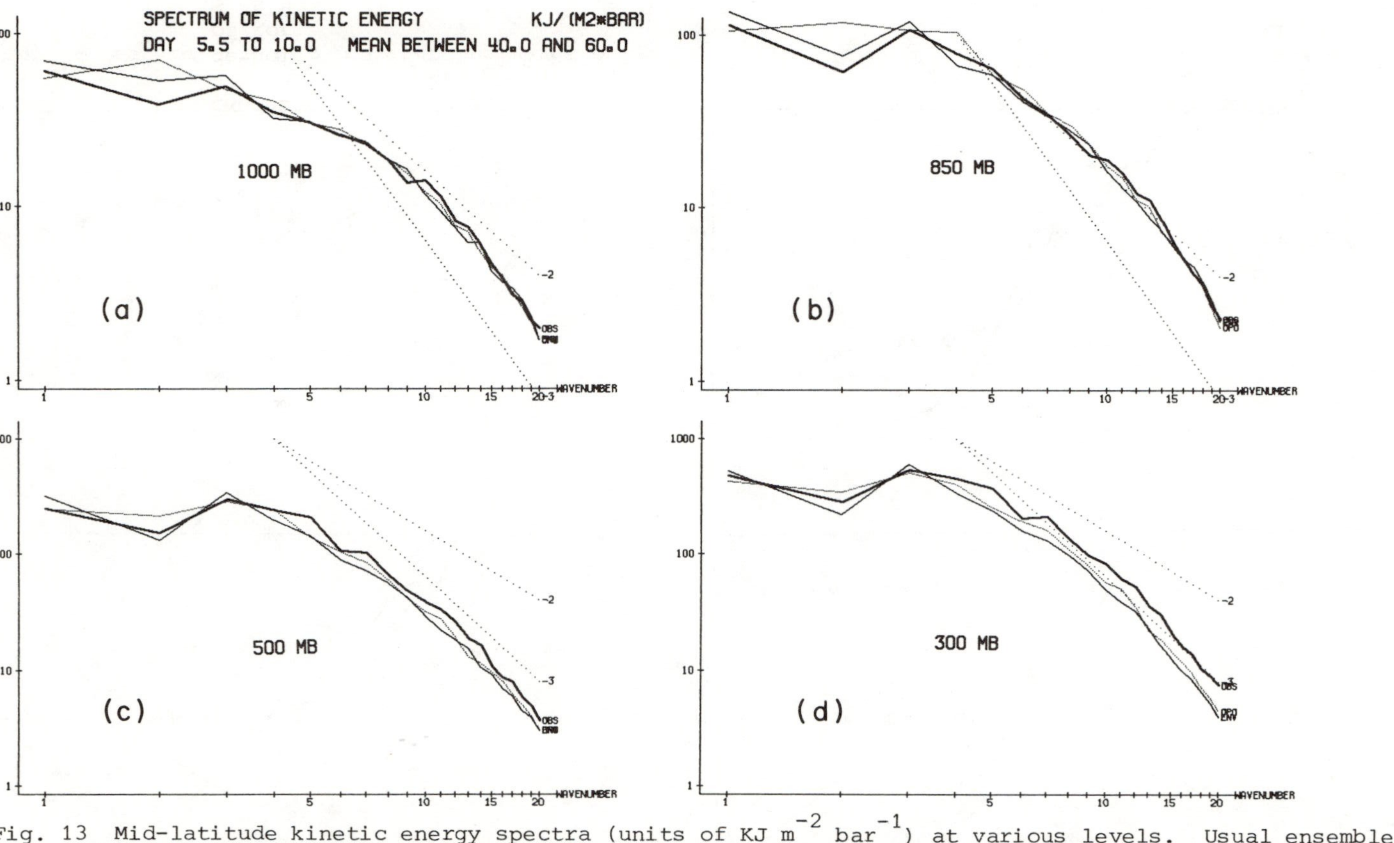

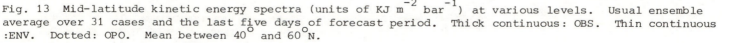

Fig. 13 Mid-latitude kinetic energy spectra (units of KJ m^{-2} bar^{-1}) at various levels. Usual ensemble average over 31 cases and the last five days of forecast period. Thick continuous: OBS. Thin continuous :ENV. Dotted: OPO. Mean between 40° and 60°N.

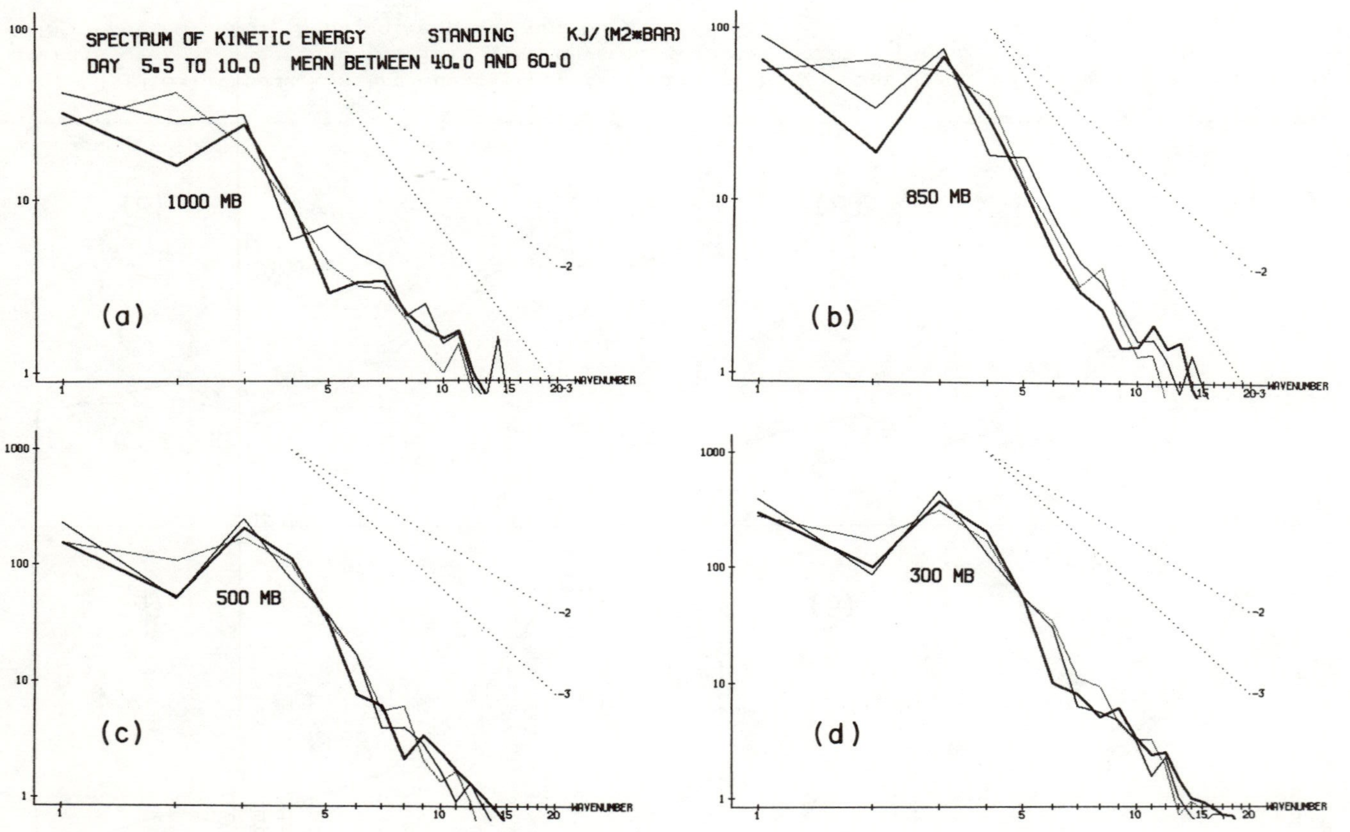

Fig. 14 As Fig. 13, but for standing waves only.

from observed values of geopotential height, or of temperature, to reach the natural variability level for that particular quantity (the "norm", see Fig. 8b to f), or any combination of these. One could then, having chosen a suitable objective measure, imagine this measure of a model's skill increasing in time, as more and more improvements are put into the model and more and more errors are eliminated ; one would probably see the forecasting skill measure progressingly levelling off as improvements become more and more hard to achieve. The asymptotic value of such a climb could be defined as the predictability of that particular model's atmosphere.

An example of such a diagram is shown in Fig. 15. The objective measure used there, devised by J. F. Geleyn at ECMWF, is described in the Appendix. An eyeball estimate of the asymptotic value of the 12-month running mean puts it at a few hours above six days. Of course, one then thinks of the real atmosphere's predictability (measured in any one of the possible objective ways) as the upper bound of all possible model atmospheres' predictability (measured in the same way, of course).

Diagrams like the one of Fig. 15 are interesting because they quantify the progress in numerical weather forecasting and they provide an estimate of an upper limit for model's fine "tuning". They also, however, pose several interesting questions about their more explicit features, like the marked seasonal cycle (why is winter objectively more predictable than summer ?) or the inter-annual variabilities more evident once the upward trend due to model changes levels out.

The black dots in correspondence to January 1981 represent, from the lower to the higher, the old operational January 1981, the January 1981 rerun with the August 1982 operational forecasting system and the January 1981 with the envelope orography. The lower dot does not coincide with the thin line intersect, as one would expect, because all value represented by black dots were obtained scoring January 1981 sets of forecasts against a set of analyses obtained using the envelope orography in the assimilating model, rather than against the operational analyses, as it is true for the values represented by the lines. This indicates the degree of sensitivity that these objective scores have with respect to the analysis used.

These results confirm that the use of an envelope-type orography can increase such an objective measure of the ECMWF's model atmosphere predictability of around 6 to 8 hours. Even having a pessimistic expectation of the improvement during the summer season (and expecting no improvement at all) one would then have raised the predictability limit for this model (grid point, 1.875° regular lat-lon grid, 15 levels in the vertical, elaborate physical parametrisation schemes) by something around 3 and 4 hours.

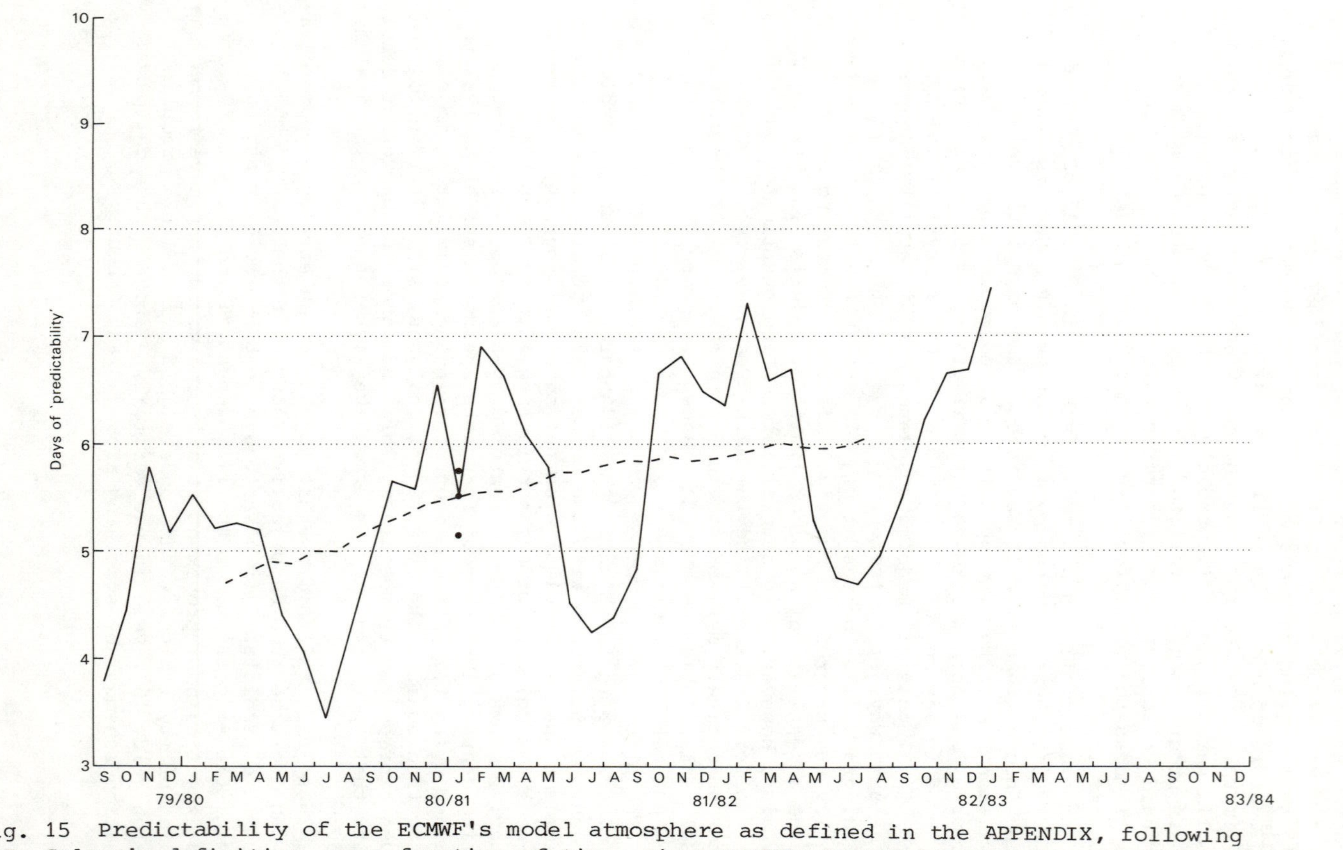

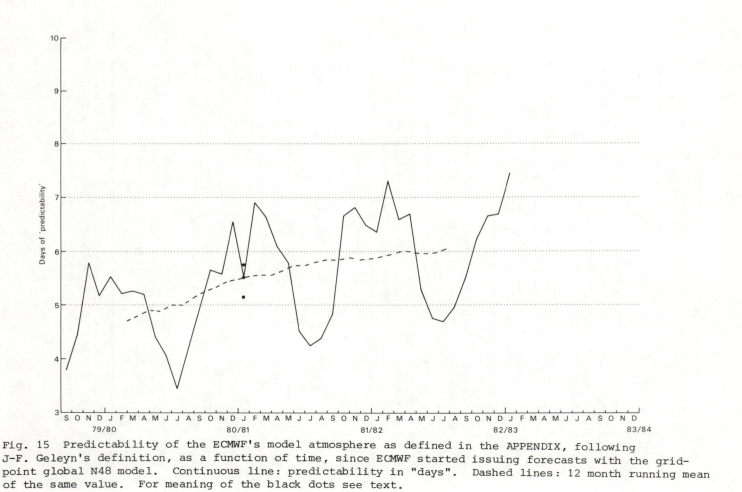

Fig. 15 Predictability of the ECMWF's model atmosphere as defined in the APPENDIX, following J-F. Geleyn's definition, as a function of time, since ECMWF started issuing forecasts with the grid-point global N48 model. Continuous line: predictability in "days". Dashed lines: 12 month running mean of the same value. For meaning of the black dots see text.

CONCLUSIONS

The use of an enhanced envelope-type orography in the ECMWF operational forecasting GCM has been shown to have benficial impact on almost all fields and diagnosed quantities, with the highest impact being shown on the zonal flow and on the ultra-long low-frequency planetary waves. Since these tend to dominate the objective measures of forecasting skill, large improvements are found in such quantities as geopotential height anomaly correlation coefficients and standard deviation form observed values.

It was also found that the considerable improvement in mean zonal quantities is consistent with the idea that the balance of zonal KE is controlled by essentially barotropic, non-linear interactions between the zonal flow and the large-scale eddies, in turn strongly dependent upon orographic forcing.

The effect of such envelope-type orography on the predictability of the ECMWF model atmosphere is to increase an objectively derived measure of it by a few hours, compared with the previously attained value of around 6 days.

AKNOWLEDGEMENTS

Several people were of great assistance with various parts of the ECMWF forecasting system : amongst others R. Strüfing with running the model, B. Norris with operational data assimilation and K. Arpe with model diagnostics.

Stimulating discussions are gratefully aknowledged with M. Wallace, J.F. Geleyn and A. Simmons. The encouragement and support of D. Burridge throughout the work was greatly appreciated.

REFERENCES

1. J. M. Wallace, S. Tibaldi and A. J. Simmons, Reduction of systematic forecast error in the ECMWF model through the introduction of an envelope orography. In the proceedings of the ECMWF Workshop on "Intercomparisons of large-scale Models used for Extended Range Forecasts", Reading 30 June - 2 July 1982, pp. 371-434.
2. W. L. Grose and B. J. Hoskins, J. Atmos. Sci., 223 (1979).
3. B. Saltzman and A. Fleisher, Tellus, XII, 374 (1960).
4. E. Oriol, Energy Budget Calculations at ECMWF part I, ECMWF TR n° 35 (1983), 113 pp.

APPENDIX

J. F. Geleyn's model atmosphere "predictability" score

Monthly means of daily Northern Hemisphere averages of the anomaly correlation (AC) and the standard deviation (FSD) of forecast geopotential height (Z) and temperature (T) errors are used to define a meausre, G, of the model's atmosphere predictability according to the following prescription :

i) a mean score $\bar{S}$ is formed by averaging the following four scores :

$$S_1 = 1 - AC_Z$$

$$S_2 = (FSD_Z / PSD_Z)^2$$

$$S_3 = 1 - AC_T$$

$$S_4 = (FSD_T / PSD_T)^2$$

Therefore :

$$\bar{S} = 1/4 \sum_{1i}^{4} S_i$$

PSD indicates the standard deviation of the error of a persistence forecast and suffices Z and T indicate geopotential height and temperature respectively.

The four scores are calculated as averages for the volume :

$$82.5^{\circ} N \div 20^{\circ} N$$

and

1000-200 mb for Z

or

850-200 mb for T

ii) Values of $\bar{S}$ are obtained for each of the first seven forecast days :

$$\bar{S}_j \quad j = 1,7$$

iii) A straight line is fitted to these points

$$\hat{S} = at$$

where t is the forecast time in days and

$$a = \frac{1}{7} \sum_{1}^{7}{}_j \frac{S_j}{t_j} \qquad t_j = 1,7 \text{ days}$$

iv) $\hat{S} = 0.5$ is taken as the limit of skillfull forecast, then predictability is defined as :

$$G = \frac{0.5}{a}$$

PREDICTABILITY AND ALMOST INTRANSITIVITY IN A BAROTROPIC BLOCKING MODEL

Richard E. Moritz
Polar Science Center
University of Washington, Seattle, WA 98105

ABSTRACT

A truncated spectral model of barotropic flow over large scale orography is perturbed with a Wiener process in order to simulate the effects, on the large scale motion, of physical processes and scales of motion neglected in the model. For a range of the control parameters, the probability distribution for the zonally-averaged part of the flow is bimodal, and the system alternates between neighborhoods of low and high zonal index equilibrium states. Probability distributions for the large scale wave components of the flow are unimodal and time dependent realizations show that the phase of the wave, relative to the large scale orography, is substantially more variable during low index episodes than are observed phases in actual atmospheric blocking events.

With the aid of a partition of variance, the predictability of the model is broken into two parts: predictability of departures from the high or low index equilibria on short time scales, and predictability of transitions between the neighborhoods of alternate equilibrium states. For the zonally averaged part of the flow, the range of predictability is determined by the distance, in phase space, between alternate equilibrium states, the persistence times in their respective domains of attraction, and the asymptotic stability properties of the equilibria. Time average statistics exhibit almost intransitivity, i.e. they depend sensitively on the initial conditions. For the zonally averaged flow, the mean value is almost intransitive. In the case of the wave components, the variance is the almost intransitive statistic.

INTRODUCTION

The predictability of the atmosphere has been defined as "the extent to which it is possible to predict it with a theoretically complete knowledge of the physical laws that govern it"[1]. If the concept of complete knowledge implies deterministic rules for atmospheric time evolution, then the only remaining source of unpredictability is error in the initial conditions. A somewhat broader definition might include amongst the physical laws nondeterministic rules whose statistical properties are known completely. We shall adopt the latter definition, since the problem of formulating perfect prediction rules, governing simultaneously phenomena as various as planetary waves, thunderstorms and gusts of wind in the surface layer, seems no more nor less irreducible than the problem of observing these phenomena at time $t = 0$. In any case, it is well known that calculations of flow predictability depend on the forecasting problem at hand, and on the prediction model used to solve the problem.

Several estimates of theoretical predictability limits are based on statistical closure hypotheses applied to two dimensional, homogeneous, isotropic turbulence, or 2DT for short.[2,3,4] For planetary wavenumbers n approximately in the range $10 < n < 20$, observational estimates of atmospheric kinetic energy spectra are not inconsistent with the shape of the spectrum in the enstrophy cascading inertial range predicted by 2DT theory.[3] However, much of the variability observed in the atmosphere occurs in the wavenumber band $n < 10$.[5] On these scales the time averaged statistics of middle latitude circulation systems are influenced decisively by the geographical distribution of heat and momentum sources and sinks, as well as by the sphericity of the earth. These factors introduce anisotropy and inhomogeneity into the statistics of the large scale flow, as illustrated by the appearance of forcing, dissipation and "beta" terms that vary with position and coordinate orientation in the equations for the time average flow.[6]

One manifestation of such effects consists in the finding that highly truncated spectral models of barotropic flow over large scale mountains can possess two or more stable equilibrium configurations, corresponding qualitatively to the so-called blocking and high zonal index patterns recognized by synoptic meteorologists.[7-10] In subsequent research, the effects of the truncated, smaller scales of motion were simulated by adding stochastic perturbations to a model that retained only three deterministic modes.[11] It was found that predictability varied with the location of the initial condition in the model phase space. In closure models of 2DT this kind of dependence is ruled out by imposing homogeneity on the statistical moments.[3]

In the present work, we extend the study of the three component model with stochastic perturbations to include time dependent and statistical properties of long realizations, generated by numerical integration. First we ask: Do the time dependent solutions display qualitative similarities to episodes of blocking in the atmosphere? Viewing the three component model as a simple prototype for planetary scale flow that is strongly anisotropic and inhomogeneous, the range of predictability is investigated, with reference to the biomodality of the steady state probability distribution and the persistence of the system near alternate equilibrium states. Numerical simulation results are also compared with results from an approximation that reduces the three component model to a single stochastic differential equation governing the zonally averaged flow.[12]

THE BAROTROPIC MODEL

Our study is based on the potential vorticity equation

$$\partial_t \nabla^2 \psi + J(\psi, \nabla^2 \psi + f_0 \frac{H}{H} + \beta y) = - f_0 k \nabla^2 (\psi - \psi^*) \quad (1)$$

governing the quasigeostrophic flow of a homogeneous fluid in a periodic beta-plane channel of length πL, width πL, and average depth H. The details of the derivations of (1) can be found in Reference 7 and will not be repeated here. $\psi(x,y,t)$ is a stream function for the geostrophic flow, f_0, is the Coriolis parameter, k is a frictional dissipation constant, ψ^* is a steady, external forcing, and J is the Jacobian operator. The channel walls $y = 0, \pi L$ are required to be flow streamlines and a rigid lid condition is imposed at the top boundary. On the lower boundary, the flow must parallel the large scale orography whose height is

$$h(x,y) = h_0 \cos\left(\frac{2x}{L}\right) \sin\left(\frac{y}{L}\right). \quad (2)$$

We shall study the three mode truncation[7]

$$\psi(x,y,t) = \psi_A(t) F_A(y) + \psi_K(t) F_K(x,y) + \psi_L(t) F_L(x,y) \quad (3)$$

and apply forcing only to the zonally averaged component of the flow

$$\psi^* = \psi_A^* F_A(y), \quad (4)$$

where

$$F_A(y) = \sqrt{2}\cos\left(\frac{y}{L}\right)$$

$$F_K(x,y) = 2\cos\left(\frac{2x}{L}\right) \sin\left(\frac{y}{L}\right) \quad (5)$$

$$F_L(x,y) = 2\sin\left(\frac{2x}{L}\right) \sin\left(\frac{y}{L}\right).$$

F_A describes a zonal average wind with maximum in the center of the channel, while F_K and F_L describe vortices centered over and in quadrature with the ridges of the orography (2), respectively. The variables t, (x,y) and ψ are nondimensionalized with f_0, L and L^2f_0, respectively. Substitution of (3) and (5) in (1), multiplication by each of the functions appearing in (5) and integration over the area of the channel then yields the following system of three coupled ordinary differential equations

$$\dot{\psi}_A = -k(\psi_A - \psi_A^*) + h_{01}\psi_L \tag{6a}$$

$$\dot{\psi}_K = -k\psi_K - b_{21}\psi_L \tag{6b}$$

$$\dot{\psi}_L = -k\psi_L + b_{21}\psi_K - h_{21}\psi_A \tag{6c}$$

where

$$\begin{aligned} b_{21} &= \alpha_{21}\psi_A - \beta_{21} \\ h_{01} &= 1.2\frac{h_0}{H} \\ \alpha_{21} &= 1.92 \\ \beta_{21} &= \frac{L}{a}\cot(\phi_0) \\ h_{21} &= 0.24\frac{H_0}{H} \end{aligned} \tag{7}$$

the dot signifies a time derivative, a is the radius of the earth, and ϕ_0 is the latitude at mid-channel. In (6) linear terms proportional to k represent frictional dissipation of vorticity. The two orographic terms describe local vorticity tendencies due to the stretching and shrinking of vortex tubes as they pass over the valleys and ridges. The b_{21} terms represent advection of relative vorticity by the zonally averged flow ψ_A, and of planetary vorticity by the wave components ψ_K and ψ_L. For a given zonal flow ψ_A, we may set $k = h_0 = 0$, so that (6b,c) can be combined in the form

$$\ddot{\psi}_L = -b_{21}^2\psi_L \,. \tag{8}$$

Solutions to (8) are Rossby wave oscillations, whose phase in the channel is fixed for the particular frequency

$$b_{21} = \alpha_{21}\psi_A - \beta_{21} = 0 \,, \tag{9}$$

as given by the strength of the zonal flow ψ_A. With nonzero h_0, (6b) and (6c) may be solved for the steady state value of ψ_L as a function of the zonal flow. As $b_{21} \rightarrow 0$, the stationary wave given by (9) is forced resonantly by the stationary orographic undulations at the same scale, and the equilibrium response blows up ($\psi_L \rightarrow -\infty$) when $k = 0$. When $k > 0$, this resonance manifests itself as an extremum in the response curve, plotted on Figure 1 in the (ψ_A,ψ_L) plane. Multiple equilibria of the full system (6) occur in parameter

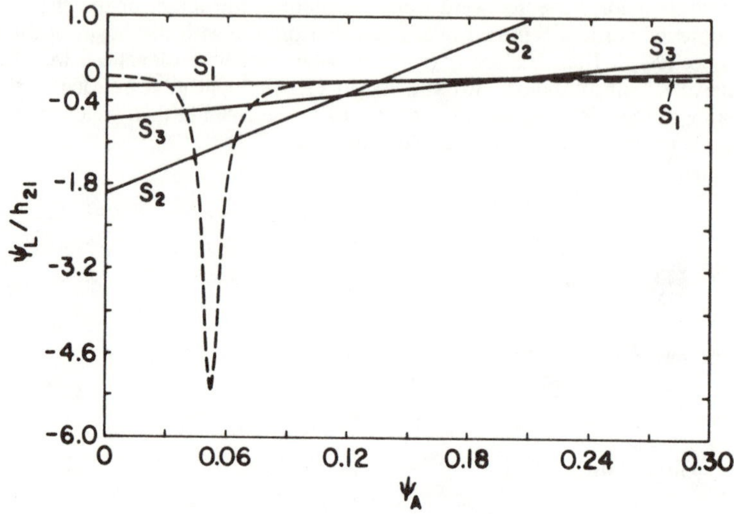

Fig. 1. Equilibrium response curve (dashed) and zonal flow equilibrium lines (solid) for cases S_1, S_2, S_3.

ranges wherein the straight lines defined by (6a), in the steady state, intersect the response curve more than once. The straight lines portray the balance among forcing, dissipation and mountain torque that determines the equilibrium zonal flow.

The equilibrium flow patterns depend on several parameters. Hence we follow earlier work[7,12] and fix $f_0 = 10^{-4}s^{-1}$, $L/a = 0.25$ and $\phi_0 = 45^0$. The remaining parameters are k, ψ_A^*, and h_0/H. The most appropriate values for dissipation and forcing in a large scale barotropic model are always subject to some uncertainty. The rough analogy between a vertically stratified, semi-infinite atmosphere and a homogeneous fluid of depth H also leaves doubt as to the best choice for h_0/H. The parameter sets $S_1 = \{k = 0.01,\ \psi_A^* = 0.20,\ h_0/H = 0.20\}$, $S_2 = \{k = 0.01,\ \psi_A^* = 0.14,\ h_0/H = 0.05\}$ and $S_3 = \{k = 0.01,\ \psi_A^* = 0.20, h_0/H = 0.10\}$, that correspond to curves plotted on Figure 1, have received detailed study in previous works.[7,12] These curves show that each of the parameter sets gives rise to three equilibria, two of which turn out to be asymptotically stable. The stable equilibria are distinguished by their different values for the zonal flow, ψ_A, that fall on either side of the intermediate, unstable state, Equilibrium streamline patterns for S_1, as given by (3), are shown elsewhere,[7] Figure 2 shows the patterns for case S_2. Differences between the two stable flows in all three cases are qualitatively similar to contrasts associated with blocking versus high index situations in the atmosphere. In the blocking state the zonal flow is relatively weak and the wave components are well developed, with a pronounced ridge to the west of the orographic ridge. The high index state has stronger zonal flow, with a weaker wave perturbation that is nearly in phase with the mountain ridge. The multiple flow equilibrium theory[7] hypothesizes that time dependent flows resembling the stable patterns in the truncated model will tend to persist, in analogy with the persistence of blocking and high index episodes.

In the truncated model (6) time dependent solutions for cases S_1, S_2, S_3, will approach one or the other stable steady state, depending on the initial condition, and time evolution then ceases. A continual sequence of transitions between neighborhoods of the alternate stable equilibria can be induced by appending a random forcing term to the right side of

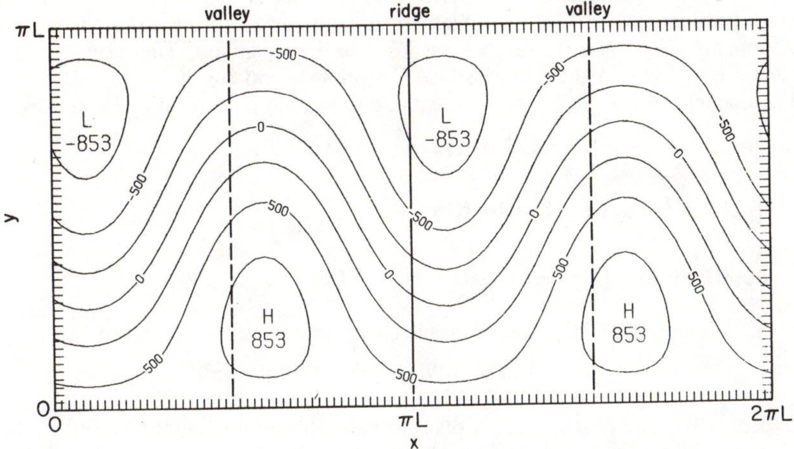

Fig. 2a. Equilibrium streamlines for blocking, case S_2. Orography in phase with solid line, contour interval 0.025.

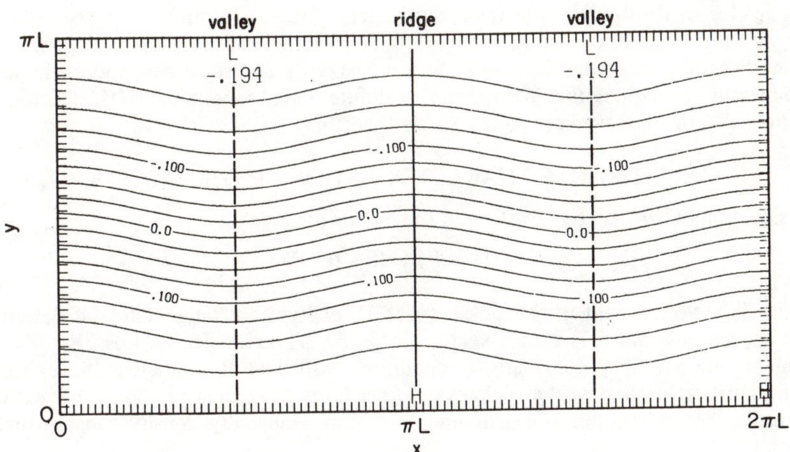

Fig. 2b. Same as 2a but for high index equilibrium.

(6)[11-14]. We shall parameterize the perturbation effects of scales of motion and physical processes not included in (6), but which occur nonetheless in the atmosphere, by adding the increments dW(t) of a Wiener process W(t) to the equations. The result is a stochastic differential equation. In the simplest case,[12] stochastic and steady forcing act only on the zonal component of flow $\psi_A(t)$, and the equations, written now in Itô form, become

$$d\psi_A(t) = \{-k(\psi_A(t) - \psi_A^*) + h_{01}\psi_L(t)\}dt + \epsilon^{1/2}dW(t) \tag{10a}$$

$$d\psi_K(t) = \{-k\psi_K(t) - b_{21}(t)\psi_L(t)\}dt \tag{10b}$$

$$d\psi_L(t) = \{-k\psi_L(t) + b_{21}(t)\psi_K(t) - h_{21}\psi_A(t)\}dt \tag{10c}$$

where the increments dW(t) are zero mean, Gaussian quantities with covariance

$$E\{dW(t)dW(t+\tau)\} = \delta(\tau)dt\ . \tag{11}$$

The expectation value E is evaluated over an ensemble of realizations of W(t). If (10) is regarded as a model for actual atmospheric variablity, then the propriety of including some random term is not really an issue.[14] The factor ϵ scales the variance of the perturbations, relative to the deterministic part of the model, and as $\epsilon \rightarrow 0$, sample realizations of (10) will follow the solutions of the deterministic model (6) with overwhelming probability.[15] Still, an important question remains: why choose the Wiener process from among all possible stochastic processes? A pragmatic advantage of this choice is that a substantial mathematical theory exists for the treatment of stochastic differential equations.[16] In this study, we adopt the attitude that more complicated stochastic forcing can be studied after the simpler cases are better understood. The physical prototype for W(t) is,, of course, the Brownian motion.

Let us note that a sample realization of (10), starting from a given initial condition $\underline{X}_0 = \{\psi_A(0),\psi_K(0),\psi_L(0)\}'$, will trace a trajectory $\underline{X}(t,\underline{X}_0)$ through the three dimensional phase space Γ that has axes $\underline{X} = \{\psi_A,\psi_K,\psi_L]'$. Vector transposes are denoted here by primes. Solution properties for the system are given a statistical description, in terms of ensembles of trajectories that correspond to different realizations of W(t), different initial conditions $\underline{X}_0$, or time averages over a single trajectory.

STATISTICAL CHARACTERIZATION OF THE PROCESS

Equations (10) may be rewritten in the vector form

$$d\underline{X}(t) = \underline{F}(\underline{X})dt + \underline{i}\epsilon^{1/2}dW(t) \tag{12}$$

where $\underline{F}$ is a vector field in the phase space Γ, everywhere tangent to the deterministic (ϵ=0) trajectories, and $\underline{i}$ is a unit vector along the ψ_A axis. The process $\underline{X}(t,\underline{X}_0)$ whose increments are given by (10) satisfies conditions sufficient to guarantee the existence of certain statistical properties that guide our procedures in evaluating the numerical simulation results.[16,17] First, there is a unique transition probability density function $p(\underline{X},t\underline{X}_0)$ such that

$$P\{\underline{X}(t) \in D \mid \underline{X}(0) = \underline{X}_0\} = \int_D p(\underline{\xi}t,\underline{X}_0)d\underline{\xi} \tag{13}$$

is the probability that a sample trajectory, starting from $\underline{X}_0$, will be found in a subset D of points in phase space, at time $t \geqslant 0$. $\underline{X}(t,\underline{X}_0)$ is a diffusion process and so ultimately visits all points in Γ, and an asymptotic probability density

$$p_\infty(\underline{X}) \equiv \lim_{t\rightarrow\infty} p(\underline{X},t,\underline{X}_0) \tag{14}$$

is established, independent of the initial condition. For a function $g(\underline{X})$, transient and asymptotic expectation values are

$$E_{\underline{X}_0,t}(g) \equiv \int_T g(\underline{X})p(\underline{X},t,\underline{X}_0)d\underline{X} \tag{15}$$

$$E_\infty(g) \equiv \int_T g(\underline{X})p_\infty(\underline{X})d\underline{X} \, . \tag{16}$$

A time average is defined by

$$\bar{g}_{\underline{X}_0,T} \equiv \frac{1}{T}\int_0^T g(\underline{X}(t,\underline{X}_0))dt \, . \tag{17}$$

The process $\underline{X}(t,\underline{X}_0)$ is transitive (ergodic) so that, with probability equal to 1

$$\lim_{T\to\infty} \bar{g}_{\underline{X}_0,T} = E_\infty(g) \, . \tag{18}$$

Thus for sufficiently large T, time average statistics computed from a single realization can be used to estimate the invariant statistical properties of the system (10). We shall adopt this estimation approach below. What is a sufficiently large value of T? Certainly T must exceed substantially all of the characteristic time scales on which $\underline{X}(t,\underline{X}_0)$ varies, for otherwise the single trajectory would, with high probability, fail to visit some regions of phase space often enough (or at all) and sampling errors $\bar{g}_{\underline{X}_0,T} - E_\infty(g)$ would be large. Time scales T_1 that characterize linear physical processes in (10) are k^{-1} (viscous dissipation), $(b_{12})^{-1}$ (Rossby wave period) and ϵ^{-1} (time scale on which the process $\psi_A(t)$ deviates from its expected value). The nonlinear terms in (10) may introduce additional time scales here denoted by T_2. For instance, numerical simulation results reported by others[12] show that the phase point $\underline{X}(t)$ typically remains in some neighborhood of a given stable equilibrium for a time $T_2 >> T_1$, before undergoing a rapid transition to a neighborhood of the other stable equilibrium state. Such behavior has been dubbed "almost intransitive"[18], and can be defined by the condition that time averages $\bar{g}_{\underline{X}_0,T}$ computed from long but finite averaging intervals T may vary greatly for different choices of the initial condition. Still according to (18) the system is transitive, and there must be some longest time scale, say T_3, such that for $T > T_3$, $\bar{g}_{\underline{X}_0,T} \simeq E_\infty(g)$. Therefore a system cannot be almost intransitive unless there exist averaging intervals that are at once "long" ($T >> T_1$) and "short" ($T << T_3$). Obviously T_2 is such a time scale.

Time evolution of the probability density $p(\underline{X},t,\underline{X}_0)$ is given by a Fokker-Planck equation

$$\partial_t p + \underline{\nabla} \cdot (\underline{F}p) = \frac{\epsilon}{2}\frac{\partial^2 p}{\partial \psi_A^2} \tag{19}$$

where the gradient operator is evaluated with respect to phase coordinates $\underline{X}$ and $p(\underline{X},0,\underline{X}_0) = \delta(\underline{X} - \underline{X}_0)$ is the initial condition. For a three component model essentially like (10), numerical solutions of (19) appeared to converge, for large t, on a bimodal distribution[11]. The occurrence of more than one probability maximum, centered on the stable equilibrium states, is another manifestation of almost intransitivity. Estimates of predictability are often judged by comparison with the asymptotic variance of the stable variables, i.e

$$\mathrm{var}(\underline{X}) \equiv E_\infty(\underline{X}'\underline{X}) - E_\infty(\underline{X}')E_\infty(\underline{X}). \tag{20}$$

However, variance alone cannot provide an adequate characterization of a bimodal distribution. A handy parametric description can be obtained by partitioning the variance into contributions due to fluctuations within the neighborhoods of blocking (B) or high index (H) equilibrium states, and due to transitions between these two neighborhoods. Formally, we consider mutually exclusive subsets B and H (to be defined later) of the phase space Γ, such that $B \cap H = \phi$ (the null set) and $B \cup H = \Gamma$. Indicator functions can be defined as follows

$$I_B(\underline{X}) \equiv \begin{cases} 0 \text{ if } \underline{X} \notin B \\ 1 \text{ if } \underline{X} \in B \end{cases}$$

$$I_H(\underline{X}) \equiv 1 - I_B(\underline{X}) \quad . \tag{21}$$

The random events $\underline{X} \in B$ and $\underline{X} \in H$ occur with probabilities

$$P\{\underline{X} \in B\} = \int_\Gamma I_B(\underline{X}) p_\infty(\underline{X}) d\underline{X}$$

$$P\{\underline{X} \in H\} = \int_\Gamma I_H(\underline{X}) p_\infty(\underline{X}) d\underline{X} \tag{22}$$

so we have the conditional probabilities $p_\infty(\underline{X}|B) = p_\infty(\underline{X}) I_B(\underline{X}) / P\{\underline{X} \in B\}$ and $p_\infty(\underline{X}|H) = p_\infty(\underline{X}) I_H(\underline{X}) / P\{\underline{X} \in H\}$. Conditional expectations $E_B(g)$ and $E_H(g)$ are defined just as in (16), but with the conditional densities in place of $p_\infty(\underline{X})$. In particular, the conditional variances are

$$\text{var}_B(\underline{X}) \equiv E_B(\underline{X}'\underline{X}) - E_B(\underline{X}')E_B(\underline{X})$$

$$\text{var}_H(\underline{X}) \equiv E_H(\underline{X}'\underline{X}) - E_H(\underline{X}')E_H(\underline{X}) \ . \tag{23}$$

It is then easy to establish, from the definitions above, that

$$\text{var}(\underline{X}) = P(\underline{X} \in B)\text{var}_B(\underline{X}) + P(\underline{X} \in H)\text{var}_H(\underline{X}) + \text{var}_T(\underline{X}) \tag{24}$$

where

$$\text{var}_T(\underline{X}) \equiv P(\underline{X} \in B)\{(E_B(\underline{X}) - E_\infty(\underline{X}))'(E_B(\underline{X} - E_\infty(\underline{X}))\}$$
$$+ P(\underline{X} \in H)\{(E_H(\underline{X}) - E_\infty(\underline{X}))'(E_H(\underline{X}) - E_\infty(\underline{X}))\} \tag{25}$$

is the transitional variance, and the first two terms on the right in (24) are called local variances. By rewriting $g(\underline{X}(t,\underline{X}_0))$ as

$$g(\underline{X}(t,\underline{X}_0)) = g(\underline{X}(t,\underline{X}_0))\{I_B(\underline{X}(t,\underline{X}_0)) + I_H(\underline{X}(t,\underline{X}_0))\} \ . \tag{26}$$

it is easily shown that the conditional expectations in B and H are approached, as $T \rightarrow \infty$, by the conditional time averages

$$\bar{g}_{B,\underline{X}_0,T} \equiv \frac{1}{T_B} \int_0^T I_B(\underline{X}(t,\underline{X}_0)) g(\underline{X}(t,X_o)) dt$$

$$\bar{g}_{H,\underline{X}_0 T} \ \frac{1}{T_H} \int_0^T I_H(\underline{X}(t,\underline{X}_0)) g(\underline{X}(t,\underline{X}_0)) dt \ . \tag{27}$$

Here T_B and T_H are the total times spent in blocking and high index sets in phase space, respectively, during the interval 0,T. Equations (27) are used below to estimate the variance partition (24) from long term numerical simulation runs. For a bimodal $p_\infty(\underline{X})$ we can, with a proper choice of B and H, parameterize two unimodal distributions using the local terms (24) and their centroids will be near the respective maxima of the bimodal $p_\infty(\underline{X})$, i.e. near the stable equilibrium states. The bimodality itself is then described by (25), whose terms are proportional to the phase space distance between the local expectations and the overall expectation value. The relevance of this partition for the predictability of (10) depends on the existence of widely disparate time scales T_1 and T_2 on which the local and transitional variances are achieved by a realization of the process. For some parameter sets (including now a choice for ϵ) we expect to see predictability of $\mathrm{var}_T(\underline{X})$ at time lags well beyond those at which the local variances are essentially unpredictable.

SIMULATION RESULTS

Realizations of (10) are obtained by numerical integration[14]. The deterministic increments $d\psi_A$, $d\psi_K$, $d\psi_L$ are computed for a small time step dt, and then zero mean, Gaussian pseudo random numbers with variance ϵdt and added to ψ_A. The parameter ϵ is varied between 1.8×10^{-1} and 4.8×10^{-5}. The latter value was adopted in[11,12] and contains the average kinetic energy of the perturbed geostrophic flow to be about twice as large as the kinetic energy in the deterministic steady states.

What are the qualitative properties of the time dependent flow? Plots for the parameter sets S_2, with $\epsilon = 2.8 \times 10^{-5}$, are shown in Figure 3. The initial value was chosen near the blocking equilibrium state, and $\psi_A(t)$ persists in a neighborhood of its low index value for about 800 time units (93 days). A transition occurs to the neighborhood of the high index equilibrium value of ψ_A at about this time, and similar transitions are evident throughout the graph. The equilibrium values of the variables are shown in Figure 3 as horizontal lines. It is obvious that time averages of $\psi_A(t)$ over, say, 100 units (12 days) tend to be quite different, depending on whether the initial value of ψ_A is greater or less than the unstable equilibrium value. It is also clear that the transitions occur on a time scale long compared to the local fluctuations of ψ_A (t) that are induced by the stochastic perturbations, and by interaction with the wave component $\psi_L(t)$. Therefore we may say that the signal $\psi_A(t)$ in Figure 3 is almost intransitive, with respect to its time mean value. One can easily distinguish blocking and high index episodes on plots of the wave components $\psi_K(t)$ and $\psi_L(t)$, also, but in this case the almost intransitivity shows up more in the variability than in the time mean. The oscillations represent Rosby waves that are excited by variations in $\psi_A(t)$ acting through the topography to influence $\psi_L(t)$. One might expect that the "frequency" $b_{21}(t) = \alpha_{21}\psi_A(t) - \beta_{21}$ of these oscillations would decrease as $\psi_A(t)$ shifts from a high index to a blocking state, and higher resolution plots of the wave components (not shown here) confirm this idea. However, the increased amplitude of the wave fluctuations in the blocking mode is somewhat surprising. Observations of blocking episodes in the atmosphere[19,20] suggest that the amplitude and, especially, the phase of the large scale blocking waves tend to persist on time scales that characterize the duration of low index situations. Therefore, the physical balances in (10) that produce multiple *equilibrium* flows (with $\epsilon = 0$) resembling blocking and high index states, fail to produce *time dependent* blocking patterns that persist for more than a fraction of a Rossby period (about two days), when only a moderate level of stochastic forcing is applied. This result appears to be new and suggests that the physics included in (10) are insufficient to provide a qualitative model for observed blocking behavior.

In Figure 4 we show frequency histograms for each flow component, computed from the S_2 simulation run discussed above. The histograms may be interpreted as estimates of the asymptotic probabilities for a single flow component, obtained by integrating $p_\infty(\underline{X})$ over all the values of the remaining two components. For the zonal average flow ψ_A, the almost intransitivity of the time mean value manifests itself as a pronounced bimodality in

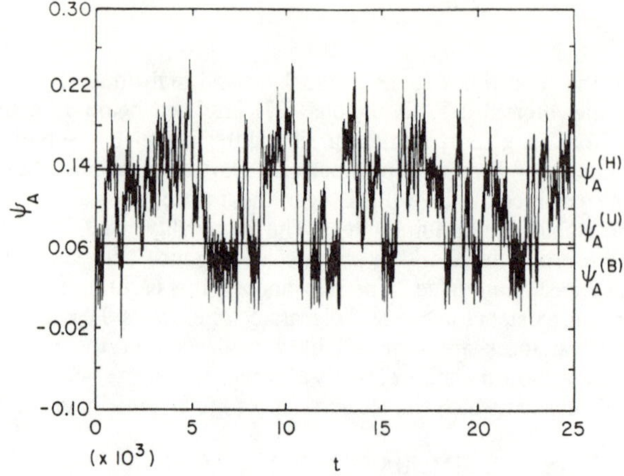

Fig. 3a. Time dependent zonal flow for case S_2, $\epsilon = 2.8 \times 10^{-4}$.

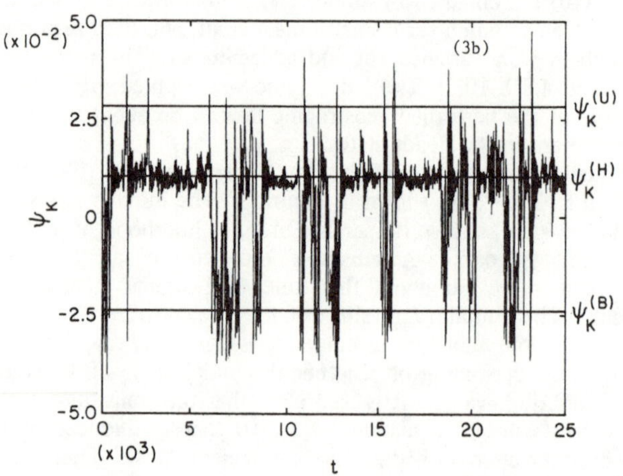

Fig. 3b. Time dependent wave components for case S_2, $\epsilon = 2.8 \times 10^{-4}$, equilibria shown as horizontal lines.

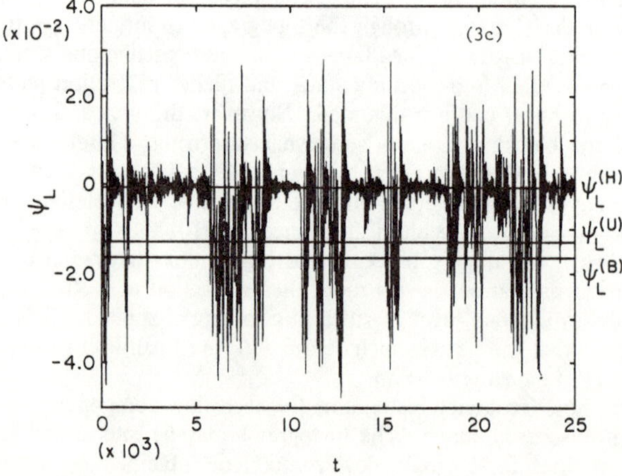

Fig. 3c. Time dependent wave components for case S_2, $\epsilon = 2.8 \times 10^{-4}$, equilibria shown as horizontal lines.

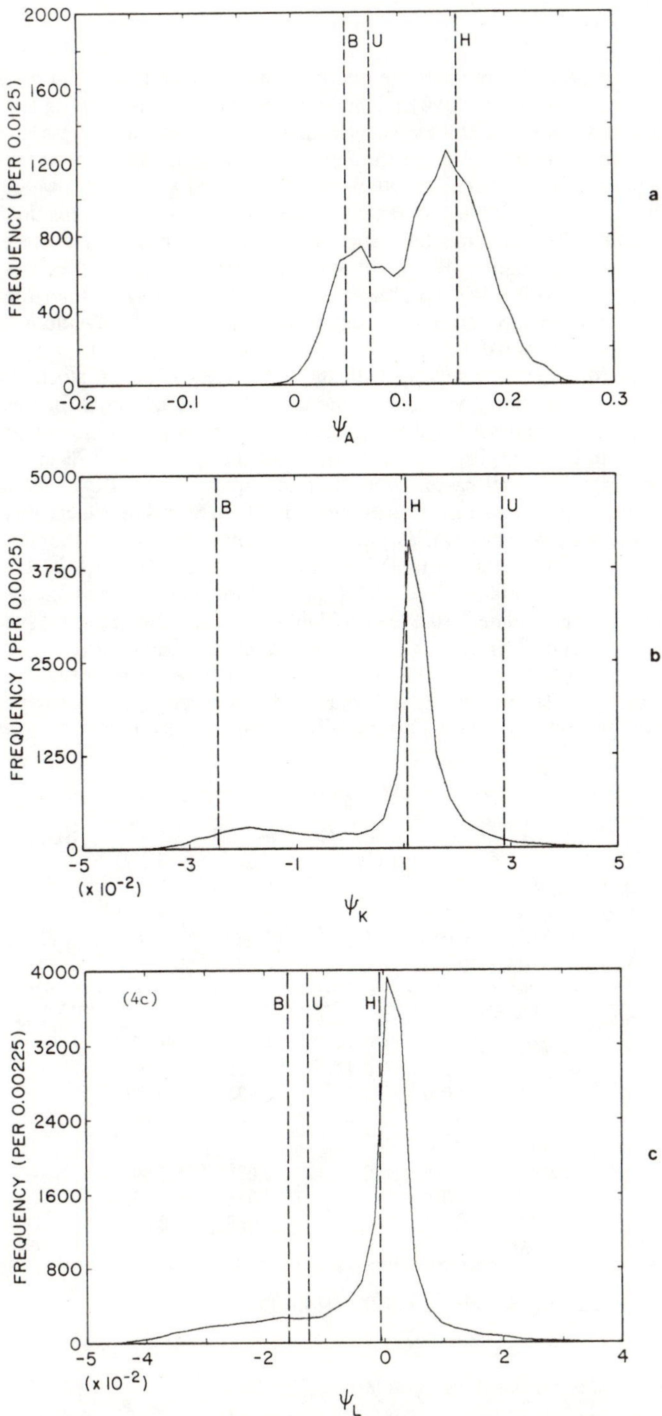

Fig. 4. Frequency histograms for S_2, $\epsilon = 2.8 \times 10^{-4}$. Equilibria shown by dashed lines.

the frequency histogram. By contrast, the distributions of ψ_K and ψ_L are dominated by single modes, and the almost intransitivity of the variance shows up as a strong left skewness.

The time series plots of Figure 3 suggest that a crude partition of variance might be established by passing a boundary plane through the unstable equilibrium value of ψ_A. Let the symbols B and H stand for the conditions $\psi_A < \psi_A^{(u)}$ and $\psi_A > \psi_A^{(u)}$, where $\psi_A^{(u)}$ is the unstable equilibrium value. Table I shows estimates for the local and transitional variances under this partition. The transitional variance dominates in the zonally averaged flow, but not so for the wave components. We might anticipate that predictability, assessed in terms of second moment quantities, will be greater at long range for $\psi_A(t)$ than for $\psi_K(t)$ and $\psi_L(t)$. Also shown in the table are local mean values and standard deviations of the phase $\theta = \tan^{-1}(\psi_L/\psi_K)$ of the wave perturbation relative to the mountain ridge at $x = 0$. Results are shown for parameter sets S_2, with three different values of stochastic forcing ϵ. In all three cases the transitional variance of the zonal flow ψ_A accounts for more than half of the total $\text{var}(\psi_A)$. By contrast, local variances in the low index regime are largest for the waves ψ_K and ψ_L. In fact, $\text{var}_B(\psi_K)$ and $\text{var}_B(\psi_L)$ contribute between 35% and 50% to their respective total variances in all cases, even after multiplication by $P(\underline{X} \in B)$ according to (24) (here the low index state has probability $\sim$0.21). These large local wave variances result mainly from wave propagation, as illustrated by the fact that the standard deviations $SD_B(\theta)$ are roughly twice as large as their high index counterparts $SD_H(\theta)$. Local standard deviations of the wave amplitude $|R| \equiv (\psi_K^2 + \psi_L^2)^{1/2}$. (not shown) are about 25% larger for "B" cases than for "H". The relative variability of the low index wave amplitude $|\bar{R}|_B/SD_B(|R|)$ ranges from 0.27 to 0.38 for cases 1, 2, and 3. Qualitatively, then, low index episodes, wherein the condition $\psi_A(t) < \psi_A^{(u)}$ persists are typified by wavy streamline patterns resembling Figure 2a, but with phases that continually propagate through the channel, so that a synoptic meteorologist would not classify the patterns as "blocking" types.

Table I. Variance partition results, S_2^*

Z	$\bar{Z}$	$\text{var}(Z)$ (10^{-3})	$\bar{Z}_B$	$\bar{Z}_H$	$\text{var}_B(Z)$ (10^{-3})	$\text{var}_H(Z)$ (10^{-3})	$\text{var}_T(Z)$ (10^{-3})	$SD_B(Z)$	$SD_H(Z)$
Case 1, T= 75,000, ϵ=2.8 X 10^{-5}									
ψ_A	0.113	2.76	0.038	0.134	0.29	1.28	1.62	--	--
ψ_K	0.006	0.17	-0.010	0.010	0.27	0.047	0.068	--	--
ψ_L	-0.005	0.12	-0.014	-0.002	0.27	0.045	0.022	--	--
θ	--	--	-128°	-8°	--	--	--	63°	30°
Case 2, T= 75,000, ϵ=1.8 X 10^{-5}									
ψ_A	0.114	2.39	0.039	0.135	0.24	1.03	1.53	--	--
ψ_K	0.005	0.17	-0.014	0.010	0.23	0.036	0.092	--	--
ψ_L	-0.005	0.10	-0.014	-0.002	0.25	0.031	0.026	--	
θ	--	--	-139°	-7°	--	--	--	52°	26°
Case 3, T=100,000, ϵ=4.8 X 10^{-5}									
ψ_A	0.116	3.67	0.032	0.140	0.49	2.02	1.99	--	--
ψ_K	0.007	0.13	-0.003	0.010	0.27	0.055	0.032	--	--
ψ_L	-0.004	0.12	-0.011	-0.002	0.27	0.058	0.013	--	--
θ	--	--	-108°	-8°	--	--	--	77°	35°

*All parameters time averages with $\underline{X}_0 = (.02,.044,-.003)$.

As the stochastic perturbation parameter ϵ decreases from 4.8 x 10^{-4} to 2.8 X 10^{-5}, the partition of zonal flow variance is changed, and a larger portion comes from $\text{var}_T(\psi_A)$. As one might expect, the bimodality of $p_\infty(\psi_A)$ (not shown) grows stronger also, with smaller scatter about the conditional averages $E_B(\psi_A)$ and $E_H(\psi_A)$. In this instance the prediction of ψ_A reduces essentially to the problem of predicting the process

$$f(t) = E_B(\psi_A)I_B(t) + E_H(\psi_A)I_H(t) \tag{28}$$

that flips from one local average to the other. For the wave components (ψ_K,ψ_L) knowledge of f(t) provides no phase information, but it does allow one to estimate the probable range of variability of wave amplitude over the persistence time scales for $I_B(t)$ and $I_H(t)$. Of course, the asymptotic probability distributions tell us nothing about such time scales. Predictability assessment requires that we consider statistics lagged in time, and their relation to the persistence times.

PREDICTABILITY

We assume that a rule is given for the time evolution of the true flow field. For example, we may choose (6) or (10). Predictions can be obtained in principle by integrating the equations, starting at some initial condition $\underline{X}_0$. For (6), exact predictions are possible only when the initial condition is known with infinite precision. In the case of (10), we have stochastic integrals[16] and, even with perfect initial data the time evolution cannot be forecast exactly. This is so because, although statistical properties of dW(t) are known, one does not know which realization of dW(t) is about to occur. Our interest here is in this latter source of uncertainty. Choices are necessary for the variables that are to be predicted and the rule, based on known information, that yields forecasts $\hat{\underline{X}}(t)$ at future instants. Moreover, the probabilistic nature of dW(t) or uncertain initial data requires that we formulate a *statistical* measure of the error, say $e(\underline{X},\hat{\underline{X}},t,\underline{X}_0)$. It is worth emphasizing that the error can be computed from different ensembles in different problems. For instance, one might compute errors from the ensemble of all dW(t), with a fixed, deterministic initial datum $\underline{X}_0$. In general, predictability assessed this way will differ from estimates based on averaging all $\underline{X}_0$ and dW(t) (or, equivalently, over an infinitely long realization of W(t) starting at an arbitrary $\underline{X}_0$).

Our purpose here is to examine, in simplest form, some of the implications of almost intransitive behavior for predictability in the prototype barotropic model. Therefore, we shall employ an approximation technique[12] that reduces the three component system (10) to a single stochastic differential equation governing the almost intransitive behavior of the zonal flow ψ_A. Furthermore, we study a simple error growth measure, the time autocorrelation function, that places a lower bound on the predictability time. Extensions to the other flow components, and consideration of more general forecast rules are deferred for later work.

Starting with (10), the approximation proceeds as follows.[12] Equation (10b) is replaced by its steady state counterpart, yielding the diagnostic relation

$$\psi_K(t) = \frac{-b_{21}(t)\psi_L(t)}{k}\,. \tag{29}$$

New variables $s \equiv kt$ and $\eta \equiv -(\psi_A - \psi_A^*) + h_{01}\psi_L/k$ are defined, so that, with the aid of (29), (10a) and (10c) become

$$d\psi_A(s) = \eta(s)ds + \frac{\epsilon^{1/2}}{k^{1/2}}dW(s) \tag{30a}$$

$$d\eta(s) = -\frac{\partial\Phi}{\partial\eta}\,ds + \frac{\epsilon^{1/2}}{k^{1/2}}dW(s) \tag{30b}$$

where

$$\phi(\eta,\psi_A) = -q(\psi_A)\eta + \left(2 + \frac{b_{21}^2}{k^2}\right)\frac{\eta^2}{2} \tag{31}$$

and

$$q(\psi_{A)} = \frac{-h_{01}h_{21}}{k^2}\psi_A - (\psi_A - \psi_A^{*})\left(1+ \frac{b_{21}^2}{k^2}\right). \tag{32}$$

For a fixed value of ψ_A (30b) is a first order stochastic differential equation, and the asymptotic, conditional transition probability density $P_\infty(\eta|\psi_{A)}$ satisfies the Fokker-Planck equation

$$\frac{\epsilon\partial^2 p_\infty}{2k\partial\eta^2} + \frac{\partial}{\partial\eta}\left(p_\infty\frac{\partial\Phi}{\partial\eta}\right) = 0. \tag{33}$$

A solution that vanishes, with its first derivative, as $|\eta| \to \infty$ is

$$p_\infty(\eta|\psi_A) = N \exp\left\{\frac{-2k}{\epsilon}\left[\left(1 + \frac{b_{21}^2}{k^2}\right)\eta^2 - q\eta\right]\right\} \tag{34}$$

where the normalization factor is

$$N = \left[\left(\pi\epsilon/k(2 + b_{21}^2/k^2)\right)^{1/2} \exp\left\{\frac{-kq}{2\epsilon(2 + b_{21}^2/k^2)}\right\}\right]^{-1}. \tag{35}$$

Conditional expectation values, averaged over the distribution p_∞, are denoted by the symbol $< >$. Averaging (30a) in this way gives

$$<d\psi_A(s)> = \frac{-\partial V}{\partial\psi_A}ds + \frac{\epsilon^{1/2}}{k^{1/2}}dW(s) \tag{37}$$

where

$$V(\psi_A) = -\int^{\psi_A}\left\{\int_{-\infty}^{\infty} p_\infty(\eta|\tilde{\psi})\eta d\eta\right\} d\tilde{\psi}$$

$$= k\left(\frac{\psi_A^2}{2} - \psi_A^{*}\psi_A\right) - \frac{k^2}{2\alpha_{21}^2}\left(k - \frac{h_{01}h_{21}}{k}\right)\log\left[\frac{k^2}{\alpha_{21}^2}\left(b + \frac{b_{21}^2}{k^2}\right)\right] - \frac{k}{\sqrt{2}\alpha_{21}}\,.$$

$$\left[\frac{\beta_{21}}{\alpha_{21}}\left(-\frac{h_{01}h_{21}}{k}\right) - k''k\Psi_A^{*}\right]\tan^{-1}(b_{21}/\sqrt{2}k)\,. \tag{38}$$

is a potential function. Figure 5 shows plots of $V(\psi_A)$ for parameter sets S_2 and S_3. If we suppose that the process $\eta(t)$ approaches its asymptotic, conditional density p_∞ on a time scale T_o much shorter than the time scale for substantial evolution of $\psi_A(t)$, we may regard (37) as a stochastic differential equation for the time dependence of $\psi_A(t)$ over increments $dt >> T_o$, that is, we have

$$d\psi_B(t) \simeq \frac{-\partial V}{\partial\psi_B}dt + \epsilon^{1/2}\, dW(t) \tag{39a}$$

where

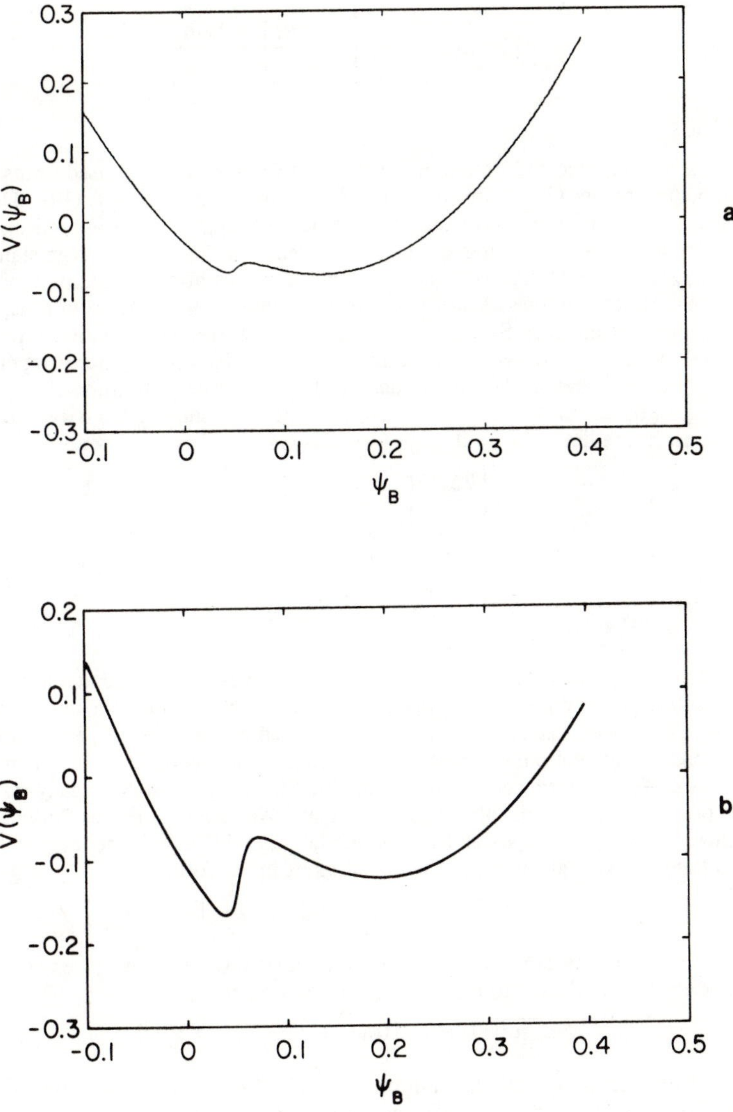

Fig. 5. Potential functions for S_2 (a) and S_3 (b).

$$\frac{\partial V}{\partial \psi_B} = \frac{-k(\psi_B - \psi_A^*)\left(1 + \frac{b_{21}^2}{k^2}\right) - \frac{h_{01}h_{21}}{k}\psi_B}{\left(2 + b_{21}^2/k^2\right)} . \tag{39b}$$

Here we have reinstated the time unit $t = s/k$ and the symbol ψ_B is used to distinguish the one-component model (39a) form the zonal flow in the original system (10). Deterministic $\epsilon = 0$ steady states of (39) occur at the extreme of $V(\psi_A)$ and correspond exactly to the equilibria of ψ_A. Essentially, the approximation requires that time average statistics of the processes $\psi_K(t)$ and $\eta(t)$ approach their asymptotic values in a time short compared to the transition times between episodes of low and high index flow. Simulation runs, for a range of parameters that includes S_2 and S_3, showed that the average transition times estimated from (39a) were in good agreement with the full model (10).[12] More importantly, this approximation facilitates analytical estimation of the average time intervals τ required for the process $\psi_B(t)$ to reach the unstable steady state $\psi_B^{(u)}$ starting from each of the stable steady states, in the limit as $\epsilon \rightarrow 0$. The formulas are

$$E\{\tau_B\} = \frac{\pi}{|V_B''V_U''|^{1/2}} \exp\left(\frac{2\Delta_B V}{\epsilon}\right) \tag{40a}$$

$$E\{\tau_H\} = \frac{\pi}{|V_B''V_U''|^{1/2}} \exp\left(\frac{2\Delta_H V}{\epsilon}\right) . \tag{40b}$$

where the double-primed quantities are second derivatives of the potential at the equilibria, and $\Delta_B V$ and $\Delta_H V$ are difference between the potentials at the unstable steady state and the respective stable ones. τ_B and τ_H may be interpreted as the average persistence times for low and high index episodes. Here the expectations are computed for an ensemble of realizations of $W(t)$, starting at the initial condition fixed on the stable steady state.

In order to study predictability, assume that (39a) governs the evolution of the true zonal flow. A simple, empirical forecast model, that illustrates the essential effects of almost intransitivity, is given by damped persistence in the form

$$\hat{\psi}_B(t + \tau) = E_\infty(\psi_B) + r(\tau)\{\psi_B(t) - E_\infty(\psi_B)\} \tag{41}$$

where we are given the present zonal flow amplitude $\psi_B(t)$, τ is the forecast interval, and the coefficient $r(\tau)$ is chosen to minimize the error variance

$$E_\infty\{e^2(\tau)\} = E_\infty\{(\hat{\psi}_B(t + \tau) - \psi_B(t + \tau))^2\} \tag{42}$$

computed from the asymptotic distribution $p_\infty(\psi_B)$. The ensemble corresponds to all realizations of $W(t)$ and all initial conditions, weighted by p_∞. Again ergodicity insures the equivalence of such expectations to an infinite time average. The variance is minimized by choosing

$$r(\tau) = \frac{E_\infty[\psi_B(t)\psi_B(t + \tau)\}}{\mathrm{var}(\psi_B)} \tag{43}$$

which parameter is simply the autocorrelation coefficient for $\psi_B(t)$. The error variance

$$E_\infty\{e^2(\tau)\} = \mathrm{var}(\psi_B)\{1 - r^2(\tau)\} \tag{44}$$

approaches the total variance as the correlation decays to zero, and a natural measure of predictive skill is

$$S(\tau) \equiv 1 - \frac{E_\infty\{e^2(\tau)\}}{\mathrm{var}(\psi_B)} = r^2(\tau)\ . \tag{45}$$

Therefore the predictability for the flow field (39a), using the forecast model (41), is determined by the rate at which the autocorrelation function decays to zero. Of course, better predictions might be made by computing the expectation value of $\psi_B(t+\tau)$ directly from (39a), averaged over the ensemble of W(t) at a fixed (given) initial condition. However, our simple procedure can place a lower bound on the predictability, and serves here to demonstrate the role of the almost intransitivity induced by the structure of $V(\psi_B)$.

Numerical solutions of (39a) were computed using the same realization of dW(t) as before, for parameter sets S_2, with ϵ as in Table I, and S_3, $\epsilon = 4.8X10^{-5}$. Due to the relevance of the autocorrelation $r(\tau)$ for our predictability problem, we present in Table II a few comparisons between the simulated series $\psi_A(t)$ and $\psi_B(t)$ that were not emphasized in Reference 12. For parameters S_2 and a range of ϵ, statistics for the two models compare favorably. In particular, the autocorrelation at lag $\tau = 25(\simeq 3\text{days})$ is essentially identical for the two models. Good agreement is likewise found at longer lags (not shown).

Case 4 is presented for purposes of demonstration, because the zonal flow in this case exhibits strong almost intransitivity (Fig. 6). Quantitatively, this effect shows up in the fact that the transitional variances account for 70% to 80% of the total variance.

Table II. Simulation results for two models

Z	$\bar{Z}$	var(Z) (10^{-3})	$\bar{Z}_B$	$\bar{Z}_H$	P(B)	P(H)	$\mathrm{var}_T(Z)$ (10^{-3})	$r(\tau=25)$
Case 1								
ψ_A	0.113	2.76	0.038	0.134	0.22	0.78	1.62	0.87
ψ_B	0.121	2.54	0.040	0.139	0.18	0.82	1.46	0.88
Case 2								
ψ_A	0.114	2.39	0.039	0.135	0.21	0.79	1.53	0.90
ψ_B	0.124	1.91	0.041	0.137	0.14	0.86	1.13	0.90
Case 3								
ψ_A	0.116	3.67	0.032	0.140	0.22	0.78	1.99	0.84
ψ_B	0.121	3.47	0.035	0.143	0.21	0.79	1.89	0.85
Case 4, S_3, T=100,000, ϵ=4.8X10^{-5}								
ψ_A	0.142	7.39	0.030	0.187	0.29	0.71	4.99	0.87
ψ_B	0.123	7.84	0.032	0.193	0.43	0.57	6.41	0.95

Although it might appear that P(B), P(H), and $\mathrm{var}_T(\psi_A)$ are not so well approximated by the single component model, detailed examination of the time series revealed that these discrepancies arise from the crudeness of the phase space partition through $\psi_A^{(u)}$, when applied to the full model (10). Thus many time intervals during which the phase point is in the domain of the low index equilibrium are counted as "H" intervals because the zonal flow exceeds the unstable value, yet the true separatrix has not been crossed. For single component model, the partition is exactly on the separatrix between the two stable states.

How is the predictability of $\psi_B(t)$, as given by $r(\tau)$, related to the phase space structures that induce almost intransitivity in the signal? Figure 7 shows the case 4 autocorrelation functions for all of the variables. Note the rapid decorrelation of ψ_K, ψ_L and especially, η, compared to the zonal flows, that is consistent with the approximation used in

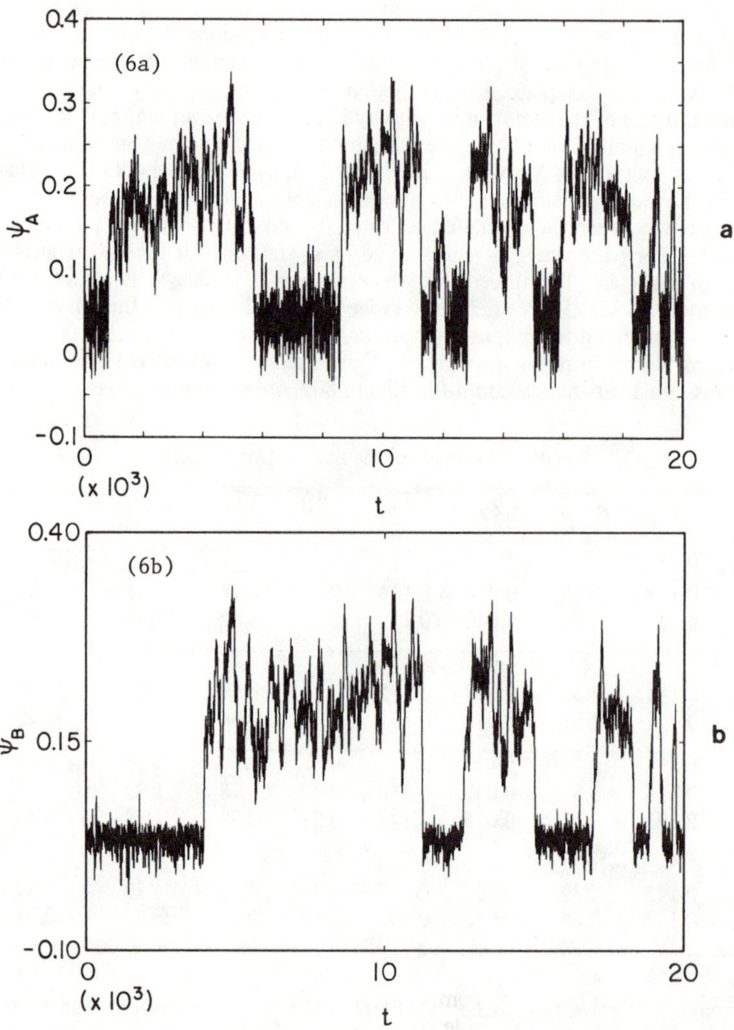

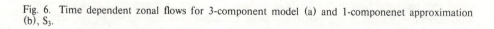

Fig. 6. Time dependent zonal flows for 3-component model (a) and 1-componenet approximation (b), S_3.

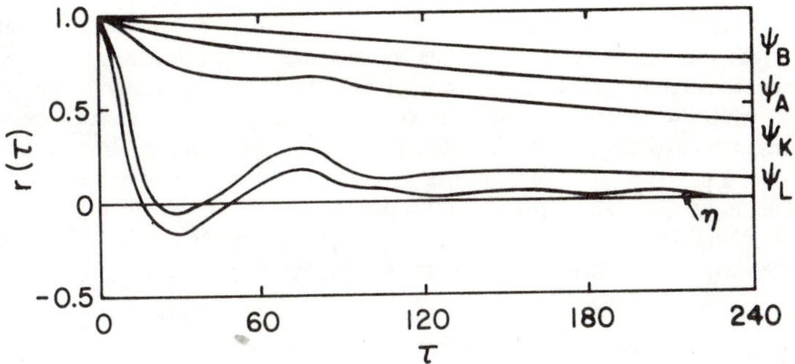

Fig. 7. Autocorrelations for S_3, $\epsilon = 4.8 \times 10^{-4}$.

(29), and in passing from (37) to (39a). The slower decorrelation of $\psi_B(t)$ can be understood in terms of the exit times (40), and a partition of the autocovariance analogous to the variance partition (25). First we write the process in the form

$$\psi_B(t) = f(t) + y(t) \tag{46}$$

where f(t) is the process that "flips" between low and high index states, as in (28), and

$$y(t) = I_B(t)\{\psi_B(t) - E_B(\psi_B)\} + I_H(t)\{\psi_B(t) - E_B(\psi_B)\} \tag{47}$$

describes departures from the conditional mean value within the domain of attraction occupied by the process at time t. When ϵ is small compared to the potential differences in (40), the process f(t) remains constant for very long time intervals because the exit times are large. During these episodes y(t) behaves like the linear Ornstein-Uhlenbeck process

$$dy(t) = -\beta_j y dt + \epsilon^{1/2} dW(t) \quad (j{=}B,H) \tag{48}$$

whose autocorrelation function is

$$\mathrm{cov}_j(\tau) \equiv E_j(y(t)y(t+\tau)) = \frac{\epsilon}{2\beta_j} \exp(-\beta_j \tau) \tag{49}$$

where $\beta_b = V_B''$ and $\beta_H = V_H''$ are autocorrelation decay constants in the low and high index basins. For small ϵ y(t) becomes uncorrelated long before a transition is likely to occur, and we may in this case consider the covariance partition

$$\mathrm{cov}(\tau) \simeq \mathrm{cp}(\tau) \equiv P(B)\mathrm{cov}_B(\tau) + P(H)\mathrm{cov}_H(\tau) + \mathrm{cov}_T(\tau) \tag{50}$$

where $\mathrm{cov}(\tau) = E_\infty(\psi_B(t)\psi_B(t+\tau))$ and $\mathrm{cov}_T(\tau) \equiv E_\infty(f(t)f(t+\tau))$. For parameter sets S_2 and S_3 the local blocking and high index decay time constants are about one and ten days, respectively. Exit times, for case 3 are $E_\infty(\tau_B) = 10.8$ days, $E_\infty(\tau_H) = 49.8$ days and for the case 4 they are $E_\infty(\tau_b) = 189.3$ days and $E_\infty(\tau_H) = 164$ days.[12] The wide disparity between local and transitional time scales suggests that the partition (50) might give an accurate breakdown of the mechanisms contributing to persistence at various lags. Table III shows numerical results, at $\tau = 45$ (5 days), and it is clear that the case the best developed almost intransitivity (case 4) is nearest to satisfying equation (50). 78% of the total variance for this case is predicted successfully by damped persistence of the process f(t) alone. In this example, the predictability $r(\tau)$ is determined essentially by the distance between stable steady states, which serve to pin down the conditional expectations in each basin, and by the exit times, which scale the autocorrelation decay of the process. In this manner, the structure induced in the phase space by the dynamical system (10) manifests itself by providing predictability that falls naturally into two categories: predictability of fluctuations away from a single steady state, as described by β, and predictability of the transitions between neighborhoods of different attractors, as described by the exit times.

Table III. Autocovariance partition results

Case	$cov(\tau=45)$ (10^{-3})	$cp(\tau=45)$ (10^{-3})	$cov_T(\tau=45)$ (10^{-3})	$cov_T(\tau=45)/var(\psi_B)$
1	2.14	1.59	1.08	0.43
3	2.67	1.77	1.08	0.31
4	7.37	6.87	6.14	0.78

CONCLUDING REMARKS

The theoretical flow model (10) and forecasting scheme (41) are arguably too simplified to be taken seriously in the context of actual atmospheric motion and operational weather prediction. In fact,it appears that time dependent flow governed by (10) fails even to simulate qualitatively the occurrence of persistent blocking events, when the behavior of the wave components is considered. Other constraints also restrict severely the range of parameters that can be considered, since the exit time must be "reasonable", the orographic amplitude something like terrestrial mountains, and the stochastic forcing at a level that can be compared with atmospheric processes varying on shorter time scales than these large scale waves. Still, the exercise of computing our crude partition of time variability and autocovariance points towards the possibility of extension to more realistic models, or to atmospheric data, as we attempt to understand the physical nature and predictability of persistent behavior in the atmospheric large scales.

ACKNOWLEDGEMENTS

The author thanks G. Holloway, A.S. Thorndike, R. Colony and A. Sutera for helpful discussions, and R. Benzi for providing an advance copy of Reference 12. This work was supported by the Division of Atmospheric Science (GARP Office), of the National Science Foundation, under grant ATM81-13252 at the University of Washington.

REFERENCES

1. P.D. Thompson, Tellus **9**, 275 (1957).
2. E.N. Lorenz, Tellus **21**, 289-307 (1969).
3. C.E. Leith, J. Atmos. Sci. **28**, 145-161 (1971).
4. C.E. Leith and R.H. Kraichnan, J. Atmos. Sci. **29**, 1041-1058 (1972).
5. B. Saltzman, Rev. Geophys. Space Phy. **8**, 289-302 (1972).
6. B. Saltzman, Met. Monogr. 8(30), 4-19 (1968).
7. J.G. Charney, J.G. Devore, J. Atmos. Sci. **36**, 1205-1216 (1979).
8. J.E. Hart, J. Atmos. Sci.**36**, 1736-1746 (1979).
9. M.K. Davey, J. Fluid Mech. **99**, 267-292 (1980).

10. J.G. Charney, J. Shukla and K.C. Mo, J. Atmos. Sci. **38**, 762-779 (1981).
11. J. Egger, J. Atmos. Sci. **38**, 2606-2628 (1981).
12. R. Benzi, A. Hanson and A. Sutera, Q.J.R. Meteor. Soc. (submitted).
13. A. Sutera, J. Atmos. Sci. **37**, 245-249 (1980).
14. R.E. Moritz and A. Sutera, Adv. Geophys. **23**, 345-383 (1981).
15. G.A. Ven'Tsel and M.I. Freidlin, Russ. Math. Surv. (Engl. Trans.) **25**, 1-55 (1970).
16. I.I. Gihman and A.V. Skorohod, Stochastic Differential Equations (Springer-Verlag, N.Y., 1972).
17. L. Arnold, Stochastic Differential Equations: Theory and Applications (Wiley, N.Y., 1974).
18. E.N. Lorenz, Met. Monogr. **8**, (30), 1-3 (1968).
19. D. Hartmann and S.J. Ghan, Mon. Wea. Rev., **108**, 114-1159 (1980).
20. R.M. Dole, Persistent Anomalies of the extratropical Northern Hemisphere wintertime circulation, Ph.D. thesis (M.I.T., Cambridge, Mass., 1982).

OBSERVATIONAL ASPECTS OF THE PREDICTABILITY OF ATMOSPHERIC BLOCKING

Anthony R. Hansen
Yale University, New Haven, CT 06511

Alfonso Sutera
Center for the Environment and Man, Inc., Hartford, CT 06120

ABSTRACT

Enstrophy and kinetic energy flux functions calculated from observed data from the winters of 1976-77 and 1978-79 are examined to compare the nonlinear cascading properties of blocking and non-blocking periods. During 1978-79 a more pronounced upscale cascade of kinetic energy as well as an upscale cascade of enstrophy from intermediate to planetary-scale wavenumbers was found during block periods as compared to nonblocking. During 1976-77, similar results appeared for a case of Rex blocking but not for the persistent, greatly amplified planetary wave pattern in January and February 1977 (as identified by Charney, et al.[1]).

The predictability time[2] based on the enstrophy flux function showed an increased predictability for the January-February 1977 event but no significant differences between the other blocking cases compared to the nonblocking sample. However, the reversal of the low wavenumber enstrophy cascade during blocking does suggest that blocking may be more persistent (due to reduced dissipation of the large-scale circulation) and therefore more predictable. Some possible implications for theoretical modelling are discussed.

INTRODUCTION

In the present study, we will attempt to identify any systematic differences in the energy and enstrophy cascading statistics of blocking events compared to nonblocking periods using data from two recent winters (1976-77 and 1978-79). It is generally accepted that the existence of persistent features such as blocking highs will allow better medium range forecast skill in numerical weather prediction models. Recently, Bengtsson[3] has shown from numerical simulations that the atmosphere is in general more predictable during blocking events than during nonblocking periods. We are interested to see if any diagnostic evidence exists for the greater predictability of blocking found by Bengtsson by qualitatively estimating a predictability time based on two dimensional turbulence theory. Although our sample is fairly small, some interesting features appear.

This report is part of a more complete set of diagnostic calculations to appear in Tellus[4].

DATA AND PROCEDURES

The data used in this study are the horizontal wind (u, v) from the twice daily operational analysis of the National Meteorological Center for the winters of 1976-77 and 1978-79. Fourier coefficients of u and v were computed for every 2.5° of latitude at the 10 mandatory levels in the troposphere with the wavenumber expansion truncated after wavenumber 18.

Following Steinberg et al.[5], the kinetic energy and enstrophy flux functions can be defined as

$$\frac{\delta F_K(m)}{\delta m} = - C_K\ (m|n,1) \tag{1}$$

and

$$\frac{\delta F_E(m)}{\delta m} = - C_E\ (m|n,1) \tag{2}$$

assuming F_K (0) = F_E (0) = 0. C_K (m|n,1) and C_E (m|n,1) represent the gain in wavenumber m kinetic energy and enstrophy, respectively, due to nonlinear, triad interactions with all possible combinations of wavenumbers n and 1. These equations were integrated from 30°N to 80°N and from 1000 mb to 100 mb.

Our primary goal is to study persistent, large-scale blocking events compared to predominantly zonal circulations which may include short duration, low amplitude, small-scale features. Therefore, we restrict ourselves to cases of stationary or slowly propagating ridges where the departure of the 500 mb height from the zonal mean, averaged over every 2.5° of latitude from 55°-80°N, exceeded 250 m for 7 days or more. Any observation not falling in a blocking period was included in the nonblocking sample.

In addition, a case of very large negative height departures occurred during January and February 1977. We will use the dates given by Charney, et al.[1] for 2 periods of persistent height departures from the climatological mean in our diagnostic calculation. Because of the existence of large amplitude features of one type or another throughout the 1976-77 winter, no nonblocking days from this winter are included in our nonblocking sample. A summary of the blocking and nonblocking days for the 2 winters is given in Table I.

Comparison of the periods of large, persistent height departures with synoptic charts indicates that these periods satisfy (for the most part) the conventional, subjective definition of blocking.

RESULTS

The results for the 1978-79 winter are given in Figure 1. Normally, enstrophy is cascaded from the longest to the shortest wavelengths[5]. This characteristic also appears in our nonblocking sample (Fig. 1a, dashed line). However, the absence of an enstrophy flux out of the lowest 5 wavenumbers during blocking and the change in sign of the enstrophy flux function near wavenumber 2 is evident

Table I Tabulation of blocking and nonblocking days from the 1978-79 winter and the 1976-77 winter determined from persistent height departures

1978-79

blocking days (90 observations)

0000 GMT 1 Dec. - 1200 GMT 7 Dec.
0000 GMT 20 Dec. - 1200 GMT 27 Dec.
1200 GMT 29 Dec. - 0000 GMT 10 Jan.
0000 GMT 14 Jan. - 1200 GMT 26 Jan.
0000 GMT 16 Feb. - 0000 GMT 22 Feb.

nonblocking days (85 observations)

0000 GMT 8 Dec. - 1200 GMT 19 Dec.
0000 GMT 28 Dec. - 1200 GMT 28 Dec.
1200 GMT 10 Jan. - 1200 GMT 13 Jan.
0000 GMT 27 Jan. - 1200 GMT 15 Feb.
1200 GMT 22 Feb. - 1200 GMT 28 Feb.

1976-77

blocking days (36 observations)

1200 GMT 11 Dec. - 0000 GMT 29 Dec.

Charney et al.[1] negative anomalies (76 observations)

30 December - 17 January
2 February - 22 February

in Fig. 1a (solid line). Also, a much larger flux of kinetic energy from the intermediate to the long waves occurred during blocking as can be seen clearly in Fig. 1b.

A comparison of the energy and enstrophy flux functions for the December 1976 case of blocking and the negative anomalies of January and February 1977 are given in Fig. 2. The December 1976 event exhibited a very striking reversal of the low wavenumber enstrophy flux function (Fig. 2a), and the same enhanced upscale kinetic energy flux as the 1978-79 blocking cases (Fig. 2b). The similarity in the behavior of Rex-type blocking in the 2 winters lends support to the significance of this result.

The January-February 1977 enstrophy flux (Fig. 2a, dashed line) is more like the 1978-79 nonblocking sample. However, notice the much lower enstrophy flux function for the January-February 1977 case in a nearly constant range from roughly wavenumbers 8 to 15. The enstrophy flux function in the same range for the December 1976 blocking case is virtually identical to the 1978-79 blocking and nonblocking samples. The kinetic energy flux for the 1977 case is dominated by exchanges amongst the lowest wavenumbers (Fig. 2b, dashed line).

It is well known that the rate of error growth in numerical

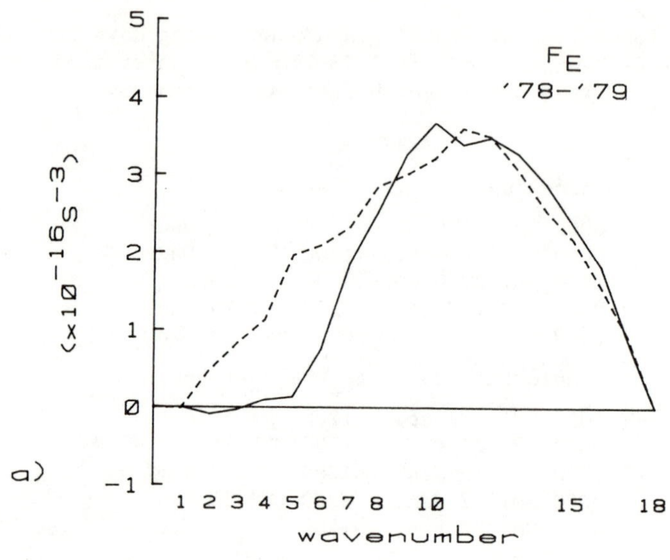

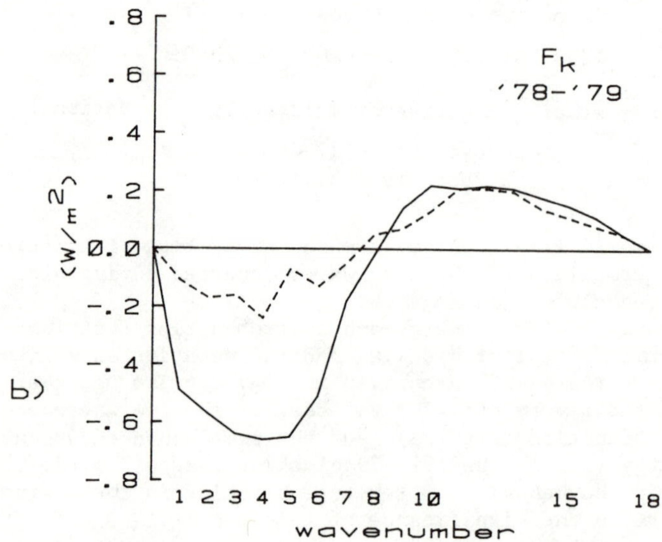

Fig. 1. The nonlinear flux functions for a) enstrophy, F_E and b) kinetic energy, F_K for blocking (solid line) and nonblocking (dashed line) for the 1978-79 winter.

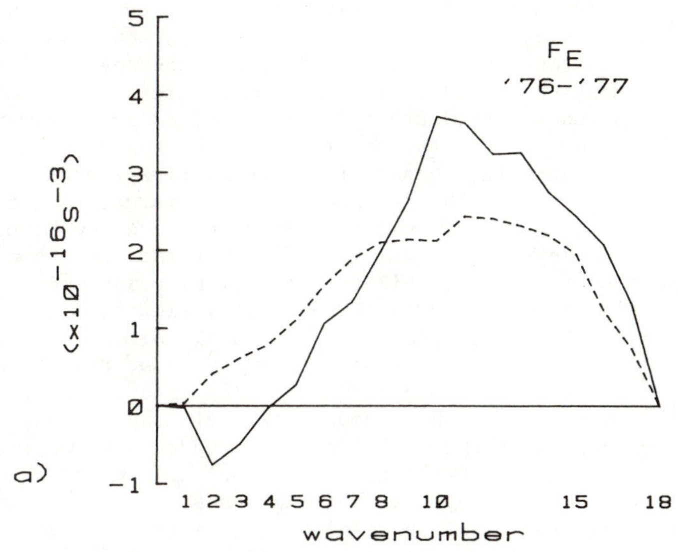

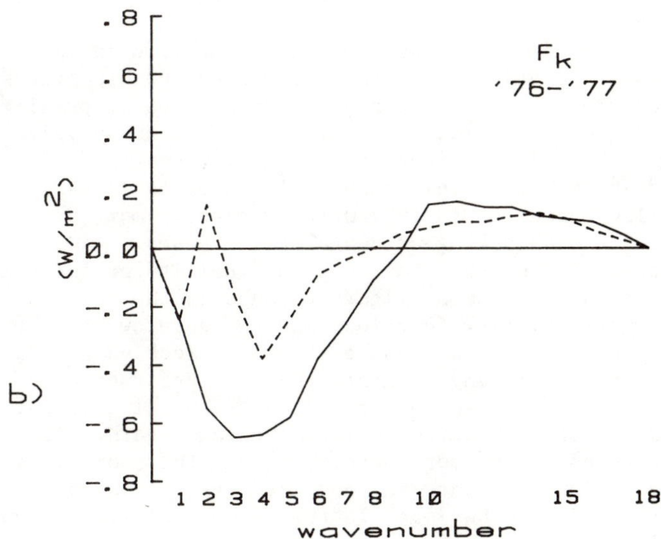

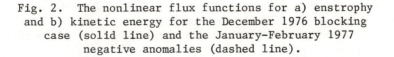

Fig. 2. The nonlinear flux functions for a) enstrophy and b) kinetic energy for the December 1976 blocking case (solid line) and the January-February 1977 negative anomalies (dashed line).

weather prediction models due to small-scale inaccuracies in the initialization can be related to the rate of downscale enstrophy cascade in a constant enstrophy flux inertial subrange of a two-dimensional turbulent fluid[6]. The characteristic time of this process can be called the predictability time and is inversely proportional to the cube root of the constant rate of enstrophy transfer to higher wavenumbers in a constant enstrophy flux inertial subrange[2,6]. The fact that the atmosphere at the larger scales behaves predominantly like a two-dimensional fluid in which the kinetic energy spectrum obeys a -3 power law [7,8,9] suggests that a qualitative estimate of the predictability time for blocking compared to nonblocking situations based on the enstrophy flux function may be made. The estimate can only be done in a qualitative way because the numerical value of the enstrophy flux function, F_E, in the constant range is dependent upon the truncation used[5]. The reason for this sensitivity is that finite amount of enstrophy is lost when the wavenumber expansion of the wind field is truncated at wavenumber 18. This will cause aliasing problems in the nonlinear enstrophy interaction in our truncated dataset that will cause changes in the shape of the enstrophy flux function at high wavenumbers. Unfortunately, the accuracy of our data at wavenumbers higher than 18 is poor so this problem cannot be alleviated. However, we will assume the values of the flux function that we have are qualitatively correct for comparison purposes.

As noted earlier, the enstrophy flux function in the constant range is nearly identical in both blocking and nonblocking situations [Figs. 1a and 2a (solid line)] suggesting no greater predictability for Rex-type blocking. However, the January-February 1977 low anomaly event exhibited a roughly 1/3 smaller enstrophy flux rate (Fig. 2a, dashed line). This result allows us to speculate that the persistent low anomalies in 1977 were inherently more predictable than either the nonblocking or Rex blocking cases.

An explanation of the greater predictability of the Rex-type blocking found by Bengtsson[2] might be found in the reversal of low wavenumber enstrophy flux function, and the absence of an enstrophy flux out of the lowest 5 wavenumbers during blocking as opposed to the normal case. This may indicate that the destruction of large-scale vorticity by down-scale cascade and eventual dissipation is greatly reduced or eliminated during blocking events. As a result, blocking patterns may be more persistent and therefore more predictable. In addition the enstrophy cascade reversal may be an indication of intermittent behavior[10] in the large-scale flow requiring no particular low wavenumber instability.

CONCLUDING REMARKS

Because of the evidently strong interaction between baroclinic cyclone-scale waves and planetary-scale waves during blocking[4,11], it would appear that theoretical models should incorporate the effects of transient, baroclinic eddies. An effort in this direction has recently been put forward by Benzi et al[12]. Using a low-order, barotropic model[13], they suggest that the inclusion of baroclinic activity as a parameterized forcing of the mean flow (as a red process), acts to mask the role of any underlying orographic instability that otherwise might be present.

ACKNOWLEDGEMENTS

This study was supported by NASA under grant NAS8-34903 at Yale University and by the National Science Foundation under grant ATM81-06034 at the Center for the Environment and Man, Inc. Computations were performed at the National Center for Atmospheric Research which is supported by the National Science Foundation.

REFERENCES

1. J.G. Charney, J. Shukla, and K.C. Mo, J. Atmos. Sci., 38, 762 (1981).
2. D.K. Lilly in Dynamic Meteorology, edited by P. Morel (D. Reidel, Dordrecht, 1970), p. 353-418.
3. L. Bengtsson, Tellus, 33, 19 (1981).
4. A. R. Hansen and A. Sutera, Tellus (in press) (1983).
5. H.L. Steinberg, A. Wiin-Nielsen, and C.H. Yang, J. Geophys. Res., 76, 8829 (1971).
6. C.E. Leith and R.H. Kraichnan, J. Atmos. Sci., 29, 1041 (1972).
7. T.-C. Chen and J.J. Tribbia, Tellus, 33, 102 (1981).
8. P.R. Julian, W.M. Washington, L. Hembree, and C. Ridley, Jr. Atmos. Sci., 27, 376 (1970).
9. A. Wiin-Nielsen, Tellus, 19, 540 (1967).
10. G.K. Batchelor, The Theory of Homogeneous Turbulence (Cambridge University Press, Cambridge, 1960).
11. A.R. Hansen and T.-C. Chen, Mon. Wea. Rev., 110, 1146 (1982).
12. R. Benzi, A.R. Hansen, and A. Sutera, Q. J. Roy. Meteor. Soc., (submitted) (1983).
13. J.G. Charney and J.G. Devore, J. Atmos. Sci., 36, 1205 (1979).

PREDICTABILITY OF A LARGE ATMOSPHERIC MODEL

J. Shukla
Laboratory for Atmospheric Sciences
NASA/Goddard Space Flight Center
Greenbelt, Maryland 20771

INTRODUCTION

The future evolution of an observed atmospheric state is not predictable beyond a few days because of the inherent instability of the large scale atmospheric flows and nonlinear interactions among motions of different space and time scales. The theoretical upper limit for deterministic prediction is mainly determined by the growth rates of the most dominant instabilities and the mechanisms for equilibration of their amplitudes. Even a small uncertainty in any one scale grows with the characteristic growth rate for that scale and nonlinear interactions among different scales of motion help spread this unpredictability to all the scales present in the flow. For a simple barotropic fluid without the beta effect and without asymmetric forcing, Lorenz[1] showed that each scale of motion has its own range of predictability which is determined by its interaction with the neighboring scales and the level of energy in that scale. Determination of the limits of predictability for atmospheric motions is of great importance for weather and climate forecasting. Limits on our ability to make accurate weather forecasts at long range can arise either due to incomplete definition of the initial state of the atmosphere or due to incomplete knowledge of the physical laws governing the future evolution of the atmospheric states.

During the last 30 years a variety of models of atmospheric flows have been used to calculate the limits of predictability. One of the standard procedures to conduct such studies is to integrate the model with two nearly identical initial conditions and then to examine the growth of the small initial difference. We would refer to such studies as the *classical predictability studies.* The first such comprehensive study was reported by Lorenz[2] in which he examined the predictability of a 28 variable atmospheric model, and showed that the doubling time for small errors was about four days. Subsequently, several large atmospheric models, generally referred to as the general circulation models (GCMs), were used to calculate the growth characteristics of small initial errors. Charney *et al.*[3] have described the results of such integrations of the then available models of Smagorinsky[4], Mintz[5] and Leith[6]. There were large differences among the estimates of the error doubling time for these three models. In each model a sinusoidal temperature error field was introduced in the initial conditions. The Leith model did not exhibit a consistent exponential growth of the initial error which underwent a transient oscillation for the first week and then levelled off at about seven days. In the Smagorinsky model, the error exhibited quasi-periodic fluctuations for the first two weeks. For very small amplitude of the initial temperature error field, the growth rate was small for the first 30 days after which an exponential growth rate of about 6-7 days was observed. The Mintz-Arakawa model results, which were considered to be more realistic because the pattern of error growth was consistent with the notion of exponential growth of a small 'linear' error field superimposed on the large scale mean flow, exhibited a doubling time of about five days. Since the root mean square difference between randomly chosen fields of Nothern Hemispheric winter temperature fields is about 8°C, a doubling time of five days for an initial perturbation of 1°C would suggest a predictability limit of about two weeks.

These experiments showed a high degree of model dependence of the results of predictability calculations. This was partly because, in the early 60's, model development was still in its infancy. If similar experiments were carried out with the three state-of-the-art GCMs available today, the predictability characteristics would be far more similar for all the models. It has been further noted by Jastrow and Halem[7] and Williamson and Kasahara[8] that even for the same model an increase in the spatial resolution results in a decrease in the error doubling time. These results suggest that a great deal of caution is required in interpreting the results of a particular GCM.

A comprehensive study of the predictability of a nine-level global GCM with orography, moist convection, radiation, and cloudiness was carried out by Smagorinsky.[9] The doubling time of the initial random error of about 0.25°C was about 2.5 days; after the initial error had reached an average value of 0.5°C, the doubling time was about 3.5 days and it took about seven days for the error to double from 1°C to 2°C. These results were far more consistent with the notion of the error growth due to hydrodynamical instabilities than those reported by Charney *et al.*[3] Smagorinsky did not find a clear model resolution dependence for predictability as found by Jastrwo and Halem[7] and Williamson and Kasahara[8]. Smagorinsky also examined predictability as a function of zonal wave number for Northern Hemisphere mid-latitudes and showed that the larger scales are more predictable. However, most of Smagorinsky's analysis concerned the growth of globally averaged error fields.

Since the growth and equilibration of the dominant instabilities are the primary determinants of the limits of predictability, and since the nature of these instabilities strongly depends upon the circulation regime, season, and the presence of quasi-stationary asymmetric forcings, it is considered desirable that the predictability characteristics of a large atmospheric model be examined separately for tropics and mid-latitudes, for winter and summer, for Northern and Southern Hemisphere, and for large and small spatial scales. In this paper, we have briefly summarized the results of such a study.

CLASSICAL PREDICTABILITY STUDIES WITH THE GLAS CLIMATE MODEL

We have carried out numerical integrations of the Goddard Laboratory for Atmospheric Sciences (GLAS) climate model[10] to determine its predictability as a function of the circulation regime (tropics and mid-latitudes of both the hemispheres), season, spatial scale (latitudinal wave number), and meteorological variable.

The two most important quantities which can summarize the results of the classical predictability studies are the error growth rate, and the maximum possible value, to be referred to as the equilibration value of the error. While comparing the predictability of two systems, it is important to note the relative values of both of these quantities. A larger growth rate does not necessarily imply a smaller predictability because larger growth rate may be accompanied with a much larger equilibration value. This is of special relevance in comparing the predictability for the winter and the summer season in the Northern Hemisphere. Based on the values of error growth rate alone, Charney *et al.*[3] had concluded that the summer season is more predictable than the winter. This conclusion is not valid because although the error growth rates are smaller in summer compared to winter, the equilibration values for error are much smaller in summer giving rise to lower predictability in summer.

Most of the earlier studies of classical predictability had examined the growth rate of global or hemisphere mean error fields. It could be argued that the examination of growth of a globally or hemispherically averaged error field is justified because the tropics and the mid-latitudes interact strongly. The drawback of this argument is that the time scale for the growth and saturation of initial observational errors in the tropics is much smaller than the time scale of interaction between the tropics and the mid-latitudes. It is, therefore, not only desirable but necessary that the error growth characteristics be examined separately for the tropics and the mid-latitudes.

The Model

The GLAS Climate Model is a global primitive equation model with a horizontal resolution of 4° latitude x 5° longtitude and in vertical levels in sigma coordinates. The model used for these numerical integrations has been described by Shukla et al.[10] It resolves the orographic features at the earth's surface reasonably well. The seasonally varying values of sea surface temperature, soil moisture, snow and sea ice are prescribed at the model grid points. There is no horizontal mixing in the model, except the one introduced by numerical finite differencing and filtering. There is no explicit parameterization for vertical mixing of momentum in the interior of the atmosphere. Vertical transport of heat and moisture is accomplished by parameterized convection. Short wave and long wave radiation fluxes are calculated every five hours and they are influenced by the model generated space-time variable cloudiness.

The Initial Conditions

Model integrations are started with the observed initial conditions obtained from the operational analysis of the National Meteorological Center (NMC). Fields of horizontal velocity (u,v), temperature (T), moisture mixing ratio (q), and pressure (p) are interpolated from the NMC model grid to the GLAS climate model grid points. The boundary conditions of sea surface temperature, snow, sea ice, and soil moisture are given by their climatological values.

The Initial Error Field

The model is first integrated with the observed initial conditions described earlier. This model integration will be generally referred to as the control run. The initial conditions are then modified by adding a random perturbation field in u and v components at all the grid points at all the levels of the model. This will be referred to as the predictability run. The grid point values of the random perturbation follows a Gaussian distribution with zero mean and standard deviation of 3 m/s in u and v components separately. The magnitude of the random perturbation in either u or v is not allowed to exceed 12 m/s at any grid point. We considered it prudent to perturb the wind field only because the variances of temperature and pressure field differ greatly between the tropics and the mid-latitude. A root mean square error of 1 mb in the pressure field for the initial random error is closer to the observational errors for the mid-latitudes but it is comparable to the observed variance for the tropics. Moreover, several earlier calculations had been done by perturbing the mass field and they did not allow a comparison of error growth for the tropics and the mid-latitudes separately.

SUMMARY OF THE RESULTS

a) The planetary scales are more predictable than the synoptic scales.

Figure 1 shows the root mean square error, averaged for six pairs of control and perturbation runs and averaged for the latitude belt 40–60°N for the 500 mb geopotential height for planetary scale wavenumbers 0-4 and synoptic scale wavenumbers 5-12. The initial conditions and the boundary conditions are for the Northern Hemisphere winter season. The dashed line is the average persistence error and the vertical bars denote the standard deviation of the error values. Although the growth rate is nearly the same for the planetary scale waves and the synoptic scale waves, the equilibration value is much larger for the planetary waves giving rise to higher predictability (about four weeks) for the planetary waves compared to the synoptic scale waves (about two weeks). The doubling time for very small errors is about 2.5 days and after the error has grown up to 25 meters, the doubling time increases to about three days. Higher predictability of the planetary scales is of special significances for predictability of space and time averages.

b) The theoretical upper limit of deterministic predictability for low latitudes is shorter than that for the middle latitudes.

(a)

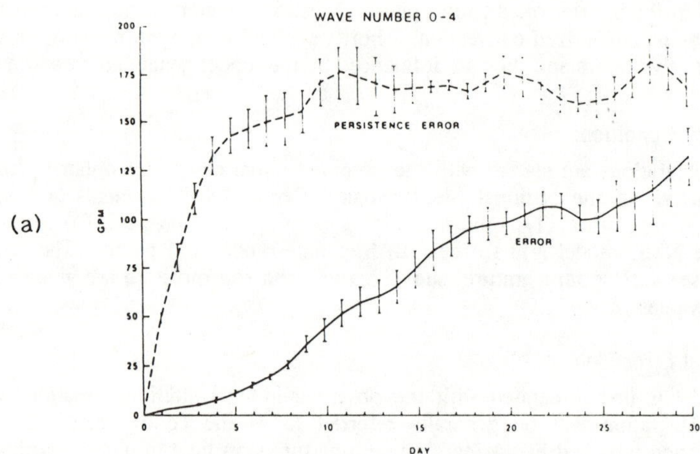

(b)

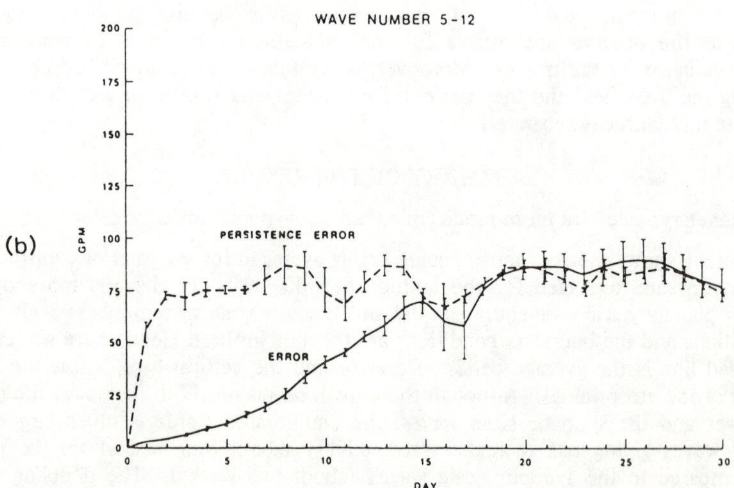

Fig. 1. Root mean square error, averaged for six pairs of control and perturbation runs and averaged for latitude belt 40°N–60°N for 500 mb geopotential height (gpm), for (a) wavenumbers 0-4, and (b) wavenumbers 5-12. Dashed line is the persistence error averaged for the three control runs. Vertical bars denote the standard deviation of the error values.

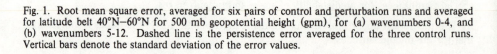

Figure 2 shows the root mean square between control and predictability runs averaged for 20° latitude belts centered at 6°N, 30°N and 58°N, respectively. The curve for 6°N has two remarkable features: the initial error growth rate is the largest, and the error equilibration value is the smallest. This suggests that the initial errors grow faster in the tropics. This could be due to the dominance of the moist-convective instabilities and inadequacy of the parameterizations of moist processes. The smallness of the equilibration value indicates smaller day-to-day variability in pressure and temperature fields in the low latitudes. This could be due to the smallness of the Coriolis parameter and lack of geostrophy. The results for the u and v components are qualitatively similar. The errors of observation in the tropics are already closer to the maximum possible value and therefore it takes only a few more days for the initial error to grow to a magnitude comparable to that between two randomly chosen maps. The above conclusions wll be valid even for an idealized case of uniform and high density observations over the globe, but in reality the situation is much worse in the tropics. The data network is sparse and the scale of the tropical disturbances is only 2000-3000 km. Based on these results it has been concluded that if the theoretical upper limit for deterministic predictability of synoptic scales is about two weeks for the mid-latitudes, it is only 5-7 days for the tropics.

c) The Northern Hemisphere winter is more predictable than summer.

Although the growth rate of the initial error is larger during the winter season compared to the summer season (Fig. 3), the equilibration value of the error is much larger in winter compared to that in summer, and therefore in winter it takes longer for the error to be comparable to the error between randomly chosen charts. This conclusion is supported by the results of operational numerical weather prediction models.

d) The Northern Hemisphere is, in general, more predictable than the Southern Hemisphere.

We have examined several pairs of control and predictability integrations for the same season for the Northern and Southern Hemispheres separately. It was found that the error equilibration value is generally higher for the Northern Hemisphere compared to the Southern Hemisphere. The error growth rates as well as the error equilibration values do not show large seasonal variation in the Southern Hemisphere. It is likely that the presence of strong zonally asymmetric forcing functions in the Northern Hemisphere enhance its predictability. The predictability of an ocean covered earth was found to be smaller than either the Northern or the Southern Hemisphere.

e) Some variables are more predictable than others.

Rainfall was found to be less predictable than circulation parameters. In the tropics, pressure and temperature are less predictable than wind velocity and, in general, all variables are less predictable in the tropics compared to the mid-latitudes.

f) Some initial conditions are more predictable than others.

Since the error growth rate depends upon the nature of the instabilities, and the growth rate of instabilities strongly depends upon the structure of the basic large scale flow, it is quite reasonable to expect a strong initial condition dependence of predictability. It is also likely that the data deficiency in certain preferred regions, for certain large-scale flow structures, might produce unusually large growth of initial error field.

g) The growth rate for spatially random initial error fields is smaller than for systematic initial errors.

We have carried out a few predictability experiments in which the random initial error field was spatially smoothed over the oceans only. This produced spatially coherent error fields over the oceans but the error remained random over the land. The error growth rate

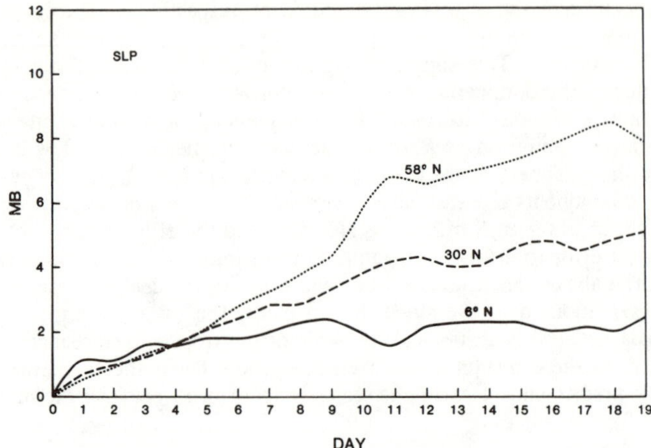

Fig. 2. Root mean square as a function of time between a summer control and a predictability run for sea level pressure (mb). Solid line, dashed line and dotted line refer to an average over 10° latitude belt centered at 6°N, 30°N and 58°N, respectively.

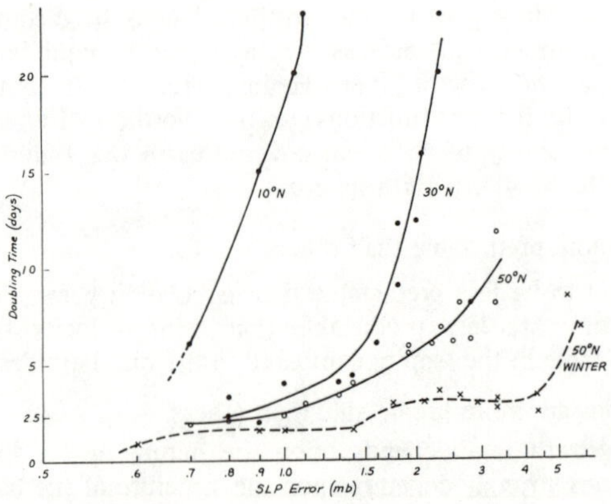

Fig. 3. Error doubling time in days along the ordinate, and magnitude of the error in mb along the abscissa for 20° latitude belt centered at 10°N, 30°N and 50°N for the summer season and at 50°N for the winter season.

in this case was larger than that for the random errors over the whole globe. This experiment was intended to simulate possible large systematic errors over the oceans. The results suggested that the lack of good data coverage over the oceans might be a possible source of error in defining the initial structure of the planetary waves and might therefore reduce the predictability at those scales. These results also suggest, albeit indirectly, that space observing systems, with uniform coverage over the whole globe to define the initial conditions, might lead to higher predictability compared to the conventional observational network with high data coverage only over the land.

h) The growth rate of the error depends upon the magnitude of the error.

Initially, small errors grow fast, but as the magnitude of the error increases, the rate at which the error grows decreases. Figure 3 is a summary of several predictability intergrations which shows that for any latitude, hemisphere, and season, the error doubling time increases with the magnitude of the error. Very large doubling times do not represent high predictability, but they only indicate that the error has grown to be so large that it cannot double again.

PREDICTABILITY OF SPACE-TIME AVERAGES

It has been shown[11] that the inherent dynamical predictability of the observed planetary waves in the atmosphere is great enough to make the errors in the predicted monthly means significantly smaller than the errors arising from small random perturbations in the initial conditions. Since space-time averages are dominated by the low-frequency planetary waves, the limit of predictability of space-time averages is not completely determined by the growth rate of the fastest growing synoptic scale instabilities, but by their interactions with the planetary scales. There is also additional predictability due to the influence of the boundary conditions at the earth's surface[12]. The boundary forcings due to anomalies of sea surface temperature, soil moisture, sea ice and snow, etc., can produce significant changes in the monthly and seasonal atmospheric circulation, and since these boundary forcings change slowly compared to the atmospheric motions, they can enhance the predictability of time averages.

The space-time averages for the tropics are potentially more predictable because the tropical planetary scale circulations are dominated by Hadley, Walker, and monsoon circulations which are intrinsicaly more stable than the mid-latitude Rossby regime. Interaction of these large scale overturnings with tropical disturbances (easterly waves, depressions, cyclones, etc.) is not strong enough to make the former unpredictable due to unpredictability of the latter. Tropical disturbances are initiated by barotropic-baroclinic instabilities, but their main energy source is latent heat of condensation. Although their growth rate is fast and they are deterministically less predictable, their amplitude equilibration is also quite rapid and they attain only moderate intensity. The intensity and geographical locations of Hadley and Walker cells is primarily determined by the boundary conditions and not by synoptic scale disturbances. It is reasonable to assume that frequency and tracks of depressions and easterly waves is primarily determined by the location and intensity of Hadley and Walker cells, and distribution of SST and soil moisture fields. It is highly unlikely that tropical disturbances will drastically alter the character of the large scale tropical circulation. This is in marked contrast to the case of mid-latitudes where interaction between synoptic scale instabilities and planetary scale circulations is sufficiently strong for baroclinically unstable disturbances to render the large scales less predictable. The mid-latitude circulation consists of baroclinic waves, long waves, and planetary waves of different wave number and frequency, so that all space and time scales are important. In contrast, the tropical circulation has a clear scale separation; the large scale Hadley and Walker cells, on the one hand, and the synoptic scale disturbances on the other. The mid-latitude atmospheric anomalies show a rapid decay of autocorrelation function, whereas tropical atmospheric anomalies persist for several months. Since the atmospheric dynamics by itself is not known to have any mechanism for long term memory, persistence of tropical sea surface temperature anomalies can change the location and intensity of Hadley and Walker cells, which can produce persistent anomalies of precipitation. Similarly, anomalies of soil moisture can be very important in determining the intensity of tropical stationary heat sources for which the maxima occur over the continents.

For favorable structures of the large scale flow, the tropical heating anomalies can produce significant changes in the mid-latitude circulation either by poleward propagation of Rossby waves or by changes in the intensity of the Hadley cell and the accompanying zonal flows which interact with the mountains and heat sources in the mid-latitudes. It is therefore likely that the tropical boundary forcings can enhance the predictability of the extratropics also. A discussion of the role of boundary conditions on monthly and seasonal predictability has been presented earlier by Shukla[13].

SUMMARY

Classical predictability studies using the GLAS climate model suggest that the short range weather forecasting is more difficult for the tropics compared to the mid-latitudes. On the other hand, the prospects for predicting monthly and seasonal averages in the tropics are better than those in the mid-latitudes because the fluctuations of time averaged tropical flows appear to be primarily determined by the slowly varying boundary forcings at the earth's surface. It is hoped that a systematic study of predictability of a variety of observed large scale atmospheric flows might provide some insight into the possible causes for why some initial conditions are more predictable than others. The task of forecasting weather would be greatly facilitated if we could identify some characteristic features of the large scale flow which determine its predictability because then we could predict the predictability of the initial state.

REFERENCES

1. E.N. Lorenz, Tellus, **21**, 289-307 (1969).
2. E.N. Lorenz, J. Atmos. Sci., **26**, 636-646 (1965).
3. J.G. Charney et al., Bull. Amer. Meteor. Soc., **47**, 220-220 (1966).
4. J. Smagorinsky. Mon. Wea. Rev., **91**, 99-164 (1963).
5. Y. Mintz, WMO-IUGG Symposium on Research and Development Aspects of Long-Range Forecasting. World Meteorological Organization, Tech. Note No.**66**, 141-155 (1964).
6. C.E. Leith, Methods in Computational Physics, Vol. 4. New York, Academic Press, 1-28 (1965).
7. R. Jastrow and M. Halem, Bull. Amer. Meteor. Soc., **51**, 490-513 (1970).
8. D.L. Williamson and A. Kasahara, J. Atmos. Sic., **28**, 1313-1324 (1971).
9. J. Smagorinsky. Bull. Amer. Meteor. Soc., **50**, 286-311 (1969).
10. J. Shukla, D. Straus, D. Randall, Y. Sud, and L. Marx, NASA Tech. Memo. 83866, 282 (1981) (NTIS No. N8218807).
11. J. Shukla, J. Atmos. Sci., **38**, 2547-2572 (1981).
12. J.G. Charney and J. Shukla, in Monsoon Dynamics, Cambridge University Press, Editors: Sir James Lighthill and R.P. Pearce, 99-109 (1981).
13. J. Shukla, in Proceedings of the seminar on "Problems and Prospect in Long Range Forecasting," ECMWF, Reading (1982).

DYNAMICAL FORECAST EXPERIMENTS WITH A BAROCLINIC QUASIGEOSTROPHIC OPEN OCEAN MODEL

Robert N. Miller and Allan R. Robinson
Center for Earth and Planetary Physics, Harvard University
Cambridge, MA 02138

ABSTRACT

We report here on a series of numerical forecast experiments using a baroclinic quasigeostrophic open ocean model. A simulation has been carried out to produce a model data set consisting of values of streamfunction and potential vorticity in four dimensions. This data set exhibits quasiturbulent characteristics similar to those of the mesoscale eddy field in the North Western Atlantic. The simulation has been carried out for several model years over many independent synoptic realizations.

Given accurate initial conditions, we can perform accurate hindcast and forecast experiments by supplying accurate streamfunction data on the entire lateral boundary and potential vorticity at inflow points. For reasonable computational parameters error can be held, e.g. to a few percent in "normalized RMS streamfunction error" over the course of a three month simulated forecast experiment.

This provides the framework for a parameter and sensitivity study whose purpose is the quantitative determination of data requirements and forecast errors. It is the baroclinic generalization of Robinson and Haidvogel's barotropic study[7] in that further experiments are performed in which the quantity and quality of initial and boundary data is degraded by the introduction of gaps and noise. In these experiments techniques for updating and optimal interpolation are tested and calibrated and the efficiency of simulated data acquisition schemes are assessed with the goal of determining which are the most efficient.

INTRODUCTION

Proper evaluation of dynamical forecasts requires sharp quantitative knowledge of the computational properties of the model being used. In this study we use a simulated data set, constructed to have time and space scales characteristic of the quasiturbulent fields observed in the MODE experiment to determine computational forecast errors and to study forecast errors resulting from incomplete data. Since our simulated data set was constructed using our forecast model, we can study computational error and error due to incomplete or noisy data in the absence of physical error. Our simulated data set also provides complete verification data, which allows exact quantitative evaluation of errors in our forecast experiments. This work is the extension of an earlier

0094-243X/84/1060457-16 $3.00

barotropic simulated data study (Robinson and Haidvogel[7]; hereafter RH).

We generated the simulated data set by using the four best fit Rossby waves[3] to the MODE data as initial and boundary conditions on a 1000 km. square region and allowing the nonlinear dynamics of the model to fill out the spectrum of the field. For our forecast experiments, we discard all but the interior 500 km. square of the domain; we also discard the first 21 months of the simulation. The remainder of the results of the simulation experiment is taken to be true ocean data, and is used to provide initial, boundary and verification data for our forecast experiments. The physical and computational model used is described in section 2. Details of the simulation are given in section 3. Section 4 contains the results of our simulated forecast experiments, and section 5 contains our summary and conclusions.

PHYSICAL AND COMPUTATIONAL MODELS

The details of the physical and computational models are presented in the model report by Miller, Robinson and Haidvogel[4] (hereafter MRH). We present a brief summary here.

The physical model employed in this study is the baroclinic quasigeostrophic vorticity equation, with dissipation parameterized by high order lateral friction and linear bottom drag. This equation is given in dimensionless form as:

$$[\frac{\partial}{\partial t} + \varepsilon J(\psi,\bullet)]\zeta + \psi_x = F \qquad (1)$$

$$\nabla^2\psi + \Gamma^2(\sigma\psi_z)_z = \zeta \qquad (2)$$

We use the scaling:

$$x = \hat{x}/d$$

$$y = \hat{y}/d$$

$$z = \hat{z}/h_T$$

$$t = \hat{t}/(\beta d)^{-1}$$

$$\psi = \hat{\psi}/(V_o d)$$

where caret denotes dimensional variables, d is the lateral length scale, V_o is the velocity scale and h_T is the vertical length scale (here chosen to be the thermocline depth). ε is the β-Rossby number $V_o/\beta d^2$. Γ^2 and σ parameterize the effects of stratification: $\Gamma^2 = f_o^2 d^2/(N_o h_T^2)$, where N_o is a typical buoyancy frequency; $\sigma(z) = N_o^2/N^2(z)$, where $N^2(z) = -(g/\rho_o)\, \partial\rho(z)/\partial z$. ρ_o is the mean density, and $\rho(z)$ is the vertical profile of density averaged in horizontal space and in time. F represents the dissipation imposed upon the model in the forms of filtering and bottom friction. The filter is a Shapiro filter, as described in MRH. Such filtering can be viewed as an approximation of a partial differential operator which describes high order lateral friction. Bottom drag is approximated by a linear drag on the vorticity at the lowest level.

The lateral boundary conditions are those introduced by Charney, Fjortoft and von Neumann[1]: values of the streamfunction are imposed on the entire boundary; vorticity is imposed on the inflow, but on the outflow is set by internal advection. The level by level implementation of these boundary conditions is described in the report on the barotropic model by Haidvogel, Robinson and Schulman[2] (hereafter HRS). The vertical boundary conditions, in the examples shown in this report, are the imposition of constant temperature at the top and bottom surfaces, which corresponds to the absence of wind forcing and bottom relief.

The numerical model is described in detail in MRH. The core of the numerical model is the finite element version of the barotropic model described in HRS. Given initial streamfunction and vorticity fields along with appropriate boundary data, the algorithm used in the barotropic code is used to compute the vorticity at each depth level for the next time step. A numerical implementation of the classic method of separation of variables is used to solve the elliptic equation (2) for the new streamfunction, using the newly calculated vorticity as the right hand side.

Testing of the numerical model on a variety of nonlinear Rossby wave problems (with and without mean advection) is reported in MRH. Figure 1 (taken from Figures 5 and 6 of MRH) shows the variation with time of the normalized level by level RMS errors for a test using an advected Rossby wave at moderate Rossby number ($\varepsilon = 1.5$). In that simulation, stratification parameters were held constant with depth at a typical value for the main thermocline. The depth structure consisted of four equally spaced levels.

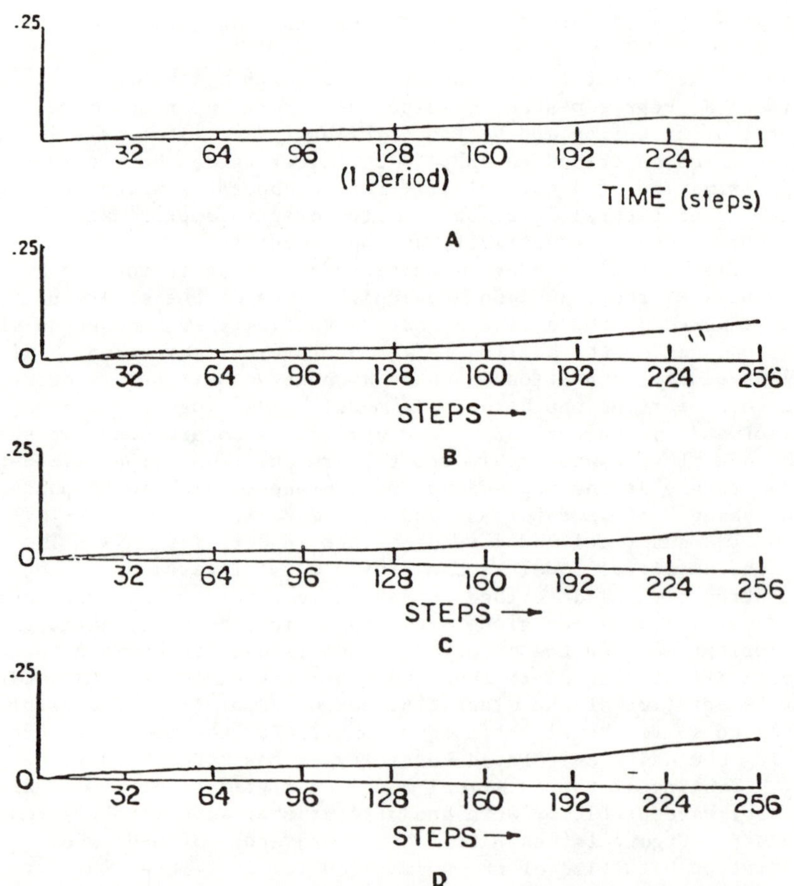

Figure 1. RMS streamfunction and vorticity errors for advected Rossby wave simulation. Values normalized by RMS of analytic expression for Rossby wave field. From a simulation with constant stratification parameters chosen to be typical of the main thermocline. A. RMS streamfunction error, top level of four level simulation. B. Bottom level. C. RMS vorticity error, top level. D. Bottom level.

THE BAROCLINIC SIMULATION

As in earlier barotropic studies, we have adopted a strategy in which our exterior numerical calculation is performed in a large computational domain (65×65×6 grid points; 1000 km square by 5 km deep) to provide initial, boundary and verification data for a smaller interior region (33×33×6 grid points; 500 km square by 5 km deep). Since the interior and exterior simulations are carried out with identical physics and discrete numerical models, the only source of forecast error is the implementation of the boundary conditions, since boundary points of the interior domain are interior points of the exterior domain. Initial and boundary conditions for the exterior simulation are calculated by evaluating the analytical expressions for the four best fit Rossby waves to the MODE-1 data[3].

The exterior simulation experiment was run with 6 levels. These levels (240 m; 476 m; 748 m; 1020 m; 2108 m; 3604 m) were chosen for optimal resolution of the maximum number of baroclinic modes of the system. These modes were determined by calculating the eigenfunctions of a discretized version of the depth operator in equation 2. In particular, the following eigenvalue problem:

$$(\sigma(z)F'(z))' = -\lambda^2F(z) \tag{3}$$

was discretized on a 36 point grid, with values of $\sigma(z)$ taken from the MODE-1 Atlas[6] and the solution to the resulting matrix problem was calculated.

Since the baroclinic flow is swifter than the barotropic, especially at the upper levels, computational stability could not be maintained with the one day time steps and fourth order filtering used in the barotropic study. Our present simulation uses six hour time steps with a fourth order filter applied each time step to the vorticity. Bottom friction with a spin-down time of approximately 1000 days (identical to that used in the barotropic study) was imposed on the lowest level of the flow. Integrated energy and enstrophy for this simulation is plotted in Figure 2. These graphs span six periods of the fastest of the linear waves, slightly over two years of model time. After five periods of nonlinear evolution, the regular structure of the initial field has given way to the quasiturbulent structure that characterizes the real ocean. Figure 3 shows the evolution of the streamfunction field at 750 m from the initial superposition of linear waves, along with objective maps of streamfunction fields from the USSR POLYMODE array. The difference in the scale between the simulated and real data maps should be noted: the real data field is 288 km square, and the simulated data field is 500 km square. The similarity between the real and simulated data fields is evidence for the relevance of this series of experiments.

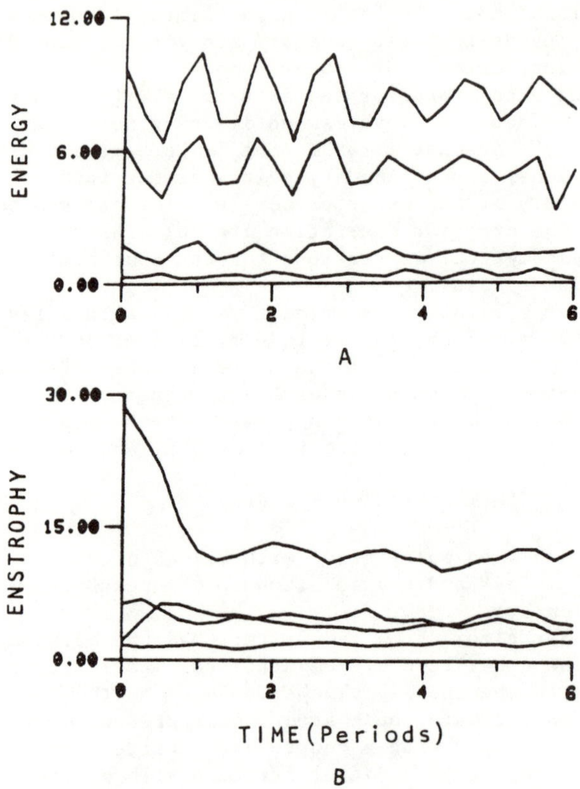

Figure 2. Integrated energy and enstrophy versus time from exterior simulation experiment. A. Level by level energy vs. time, levels 1,2,3,6. Level 1 is the most energetic, followed by 2, 3 and 6. B. Level by level enstrophy vs. time, levels 1, 2, 3 and 6. Enstrophy decreases with increasing depth at time 0. At the end of the simulation, level 3 has greater enstrophy than level 2.

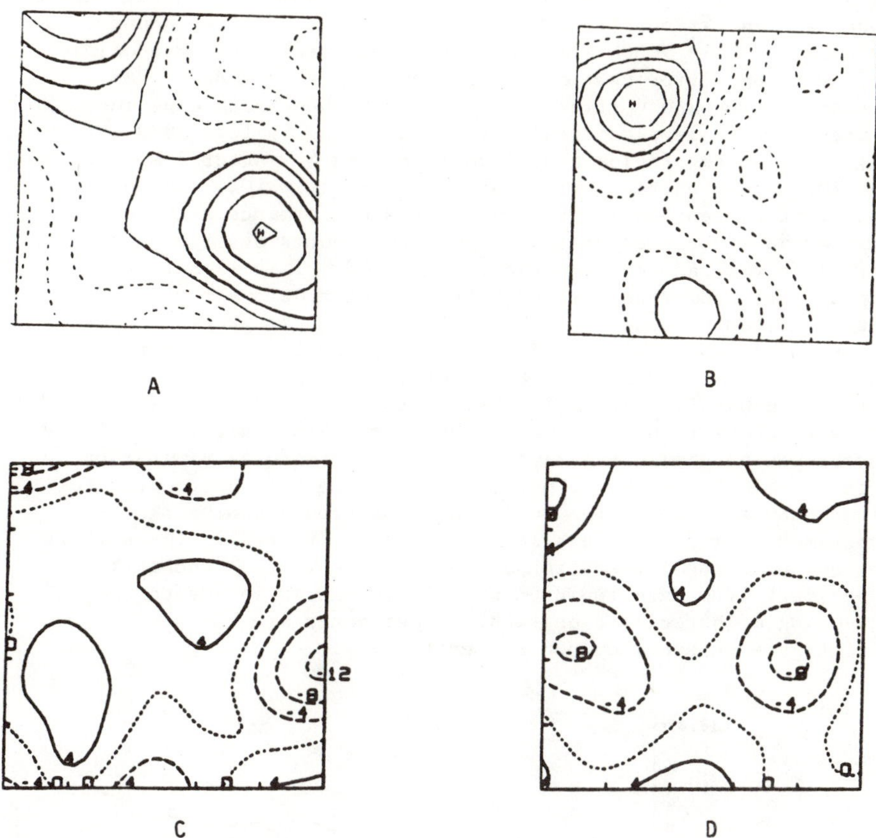

Figure 3. Simulated and observed streamfunction fields. A,B. Objectively analyzed streamfunction fields at 700 m from USSR POLYMODE array. Fields are 288 km square, separated in time by 24 days. Dashed lines are negative contours. Solid lines are 0 and positive contours. C,D. Simulated streamfunction fields at 750 m five and one half and five and three quarters period after start of simulation, spanning 32 days of model time. Fields are 500 km square. Dashed lines are negative contours. The zero contour is a dotted line. Solid lines are positive contours. All subsequent contour plots are of this form.

RESULTS OF FORECAST EXPERIMENTS

We now turn to the use of the model and simulated data set described above for a parameter and sensitivity study whose purpose is the quantitative determination of data requirements and forecast errors. This is the baroclinic generalization of RH. In our initial (benchmark) experiment, complete initial and boundary data are provided to the model. This is a "nowcast" rather than a forecast experiment, since new data is provided on the boundary at each time. This benchmark experiment determines the irreducible computational error that can be achieved by this model. RMS streamfunction and vorticity errors in the benchmark experiment at representative levels normalized by the level by level RMS amplitude of the fields are shown in Figure 4. After 3 months, the RMS streamfunction errors are approximately 10%. At this point, it is reasonable to compare this performance with the similar barotropic experiment. The experiments described in RH were done with a β-Rossby number $\varepsilon = 1.48$. This is not an appropriate comparison, since this simulated data set was characterized by speeds of the order of twice as great as those reported in RH. This is due to the fact that we use four waves instead of two, and also to the fact that the amplitude of the baroclinic waves, which were absent from the barotropic study, is quite large near the maximum of the first baroclinic mode. A new barotropic benchmark experiment was performed for a simulation with $\varepsilon = 3$ by Dr. E. S. Hertel with a data set constructed for other purposes. RMS streamfunction error from a benchmark experiment using that data set is noted in Figure 4. The authors believe that the degradation of performance in the benchmark experiment from barotropic to baroclinic models is due to the inclusion of baroclinic physics. Experiments are planned to investigate this phenomenon systematically.

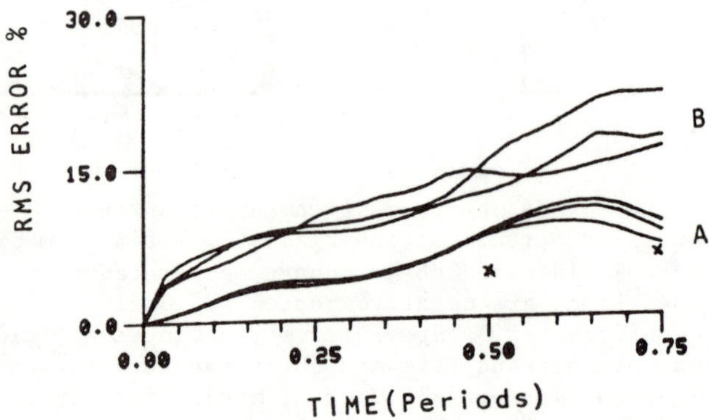

Figure 4. RMS errors of benchmark experiments. A. RMS streamfunction errors, levels 2, 3, 5. Errors decrease with increasing depth. B. RMS vorticity errors levels 2, 3, 5. At the beginning, errors decrease with increasing depth. At the end, level 5 has the largest error, followed by level 3 and level 2. "x" denotes RMS streamfunction error of barotropic benchmark experiment, $\varepsilon = 3$, courtesy of Dr. E. S. Hertel.

Figure 5 shows the comparisons of fields from the benchmark experiments. After 90 days of forecasting, the streamfunction and vorticity fields are nearly identical to the corresponding verification fields. As expected, the errors in the field are primarily concentrated in the large scales, and more near the boundary than in the interior.

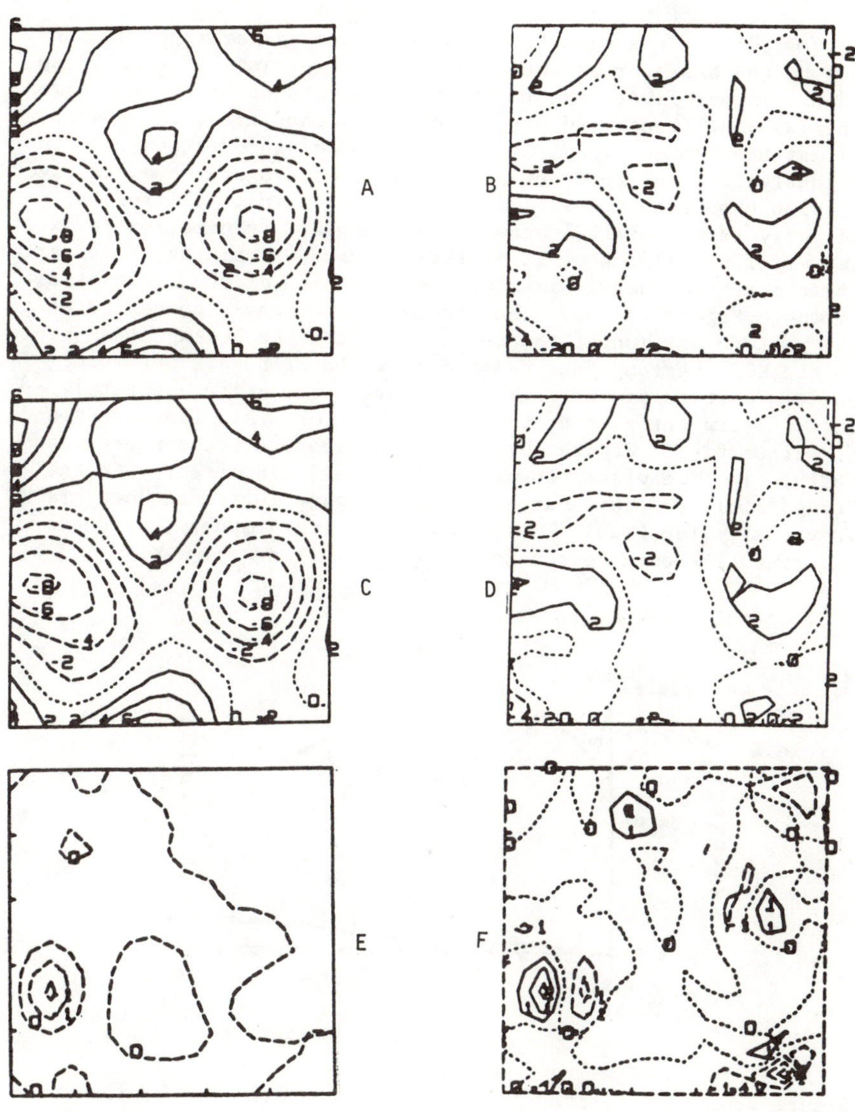

Figure 5. Results of benchmark experiment. Fields at 3/4 period, approximately 3 months of model time, at 750 m. A. Predicted streamfunction. Contour interval = 2. B. Predicted vorticity. Contour interval = 2. C. True streamfunction field. Contour interval = 2. D. True vorticity field. Coutour interval = 2. E. Streamfunction error field. Contour interval = .4. F. Vorticity error field. Contour interval = .5.

Guided by the results of the benchmark experiment, we turn now to the behavior of the model under conditions of incomplete data. In our first experiment, the model is provided with zero initial conditions, and perfect boundary conditions in order to illuminate the nature of the propagation of information from the boundary into the interior. Normalized errors are shown in Figure 6, with the benchmark as a comparison. The initial error is slightly less that 100% because the boundary information is exact. The error oscillates as it drifts downward, remaining at 70% after three months of model time. But examination of the field itself, shown in Figure 7, shows the error to be strongly concentrated in the western portion of the region. This is to be expected, since, in the MODE region, features are known to propagate westward. The eastern portion of the region is represented quite accurately -- note the low entering at the right in Figure 7B. The corresponding error field, Figure 7D, shows that feature to be reproduced faithfully. The visual comparison of the field directly to the "true" simulated date can be seen in Figure 10, since "benchmark" is visually identical. These results are preliminary, as these experiments should be run for longer duration.

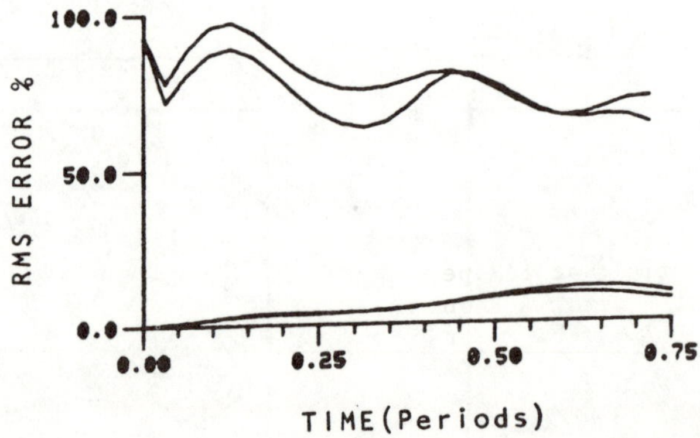

Figure 6. RMS streamfunction errors from zero initial condition experiment, with benchmark shown as the two lower curves for comparison. Curves with less error at the end are for level 5; others are for level 2.

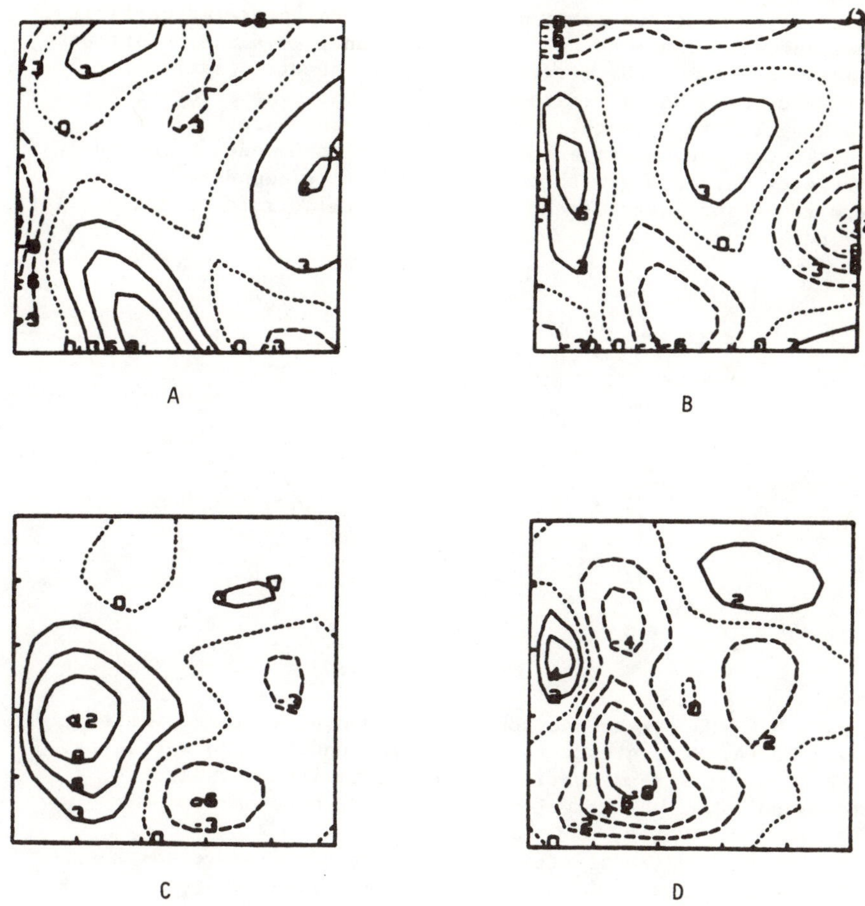

Figure 7. Fields at 750 m, from zero initial condition experiment.
A. Predicted field at 1/4 period. Contour interval = 3.
B. Predicted field at 1/2 period. Contour interval = 3.
C. Error at 1/4 period. Contour interval = 3.
D. Error at 1/2 period. Contour interval = 2.

In our next set of experiments, we examine the importance of providing accurate boundary information. In our first experiment, a true forecast experiment, we hold both vorticity and streamfunction constant on the boundaries. In this case, the forecast quality deteriorates rapidly. The normalized RMS error reaches 100% in approximately one week. Providing exact vorticity information while holding streamfunction constant on the boundary results in perceptible but insignificant improvement. Providing exact streamfunction data, however, results in dramatic improvement. One week into the experiment, when the persistence experiment shows 100% error, the experiment with exact streamfunction and persistent vorticity shows an error of 7%, reaching 23% in a month. This bears out the barotropic experience: the relative insensitivity of the model to boundary vorticity data along with the sensitivity to boundary streamfunction data. These results are illustrated graphically in Figure 8.

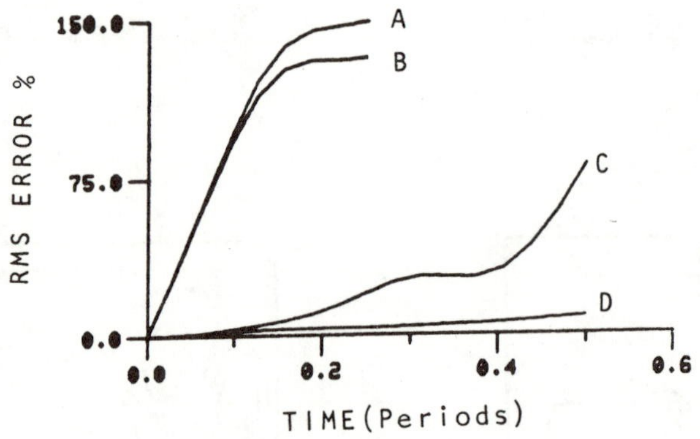

Figure 8. RMS streamfunction errors from persistence series, level 2. A. Persistence in both streamfunction and vorticity. B. Persistent streamfunction, exact vorticity. C. Exact streamfunction, persistent vorticity. D. Benchmark.

In the barotropic study, it was further discovered that the error could be controlled by the time interval between successive updates of the individual points on the boundary. To this end, we performed an experiment in which streamfunction and vorticity on the boundary are updated every four days. As in the barotropic study, we refer to this as the "P4" or four-day persistence experiment. Results of this experiment are shown in Figure 9. From Figure 9 it can be seen that the error at a given time is controlled by the time elapsed since the most recent update. Examination of Figure 9 shows that the minima and maxima of the P4 error curve creep upward slowly, at approximately the same rate as the growth rate of the error in the benchmark experiment. It should be noted here that the error statistics were not computed at every time step, and thus the particular structure of the P4 error curve derives in part from the phasing of the computation relative to the updating of the boundary values.

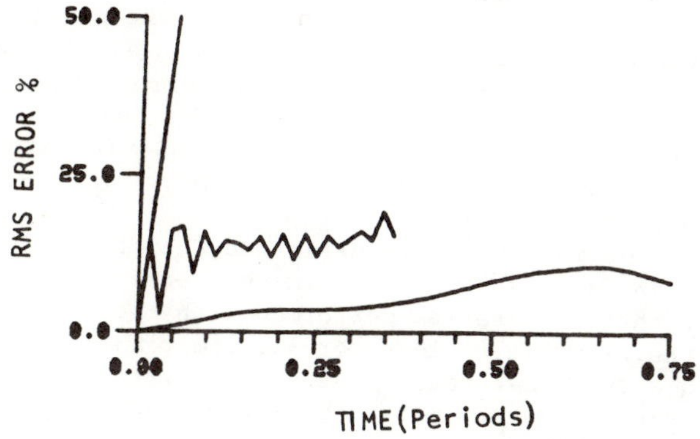

Figure 9. RMS streamfunction errors at 750 m from updating experiment. Top curve: persistence. Middle curve: boundary updating every four days (P4). Lowest curve: benchmark.

SUMMARY AND CONCLUSIONS

We have presented a series of forecast experiments using a baroclinic quasigeostrophic open ocean model with simulated data. Our simulated data set was constructed to have scales and phases typical of the MODE region and to mimic the quasiturbulent nature of the observed field.

Dynamical forecasting errors can be ascribed to three sources: 1) physical errors related to inadequate dynamical representation of resolved and/or unresolved scales of motion; 2) computational errors which arise from the method by which the model equations and boundary conditions are approximated on the computer; 3) observational errors associated with inaccurate or incomplete boundary, initial and verification data. This paper, like its barotropic predecessor addresses the question of determining quantitative limits on computational errors, and on the effect of systematically degraded data on forecast errors.

As in the barotropic case, the only source of computational error is the implementation of the open boundary condition, since the simulated data set was generated by the same model with the same physics. Thus the only difference between interior and exterior simulations arises from the fact that the boundaries of the interior simulation are ordinary interior points in the exterior simulation. In that case where exact initial data is supplied to the scheme, and exact boundary data is supplied at each time step (the benchmark experiment), the error can be held to approximately 10% over the course of a three month simulation.

An experiment with zero initial condition showed that the

normalized RMS streamfunction error decreased only gradually as new information propagated into the field. In just over three months, the error was just under 70%, with an oscillation superimposed on the slow decrease of the error. The spatial distribution of the error was highly inhomogeneous: the fields themselves showed the expected concentration of error near the western boundary of the region due to the westward phase propagation typical of features in the MODE region. Despite the large RMS errors, features near the eastern boundary of the field are reproduced quite faithfully.

In our other experiments, we investigated the effects of incomplete data upon forecast quality. In the persistence calculation, i.e. that calculation performed with the boundary values fixed at their initial values for all time, the error reached 100% in a week. Improvement was not significant when exact vorticity information was provided at the boundary each step, with the streamfunction held constant. Providing exact streamfunction information with persistent vorticity resulted in dramatic improvement, with the error not reaching 10% for just over two weeks, and not reaching 50% for nearly two months. This bears out the barotropic experience that the method is more sensitive to errors in streamfunction than it is to errors in vorticity.

An updating experiment, with the boundary values updated every four days, was performed, following the barotropic experience that the error was controlled by the frequency of boundary updating and was not sensitive to the sequence in which points on the boundary were updated. In our updating experiment, we found that in approximately one month, the error could be controlled at a level near 15% and the graphical representation of the field was quite faithful, with most of the details well represented. The error at that point appears to have a box mode structure. A summary of forecast fields after one month with their associated error fields appears in Figure 10.

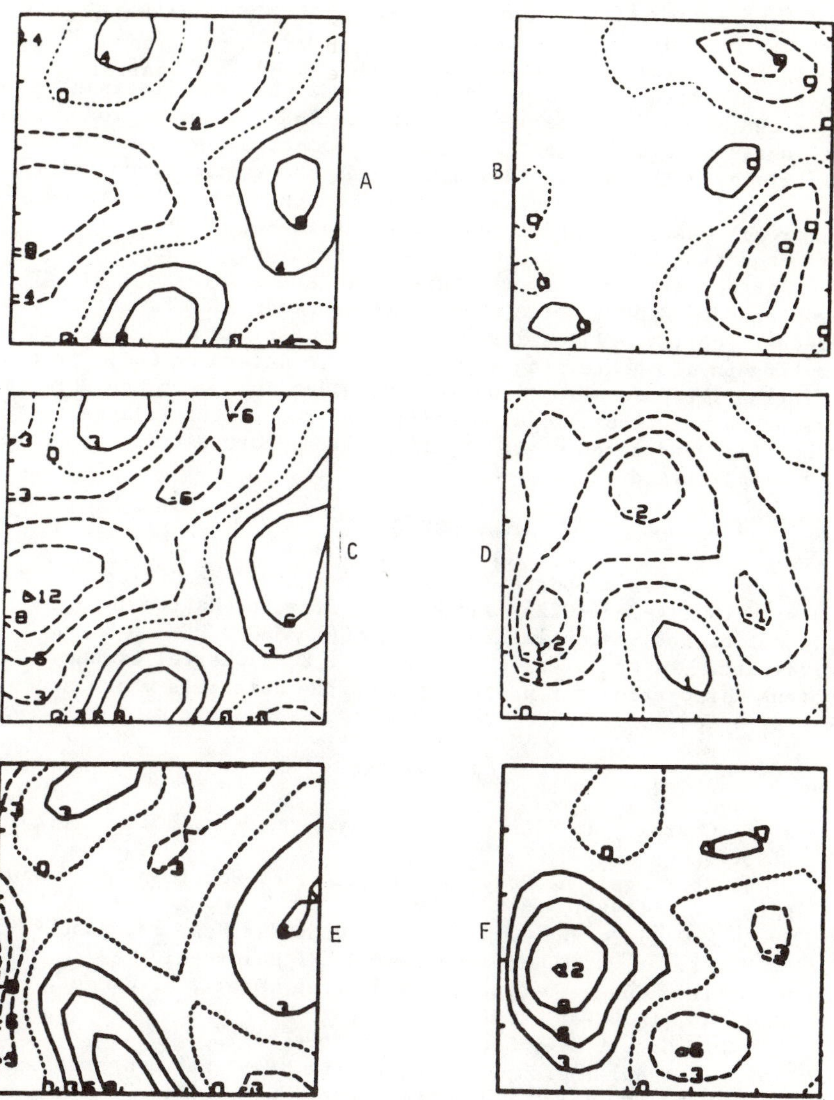

Figure 10. A summary of 30 day forecast experiments. Streamfunction fields and their asssociated error fields shown at the 750 m level. A. Benchmark. Contour interval = 4. B. Error in benchmark simulation. Contour interval = .2. C. Boundary values updated every 4 days (P4). Contour interval = 3. D. Error field associated with C. Contour interval = .4. E. Zero initial condition experiment. Contour interval = 3. F. Error associated with E. Contour interval = 3.

This series of dynamical forecast experiments with simulated data has provided valuable quantitative information about the model, which will be of major importance to our modeling experiments with real data. We have found that computational error can be held to the order of 10% in RMS streamfunction over the course of a three month forecast. Comparison with barotropic simulations with comparable scale speeds shows significant effect due to inclusion of baroclinic physics in the model. As in the barotropic case, the model was more sensitive to changes in streamfunction than it was to degraded vorticity data.

Further experiments with this model and simulated data set are planned. Random noise will be introduced, and its effect upon forecast quality will be determined. Simulations with different sampling in depth are planned, in order to determine the effect of differing depth sampling schemes. Finally, following the work by Tu[8] on the barotropic model, statistical quantities will be calculated, and mixed statistical/dynamical forecast experiments will be performed.

ACKNOWLEDGMENTS

This research was supported by a contract from the Office of Naval Research (N0014-75-C-0225) and a grant from the National Aeronautics and Space Administration (NASA-NSG-5228) to Harvard University. We are pleased to thank Dr. E. S. Hertel of the Proteus Corporation for preparation of the data sets which made this study possible.

REFERENCES

1. J. G. Charney, R. Fjortoft and J. von Neumann, Tellus 2, 237 (1950).
2. D. B. Haidvogel, A. R. Robinson and E. E. Schulman, J. Comput. Phys. 34, 1(1980).
3. J. C. McWillams and G. R. Flierl, Deep-Sea Res. 23, 285(1976).
4. R. N. Miller, A. R. Robinson and D. B. Haidvogel, "A baroclinic quasigeostrophic open ocean model", J. Comput. Phys. 49, In press. (1983).
5. The MODE Group, Deep-Sea Res. 25, 859(1978).
6. The MODE-1 Atlas Group, "Atlas of the Mid-Ocean Dynamics experiment (MODE-1)", MIT, Cambridge, Mass., 1977.
7. A. R. Robinson and D. B. Haidvogel, J. Phys. Oceanogr. 10, 1909(1980).
8. K-S Tu, A Combined Statistical and Dynamical Approach to Regional Forecast Modeling of Open Ocean Currents. Harvard University Reports in Oceanography and Metererology #13. Harvard University, Cambridge, Mass. 1981.

SENSITIVITY STUDIES WITH AN UPPER OCEAN PREDICTION MODEL

Russell L. Elsberry and David Adamec
Naval Postgraduate School, Monterey, California 93940

ABSTRACT

A series of sensitivity studies is carried out to demonstrate the relative importance of initial temperature profile errors and atmospheric forcing errors on upper ocean thermal structure prediction. Given a proper data assimilation technique, the sensitivity of the model to initial temperature profile errors is minimized. The seasonal dependence in the model sensitivity is due to variations in the magnitude and character of the atmospheric forcing as well as in the pre-existing ocean thermal structure profile. The most important atmospheric forcing variable is the wind speed during both winter and summer. The conclusion of these studies is that the accuracy and predictability of the upper ocean thermal structure is intimately linked to the ability to specify the atmospheric forcing, and thus to the atmospheric predictability.

INTRODUCTION

First generation ocean prediction systems have only recently been developed.[1,2,3] Thus far, these systems have attempted to predict the upper ocean thermal structure changes on time scales of 72 h as an initial value problem. Therefore, the focus of these models is quite different than the systems described by A. Robinson and by J. Miller and A. Robinson in the previous talks in this session. In this paper, the focus will be on the diurnal and synoptic time scale response of the upper ocean to atmospheric forcing. Discussion will also be limited to a single location--Ocean Weather Ship Papa at 50°N, 145°W. Three sources of error in the ocean prediction system can be treated. First, the numerical and physical aspects in the model may be in error. For example, the version of the one-dimensional, bulk oceanic mixed layer model of Garwood[4] used in this work does not treat advective effects or include salinity processes. Nevertheless, the predictions of this simple model will be used as a control. Second, the representation of the initial thermal structure profiles will be imperfect, and will lead to error growth even though the prediction is assumed to be perfect. Finally, an imperfect knowledge of the atmospheric forcing variables will also be an important source of error growth in the upper ocean prediction system.

The ocean model and data will be described in the next section. The primary purpose of that section will be to demonstrate that a model without horizontal or vertical advection has some credibility in the region of OWS Papa. In the following section, the sensitivity of the model to various sources of error in the initial temperature profiles will be described. These results are an extension of the data assimilation study of Elsberry and Warrenfeltz.[5] Finally,

0094-243X/84/1060473-15 $3.00

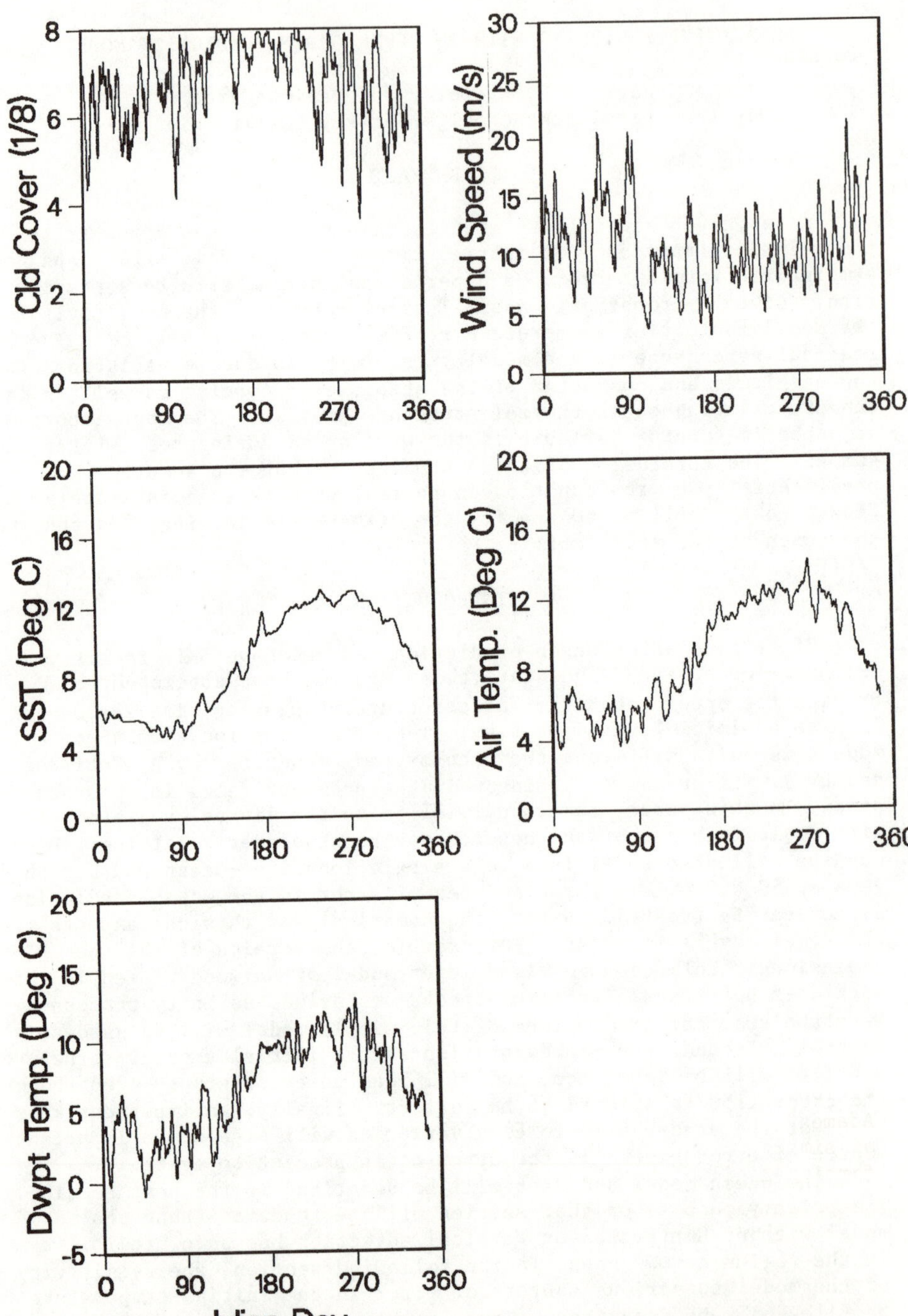

Fig. 1 Atmospheric forcing variables at OWS Papa during 1959 necessary for the ocean prediction model. A four-day running mean has been applied for illustration purposes only.

the sensitivity of the ocean model to errors in the atmospheric forcing variables will be described.

OCEAN MODEL

The one-dimensional, oceanic mixed-layer model of Garwood [4] is used in these studies. This vertically integrated or bulk model utilizes an entrainment hypothesis that is dependent on the horizontal and vertical components of turbulent energy. Part of the wind-generated turbulent kinetic energy that increases potential energy by deepening the mixed layer is dependent on stability. This results in modulation of the entrainment rate by the diurnal heating/cooling cycle. Another feature of the model is the planetary influence on the dissipation time scale for turbulence, which enhances dissipation for deeper mixed layers.

The atmospheric forcing for the model is calculated from 3-h observations of wind speed, cloud cover, sea surface temperatue, air temperature and dew point. Surface fluxes of buoyancy and momentum are calculated each 3 h using the bulk aerodynamic formulae, and the fluxes are then interpolated to the 1 h time step of the model. Garwood and Adamec[6] have simulated 17 y of mixed layer evolutions at OWS Papa. An example of the atmospheric forcing variables during one of these 17 y is shown in Fig. 1. The seasonal and synoptic variability in the wind speed and cloud cover are well marked. The sea-surface, air and dew point temperatures have corresponding seasonal variations, and the latter two variables also have a synoptic time scale variability.

Each of the 17 annual simulations is begun from an idealized temperature profile on 1 January. The temperature profile is resolved to 200 m at 1 m intervals. No additional temperature profiles are provided the model during the annual simulations. Thus it is noteworthy that the predicted isothermal layer depth in Fig. 2 agrees as well as it does with the observed mixed layer depth (defined to be the depth at which the temperature is 0.2^{o}C below the surface temperature). Notice in particular that the synoptic deepening events during the summer and autumn are well represented in the simulation. There is a tendency for the predicted layer depth to be too shallow during the summer. Thus the model-predicted temperatures in Fig. 3 are too high by about 1°C during this period. There is again a marked synoptic time scale signal in both the simulations and in the observations. Other examples of such simulations are given in Garwood and Adamec[6], and other authors[1,2] have demonstrated the credibility of similar ocean prediction models, given the excellent atmospheric forcing data at the ocean weather ship locations.

EFFECT OF INITIAL TEMPERATURE PROFILE ERRORS

An integral part of a prediction system is a technique for providing the best possible initial temperature profile. An additional requirement is that the profile should be consistent with the vertical structure of the model. Elsberry and Warrenfeltz[5] have

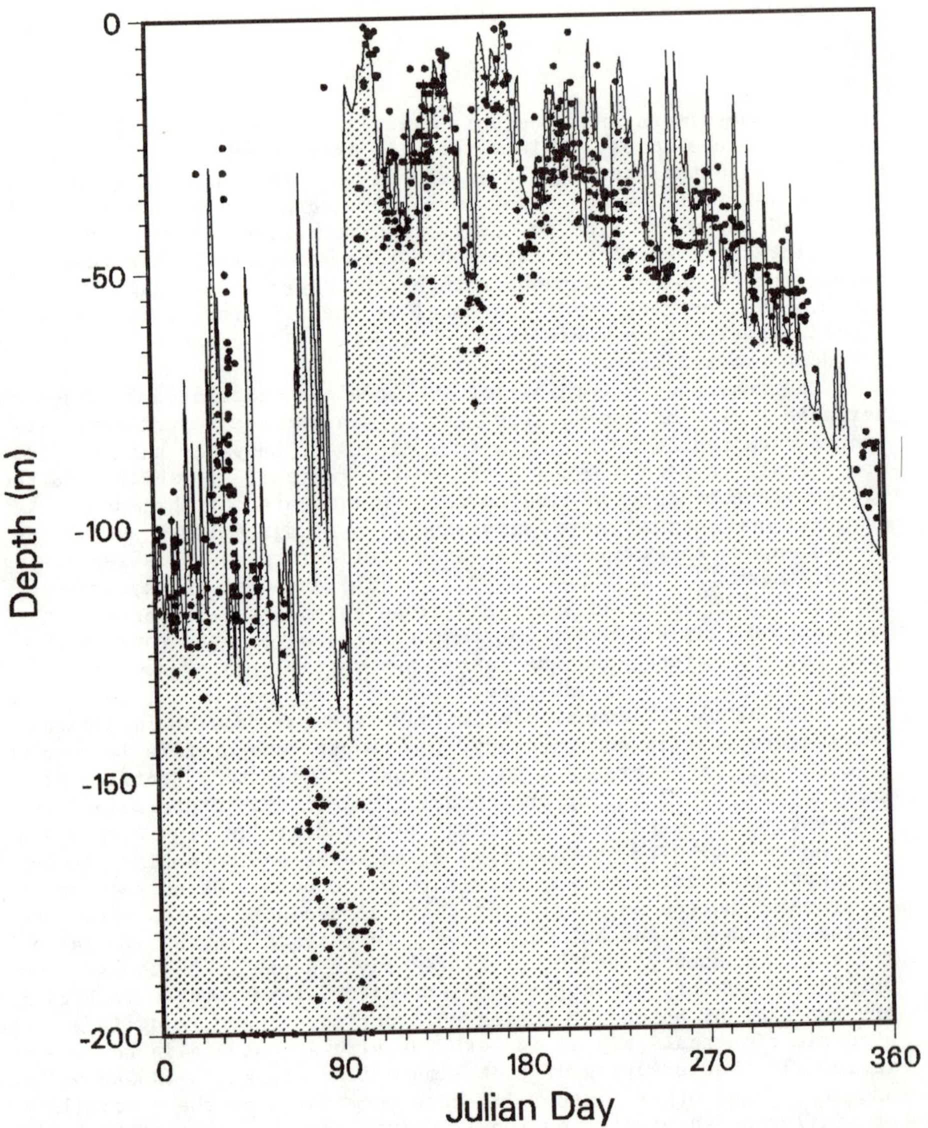

Fig. 2 Maximum mixed layer depth predicted (solid) during each day of 1959 and observed depths (dots) at OWS Papa.

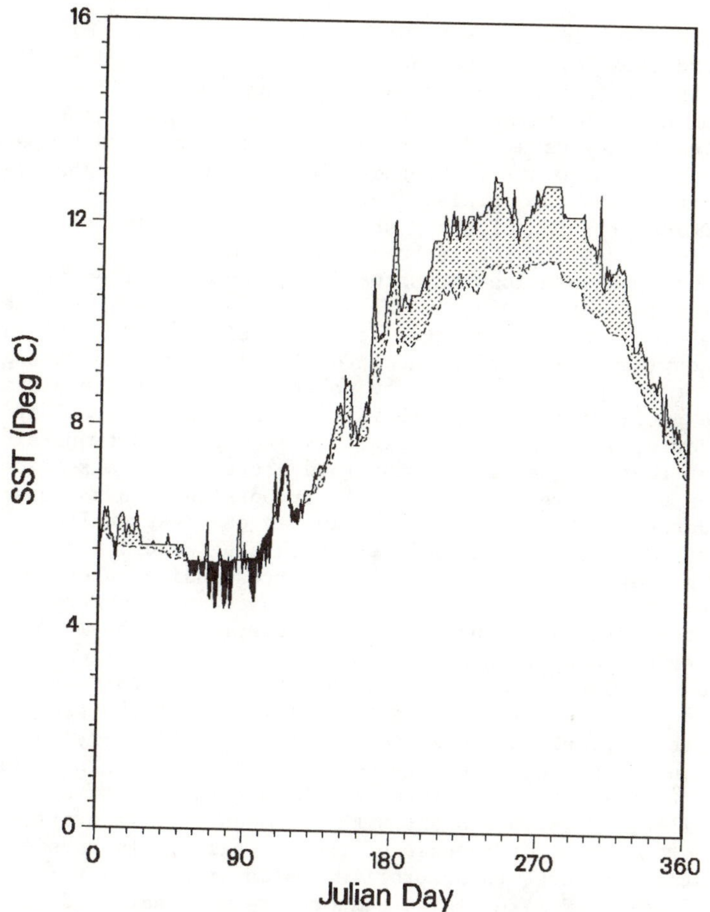

Fig. 3 Mixed layer temperature predictions (solid) and observations (dashed) during 1959 at OWS P.

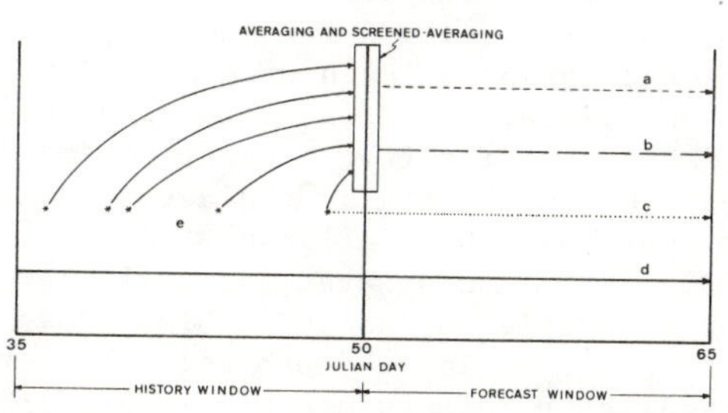

Fig. 4 Forecast paths for study of effect of initial temperature profile errors at the time indicated by asterisks (e). Forecasts are based on the screened-average of observations (a), the average of all observations (b), the last available profile only (c) and the control run (d).

described a data assimilation technique which satisfies these requirements for the ocean mixed layer model. They demonstrated that the data assimilation technique can result in improved predictions in tests in which purely random errors were added to model-generated initial profiles. They used different mixed layer depth and temperature errors in summer and winter, but they did not test the technique over a range of errors in the initial profiles.

A short description of the Elsberry and Warrenfeltz tests will be given here to illustrate the technique. As indicated in Fig. 4, each of the 17 annual simulations of Garwood and Adamec[6] serves as a control. During the late winter and also during midsummer, a series of temperature profiles are extracted at random times during the 15-d history window, and 3-h temperature profiles are retained during the 15-d forecast period to serve as the control. Each of the history temperature profiles becomes a simulated "observation" by addition of a Gaussian error distribution in mixed layer depth and temperature and in the upper thermocline slope. The random error with a zero mean added to the mixed layer temperature is proportional to the standard deviation during the history window, plus an observational error of 0.25°C. Random errors in mixed layer depth are 7 and 3 m for the winter and summer cases, respectively. Corresponding values of the slope random errors are 0.025 and 0.02. Adamec and Elsberry[7] have extended these experiments by varying systematically the magnitude of the random errors and by shifting the Gaussian distribution to have either a positive or a negative bias rather than a zero mean.

A standard of comparison (see Fig. 4) is to use the predictions from the last available of these "observations." Elsberry and Warrenfeltz used the model to advance all of the available profiles to the initial time and then combined them by a simple average into a single profile that is consistent with the model. This profile will also have a reduced error because of the averaging. As an additional quality control feature, those profiles which exceed 1.5 times the standard deviation of mixed layer temperature were screened out prior to averaging. Elsberry and Warrenfeltz[5] show examples of the mixed layer depth and temperature errors in individual years due to the initial temperature profile errors.

The mixed layer depth and temperature error growth relative to the control is shown in Figs. 5 and 6 for the 17-y ensemble of winter and summer 15-d forecast periods. Even with an error factor of zero, which means that no additional error is included beyond the extraction of an idealized "observation"from a detailed, model-generated temperature profile, the superiority of the averaged and screened-averaged approaches over the last-available strategy is clear during the winter period (Fig. 5). Adding systematically larger random errors, or including a positive or negative bias, does not change the relationship between the mixed layer depth errors of the three approaches. The mixed layer temperature errors are larger as the error factor is increased, with the largest errors occurring if there is a negative bias in the observations which adds to the tendency to have a negative bias in the model predictions with a random error. With the additional quality control feature of screening the initial

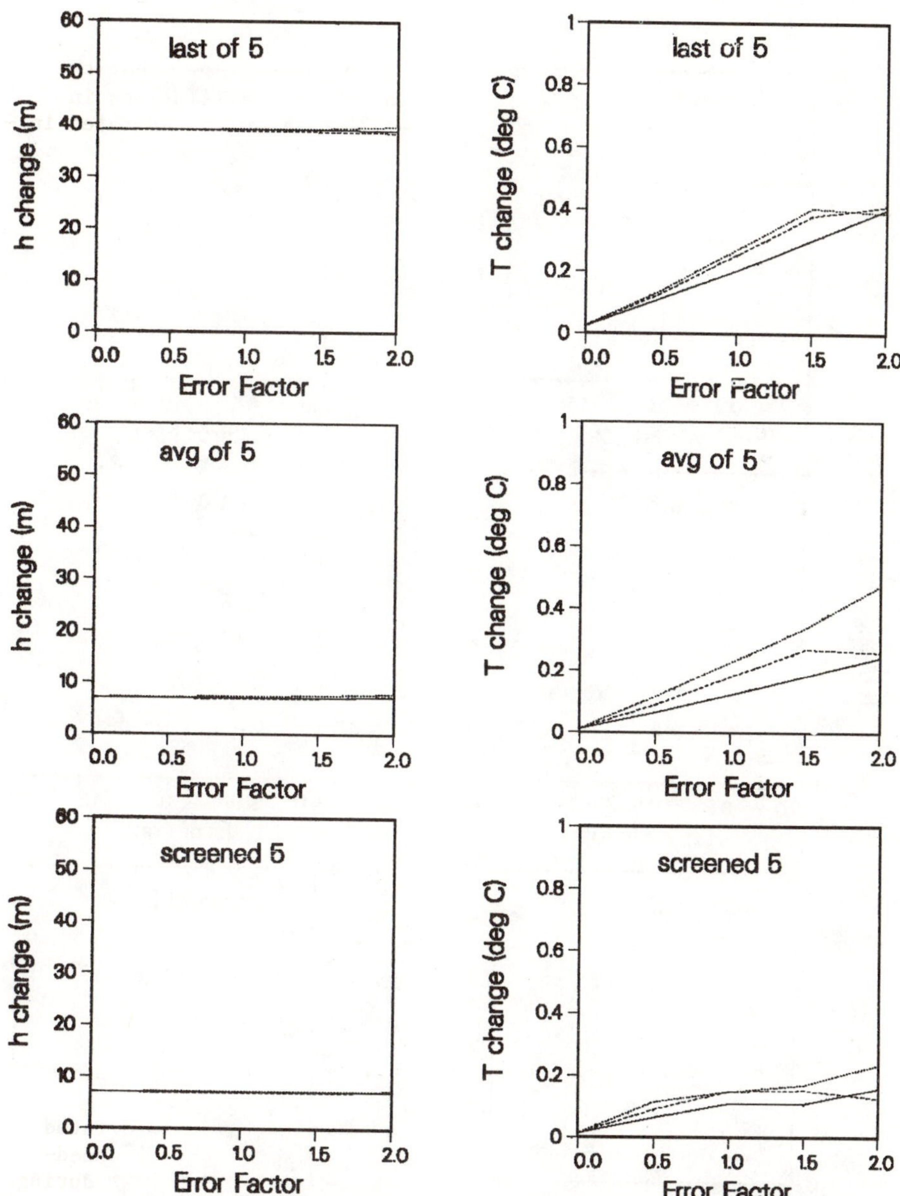

Fig. 5 Root mean square deviations of the predicted mixed layer depths (h) and temperatures (T) during the winter 15-day predictions due to imposed errors on the set of 5 initial profiles. An error factor of 1.0 indicates a Gaussian error equal to one standard deviation. Random errors with a zero mean (solid), a positive bias (dotted) and a negative bias (dashed) in depth, temperature and thermocline slope have been added. Last of 5 (top), average of 5 (middle) and screened average of 5 refer to the forecast paths c, b and a in Fig. 4.

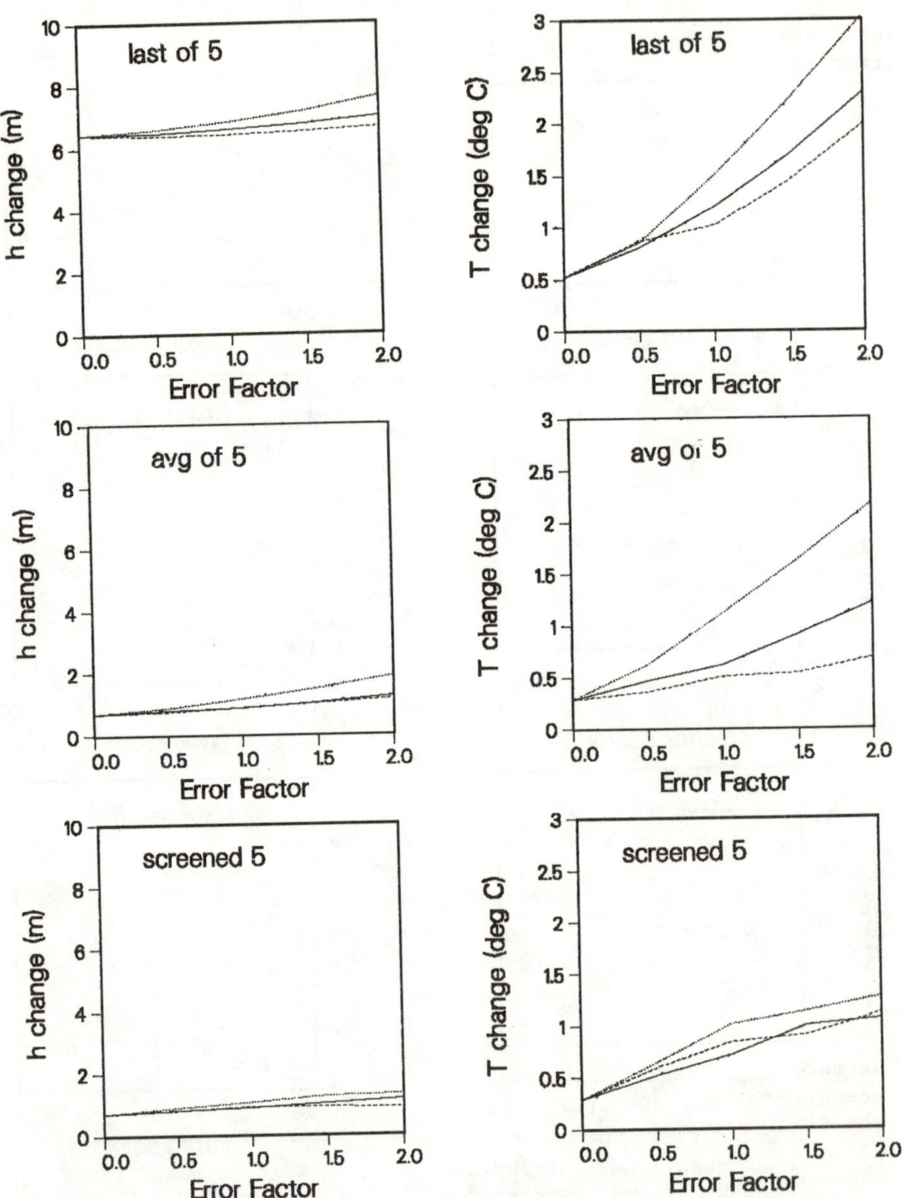

Fig. 6 As in Fig. 5, except during summer.

profiles, the error growth is slightly smaller. During the summer (Fig. 6), the superiority of the assimilation techniques is evident for depth errors, but not for temperature errors until rather large error factors are used. The error growth with increasing error factor is again smallest if the screened-averaged approach is used. In particular, this approach is nearly as effective in handling very large random errors with a zero mean as it is with small random errors. There is a residual error even with the best strategy for specifying the initial profile. This is a reflection of the contribution to predictability due to uncertainty in the initial temperature profile information.

SENSITIVITY TO ATMOSPHERIC FORCING VARIABLES

One can see from Fig. 1 that the fluctuations of the atmospheric forcing variables is a function of season. Table I is a listing of the 17-y ensemble standard deviations of each of these variables during the late winter and midsummer periods. The natural variability is larger during winter than during summer, except for sea-surface temperature. It will be assumed below that the errors in observing these variables will be proportional to these standard deviations. A Gaussian distribution of the errors with a zero mean or with a positive (negative) mean equal to the standard deviations will be added to each variable one at a time. A lower bound of zero wind speed and cloud cover and an upper bound of 1.0 on cloud amount are imposed. The sensitivity of the ocean model to each atmospheric forcing variable can then be estimated by comparing the prediction with the modified forcing to the control run based on "perfect" forcing.

Examples of the departures from the control run during an individual integration are shown in Fig. 7. The addition of a random error with a zero mean introduces departures which tend to cancel out over the 15-d period. However, if there is a bias in the forcing, the errors tend to accumulate. For example, the temperature error (Fig. 7d) due to a negative bias in wind speed contributes to a generally increasing warm bias. This is accompanied by a systematic shallow bias in the mixed layer depth (Fig. 7c). By contrast, a positive bias in wind speed leads to a deeper and cooler mixed layer. These positive or negative biases add or subtract to the synoptic fluctuations, so there is an enhancement of synoptic time scale response in the ocean. A negative bias in cloud cover leads to a warm bias (Fig. 7b) as more insolation will be available for absorption in the upper ocean. A positive bias in cloud cover has almost no effect because the cloud cover is already near 1.0 at OWS Papa during the summer (see Fig. 1). The next largest contribution to temperature error arises from dew point errors (Fig. 7f). One can see again that a bias in the atmospheric forcing variable is much more detrimental to the ocean prediction than is a truly random error with a zero mean. A positive bias in dew point decreases the surface evaporative heat flux and thus leads to less cooling of the upper ocean. Adamec and Elsberry[7] also show examples of the departures during the winter periods, when the most sensitive ocean variable is the mixed layer depth.

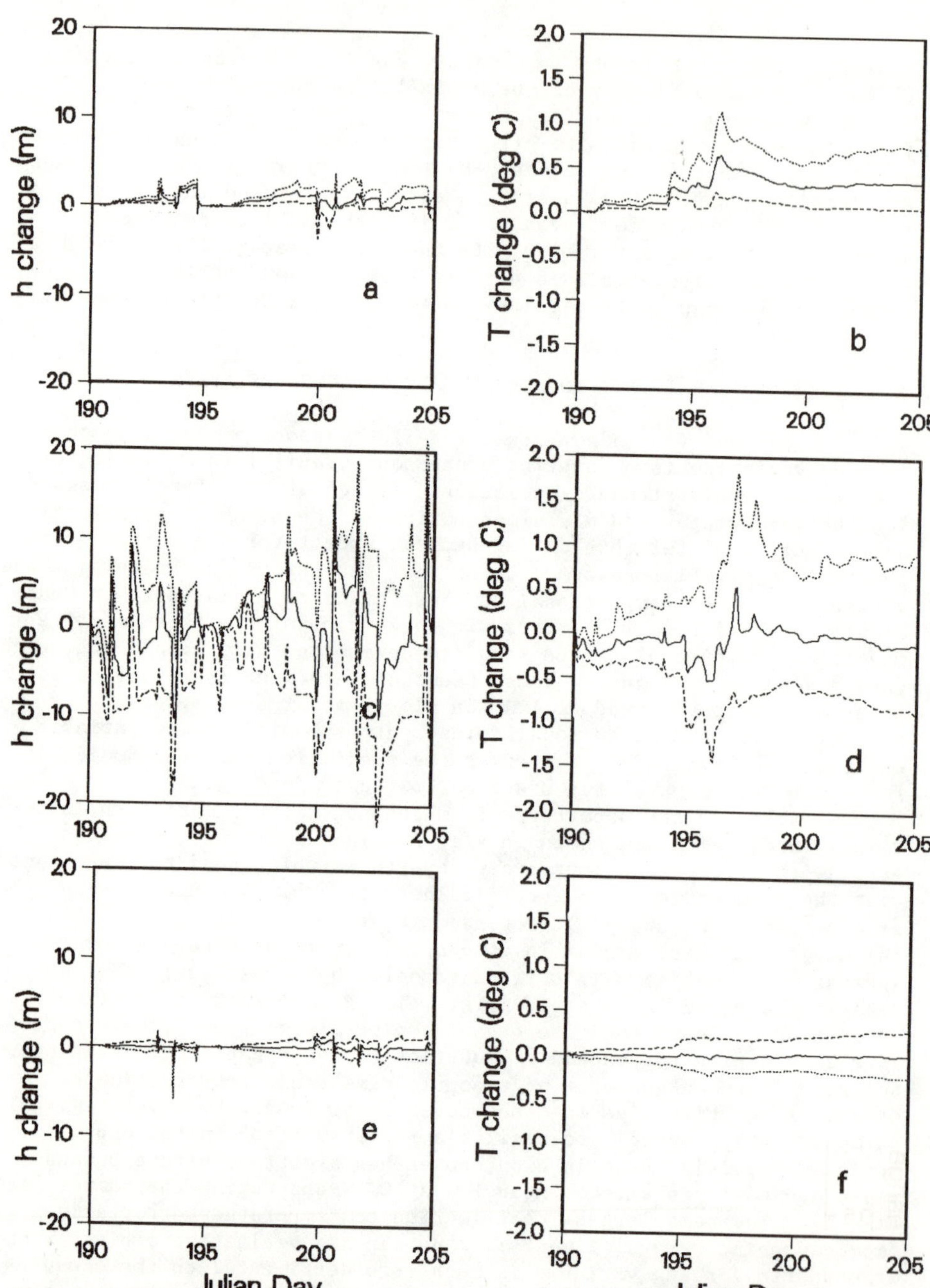

Fig. 7 Depth and temperature deviations from the control for summer 1965 predictions upon varying cloud cover (a,b), wind speed (c,d) and dewpoint temperature each 3 h by adding a Gaussian error with a positive (dashed), zero (solid) or negative (dotted) mean.

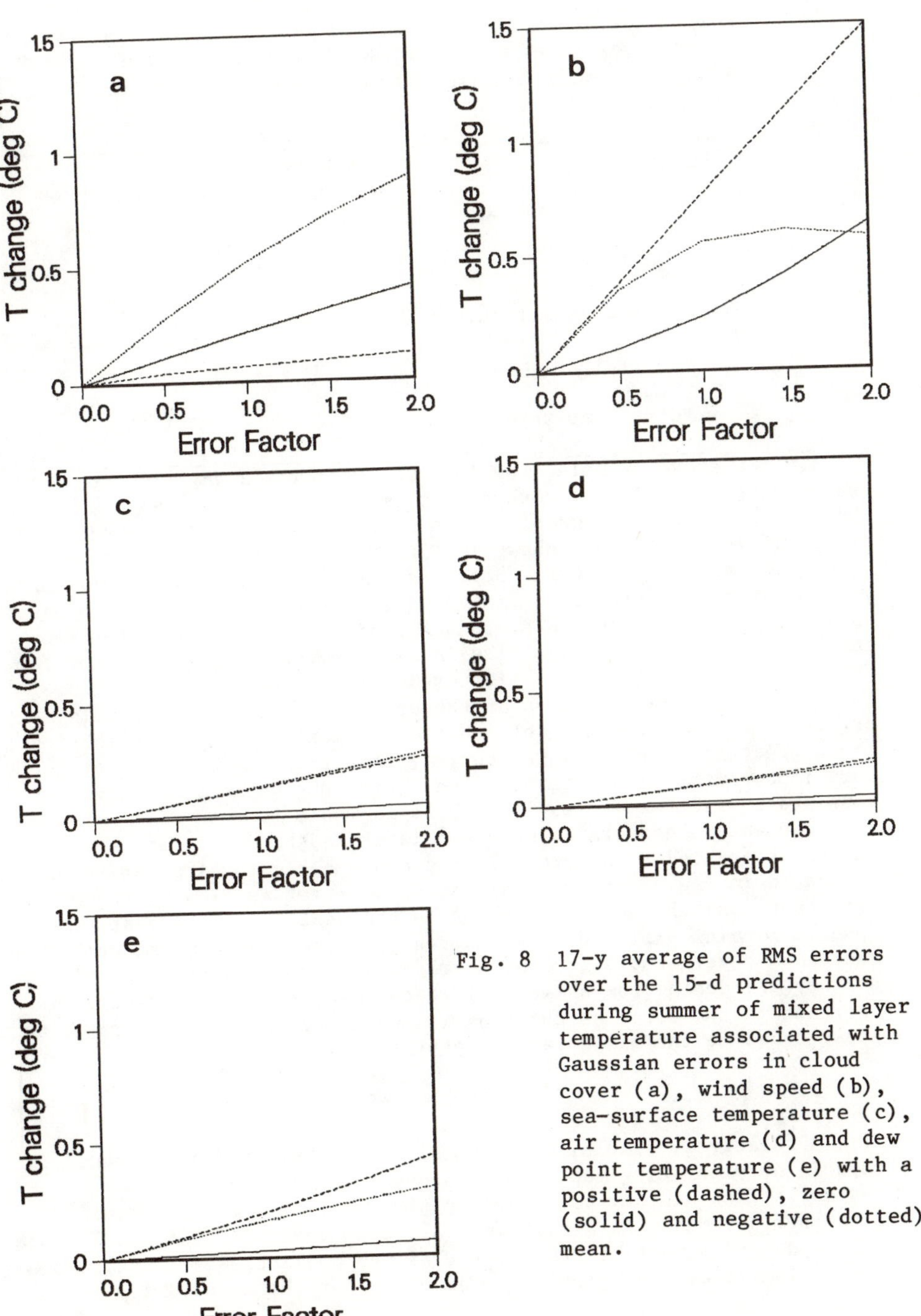

Fig. 8 17-y average of RMS errors over the 15-d predictions during summer of mixed layer temperature associated with Gaussian errors in cloud cover (a), wind speed (b), sea-surface temperature (c), air temperature (d) and dew point temperature (e) with a positive (dashed), zero (solid) and negative (dotted) mean.

Table I. Ensemble average of the standard deviations of atmospheric forcing variables within 15-d periods during winter (Days 50-65) and summer (Days 190-205) based on 17-y record at OWS Papa.

Variable	Winter	Summer
Cloud Cover (Eighths)	2.0	1.0
Wind Speed (m/s)	5.31	3.22
Sea Surface Temperature (^{o}C)	0.24	0.50
Air Temperature (^{o}C)	1.21	0.81
Dew Point Temperature (^{o}C)	2.44	1.19

In the above example, the imposed error was always one standard deviation (see Table 1). Other error magnitudes were also used, and the 17-y ensemble average of the resulting mixed layer temperature errors during summer are shown in Fig. 8. The largest errors occur with wind speed errors (Fig. 8b). As was the case in the individual year example, positive or negative biases in the atmospheric forcing variable are more detrimental than a random error with a zero mean. In the case with a positive bias in cloud cover (Fig. 8a), error factors of 1.5 or 2.0 have to be imposed before a net effect is shown, but this only occurs because of the upper bound on this variable. Errors in dew point (Fig. 8e) provide the next largest contribution to ocean prediction errors. Sea-surface temperature errors are slightly more improtant than air temperature during the summer, which is consistent with the variability shown in Table 1.

The ensemble averages of mixed layer depth errors during the winter associated with errors in the atmospheric forcing variables are presented in Fig. 9. Here the wind speed error is clearly the most important, and the dew point error is the next most important atmospheric forcing variable. By contrast, the sea-surface temperature or the cloud cover errors are not nearly so important during the winter as they are during the summer. In each case, there is a systematic increase in the error as the bias is increased, although there is an upper bound on the mixed layer depth errors.

CONCLUSIONS

A relatively simple, one-dimensional ocean mixed layer model is capable of simulating much of the upper ocean thermal structure variability on diurnal and synoptic time scales, provided excellent atmospheric forcing is available. Although there is additional variability due to advective processes that are not included in this model, these simulations can provide a guideline for the sensitivity of an

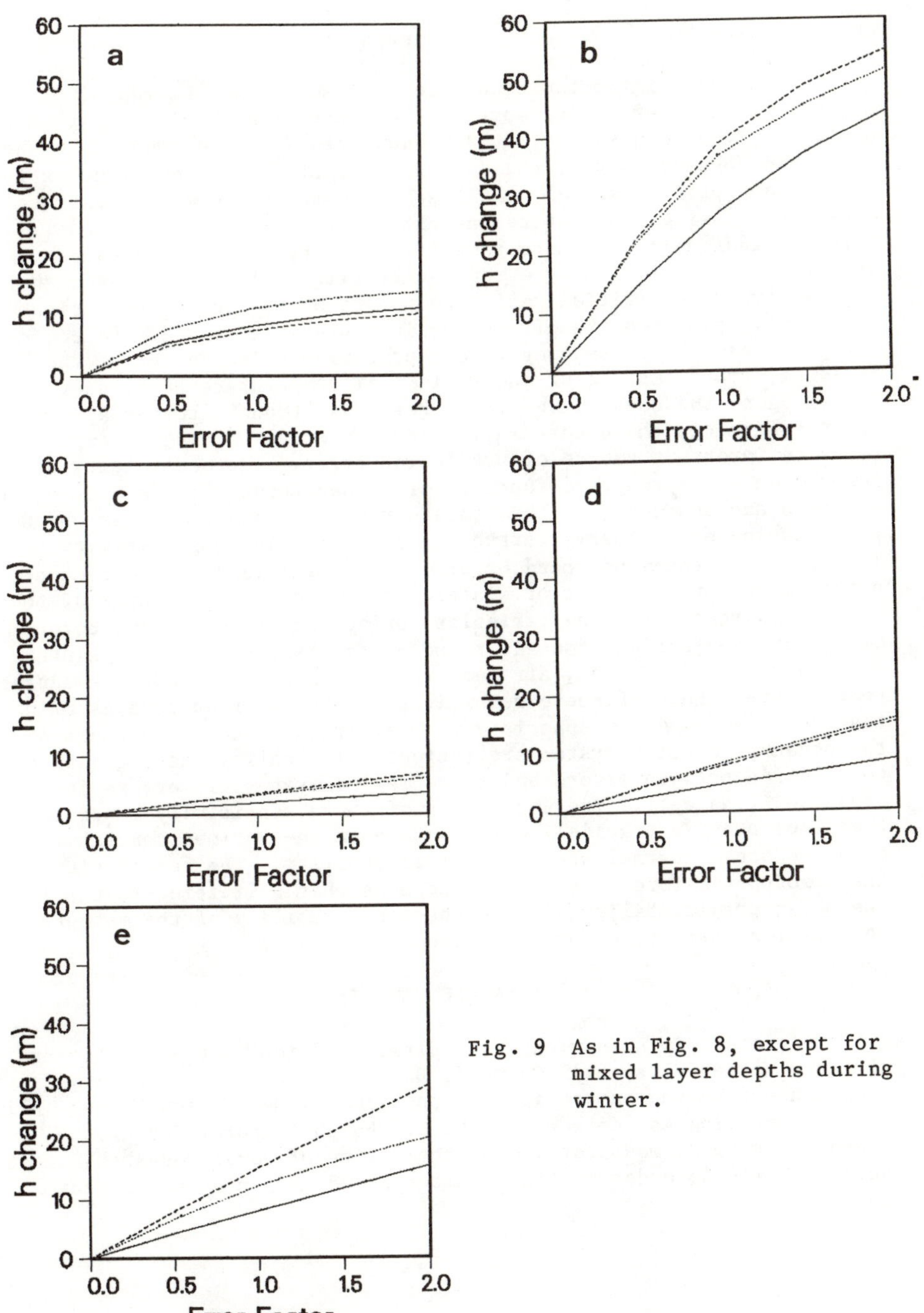

Fig. 9 As in Fig. 8, except for mixed layer depths during winter.

ocean prediction system to initial temperature profile errors or to atmospheric forcing errors.

The data assimilation technique of Elsberry and Warrenfeltz[5] appears to be an effective approach for improving the initial tempera-profile. Even when fairly large random errors are introduced in the simulated "observations", the data assimilation technique results in predictions with relatively small errors compared to an alternative strategy of using the most recent profile to initialize the model. This conclusion is even more true if a quality control procedure is used to screen out clearly erroneous profiles. Thus it appears that a good analysis and initialization component of the ocean prediction system will minimize the effects of incomplete and inaccurate ocean data sources. These studies give a preliminary indication of the residual errors that occur due to initial temperature profile errors.

Because the ocean mixed layer responds strongly to the atmospheric forcing, the uncertainty in the external forcing is expected to be an important source of prediction error. The sensitivity of the ocean model is strongly a function of season because it depends on the strength and character of the atmospheric forcing, as well as on the pre-existing ocean thermal structure profile. These sensitivity tests indicate that the wind speed error must be minimized for correct ocean predictions during summer or winter. The order of importance of the other atmospheric forcing variables during summer is: cloud cover, dew point, sea-surface and air temperature. However, during winter the order is: dew point, air temperature, cloud cover and sea-surface temperature. Each of these sensitivity tests has been done assuming that the remaining atmospheric variables or the other components of the ocean prediction system are perfect. In reality there may be a correlation between errors and there may be systematic errors in the ocean model as well. The usefulness of these sensitivity tests is that they provide a guideline for assessing the various contributions to upper ocean thermal structure predictability. The sensitivity to the atmospheric forcing strongly suggests that a limiting factor in the ocean predictability will be the predictability of the meteorological forecast system over the oceans.

ACKNOWLEDGEMENTS

We gratefully acknowledge the direct and indirect contributions of Larry Warrenfeltz, Bill Garwood, Bob Haney and Pat Gallacher. Assistance with the manuscript was provided by Ms. Marion Marks. All of the computing was done at the W. R. Church Computer Center. Financial support was provided by the Office of Naval Research, Ocean Science Division under contract number NR 083-275 program element 61153N.

REFERENCES

1. R. L. Elsberry, R. W. Garwood, Jr., Bull. Amer. Meteor. Soc., 61, 1556 (1980).
2. R. M. Clancy, P. Martin, Bull. Amer. Meteor. Soc., 62, 770 (1981).
3. A. Warn-Varnas, M. Clancy, M. L. Morriss, P. Martin, S. Horton, Studies of Large-scale Thermal Variability with a Synoptic Mixed-layer Model (this volume).
4. R. W. Garwood, Jr., J. Phys. Oceanogr., 7, 455 (1977).
5. R. L. Elsberry, L. L. Warrenfeltz, J. Phys. Oceanogr., 12, 839 (1982).
6. R. W. Garwood, Jr., D. Adamec, Naval Postgraduate School Technical Report NPS68-82-006, Monterey, CA, 37 pp (1982).
7. D. Adamec, R. L. Elsberry, J. Phys. Oceanogr., in review (1983).

PRELIMINARY RESULTS FROM A NUMERICAL STUDY OF THE NEW ENGLAND SEAMOUNT CHAIN INFLUENCE ON THE GULF STREAM

Harley E. Hurlburt and J. Dana Thompson
Dynamical Ocean Forecasting Branch
Naval Ocean Research and Development Activity, Code 324
NSTL Station, MS 39529

ABSTRACT

The Dynamical Ocean Forecasting Branch of NORDA is tasked with developing an ocean forecasting capability for the Navy using dynamical ocean models. In addition to model development, performance of this mission requires studies of ocean dynamics, ocean predictability, data assimilation, and oceanic prediction. The Gulf Stream is an important part of the system to be predicted. Approximately 1000 km east of Cape Hatteras, the Gulf Stream crosses the New England Seamount Chain. Observations presented by Richardson[1] suggest that the New England Seamount Chain has a marked influence on the Gulf Stream. If so, this substantially complicates the Gulf Stream prediction problem. Some alternatives are to a) resolve the seamounts, b) parametrize the seamounts, or c) introduce data downstream to correct the field.

In this study we use a simple two-layer primitive-equation model to study the influence of the New England Seamount Chain on the Gulf Stream. The model domain extends from Cape Hatteras to the Grand Banks and has closed boundaries except for inflow and outflow ports for the Gulf Stream at the eastern and western ends. The Gulf Stream inflow to the domain is confined entirely to the upper layer and the seamounts are confined entirely to the lower layer. Thus, this study departs from the usual design for flow over topography where the current would be forced to impinge directly on the topography. In this investigation, any energy in the lower layer must come from the upper layer, and any influence of the seamount chain on the upper layer current must be a back interaction, a rather circuitous influence.

Results from several situations are compared: 1) reduced gravity, 2) flat bottom, 3) a ridge in the location of the seamount chain, 4) a small amplitude seamount chain which obeys the quasi-geostrophic approximation, and 5), a large amplitude seamount chain permitted in the primitive-equation model. Only in experiments 2) and 5) does the character of the model Gulf Stream differ substantially from the preceding experiment. Introduction of a second active layer results in smaller and weaker eddies. Introduction of the large-amplitude seamounts changes the mean path of the Gulf Stream, the amplitude of the meanders, and the distribution of the eddies in the vicinity of the seamounts. Prolific generation of warm core rings occurs immediately upstream of the seamount chain.

INTRODUCTION

The late 80's and the 90's should be to numerical ocean prediction what the late 50's and the 60's were to numerical weather prediction, a time when the essential elements came together to produce a skillful and useful forecasting capability. These elements are 1) adequate computing power, 2) adequate data input, and 3) numerical models which have been adequately tested and understood. Class 7 computers (32-128M words and ~1 gigaflop) are required for efficient eddy-resolving models of major ocean basins. The most promising new sources of data are satellite altimeters for sea surface elevation and satellite scattermeters for surface wind stress. The potential for ocean forecasting within the next decade is a subject explored by Hurlburt.[2]

The Dynamical Ocean Forecasting Branch at the Naval Ocean Research and Development Activity is charged with the task of helping the U.S. Navy develop a forecasting capability for the ocean circulation, including meandering current systems, eddies, and frontal positions. Three types of model application studies are especially useful to accomplish this mission; 1) phenomenological, 2) predictability, and 3) prediction. Phenomenological studies focus on simulating and understanding the dynamics of observed phenomena. Exploration of the model parameter space and testing of model sensitivity are also important.

Predictability studies investigate the potential for predicting the state of the system based at least partly on knowledge of another earlier state, particularly when the prediction may be sensitive to small errors in any aspect of the prediction system. Prediction studies evaluate the skill of actual forecasts and investigate methods of data assimilation and forecast verification.

In developing an ocean prediction capability for the U.S. Navy we are primarily at the phenomenological level investigating simulation skill and ocean dynamics. We are attempting to demonstrate the feasibility of developing ocean models which are computationally tractable, able to simulate the phenomena of interest, and which have the possibility of true predictive skill given the expected data input. Although not predictability studies in the strictest sense, they are of great value in developing an ocean prediction capability. Simulation skill is especially important in ocean prediction because of an extra burden placed on the forecast model not done in meteorology. This burden is subsurface forecasts of the ocean based primarily on data input at the surface. Thus, we are confronted by a spatial prediction problem as well as a temporal one.

The meandering Gulf Stream and the associated eddies and eddy shedding pose some of the most important prediction problems in the ocean. This paper is focussed on the influence of the New England Seamount Chain on the Gulf Stream. This seamount chain intersects the Gulf Stream approximately 1000 km east of Cape Hatteras. Figure 1 from Legeckis (personal communication) and Figure 2 from Richardson[1] both indicate that the seamount chain has a marked influence on the Gulf Stream. This substantially complicates the Gulf Stream prediction problem because of the difficulty in resolving the seamounts or in quasi-geostrophic models even of including them at full amplitude. The difficulty in resolving the seamounts is illustrated in Figure 3 taken from Richardson.[1] The diameter of the individual seamounts is comparable to the radius of deformation. It is clear that any ocean prediction model for the Gulf Stream must confront the predicament imposed by the seamounts. Some possible responses are 1) to resolve the seamounts, 2) to parameterize the seamounts or their effect, and 3) to introduce data downstream to correct the field and forego prediction in the immediate vicinity of the seamounts.

THE MODEL DESIGN AND PARAMETERS

In this study a simple two-layer primitive equation model is used to investigate the influence of the New England Seamount Chain on the Gulf Stream. The results from two possible simplifications of the seamount chain, a ridge and small amplitude seamounts, are compared to those produced by a large amplitude chain. The model domain extends from Cape Hatteras to the Grand Banks as shown in Figure 2. It is rotated 28° counterclockwise from zonal. To avoid undue problems with open ocean boundaries, the domain is closed except for inflow and outflow ports for the Gulf Stream at the western and eastern ends. The inflow port is positioned to represent the Gulf Stream as it leaves Cape Hatteras and the outflow port is approximately in the mean position of the stream as it passes south of the Grand Banks (see Fig. 2).

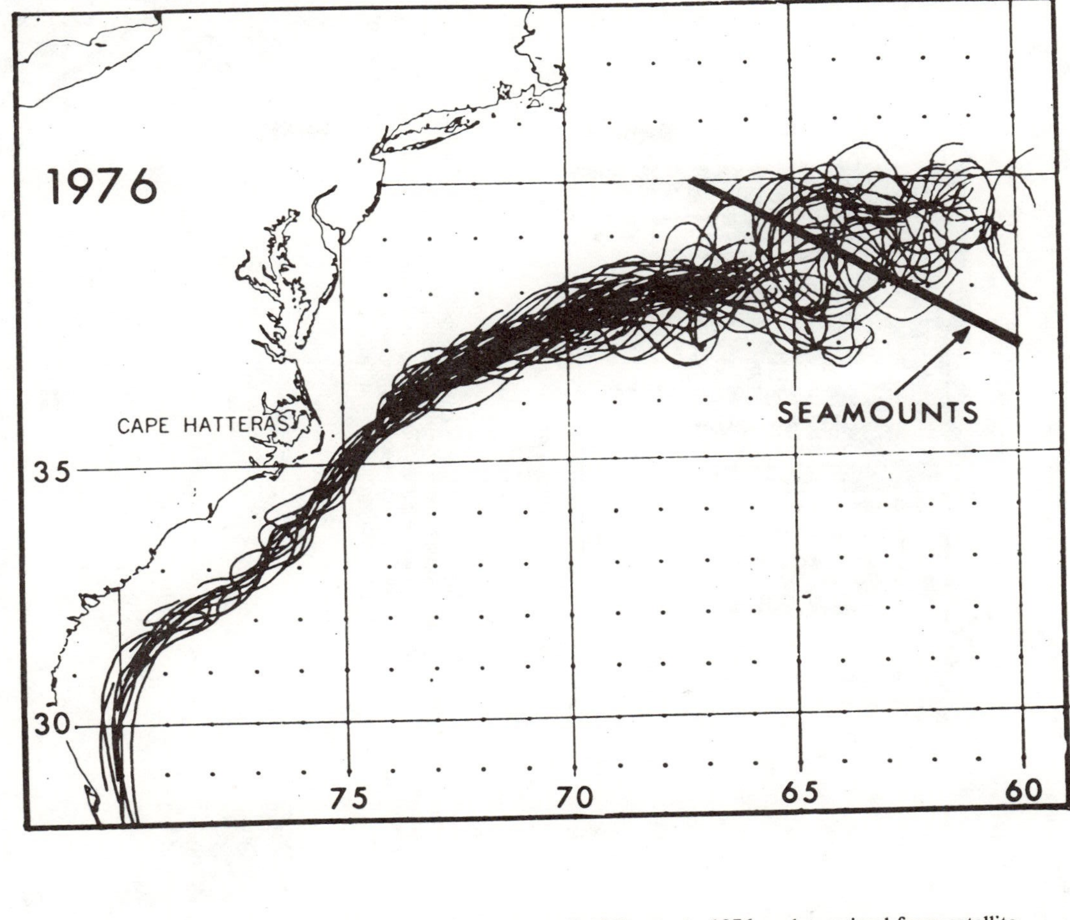

Fig. 1. Positions of the northwest boundary of the Gulf Stream in 1976 as determined from satellite IR (from Richard Legeckis, personal communication).

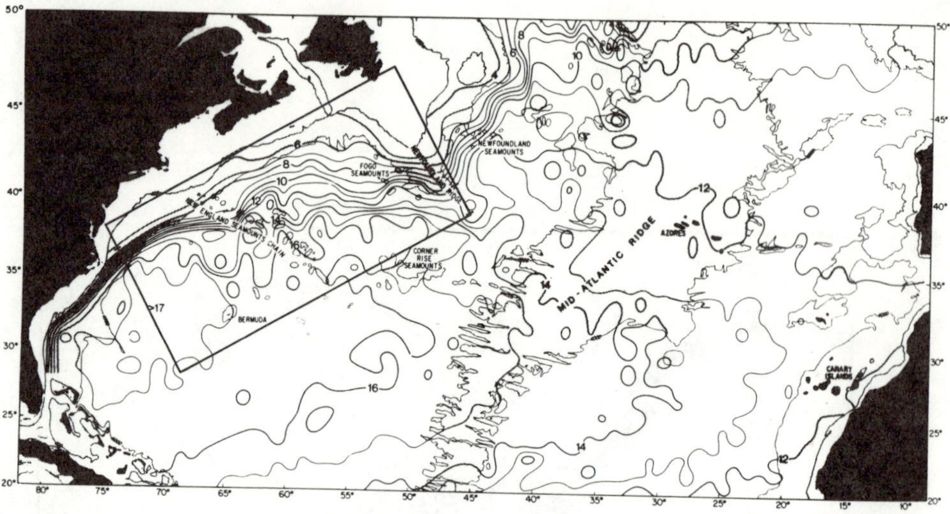

Fig. 2. The model domain superimposed on a map of the mean temperature at 450 m depth based on 1° square averages of the NODC data (from Richardson[1]). East of the New England Seamount chain there is a marked spreading of the isotherms, an indication of the large amplitude meaners in this region. Inflow to the model domain is through a port just off Cape Hatteras compensated by outflow through a port near the Newfoundland Ridge. Lateral boundaries are otherwise closed.

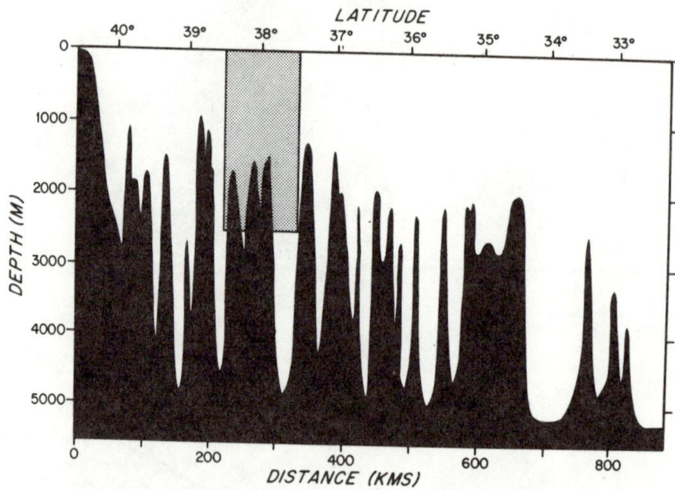

Fig. 3. "A north-south section showing the projected depth profile along the New England Seamounts. The Gulf Stream is shown schematically by the stippled region, which is 110 km wide and 2.5 km deep. The seamounts extend from Georges Bank southeastward to the Sohm Abyssal Plain near 34°N, 56°W; thus this projection emphasizes the close spacing of seamounts" (from Richardson[1]).

The flow of the Gulf Stream is confined to the upper layer as it enters the model domain through the western port. Thus, this study departs from the usual design for flow over topography where the flow would be forced to impinge directly on the topography. Here, any energy in the lower layer must come from the upper layer and any influence of the seamount chain on the current in the upper layer must be a back interaction, a rather circuitous influence.

All but one of the numerical experiments discussed here use a model with two active layers, the minimum to allow baroclinic instability and the coexistence of topography and the pycnocline. One experiment uses a reduced gravity model with an active upper layer and a lower layer which is infinitely deep and at rest. The pycnocline is represented by an immiscible interface between two layers with a prescribed density contrast.

The models are primitive equation on a $\beta-$ plane and retain a free surface. Using common approximations the two-layer model equations are

$$\frac{\partial \vec{V}_i}{\partial t} + \left(\nabla \cdot \vec{V}_i + \vec{V}_i \cdot \nabla\right)\vec{v}_i + \hat{k} \times f\vec{V}_i = -h_i \nabla p_i + \left(\vec{\tau}_i + \vec{\tau}_{i+1}\right) / \rho + A\nabla^2\vec{V}_i$$

$$\frac{\partial h_i}{\partial t} + \nabla \cdot \vec{V}_i = 0$$

where i=1 for the upper layer and 2 for the lower layer and

$$\nabla = \frac{\partial}{\partial x}\hat{i} + \frac{\partial}{\partial y}\hat{j} \, , \; p_1 = g\eta_1 \, , \; p_2 = p_1 - g'(h_1 - H_1) \, ,$$

$$\vec{V}_i = h_i\vec{v} = h_i(u_i\hat{i} + v_i\hat{j}) \, , \; g' = g(\rho_2 - \rho_1) / \rho \, , \; f = f_o + \beta_x(x - x_o) + \beta_y(y - y_o) \, ,$$

$$\beta_x = \beta \sin\alpha \, , \; \beta_y = \beta \cos\alpha \, , \; \vec{\tau} = \tau_i^x\hat{i} + \tau_i^y\hat{j} \; .$$

Symbol definitions are common in oceanography and are listed in the Appendix. In the reduced gravity model the lower layer momentum equation is $g\nabla\eta_1 = g'\nabla h_1$.

The numerical experiments have been driven from rest by prescribed inflow through a port in the western boundary compensated by outflow through a port in the eastern boundary. Except at the ports the boundaries are rigid and the no-slip condition is used. $\vec{V}$ is prescribed at the western (inflow) port using a parabolic inflow profile. At the eastern (outflow) port the normal flow is self determined and the tangential flow is free slip.[3]

The model equations were integrated using a semi-implicit numerical scheme where the external and internal gravity waves were treated implicitly.[4] This allows a time step much larger than possible in the corresponding explicit free-surface model and substantially longer than possible in a primitive-equation rigid-lid model. The time step limitation is more stringent than that for a quasi-geostrophic model only because of a stability criterion imposed by the numerical scheme for the Coriolis force which by itself is $\Delta t \leqslant 1/|f|$. Hurlburt and Thompson[3] discuss the numerical formulation of the models in detail. To the best of our knowledge the investigations of the Gulf of Mexico by Hurlburt and Thompson[3,5] represent the first application of the "semi-implicit method" to long-term oceanographic integrations.

Table 1 presents the parameters for the two-active-layer model. The reduced gravity model uses the same parameters when applicable except that $g' = 2.5$ cm/s^2. These parameters imply a maximum upper layer inflow velocity of about 120 cm/s and an internal radius of deformation of $\lambda = c/f \simeq 40$ km, where c is the internal gravity wave speed.

TABLE 1
Model parameters

A	$3x10^6 cm^2 s-1$	β	$2x10^{-13} cm^{-1} s^{-1}$
f_o	$6.73x10^{-5} s^{-1}$	ρ	1 gm cm^{-3}
g	980 cm s^{-2}	$\vec{\tau}_i$	0
g'	2.27 cm s^{-2}	$\Delta x, \Delta y$	20 km*
H_1	600 m	Δt	4800 s
H_2	4400 m		

Domain Size, x_L by y_L	2500 x 1280 km
Western Port Width, L_w	100 km
Eastern Port Width, L_e	200 km
Center of western port at y_w	1030 km
Center of eastern port at y_e	100 km
Upper Layer Inflow Transport	50 x 10^6 $m^3 s^{-1}$(50 Sv)
Lower Layer Inflow Transport	0
Angle of inflow from x-axis, θ_I	0°
Angle of x-axis from zonal, α	28°
Inflow spin-up time constant	90 days

*for a given variable

MODEL RESULTS

The five numerical experiments listed in Table 2 were integrated from rest for seven years, ~4 years to statistical equilibrium, then three additional years. All five use the applicable parameters from Table 1, except that the reduced gravity experiment used $g' = 2.5 cm/s^2$. The last four are identical except for the bottom topography. The topography for Experiments 3-5 is shown in Figure 4. Figure 4a shows a Gaussian ridge with an amplitude of 2500 m and an e-folding half-wide scale of 50 km. Figure 4b shows a chain of Gaussian seamounts with an amplitude of 2500m for Experiment 4 and 500 m for Experiment 5. The seamounts are spaced 200 km apart and have an e-folding radius of 50 km. The topography drops nearly to the abyssal plain between the seamounts. The separation and the diameter of the seamounts is substantially greater than that for the real seamounts, a compromise dictated by the affordable grid resolution and the desired domain size. The diameter (radius) of the real (model) seamounts is comparable to the radius of deformation.

TABLE 2
Numerical Experiments

Exp. #	Description	Topographic amplitude as a fraction of the depth of the abyssal plain
1	Reduced gravity	-
2	Two-layer flat bottom	0
3	Ridge topography (Fig. 4a)	.5
4	Large amplitude seamounts (Fig. 4b)	.5
5	Small-amplitude seamounts (Fig. 4b)	.1

Figure 5 shows the pycnocline height anomaly (PHA) for the reduced gravity experiment at day 1620. The PHA is the deviation of the interface between the layers from its initial uniform elevation. This is not a realistic simulation of the Gulf Stream because the scale of the eddies and the meanders is too large and the evolution of the features is too slow.

The diameter of the major cyclonic vortices in the reduced gravity experiment is ~300 km between speed maxima ~350 km e-folding, and ~500 km total, 2 and 3 times larger than observed.[6,7,8] Despite their size, the radius of the speed maximum of these rings is less than the value $r = (v_c/\beta)^{1/2}$ found for the eddies shed from the Loop Current.[3,5] For the eddies from the reduced gravity Gulf Stream experiment, $r \simeq 200$ km where v_c is the typical current speed at the radius of the speed maximum and $v_c \simeq 80$ cm/s. Let L = the radius of the speed maximum. Then, the β Rossby number, $R_B = v_c/\beta L^2 = (r/L)^2 = 1$ for the Gulf of Mexico, and is $\simeq 1.8$ for the reduced gravity Gulf Stream rings, much less than the value $R_B = 10$ to 15 for observed Gulf Stream rings.[9]

In the reduced gravity experiment we also find 2 to 3 wavelengths in the major meanders covering the 2700 km distance between the inflow and outflow ports of the model. The elevation of the pycnocline changes approximately 500 m across the stream and the amplitude of the ring in Figure 5 is 600 m. Six months later a ring is shed from the meander near the center of the basin. However, in some cases the time interval for the detachment of cyclonic (cold core) rings anywhere in the model domain exceeded three years. No detachment of anticyclonic (warm core) rings was observed. Two of the cyclonic rings were observed to move WNW at 3.2 and 4.2 km/day until they encountered the western boundary.

The reduced gravity model consists solely of a single internal vertical mode and no transfer of energy from the upper ocean to the deep ocean is permitted. Unlike the eastern Gulf of Mexico,[3,5] and the eastern Caribbean,[10] many features of the flow are substantially altered when the barotropic mode is added and an exchange of energy is permitted between the shallow and deep flow. This is illustrated in Figure 6 by the upper layer pressure at day 1500 for Experiment 4 with the large amplitude seamount chain. The axis of the seamount chain is indicated by the solid line at x = 1000 km. Immediately obvious is the approximately 2-fold decrease in the wavelength of the meanders and the decrease in the size of the eddies. The total diameter of the eddies in Experiments 2-5 is typically 200 km with a few around 300 km, more in line with observations. The β Rossby number for one of the larger eddies was measured at 7.5.

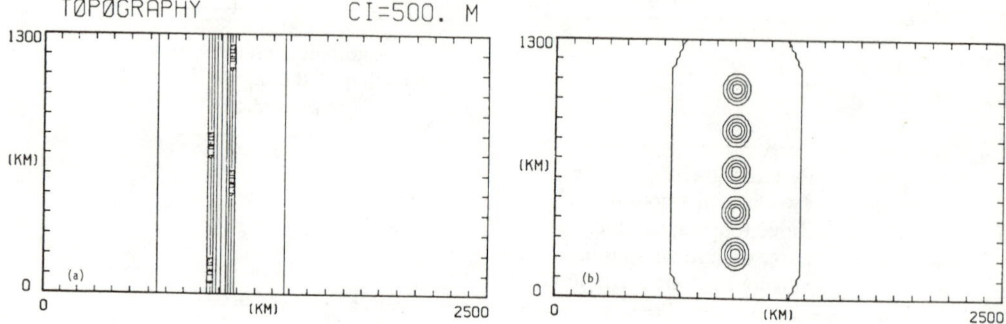

Fig. 4. Bottom topography for (a) Experiment 3 with a ridge and (b) Experiments 4 and 5 with a seamount chain. The amplitude of the topography is 2500 m for Experiments 3 and 4 and 500 m for Experiment 5. The shape is Gaussian with an e-folding scale of 50 km. The separation of the seamounts is 200 km. The contour interval is 500 m.

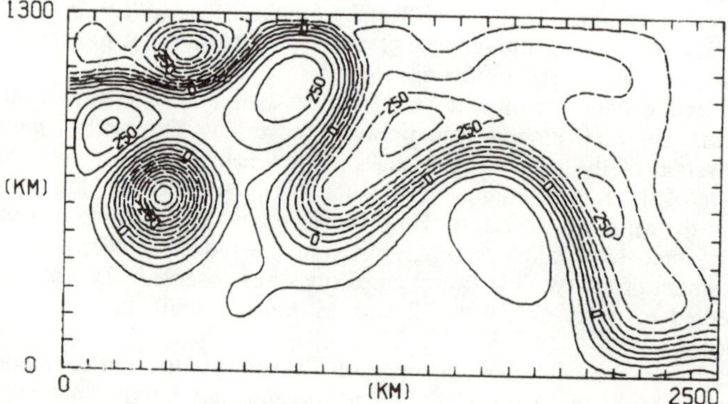

Fig. 5. Pycnocline height anomaly in meters at day 1620 for the reduced gravity experiment, Experiment 1. Dashed contours denote shallowing of the layer interface, upper layer thickness less than the initial uniform value. The contour interval is 50 m.

Occasional bifurcation of the stream is a phenomenon found in the two-layer experiments, but not in the reduced gravity. In most cases this consisted of two branches of the stream enclosing a counter-rotating vortex pair, but in Experiment 4 with the large amplitude seamount chain sometimes a larger number of eddies was enclosed. This occurred a few days after the time of Figure 6.

Figure 6 suggests that the seamount chain is a region of intense eddy activity, an observation confirmed by a map of the RMS fluctuations in p_1 (not shown). Figure 7 shows Richardson's[1] interpretation of a NOAA 5 satellite infrared image of the Gulf Stream where the Gulf Stream crossed the New England Seamount Chain. Both the model and the satellite image show northward anticyclonic meanders over and just downstream of the seamount chain. Both also show a southward cyclonic meander immediately upstream of the seamount chain which Richardson calls a "ring meander". Both show comparable scales for the meanders and the eddies, including the total N-S amplitude of the meandering in the vicinity of the seamount chain.

Table 3 compares the frequency of ring generation in three of the numerical experiments. The frequency is dramatically increased when the barotropic mode and the possibility of energy exchange between the layers is added. Particularly notable is the effect of the seamount chain on the generation of warm core rings. In Experiment 4, 6 out of 8 warm core rings generated in the model domain in three years formed immediately upstream of the seamount chain. Furthermore, the rings which formed in the vicinity of the seamount chan tended to be larger and stronger than those found in the other 2-layer experiments.

Figure 8 shows a two-year mean (years 6 and 7) of the upper and lower layer pressure for Experiments 2-5. For Experiments 3-5 a solid line marks the axis of the topography at $x = 1000$ km. In all cases there are standing waves in the upper layer pressure and some nonlinear recirculation in the western part of the basin which are least partly due to the artificial boundaries of the model domain, a problem not considered here but one which deserves scrutiny.

TABLE 3
Number of rings generated in 3 years (years 5-7)

Exp. #	Description	#Cold core	#Warm core	Total
1	Reduced gravity whole domain	3	0	3
2	Two-layer flat bottom whole domain	4	7	11
4	Large amplitude seamounts whole domain	6	8	14
4	Large amplitude seamounts just west of seamount chain	3	6	9

Of primary interest is the effect of the different topographies on the mean upper and lower layer pressure. Recall that inflow through the western boundary is confined entirely to the upper layer and the topography entirely to the lower layer. Thus, any energy in the lower layer came from the upper layer and any influence of the topography on the upper layer flow is a back interaction from the lower layer. Figure 8a,b is for the flat-bottom experiment. Figure 8c (ridge topography) and Figure 8g (small amplitude seamounts) show only small effects of the topography on the mean upper layer pressure. In contrast Figure 8e (mean p_1 for the large amplitude seamounts) shows a large amplitude meander in

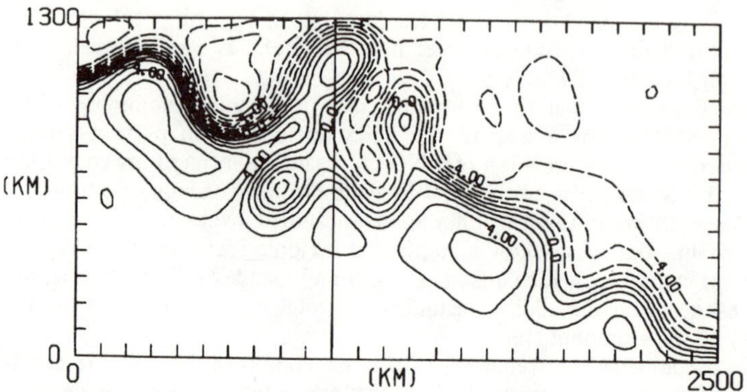

Fig. 6. Density-normalized pressure field in the upper layer (p_1) at day 2500 for Experiment 4 with the large amplitude seamount chain. The axis of the seamount chain at x=1000 km is marked with a solid line. The contour interval is 1 m^2/sec^2.

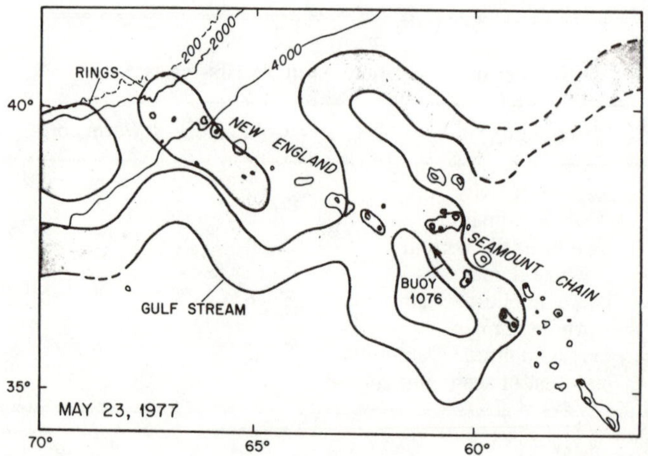

Fig. 7. "Interpretation of a NOAA 5 satellite infrared image of the Gulf Stream region, 23 May 1977, as buoy 1076 (speed 45 cm/s) began looping near the New England Seamounts. The Gulf Stream formed a large, 350 km diameter ring meander partially overlying the seamounts and another meander north of the seamounts. The total north-south extent of the Gulf Stream near the seamounts is 700 km." (from Richardson[1]).

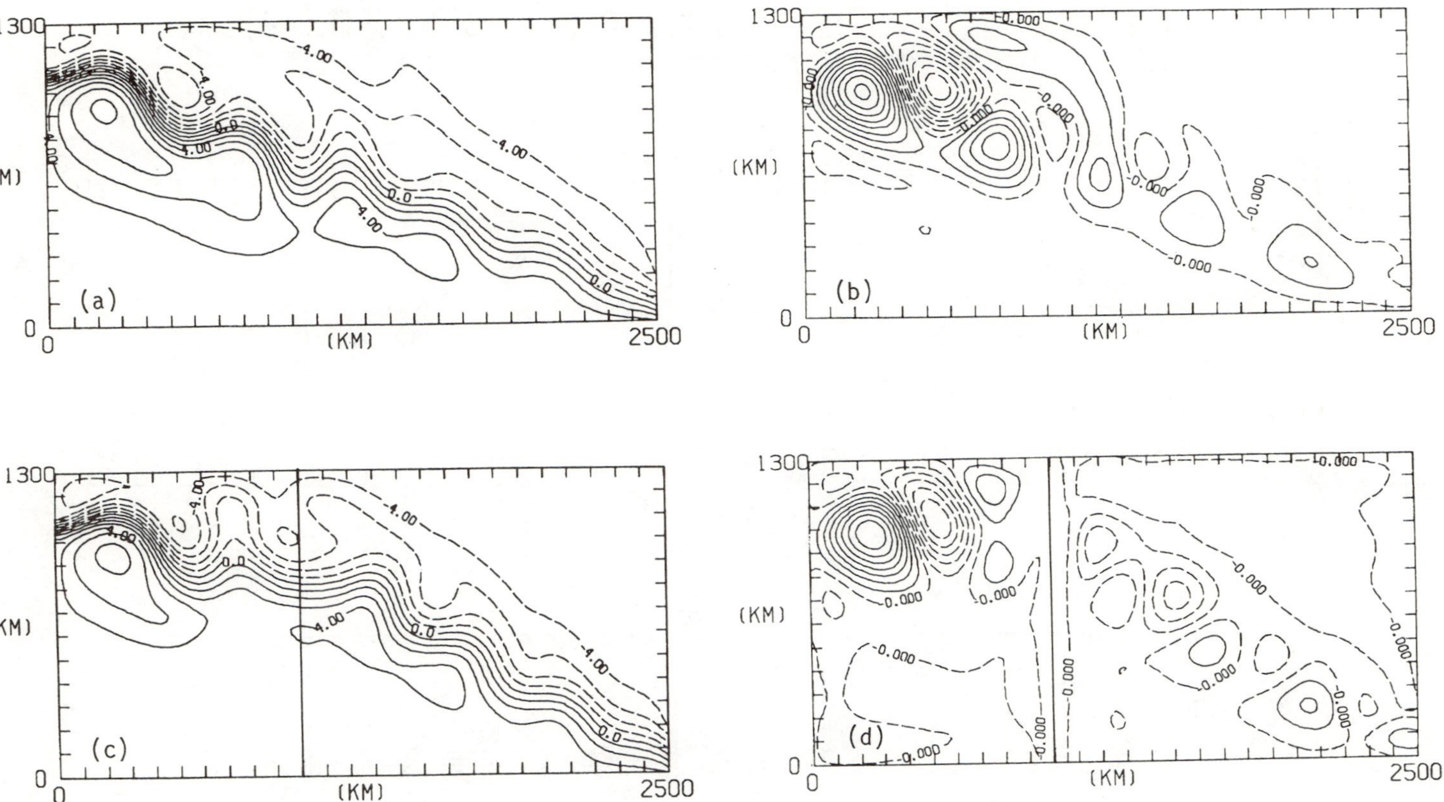

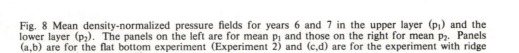

Fig. 8 Mean density-normalized pressure fields for years 6 and 7 in the upper layer (p_1) and the lower layer (p_2). The panels on the left are for mean p_1 and those on the right for mean p_2. Panels (a,b) are for the flat bottom experiment (Experiment 2) and (c,d) are for the experiment with ridge topography (Experiment 3).

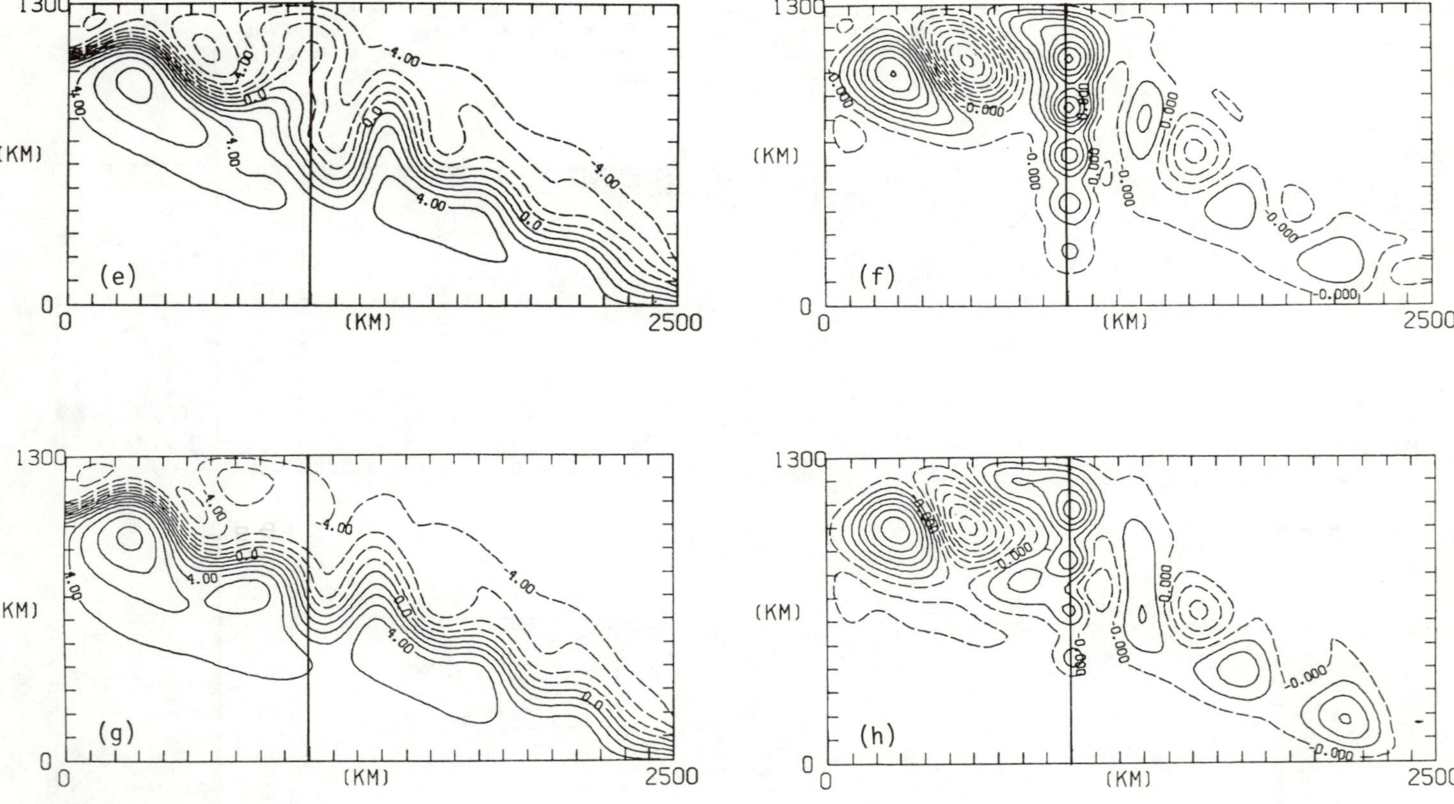

Fig. 8 (continued) Panels (e,f) are for the experiment with the large amplitude seamount chain (Experiment 4) and panels (g,h) for the experiment with the small amplitude seamount chain (Experiment 5). A solid line marks the axis of the topography in Experiments 3-5. The contour interval is 1 m^2/s^2 for for mean p_1 and .2 m^2/s^2 for mean p_2.

the vicinity of the seamount chain. These results are consistent with the patterns in the lower layer pressure field (p_2). In the case of the ridge there are no closed f/h contours and little deep flow in its vicinity. In contrast there is relatively strong deep flow around the seamounts and the seamount chain and stronger flow around the tall seamounts than around the short ones.

There is a barotropic phase relationship between the standing eddies in the lower layer and standing waves in the upper layer except in the vicinity of the seamount chain where there is almost a 90° phase shift. As noted earlier, the flow in the vicinity of the seamount chain is exceptionally time dependent. Some insight into this behavior can be obtained from a kinematic analysis[3] used in explaining how deep southward flow along the West Florida Shelf could prevent northward penetration of the Loop Current into the Gulf of Mexico.

The continuity equation for the upper layer can be written as

$$h_1t + h_1\nabla \cdot \vec{v}_1 + \vec{v}_1 \cdot \nabla h_1 = 0$$

Since

$$\vec{v}_1 \cdot \nabla h_1 \simeq \vec{v}_{1g} \cdot \nabla h_1 = \vec{v}_{2g} \cdot \nabla h_1$$

(where $\vec{v}_{ig}$ is the geostrophic velocity component in layer i) there is a strong tendency for advection of the upper layer current by the lower layer current where they intersect at large angles, a situation likely to occur in the vicinity of the seamount chain and consistent with a mean northward (southward) exclusion of the stream west (east) of the seamount chain.

Additional insight can be gained from potential vorticity. Either cyclonic or anticyclonic flow around the seamounts could conserve potential vorticity. Hurlburt and Thompson[3] found the deep flow in the Gulf of Mexico following a cyclonic path around the high topography. Here it is anticyclonic. It is useful to note that when flow encounters a rise (depression) anticyclonic (cyclonic) relative vorticity is generated in order to conserve potential vorticity, $(\zeta + f)/h$ = constant. Alternately, a source of anticyclonic (cyclonic) vorticity over a seamount will cause flow around the seamount to seek a higher (lower) elevation $d\,ln h/dt = d\,ln(\zeta + f)/dt$). These observations are consistent with anticyclonic flow around the seamounts under the Gulf Stream, the exceptionally time dependent flow in that region, and the baroclinic phase relation in the mean near the seamount chain.

SUMMARY AND CONCLUSIONS

This paper has presented some preliminary results from a numerical study of the New England Seamount Chain influence on the Gulf Stream. These results indicate that the seamount chain has a substantial influence on the Gulf Stream even when the stream does not directly impinge on the topography. The model exhibits particularly intense eddy activity in the vicinity of the seamount chain in accord with observations (Figs. 1 and 7). Prolific generation of warm core rings occurs immediately upstream of the seamount chain with a frequency of approximately 2/year.

There is also a large amplitude meander in the mean with a ridge immediately upstream of the seamount chain and a trough followed by a ridge downstream. This pattern is approximately 90° out of phase with the weaker standing waves found in the flat bottom cases, waves which may have a significant influence on the results for the seamount chain. This point bears further investigation particularly since the standing wave pattern is probably influenced by the artificial boundaries of the model domain. The model shows the seamount influence confined to within a few hundred kilometers of the chain, contrary to the extended downstream influence indicated by the observations in Figures 1 and 2. This discrepancy may result from the artificial constraint on the location of the outflow, another hypothesis which bears testing.

Both the observations and the model indicate that the New England Seamount Chain needs to be accounted for in an eddy-resolving forecast model covering the region. Attempts to parameterize the hard to resolve seamounts using a ridge or small amplitude seamounts obeying the quasi-geostrophic approximation were not successful. These topographic configurations demonstrated much less influence on the model Gulf Stream than the large amplitude seamounts. Even the large amplitude seamounts used in this study are a fairly severe parameterization with seamount diameter and separation roughly three times too large.

The results from a reduced gravity model were quite unrealistic; exhibiting time and space scales for the eddies and meanders which are too large. The addition of the barotropic mode which allowed energy transfer to the deep water and the possibility of baroclinic instability brought the scales more in line with observations. The nature of the energy transfer to the deep water has not yet been determined, pending application of the eddy-mean energetics for this model.[5]

ACKNOWLEDGEMENTS

Dr. Daniel Moore of Imperial College, London provided the fast vectorized Helmholtz solver for the models. Some of the graphics software was supplied by the National Center for Atmospheric Research, which is sponsored by the National Science Foundation. Computations were performed on the two-pipeline Texas Instruments Advanced Scientific Computer at the Naval Research Laboratory in Washington, D.C. We thank Charles Parker for the original typing of the manuscript and Cynthia Seay for help in preparing the figures.

REFERENCES

1. P.L. Richardson, J. Phys. Oceanogr., **11**, 999-1010 (1981).
2. H.E. Hurlburt, submitted to Marine Geodesy (1983).
3. H.E. Hurlburt and J.D. Thompson, J. Phys. Oceanogr., **10**, 1611-1651 (1980).
4. M. Kwizak and A.J. Robert, Mon. Wea. Rev., **99**, 32-36 (1971).
5. H.E. Hurlburt and J.D. Thompson, in Hydrodynamics of Semi-enclosed Seas, J.C.J. Nihoul, Ed., (Elsevier Scientific Publishing Company, 243-297, 1982).
6. D.J. Lai and P.L. Richardson, J. Phys. Oceanogr., **7**, 670-683 (1977).
7. D.E. Hagan, D.B. Olson, J.E. Schmitz and A.C. Vastano, J. Phys. Oceanogr., **8**, 997-1008 (1978).
8. A.C. Vastano, J.E. Schmitz and D.E. Hagan, J. Phys. Oceanogr., **10**, 493-513 (1980).
9. J.C. McWilliams and G.R. Flierl, J. Phys. Oceanogr., **9**, 1155-1182 (1979).
10. G.W. Heburn, T.H. Kinder, J.H. Allender and H.E. Hurlburt, in Hydrodynamics of Semi-enclosed Seas. J.C.J. Nihoul, Ed., (Elseiver Scientific Publishing Company, 299-327, 1982).

APPENDIX – LIST OF SYMBOLS

Symbol	Definition
A	horizontal eddy viscosity
c	internal gravity wave speed
f, f_o	Coriolis parameter; f_o taken at southwest corner
g	acceleration due to gravity
g'	reduced gravity, $g(\rho^2 - \rho_1)\rho$
$H_1, H_2(x,y)$	initial thickness of the layers
h_1, h_2	instantaneous local thickness of the layers
L	radius of the speed maximum in an eddy
p_1	upper layer density-normalized pressure, $g\eta_1$
p_2	lower layer density-normalized pressure, $g\eta_1 - g'(h_1 - h_2)$
R_B	beta Rossby number, $v_c/(\beta L^2)$
r	eddy radius when $R_B = 1$
t	time
Δt	time increment in the numerical integration
u_1, u_2, v_1, v_2	x and y-directed components of current velocity
v_c	current speed at the radius of the speed maximum in an eddy
$\vec{v}_{ig}$	geostrophic velocity component in layer i
$\vec{V}_1, \vec{V}_2$	$h_1 v_1, h_2 \vec{v}_2$
x, y, z	right-handed, tangent-plane Cartesian coordinates with a z positive upward
x_L, y_L	dimensions of the model domain
$\Delta x, \Delta y$	horizontal grid increments
α	counterclockwise angle of rotation for the basin. $\alpha = 0$ would imply an eastward positive x-axis
β	differential rotation, $\beta_x = \partial f/\partial x, \beta_y = \partial f/\partial y$
ζ	relative vorticity, $v_x - u_y$
η_1	free surface anomaly; height of the free surface above its initial uniform elevation; $\eta_1 = h_1 + h_2 - H_1 - h_2$
η_2	$\eta_2 = H_1 + \eta_1 - h_1 = h_2 - H_2 = -PHA$
θ_I	angle of inflow with respect to the positive x-axis
λ	internal radius of deformation
ρ, ρ_1, ρ_2	densities of sea water
τ_i^x, τ_i^y	x and y directed tangential stresses at the top (i) and bottom (i+1) of layer i

AN OBSERVING NET FOR OCEANIC PREDICTION

Walter Munk
Scripps Institution of Oceanography, La Jolla Ca 92093

I will discuss the oceanographic measurements that might become available for assimilation into ocean prediction models.

The ocean situation is very different from the atmospheric situation. There has never been a routine collection network of ocean stations, with the exception of very few weather ships. Since the days of the Challenger expedition a century ago, most of the observations consisted of hydrographic casts, using Nansen bottles. From these one obtains temperature and salinity as a function of pressure at rather widely spaced depths; one then calculates the density field and infers the velocity field using geostrophy.

The important consideration is that such measurements are not taken simultaneously. Usually a "section" is occupied in one year, another section in another year, etc. (We still depend heavily on the sections occupied during the International Geophysical Year.) There have been very few repeated sections. We now know that the analysis of these sections under the assumption of a steady state circulation leads to serious problems in connection with mass conservation.

The situation became even worse with the realization, about 20 years ago, of a very active mesoscale circulation with typical dimensions of 100 km and 100 days. The ocean mesoscale corresponds to atmospheric weather. Ships making standard casts are incapable of keeping up with the average changing ocean weather. (The speed of oceanographic vessel remains near 10 knots, as in the days of the Challenger.)

There have, of course, been very considerable modern improvements in making ocean measurements. XBT's and particularly their airborne sisters, the AXBT's, can do a creditable job of keeping up with mesoscale developments. Drifters, both surface and submerged, reporting via satellites are the equivalent of weather balloons. They give very interesting information, though we have not become quite accustomed to digest the Lagrangian representation of ocean circulation.

An even more important development took place with the deployment of ocean observing satellites, particularly SEASAT. With proper orbits these satellites can for the first time sample the ocean surface with mesoscale resolution on a global scale. By using altimeters and scatterometers one obtains (in principle) the three components of surface stress. With this stress vector known it is possible to extrapolate ocean motion into the interior, using some fundamental conservation principles.

However, sea *surface* observations by themselves will be no more acceptable than depending on sea level pressure maps for describing the atmosphere. One needs direct measurements of the interior ocean, just as one needs measurements aloft. Our group has proposed that measurements of travel time between distributed acoustic sources and receivers can be used to obtain three-dimensional maps of density and flow fields using methods of analysis analogous to medical tomography. I envision an ocean observing system that depends heavily upon surface stress measurements from satellites and interior measurements using ocean acoustic tomography. (Observations in critical areas could be augmented by special flights dropping AXBT's.) The requirements of such a system appear to be within practical bounds. Measurements could be readily assimilated into numerical prediction models. There is a fascinating question concerning the trade-off between the potency of the numerical modelling and the requirement for oceanographic hardware.

ON PREDICTABILITY OF DEEP-WATER WAVES

M.Y. Su and A.W. Green
Naval Ocean Research and Development Activity,
NSTL Station, MS 39529

ABSTRACT

The classical theory of linear surface gravity waves has little utility in describing the natural wave evolution in which strong and weak wave interactions occur. Recent theoretical work, results of analyses of laboratory experiments and observations of oceanic storm seas show that steep waves (steepness ak > 0.1) are subject to at least two types of rapidly growing instabilities. Evidence is presented that these distinct nonlinear instabilities, one a 2-dimensional instability and the other 3-dimensional, are at times strongly coupled, so that the growth of one type triggers rapid growth of the other. Further, these instabilities appear to be sources for two types of bifurcated wave states. Experimental results have demonstrated that 3-dimensional bifurcations occur in steep wave trains; one bifurcation (symmetric pattern) propagates along the direction of the original carrier and the second (skew pattern) propagates obliquely to the carrier. A second feature of wave evolution observed in experiments is the irreversible transfer of energy from the original carrier to lower frequency side bands; this "red shift" is also characteristic of natural waves with steepness ak $\gtrsim 0.1$. Current theoretical computations have predicted cyclical reversible exchanges among the carrier and side bands. Future developments in the theory of surface waves must encompass accurate representations of the nonlinearly superposed instabilities, frequency downshift and turbulent dissipation through wave breaking.

INTRODUCTION

The prediction of oceanic wave statistics is in a somewhat different category of fluid dynamic prediction than the deterministic meteorological and oceanic circulation predictability. The small temporal and spatial scales of random surface gravity waves preclude the use of numerical models to develop the statistics, even over very limited regions; further, the dynamical processes that govern the exchange of momentum and energy between the wind and the waves are not yet understood sufficiently to construct accurate parametric representations of the dynamics. As a consequence of these limitations, present ocean wave prediction models are based on simple kinematics of wave propagation and semi-empirical relationships of wave statistics with the wind stress, duration and fetch. Several of these models have been in use over a moderate range of conditions, but none include the features required to reliably predict the occurrences of extreme wave conditions. The extreme waves or wave groups are critical hazards to ships or offshore structures.

Extreme waves are highly nonlinear in the sense that they result from the interactions of steep waves with differing frequencies, the durations of these interactions are only tens of wave periods and the final composite of wave frequencies and amplitudes is significantly different from that in the initial state. Wave breaking is probably the most intense of the strongly nonlinear processes. Recent experimental and theoretical results have given some evidence that steep deep water waves are subject to 2- and 3-dimensional instabilities that have features similar to extreme ocean waves. Storm seas frequently meet the conditions for onset of nonlinear wave instabilities, consequently, an improved understanding of these instabilities may lead to better predictions of extreme wave groups.

WAVE INSTABILITIES

Steep, deep water waves are not hydrodynamically stable. Benjamin and Feir[1] demonstrated experimentally and theoretically that a train of Stokes waves with small but finite steepness (the amplitude times the wave number) are subject to instability when the train is modulated by infinitesimal waves. Longuet-Higgins[2] considerably extended the theoretical analysis to include a wide range of steepness with infinitesimal modes of perturbation. This "modulational" instability is characterized by the growth of upper and lower sidebands that are phase-locked with the primary wave train. The incipient form of this instability for small initial steepness is also a type of weak resonance interaction. The most unstable sideband modes are offset from the primary frequency by a difference proportional to the steepness of the primary waves. The instabilities are observed to cause intense modulations that eventually exceed the amplitude of the primary waves.

a. 2-Dimensional Instabilities

Numerical models of 2-dimensional irrotational flow with nonlinear boundary conditions give results that are consistent with earlier analyses of the Benjamin-Feir instability, which was restricted to small amplitude perturbations; however, the numerical models were able to show a more curious phenomenon, wave-form recurrence. The cycle of recurrence includes the growth of the sideband instabilities accompanied by intense modulations; eventually, the modulations subside and the wave forms return to the initial state.[3] This type of recurrence is quite similar to that noted in numerical solutions to the nonlinear Schrodinger equation.[4] The irrotational models have given results that differ significantly from the observations; no exact recurrence has been observed[5-7]. In the experimental cases with long wave trains in which the steepness exceeds about 0.1, the lower frequency sideband instability steadily increases amplitude as the primary waves and the lower sideband diminishes. Within a few tens of wave periods the initial carrier appears to transfer the bulk of its energy to the lower sideband irreversibly, and the total wave energy and action (energy/frequency, to first order) decrease. The obvious differences between the numerical and laboratory experiments were the presence of turbulent and viscous dissipation. Secondarily, experimental waves were good approximations of Stokes waves, but not exact; the wave trains also had finite space-time extent. Of these, dissipation due to breaking appears to play the dominant role in the observed irreversible red shift. The transition to groups and frequency red shift are accentuated in wave packets.

For weakly nonlinear cases Benny[8] showed that the wave equation could be transformed to a nonlinear Schrodinger equation (NSE). Zakharov and Shabat[9] presented an analytical solution of the NSE for an initial value problem applied to the evolution of a wave packet. Their solution predicted a series of envelope solitons followed by a small oscillating tail of higher frequency waves. For $a_o k_o \leqslant 0.1$, the theoretical prediction agrees well with experimental observation. For larger $a_o k_o > 0.1$, Su[7] has found that some of the leading envelope solitons have carrier frequencies that correspond to the lower sideband frequency associated with the modulational instability of the initial wave packet. It appears that the evolution of a packet of waves of moderate to high steepness $0.1 \leqslant a_o k_o < 0.3$ leads to formation of more stable lower-frequency envelope solitons with a nonlinear energy/momentum transfer from higher to lower frequency (and wavenumber) modes. A typical example of wave packet evolution is shown in Figure 1, where the initial packet consisting of 20 waves with $a_o k_o = 0.22$. After the packet has traversed about 100 wavelengths, five envelope solitons emerge. The first two have the carrier frequency associated with the lower side-band of the modulational instability with $a_o k_o = 0.22$. The carrier frequencies of the three trailing packets are at the original carrier frequency.

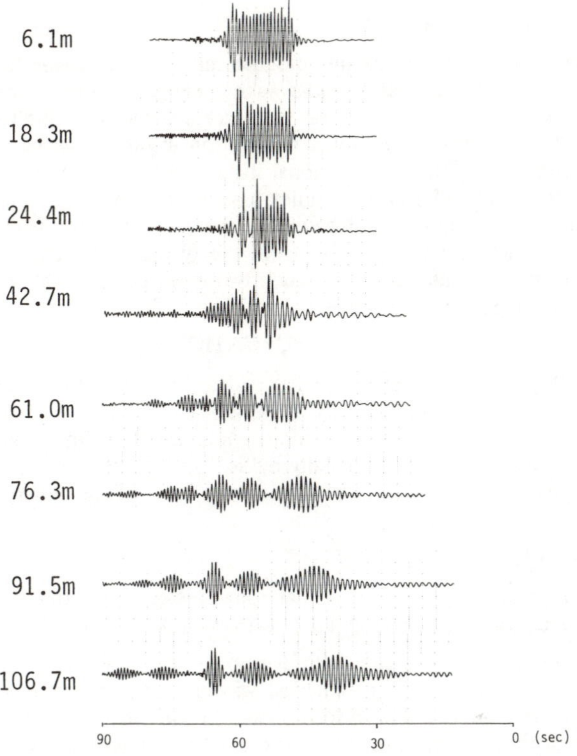

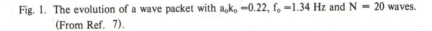

Fig. 1. The evolution of a wave packet with $a_o k_o = 0.22$, $f_o = 1.34$ Hz and N = 20 waves. (From Ref. 7).

b. 3-Dimensional Wave Instability

Recently, a 3-dimensional wave instability was discovered. This new instability arises from skew perturbations, not colinear modulations[11,12] The perturbation modes that grow most rapidly have significant components normal to the wave vector of the initial unperturbed waves. The most unstable modes have wavelengths twice that of the primary wave, and these subharmonic modes have the same phase speeds as the primary wave. Hence, the most rapidly growing modes are phase-locked to the unperturbed waves. This 3-dimensional instability exists for $a_o k_o > 0$, but its growth rate is smaller than the 2-dimensional instability when $a_o k_o < 0.3$, and exceeds that of the 2-dimensional modulation instability when $a_o k_o > 0.3$.

When $a_o k_o > 0.41$, a superharmonic 2-dimensional instability[2] occurs but this instability is still weaker than the 3-dimensional instability modes. Further, in a practical sense, we expect that these instabilities at very high steepness ($a_o k_o > 0.4$) are not important in dominant oceanic waves where $a_o k_o > 0.2$ is very rare. At lower values of $a_o k_o$, that occur in storm seas, it appears that 2- and 3-dimensional instabilities interact nonlinearly.

c. Coupling Between the 2-D and 3-D Instabilities

During the stages of maximum modulation of the wave train packets with $0.12 < a_o k_o < 0.16$, the amplitudes of some waves may grow to more than double the initial value; consequently, the local steepness increases accordingly. The locally higher wave steepness induced by the two-dimensional modulation instability appears to accelerate the growth of the contemporary three-dimensional instability. Observational evidence for such coupling between the 2-D and 3-D instabilities have been provided by Su and Green[13], but no theoretical analyses of these coupled nonlinear instabilities are available at this time. The stability analysis would have to differ from the typical "monochromatic perturbation" method, since the model would have to include two types of perturbations with amplitudes comparable with the unperturbed waves.

WAVE BIFURCATIONS

The possibility of 2-dimensional bifurcations of very steep Stokes waves with $a_o k_o \simeq 0.44$ was first proved numerically by Chen and Saffman[22], but it has not been confirmed experimentally. Recently, two more complicated 3-dimensional bifurcations of Stokes waves were discovered experimentally by Su[14] and analyzed theoretically by Saffman and Yuen[15] and Meiron et al.[16] We shall now describe some of the experimentally observed characteristics of the bifurcations.

a. Skew Bifurcation

The skew bifurcation of 2-dimensional long-crested wave trains occurs most frequently in our laboratory experiments when $0.16 \lesssim a_o k_o \lesssim 0.18$. A typical example of skew bifurcation ($a_o k_o = 0.17$) is shown in Figure 2, which is a photograph of the wave field taken from the center of the mechanical wavemaker. The areas with lighter toned areas of the waves are associated with lower wave heights in the bifurcated pattern. The average direction of these lighter-tone bands (skew bifurcations) is between 15° to 20° away from the original wave direction, and the average distance between two consecutive bands is approximately 3 times greater than the original unpertubed wavelengths. The interaction of the right- and left-moving bifurcations creates compact, diamond-shaped wave packets. The skew bifurcated waves are subject to Benjamin-Feir type instability that causes the wave envelope modulations that appear in the upper part of Figure 2. The skew wave pattern has a very slow forward velocity that is about 1/50 of the phase speed of the original waves. The observed skew bifurcations are qualitatively similar to the patterns computed by Saffman and Yuen[15].

b. Symmetric Bifurcations

For $0.25 \lesssim a_o k_o \lesssim 0.35$, following the occurrence of unstable 3-dimensional perturbations, the long-crested wave trains are observed to undergo a rapid transition into regular symmetric, 3-dimensional wave forms. A typical example with $a_o k_o = 0.32$ is shown in Figure 3. The crestwise length of these crescent-shaped bifurcated waves is about 0.8 of the wave length of the original waves. There is also a one-half wave shift crestwise between two consecutive rows of crescent waves. Hence, a wave record measured at a fixed point in this stage of wave bifurcation exhibits an alternating series of high and low waves. Near the center of each crescent wave, rapid acceleration at the water surface causing plunging and spilling breaking waves with coincident air entrainment. The crescent-shaped pattern persists for 6-8 wavelengths and then divides into two parts consisting of oblique wave groups that propagate at an angle of about 30° away from the direction of the primary waves and 2-dimensional waves that have modulation patterns characteristic of the Benjamin-Feir instability. These 2-dimensional waves, that propagate in the direction of the original wave train, undergo a frequency downshift that causes the bulk of the wave energy

Fig. 2. A typical photographic image of skew bifurcation with $a_o k_o$ =0.17 (From Ref. 14).

Fig. 3. A typical photographic image of symmetric bifurcation with $a_o k_o$ = 0.32 (From Ref. 14).

to be concentrated at a frequency band as much as 25% lower than the primary waves near the wavemaker.

These experimental results are in qualitative agreement with the theoretical computations[15,16], but the latter stages of evolution in which fragmentation of the patterns and the frequency downshift occurs exceed the scope of present theory.

COMPARISONS WITH OCEANIC WAVES

Results of the work that we have described in the preceding sections have encouraged us to reconsider the seemingly chaotic oceanic waves. The relatively clear views of instabilities and bifurcations that we have obtained allow us to gain some new insights about some types of apparently random events commonly observed by hearty souls during high seas.

First we note that most waves in growing seas are short-crested; this is also a characteristic of 3-dimensional instabilities and bifurcating waves. Both phenomena have been found to lead to wave breaking in the laboratory, and we surmise that similar events occur at sea. These instabilities and bifurcations will create local "order" from the surrounding chaos, and will cause the spreading of wave energy as we have observed in the secondary waves radiated from the bifurcations. We have noted that modulational instabilities create wave groups in which the primary and sideband components are phase-locked, and as the instability proceeds the peak waves in the groups may greatly exceed the initial wave amplitudes. Due to the relatively rapid growth of the modulations, we could expect that many of the largest waves in storm seas could be the results of this type of instability. In natural conditions we shall also find occurrences of linear wave superposition, but we would not expect to frequent repetitions of coherent groups with a particular shape.

An additional insight to an important feature of growing seas has been gained by the experiments, namely the red shift associated with the advanced stages of large amplitude modulations where energy is irreversibly extracted from the primary wave by the lower sideband modulation. The laboratory and natural waves are subject to breaking and turbulent dissipation that prohibit reversibility. Further, we have noted that the amount of the red shift is proportional to the initial wave steepness.

We have recently conducted statistical analyses of extensive records of oceanic storm waves in order to determine whether some of our ideas about the role of strong nonlinearity and wave instabilities could be confirmed. The records are comprised of nearly 600, 20-minute segments in which the significant wave height ($H_s = H_{1/3}$) exceeded 2m[17,18]. So far we have found two types of evidence that indicate the effects of nonlinearity. The first characteristic of nonlinearity that we noted was in the joint distribution of heights and periods for the individual waves; the data were obtained using the "zero-up-crossing" method. For cases of narrow spectral peaks, where "spectral peakedness" (Q_p) is defined as in Goda[19], with $Q_p \geqslant 3$ the joint distribution of heights and periods in bimodal (Fig. 4), and the associated marginal distribution of wave periods is markedly skewed toward longer periods (Fig. 5). At present the available theories of these distributions are based on assumptions that the distributions results of linear, narrow-band Gaussian process[20,21]; these theories do not predict the observed bimodality of the joint distribution nor the skewness of the marginal distribution for highly peaked spectra.

The second type of evidence arising from our analysis is the occurrence of wave groups of characteristic shape. By setting a threshold to trap the highest waves in the record, and then analyzing the distributions of wave heights, periods, and steepness around the highest waves, we were able to statistically characterize the groups of extreme waves. The typical wave group found by this method contained three waves. For two or more wave periods away from the extreme waves, the heights of the surrounding waves appeared to have little correlation with the extreme waves. The incidence of the higher peaks with other high peaks in sequence is higher than expected in a purely narrow-band Gaussian process. From this evidence and other future measurements we hope to show that these

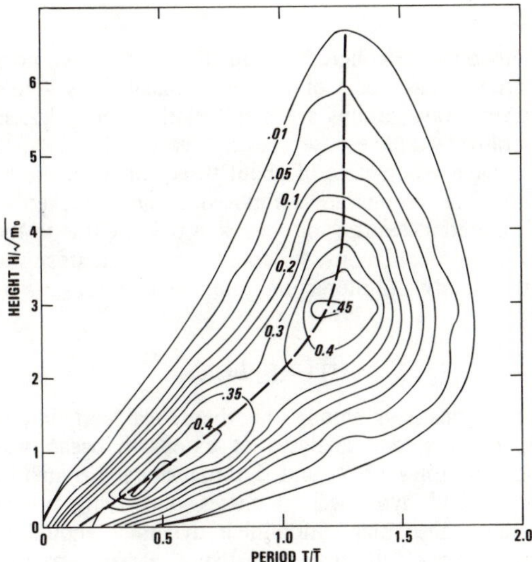

Fig. 4. The joint distribution of heights and periods P(H,T) for wave records with $Q_p \geqslant 3.0$ (From Ref. 17).

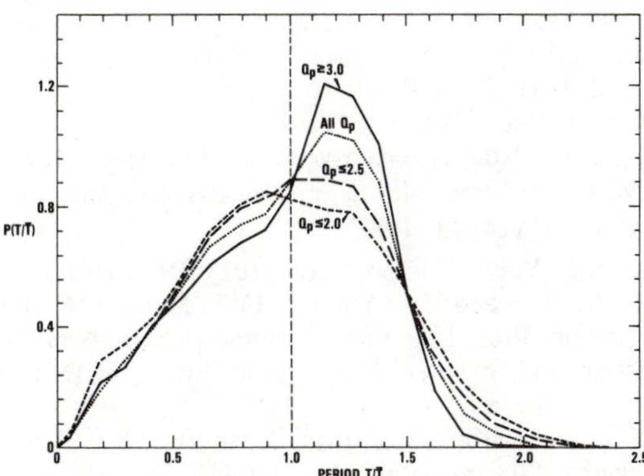

Fig. 5. The marginal distribution of wave periods P(T) for various ranges of Q_p. (From Ref. 17).

extreme waves are components of coherent groups that result from nonlinear wave instabilities. Results from our measurements of nonlinear instabilities give some circumstantial evidence that the extreme wave groups are manifestations of 2-dimensional modulational instability. In the laboratory we have observed that waves with $a_o k_o = 0.15$, a typical value for intense storm seas, have group sizes of about three waves in the period of most active modulational instability. The groups are composed of phase-locked components, and we suggest that the locally coherent features of oceanic waves are due to the same causes. Due to the rapid growth of these modulations, we could expect that steep waves in a random sea would have a high probability of interaction long enough to grow to extremes via the modulational instability.

FUTURE WORK

Through the intercomparison among the theoretical computations, results of controlled laboratory experiments, and analyses of extensive ocean wave records, we are attempting to sort out the roles of 2- and 3 dimensional instabilities, two kinds of 3-dimensional bifurcations, and dynamical processes underlying predominant features of storm waves. Ultimately, these highly nonlinear dynamics should be incorporated into operational wave forecasts models in terms of statistical parameterizations. Obviously, considerably more in-depth research of the kind reported herein is needed to reach this important goal.

REFERENCES

1. T.B. Benjamin and J.E. Feir, J. Fluid Mech. **27**, 417 (1967).
2. M.S. Longuet-Higgins, Proc. R. Soc. Lond. **A360**, 489 (1988).
3. B. Fornberg, private communication (1982).
4. H.C. Yuen and B.M. Lake, Ann. Review, Fluid Mech. **12**, 303 (1980).
5. B.M. Lake, H.C. Yuen, H. Rungaldier and W. Ferguson, J. Fluid Mech.**83**, 49 (1977).
6. W.K. Melville, J. Fluid Mech. **115**, 165 (1982).
7. M.Y. Su, Phys. Fluids **25**, 2167 (1982).
8. D.J. Benny, J. Fluid Mech. **14**, 577 (1962).
9. V.E. Zakharov and A.B. Shabat, Sov. Phys. JETP **34**, 62 (1972).
10. H.C. Yuen and B.M. Lake, Phys. Fluids **18**, 956 (1975).
11. J.M. McLean, J. Fluid Mech. **114**, 315 (1982).
12. M.Y. Su, M. Bergin, P. Marler and R. Myrick, J. Fluid Mech. **124**, 45 (1982).
13. M.Y. Su and A.W. Green, Proc. 14th Sym. Naval Hydrodynamics, 1982 (in press).
14. M.Y. Su, J. Fluid Mech. **124**, 73 (1982).
15. P.G. Saffman and H.C. Yuen, J. Fluid Mech. **101**, 797 (1980).
16. D.I. Meiron, P.G. Saffman and H.C. Yuen, J. Fluid Mech. **124**, 109 (1982).
17. M.Y. Su and M. Bergin, Proc. 1983 Buoy Technology Sym. New Orleans (1983).
18. M.Y. Su, M. Bergin and S. Bales, Proc. Ocean Structural Dynamics Sym. '82, **118** (1982).
19. Y. Goda, Rept. Port and Harbour Res. Inst. Japan **9**(3), 3 (1970).
20. M.S. Longuet-Higgins, J. Geoph. Res. **80**, 2688 (1975).
21. A. Cavanie, M. Arhan and Ezraty, BOSS '76, 354 (1976).
22. B. Chen and P.G. Saffman, Stud. Appl. Math., **62**, 1 (1980).

STUDIES OF LARGE-SCALE THERMAL VARIABILITY WITH A SYNOPTIC MIXED-LAYER MODEL

Alex Warn-Varnas, Mike Clancy, Mary Lou Morris, Paul Martin, Shelley Horton

Naval Ocean Research and Development Activity, NSTL Station, MS 39529

INTRODUCTION

In recent years large-scale thermal variability in the Central North Pacific Ocean has been studied by many investigators. Most of the studies were concerned with the explanation of sea surface temperature anomalies. Causes of these anomalies were sought in air-sea heat exchange, atmospherically forced vertical mixing, and advection by geostrophic and wind-driven currents. The contributions of these various mechanisms to the heat budget of the upper ocean were estimated in various ways. Clark[1] found correlation between the patterns of the observed anomalies and air-sea heat exchange. Camp and Elsberry[2] simulated upper-ocean response to observed atmospheric forcing with a one-dimensional bulk mixed-layer model, and demonstrated the importance of strong forcing events and vertical mixing processes. Barnett[3] analyzed AXBT data and found that the seasonal cycle is confined largely to the upper 100 m, and that approximately 90-95% of the variance in the seasonal change of heat storage in the region can be accounted for by air-sea heat exchange and vertical mixing. Barnett also pointed out that there is no agreement on the principal mechanisms responsible for anomaly development. He felt that some of the apparent uncertainties may be due to the importance of different physics in different parts of the ocean.

Haney et al.[4] and Haney[5] studied the development of large-scale thermal anomalies in the North Pacific with a three-dimensional numerical model. They found that vertical mixing and horizontal advection of mean temperature by anomalous surface Ekman currents explained a large fraction of the observed temperature anomalies in the upper ocean. The effects of anomalous horizontal advection were confined primarily to the upper 50 meters, while the effects of vertical mixing extended down to 125 m. Anomalous surface heating improved the simulation and was important for the development of a shallow warm anomaly to the east of the large scale cold anomaly in the Central North Pacific in the fall of 1976. It was found that the temperature anomalies were not simply due to an anomalous initial state.

Our study deals with ocean forecasting from an operational point of view, which involves the predictability of the state of the ocean on the daily to seasonal time scale with numerical models that rely on operational (i.e., real-time) data bases for initial conditions and surface forcing. Our modeling work differs from previous studies in the sense of approach, model used, and sensitivity studies performed.

Our work will focus on predictability and verification studies with a limited area version of the NORDA Thermodynamic Ocean Prediction Systems (TOPS) model that is used operationally at the U.S.Navy's Fleet Numerical Oceanography Center (FNOC). This model can be described as an "NX1-D" or synoptic mixed-layer model.[6,7] It contains a detailed treatment of thermodynamics and mixed-layer physics, and its primary objective is the prediction of changes in the thermal structure of the upper ocean. For the studies reported here, the model is forced by surface flux fields predicted by FNOC atmospheric models and initialized from the FNOC Ocean Thermal Structure (OTS) analysis, which was an operational Navy product during the period of interest.

The Central North Pacific during the fall of 1976 was chosen for the present study. This choice was made on the basis of available data associated with the NORPAX experiment and the existence of the previously discussed experimental and modeling work.

Our verification studies involve the comparison of model results with the daily OTS analysis and the monthly NORPAX analysis of XBT data obtained by the TRANSPAC ships-of-opportunity program.[8] In this way we hope to gain insight into the validity of the model as well as the OTS analysis itself. Comparison with the results of Haney[5] is also done together with an investigation of the effects of model physics and initial conditions on the simulated anomalies.

Our initial state studies involve a pair of initial states obtained by considering a time average and deviation of the OTS analysis over a chosen time period. The differences in the resulting solutions in space and time are considered.

DESCRIPTION OF THE MODEL

(a) Basic Equations

The basic equations of the model express conservation of temperature, salinity, and momentum in the upper ocean and take the form

$$\frac{\partial \bar{T}}{\partial t} = \frac{\partial}{\partial z}\left(\overline{-w'T'} + \kappa \frac{\partial \bar{T}}{\partial z}\right) + \frac{1}{\rho_w c}\frac{\partial \bar{F}}{\partial z} - \frac{\partial}{\partial x}(u_a\bar{T}) - \frac{\partial}{\partial y}(v_a\bar{T}) - \frac{\partial}{\partial z}(w_a\bar{T}) + A\left(\frac{\partial^2 \bar{T}}{\partial x^2} + \frac{\partial^2 \bar{T}}{\partial y^2}\right) \tag{1}$$

$$\frac{\partial \bar{S}}{\partial t} = \frac{\partial}{\partial z}\left(\overline{-w'S'} + \kappa \frac{\partial \bar{S}}{\partial z}\right) - \frac{\partial}{\partial x}(u_a\bar{S}) - \frac{\partial}{\partial y}(v_a\bar{S}) - \frac{\partial}{\partial z}(w_a\bar{S}) + A\left(\frac{\partial^2 \bar{S}}{\partial x^2} + \frac{\partial^2 \bar{S}}{\partial y^2}\right) \tag{2}$$

$$\frac{\partial \bar{u}}{\partial t} = f\bar{v} + \frac{\partial}{\partial z}\left(\overline{-w'u'} + \nu \frac{\partial \bar{u}}{\partial z}\right) - D\bar{u} \tag{3}$$

$$\frac{\partial \bar{v}}{\partial t} = -f\bar{u} + \frac{\partial}{\partial z}\left(\overline{-w'v'} + \nu \frac{\partial \bar{v}}{\partial z}\right) - D\bar{v} \tag{4}$$

where T is the temperature, S the salinity, u and v the x- and y-components of the current velocity (the x and y horizontal coordinates are defined relative to the grid), w the z-component of the current velocity, F the downward flux of solar radiation, ρ_w a reference density, c the specific heat of seawater, D a damping coefficient, ν a diffusion coefficient, f the Coriolis parameter, A the horizontal eddy diffusion coefficient, t the time, and z the vertical coordinate (positive upward from the level sea surface). Ensemble means are denoted by an overbar and primes indicate departure from these means. The quantities u_a, v_a, and w_a are the x-, y-, and z- components of an advection current, which will be defined subsequently.

The advective terms are retained in the temperature and salinity equations and neglected in the momentum equations on the basis of scale analysis.[9] Such an analysis shows that the advective terms in the thermal energy equations are of order unity, while the advective terms in the momentum equations are of the order of the Rossby Number. Since the Rossby Number is very small in most regions of the open ocean, the advective terms are dropped in the momentum equations.

Because there are no horizontal pressure gradient terms in (3) and (4), u and v represent the wind-drift component of the current. Neglect of horizontal pressure gradients here is motivated by the fact that geostrophic currents generally do not play an important role inside the mixed-layer region which is the issue of most concern in this study.

The terms involving the damping coefficient D in (3) and (4) represent the drag force caused by the radiation stress at the base of the mixed layer associated with the propagation of internal wave energy downward and away from the wind-forced region[10,11] The terms involving ν in Equations (1)-(4) account for very weak "background" eddy diffusion (due to intermittent breaking of internal waves, for example) that exists below the mixed layer. We take $D=0.1$ day^{-1} and $\nu = 0.1$ cm^{2}s^{-1} and note that these values are within the range of estimates for these quantities.

The Level-2 turbulence closure theory of Mellor and Yamada[12] is used to parameterize the vertical eddy fluxes of temperature, salinity, and momentum. This turbulence model has been described in a number of papers[13,14] and will not be presented here. Its energetics are essentially the same as those of Pollard et al.[15] and Thompson[16], with the increase in potential energy during mixed-layer deepening due to the buoyancy flux at the layer base balanced locally by mean flow shear generation minus viscous dissipation of turbulent kinetic energy.

The horizontal eddy diffusion coefficient A is simply taken to be equal to a constant value of 10^7cm^{2}s^{-1}, and the divergence of the solar radiation flux is based on the data of Jerlov[17] for seawater optical type IA.

(b) Grid

A vertically stretched grid of 17 points, extending from the level sea surface to 500 m depth, is used in the model. The horizontal grid, on which $\overline{T}$, $\overline{S}$, $\overline{u}$, $\overline{v}$, and w_a are defined, is a rectangular subset of the standard FNOC 63 x 63 Northern Hemisphere Polar Stereographic Grid, and is shown in Figure 1.

(c) Calculation of the Advection Current

The current used to advect the temperature and salinity is given by

$$u_a=u_i + u_g^* \ , \ v_a=v_i + v_g^* \ , \ w_a=w_i \tag{5}$$

where u_i and v_i are the x- and y-components of the instantaneous wind-drift current, w_i is the vertical component of the current resulting from he divergence of u_i and v_i, and u_g^* are v_g^* are the components of a divergence-free geostrophic current. The geostrophic current is set to zero for all calculations presented here.

(d) Initial Conditions

The initial temperature for the model is provided by fields produced by the FNOC OTS Analysis System. The OTS analysis is based on the Fields-by-Information-Blending methodology[18] and was used at FNOC in the mid-1970's to produce daily objective analyses of ocean thermal structure in the upper 400 m.

(e) Boundary Conditions

The lower boundary conditions for the basic equations are provided by simply holding the initial temperature, salinity, and momentum at the lower boundary of the model constant throughout the integration. The upper boundary conditions are supplied by surface fluxes of heat, moisture, and momentum.

STUDIES AND DISCUSSION

The region of study, located in the North Central Pacific Ocean for 30°N to 50°N and 130°W to 180°W will be referred to as the TRANSPAC region. The period of study is fall 1976. The model described in the preceding section is initialized at 0000 GMT 29 October 1976 from the OTS analysis and forced for 60 days by the FNOC latent and sensible heat fluxes, back radiation, solar radiation, precipitation, evaporation, and wind stress.

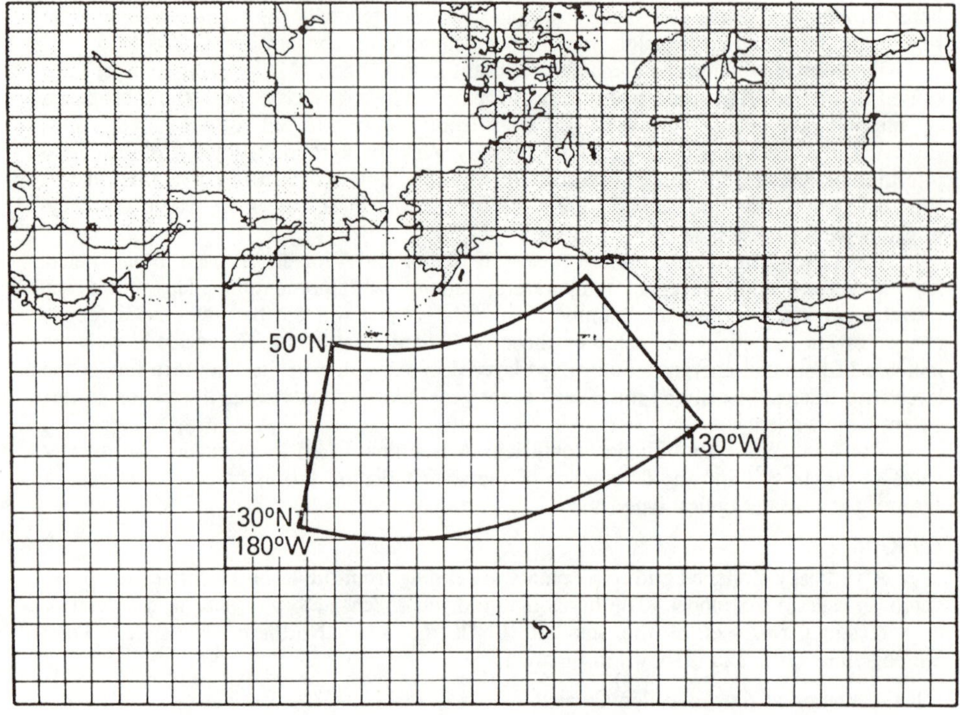

Fig. 1. Subsection of the FNOC 63x63 Northern Hemisphere Polar Stereographic Grid showing the model domain (rectangle) and the TRANSPAC region (sub-area inside rectangle).

Our studies of the resulting model predictions are roughly divided into heat content calculations, comparisons with the OTS and NORPAX ocean thermal analyses, and variations in initial conditions.

(a) Heat Content Calculations

Elsberry et al.[19] and Budd[20] noted the presence of a large-scale bias in the FNOC heat fluxes in the TRANSPAC region. They examined the accuracy of the heat fluxes through a comparison with observed upper-ocean heat content changes derived from the TRANSPAC XBT data and found that the FNOC net surface heat (long-wave + latent + sensible) flux produced too much cooling. We found similar results and pursued the problem by performing several studies. In one study we compared the monthly averaged FNOC net surface heat flux against the monthly averaged NORPAX surface heat flux calculated by N. Clark from surface ship observations (private communication). Some of the results are shown in Figure 2. The monthly averaged difference of the net surface heat fluxes over the TRANSPAC region varied from 140 ly day^{-1} to 240 ly day^{-1} from September 1976 to January 1977. This yielded an average cooling of 169 ly day^{-1} for the FNOC fluxes relative to the Clark results, which is in good agreement with the heat flux correction derived by Budd[20] from the NORPAX XBT data.

Subsequently, we corrected the FNOC surface heat flux by simply reducing the surface latent heat loss by 150 ly day^{-1} at all points.

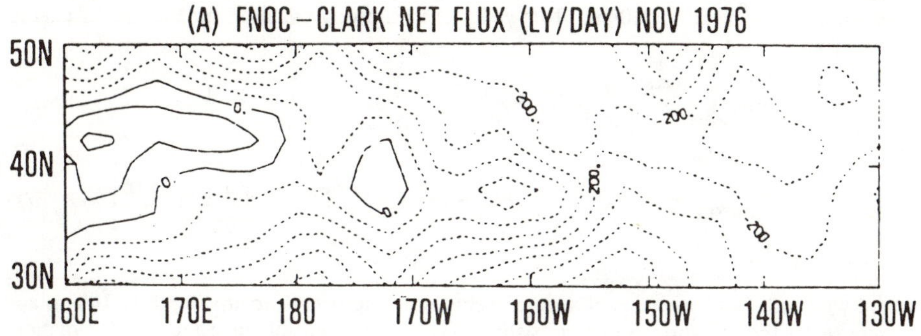

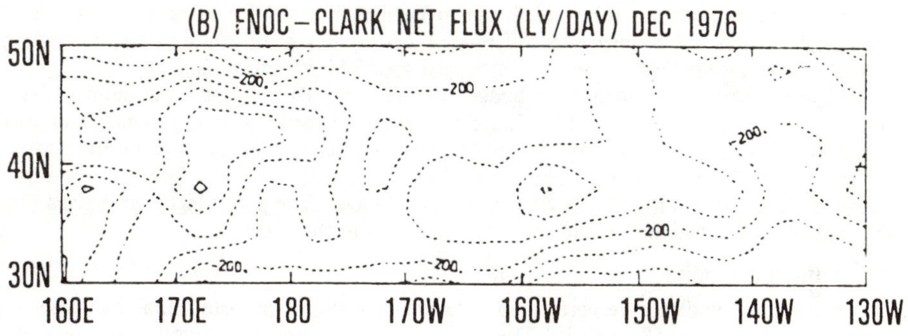

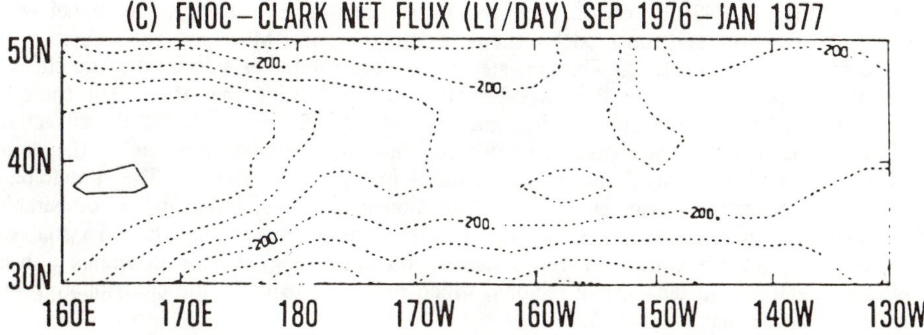

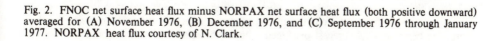

Fig. 2. FNOC net surface heat flux minus NORPAX net surface heat flux (both positive downward) averaged for (A) November 1976, (B) December 1976, and (C) September 1976 through January 1977. NORPAX heat flux courtesy of N. Clark.

Using the results of numerical simulations and the temperature equation, Equation (1), we have calculated the heat budget of the upper 100 m of the ocean. We obtained the equation for the rate of change of heat storage by multiplying Equation (1) by ρc and integrating over a depth of 100m. This yields the following equation for the rate of change of heat content

$$\frac{\partial q}{\partial t} = - \rho c \int_{z=o}^{z=100m} \left[\frac{\partial}{\partial x} \left(u_a T \right) + \frac{\partial}{\partial y} \left(v_a T \right) \right] dz$$

$$- \rho c w_{100} T_{100} + S_F + \left(\nu \frac{\partial T}{\partial z} \right)^{z=100m} + \rho c A \int_{z=o}^{z=100m} \left[\frac{\partial^2 T}{\partial x^2} + \frac{\partial^2 T}{\partial y^2} \right] dz \qquad (6)$$

where w_{100} and T_{100} are values of vertical velocity and temperature at a depth of 100 m and S_F represents the net surface flux consisting of latent and sensible heat flux, back radiation, and solar heating at the surface (it is assumed that all incident solar radiation is absorbed over the integration length of 100 m and that all turbulent fluxes are zero at 100 m). The last two terms in Equation (6) were found to be negligible in the subsequent balance of terms analysis and will be neglected. The heat budget terms were integrated horizontally over the entire TRANSPAC region and averaged over 24 hour intervals.

Figure 3 shows the total heat budget calculation. The largest contribution of the advective terms is around Day 15. The differences between the rate of change of heat storage and surface flux curves are due to the contribution of the advective terms. On the average, over the 60 day period, atmospheric forcing contributes about 80% and wind-drift advection about 20% to the rate of change of heat storage. The percentages can have a possible variation of 10 percent due to the inherent errors in the analysis.

(b) Verification Studies

Verification studies were performed by comparing model, persistence, and climatological forecasts with the OTS analysis. The OTS analysis, in turn, was monthly averaged and compared with the monthly averaged NORPAX XBT analysis.

Figure 4 shows root-mean-square (RMS) sea surface temperature (SST) forecast errors for the TRANSPAC region produced by persistence, climatology, the model with wind-drift advection, and the model without advection. The RMS errors are defined relative to the daily OTS analysis. The persistence forecast, defined as a forecast of no change from the initial state, shows the largest RMS error, indicating that the ocean changes significantly from the initial state as time goes by. The figure also shows that the advective version of the model gives a smaller RMS error than the non-advective version (i.e., one which neglects all terms involving u_a, v_a, w_a and A in Equations (1)-(4)). The climatological and non-advective model curves cross occasionally in time, indicating a comparable RMS error for these forecasts. However, it must be remembered that the OTS analysis against which we are comparing relies heavily on climatology.[21] This generates a bias towards climatology in data-sparse regions, which may tend to make the climatology forecast appear better than is justified.

Relative to persistence, the model exhibited skill in forecasting the temperature change over a period of 60 days. The advective model's RMS error was about a quarter of the persistence error at 60 days. Thus the model could be used to represent the state of the ocean for regions where the ocean thermal data is sparse.

In Figure 5 we show pattern correlations between forecast and analyzed changes in SST from the initial state. Again, the advective version of the model gave the best results. This suggests that there is a signature of the advective effects in the OTS analysis. Furthermore, the climatological forecast produced a higher correlation than the non-advective ver-

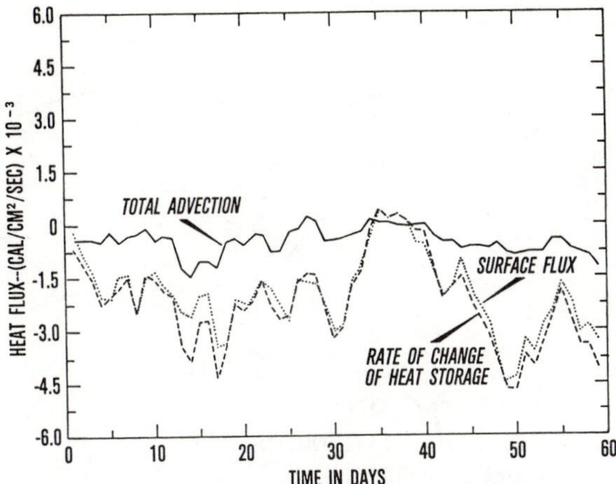

Fig. 3. An analysis of the rate of change of heat storage for the model simulation with wind-drift advection.

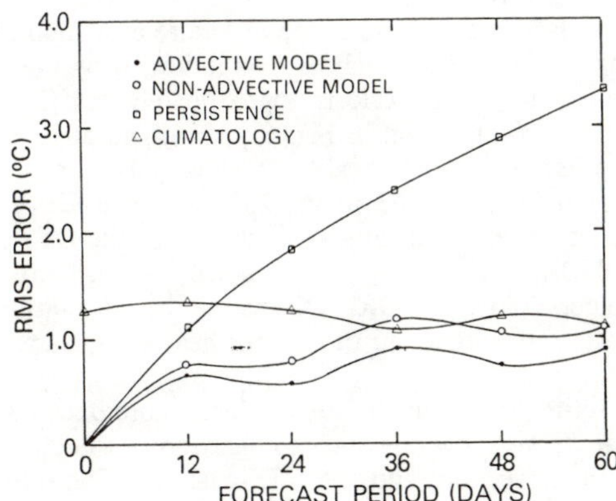

Fig. 4. RMS SST forecast errors for the TRANSPAC region as a function of forecast time. The OTS analysis was used for verification.

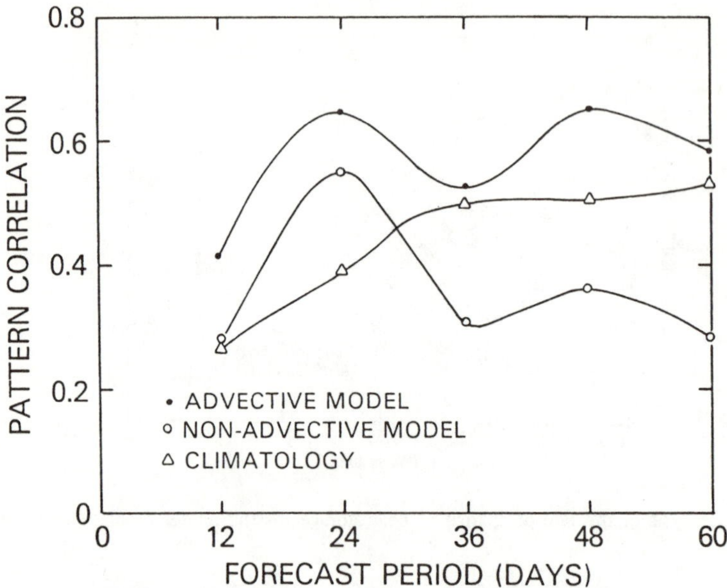

Fig. 5. Pattern correlations between forecast and analyzed changes in SST from the initial state. The OTS analysis provided the analyzed changes.

sion of the model. With regard to the correlation between climatology and the OTS analysis, we again suspect that the results may appear better than is justified because of the bias of OTS towards climatology.

It was also of interest to compare the OTS analysis against the monthly averaged NORPAX analysis. We made such a comparison in Figure 6 by monthly averaging the OTS analysis and subtracting the NORPAX analysis. As indicated by the figure, this difference in temperature on the monthly time scale is approximately $\pm 1°C$. A comparison of the difference patterns at 2.5 m (not shown in Fig. 6), 10 m and 50 m shows a weak correlation between these depths. This upper region of about 50 m roughly defines the mixed layer. As we go deeper, to depths of 162 m and 262 m, the structure of the difference patterns becomes simpler, and there is an overall tendency for the OTS analysis to give lower temperatures than the NORPAX analysis. In the upper region, there is a tendency for the OTS analysis to be colder than the NORPAX analysis in the northwestern quadrant and warmer in the southwestern quadrant. In the far northeastern quadrant, the OTS analysis is warmer.

At this point we return to the model results for a more detailed comparison with the OTS analysis. In Figure 7, we show one such comparison of the difference between model prediction with wind-drift advection, and the OTS analysis. The differences at each point are averaged in time over 12-day intervals. At the 2.5 m level, the point of reference for SST, the magnitude of the differences is plus or minus one degree. In the far eastern and northwestern regions the model results are colder than the OTS analysis. There is a tendency for coherence of the difference patterns in the upper region. At a 102 m there is an average tendency for the model results to be colder than the OTS analysis.

(c) Anomaly Calculations and Studies

Another study performed was the calculation of temperature anomalies (i.e., deviations from climatology) from the model predictions. These anomalies were compared with

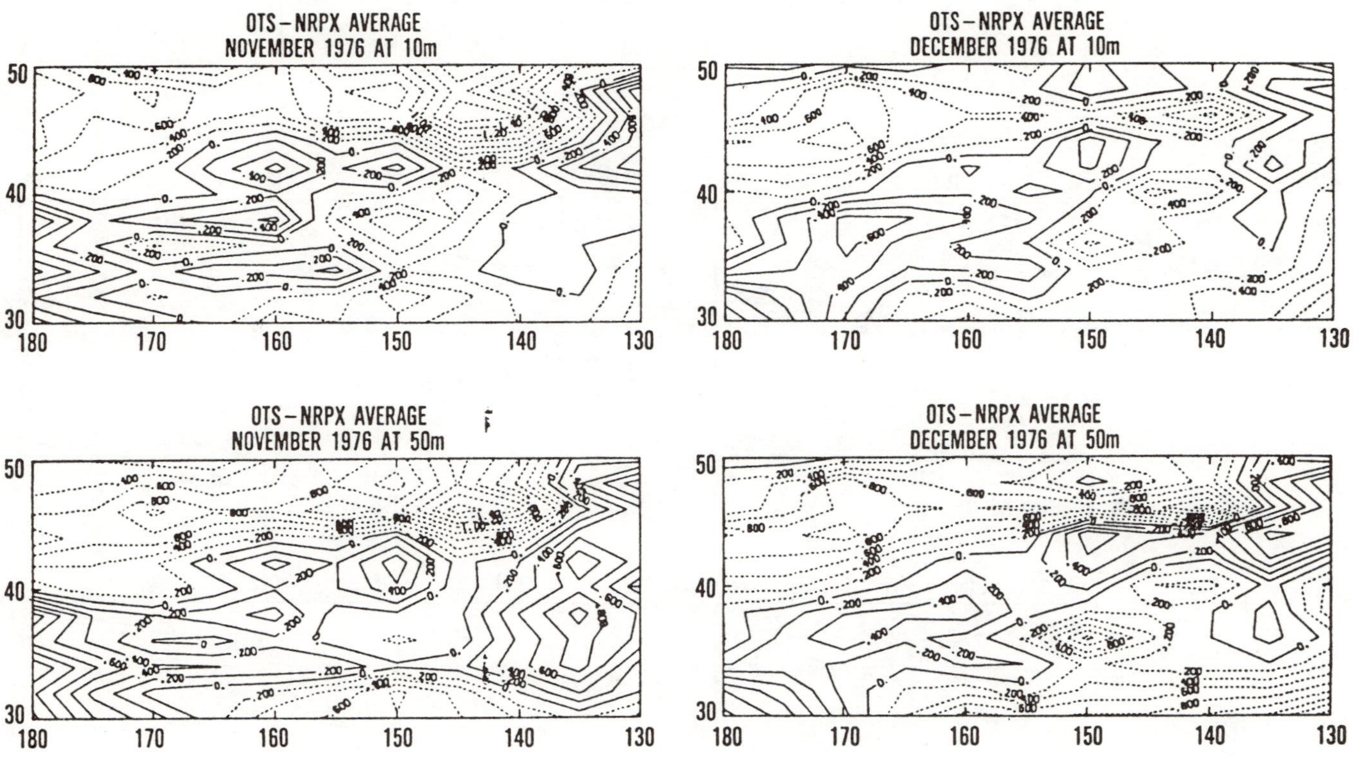

Fig. 6a. Differences, in °C, between the monthly averaged OTS analysis and the monthly averaged NORPAX analysis.

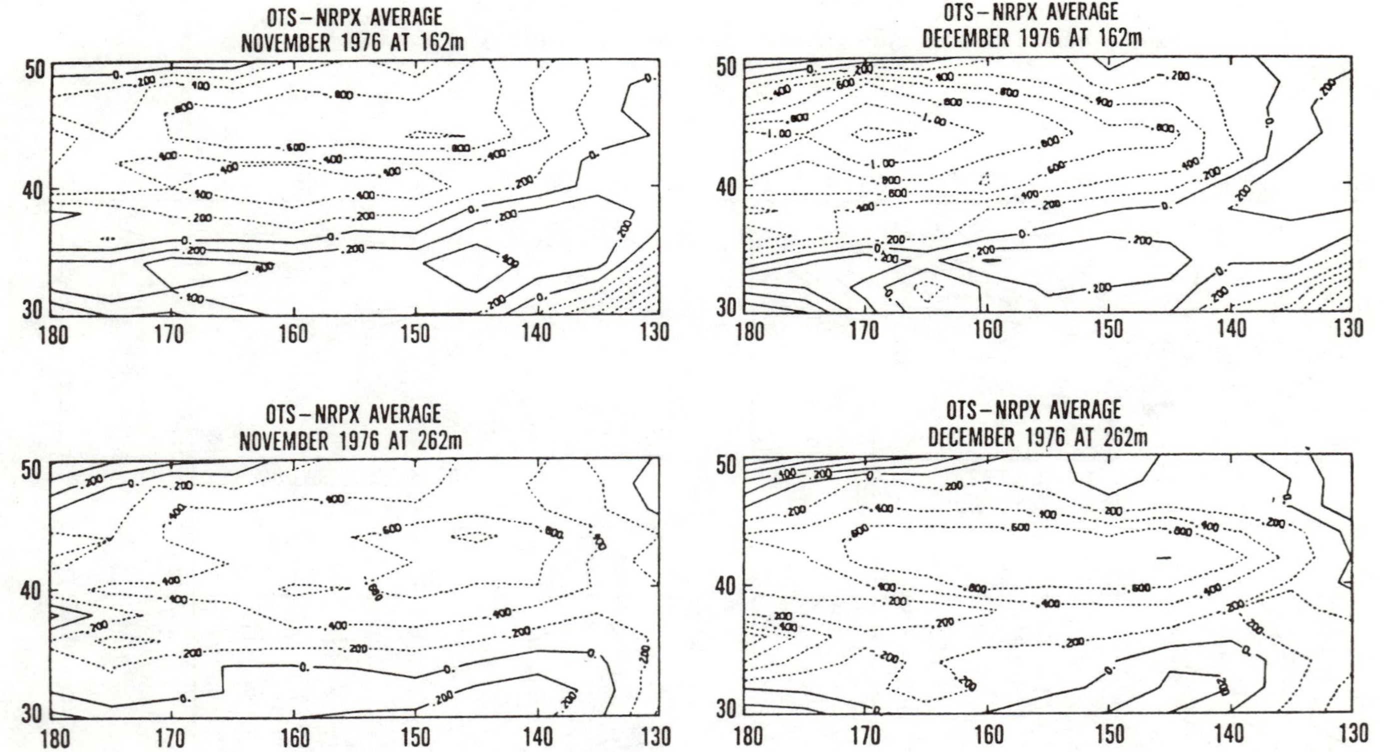

Fig. 6*b* Differences, in °C, between the monthly averaged OTS analysis and the monthly averaged NORPAX analysis.

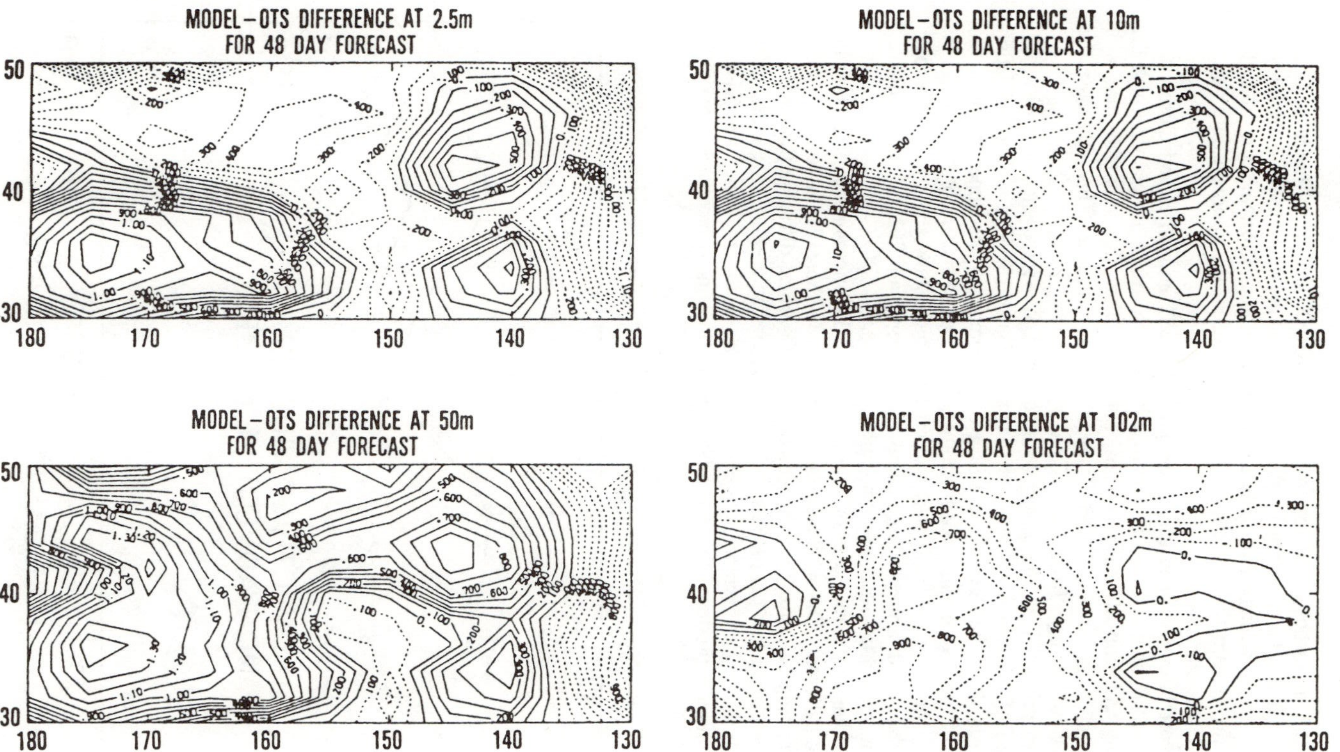

Fig. 7. Differences between the model simulations and the OTS analysis. Each result is averaged over a 12 day interval centered at Day 48. The units are °C.

the monthly averaged anomalies calculated from the NORPAX analysis. The development of these anomalies can be characterized by the formation of an intensely cold anomaly in the upper 100 m over the western two-thids of the region and the formation of a shallow warm anomaly in the eastern third of the region. In all cases, the anomalies are defined relative to the NORPAX climatology.

The pattern correlation between the model-predicted SST anomalies and those calculated from the OTS analysis are shown in Figure 8. The correlations are generaly high for both the advective and non-advective models, indicating that the anomalies are well predicted. The correlations show, on the average, a decrease in time. This is because the model simulations slowly diverge from the OTS analysis in time.

The anomalies predicted by the advective model at Day 48 are shown in Figure 9. The corresponding OTS analysis anomalies are shown in Figure 10. The model simulation and the OTS analysis show the development of a cold anomaly in the upper layers of the western two-thirds of the region and a warm anomaly in the eastern third of the region. At 162 m and 262 m, there are cold and warm anomalies which show some resemblance to the NORPAX analysis and Haney's simulation[5]. At a depth of 262 m, the NORPAX analysis shows a widespread but weak warm anomaly.

At the depths below the mixed layer, the patterns of the anomalies closely resemble the initial anomaly patterns. The model-predicted changes in these depths can be caused only by Ekman pumping or diffusion. The horizontal and vertical diffusive time scales in this region are far too long for this process to have an appreciable effect on the anomalies over a period of 60 days. Furthermore, consistent with the results of Haney and Risch[22], the *net* Ekman pumping produced by the FNOC synoptic winds during this period is insignificant.

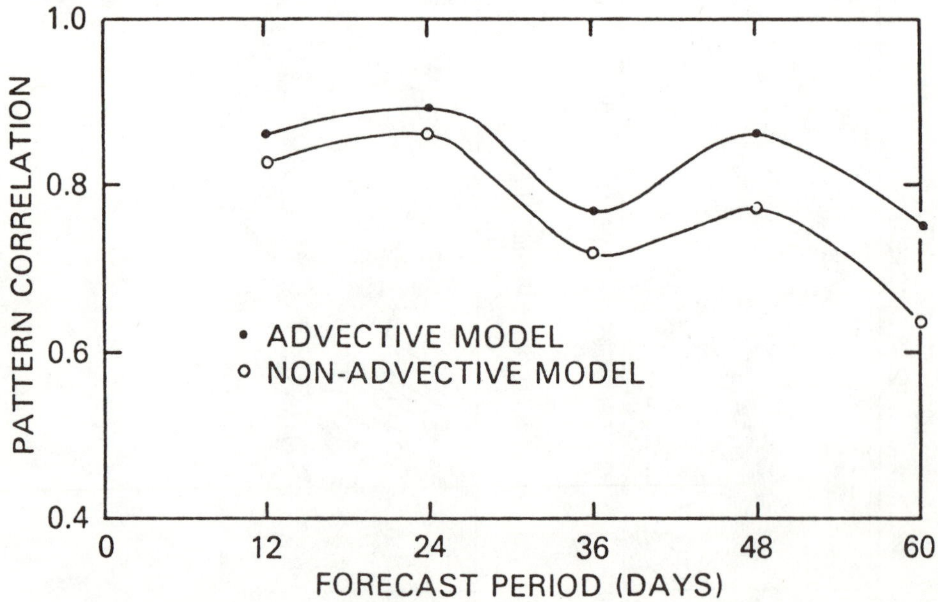

Fig. 8. Pattern correlation between predicted SST anomalies and anomalies calculated from the OTS analysis for the TRANSPAC region.

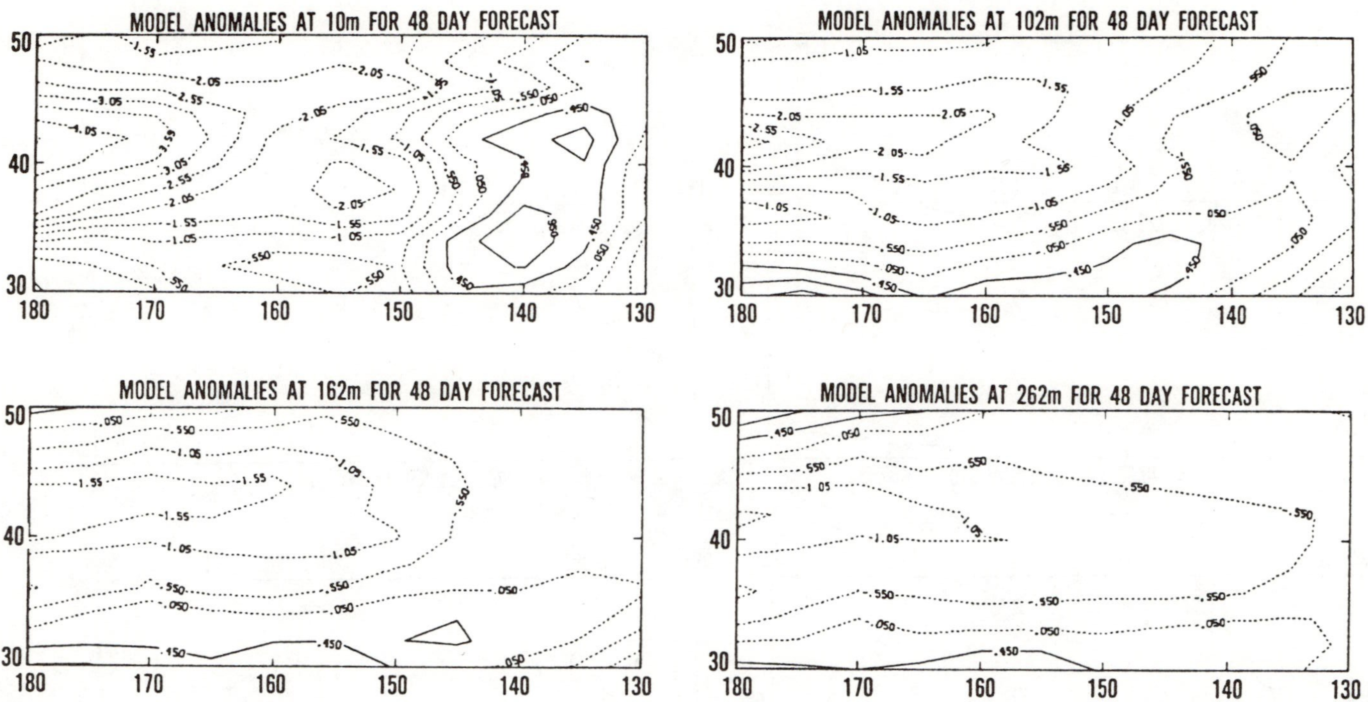

Fig. 9. Anomalies predicted at Day 48 by the model with wind-drift advection. The contour interval is 0.5°C.

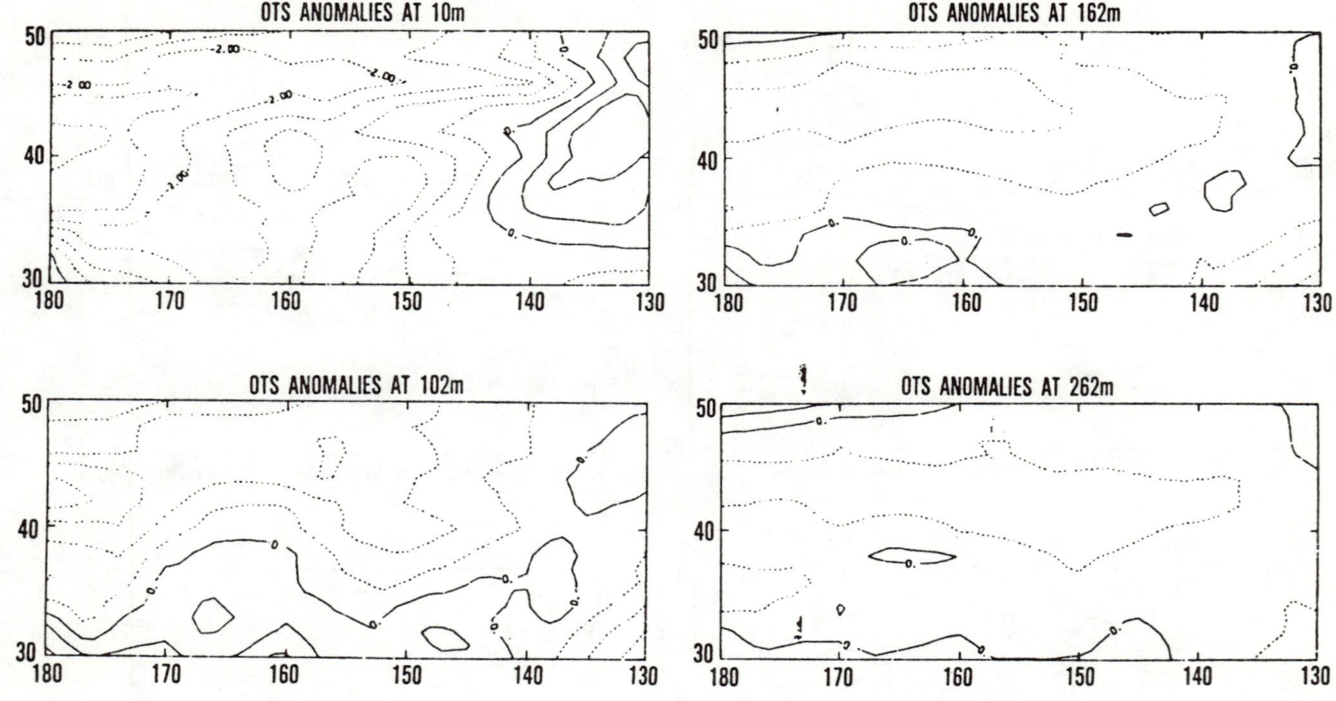

Fig. 10. Anomalies calculated at Day 48 from the OTS analysis. The contour interval is 0.5°C.

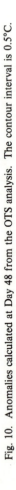

The previous analysis suggested that the model-simulated anomalies at the deeper depths are strongly dependent on the initial conditions from which the simulation is started. We studied the dependence of the anomalies on the initial conditions by initializing the model from the monthly averaged NORPAX analysis instead of the OTS analysis. The results of the model simulation with wind-drift advection and an ambient viscosity of 1 cm^2/sec, ten times larger than usual and comparable to 1.5 cm^2/sec used by Haney[9], are shown in Figure 11. The anomaly in the upper layers is weaker than its analogue shown in Figure 9. At the lower depths there is a dominant warm anomaly. The trend of a cold anomaly in the surface layers and a warm anomaly at depth agrees with the NORPAX analysis. At depths below the mixed layer, the patterns of the anomalies closely resemble as before, the initial anomaly pattern indicating that the increased ambient viscosity or 1 cm^2/sec does not produce an anomaly structure.

As shown in Figure 8, the anomalies obtained with the model with wind-drift advection show a higher correlation with the anomalies calculated from the OTS analysis than do those predicted by the non-advective model. An analysis of the anomalies with and without advection at depths of 2.5 m, 10 m, and 50 m shows that the simulation with advection leads to colder anomalies in the western two-thirds of the region. This agrees with Haney's observation[5] that anomalous horizontal advection produces a widespread pattern of cooling in the upper layers and plays no significant role in the deeper layers. The anomalous current consists of a strong Ekman flow from the north in the western two-thirds of the region which contributes to the cold anomaly formation by advecting the cold northern surface water southward. In the eastern third of the region the anomalies horizontal advection is weak and directed from the south.

(d) Initial State Studies

We investigated the sensitivity of model solutions to different initial conditions. Two model simulations were performed which were identical except for their initial conditions. To derive a pair of initial states, we considered the daily OTS analysis over a 14-day interval from 26 October to 8 November 1976, and computed a 14-day average of the OTS analysis and a standard deviation. The averaged OTS analysis and the standard deviation are shown in Figure 12 for two 14-day time intervals located at the beginning and end of the 60 day period. Note that the standard deviation does not exceed 1°C.

From the 14-day interval we chose two initial conditions for our model simulations. Initial condition A was the 14-day average. Initial condition B was OTS analysis at 0000 GMT, 26 October 1976.

The differences in time and space between two solutions are shown in Figure 13. Each of the solutions is averaged over the 12 days surrounding the indicated time of forecast. The difference between solutions has a high degree of correlation in magnitude and shape. At Day 48 the difference patterns correlate as a function of depth in the upper layers and decrease in correlation as the bottom of the mixed layer is approached.

The predominant signature of the difference between the solutions is the propagation of approximately the same difference patterns in time at all depths. This situation occurs because this is an externally forced problem with the evolution of the model prediction controlled by the atmospheric forcing. The problem is approximately linear in relation to the forcing, and the differences between two initial conditions propagate in time in response to it. This situation can be contrasted with the internally forced meteorological forecast problem in which small perturbations to the initial state will grow.

CONCLUSION

We performed model simulations of upper-ocean thermal structure in the TRANSPAC region of the Central North Pacific within the framework of operational ocean prediction. Limited-area versions of the FNOC TOPS model were used to predict changes

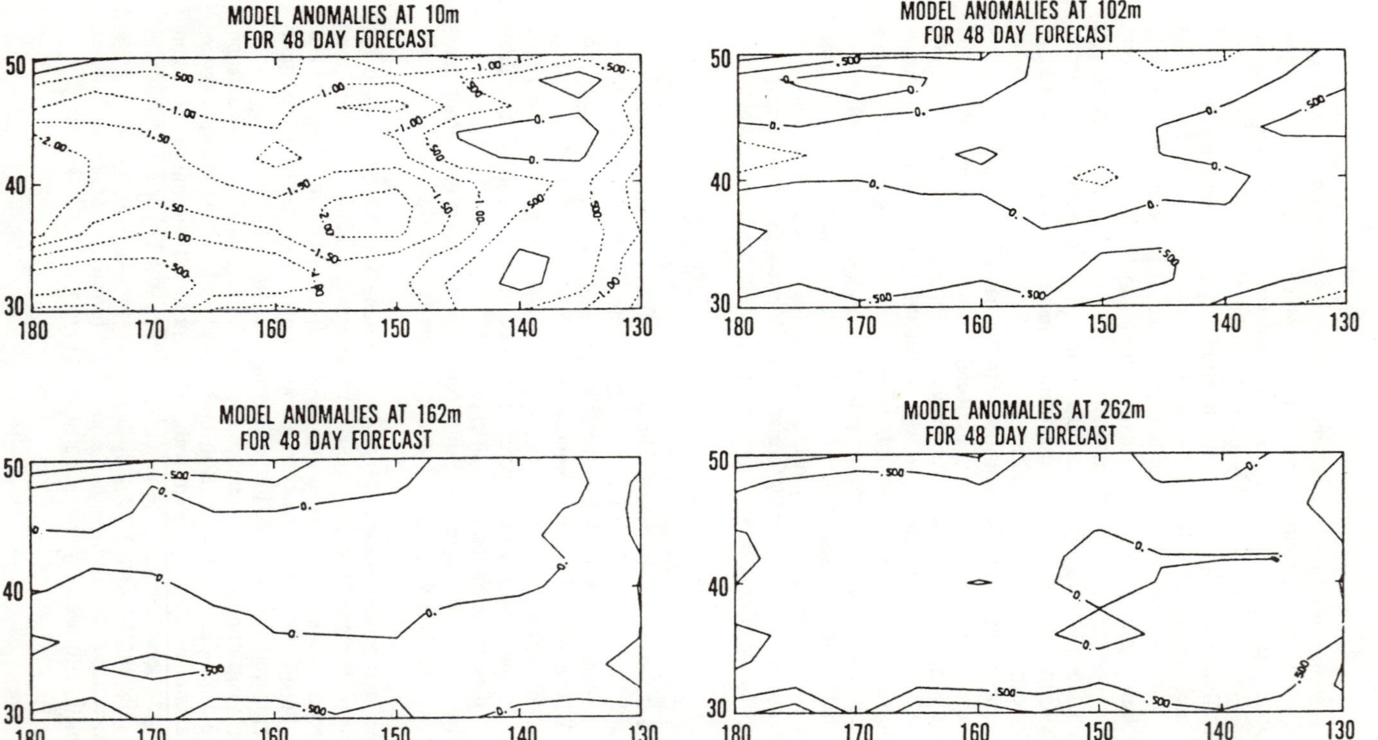

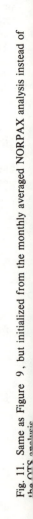

Fig. 11. Same as Figure 9, but initialized from the monthly averaged NORPAX analysis instead of the OTS analysis

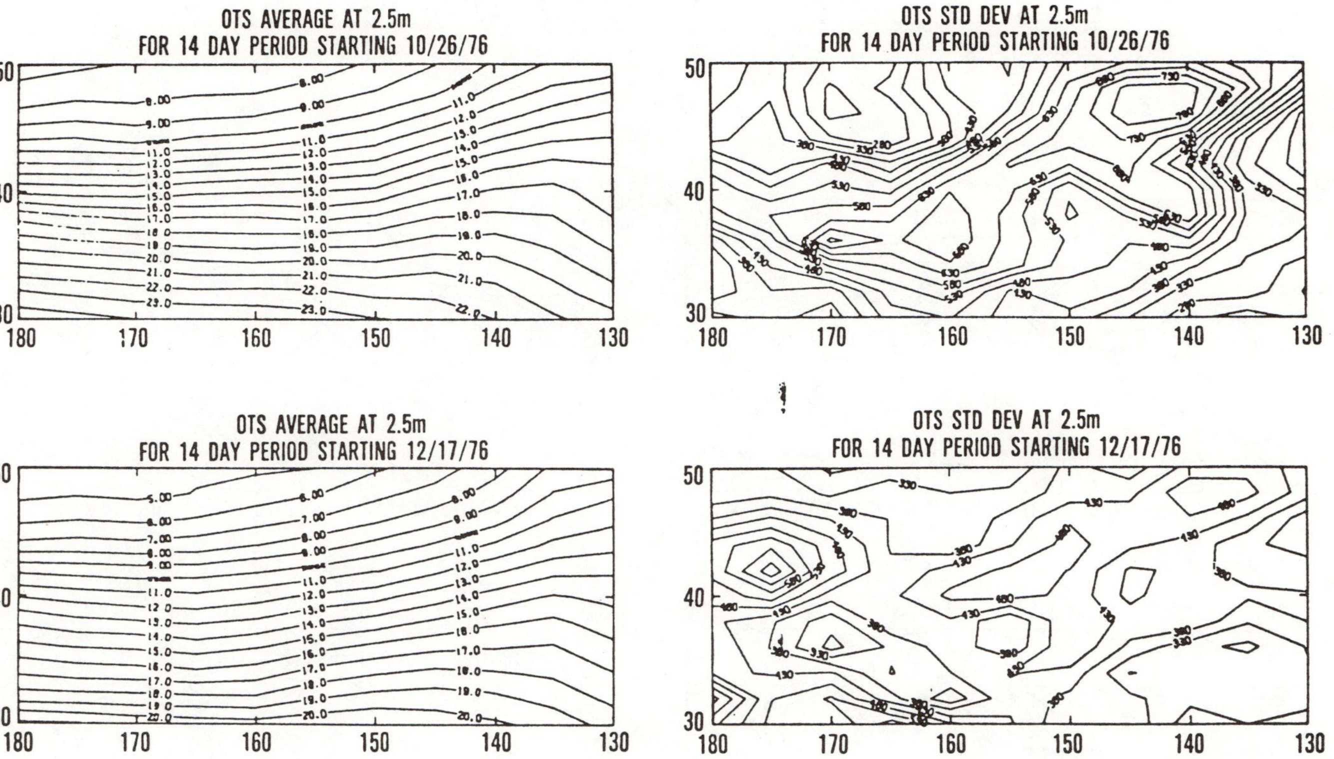

Fig. 12. Fourteen-day time averages of the OTS analysis and the associated standard deviations. Units are in °C.

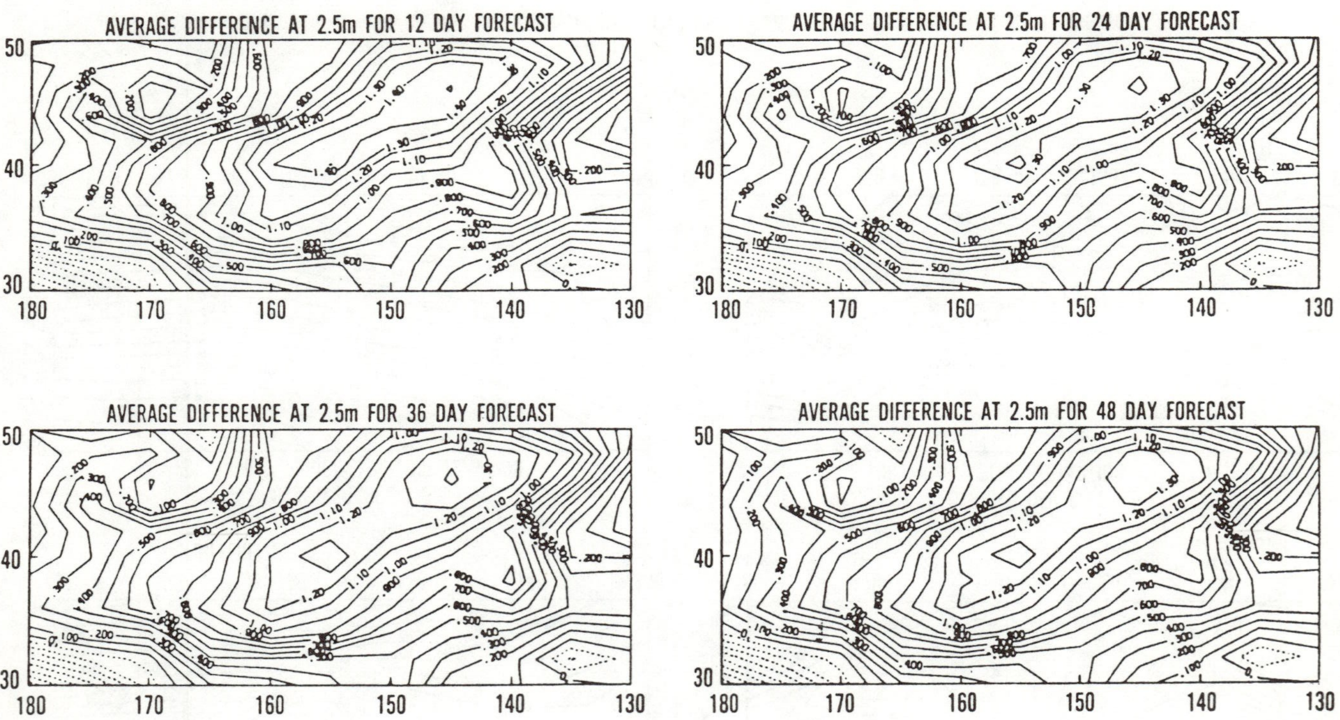

Fig. 13a. Differences between two simulations started from different initial conditions at a number of depths and times. Each solution is averaged over the 12 days surrounding the indicated time of forecast. Units are °C.

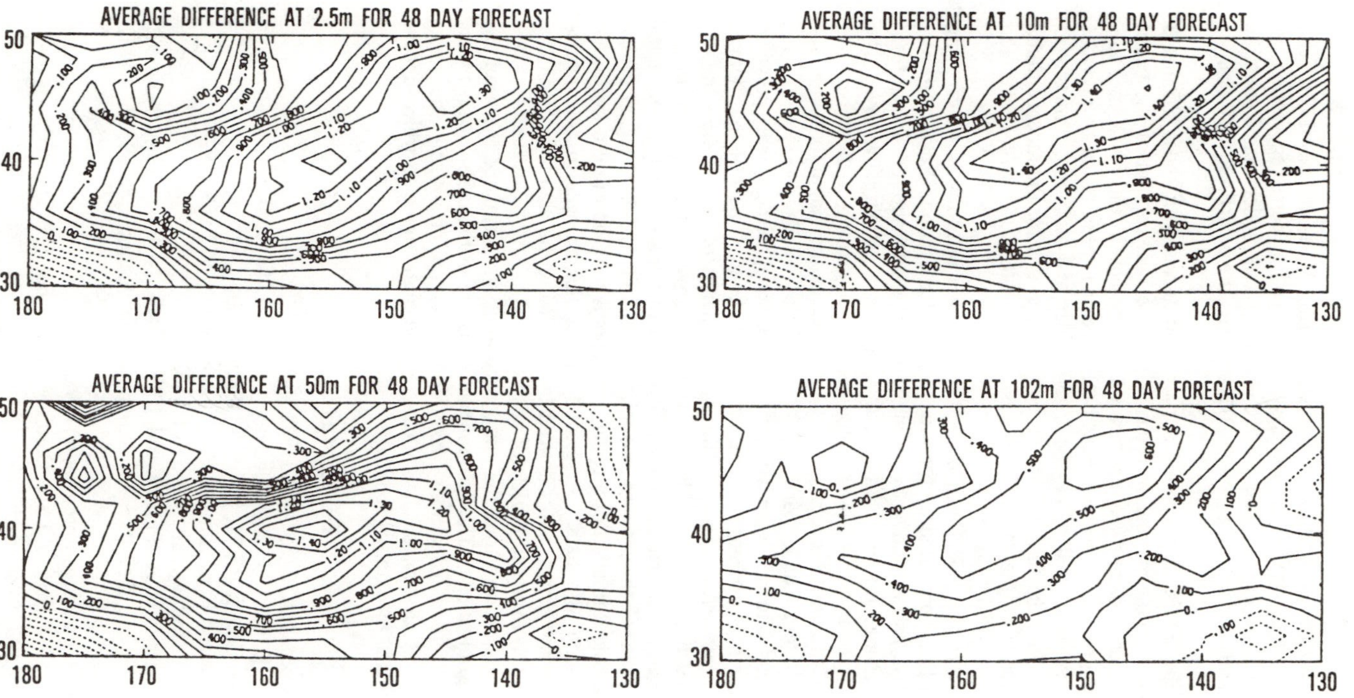

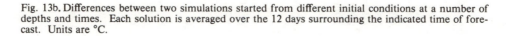

Fig. 13b. Differences between two simulations started from different initial conditions at a number of depths and times. Each solution is averaged over the 12 days surrounding the indicated time of forecast. Units are °C.

in upper-ocean thermal structure during the period September 1976 to January 1977. The models were initialized from the FNOC OTS analysis and forced by corrected versions of the FNOC atmospheric surface fluxes. The predictive skill of the models was studied by comparing the model simulations with the FNOC OTS analysis, the NORPAX XBT analysis and previous work.

A comparison of the FNOC atmospheric surface heat fluxes with those calculated independently from ship observations and with changes in the heat content of the ocean determined from the OTS analysis showed a significant cooling bias in the FNOC net surface heat flux. A similar problem with the FNOC heat fluxes has been noted by previous investigators. The bias was estimated to be about 150 ly/day and attributed primarily to excessive latent heat loss. In order to diminish this bias the heat flux for the model simulations was modified by reducing the latent heat loss by a constant 150 ly/day over the modeling region.

Predictions produced by the model with wind-drift advection agreed with the verifying OTS analysis better than those produced by the model without advection. Relative to persistence, both the advective and non-advective versions of the model showed skill in forecasting the temperature change for the TRANSPAC area over a period of 60 days. The advective model's RMS error at 60 days was approximately one quarter of the persistence error. This implied that the model could provide upper-ocean thermal information for regions where the ocean thermal data is sparse.

Our heat content and anomaly calculations indicated, in agreement with previous investigators[3,5] that air-sea heat exchange and vertical mixing control most of the thermal variability of the upper ocean. An analysis of the rate of change of heat storage showed that air-sea exchange contributed about 80% and wind-drift advection about 20% to the total heat budget. Anomalous Ekman currents tended to advect the colder northern temperatures in the upper layers of the western two-thirds of the TRANSPAC region southward. This contributed to the formation of a upper-layer anomaly in the western two-thirds of the region, in agreement with data. At greater depths the changes in the model-simulated anomalies were very small, as evidenced by initializing the simulation with the OTS analysis and the monthly averaged NORPAX analysis.

In our initial state studies, we varied the initial conditions and examined the resulting differences in the model simulations. The predominant signature of the difference between solutions due to different initial conditions was the propagation of roughly the same difference pattern in time of all depths. This illustrated the high degree to which upper-ocean variability is dominated by external forcing.

REFERENCES

1. N. Clark, J. Phys. Oceanogr., **2**, 391-404 (1972).
2. N.T. Camp and R.L. Elsberry, J. Phys. Oceanogr., **8**, 215-224 (1978).
3. T.P. Barnett, J. Phys. Oceanogr., **11**, 887-906 (1981).
4. R.L. Haney, S. Shiver and K.H. Hunt, J. Phys. Oceanogr., **8**, 952-969 (1978).
5. R.L. Haney, J. Phys. Oceanogr., **10**, 541-556 (1980).
6. R.M. Clancy, P.J. Martin, S.A. Piacsek and K.D. Pollak, NORDA Tech. Note 92, Naval Ocean Research and Development Activity, NSTL Station, MS 39529, (1981) p. 66.
7. R.M. Clancy and K.D. Pollak, (in preparation, 1983).
8. W.B. White and R.L. Bernstein, J. Phys. Oceanog., **9**, 592-606 (1979).
9. R.L. Haney, J. Phys. Oceanogr., **4**, 145-167 (1974).
10. R.T. Pollard and R.C. Millard, Deep Sea Res., **17**, 813-821 (1970).
11. P.P. Niiler and E.B. Kraus, in Modelling and Prediction of the Upper Layers of the Ocean, edited by E.B. Kraus, (Pergamon, N.Y., 1977) Chap. 10.

12. G.L. Mellor and T. Yamada, J. Atmos. Sci., **31**, 1791-1806. (1974).
13. G. L. Mellor and P.A. Durbin, J. Phys. Oceanogr., **5**, 718-725 (1975).
14. R.M. Clancy and P.J. Martin. Bull. Am. Meteorolo. Soc., **62**, 770-784 (1981); R.M. Clancy, Mon. Wea. Rev., Mon. Wea. rev., **109**, 1807-1809 (1981).
15. R.T. Pollard, P.B. Rhines and R.O. Thompson, Geophys. Fluid Dyn., **3**, 381-404 (1973).
16. R.O. Thompson, J. Phys. Oceanogr., **6**, 496-503 (1976).
17. N.G. Jerlov, in Optical Oceanography (Elsevier, N.Y., 1968) p. 352.
18. M.M. Holl and B.R. Mendenhall, Tech. Rept. M-167. Meteorology International, Inc., Monterey, Calif., (1971) p. 71.
19. R.L. Elsberry, P.C. Gallacher and R.W. Garwood, Tech. Rept. NPS 63-79-003, Naval Postgraduate School, Monterey, Calif., (1979) p. 30.
20. B.W. Budd, Prediction of the Spring Transition and Related Sea-Surface Temperature Anomalies. Master's Thesis. Naval Postgraduate School, Monterey, Calif., (1981) p. 95.
21. B.R. Mendenhall, M.J. Cuming and M.M. Holl, Tech. Rept. M-232, Meteorology International, Inc., Monterey, Calif., (1978) p. 110.
22. R.L. Haney and M.S. Risch. Bull. Am. Meteorol. Soc., **60**, 1254 (1979).

HARMONICS RESONANCE AND CHAOS IN THE EQUATORIAL WAVEGUIDE

P. Ripa
C.I.C.E.S.E.
Ensenada, B.C.N., México

ABSTRACT

The purpose of this note is to report some preliminary results of (apparently) chaotic solutions found in the study of *weak* interactions among equatorial waves.[6] Chaos is usually related to models in the classical turbulence limit, strong interactions (e.g., see Lee[3]), it is interesting to find it also in the opposite limit, weak interactions. Meiss[5] has also reported chaotic evolution for a system with a few resonant waves. The important questions of how is the chaos/order picture changed (1) by adding more modes to the model, and/or (2) for intermediate energies (such that neither the weak - nor the strong - interactions approximation is applicable), will not be addressed here.[2]

THE MODEL

We work with the shallow water equations in the unbounded equatorial β-plane. The position-dependent part of the linear waves span a complete basis,[4] which is used to expand the dynamical fields at any time. The state of the system is then fully described by the set of expansion amplitudes, $Z_a(t)$; the label "a" groups the zonal wavenumber, k, the meridional quantum number, n, the vertical separation constant, c (not to be confused with a phase-speed), and a discrete variable (which may take, at most, three values). The evolution of the system is controlled by

$$\partial_t Z_a(t) + i\omega_a Z_a(t) = \tfrac{1}{2} \tilde{\sum_{bc}} \sigma_a{}^{bc} Z_b^*(t) Z_c^*(t) + F_a \,, \tag{1}$$

where the ω_a (free frequencies) and $\tau_a{}^{bc}$ (coupling coefficients) are real, and F_a represents forcing plus dissipation projected into mode "a". The tilde over the summation sign means that, in addition to a regular sum over the discrete labels, there is an integral in k_b, or in k_c, subject to

$$k_a + k_b + k_c = 0. \tag{2}$$

Hereafter we neglect forcing and dissipation by setting $F_a = 0$. The solutions of (1) are constrained by the conservation of total energy and pseudomomentum, *viz.*

$$E^{(2)} + 0(Z^3) = \text{const.},\ P^{(2)} + 0(Z^3) = \text{const.}, \tag{3}$$

whose quadratic parts have a diagonal representation in phase space

$$E^{(2)} = \tfrac{1}{2} \tilde{\sum_a} |Z_a|^2 \,,\ P^{(2)} = \tfrac{1}{2} \tilde{\sum_a} s_a |Z_a|^2 \,, \tag{4}$$

where

$$s_a = k_a/\omega_a \tag{5}$$

is the *slowness*[1] of mode "a".

Equations (1) through (5) represent quite general results: they are found in fluid systems with (at least) one homogeneous coordinate. Invariance under translations along that coordinate is linked to pseudomomentum *and* (regular) momentum conservation; transversal boundaries, which break this symmetry, are sources and sinks of these momenta.

STRONG INTERACTIONS

Formally, this corresponds to the limit $\omega_a \to 0$ in (1). For instance, consider quasi-geostrophic flow in flat bottom ocean: the expansion modes represent (planetary) Rossby waves and $-\beta P^{(2)}$ is no more than total enstrophy (variance of potential vorticity) because $-\beta s$ is equal to squared wavenumber. Similarly, the uniform potential vorticity flow discussed by William Blumen in this meeting, could be taken as bounded by rigid planes with slopes equal to $\pm \epsilon$ ($\epsilon << 1$): the expansion basis is made up of topographic Rossby waves with $\omega \propto \epsilon$, and $\epsilon P^{(2)}$ is proportional to the second integral of motion presented by Blumen. The strong interactions limit corresponds to making $\beta \to 0$ or $\epsilon \to 0$ in these two problems. In both cases the $0(Z^3)$ terms in (3) vanish identically (a property with many important consequences) and Fjortoft's theorem follows: any cascade of energy in s-space must be balanced by a decascade (the slowness may be replaced by βs, which is independent of β, or ϵs, which is independent of ϵ, respectively). In the equatorial β-plane, the $0(Z^3)$ terms in (13) do not vanish and Fjortoft's theorem is (approximately) valid only if those terms are negligible with respect to $E^{(2)}$ and $P^{(2)}$. (In this case, s may have both signs: a certain ordering in s does not translate into the same ordering in the phase-speed s^{-1}.)

WEAK INTERACTIONS

If the main balance in (1) is between the two terms in the left-hand side, the more important nonlinear terms correspond to resonant triads, i.e., those trios of interacting modes that, in addition to (2), satisfy

$$\omega_a + \omega_b + \omega_c = 0. \tag{6}$$

An important consequence of energy-pseudomomentum conservation is a factorization of the coupling coefficients of a *resonant triad*, viz

$$\sigma_a{}^{bc}/\omega_a = \sigma_b{}^{ca}/\omega_b = \sigma_c{}^{ab}/\omega_c \overset{\text{def}}{=} \gamma_{abc}. \tag{7}$$

Now, let us "approximate" (1) by taking a discrete set of modes resonantly coupled. Defining new amplitudes $X_a(t)$ by

$$Z_a(t) = X_a(t)\exp(-i\omega_a t), \tag{8}$$

we get

$$\dot{X}_a = \tfrac{1}{2}\omega_a \sum_{bc} \gamma_{abc} X_b^* X_c^*, \tag{9}$$

where the coupling coefficients have been written using (7), and there is no tilde over the sum symbol because it is a regular summation [over trios that satisfy (2) and (6)]. The solutions of (9) are constrained by

$$E^{(2)}\text{=const., } P^{(2)}\text{=const., } H\text{=const.,} \tag{10 a,b,c}$$

where

$$H = \frac{1}{3!} \sum_{abc} \gamma_{abc} \mathrm{Im}(X_a X_b X_c) \tag{11}$$

is the Hamiltonian. Notice the difference between (3) and (10a,b).

If $X_a(t)$ is a solution of (9), then $\lambda X_a(\lambda t)$ is also a solution for any real λ. The same invariance is found in the strong interactions limit, replacing X_a by Z_a, modeled by (1) with $\omega_a=0$ (and $F_a=0$). Mathematically, this reflects the fact both are asymptotic approximations in the limit $E^{(2)}\rightarrow 0$ (weak) or $E^{(2)}\rightarrow \infty$ (strong). Practically, this means that either the nonlinear time scale or the value of $E^{(2)}$ are unimportant.

A single Rossby wave is an exact nonlinear solution of the quasi-geostrophic equations: this may be derived as a consequence of the lack of $0(Z^3)$ terms in (3). An equatorial wave is not a nonlinear solution (cubic energy and pseudomomentum do not vanish): if initially there is only one wave, say

$$Z_a(0)=0 \quad \text{for} \quad a \neq 1, \tag{12}$$

then all harmonic modes, k/k_1 = integer, are in principal generated. We now consider the possibility that a subset of the harmonics are in resonance, namely

$$\omega_j = j\omega_1, \ k_j = jk_1 \ , \tag{13}$$

for certain integer j's. The following results are valid for a one-layer model.

KINEMATICS

Many equatorial *Rossby* waves resonate with the second harmonic [$j = 1,2$ in (13)]. There are also cases with higher harmonics in resonance [e.g. $j = 1,2,4$ in (13). See inset in Figure 1]. Resonant chains of Rossby waves must have a finite number of components, because these modes have bounded $|\omega|$. Resonant *inertia-gravity* harmonics, on the other hand are found in infinite chains: all integer values of j in (13), except zero and multiples of 3. Finally, *Kelvin* waves constitute a trivial case of resonant harmonics: equation (13) is satisfied for all j, because these modes are non-dispersive.

Despite that both have infinitely many components, there is a very important difference between the chains of inertia-gravity and Kelvin waves: in the latter case, the interaction coefficient γ is the same for all triads because these modes have all the same meridional structure, whereas in the former case γ is different for each triad.

DYNAMICS

Resonant harmonics have the same value of s and therefore (10a) and (10b) do not represent independent constraints on the evolution of the system (viz. $P^{(2)} \equiv s\, E^{(2)}$). An M-wave system has M degrees of freedom; since it has (at least) two integrals of motion, it follows that the two-harmonic problem is integrable. For instance, the solution corresponding to the initial condition (12) is [assuming $X_1(0)$ real]

$$X_1(t) = X_1(0) \operatorname{sech} (\mu t) \tag{14}$$

$$X_2(t) = X_1(0) \tanh (\mu t)$$

where

$$\mu = \gamma\omega_1 X_1(0) \tag{15}$$

with $\gamma = \gamma(j)$ = 1,1,-2). The relative energies of the first and second harmonic $X_j^2(t)/X_1^2(0)$ $j = 1,2$ are shown in curves a_1 and a_2 in Figure 1, as a function of μt.

Probably the other extreme case of harmonics resonance is that of the Kelvin modes: all interacting triads are in resonance and with the same value of γ. This system is also

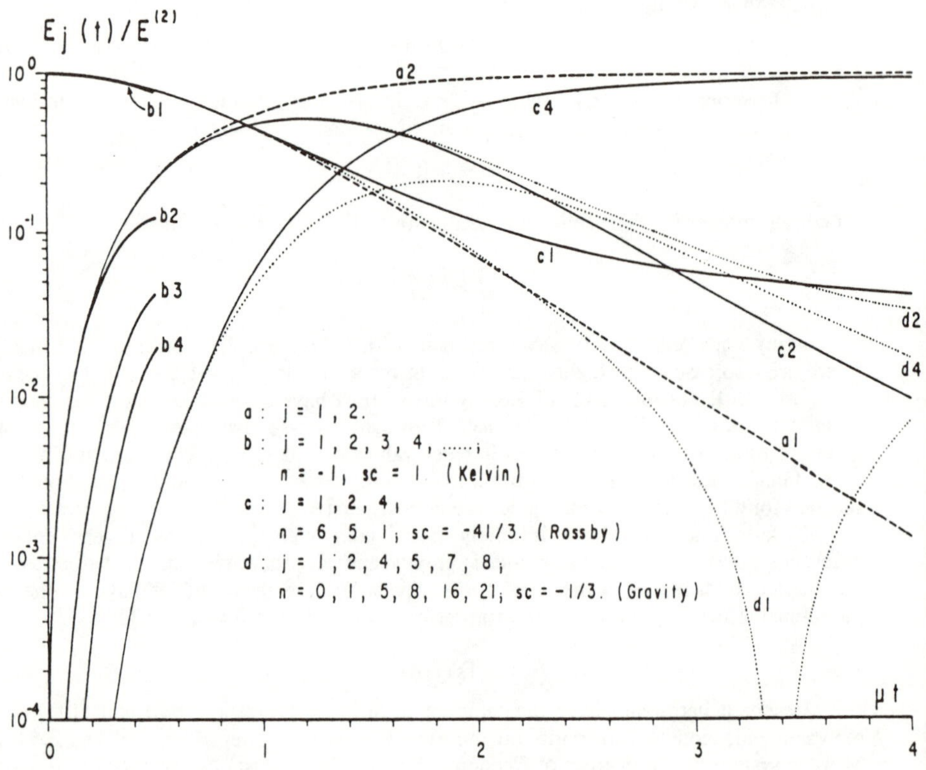

Fig. 1. Evolution of a wave (j = 1) that interacts resonantly with harmonics (j > 1). The energies of the harmonics with j < 5 are shown, relative to the total quadratic energy $E^{(2)}$ (which is a constant of motion). The nonlinear inverse time scale, μ, is defined in (15). (a): Only the fundamental wave and the second harmonic interact resonantly the solution is given by (14). (b): The Kelvin modes problem (all interacting triads are resonant and with the same interaction coefficient). Neglect of the off-resonant excitation of non-Kelvin mode makes the solution invalid for $\mu t > ½$ (see Ripa, 1982). (c): First, second and fourth harmonics interact resonantly. (d): A chain of inertia-gravity modes that interact resonantly, truncated down to the first six components.

integrable. The evolution of the relative energies of the first four harmonics are shown as curves b_j(j = 1,...,4) in Figure 1, again for the initial condition as in (12). High harmonics gain energy coherently (they do so from any initial condition indeed) building up a front at $\mu t = \frac{1}{2}$. The solution of (9) is meaningless for $\mu t > \frac{1}{2}$: the approximation of considering only resonant interactions breaks down quickly in this case.

Going back to a few harmonics case, curves c_j in Figure 1 show the evolution of the relative energies for one of the Rossby cases in which the first, second, and fourth harmonics resonate. All the energy goes ultimately to the shortest wave but as a *power* μt, instead of an *exponential* as in (14). (Notice how curve c_1 levels off in the logarithmic scale of Figure 1.)

Last but not least, we consider a chain of resonant harmonics of inertia-gravity waves: all harmonics with j different from zero or a multiple of 3 are resonantly coupled but, unlike the case of the Kelvin modes, the interaction coefficient γ varies considerably (both in magnitude and sign) among the different triads. The system (9) must be integrated numerically in this case, and the infinite chain has to be truncated to a finite number of components, say M; two different truncations are reported here: M = 5 (j = 1,2,4,5,7) and M = 6 (same as before plus j=8). The evolution of the relative energies of the first three resonant harmonics are shown in Figure 1, curves d_j (j = 1,2,4), calculated with the M = 6 truncation. (M = 5 and M = 6 give practically the same results in Figure 1 up to $\mu t \doteq 3$.) Curves d_j present a behavior quite different from the other three cases (e.g., X_1 goes through a zero at $\mu t \doteq 3.4$); in order to clarify this, the M = 5 and M = 6 were integrated for a longer time, as discussed next. (We are *not* assuming the results for M = 6 and M>6 to be equivalent for, say, $\mu t \leqslant 50$. In fact, they are very likely quite dissimilar, as the difference between the M = 5 and M = 6 cases hints: the conclusions here are only valid for the truncated systems.)

The normalized amplitudes, $X_j(t)/[E^{(2)}]^{\frac{1}{2}}$, are shown in Figure 2, the orbits over energy partition diagrams (for selected triads) in Figure 3, and variance spectra of the X_j in Figure 4; in each one, (a) corresponds to the M = 5 truncation and (b) to M = 6.

With only five components, Figure 2(a), the system rapidly goes into one of the solutions of the three-wave problem for the triad (j=1,4,5,), the other two components (j=2,7) have an appreciable energy only for a short initial transient. (The presence of component "2", however, is essential for the transfer of energy from "1" to "4," *via* the successive generations "1" + "1" → "2" and "2" + "2" → "4".) After the initial transient there is catalytic energy exchange between "1" and "5" (in a time scale determined by the energy of "4") as it can be seen in the (1,4,5) triangle of Figure 3(a). (A point near, say, the vertex E_1, in that triangle means that $E_1 >> E_4 + E_5$; the opposite inequality holds near the side E_4 E_5.) The orbit on the triangle (2,5,7,) is "most of the time" at the vertex E_5, with sudden excursions to the E_2 E_7 side each time X_5 goes through a zero. Surprisingly, the same asymptotic solution was reached starting from many "random" initial conditions [constrained to real $X_j(0)$, which implies $H \equiv 0$], i.e., this is a "stable limit cycle" of the system.

With one more component, Figures 2(b) and 3(b), the solution looks, quite unexpectedly, "chaotic." (Once again, a similar behavior was found starting from several "random" initial conditions, constrained to real X_j). The triangles on Figure 3(b) are "filled up" by the trajectory in a longer elapsed time, i.e., the evolution looks "ergodic." (This result should be taken with some caution, because the orbit is unstable, and the unavoidable numerical errors make it irreversible after $\mu t \doteq 50$, even though energy is conserved better than one part in 10^{10}.)

A last piece of evidence of the difference between the truncations to five or six components is presented in Figure 4 in the form of variance spectra of the amplitude X_j. Even though similar spectra are obtained from the records in Figure 2, those shown here were calculated from a run with "random" initial condition (to avoid transient effects) and with a

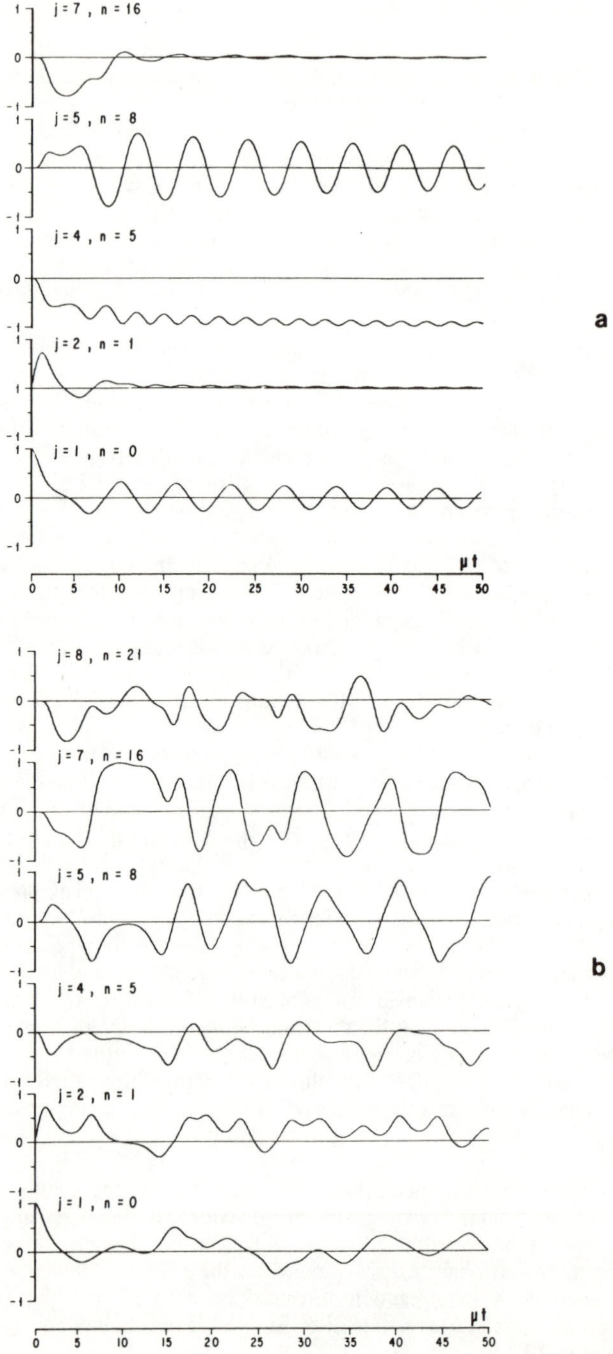

Fig. 2. Evolution of the envlopes of the resonant harmonics which have the mixed Rossby-gravity wave as fundamental. The ordinates are $X_j(t)/E^{(2)1/2}$ (a): The system is truncated to *five* components; the solution is regular and stable. (b): The system is truncated to *six* components; the solution is irregular, aperiodic, and unstable.

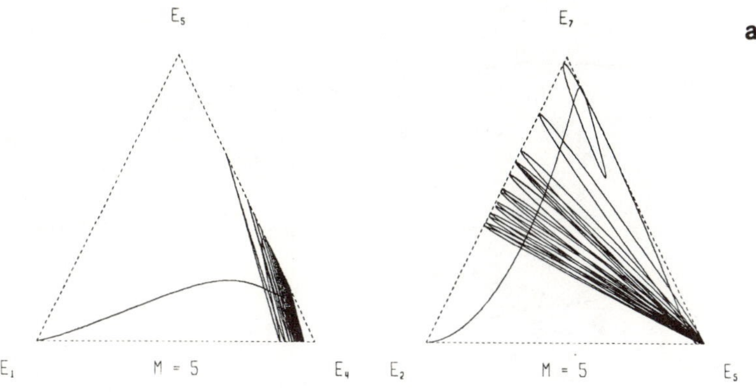

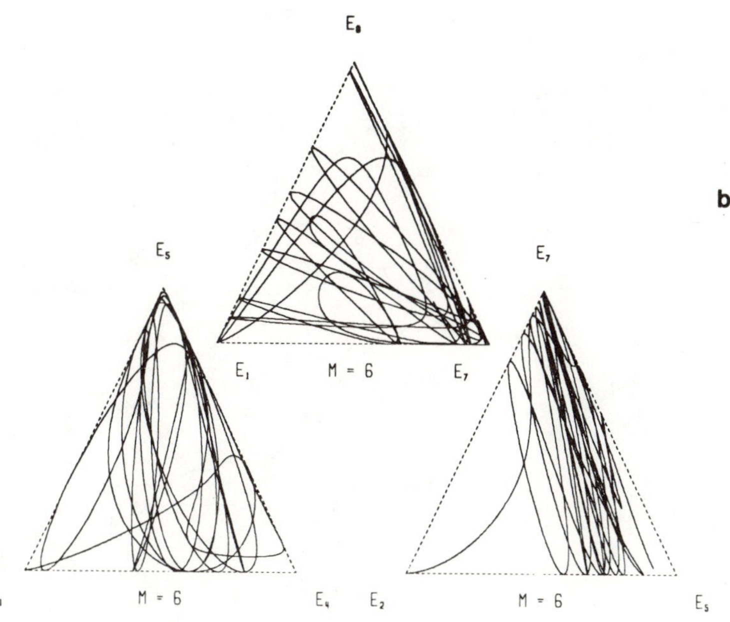

Fig. 3. Trajectories of the system with (a): five components, and (b): six components, in an energy partition diagram for selected triads.

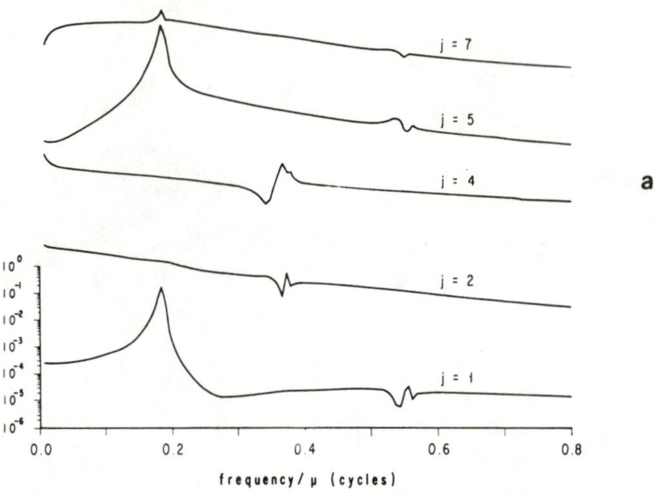

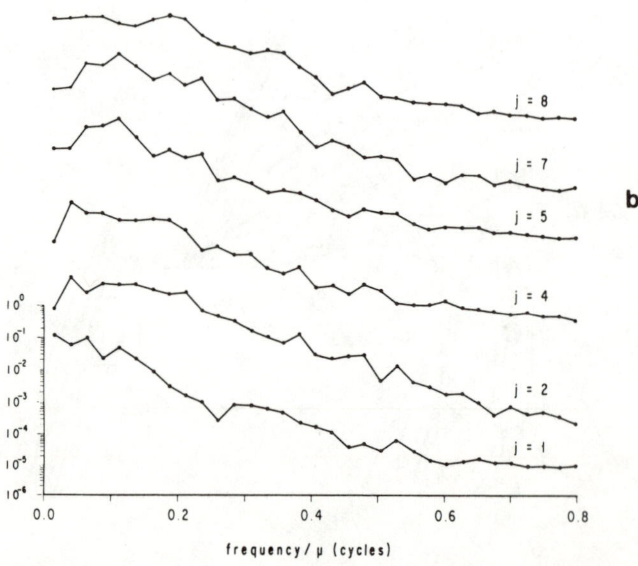

Fig. 4. Variance spectra of the envelopes X_j of the problem for (a): five components, and (b): six components.

length of μt = 163.68 (to increase frequency resolution); the spectra in Figure 4(b) represent averages of four consecutive bands because otherwise the curves were too "noisy" [no band average was done for Figure 4(a)].

The most outstanding feature in Figure 4(a) are the peaks at the energy-exchange frequency for components "1" and "5" (the second and third harmonics of that nonlinear frequency also show up in some of the curves). No significant peaks are found in Figure 4(b).

In summary, two different few-component truncations of an infinite chain of unevenly coupled resonant harmonics exhibit quite distinct behavior: with five modes the system always goes into one of the periodic solutions of the three-wave problem; the addition of one more component results in a chaotic evolution. (The total energy, which is conserved, only fixes the nonlinear time scale and therefore its value is not related to the onset of chaos.)

ACKNOWLEDGMENTS

I am grateful to S. Ramos and M. Noriega for drafting the figures. I would also like to express my appreciation to G. Holloway and B.J. West for making it possible for me to attend such a fruitful meeting.

REFERENCES

1. W.D. Hayes, in Nonlinear Waves, Leibovich and Seebass (Eds.). Cornell U. Press. (1974).
2. G. Holloway and M.C. Hendershott, J. Fluid Mech., *82* 747-765, (1977).
3. J. Lee, J. Fluid Mech., **120**, 155-183 (1982).
4. T. Matsuno, J. Meteor. Soc. Japan, **48**, 25-43 (1966).
5. J.D. Meiss, in Nonlinear Properties of Internal Waves, B.J. West (Ed.). A.I.P. Proceedings. **76** (1981).
6. P. Ripa, J. Phys. Ocean., in press (1983).

COEXISTENCE OF STOCHASTICITY AND DETERMINISM IN SIMPLE TRUNCATED MODELS OF BAROTROPIC FLOW IN β- PLANES

A. Navarra and A. Speranza
C.N.R. - FISBAT, Via de' Castagnoli 1, 40126
Bologna, Italy

ABSTRACT

The structure of the phase space of a simple barotropic truncated spectral model in a beta-plane is investigated and its deformation under a perturbation is analyzed. The perturbation is obtained by increasing the order of truncation and it does not seem to affect in a significant way the overall structure of the phase space. Aperiodic motions dominate the scene both in the perturbed and in the unperturbed case.

THE GENERAL PROBLEM

The northern-hemisphere, extratropical circulation is known to undergo frequent sudden transition between states of strongly zonal circulation, in which the large scale fluctuation field is dominated by synoptic scale traveling baroclinic waves, and states of pronounced meridional flow associated with standing planetary waves (see Dole and Gordon[4] for a recent exhaustive analysis of the statistical properties of the wintertime circulation).

A considerable amount of literature has recently appeared concerning the observational and theoretical properties of such states; we will not comment here upon the numerous concepts and problems arising in such a context, for a partial review see Speranza[12]. We want to call attention, however, to the fact that the operational forecast experiment has brought to light one of the quantities that, among others, varies considerably in correspondence with changes in the type of circulation, i.e. the numerical predictability of the atmosphere itself. Figure 1 shows the variation of the forecast time of the model of ECMWF.

It is clear that, apart from a systematic variation, due to bettering the model in time and a marked seasonal variation, there are strong winter fluctuations of the forecast time. Such fluctuations are associated with different types of circulation. For example the absolute maximum of predictability of February 1982 is associated with a very strong double (Atlantic and Pacific) blocking.

The general impression is that the atmospheric behavior is a rather even mixture of "stochasticity" and "determinism". Although this admittedly rough nomenclature should be made more precise we will take for granted the intuitive notion that the phase space of the atmospheric system must be non-homogeneous in its properties and address the question: are the available prototype models of atmospheric circulation able to reproduce such a mixed statistical behavior? If the answer is affirmative we might further ask: is it possible to describe at last the coarse structure of the phase space in terms of a limited number of parameters and variables?

In strictly meteorological terms, the hope is that the existence of different "equilibria" of the circulation not revealed by objective analysis of physical fields (see again Dole and Gordon[4]), may be detected by means of analysis of statistical properties like the mentioned "predictability". For example, simple (hamiltonian) mechanical systems displaying a mixed stochastic-deterministic behavior ("islands of predictability in a stochastic sea") are often characterized by an association between determinism (enhanced predictability) and equilibrium solution of minimal, unperturbed versions of the same system. It is clear, however, that the meteorological problem may be considerably more complicated. For example, in the truncated model of Reinhold and Pierrehumbert no association is found between persistent (localized in phase space) solution of the complete model and equilibria (fixed

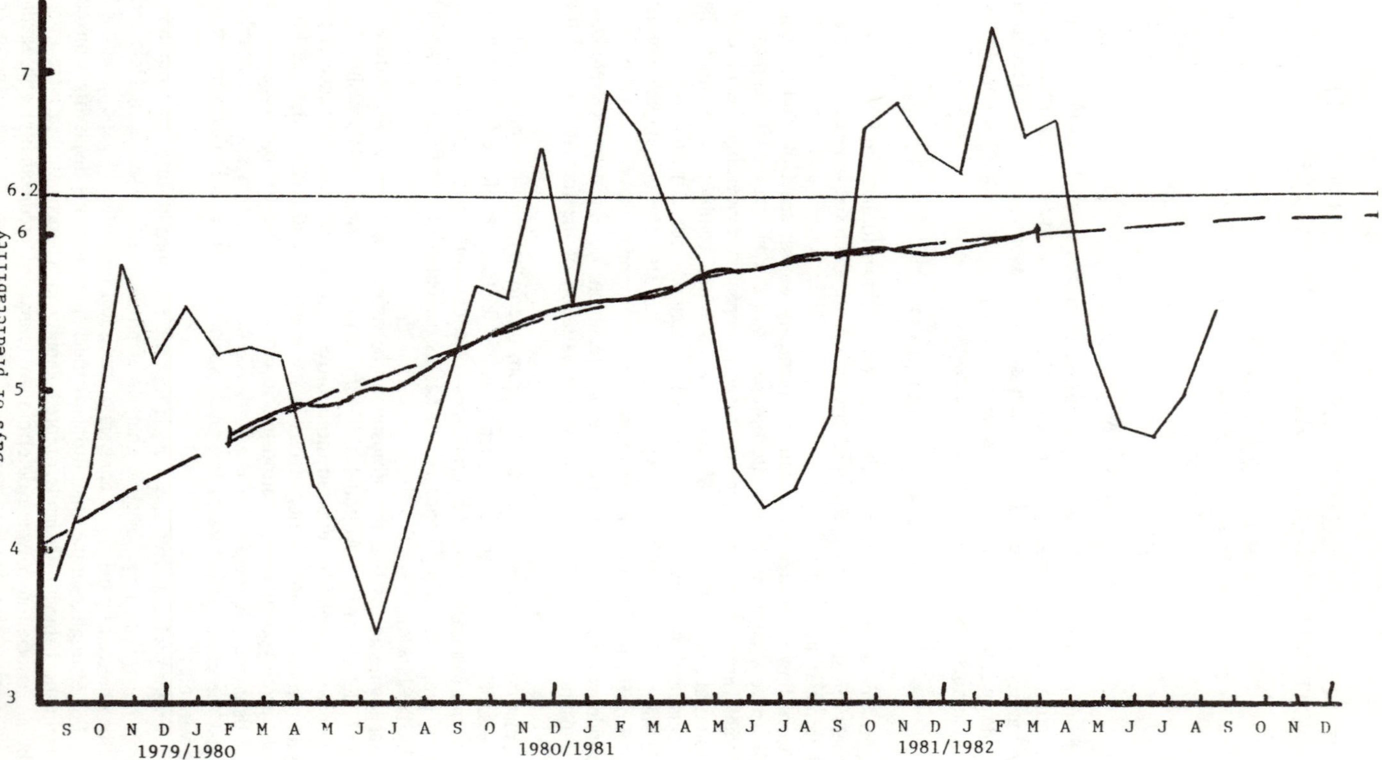

Fig. 1. Predictability of ECMWF model, adapted from Tibaldi.[14] The curve shows the trend due to improvements in the model, it is empirically asymptotic to 6.2 days (see text).

points) of the version without the higher zonal harmonic (representing baroclinic instability).

It goes without saying that the problem of modeling the coarse structure of phase space with a limited number of parameters and variables is intimately connected with adequately representing the non-resolved ("truncated") components of the fields defining the state of the system. In the meteorological context, different representations have been used: from the classical "parametrization", to the reduction of non-resolved components to the stochastic noise forcing the explicitly resolved components (in the literature concerning blocking, for example, see Egger[6] and Benzi et al.[2]. It must be remembered that the problem of representation of the fields, or, in other words, the choice of the functional basis of expansion, can be critical with respect to the very existence of an adequate truncation permitting representation of both the coarse structure of the phase-space and the transition probability among different basins. In fact, in view of the regional nature of blocking which still escapes adequate representation (see Malguzzi and Speranza[9]), and the difficulties encountered in producing statistically realistic models of atmospheric circulation, it seems probable that some profound problem of representation is preventing us from really improving our ability to model the essential mechanisms involved in the production and maintenance of anomolies. After all, the almost universally used basis of expansion is the series of eigenfunctions of the Laplace operator which has no particular meaning in the context of the non-linear Navier-Stokes equations. It therefore seems wise for the time being to leave the realism of representation aside in addressing our question concerning the coexistence of stochasticity and determinism.

The simplest mechanical system displaying a mixture of deterministic and stochastic behavior are systems subject to the KAM (Kolmogorov, Arnold, Moser) theorem. This theorem concerns the perturbation problem of an integrable conservative system. Under the perturbation, the ordinate structure of the phase space of the basic system, usually divided by a conserved quantity in invariant subspaces, breaks down. The invariant subspaces corresponding to periodic orbits are completely destroyed and a very turbulent area appears in their place. The whole process results in a substantial dishomogeneity of the phase space[1].

A meteorological analogue that we find appropriate is the inviscid barotropic equation

$$\frac{\partial}{\partial t}\nabla^2\psi + J(\psi,\nabla^2\psi + f) = 0\,. \tag{1}$$

We will consider the channel version, $f=\beta y$, $\psi = 0$, at $y=0$, πL where L is the width of the channel. Over projection on the simple harmonic base

$$L^2\nabla^2 F_i = -a_i F_i$$

$$\psi = \sum_{i=1}^{N} \psi_i(t) F_i(x,y) \tag{2}$$

it gives a sequence of truncated problems

$$(-a_l)^2\dot{\psi}_l + \sum_{k,j} c_{jkl}(-a_k)^2\psi_k\psi_j + \sum_j \beta\, d_{jl}\psi_j = 0 \quad l = 1,\dots,N$$

$$c_{ijk} = \int_0^{2\pi} dx \int_0^{\pi L} J(F_i,F_J)F_k dy\,; \quad d_{ij} = \int_0^{2\pi} dx \int_0^{\pi L} \frac{\partial F}{\partial x} iF_j dy \tag{3}$$

In particular, in view of any future study of the effect of asymmetric forcing, we select the sequence of eigenfunction used by Charney and Devore[3]

$$F_1 = \frac{\cos(y)}{\pi}$$

$$F_{2j} = \sqrt{\frac{2}{\pi}}\sin((2j-1)y)\sin(2x) \qquad (4)$$

$$F_{2j+1} = \sqrt{\frac{2}{\pi}}\sin(2j-1)y)\cos(2x)$$

The non vanishing c_{ijk} and d_{ij} become then

$$c_{1,2j,2k+1} = \frac{4}{\pi^2}\left[\frac{1}{1-(2_j-2k)^2} - \frac{1}{1-(2_j+2k-2)^2}\right] ; d_{2i,2j+1} = 2 \qquad (5)$$

The system also admits two quadratic invariants: the energy E and the enstrophy V

$$E = \sum_{i=1}^{N} (a_i)^{1/2}|\psi_i|^2 \; ; \; V = \sum_{i=1}^{N} a_i|\psi_i|^2$$

The orbits are bound to stay on the intersection of the ellipsoids defined by E and V.

The analogy we have in mind consists of the minimum truncation (N=3, in Eq. 3) known to be integrable Lorenz,[7] Platzman,[11,12] Galin,[7] Dutton,[5] Pellacani and Lupini,[9] and if the action of additional modes can be assimilated to a perturbative forcing then chaos could be generated near resonant (periodic) orbits by a mechanism similar to that operating in KAM systems. Our aim is therefore to study, starting from the minimal, completely integrable system, the effect of the successive addition of modes in terms of the distribution of predictability in the phase space. To this purpose, we begin with a rapid summary of the known properties of the minimal system (Section 2) and proceed to discuss the "perturbative" approach by the addition of one mode (the non-linear coupling coefficient of the additional mode is scaled with a parameter ϵ varying from 0, corresponding to no coupling, to 1, corresponding to complete coupling) in terms of the predictability in the phase space (Section 3). Some temptative conclusion is finally given.

THE MINIMAL PROBLEM

The minimal non-linear problem in the sequence (4) is obtained for n=2

$$-a_1\mathring{\psi}_1 + C_{251}(a_2 - a_5)\psi_2\psi_5 + C_{341}(a_3 - a_4)\psi_3\psi_4 = 0$$

$$-a_2\mathring{\psi}_2 + C_{312}(a_3 - a_1)\psi_3\psi_1 + C_{512}(a_5 - a_1)\psi_5\psi_1 - 2\beta\psi_3 = 0$$

$$-a_3\mathring{\psi}_3 + C_{123}(a_1 - a_2)\psi_1\psi_2 + C_{413}(a_4 - a_1)\psi_4\psi_1 + 2\beta\psi_2 = 0$$

$$-a_4\mathring{\psi}_4 + C_{514}(a_5 - a_1)\psi_5\psi_1 + C_{134}(a_1 - a_3)\psi_1\psi_3 - 2\beta\psi_5 = 0$$

$$-a_5\mathring{\psi}_5 + C_{145}(a_1 - a_4)\psi_1\psi_4 + C_{125}(a_1 - a_2)\psi_1\psi_2 + 2\beta\psi_4 = 0 \qquad (6)$$

The system (6) can equivalently be written in terms of complex variables $y_1 = \psi_1$, $y_2 = \psi_2 + i\psi_3, y_3 = \psi_4 + i\psi_5$

$$\mathring{y}_1 - \alpha_1 \mathrm{Im}(y_2 y_3^*) = 0$$

$$\mathring{y}_2 + i\gamma_1 y_1 y_2 + i\gamma_2 y_1 y_3 + iB' y_2 = 0 \tag{7}$$

$$\mathring{y}_3 + i\epsilon_1 y_1 y_3 + i\epsilon_2 y_1 y_2 + iB'' y_3 = 0$$

where the coefficients in greek letters are constants depending on the c's and the B′, B″ depends on the d's and the beta factor.

In terms of the amplitudes and phases $A_1 = y_1, A_i = |y_i|\ \theta_i = \arctan \dfrac{\mathrm{Im}y_i}{\mathrm{Re}y_i}$; i=2,3

$$\mathring{A}_1 - \alpha_1 A_2 A_3 \sin(\theta_2 - \theta_3) = 0$$

$$\mathring{A}_2 - \gamma_2 A_1 A_3 \sin(\theta_3 - \theta_2) = 0$$

$$\mathring{A}_3 - \epsilon_2 A_1 A_2 \sin(\theta_2 - \theta_3) = 0$$

$$\mathring{\theta}_2 + \gamma_1 A_1 + \gamma_2 \frac{A_1 A_3}{A_2} \cos(\theta_3 - \theta_2) + B' = 0$$

$$\mathring{\theta}_3 + \epsilon_1 A_1 + \epsilon_2 \frac{A_1 A_2}{A_3} \cos(\theta_3 - \theta_2) + B'' = 0\,. \tag{8}$$

The existence of two separating invariants (energy and enstrophy) in the three-dimensional space of amplitudes A_i guarantees integrability: the A(t) have, in fact, been proved by Dutton[5] to be elliptic functions of time. The phase space of the amplitudes is shown in Figure 2. The periodicity of the total solution depends then entirely on the nature of the time evolution of the phases: for the motion to be periodic it is necessary that θ_2 and θ_3 both be periodic and their period commensurable. Due to the dispersion introduced by the beta-effect this is usually not the case (Lupini and Pellacani).[9] There are isolated periodic orbits, however. Their existence is well illustrated by the two fixed points of the phase space of amplitudes $(0,1/\sqrt{13},0)$ and $(0,0,1/\sqrt{5})$ (in Figure 2 they are the poles of the ellipsoid in the horizontal plane) which are periodic in phase-space θ_2, θ_3. These cyclic solutions are obviously "bare" Rossby waves which are known to be exact solutions of (1), the inviscid barotropic equation.

The presence in the minimal system of both periodic and quasi-periodic behavior makes it a possible candidate for the production of a statistically non-uniform phase-space of higher order of truncation.

Figure 3 shows the time evolution of the two phases are in the neighborhood of the cyclic orbits corresponding to pure Rossby waves. On this plot the solution corresponding to the Rossby waves themselves would appear as a straight line. The drift in phase which does not allow orbits to close can be clearly seen. It must be noted that non-periodicity is dense near the cyclic points since the transition to the cyclic orbit is marked by the vanishing of amplitude and of the frequency of one of the two Rossby waves.

THE ADDITION OF ONE DEGREE OF FREEDOM AS A PERTURBATION PROBLEM

The addition of one mode (N=3, in Eq. 3) modifies the equation by including new non-linear coupling. For example, to illustrate the basic idea, the equation for the zonal

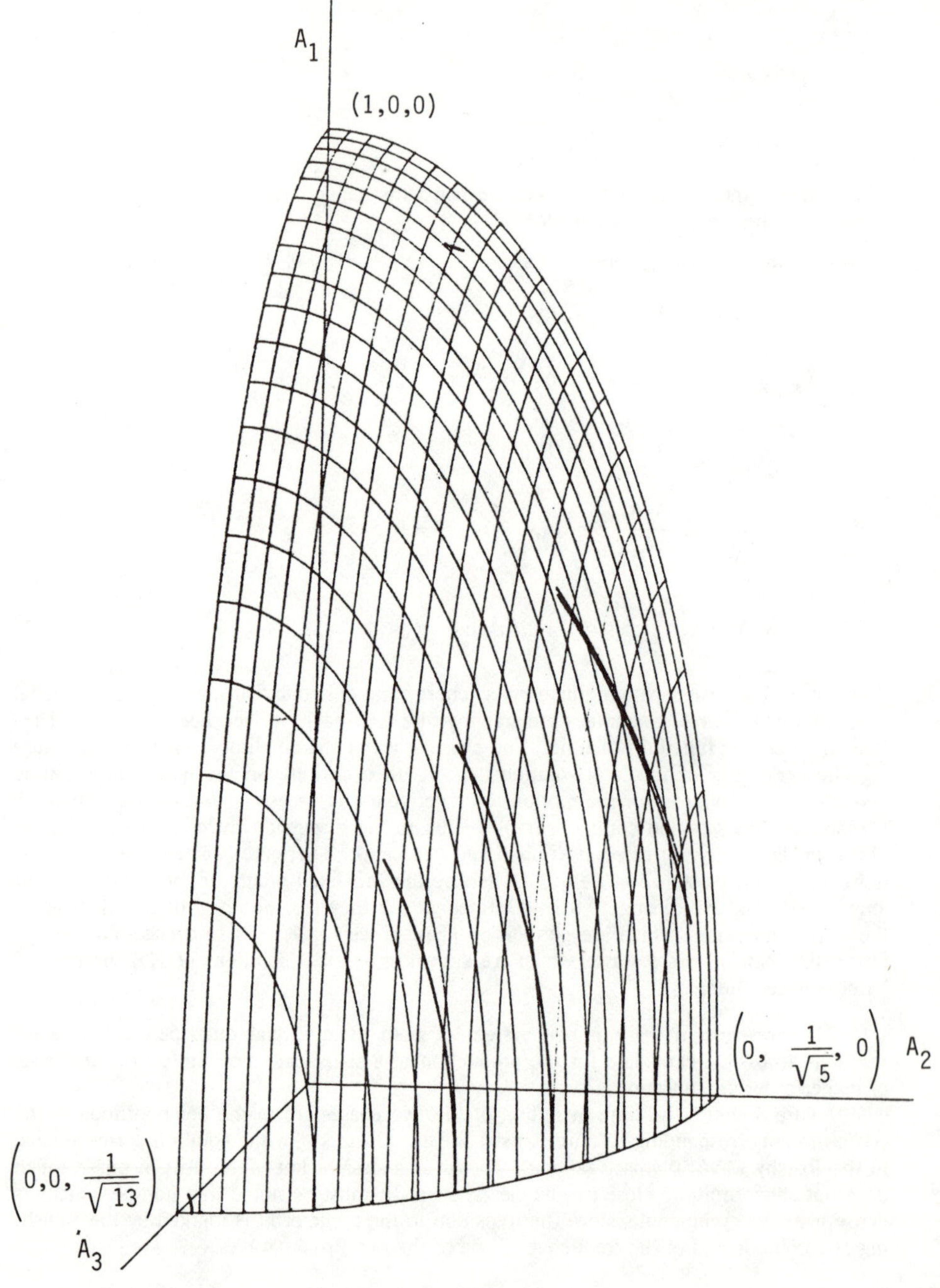

Fig. 2. A few orbits of system (8) (thick curves) shown on the energy ellipsoid.

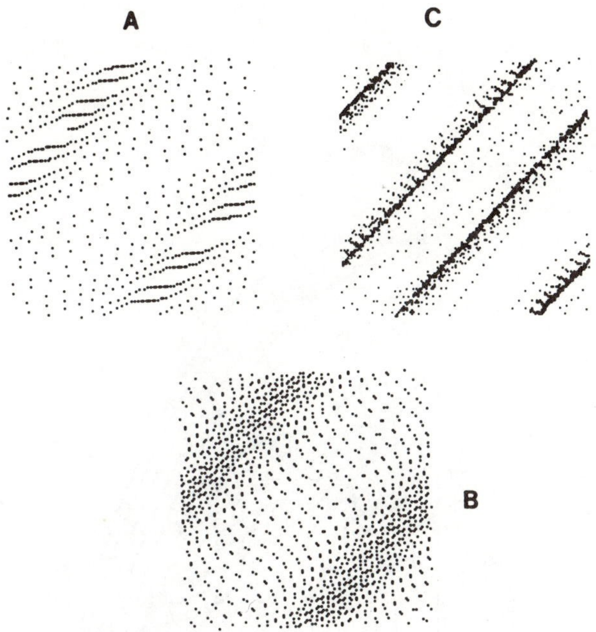

Fig. 3 Phase plots (θ_2,θ_3), for three different orbits. (A) is in the A_1A_3 plane, but away from the A_2A_3 plane. ψ_1 is 0.1. (C) has $\psi_1 = 0.00001$, and therefore is closer to $(0,0,1/\sqrt{13})$. (B) has $\psi_1 = 0.00001$ and is closer to $(0,1/\sqrt{5},0)$.

component becomes

$$
\begin{aligned}
&-a_1\mathring{\psi}_1 + C_{341}(a_3 - a_4)\psi_3\psi_4 + C_{251}(a_2 - a_5)\psi_2\psi_5 \\
&\quad + \epsilon C_{361}(a_3 - a_6)\psi_3\psi_6 + \epsilon C_{561}(a_5 - a_6)\psi_5\psi_6 \\
&\quad + \epsilon C_{271}(a_2 - a_7)\psi_2\psi_7 + \epsilon C_{471}(a_4 - a_7)\psi_4\psi_7 \\
&\quad = 0 .
\end{aligned}
\tag{9}
$$

The other equations are modified accordingly. The total number of equations to be considered must also be risen to 7. We have inserted a coupling coefficient ϵ which, varied between 0 and 1, allows the system to change continuously from the minimal to that with an extra mode. In particular, if $\epsilon < 1$, the additional mode acts as a "perturbation" on the motion of the minimal system.

Hence the two quadratic invariants E and V are no longer separating in the four-dimensional space of amplitudes of the full problem (9), non-periodicity now appears also in the evolution of A_i's. As an example, Figure 4 shows the projection of some orbits on the E=1 ellipsoid of the minimal system. Comparison with Figure 2 illustrates the drastic action of non-linear coupling with the new mode.

That the statistical properties of the system are not completely uniform in the phase-space is again illustrated by the existence, also in (9), of exactly periodic orbits, corresponding to isolated Rossby waves, being exact solutions of Equation 1 they will exist at any

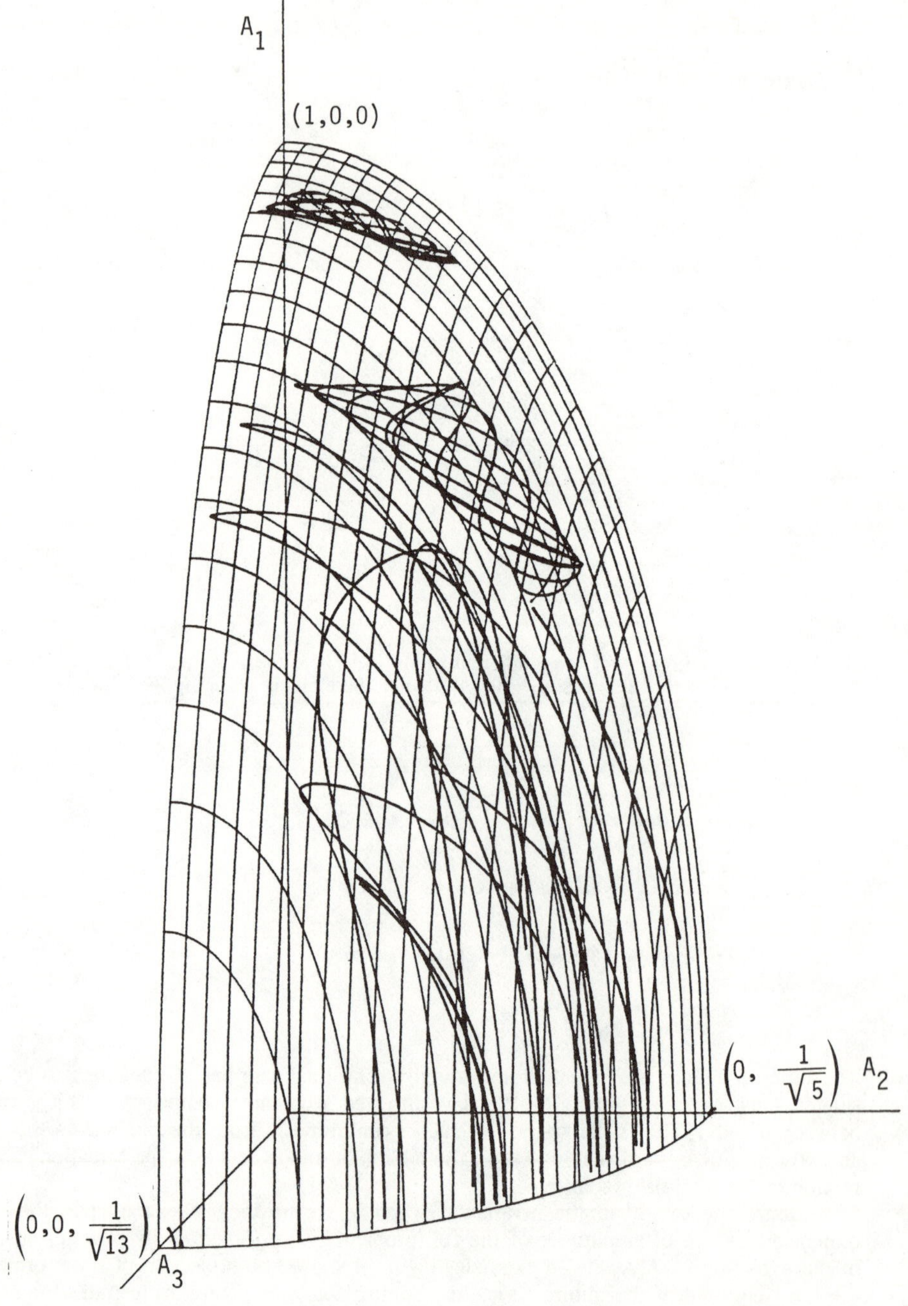

Fig. 4. Orbits for the system (9). Initial conditions are the same as in Figure 2.

order of truncation. As a preliminary step in the application of a KAM strategy we have analyzed the neighborhood of these periodic solutions.

Figures 5, 6, and 7, are to be compared with Figure 3, and illustrate the influence of the additional mode (added as a perturbation and with full interaction) on the phases of the minimal space in the neighborhood of the pure Rossby wave solutions and far from there. Even if a quantitative analysis was not yet performed, visual inspection does not show any qualitative difference among the different behaviors: perturbation (= 0.1) by the additional mode induces very slight deformation of the orbits everywhere and cause the phase evolution to become totally irregular; in the limit (=1) of complete coupling the fully periodic (pure rossby wave) solutions seem to be completely isolated in a "sea" of irregular orbits.

CONCLUSION AND FUTURE WORK

It is obviously not possible to deduce general conclusions from the very preliminary work we have performed. It does seem possible, however, that non-homogeneous statistical behavior is produced by non-linearity, and that opportunity chosen minimal systems may approximately define the coarse structure of the phase space.

In order to check the above hypothesis it is necessary to investigate systematically the neighborhood of all completely periodic orbits of the minimal system and to understand whether or not the action of truncated modes can really be assimilated to that of perturbative forcing.

REFERENCES

1. V.I. Arnold, Mathematical methods of Classical Mechanics, Springer, (1978).
2. R. Benzi, A.R. Hansen and A. Sutera. On stochastic perturbations of simple blocking models, in preparation.
3. J. Charney and J. Devore, J. Atmos. Sci., **36**, 1205, (1979).
4. R.M. Dole and N.D. Gordon, preprint, submitted to MWR.
5. J.A. Dutton, J. Atmos. Sci., **33**, 2606, (1981).
6. J. Egger, J. Atmos. Sci., **38**, 2602, (1981).
7. M.B. Galin, Izv. Atm. Ocean. Phys., **10**, 455, (1975).
8. E.N. Lorenz, Tellus, **12**, 243 (1960).
9. R. Lupini and C. Pellacani. On forced and unforced models of atmospheric flows, submitted to Tellus.
10. P. Malguzzi and A. Speranza, J. Atmos. Sci., **38**, 1939, (1981).
11. A. Platzman, J. Atmos. Sci., **19**, 313, (1963).
12. A. Platzman, J. Meteor., **17**, 365, (1960).
13. A. Speranza, Deterministic and Statistical Properties of the Westerlies, PAGEOPH, (1983).
14. S. Tibaldi, see article in this volume.

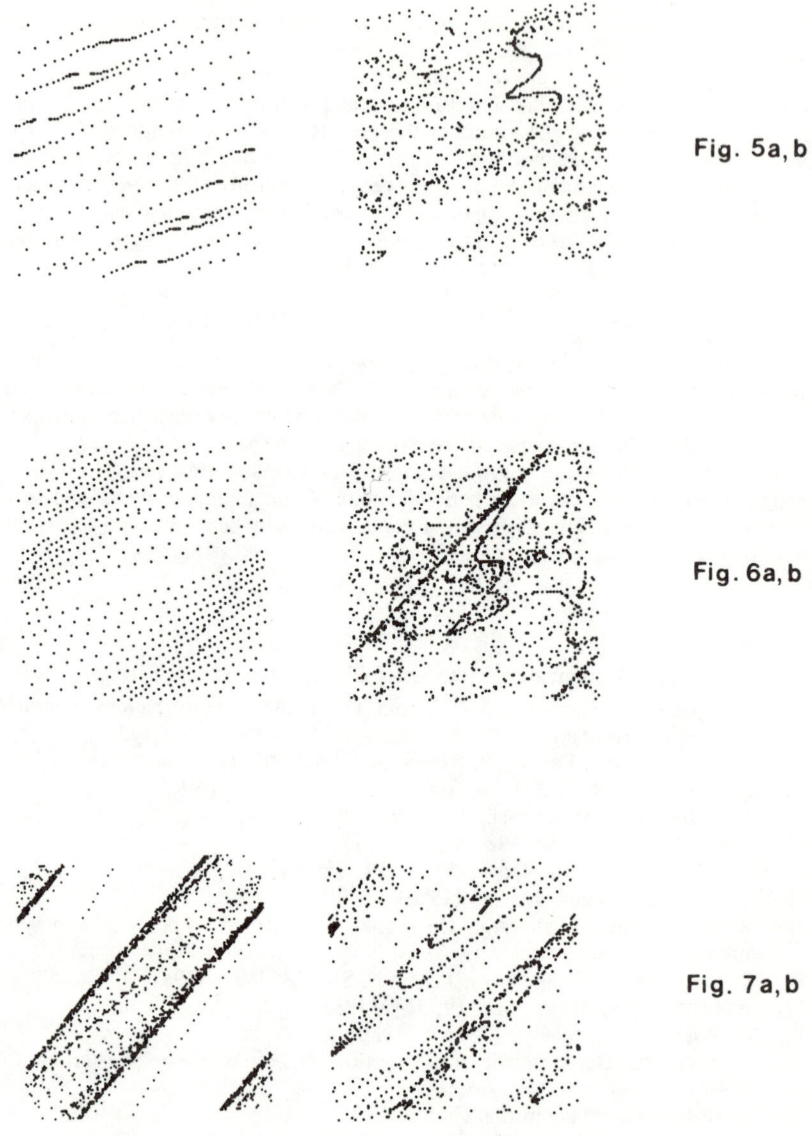

Fig. 5. Phase plots for the perturbed system. Figures 5, 6, 7 have the same initial condition as A,B,C in Figure 3. On the left $\epsilon = 0.1$ and on the right $\epsilon = 0$.

LINEAR PREDICTABILITY: EFFECTS OF STATIONARY FORCING

John O. Roads
and
Richard C.J. Somerville

Climate Research Group, Scripps Institution of Oceanography
University of California, San Diego, La Jolla, California 92093

ABSTRACT

Large-scale atmospheric predictability is examined in a series of simulations and real-data numerical weather forecasts with an extremely idealized system: a linear quasi-geostrophic barotropic low-order model with prescribed stationary forcing. We find that forecast skill depends crucially on the specification of the stationary forcing. A lack of stationary forcing leads to spurious westward propagation of the ultralong waves.

Forecasts made with stationary forcings derived from climatological data are superior to those using forcings inferred from observations immediately preceding the forecast period. Interhemispheric differences in forecast skill are examined. The model error is compared to errors of a simple persistence-damped-to-climatology scheme and to errors of a modern general circulation model.

INTRODUCTION

Predictability theory[1] demonstrates that we must expect errors to occur first in the small scales, which, by nonlinear interactions, eventually infect intermediate scales and ultimately reach the largest scales. However, the errors in operational numerical weather prediction often occur in the ultralong waves rather than the intermediate scales and also appear on a much shorter time scale than anticipated theoretically, i.e., days instead of weeks.[2]

Recent work[3,4] indicates that one reason for errors immediately appearing preferentially in the planetary waves is related to improper representation of the ultralong Rossby waves. For example, a hemispheric model necessarily contains an incomplete set of modes because of the equatorial boundary conditions. One important conclusion which may be drawn from the results of Daley et al.[4] is that the basic dynamical processes which underlie some of these effects on predictability are mainly barotropic in nature. This result offers the hope that theoretical insight to be gained from relatively simple models may prove to be of real practical utility.

The research described below is motivated by the need to understand how errors appear in the ultralong waves of numerical weather prediction models and how these errors might be alleviated. Efforts to improve the skill of numerical forecasts of planetary waves are especially important, because these very energetic and very large-scale phenomena are the basis for our expectations of progress in extended-range weather forecasting.

In the work described in this paper, we have employed a particularly simple model (linear, quasi-geostrophic and barotropic) in simulations and real-data forecasts. Our aim has been to explore in an idealized context several effects, particularly those due to stationary forcing, which may explain some of the discrepancies between predictability theory and the performance of current numerical weather prediction models. We are especially interested to learn the degree to which some of these discrepancies may be accounted for by linear theory.

ANALYSIS

The model described in more detail in Appendix I is based upon the linear quasi-geostrophic vorticity equation:

$$\frac{\partial}{\partial t}\{\nabla^2\hat{\psi}-\lambda^2\hat{\psi}\} = -J(\overline{\psi},\nabla^2\hat{\psi}) - J(\hat{\psi},\nabla^2\overline{\psi}+\text{siny})$$

$$-k\nabla^2\hat{\psi}+\hat{F}\ .$$

Let $\hat{\psi}$ be expressed as a finite sum of associated Legendre polynomials,

$$\hat{\psi} = \sum_{n=0}^{\infty}\sum_{m=-n}^{n} \hat{\psi}_n^m P_n^m(\text{siny})e^{im\lambda}$$

and for illustrative purposes let

$$\overline{\psi} = \overline{\psi}_1 P_1(\text{siny})\ .$$

That is, let the zonal mean state describe solid-body rotation only. The model actually used includes a more accurate description of the zonal wind. In the case of solid-body rotation, the model equations reduce, for a single wave, to

$$\frac{\partial}{\partial t}\{-n(n+1)-\lambda^2\}\hat{\psi} = \hat{\psi}(im((2-n(n+1))\overline{\psi}_1 - 1)$$

$$+ kn(n+1)) + \hat{F}\ .$$

For brevity, we omit the indices. The solution to this equation, if $\hat{F}$ is assumed independent of time, is

$$\hat{\psi} = \hat{\psi}(0)e^{ct} - \frac{\hat{F}}{b}(1-e^{ct})$$

where

$$\begin{aligned}
\hat{\psi}(0) &= \hat{\psi}(x,y,t=0) \\
a &= -(n(n+1)+\lambda^2) \\
b &= [im([2-n(n+1)]\overline{\psi}_1-1) + kn(n+1)] \\
c &= b/a = -c_r + ic_i \\
c_r &= \frac{kn(n+1)}{(n(n+1)+\lambda^2)} \\
c_i &= \frac{-m([2-n(n+1)]\overline{\psi}_1-1)}{(n(n+1)+\lambda^2)}
\end{aligned}$$

The stationary wave in the above system is simply

$$\hat{\psi}_s = -\hat{F}/b$$

so the solution can also be written as

$$\hat{\psi} = \hat{\psi}_s + (\hat{\psi}(0)-\hat{\psi}_s)e^{ct}$$

That is, the solution is composed of a stationary part plus a damped transient wave whose amplitude is the initial value minus the stationary part.

Suppose that $\tilde{\psi}$ is the solution of a system differing from the above only in parameter values or initial condition:

$$\tilde{\psi} = \tilde{\psi}_s + (\tilde{\psi}(0) - \tilde{\psi}_s)e^{\tilde{c}t} \ .$$

The difference or error between the two models would then behave as

$$\tilde{\psi} - \hat{\psi} = (\tilde{\psi}_s - \hat{\psi}_s) + (\tilde{\psi}(0) - \tilde{\psi}_s)e^{\tilde{c}t} - (\hat{\psi}(0) - \hat{\psi}_s)e^{ct}$$

Let us now examine how different terms of the solution affect the error growth.
Example 1. Difference is in initial conditions only.

In this case,

$$|\tilde{\psi} - \hat{\psi}| = |\tilde{\psi}(0) - \hat{\psi}(0)|e^{-c_r t}$$

Thus the initial error will decay in time as both models approach the stationary state.
Example 2. Difference is in frictional component only.

$$|\tilde{\psi} - \hat{\psi}| = |\hat{\psi}(0) - \hat{\psi}_s|(e^{-2\tilde{c}_r t} + e^{-2c_r t} - 2e^{-(c_r + \tilde{c}_r)t})^{1/2} \ .$$

At $t\sim 0$, $|\tilde{\psi} - \hat{\psi}| \sim |\hat{\psi}(0) - \hat{\psi}_s \, |t|\tilde{c}_r - c_r|$. Thus the error grows linearly in time.

Example 3. Difference is in wave propagation characteristics only.

$$|\tilde{\psi} - \hat{\psi}| = |\hat{\psi}(0) - \hat{\psi}_s|e^{-c_r t}(2 - e^{it(c_i - \tilde{c}_i)} - e^{-it(c_i - \tilde{c}_i)})^{1/2} \ .$$

At $t\sim 0$, $|\tilde{\psi} - \hat{\psi}| \sim |\hat{\psi}(0) - \hat{\psi}_s| \, t|c_i - \tilde{c}_i|$. Thus the error grows linearly in time.
Example 4. Difference is in stationary components only.

$$|\tilde{\psi} - \hat{\psi}| = |\tilde{\psi}_s - \hat{\psi}_s|(1 + e^{-2c_r t} - 2e^{-c_r t}\cos c_i t)^{1/2} \ .$$

At $t\sim 0$, $|\tilde{\psi} - \hat{\psi}| \sim |\tilde{\psi}_s - \hat{\psi}_s| t \ (c_r^2 + c_i^2)^{1/2}$. Thus the error grows linearly in time.

A characteristic of Examples 2 and 3 is that initial error growth rate is proportional to t, with the proportionality constant dependent upon the weighted spectrum of the transient wave (initial state minus stationary state) times the difference in frictional components or in phase speeds. In Example 4, the weighted spectrum depends upon the difference in the stationary wave component as well as the friction and wave speed characteristics. Since the synoptic and stationary wave spectrum is weighted toward the largest planetary waves and since these waves potentially have the largest phase speeds, we can expect a large error response in these modes.

NUMERICAL RESULTS

Figure 1 shows numerical results from the quasi-geostrophic model. We carried out a predictability experiment by comparing the results of two models differing only in the stationary forcing. The initial state was taken from the FGGE (First GARP Global Experiment) observations for 15 January 1979 at 00Z, and the stationary state was taken as the time average of the period 15 to 22 January. The constant (in time only) zonal wind was taken from the zonal wind of the initial state. Figure 1 shows the RMS area-weighted error in only zonal waves 1, 2, and 3. Most of the error occurs in waves 1-3 with only a small contribution in the additional waves (see Appendix 1). As expected, the errors are twice as large for doubled stationary forcing. This predictability experiment shows that knowledge of the stationary forcings leads to much smaller errors even for time scales as short as a few days.

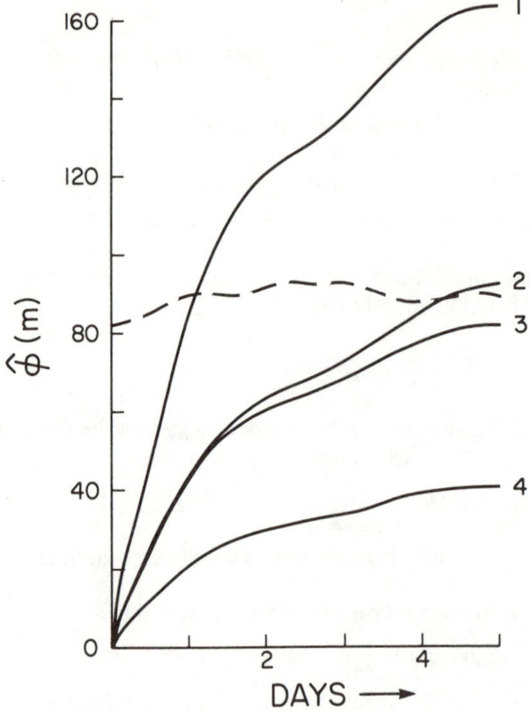

Fig. 1. Time variation of global RMS 500 mb height differences between pairs of model integrations, low-pass filtered to include only zonal wavenumbers 1-3. The model is a linear quasi-geostrophic barotropic model.

Case 3 is the difference between two models, one including and one excluding stationary forcing. Case 2 is the same as Case 3, except that both models have one order of magnitude less friction. Case 1 is the same as Case 3, except that the stationary forcing is two times as large. Case 4 is the same as Case 3 except that the stationary forcing is half as large. The dashed line is the RMS value of 15-20 January 1979 FGGE 500 mb data. The time-averaged value of this field is used for the stationary forcing in the model.

The next question we asked was whether knowledge of the time-averaged state would improve a forecast when verifying against real data. The answer is shown in Figure 2. Note that in these forecast experiments the largest errors occur for a stationary forcing of twice the observed time-averaged state. The second worst forecast occurs for no stationary forcing; the next worst forecast is obtained when the forcing is halved. The best forecast is found when the actual time-averaged forcing is used. Thus an appropriate stationary forcing is required in order to obtain a good forecast.

Harmonic dials of the various spherical harmonic components can be constructed and are shown in Figure 3 for the observed data, for the forecasts with no stationary forcing and for the forecasts with stationary forcing. Also shown are observed data for a subsequent time period (25-30 January). Forecasts with no stationary forcing yield harmonic dials corresponding to spurious westward propagation, whereas the ones with stationary forcing are fixed in a phase region which resembles that of the actual observed data. It is probable that the erroneous westward propagation of planetary waves in early operational barotropic numerical weather prediction models was due in large measure to the lack of a realistic stationary forcing (Thompson[5] p. 156). However, note that for the subsequent time period

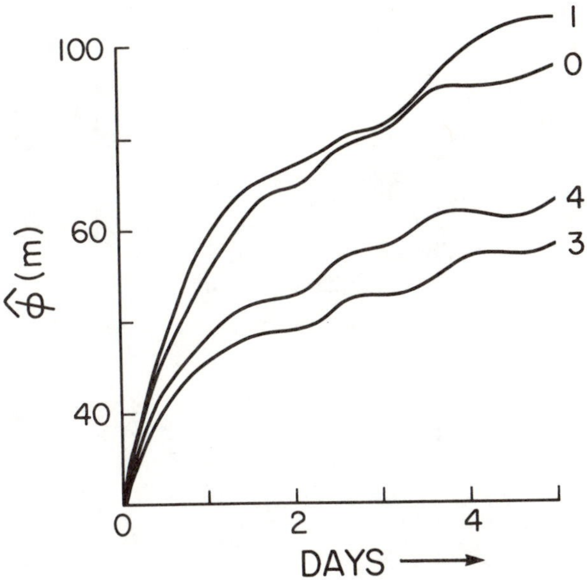

Fig. 2. Time variation of global RMS 500 mb height differences between model integrations and verifying data. The models are the same linear quasi-geostrophic barotropic models as in Figure 1. The verifying data is the 15-20 January 1979 FGGE data as analyzed by GLAS.

Case 0 is for a model with no stationary forcing. Case 3 is for a model with stationary forcing taken from the time-averaged data shown as the dashed line in Figure 1. Case 1 is for a model with doubled stationary forcing. Case 4 is for a model with halved stationary forcing.

the phase of the observed stationary component seems to change. In fact, knowledge of the stationary forcing for one time period often was not helpful for forecasting subsequent time periods in our experiments.

Various experiments were then performed to see if the model could be improved to give a smaller forecast error. These are shown in Figure 4. Figure 4a shows the result for 15 January initial data, 4b for 17 January initial data and 4c for 25 January initial data. Examining Figure 4a first, we note that the worst forecast occurs for low friction and no stationary forcing, next was for friction increased by an order of magnitude, next was for λ^2 increased from 4 to 40, next was persistence, next was low friction with stationary forcing, and the best forecast was for stationary forcing and large friction. In fact, simply using as a forecast the time-averaged state gives the best forecast after about 2 days. Similar results occur for the 17 and 25 January cases. It would thus seem that knowledge of the time-averaged state should give a better forecast, but, as shown in Figure 4c, there is no way of constructing this from the previous week's activity. For the 25 January case, the worst forecast was either the 15-22 January climatology or the model using this time average to derive the stationary forcing.

However, if a long-term average state is used to infer the stationary forcings, then better forecasts result. For this test we obtained a 30-year average of observations of the 5-day mean data for the period 15-20 January and used this average to obtain the stationary forcing for the forecast with the January 15 initial state. Only data for the region 20-90°N were available, and hence the stationary forcing was replaced by the forcing which would produce the initial state south of 20°N. The results are shown in Figure 5 for the Northern

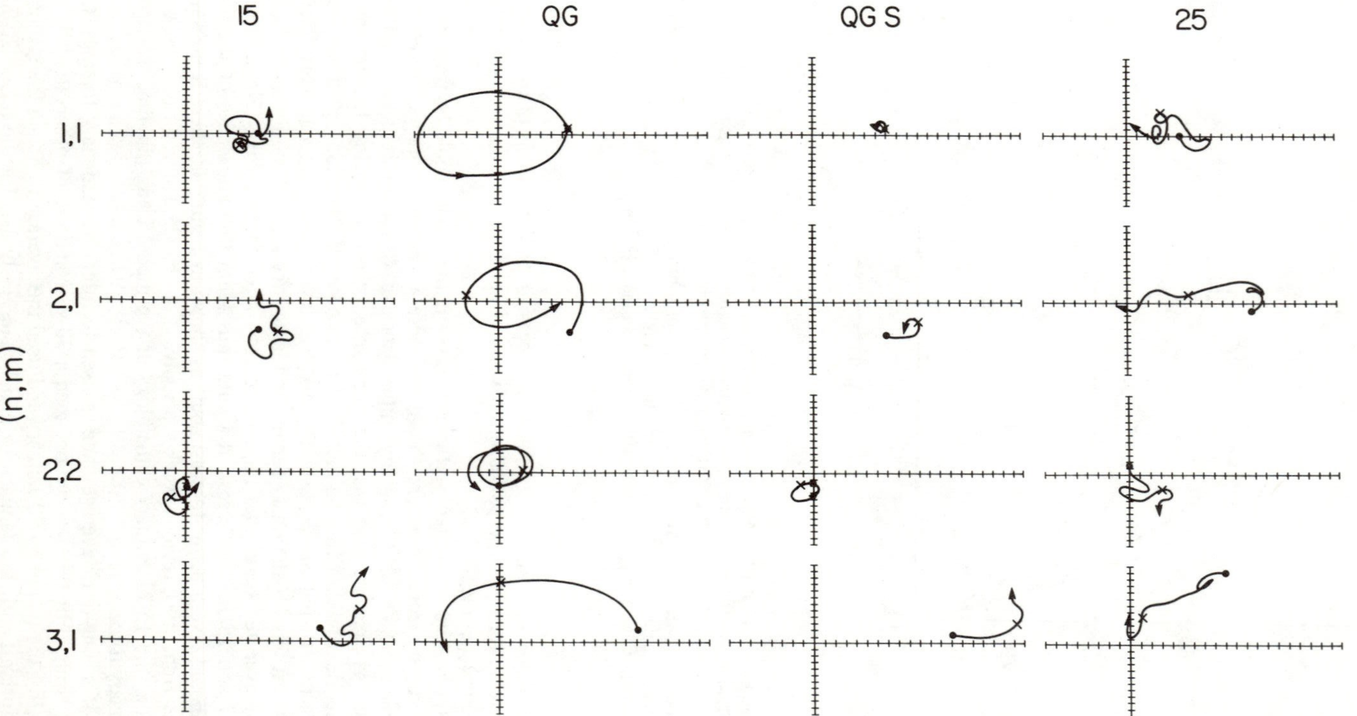

Fig. 3. Harmonic dials of various spherical harmonic components, P_1^1, P_2^1, P_2^2, P_3^1. In each figure the ordinate is the imaginary component and the abscissa is the real component, so counterclockwise movement occurs for westward propagating waves. On each dial a dot denotes the initial value, x the value at day 3 and the arrowhead the value at day 5. Four dials are shown for each wave. "15" is the verifying data starting on 15 Janurary 1979. QG is the model without stationary forcing, QGS is the

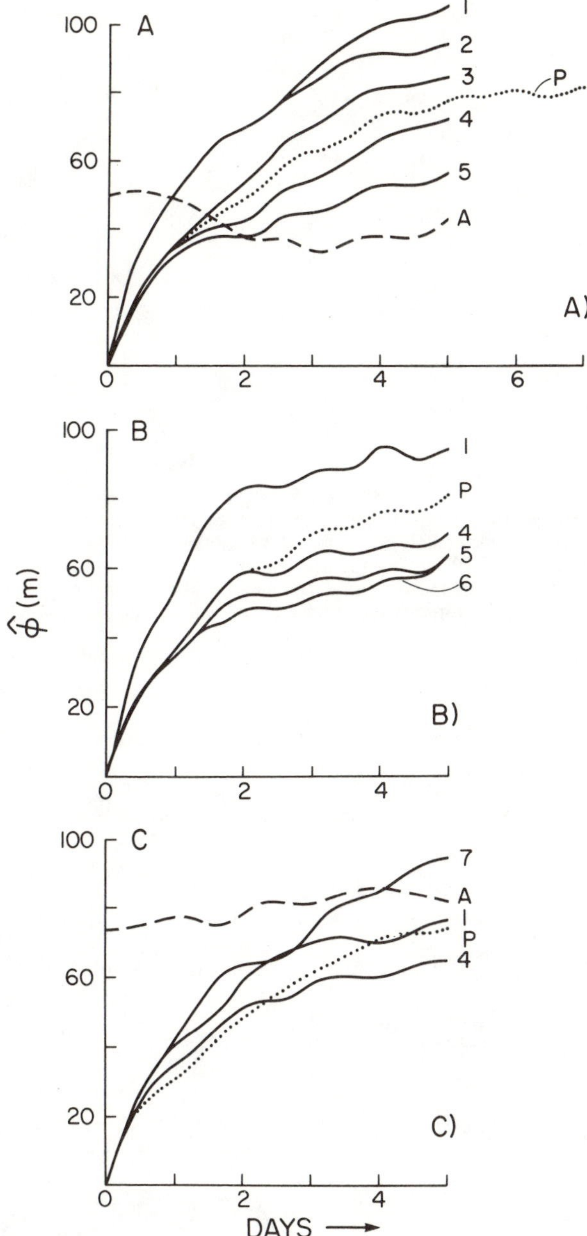

Fig. 4. Time variation of global RMS 500 mb height difference between forecasts and verifying data. Case 1 is the model with no stationary forcing and small friction. Case 2 is the model with no stationary forcing and large friction. Case 3 is the model of Case 1 with the barotropic divergence correction increased by a factor of 10. Case 4 is the model of Case 1 with stationary forcing added. Case 5 is the model of Case 2 with stationary forcing added. Case 6 is the model of Case 5 with the mean zonal wind speed increased by a factor of 2. Case 7 is the model of Case 1 with stationary forcing appropriate to a period 10 days earlier added. Case P denotes the error of a persistence forecast. Case A denotes the RMS difference between the verifying FGGE data and the value of the FGGE data, time-averaged over the period 15-22 January 1979. The three figures present ensembles of forecasts from three different initial states: (a) 15 January 1979, (b) 17 January 1979, (c) 25 January 1979.

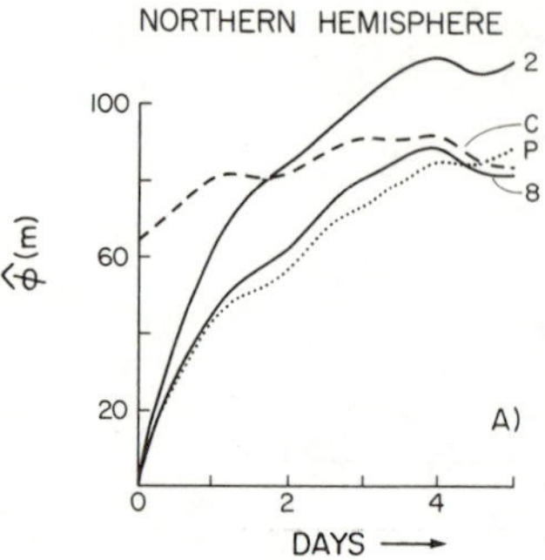

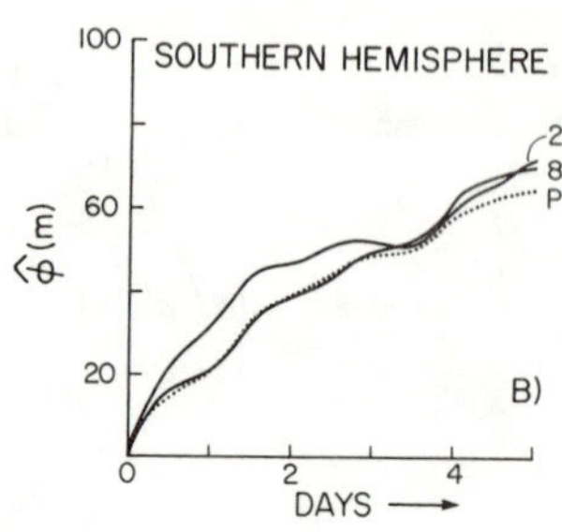

Fig. 5. Time variation of hemispheric RMS 500 mb height differences betwen forecasts and verifying data, in both hemispheres, for the initial state of 15 January 1979.

Case C is the error of a climatological forecast, where the climatology is the 30-year average of observations for the period 15-20 January for latitudes north of 20°N and the initial data elsewhere. Case 8 is the error of the forecast made by a model using the above climatology as its stationary forcing. Case 2 is the error of the forecast made by a model with no stationary forcing. Case P is the error of a persistence forecast.

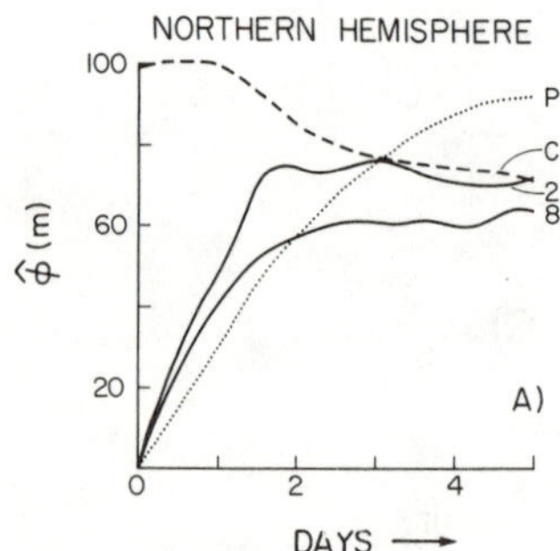

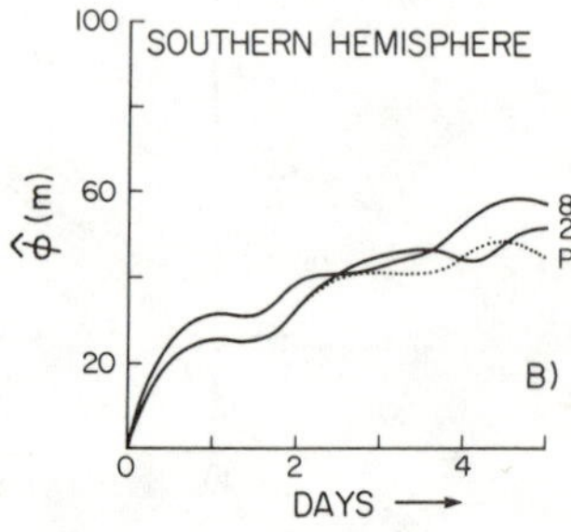

Fig. 6. Same as Figure 5, but for the initial state of 25 January 1979.

and Southern Hemispheres. It is clearly better to use observed climatology rather than no climatology, at least in the Northern Hemisphere. For the 25 January example, shown in Figure 6, the forecasts are markedly better than simple persistence. For this case, the 5-day climatological mean state for 25-30 January was used.

PERSISTENCE DAMPED TO CLIMATOLOGY

Since a linear, barotropic, quasi-geostrophic model is a gross approximation to the actual governing equations, we wished to know how well the model results would compare to those of the Goddard Laboratory for Atmospheric Sciences (GLAS) general circulation model (GCM) and also to those of even simpler models. The simpler models that we wished to test are versions of a persistence-damped-to-climatology scheme.

Earlier, the linear barotropic model was shown to have the solution

$$\hat{\psi}(t) = \hat{\psi}_s + (\hat{\psi}(0) - \hat{\psi}_s)e^{ct}$$

If c is assumed real and negative then we have Scheme I: $\hat{\psi}(t) = \hat{\psi}_s + (\hat{\psi}(0) - \hat{\psi}_s)e^{-|c|t}$, a simple version of persistence damped to climatology. That is, $\hat{\psi}(t) = \hat{\psi}(0)$, which is persistence, if $c = 0$. On the other hand, $\hat{\psi}(t) = \hat{\psi}_s$, which is climatology, if $c \rightarrow \infty$. Thus persistence initially dominates, and then after a sufficiently long time climatology dominates. For c complex, then the solution also includes a propagating wave. This is Scheme II:

$$\hat{\psi}(t)e^{ikx} = (\hat{\psi}_s + (\hat{\psi}(0) - \hat{\psi}_s)e^{-c_r t}e^{ic_i t})e^{ikx}$$

For $c_i > 0$ the solution includes a westward propagating wave and for $c_i < 0$ an eastward propagating wave.

Scheme I is shown in Figures 7a and 7b for the 15 and 25 January cases. Note that for $\frac{1}{|c|} \sim .5$ day, the solution is damped too quickly to the climatology, which results in a bad forecast for the first two days. On the other hand for $\frac{1}{|c|} = 32$ days, the solution is not damped to climatology sufficiently quickly, and we have essentially a persistence forecast which is bad after a few days. The best forecasts occurred for $\frac{1}{|c|} = 8$ days.

Scheme II is shown in Figures 7c and 7d for the case in which the real damping coefficient is $\frac{-1}{c_r} \sim 8$ days. Note that c_i negative gives a worse forecast than c_i positive, indicating the importance of westward-propagating waves. Also note that again $\frac{1}{c_i} \sim 8$ days gives the best solution. This is a fairly slow westward-propagating wave (period $= 16\,\pi$ days).

COMPARISON OF MODELS

Comparing now the solutions for the various models in Figure 8, one notes that the worst forecast for the Northern Hemisphere is the quasi-geostrophic model without stationary forcing, and next is the quasi-geostrophic model with stationary forcing. Somewhat surprisingly, persistence-damped-to-climatology via Scheme II is better than either quasi-geostrophic model, and the general circulation model is best, but not by a large margin. In the Southern Hemisphere, the models all show comparable skill. Of course, this case may not be typical.

A final question is how well these models would do if verified against an independent objective analysis scheme. In all preceding experiments, the verification was the GLAS GCM objective analysis. An independent objective analysis was available from the National

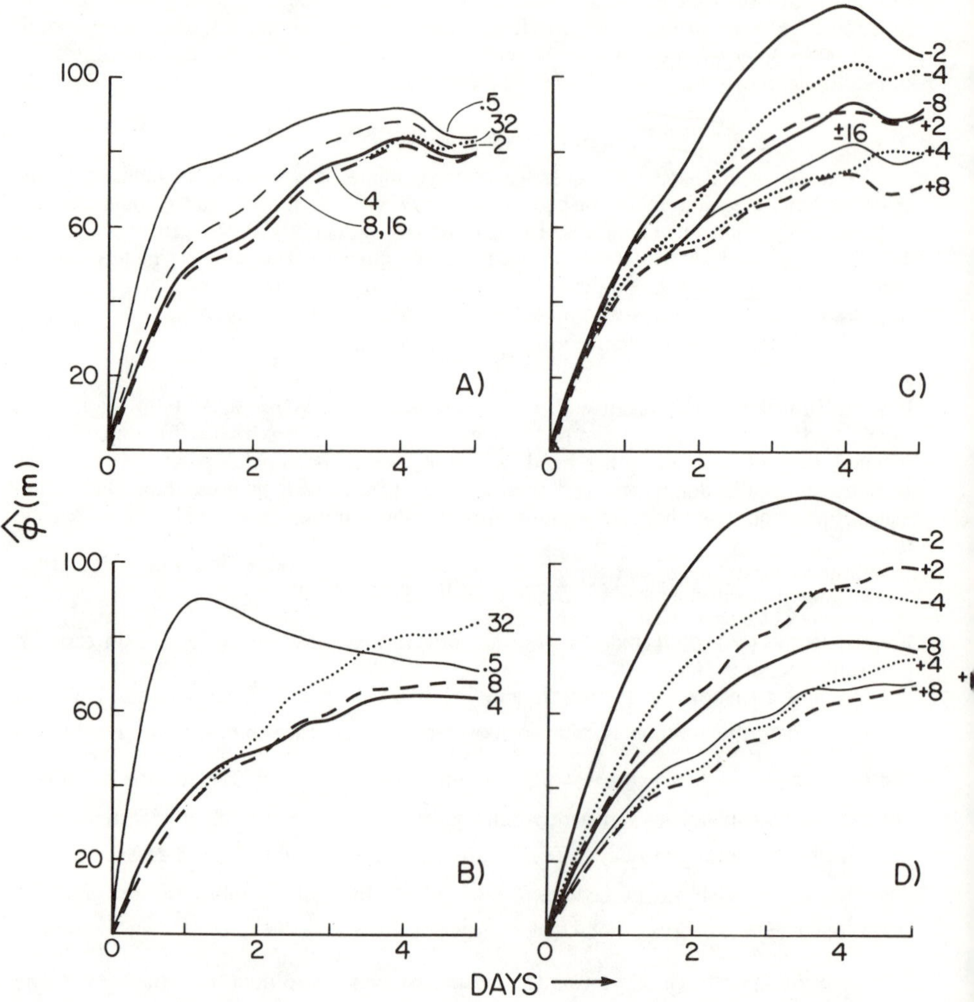

Fig. 7. Time variation of Northern Hemisphere RMS 500 mb height errors.

Forecasts from 2 initial states are shown: 15 January 1979 (Figs. 7a and 7c); and 25 January 1979 (Figs. 7b and 7d). Forecasts made by persistence damped to climatology are shown in Figures 7a and 7b. The numbers on each curve refer to the damping constant (e-folding time) in days. In Figures 7c and 7d, the above forecast with a damping constant of 8 days has been modified by the addition of a propagating wave. The numbers on each curve are the inverse wave propagation frequencies in days, with negative numbers indicating eastward propagating waves.

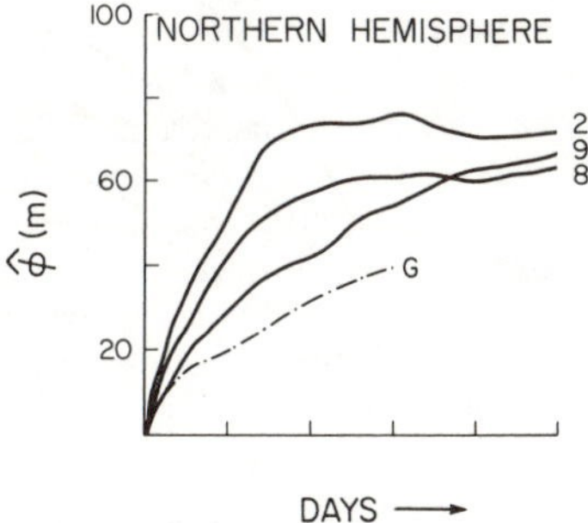

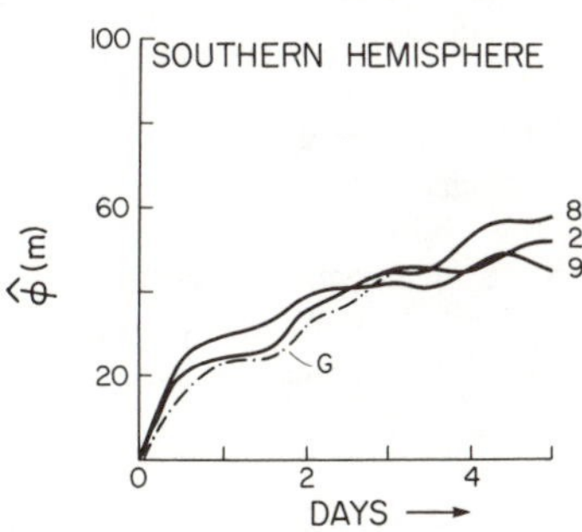

Fig. 8. Time variation of hemispheric RMS 500 md height differences between forecasts and verifying data, in both hemispheres, for the initial state of 25 January 1979.

Case 2 is the forecast made by a model with no stationary forcing. Case 8 is the forecast made by a model with stationary forcing derived from climatology. Case 9 is the persistence forecast damped to climatology and modified by the inclusion of a westward propagating wave with an inverse propagation frequency of 8 days. Case G is the forecast made by the GLAS general circulation model. All of these forecasts are verified against analyses produced by the GLAS objective analysis scheme.

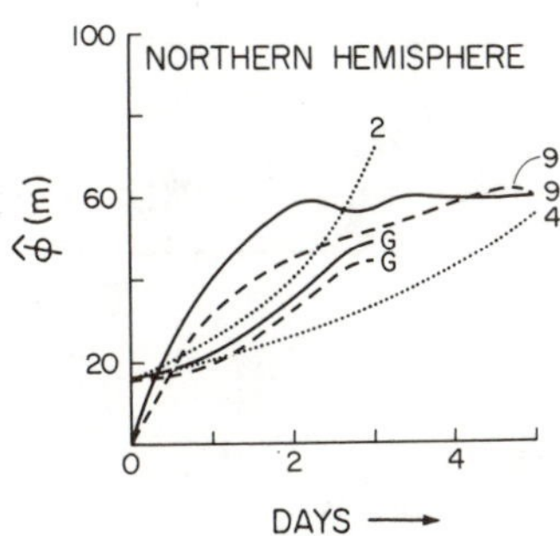

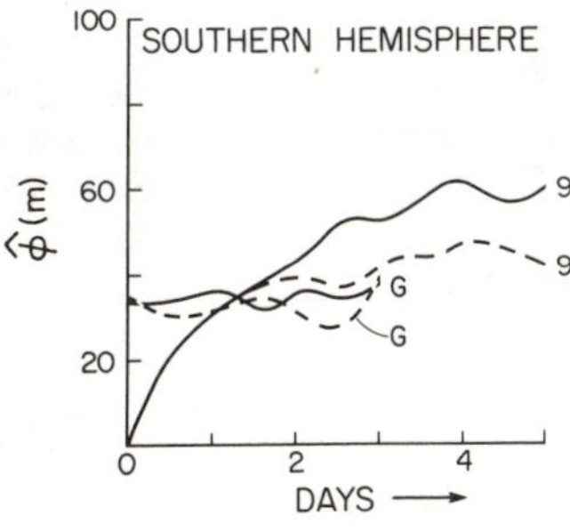

Fig. 9. Time variation of hemispheric RMS 500 mb height differences between forecasts and verifying data, for 2 initial states and 2 forecast techniques, in both hemispheres.

Case 9 is a persistence forecast damped to climatology and modified by the inclusion of a westward propagating wave with an inverse propagation frequency of 8 days. Case G is the forecast made by the GLAS general circulation model. Solid lines refer to the initial state of 17 January 1979. Dashed lines refer to the initial state of 25 January 1979. The verification data for these forecasts are analyses produced by the NMC objective analysis scheme. Also shown in dotted lines are exponential growth rates labeled with the e-folding times in days.

Meteorological Center (NMC). Figure 9 shows forecast results verified against this analysis. The Southern Hemisphere error is so large that essentially no error increase with time is found in the GCM results. The Scheme II forecast in the Southern Hemisphere is initially better for about 1 day and then remains comparable in error to that of the GCM for the 25 January case and becomes worse for the 17 January case. In the Northern Hemisphere the GCM error is much less initially, due to the two objective analysis schemes being more in agreement, and a much weaker error growth occurs. In fact, only for this GCM forecast does the error growth resemble the exponential predictability curve expected theoretically, rather than the quasi-linear error growth curve of our linear barotropic model.

CONCLUSION

The results achieved to date with the linear barotropic quasi-geostrophic model answer several questions but raise others. They make clear that an accurate treatment of stationary forcing is essential to skillful large-scale forecasts, and may in fact partially explain the spurious retrogression of planetary waves in early operational barotropic forecasts. However, these results are somewhat puzzling in view of the demonstrated skill of physically simple primitive-equation models[2]. It is somewhat paradoxical that the linear barotropic model seems so sensitive to stationary forcing, while the primitive-equation model is relatively insensitive to diabatic forcing and parameterized friction.

ACKNOWLEDGMENTS

This research has been supported by the National Aeronautics and Space Administration under Grant No. NAG5-236. JOR was supported in part by National Science Foundation Grant No. ATM82-10160. We are grateful to Dr. Wayman Baker for data and helpful advice. Ms. Beth Chertock and Drs. John Horel and Geoffrey Vallis provided generous assistance. T. Barnett and J. Namias gave helpful advice.

REFERENCES

1. E. N. Lorenz, Tellus, **21**, 289 (1969).
2. R. C. J. Somerville, J. Atmos. Sci., **37**, 1141 (1980).
3 J. O. Roads and R. C. J. Somerville, J. Atmos. Sci., **39**, 745 (1982).
4 R. Daley, J. Tribbia, and D. L. Williamson, Mon. Wea. Rev., **109**, 1836 (1981).
5. P. D. Thompson, *Numerical Weather Analysis and Prediction*, (MacMillan, N.Y., 1961), 170 pp.

APPENDIX I. MODEL

The quasi-geostrophic model written in terms of stream function is

$$\frac{\partial}{\partial t}\{\nabla^2\psi-\frac{\lambda^2}{a^2}\psi\} = -J(\psi,\nabla^2\psi+f) -k\nabla^2\psi+F$$

Linearizing around a zonal mean state yields

$$\frac{\partial}{\partial t}\{\nabla^2\hat{\psi}-\frac{\lambda^2}{a^2}\hat{\psi}\} = -J(\bar{\psi},\nabla^2\hat{\psi})-J(\hat{\psi},\nabla^2\bar{\psi}+f)$$

$$-k\nabla^2\hat{\psi} + \hat{F} \ .$$

The parameters governing these equations are λ^2, $\bar{\psi}$, k, $\hat{F}$, f, and a.

Here

λ^2 is the barotropic divergence correction = constant.

$\bar{\psi}$ is the zonal mean 500 mb stream function = $\bar{\psi}(y)$.

k is the friction coefficient = constant.

$\hat{F}$ is the stationary forcing = $\hat{F}(x,y)$.

f is the Coriolis parameter = $2\ \Omega \sin y$.

a is the radius of the earth = constant.

$\hat{\psi}$ is the deviation of the stream function ψ from $\bar{\psi}$.

In the text, a dimensionless form of this equation is employed; lengths have been scaled by (a) and times by $(2\Omega)^{-1}$. We assume physically realistic parameter ranges, e.g., $\frac{4}{a^2} \leqslant \lambda^2 \leqslant \frac{40}{a^2}$; $\frac{1}{80} < k < \frac{1}{8}$ day^{-1}; $-2 < -\frac{1}{a}\frac{\partial \bar{\psi}}{\partial \theta} < 27$ m/s. $\hat{F}$ is determined such that the space- and time-averaged geopotential heights have RMS values of about 60 m.

In addition to the linearization assumption, two major approximations are used to solve this equation. The first is concerned with the number of waves allowed in the system. Due to the dominance of the planetary modes, linear models retaining only a few zonal waves are adequate to describe the geopotential variance. For example, Figure 10 shows a typical comparison between the GLAS GCM and the linear quasi-geostrophic model for the first few waves. About half of the difference is due to wave number 1. Waves 1 to 3 contain about 80% of the total error contributed by waves 1-6 and describe all major variations.

The next major assumption is that the stream function can be related to the geopotential via the simple relationship $\phi = f_0\psi$, where f_0 is a constant. This assumption was tested for several cases by using the linear balance relationship $\nabla^2\phi = \nabla \cdot f \nabla \psi$ to convert geopotentials to stream function. No significant difference in large-scale error was found although there were some differences in the location of the errors. Thus for simplicity the relationship $\phi = f_0\psi$ was used. However, the zonal wind, around which the models are linearized, was taken from the actual 500 mb zonal wind.

In a linear model like ours, there are various ways to implement a stationary forcing. One is simply to subtract from the initial state the stationary wave, run the time dependent linear model with no forcing and then add the stationary wave to the solution at the end of the run. Another way, used in this paper, is to solve for the forcing from a steady state version of the model and then solve for the complete solution using the prescribed forcing. That is, given a stationary wave $\hat{\psi}_s$, the steady state model is

$$A\hat{\psi}_s = \hat{F}$$

where A is the matrix describing the linear operator. Since A and $\hat{\psi}_s$ are known, $\hat{F}$ is determined.

APPENDIX II. MEASURE OF ERROR

The measure of error used in this paper is an areally-weighted average of the geopotential height deviations on a latitude-longitude grid. That is, the global measure $\hat{\phi}$ is

$$\hat{\phi} = \left\{ \frac{\int_{-\pi/2}^{\pi/2}\int_0^{2\pi} (\tilde{\phi}(x,y) - f_0\hat{\psi}(x,y))^2 \cos y\, dx\, dy}{\int_{-\pi/2}^{\pi/2}\int_0^{2\pi} \cos y\, dx\, dy} \right\}^{1/2},$$

where $\tilde{\phi}$ is the observed geopotential height and $\hat{\psi}$ is the forecast streamfunction.

The spectral quantities are first transformed to the grid before the error is measured. For most of the calculations, a transform involving only the first few zonal wave numbers (1, 2, 3) is sufficient to capture most of the error. Omitting the zonal mean does not significantly affect the error field.

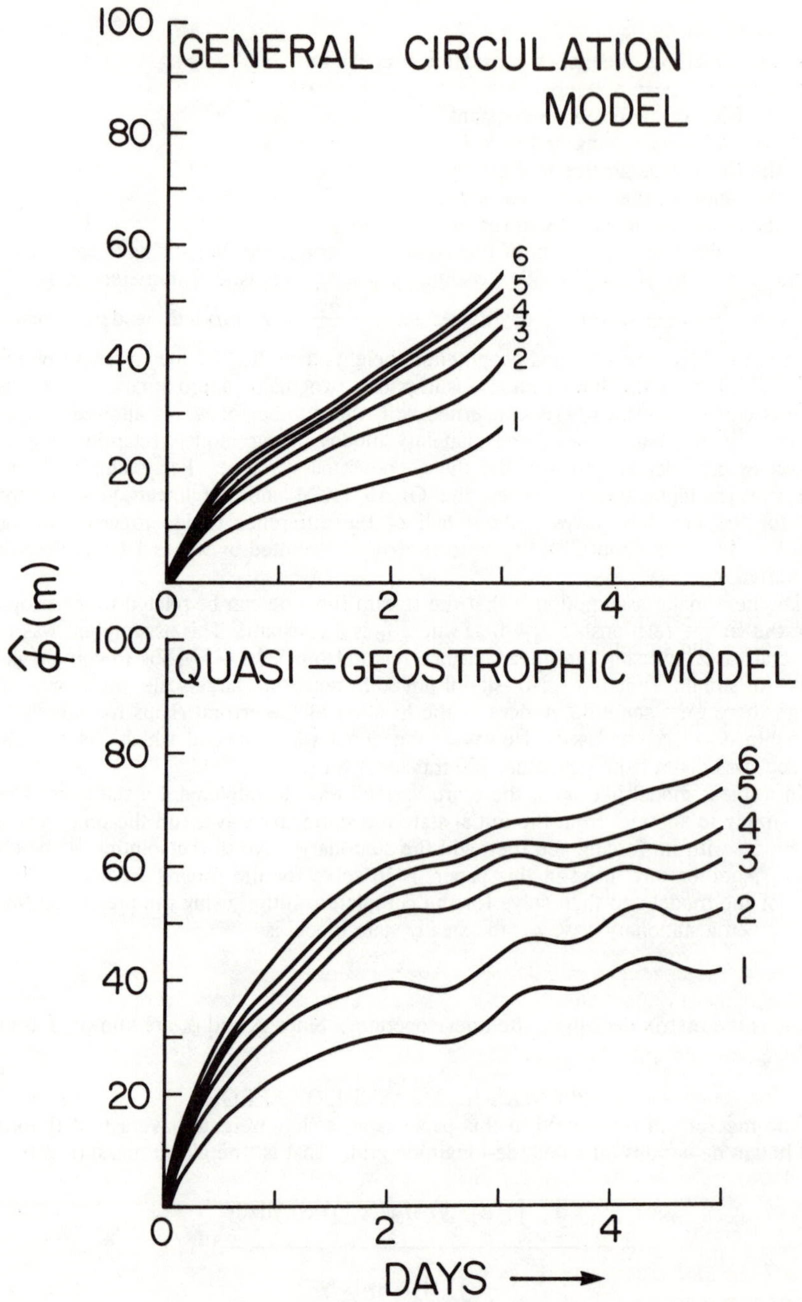

Fig. 10. Time variation of global RMS 500 mb height differences between model forecasts and the GLAS FGGE analyses for the initial state of 17 January 1979. The cases are labeled with the truncation zonal wavenumber. For example, Case 3 includes wavenumbers 1, 2, and 3.

THE IMPORTANCE OF LARGE SCALES OF TURBULENCE FOR THE PREDICTABILITY OF THE TURBULENT ENERGY DECAY

J.A. Domaradzki† and G.L. Mellor
Geophysical Fluid Dynamics Program
Princeton University, Princeton, N.J. 08540

ABSTRACT

A simple eddy viscosity closure of the Kármán-Howarth equation is introduced. Within the framework of this closure scheme it is shown that the turbulent energy decay is strongly influenced by the assumptions regarding behavior of the double velocity correlation functions for large separation distances where experimental data do not exist.

INTRODUCTION

The experimental results on the time decay of the turbulent energy in isotropic turbulence for moderate internal Reynolds numbers ($R_\lambda \sim 10^2$) are usually represented by power law[1,2]

$$\overline{u^2} \propto (t-t_0)^{-a} \ . \tag{1}$$

However, the experimental values of the exponent a vary between 1 and 1.4 for different experiments[2,3] and sometimes it is possible to fit the same data with two different curves by changing the exponent a and choosing appropriate virtual time origin t_0 as reported in Reference 3. On theoretical grounds the decay laws with $a=10/7$ and $a=6/5$ were derived by Kolmogoroff[4] and Saffman[5], respectively. They assumed self preservation of the correlation functions outside the viscous range. Additionally, Kolmogoroff assumed that the Loitsyanskii integral is finite and invariant, implying that as $r\rightarrow\infty$ then, at least, $f(r,t)=0(r^{-5})$, where f is the double velocity correlation function, whereas Saffman assumed that $f(r,t)=0(r^{-3})$ as $r\rightarrow\infty$. These results suggest that the turbulent energy decay rate is closely related to the behavior of the large scales of turbulence. In the present paper, we support this conjecture, using a simple scheme to close the Kármán-Howarth equation, which is then solved with different assumptions regarding large scales of turbulence.

Since for these scales the experimental data on double velocity correlation functions do not exist, the degree of arbitrariness in our assumptions results in the degree of unpredictability of the turbulent energy decay. Therefore, this investigation stresses these aspects of predictability which are related to our limited knowledge of large scales rather than small scales; the latter question being addressed extensively elsewhere in these proceedings.

CLOSURE ASSUMPTION

The Kármán-Howarth equation[1,2] for isotropic turbulence describes the time evolution of the double velocity correlation function f in terms of the unknown triple correlations k

$$\partial_t(\overline{u^2}f)=r^{-4}\partial_r(r^4(\overline{u^2})^{3/2}k)+(2\nu/r^4)\partial_r(r^4\partial_r\overline{u^2}f) \tag{2}$$

where ν is the kinetic viscosity of the fluid.

To those equation (2) we use a simple eddy viscosity assumption

$$(\overline{u^2})^{3/2}k=2A(r,t)\partial_r\overline{u^2}f \tag{3}$$

†On leave from Institute of Geophysics, Warsaw University, Poland.

specifying the scalar function A as follows

$$A=\gamma\epsilon^{1/3}r^{4/3} \tag{4}$$

where ϵ is the energy dissipation rate and γ is constant.

For large internal Reynolds numbers ($R_\lambda \gtrsim 10^3$) expressions (3) and (4) give proper relation between the double and triple correlation functions in inertial subrange of the Kolmogoroff 1941 theory with $.07 \lesssim \gamma \lesssim 08$, whereas for moderate Reynolds numbers ($R_\lambda \sim 10^2$) relation (4) with γ=.06 is derived from the experimental data of Stewart and Townsend[6]. More detailed description of the closure is given in Reference 7.

In what follows we will focus our attention on a case of laboratory, grid generated turbulence, which is generally characterized by moderate Reynolds numbers; we therefore use γ=.06 in our calculations.

The Kármán-Howarth equation nondimensionalized on the grid mesh size M and the kinematic viscosity ν is

$$\partial_t F=(2/r^4)\partial_r(r^4\partial_r F)+\gamma(15F(0,t)/\lambda(t)^2)^{1/3}(2/r^4)\partial_r(r^{16/3}\partial_r F) \tag{5}$$

where $F(r,t)=\overline{u^2}f(r,t)$ and λ is the nondimensional Taylor microscale. The internal Reynolds number $R_\lambda=\lambda(\overline{u^2})^{1/2}$. the boundary conditions imposed on the function F are

$$\partial_r F(r,t)|_{r=0}=0 \tag{6a}$$

$$F(r,t) \rightarrow 0 \text{ as } r\rightarrow\infty \tag{6b}$$

THE INITIAL PERIOD OF DECAY LAW

A notion of the initial period of decay law was introduced by Batchelor[8] and corresponds to the value of the exponent a=1 in formula (1). This decay law may be easily derived from the Kármán-Howarth equation (2) assuming the self preservation of the double and triple correlation functions[1,2]. To investigate what restriction on the behavior of the correlation functions are imposed by this assumption let

$$F(r,t)=\sqrt{\epsilon}\Phi(\eta,t) \tag{7}$$

where $\eta\equiv r/\eta_K=r\epsilon^{1/4}$ is distance measured in the Kolmogoroff length scale units. Equation (5) now becomes

$$(1/2\sqrt{\epsilon})\partial_t\Phi=\eta^{-4}\partial_\eta(\eta^4(1+\gamma\eta^{4/3})\partial\eta\Phi) \tag{8}$$
$$+E(\eta\partial_\eta\Phi/2+\Phi)$$

The conditions of self similarity are $\Phi=\Phi(\eta)$, implying $\partial_t\Phi=0$, and E=const., implying $\overline{u^2}\propto t^{-1}$ i.e. the initial period of decay law.

Equation (8) with $\partial_t\Phi=0$ and with boundary condition (6) may be solved analytically in two limiting cases:

1. If $\gamma\eta^{4/3} << 1$ i.e. the viscous term is large as compared with the inertial term, we get

$$\Phi(\eta)=(6\Phi(0)/E)(\eta^{-2} - 2E^{-1/2}\eta^{-3}\exp(-E\eta^2/4)\mathrm{erfi}(\sqrt{E/4}\eta)) \tag{9}$$

where

$$\mathrm{erfi}(x) = \int_0^x e^{t^2}dt \ . \tag{10}$$

2. If $\gamma\eta^{4/3} >> 1$ i.e. the viscous term is small as compared with the inertial term, we get

$$\Phi(\eta)=F(3,7.5;z) \tag{11}$$

where F is the confluent hypergeometric function and $z=-.75E\eta^{2/3}/\gamma$.

The asymptotic expansions of formulas (9) and (10) for large values of η give immediately

$$\Phi(\eta) = 0(\eta^{-2}) \text{ as } \eta\to\infty \tag{12}$$

which means, that for the initial period of decay law to be valid the double velocity correlation function must approach zero as r^{-2} for large values of r.

THE INFLUENCE OF THE STRUCTURE OF LARGE SCALES OF TURBULENCE ON THE TURBULENT ENERGY DECAY

The survey of a variety of assumptions related to the behavior of the correlation functions for large r is given by Monin and Yaglom[2]. There is in general no rational basis for making a choice among them, but all of them lead to the decay of the double correlation functions at infinity more rapid than r^{-2}.

To investigate the influence of the structure of large scales of turbulence on the turbulent energy decay we used two different assumptions about the behavior of the double correlation functions at infinity. The first one is consistent with the properties of the self similar solution discussed in the previous section; the other one is based on the Batchelor and Proudman[9] result that the double correlation functions behave as r^{-6} at large r in isotropic turbulence. The numerical solutions of the Equation (5) for both assumptions are shown in Figures 1, 2 and 3, where they are compared with the experimental results of Comte-Bellot and Corrsin[10].

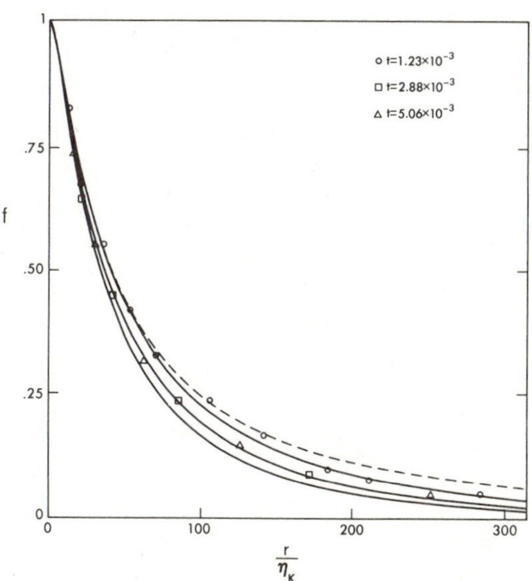

Fig. 1. The comparison between the experimental values of the correlation function and the theoretical predictions. The descending curves are, respectively, the self similar solution (dashed line), the initial condition (at $t=1.23x10^{-3}$) decaying as r^{-6} at infinity and the correlation functions calculated from this initial condition at the subsequent times $t=2.88x10^{-3}$ and $t=5.06x10^{-3}$.

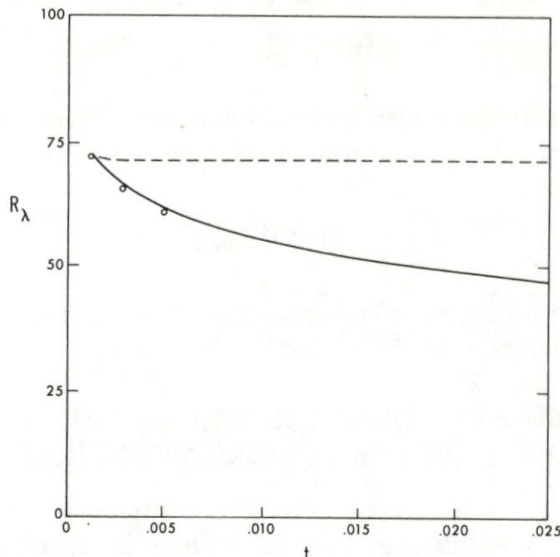

Fig. 2. Time dependence of the internal Reynolds number for the self similar (dashed line) and the r^{-6} (solid line) initial conditions.

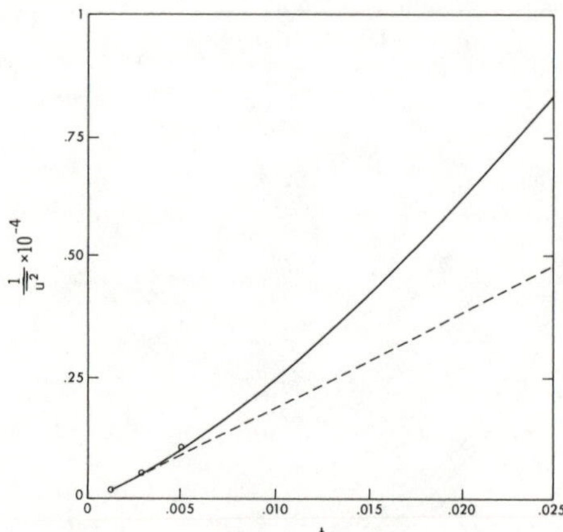

Fig. 3. The energy decay curves for the self similar (dashed line) and the r^{-6} (solid line) initial conditions.

The initial shapes of correlation functions were chosen by fitting them to the experimental values of the measured correlation functions at dimensionless time $t=1.23\times10^{-3}$ and extending them for large r (up to $r\approx 2\times10^{3}\eta_K$) according to our assumptions. With both assumptions considered it was possible to obtain a reasonable fit in the experimental range. However, differences in the large r behavior of the correlation functions beyond the experimental range lead to differences between predictions, growing in time. The experimental data clearly favor Batchelor and Proudman's result.

DISCUSSION

Measurements of double correlation functions in isotropic turbulence have limited range and do not provide information about large scales of turbulence, which must be modeled in somewhat arbitrary manner. This arbitrariness may lead to significant limitations of the predictability of turbulence properties. In a simple example presented it was possible to choose the preferred continuation of the correlation functions from the experimental range into the large r range on the basis of the carefully theoretical analysis by Batchelor and Proudman[9]. If, however, this procedure cannot be carried out for more complicated fluid problems, this example can give, at least qualitative, insight into the importance of modeling the large scales of turbulence, which are beyond the experimental range.

REFERENCES

1. J.O. Hinze, Turbulence (McGraw-Hill, New York, 1959).
2. A.S. Monin and A.M. Yaglom, Statistical Fluid Mechanics: Mechanics of Turbulence (MIT Press, Cambridge; Vol. 1., 1971; Vol. 2, 1975).
3. T.D. Dickey and G.L. Mellor, J. Fluid Mech., **99**, 13 (1980).
4. A.N. Kolmogoroff, C.R. Acad. Sci. USSR **30**, 301 (1941).
5. P.G. Saffman, Phys. Fluids **10**, 1349 (1967).
6. R.W. Stewart and A.A. Townsend, Roy. Soc. Phil. Trans. **A243**, 359 (1951).
7. J.A. Domaradzki and G.L. Mellor (submitted for publication to J. Fluid Mech.)
8. G.K. Batchelor, The Theory of Homogeneous Turbulence, (Cambridge 1953).
9. G.K. Batchelor and I. Proudman, Roy. Soc. Phil. Trans. **A248**, 369, (1956).
10. G. Comte-Bellot and S. Corrsin, J. Fluid Mech, **48**, 273 (1971).

APPLICATIONS OF ENTROPY TO PREDICTABILITY THEORY

G. F. Carnevale
Center for Studies of Nonlinear Dynamics
La Jolla Institute, La Jolla, California 92038

G. K. Vallis
Scripps Institution of Oceanography, La Jolla, California 92093

ABSTRACT

Information theory provides a prescription for quantifying in a statistical sense the concept of knowledge. This can be used to construct a measure of predictability defined in terms of ensemble averaged kinetic and error energy spectra. As a theoretical tool it offers certain advantages over other predictability measures such as error energy. Recent applications of this concept to the standard predictability experiment have pointed out certain ambiguities, difficulties and curiosities. A sequence of barotropic predictability experiments that illustrates and elucidates some of these points is presented.

INTRODUCTION

This paper is directed toward two aspects of fluid predictability. First we consider the role of entropy, the second law of thermodynamics and H-theorems in fluid dynamics. Second we consider the predictability of two-dimensional flow, the effects of forcing and viscosity and the role of entropy as an indicator of information degradation as a forecast proceeds. The discussion of the second law and the relationship to the H-theorems will be partly pedagogical; but as there is some controversy in the literature concerning their relationship with each other, if any, we feel some statement of our perspective is warranted.

Entropy as defined by Gibbs is the negative of the quantity that measures information in modern information theory. It may be written as

$$S_G = -\int P \ln P \prod_i dy_i \, , \tag{1}$$

where $P\{y\}$ is the probability density in the phase space defined by the set of f degrees of freedom or co-ordinates y_i. That the Gibbs entropy is a natural measure of information contained in any probability distribution is shown in a number of works on information theory or statistical mechanics[1-3]. For Liouvillian systems S_G has a very important property. A system is Liouvillian if the motion of points in the coordinate space is incompressible, that is, if

$$\sum_i \frac{\partial \dot{y}_i}{\partial y_i} = 0. \tag{2}$$

In this situation, S_G is a constant of the motion — evolution does not destroy information. This result leads to the second law of thermodynamics. To see this we must first introduce the notion of experimental entropy.

It is usually quite impossible to follow the detailed evolution of $P\{y\}$. Progress is made by considering some restricted aspects of the system under consideration, and so we must infer a probability distribution which reflects a knowledge of these aspects. This distribution is the one which maximizes the integral (1) subject to the constraints of the given knowledge. The actual maximum value achieved is defined to be the experimental entropy, S_e. The constraints can be put in the following form

$$\int Q_\alpha P \prod_i dy_i = <Q_\alpha>, \tag{3}$$

where the Q_α are representations in terms of the $\{y_i\}$ of the quantities about which we have some knowledge, and the $<Q_\alpha>$ are the measured values of these quantities. In classical thermodynamics these often are thought of as 'macroscopic' variables, e.g. temperature (or energy), pressure, volume, etc. However, this distinction between microscopic and macroscopic variables is somewhat arbitrary, as will be seen in the applications made below. Using Lagrange multipliers leads to the well-known canonical distribution for the probability distribution, and a *formula* for the experimental entropy in terms of the $\{Q_\alpha\}$. In the sense of Tolman[2] one might interpret experimental entropy as a coarse grained entropy.

As stated by, for example, Jaynes[4], the second law is "The experimental entropy cannot decrease in a reproducible adiabatic process that starts from a state of thermal equilibrium." This may be *proved* as follows[4]. In a state of thermal equilibrium (i.e. one in which only constants of the motion are known) the information entropy is maximal and must be set equal to the experimental entropy. A measurement or other specification of the set $\{Q_\alpha\}$ defines the initial state. This is the only information which goes into defining the initial probability distribution, and so $S_e = S_G$ at $t = 0$. If the system is Liouvillian, the probability density is conserved and the information entropy,S_G, stays constant. At a later time t, we suppose the "macroscopic" parameters $\{Q_\alpha\}$ have undergone change to new values $\{Q'_\alpha\}$. But since S_e at the later time is defined as the maximum of the integral (1) under the new constraints, we must have $S_e(t\neq 0) \geqslant S_G = S_e(t=0)$. In other words, Gibbs entropy is a measure of phase volume of probable states. This volume is conserved by Liouville's theorem. But we measure (in the experimental entropy) the maximum possible volume subject to the given knowledge. Hence at a later time this must be as large or larger than the original volume. This theorem of statistical mechanics is somewhat more comprehensive than the conventional second law of thermodynamics, in that it generalizes to describe the evolution between nonequilibrium states. Strictly speaking, in thermodynamics entropy is defined only in states of thermal equilibrium. This statistical mechanics approach can be used to extend the definition outside of equilibrium, provided the $\{Q_\alpha\}$ can still be defined.

The power in the second law is its universal applicability (all Liouvillian systems). Note, however, that the second law actually makes no attempt to describe the evolution from initial to final experimental entropy S_e. The canonical equilibrium state corresponds to a knowledge of only the values of the constants of motion which corresponds to a subset of the $\{Q_\alpha\}$. Other states, for which we measure more of the $\{Q_\alpha\}$, other than just constants of the motion, must have a lower entropy. But the second law does *not* tell us in what manner the system evolves toward the maximum entropy state, or indeed if it does at all. Given knowledge only of the constants of the motion, we must certainly *assume* a maximum entropy state. If the system is mixing, then we hypothesize that the phase space of the system is filled out and hence the experimental entropy of the system increases as it approaches equilibrium. Thus we *postulate* evolution toward maximum experimental entropy, but we cannot prove it unless we have a Boltzmann-like H-theorem for the system.

The Boltzmann-like H-Theorems attempt to go further than the second law by showing that for adiabatic conditions the experimental entropy increases monotonically toward its equilibrium value. However, to accomplish this requires *ad hoc* assumptions, and so such theorems cannot claim the fundamental status clearly held by the second law.

In statistical fluid studies the specified information usually takes the form of average fields, $<y_i>$, and second order correlations, $Y_{ij} = <y_i y_j>$. For simplicity we restrict this discussion to the case $<y_i>=0$. Under these conditions the prescription for experimental entropy results in[6]

$$S = \tfrac{1}{2}\ln|Y_{ij}|, \tag{4}$$

where $|Y|$ is the determinant of the second order correlation matrix — actually, in writing (4) we have neglected some additive constants which are of no consequence here.

Jaynes[4,5] has pointed out that in general the Boltzmann and Gibbs entropies cannot agree even in equilibrium. This follows because even in equilibrium the probability distribution need not factorize. The Boltzmann-like entropy we use here is subject to the same criticism unless we restrict ourselves to using only quadratic constants of motion to determine the canonical equilibrium. Under this restriction, which is appropriate to many fluid problems, the Boltzmann and Gibbs entropies are identical in equilibrium. If additional information in the form of higher order moments is available, then (4) would not be appropriate; the work here is based entirely on spectra and hence (4) will suffice.

For example, a two-dimensional flow with doubly periodic boundary conditions is represented entirely in terms of the Fourier amplitude of the streamfunction, $\psi_{\vec{k}}$. In enumerating independent degrees of freedom the Hermiticity condition,

$$\psi_{\vec{k}} = \psi_{-\vec{k}}^{*}, \tag{5}$$

must be taken into account. A convenient definition of the set of independent variables lists the real and imaginary parts of the Fourier amplitudes for a restricted set of wavevectors, $\{\vec{k}\}$. According to (5), this set of wavevectors is such that if a particular $\vec{k}$ is included then $-\vec{k}$ is not — furthermore, the zero wavevector is noninteracting in the case of periodic boundary conditions and is ignored in this work. The wavevector set considered below is appropriate to the numerical simulation of the flow, and so it is discrete and finite (finite resolution implies an upper wavenumber cutoff).

To investigate homogeneous turbulence the appropriate ensemble is such that

$$<\psi_{\vec{k}}\,\psi_{\vec{p}}> = \delta_{\vec{k},-\vec{p}}<|\psi_{\vec{k}}|^2>. \tag{6}$$

Thus the matrix Y_{ij} reduces to block diagonal form with blocks given by

$$\begin{bmatrix} <\psi_{\vec{k}}'\psi_{\vec{k}}'> & <\psi_{\vec{k}}'\psi_{\vec{k}}''> \\ <\psi_{\vec{k}}''\psi_{\vec{k}}'> & <\psi_{\vec{k}}''\psi_{\vec{k}}''> \end{bmatrix}.$$

The determinant is then the product of the determinants of these blocks labeled by $\vec{k}$. Furthermore, the vanishing of (6) when $\vec{k}=\vec{p}$ implies

$$<\psi_{\vec{k}}'\psi_{\vec{k}}''> = 0, \tag{7}$$

and

$$<\psi_{\vec{k}}'\psi_{\vec{k}}'> = <\psi_{\vec{k}}''\psi_{\vec{k}}''> = \tfrac{1}{2}U_{\vec{k}}/k^2, \tag{8}$$

where prime and double prime denote real and imaginary parts and where $\frac{1}{2}U_{\vec{k}}$ is the kinetic energy per mode. The entropy then assumes the following simple form

$$S = \tfrac{1}{2}\sum_{\vec{k}\neq 0} \ln(U_{\vec{k}}/k^2). \tag{9}$$

Thus S is the experimental entropy, given knowledge of the energy in each mode[6-8]. Note that the differentiation between microscopic and macroscopic variables is irrelevant here.

The analogs of Boltzmann's equation in fluid dynamics are the second-order Markovian closure equations. These equations estimate the rate of change of $U_{\vec{k}}$ given only its instantaneous value. They imply that for unforced, inviscid flow the entropy (9) evolves according to[6]

$$\frac{dS}{dt} = \frac{1}{3} \sum_{\substack{\vec{k}\,\vec{p}\,\vec{q} \\ \vec{k}+\vec{p}+\vec{q}=0}} \theta_{\vec{k}\,\vec{p}\,\vec{q}} \frac{\sin^2\alpha}{k^2} U_{\vec{k}} U_{\vec{p}} U_{\vec{q}} \left[\frac{(p^2-q^2)}{U_{\vec{k}}} + \frac{(k^2-q^2)}{U_{\vec{p}}} + \frac{(p^2-k^2)}{U_{\vec{q}}}\right]^2, \tag{10}$$

where α is the angle between $\vec{p}$ and $\vec{q}$. $\theta_{\vec{k}\,\vec{p}\,\vec{q}}$, called the the triad relaxation time scale, can be modeled consistently in several ways; for our purposes we need note only that it is positive and symmetric under permutations of the triad ($\vec{k}$, $\vec{p}$ and $\vec{q}$). The right hand side of (10) is manifestly non-negative. Thus according to (10) the entropy increases monotonically except in the canonical equilibrium state

$$U_{\vec{k}}^{eq} = \frac{1}{a+bk^2} \tag{11}$$

(a state of detailed balance)[9]. The Lagrange multipliers a and b are determined by the values of the total energy and enstrophy (which are the important dynamical invariants for 2-D flow). Now (11) is also the maximum entropy state subject only to knowledge of the total energy and enstrophy in the system. Thus the H-theorem (10) implies monotonic relaxation toward a maximum value of the experimental entropy. This is the direct analog of Boltzmann's H-theorem.

In predictability experiments we consider two flows represented by streamfunctions ψ and ϕ simultaneously. In the statistical analysis of this problem, the statistics of the two fields are taken to be identical[10,11]; that is,

$$\langle \psi_{\vec{k}} \psi_{\vec{k}}^* \rangle = \langle \phi_{\vec{k}} \phi_{\vec{k}}^* \rangle = U_{\vec{k}}/k^2. \tag{12}$$

Spatial homogeneity is assumed throughout. The predictability problem is concerned with how correlated these fields remain if initially not identical. This correlation is measured by

$$\langle \psi_{\vec{k}} \phi_{\vec{k}}^* \rangle = W_{\vec{k}}/k^2. \tag{13}$$

Furthermore, it is assumed that there is an equivalence in ψ and ϕ in the form of a reflection symmetry,

$$\langle \psi(\vec{x}+\vec{r})\phi(\vec{x}) \rangle = \langle \phi(\vec{x}+\vec{r})\psi(\vec{x}) \rangle,$$

which implies that $W_{\vec{k}}$ is real. In order to define a measure of predictability, or rather lack of predictability, based on information theory measures, we include both ϕ and ψ in the list of coordinates {y}. Then the prescription (4) yields

$$S = \tfrac{1}{2} \sum_{\vec{k}\neq 0} \ln[(U_{\vec{k}}^2 - W_{\vec{k}}^2)/k^4]. \tag{14}$$

By using the relationships dictated by homogeneity this predictability entropy can be rewritten (except for an additive constant, $2n\,ln2$ where n is the number of wavevectors in the summations) as

$$S = \tfrac{1}{2} \sum_{\vec{k}\neq 0} \ln\langle |\psi_{\vec{k}} - \phi_{\vec{k}}|^2 \rangle + \ln\langle |\psi_{\vec{k}} + \phi_{\vec{k}}|^2 \rangle. \tag{15}$$

The analog of Boltzmann's equation for (15) has been recorded elsewhere[12] and, for present purposes we need merely note that it implies monotonic evolution of this entropy toward its canonical equilibrium value. The equilibrium state has the energy spectra given by (11) and no cross-correlation, $W_{\vec{k}}=0$.

APPLICATIONS

The entropy prescription as suggested by information theory and Markovian closure does not translate immediately into an operational procedure for quantifying predictability. The difficulty is that the prescription requires an ensemble average and this is quite impractical for high resolution applications. If ensemble averaged quantities are simply replaced by their single realization values, a catastrophic contradiction ensues. Specifically the replacement of

$$S = \tfrac{1}{2} \ln|\langle y_i y_j \rangle| \tag{16}$$

by

$$S = \tfrac{1}{2} \ln|y_i y_j| \tag{17}$$

yields $-\infty$ because $|y_i y_j| = 0$. That is, if the number of degrees of freedom, f, is greater than one, then the matrix whose elements are the *unaveraged* products $y_i y_j$ has vanishing determinant. For example, imagine the case f=2; the determinant is $y_1^2 y_2^2 - (y_1 y_2)^2 = 0$. This does not mean the situation is hopeless; it does, however, mean that some *ad hoc* modification specific to the system under investigation is necessary to proceed. One possibility, which has been applied to systems where spatial homogeneity is a reasonable assumption for the statistics, is to begin with the expression (9) and simply substitute the instantaneous single realization energy spectrum in place of the ensemble average spectrum[7,8]. The single realization prescription then is

$$S_s = \tfrac{1}{2} \sum_{\vec{k} \neq 0} \ln|\psi_{\vec{k}}|^2, \tag{18}$$

where the summation includes *all* simulated wavevectors.

Since this prescription is based on a single realization it must be expected to show fluctuations in time even in states where the energy spectrum is nearly stationary. However the time average and the standard deviation about the mean of this single realization entropy can be accurately predicted by canonical statistics. That is, the statistics are computed using the canonical distribution function:

$$P(\{y\}) = \prod_{\{\vec{k}\}} \eta_{\vec{k}} \exp[-(a+bk^2)k^2|\psi_{\vec{k}}|^2], \tag{19}$$

where the normalization coefficient, $\eta_{\vec{k}}$, is given by

$$\eta_{\vec{k}} = k^2(a+bk^2)\pi^{-1}, \tag{20}$$

and the product runs over the set of wavevectors which includes only one of each of the pairs, $(\vec{k},-\vec{k})$. The equilibrium entropy (9) takes the value

$$S = -\tfrac{1}{2} \sum_{\vec{k} \neq 0} \ln(a+bk^2)k^2. \tag{21}$$

The mean and standard deviation of the single realization entropy are calculated using (19) and the following definite integrals:

$$\int_0^\infty dz\, e^{-\lambda z} \ln z = -(C+\ln\lambda)\lambda^{-1},$$

and

$$\int_0^\infty dz\, e^{-\lambda z}(\ln z)^2 = [\frac{\pi^2}{6} + (C+\ln\lambda)^2]\lambda^{-1},$$

where C (~ 0.5772157) is Euler's constant. The expectation value of the single realization entropy is given by

$$<S_s> = S - \tfrac{1}{2}fC. \qquad (22)$$

The expectation of the size of the fluctuations about this mean is given by the standard deviation

$$\sigma^2 = <S_s^2> - <S_s>^2 = f\pi^2/24. \qquad (23)$$

Note that (19), a multivariate Gaussian distribution, has rms fluctuations of the modal energies equal to the modal energies while the relative rms fluctuation in the total energy is inversely proportional to $\sqrt{f}$. Even though the fluctuations in the single realization modal energies are not small, the relative rms difference between $\ln|\psi_{\bar{k}}|^2$ and $\ln<|\psi_{\bar{k}}|^2>$ is inversely proportional to lnf.

It is often useful in analyzing turbulent flow to construct an isotropic energy spectrum from the single realization data. This spectrum is actually an average of modal energies. For situations in which anisotropy is not anticipated as being a significant feature, it would seem practical to replace the ensemble average in the prescription by the average over a wavenumber ring. There is, however, a slight ambiguity associated with doing this. Namely the averages can be taken for the modulus squared streamfunctions, the modal energies, or the modal enstrophies, etc. In a discrete wavevector representation the numerical results will differ depending on the choice made. However the qualitative behaviors of the entropies arrived at by these different prescriptions are the same. In what follows the average over the ring will be performed over the stream function modulus squared. To the extent that this is a reasonable estimate for the ensemble average, this "pseudo-ensemble" entropy, which shall be denoted by S_p, should approximate S^{eq} at long times in the inviscid simulation. Also it is expected that the behavior of S_p will be much smoother than the behavior of S_s.

The simplest possible test of these ideas and a good check on the numerics is to simulate the approach to inviscid equilibrium. For simplicity the initial condition spectrum is chosen so that the ratio of enstrophy to energy is such that the Lagrange multiplier b vanishes; that is, the expected equilibrium is energy equipartition. The value of the Lagrange multiplier a is then calculated from the total energy (i.e., $a=f/2E_T$). The dealiased spectral code used here retains a total of n=3128 wavevectors; this represents f=n degrees of freedom, since there is only the one field ψ. Quantities such as the sum of the logarithms of wavenumber squared over all retained modes can be calculated numerically. Thus for a total energy of 0.3938×10^{-3} and enstrophy to energy ratio of 499.24 (chosen to make b=0), the predictions are given in Table I. Relevant time scales are provided by the local eddy turnover time defined by

$$\tau_e(k)=(k^3E(k))^{-1/2}. \qquad (24)$$

The smallest such time scale in this run is roughly 0.5 of our time units. The initial energy spectrum is shown in Figure (1b). The simulation was carried out for approximately 80 of the fastest turnover times. Approximate energy equilibration of wavenumber k is found to occur after roughly a time period $\tau_e(k)$ calculated with the equilibrium spectrum. The final energy spectrum is shown on Figure (1c). The behavior of S_s and S_p during this equilibration are shown in Figure (1a). Agreement with the predictions in Table I is excellent. For

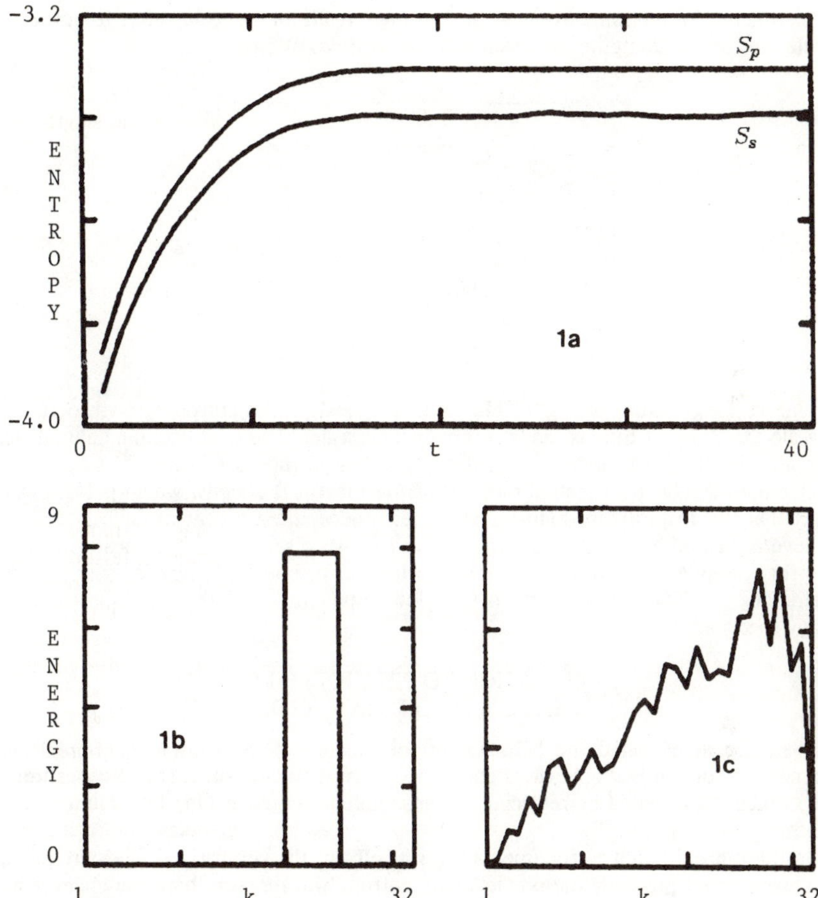

Fig. 1. Evolution to canonical equilibrium.

(1a) Entropy measures S_s and S_p, based on unaveraged and wavenumber-ring averaged modal energies (see text). The entropies are measured in units of 10^4. In the time units used here the fastest eddy turnover time is roughly 0.5.

(1b) Initial energy spectrum (in units of 10^{-5}). This is the one dimensional spectrum produced by integrating the modal energy over a wavenumber ring.

(1c) Final energy spectrum (in units of 10^{-5} at time t=40). The fall off of the spectrum at the highest wavenumbers is due to the use of an octagonal truncation, used for dealiasing purposes, and does not indicate a failure in equipartioning.

the experimental results,time averages of the single realization entropy after it has begun to fluctuate are used to compute the mean and standard deviation.

Table I			
Theoretical Predictions		Experimental Results	
S^{eq}	-33004	max S_p	-33028
$<S_s>$	-33907	mean S_s	-33947
σ	36	σ	29

One striking feature of Figure (1a) is that almost all the entropy growth occurs on a time scale the order of the fast eddy turnover time scale. The overwhelming abundance of modes are fast (high k) modes, and entropy weights all modes equally. The fact that the slowest modes are far from equilibrium is for the most part a signal which is lost owing to the fact that the majority are high wavenumber modes and are near canonical equilibrium after several fast-eddy turnover times. In the following sections the comparison between total error energy and entropy shows that the total error energy suffers from this defect to the same degree. The only recourse is to follow the evolution of the error spectrum or the "entropy spectrum".

APPLICATION TO PREDICTABILITY: (A) UNFORCED, INVISCID

We have investigated the behavior of the analogs of both S_s and S_p in predictability experiments. For the isotropic cases considered here it will be sufficient to concentrate on just S_p, which is calculated by replacing the ensemble averages in (15) by wavenumber ring averages.

In these predictability experiments the statistics of the comparison fields are identical. As the experiment proceeds the predictability entropy will measure both the changes in the evolving autocorrelations as well as cross-correlations. In the experiments described below, only stationary energy spectra are used so as to focus attention on the evolution of the error field. The first experiment has an initial condition such that the energy spectra of both ψ and ϕ are the same as the approximate energy equipartition exhibited in Figure (1c). The error energy is taken as maximal for wavenumbers above 25 and zero below. In an unforced inviscid simulation the entropy must tend monotonically toward its maximum value while the error energy tends to become total at all wavenumbers. The behavior of S_p and the total error energy during this process is shown in Figure (2a). Figure (2b) shows the the evolution of the relative error energy spectrum. It is quite natural also to inquire about the behavior of the entropy spectrum. Here it is more convenient to display the entropy deficit in each wavenumber band. For this experiment this deficit is computed according to

$$I(k) = \ln(U_{\bar{k}}^{eq})^2 - \ln(U_{\bar{k}}^2 - W_{\bar{k}}^2),$$

where the spectra are experimentally evaluated as isotropic wavenumber band averages. This definition of entropy deficit is somewhat analogous to the information theory quantity called the surprisal[13]. Strictly speaking the surprisal is a function on the domain of all possible values of some random variable. This function gives a quantitative measure of how

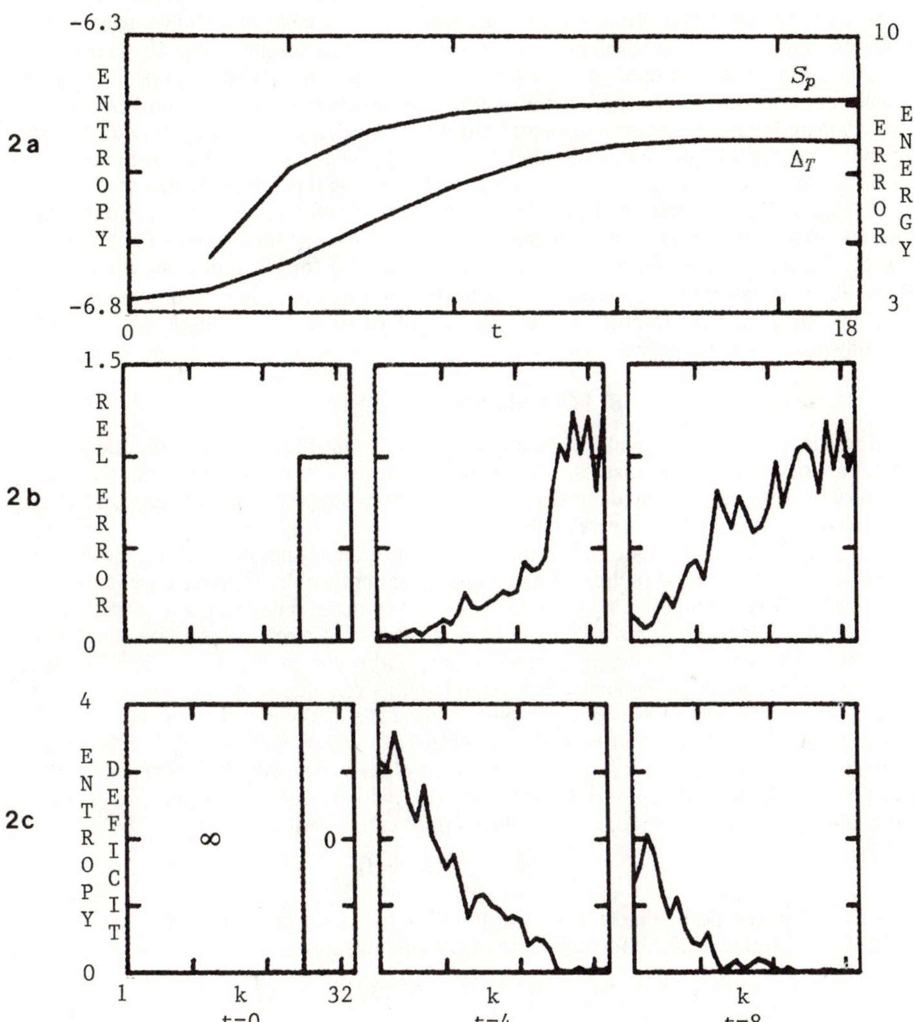

Fig. 2. Predictability: unforced, inviscid.

(2a) The evolution of the total error energy in units of 10^{-4}) and predictability entropy (in units of 10^{4}). The smallest eddy-turnover time is roughly 0.5.

(2b) Relative error spectra, $\Delta(k)/E(k)^{eq}$, at times t=0,4 & 8.

(2c) Entropy deficit (see eq. (25)) at t=0,4 & 8. The graphs for t=0 are schematic.

surprised one should be by a particular value resulting from a measurement of the random variable. In the applications made here one attempts to quantify the surprise in obtaining, at a given time, the simulated spectra of $U_{\bar{k}}$ and $W_{\bar{k}}$. Certainly one should register no surprise in obtaining the canonical equilibrium spectra, so this baseline must be subtracted. The subtraction procedure used in Equation (25) is, in fact, not the only possibility. For example, in chemical physics applications, Levine and coworkers[14] have found it appropriate to compare measured values and canonically expected values by taking the ratios of the corresponding inferred probability distributions. For the work at hand we find the spectra (25) to be the most meaningful and shall simply refer to it as the entropy deficit spectra.

In Figure (2c) the entropy deficit spectra at three different times in the evolution are displayed. Where the deficit is large there is more "information" than could be anticipated. Here, for convenience, we used the spectrum shown in (1c) for $U_{\bar{k}}$ in defining the plotted deficits and relative errors; however, the spectrum is not exactly steady. An alternative is to use the instantaneous spectra, which might avoid some of the apparent discrepancies which now appear in the graphs.

(B) FORCED AND VISCOUS

The addition of forcing and dissipation is, of course, incompatible with the formalism of equilibrium statistical mechanics. However, the success of nonequilibrium statistical mechanics in using an entropy defined out of equilibrium suggests that extensions to non-conservative systems may also be valuable.

An immediate consequence of introducing dissipation is the violation of Liouville's theorem — no longer is the motion in phase space incompressible. Furthermore, although a statistically sharp external forcing does not destroy incompressibility in phase space, a random external forcing will. Thus when considering the forced viscid problem we must give up the notion of an incompressible spreading out of the ensemble phase points through convoluted, filamentous streaming. One should then even question the validity of one choice of dynamical variables over any other. Noting the shaky ground, we shall adopt a pragmatic stance and proceed to examine the effect of random forcing and dissipation on the behavior of the information theory quantities already defined. For simplicity, only Gaussianly distributed white-noise (in time) forcing will be considered here. In a convenient abstract notation, the primitive field equations can be written as

$$\dot{y}_i = \Omega_i(y) - \nu_i y_i + f_i, \tag{26}$$

where Ω represents the nonlinear advective interactions, ν_i viscous dissipation, and f_i stochastic external forcing. If the forcing variance is written as

$$<f_i(t)f_j(t')> = \mu_{ij}\delta(t-t'), \tag{27}$$

where by definition μ_{ij} must be a nonnegative matrix, then the probability distribution of the y's averaged over the forcing ensemble is given by

$$\frac{\partial P}{\partial t} + \sum_i \frac{\partial}{\partial y_i}(\Omega_i(y)P) - \sum_i \nu_i \frac{\partial}{\partial y_i}(y_i P) - \sum_{ij} \mu_{ij} \frac{\partial^2}{\partial y_i \partial y_j} P = 0. \tag{28}$$

Equation (28) is derived from the conservation of total probability. The Gibbs entropy, however, is no longer a constant of the motion; its evolution is given by

$$\frac{d}{dt}\left(-\int P \ln P\right) = -\sum_i \nu_i + \sum_{ij} \mu_{ij} \int \frac{\partial P}{\partial y_i}\frac{\partial P}{\partial y_j} P^{-1} \prod_k dy_k. \tag{29}$$

The dissipation contributes a negative term which corresponds to a contraction of probability volume in phase space — viscosity tends to drive each member of an ensemble to the state of $y_i=0$, which tends to make the probability distribution infinitely sharp. The forcing contributes a positive term which implies a divergence in phase space — random forcing causes the probability density to diffuse, which tends to broaden the distribution.

We shall now examine the effects of forcing and viscosity on the Markovian closure predictions. For concreteness only two dimensional flow is considered in the rest of the equations — the generalization is straightforward. The energy spectrum evolves according to

$$\frac{dU_{\vec{k}}}{dt} = 2 \sum_{\substack{\vec{p}\vec{q} \\ \vec{k}+\vec{p}+\vec{q}=0}} \theta_{\vec{k}\vec{p}\vec{q}} \frac{\sin^2\alpha}{k^2} (p^2-q^2)(k^2-q^2)[U_{\vec{p}}U_{\vec{q}} - U_{\vec{k}}U_{\vec{q}}] -2\nu_k U_{\vec{k}} + F_{\vec{k}}, \tag{30}$$

and the error energy, $\Delta_{\vec{k}} = U_{\vec{k}} - W_{\vec{k}}$, evolves according to[10,11]

$$\frac{d\Delta_{\vec{k}}}{dt} = 2 \sum_{\substack{\vec{p}\vec{q} \\ \vec{k}+\vec{p}+\vec{q}=0}} \theta_{\vec{k}\vec{p}\vec{q}} \frac{\sin^2\alpha}{k^2} (p^2-q^2)(k^2-q^2)[\Delta_{\vec{p}}U_{\vec{q}} - U_{\vec{q}}\Delta_{\vec{k}} + W_{\vec{p}}\Delta_{\vec{q}}] -2\nu_k\Delta_{\vec{k}} + (F_{\vec{k}} - R_{\vec{k}}), \tag{31}$$

where F is the forcing variance and R is the forcing cross correlation. From these follows the Boltzmann type H-Theorem which takes the form:

$$\frac{dS}{dt} = NL - 2\sum_{\vec{k}}\nu_k + \sum_{\vec{k}} \frac{U_{\vec{k}}F_{\vec{k}} - W_{\vec{k}}R_{\vec{k}}}{U_{\vec{k}}^2 - W_{\vec{k}}^2}, \tag{32}$$

where NL is short hand for the complicated nonlinear terms. NL is written out elsewhere, here all we need note is that NL is nonnegative and vanishes only when $U_{\vec{k}} = U_{\vec{k}}^{eq}$ and $\Delta_{\vec{k}} = U_{\vec{k}}^{eq}$ or $\Delta_{\vec{k}}=0$. Note that the explicit contribution of the viscosity is precisely the same as for the Gibbs entropy; the contribution of the random forcing can be simply obtained from (29) by replacing the exact probability distribution by the coarse grained Gaussian approximate which is used in the definition of the experimental entropy. By the Schwartz inequality the contribution of the random forcing at each wavevector is positive. However, note that the partial contribution due to cross-correlations is negative (for positive streamfunction cross correlation). This makes intuitive sense — the more correlated the forcing the less it can destroy the predictability of the system, or, in other words, the more we know about the forcing the greater is our predictive capability.

For simplicity the forcing is taken to be perfectly correlated, $R_{\vec{k}} = F_{\vec{k}}$, that is, here attention is focused on the error increase due only to imprecise knowledge of the streamfunction. An interesting feature of this situation is that the error can never become complete. To see this substitute $\Delta_{\vec{k}} = U_{\vec{k}}$ into (31) and sum over all $\vec{k}$. Thus the total error *decreases* according to

$$\frac{d}{dt}\sum_{\vec{k}}\Delta_{\vec{k}} = -2\sum_{\vec{k}}\nu_k U_{\vec{k}}. \tag{33}$$

This implies that the completely decorrelated state must become correlated, and the incompletely correlated state cannot approach the state of complete correlation too closely.

To examine the effects of forcing and viscosity as cleanly as possible, we shall consider a forced viscid state with precisely the same energy spectra as the thermal equilibrium state. Here the net transfer of energy by nonlinear interactions vanishes and forcing in each mode is balanced exactly by linear dissipation. Thompson[15] showed that such a steady state exists when the forcing is given by

$$F_{\vec{k}} = 2\nu_k U_{\vec{k}}^{eq}, \tag{34}$$

with $U_{\vec{k}}^{eq}$ defined above. It is not clear to us whether this state is stable against arbitrary perturbations. From the closure we obtain

$$\sum_{\vec{k}} U_{\vec{k}}^2 \frac{d}{dt}(f_{\vec{k}}^2) = -2\sum_{\vec{k}} \frac{\nu_k f_{\vec{k}}^2}{U_{\vec{k}}^{eq} U_{\vec{k}}} \tag{35}$$

$$-\frac{1}{3} \sum_{\substack{\vec{k}\vec{p}\vec{q} \\ \vec{k}+\vec{p}+\vec{q}=0}} \theta_{\vec{k}\vec{p}\vec{q}} \frac{\sin^2\alpha}{k^2} U_{\vec{k}} U_{\vec{p}} U_{\vec{q}} \left[(p^2-q^2)f_{\vec{k}} + (k^2-q^2)f_{\vec{p}} + (p^2-k^2)f_{\vec{q}}\right]^2 .$$

where α is the angle between $\vec{p}$ and $\vec{q}$ and

$$f_{\vec{k}} = \frac{\delta U_{\vec{k}}}{U_{\vec{k}}^{eq} U_{\vec{k}}}, \tag{36}$$

or in terms of $\delta U/(UU^{eq})$ we have the inequality

$$\sum_{\vec{k}} \left(1 + \frac{\delta U}{U^{eq}} + \left(\frac{\delta U}{U^{eq}}\right)^2\right) \frac{d}{dt}(\delta U)^2 \leqslant 0. \tag{37}$$

This does not prove that all perturbations must decay; however, it does show that small perturbations must decay (i.e., linear stability).

Now note that in the Thompson state, if the error is complete (i.e., $\Delta_{\vec{k}} = U_{\vec{k}}^{eq}$ for all $\vec{k}$), then we have

$$\dot{\Delta}_{\vec{k}} = -2\nu_k U_{\vec{k}}^{eq}. \tag{38}$$

This is a much stronger result than (33); here the correlations in the forcing keeps *each* mode from becoming totally uncorrelated; or if uncorrelated it causes it to become correlated. A possible steady state of this situation is $\Delta_{\vec{k}}=0$ for all $\vec{k}$. We do not at this time have an analytic determination of the conditions necessary to achieve this state of complete correlation.

Among the several test cases we are considering is the case in which the streamfunctions are initially completely uncorrelated. In Figure 3 we display our results obtained so far. Both the error entropy and total error energy are decaying more or less monotonically. This may suggest asymptotic evolution to a state of complete correlation.

SUMMARY AND FUTURE WORK.

This paper has been concerned with applying a few of the ideas of statistical mechanics to the "predictability problem" of fluid mechanics. In order to keep the fluid dynamical problem as simple as possible only two of the simplest possible turbulent states have been examined — an unforced inviscid state of thermal equilibrium and a forced viscid "Thompson" state of precisely the same energy spectra.

We have first shown that the maximum entropy state is indeed approached by unforced, inviscid flow and that the energy spectra and entropy of this state may be predicted by the methods of equilibrium statistical mechanics with some degree of accuracy. Next we performed some predictability experiments on the thermal equilibrium state. The final entropy of the two field system is accurately predicted. Furthermore, the H-theorems following from the closure theories tell us that this state is monotonically approached, implying a monotonic loss of predictability which is intuitively very satisfying.

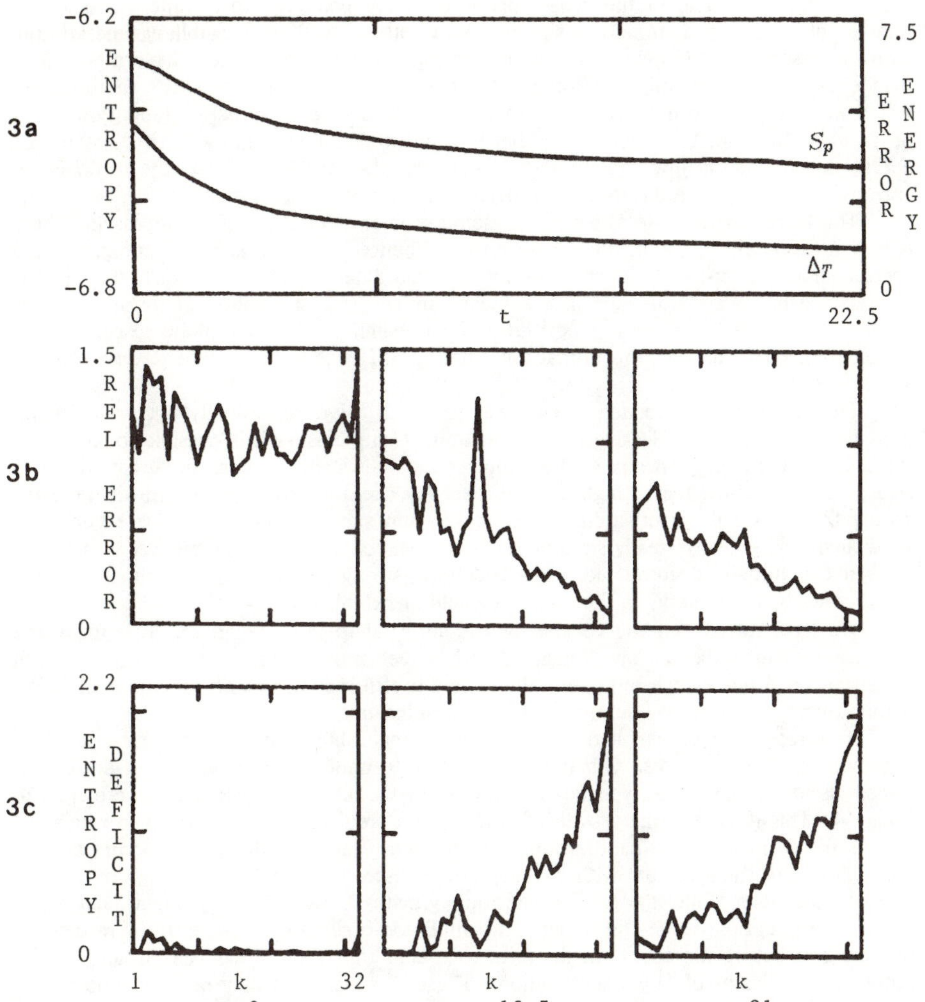

Fig. 3. Predictability: forced, viscid (Thompson state).

(3a) The evolution of the total error energy (in units of 10^{-4}) and predictability entropy (in units of 10^{4}). The smallest eddy-turnover time is roughly 0.5.

(3b) Relative error spectra, $\Delta(k)/E(k)^{eq}$, at times t=0,10.5 & 21.

(3c) Entropy deficit $(\ln E^2(k) - \ln(E^2(k) - \Delta^2(k))$ at t=0,10.5 & 21.

However, there are some rather disquieting aspects involved with entropy. First, it is not a "rugged" variable in the sense that it does not converge to a finite value as the number of modes in a numerical simulation increases. And, as a corollary, because the entropy prescription (9) gives equal weight to high wavenumbers, these have more effect than energetically they might deserve. Hence in terms of a predictability experiment in which we consider two initially almost identical fields diverging, initial entropy growth will be rapid as high wavenumber modes become decorrelated very quickly. If most of the energy is contained at low wavenumbers (as in typical atmospheric flows) the *usefulness* of entropy as quantitative indicator of flow decorrelation must be questionable.

The forced viscid state also presents some problems for the entropy approach. There is now no Liouville theorem for the system, and hence any information is ambiguous and coordinate dependent. However, consider the coordinate system for which the viscous terms are a linear addition. Then any linear similarity transformation of dynamical variables will still preserve the rate of evolution of probability volume in phase space, and the viscous terms will cause contraction at the same rate. Hence to this extent we may proceed as before.

The addition of correlated forcing was found to drive two initially uncorrelated systems toward a state of at least partial correlation. Using closure it is possible to show that the shape of the energy spectra of the Thompson state is stable to small perturbations. It is also possible to show that a high enough viscosity can always damp the growth the error arising from a small perturbation to the equilibrium state. However, we were unable to show analytically if any steady state of partial correlation can exist. Numerical simulations showed two initially decorrelated states becoming strongly correlated with time, as measured either by total energy in the difference field or by entropy.

We have introduced the concept of the surprisal to fluid dynamics. Essentially the surprisal measures the entropy deficit in each waveband. Although the entropy in each waveband need not evolve monotonically, it is a useful measure of one's surplus of information over that given by assuming canonical equilibrium.

If entropy is to be useful then it must have applicability outside the narrow range of canonical equilibrium states. Our next task will be to understand the growth of entropy in forced-viscid inertial range type flows. The problems here are illustrated by reference to Figure 4. This illustrates the growth of entropy in a predictability experiment in which the fluid is forced in a narrow band around wavenumber 7 and the dissipation occurs both at the high end of the spectrum with a fourth power viscosity and with a Rayleigh drag (after Vallis[16], see also Holloway[17]). The entropy growth is initially very rapid, apparently overshooting its final value and undergoing amplitude oscillations. The entropy reaches its first maxima before the energy in the difference field (not shown) has peaked, and here is not a good indicator of flow decorrelation. These disadvantages may be overcome by the use of entropy spectra or surprisals. There is no H-theorem for each surprisal, and hence the theoretical advantages over simply using the error energy spectra would seem moot. On the other hand, studying the entropy evolution may help to disentangle the complicated effects of viscosity, forcing, and nonlinearity. A full understanding will probably require a combination of explicit simulations, closure integrations and analysis.

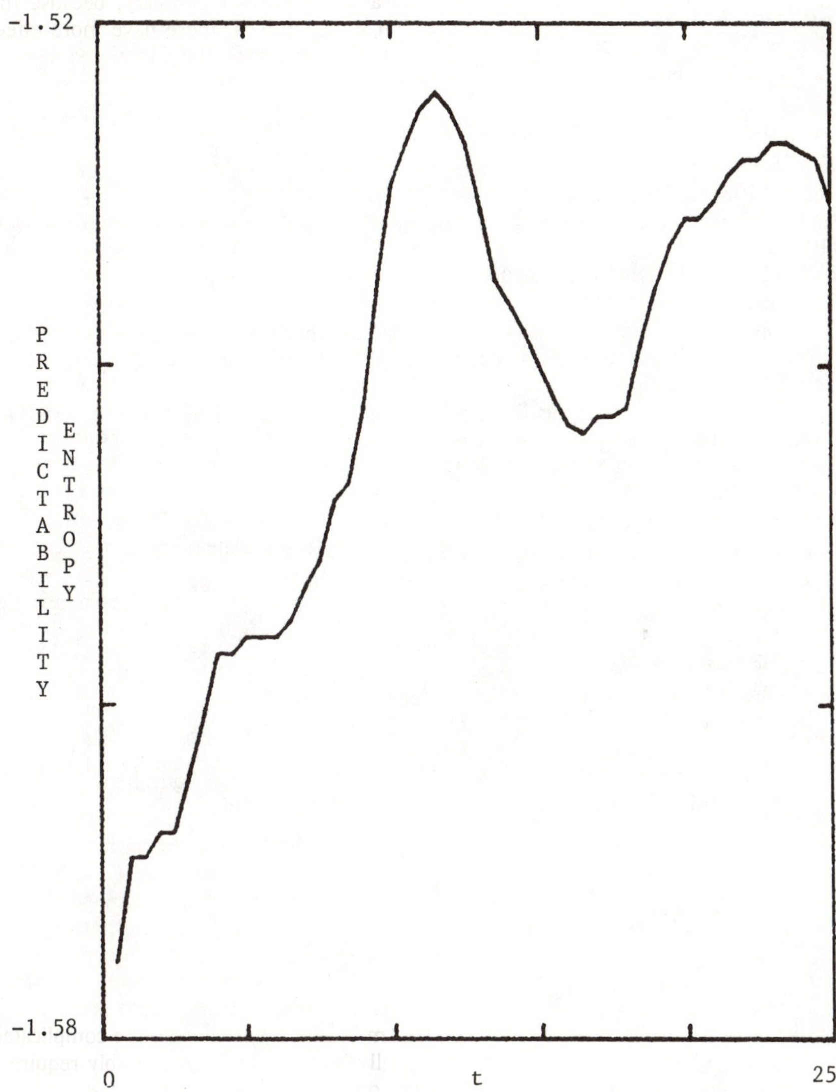

Fig. 4. Predictability: forced, viscid, nonequilibrium.

The entropy (in units of 10^4) varies widely as nonlinear, viscous and forcing terms struggle for balance. The mean eddy turnover time based on the r.m.s. velocity is roughly 0.5.

ACKNOWLEDGEMENTS

We thank Michael Tabor for interesting suggestions and specifically for bringing to our attention the use of surprisals in statistical physics. This work is supported in part by the National Science Foundation (ATM 82-10160).

REFERENCES

1. A. I. Khinchin, "Mathematical Foundations of Information Theory." (Dover Publications), (1957).
2. R. C. Tolman, "The Principles of Statistical Mechanics." (Oxford University Press), (1938).
3. E. T. Jaynes, Phys. Rev., **106**, 620 (1957).
4. E. T. Jaynes, Amer. J. Phys., **33**, 391 (1965).
5. E. T. Jaynes, "Information theory and statistical mechanics", in *Statistical Physics*, (1962 Brandeis Theoretical Physics Lectures, Vol. 3), ed. K.W. Ford (Benjamin Inc., New York), (1963).
6. G. F. Carnevale, U. Frisch and R. Salmon. J. Phys. A: Math. Gen., **14**, 1701 (1981).
7. R. Betchov, Phys. Fluids, **7**, 1160, (1964).
8. G. F. Carnevale, J. Fluid Mech., **122**, 143 (1982).
9. R. H. Kraichnan, Phys. Fluids, **10**, 1417 (1967).
10. R. H. Kraichnan, Phys. Fluids, **13**, 569 (1970).
11. C. E. Leith and R. H. Kraichnan, J. Atmos. Sci., **29**, 1041, (1972).
12. G. F. Carnevale and G. Holloway, J. Fluid Mech., **116**, 115, (1982).
13. M. Tribus, "Thermostatics and Thermodynamics", (Van Nostrand), (1961).
14. R. B. Bernstein and R. D. Levine, J. Chem. Phys., **57**, 434, (1972); R. D. Levine and R. B. Bernstein, Acc. Chem. Res., **7**, 393,(1974); I. Procaccia, Y. Shimoni and R. D. Levine, J. Chem. Phys., **65**, 3284 (1976).
15. P. D. Thompson, J. Fluid Mech., **55**, 711 (1972).
16. G. Vallis, J. Atmos. Sci., **40**, 10, (1983).
17. G. Holloway, "Effects of Planetary Wave Propagation and Finite Equivalent Depth on the Predictability of Atmospheres." J. Atmos. Sci., **40**, 314 (1983).

CONTRARY ROLES OF PLANETARY WAVE PROPAGATION IN ATMOSPHERIC PREDICTABILITY

Greg Holloway
Institute of Ocean Sciences
Sidney, BC, Canada V8L 4B2

ABSTRACT

Depending upon how one interprets "predictability," planetary wave propagation may play opposing roles. In one interpretation, a difference field between two evolving flow realizations is considered. Initially confined to small scales, differences grow in scale and in amplitude until the two realizations become nearly as different as two randomly selected realizations. Predictability is thereby lost. Studies of this process in cases of barotropic motion show that planetary wave propagation inhibits the growth of difference fields, thereby *enhancing* predictability.

A alternative view is suggested by studies of uniformly rotating, decaying barotropic turbulence in which one observes the spontaneous formation of relatively isolated, long-lived vortices. The persistence of these vortices suggests a useful predictability for times longer than those times implied by measures of difference fields. However, when a mean potential vorticity gradient is imposed, planetary wave propagation may effectively prevent the formation of such isolated vortices, thereby *preventing* a kind of predictability associated with long-lived flow features.

In each case, various points of controversy are discussed.

WAVES MAY ENHANCE PREDICTABILITY

Among the many processes which affect the predictability of atmospheres and oceans, the propagation of planetary, or Rossby, waves has recently attracted a good deal of attention. Idealizing atmospheric dynamics in terms of turbulent interactions among many scales of motion, Leith[1] suggested that inclusion of Rossby wave propagation might significantly affect rates of growth of error due to initial uncertainty of observations. Subsequent numerical investigations[2-4] as well as statistical closure theory[5] have shown that, indeed, for the case of barotropic motion, wave propagation significantly inhibits error growth and thus enhances predictability.

For baroclinic motion, the situation is less clear. Numerical simulations[3,6] of layered flow in the $\beta-$ plane approximation show that predictability characteristics are not significantly affected by increasing β. At least in part, this may be due to the role of β modifying the processes of release of mean potential energy into eddy kinetic energy[7]. However, for the remainder of this note, we return to the strictly barotropic case. Even in the barotropic case, some interesting contradictions will appear.

Using conventional measures of error such as kinetic energy of the difference field, several studies [2-5] have shown clearly that *increasing* β gives *increased* predictability, i.e. slower growth of difference kinetic energy. Although such increase of predictability is apparent [2-5] in a overall sense, there are discrepancies in certain details.

Decomposing total error energy (difference kinetic energy) into its wavenumber spectrum, Vallis[3,7] observes from numerical experiments that the relative error in the gravest modes increases when β is present. That is, the very longest scales of motion ($k = 1$) in Vallis' experiments are *less* predictable when β is present. Relative error here refers to the quotient of error energy divided by the average of the energy of each realization.

Numerical experiments similar to those of Vallis have been performed by Holloway[4] for the case of equivalent barotropic flow, i.e. one-layer flow including the effect of a free surface (finite deformation radius). However, even in the case of rigid surface (infinite deformation radius), Holloway's experiments do not seem to exhibit consistently the anomalous loss of predictability at k = 1.

One possible explanation may be that numerical simulations, as direct realizations of turbulent flows, will exhibit some degree of variability among experiments. Especially at very low wavenumbers, few degrees of freedom will be included in any wavenumber band-averaged quantity. The statistical reliability of a few numerical experiments may be questioned. An alternative explanation as advanced by Vallis is that his calculation[3] included a fairly large ground friction, resulting in smaller energy at small k as compared with Holloway[4]. Thus, for a given amount of error energy, larger relative error (i.e. normalized by flow energy) would be reported by Vallis.

Another discrepancy occurs with regard to the anisotropy of error growth at small k Vallis[3,7] observes that predictability of the *zonal* flow is increased when β is present. This may be due to production of strong, steady zonal jets on the $\beta-$ plane[8]. Holloway finds the opposite, that error more readily invades the zonal flow so that it is *meridional* flow at small k which is more predictable. The latter result could be explained by arguing that the higher Rossby wave frequencies associated with meridional flows will tend to decouple these components from the error invading interactions. Again, questions of statistical significance as well as the shape of energy spectrum at small k as dependent on ground friction both appear as possible explanations for these differences.

The problem of statistical reliability of numerical experiments can be avoided by examining these questions from the view of statistical closure theory. (Nonetheless, uncertainty is conserved with the assumption of dubious[18] closure hypotheses!) In their Figure 2, Litherland and Holloway[5] show no anomalous loss of predictability at k = 1. However, for this calculation the assumed energy spectrum is more like Holloway's[4] than Vallis'[3]. An examination of closure theoretical results when the energy spectrum is depressed at small k remains to be done.

Anisotropy of error evolution is also calculated by closure theory and is seen in Figure 1. Shown in this figure is the average rate of change of error. A "lazy 8" contour at small k shows error invading along the k_y - axis (zonal flow) while avoiding the k_x- axis (meridional flow).

Although there are minor points of discrepancy, the main result[2-5] seems quite clear. For barotropic flows and using conventional predictability measures such as kinetic energy of the difference field, a larger tendency for *wave propagation* (viz. larger β) *enhances predictability* by suppressing the rates of growth of difference or error fields. I emphasize this agreement "in the main" because, in the next section, we will see just a contrary indication.

WAVES MAY DEGRADE PREDICTABILITY

Such conventional predictability measures as error energy can be misleading, especially in flows which exhibit a tendency toward intense, localized features. Anthes[9] provides a graphic example in which a modest displacement error in the location of a cyclone yields quite large error kinetic energy, suggesting almost total loss of predictability, whereas practically such a forecast could be considered useful. There would seem to be a need which, to my knowledge, has not been satisfied quantitatively; that is to describe predictability in terms of the persistence of recognizable flow "structures" or features. This question can be divided into two parts: (1) the forecast that a feature simply will persist, and (2) a forecast for the space-time trajectory of the feature.

The existence and stability of isolated features in two-dimensional flows have been the objects of several studies[10-12] which have assumed a particular flow structure as an initial condition. However, we may be especially interested in situations in which more-or-less

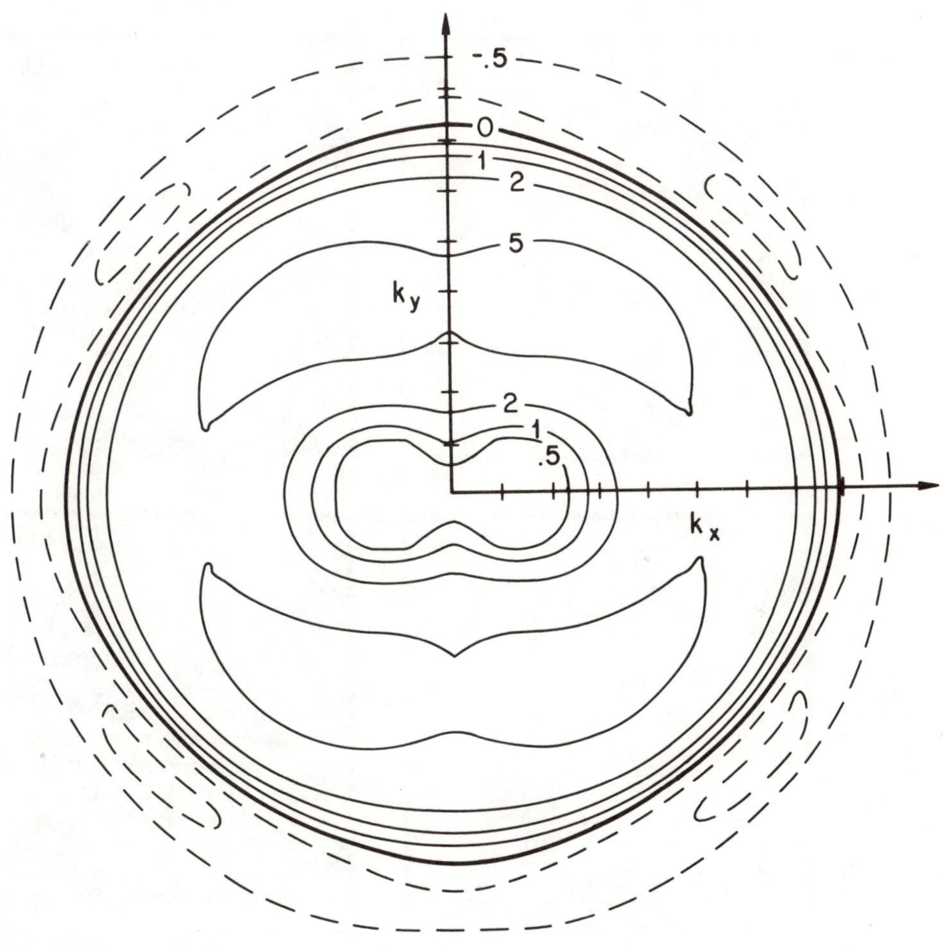

Fig. 1. Contours of error energy change in k_x,k_y are shown. β is such that $k_\beta \equiv \beta/\zeta_{rms} \approx 2$ where ζ_{rms} is r.m.s. vorticity. Rigid surface (infinite deformation radius) is assumed.

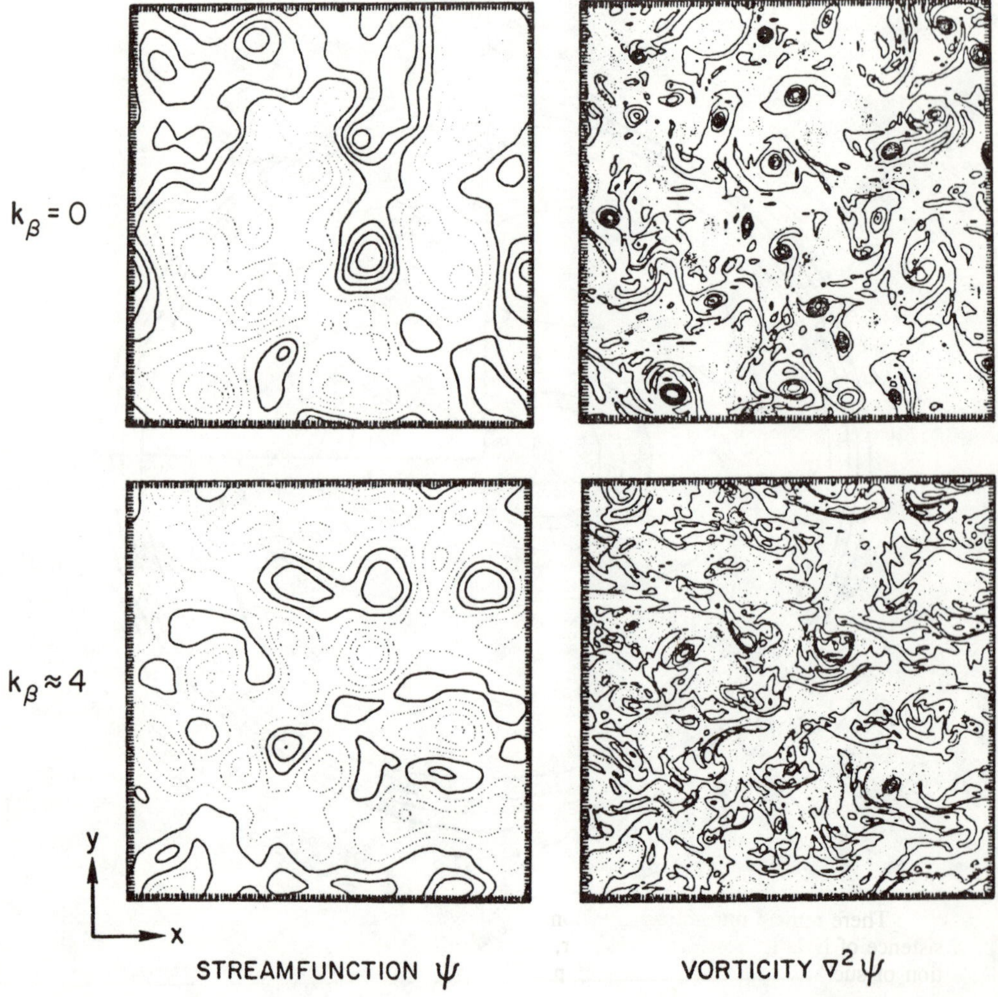

Fig. 2. Fields of streamfunction and of vorticity are shown at comparable stages of decay in two cases. Top panels, $k_\beta = 0$; bottom panels, $k_\beta \approx 4$. A rigid surface is assumed. Resolution corresponds to 128 x 128 points.

isolated features appear spontaneously and persist. An example is the formation of long-lived, axisymmetric vortices in decaying two-dimensional turbulence[13] ($\beta = 0$). See particularly the extensive investigation by McWilliams[14] in which individual vortices survive for times quite long compared with conventional predictability time for two-dimensional turbulence[1]. It is not known, however, to what extent vortex trajectories are predictable.

Recognizing that there is a kind of long term predictability given by the persistence of isolated vortices, we inquire to what extent wave propagation ($\beta \neq 0$) affects this formation and persistence. On the one hand, a number of persistent solutions[10-12] are known on the β plane; on the other hand, we may expect dephasing associated with wave propagation to interfere with the formation process. Indeed, the role of β in preventing formation of isolated vortices is seen in Figure 2 from an early study by Holloway[15]. In terms of formation of persistent features, wave propagation here *degrades* predictability.

To quantify the role of β in preventing vortex formation, I have performed a number of numerical simulations each from initial conditions consisting of a broad spectrum, random phase vorticity field. A mean energetic wavenumber characterizing the vorticity field is initially near $\bar{k} \approx 5$, decreasing during the decay of the flow toward $\bar{k} \approx 2$ to 3. The method of simulation is dealiased pseudospectral[16] and resolution corresponds to 128 x 128 grid points. The flow decays in the presence of a high wavenumber (biharmonic) dissipation operator. For details see Holloway[4].

Adopting vorticity kurtosis $Q = \overline{\zeta^4} / (\overline{\zeta^2})^2$, where ζ is vorticity and overbars denote areal average, as a measure of tendency toward isolated vortices, we see time evolution of Q plotted in Figure 3. Five values of β are taken corresponding to $k_\beta \equiv \beta / \zeta_{rms} \approx 0, .5, 1, 2, 4$, where ζ_{rms} is the initial r.m.s. vorticity. Whereas it might be expected that β is effective in suppressing Q when $k_\beta \geqslant \bar{k}$, we observe a result that, apparently, β is effective when $k_\beta \geqslant 1$. Thus, β prevents isolated vortex formation if only the longest resolved waves feel the influence of β significantly. The observed scales of vortices which form when $\beta = 0$ are a good deal smaller than the longest resolved waves, viz. Figure 2.

It is unclear why β should be so effective. For $\beta = 0$, vortices are seen to form although large scale gradients of *relative* vorticity are present. Recall too that isolated vortices of particular forms, if present initially, many persist[10-12] at larger β.

It is also curious that dissipation, acting only at very small scales, must be essential. Although the *form* of dissipation may not be crucial[14], the presence of some nonconservative effect is crucial; otherwise an equation of motion which is just advection of vorticity should conserve Q as an integral of the motion. (In finite numerical representation, such ideal integrals of the motion may not be preserved. Q will only be quasi-conserved. However, for finitely resolved, non-dissipative, two-dimensional flow, we expect[17] flow statistics to approach at long times a maximum entropy solution giving $Q \approx 3$. I have performed some experiments with $\beta = 0$ and including small scale dissipation. Q grows as expected. At later time, dissipation is set identically to zero. Q growth then ceases although relaxation toward $Q \approx 3$ was not observed during the limited duration of the experiment.)

There remain unresolved questions concerning the role of β in the formation and persistence of isolated vortices. However, the clear result is that larger β prevents the formation of such vortices from a random phase initial vorticity field. To the extent that spontaneous formation of long-lived features may provide some useful kind of predictability, we have an example where *wave propagation degrades predictability.* This difference "in the main" from the previous section depends upon how one formulates the question of predictability.

ACKNOWLEDGMENTS

I am grateful for stimulating discussions with Jackson Herring, Jim McWilliams and Geoff Vallis. Figure 1 is taken from calculations by John Litherland. Figures 2 and 3 are

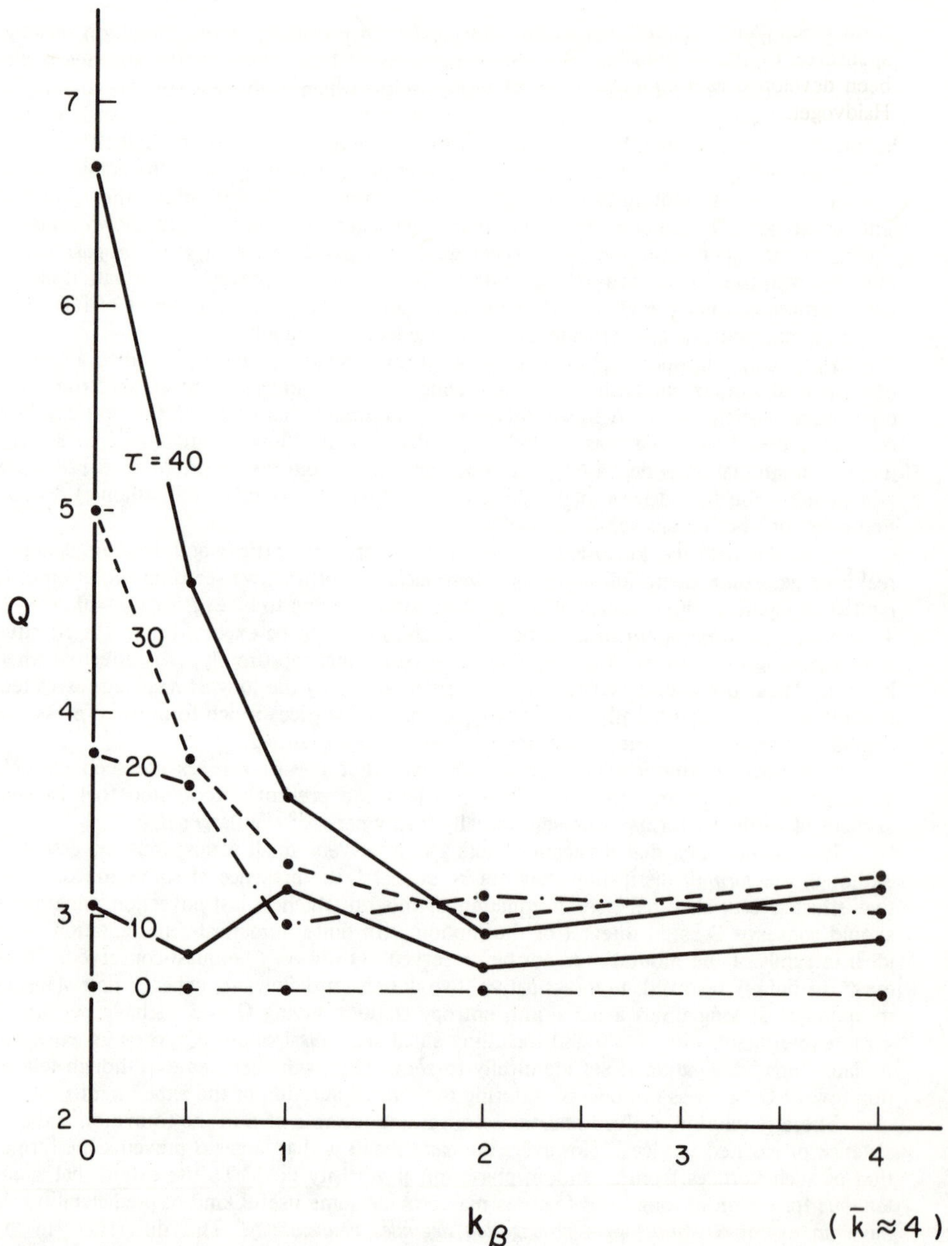

Fig. 3. Evolution in time of vorticity kurtosis $Q \equiv \overline{\zeta^4}/(\overline{\zeta^2})^2$ is plotted as a function of k_β. Curves are labeled by nominal non-dimensional time $\tau \approx t\zeta_{rms}$. Each experiment was begun from the same random phase vorticity field with $Q = 2.64 \approx 3$. Only for $k_\beta \leqslant 1$ is significant growth of $Q > 3$ evident.

from computations performed at the National Center for Atmospheric Research which is sponsored by the National Science Foundation. Programs for numerical simulation have been developed as a community effort including participation of Michael Davey and Dale Haidvogel.

REFERENCES

1. C.E. Leith, J. Atmos. Sci., **28**, 145 (1971).
2. C. Basdevant, B. Legras, R. Sadourny and M. Beland, J. Atmos. Sci., **38**, 2305 (1981).
3. G.K. Vallis, J. Atmos. Sci., **40**, 10 (1983).
4. G. Holloway, J. Atmos. Sci., **40**, 314 (1983).
5. J. Litherland and G. Holloway, in these proceedings (1983).
6. J.C. McWilliams and J.H.S. Chow, J. Phys. Oceanogr., **11**, 921 (1981).
7. G.K. Vallis, in these proceedings (1983).
8. R.B. Rhines, J. Fluid Mech., **69**, 417 (1975).
9. R. Anthes, in these proceedings (1983).
10. G.R. Flierl, B.D. Larichev, J.C. McWilliams and G.M. Reznik, Dyn. Atmos. Oceans, **5**, 1 (1980).
11. P.M. Rizzoli and M.C. Hendershott, Dyn. Atmos. Oceans, **4**, 247 (1980).
12. J.C. McWilliams and G. Flierl, J. Phys. Oceanogr., **9**, 1155 (1979).
13. B. Fornberg, J. Comput. Phys., **25**, 1 (1977).
14. J.C. McWilliams, in these proceedings (1983).
15. G. Holloway, PhD. thesis, University of California at San Diego (1976).
16. S.A. Orszag, Stud. Appl. Math., **50**, 297 (1971).
17. G. Holloway and S.S. Kristmannsson, J. Fluid Mech., submitted (1983).
18. J.R. Herring, in these proceedings (1983).

PREDICTABILITY OF THE MARINE PLANKTONIC ECOSYSTEM

Kenneth Denman
Institute of Ocean Sciences
Sidney, B.C., Canada VBL 4B2

Predictability studies in geophysical fluid dynamics have focused on the study of fluid motions in the context of nonlinear dissipative systems[1]. In particular, studies of low order dynamic systems have been used to develop our analytical tools and intuition for the behavior of more complex nonlinear systems[2]. Lorenz[3] demonstrated how a simple 3 component system could possess attractor states: the system was attracted to certain domains of state variables, but the cyclic behavior was aperiodic and not predictable. In an analogous manner, May[4] showed how the simplest nonlinear difference equations describing the growth of biological populations can exhibit behavior with cycles of any period or even aperiodic but bounded cycles, the so called *chaotic* regime. Coupled nonlinear equations describing predator-prey interactions can develop spatial pattern through a process of dissipative instability[5], which has been invoked as a possible mechanism for development of patchiness in oceanic plankton distributions[6]. Prediction of ecosystem behavior then is very similar to prediction of geophysical fluid motion: both involve noisy, nonconservative, nonlinear systems with many degrees of freedom.

The previous studies[1-5] have not considered the complexity and predictability of a marine ecosystem that is subjected to advection by oceanic motions with a wide range of time and space scales. The problem of biological prediction in the sea has been considered previously[7,8], but under the assumption that all unsteady currents could be parameterized by a scale dependent horizontal turbulent eddy diffusion coefficient. Mesoscale eddies cannot be represented by such a coefficient and we do not yet know whether simple parameterizations will be possible[10]. We do expect however that the time scale for the enstrophy cascade associated with mesoscale eddies (~10 days) is comparable with time scales for the doubling of planktonic biomass due to growth and reproduction (~1 day for phytoplankton to ~1 month for zooplankton), indicating that eddy motions should affect the formation of biological spatial pattern.

In the terminology of tracers in turbulent flow, phytoplankton (the photosynthesizers of the sea) are passive but non-conservative scalar tracers: they grow, die, and are eaten by zooplankton and fish. Because the time scale for their growth is comparable with the time scale for the enstrophy cascade associated with quasi two dimensional mesoscale eddies, one would expect intuitively the spectrum of their spatial variance to differ from that for a conserved scalar. Observations of ocean color from satellite, from which are inferred phytoplankton pigment concentrations, give roughly a -3 power law spectrum for the spatial pattern[9]. However, this is inconsistent with present theories[12] for passive scalar variance in two dimensional turbulent flow regardless whether the scalar is conserved or not.

To study the effects of oceanic mesoscale eddies on the spatial patterns and the predictability of the planktonic ecosystem, we have initiated numerical simulations with a two dimensional turbulence model. The models are pseudospectral and doubly periodic and have already been used for tracer studies[10]. In its simplest form, the conservation equation for phytoplankton biomass P is

$$\frac{\partial P}{\partial t}+J(\Psi,P)=rP+D_pP \tag{1}$$

where r is the net growth rate of the phytoplankton, D_p is a linear parameterization of subgrid scale diffusion, and $J(\Psi, P)$ is a Jacobian representing the advective distortion of the scalar field by the vorticity field (Ψ = streamfunction) which is simulated from solution of a similar conservation equation. Taking the diffusive and advective terms as the familiar K $\nabla^2 P$ and $u.\nabla$ P (where K is a coefficient for subgrid scale diffusion), we have an equation which has been considered previously in the same context[11]. However, provided that the growth rate r is a function of time only, a substitution in Equation (1) of the form

$$Y(t)=P(t)\exp(-\int_0^t r(t')dt') \quad (2)$$

yields an equation for Y(t) which is just the conservation equation for a passive *conserved* scalar. In words, if the growth rate r is only a function of time but is constant in space, the biological and physical systems are coupled only linearly, and the growth of the biological scalar does not modify the shape of its variance spectrum from that of a purely conserved scalar tracer.

Thus, the complexity of the system will be increased by the existence of biological growth rates only if they are spatially varying. Spatial pattern in the net growth rate can result from spatially variable nutrient supply, temperature, grazing by zooplankton, etc. As an example, the growth term rP in (1) can be written[7,8] as

$$rP = PP_m\exp(-bI)(1-\exp(-aI)) + R_m(1-\exp(-c(P-P_o))) \quad (3)$$

where P_m, R_m, P_o, a, b, c are all "constants." The first term on the right represents the growth rate dependence on incident light I, and the second term represents the rate of grazing by zooplankton which depends on the abundance of phytoplankton. These functional relationships are nonlinear. We have omitted dependence on temperature and nutrient supply. The "constants" themselves vary in time, adapting to the other variables. Thus, both the nonlinear coupling between the biological and the physical dynamics and the time varying "constants" may cause biological variables to be even less predictable than physical variables. Specification both of the form of these functional relationships and of realistic autocorrelation functions in time and coefficients is necessary before we can evaluate their role in modifying the spatial pattern of phytoplankton from that expected for conserved tracers.

ACKNOWLEDGMENT

Encouragement and helpful conversations with G. Holloway are gratefully acknowledged.

REFERENCES

1. O.E. Lanford, Ann. Rev. Fluid Mech. **14**, 347 (1982).
2. G. Nicolis, p. 185, in Nonlinear Phenomena in Physics and Biology edited by R. Enns, W. Jones, R. Miura, and S. Rangnekar (Plenum, N.Y., 1981).
3. E.N. Lorenz, J. Atmos. Sci. **20**, 130 (1963).
4. R.M. May, Science **186**, 645 (1974).
5. L.A. Segel and J.L. Jackson, J. Theor. Biol. **37**, 545 (1972)
6. S.A. Levin and L.A. Segel, Nature **259**, 659 (1976).
7. K.L. Denman and T. Platt, p. 251, in Modelling and Prediction of the Upper Layers of the Ocean edited by E. Kraus (Pergamon, Oxford, 1977).
8. T. Platt, K.L. Denman, and A.D. Jassby, p. 807, in The Sea, Vol. 6 edited by E. D. Goldberg (John Wiley, N.Y., 1977).
9. J.F. Gower, K.L. Denman, and R.J. Holyer, Nature **288**, 157 (1980).
10. G. Holloway and S.S. Kristmannsson, submitted to J. Fluid Mech.
11. K.L. Denman, A. Okubo, and T. Platt, Limnol. Oceanogr. **22**, 1033 (1977).
12. M. Lesieur, J. Sommeria and G. Holloway, Comptes-Rendu, **II-20** (1981).

PREDICTABILITY IN THE WAVENUMBER-FREQUENCY DOMAIN[1]

Joseph J. Tribbia
National Center for Atmospheric Research[2]
Boulder, Colorado 80307

INTRODUCTION

Predictability has historically been defined in terms of the time it takes for two nearly identical initial states to depart from each other to the extent that the difference between the two states is that which might be expected from the random selection of two states from the climatological ensemble. This time, denoted the predictability time, can differ for differing dependent variables within the system. The rationale behind space-time filtering is to maximize the predictability time for dynamically interesting variables through the use of specified linear transformations, Fourier decomposition in space and band-pass frequency decomposition in time.

The mechanism through which spatial filtering acts to increase the predictability time for large scales has been understood since the earliest predictability experiments of Leith and Kraichnan[1], i.e. large scales are contaminated primarily through the inverse cascade of error energy which takes a longer time for larger scales. The mechanism through which temporal filtering increases the predictability time for low-pass filtered fields is rather different and may be envisioned as follows: rapid phase decorrelation of the two initial states are prohibited by the filter and thus the primary exponential growth of error energy is the primary mechanism leading to loss of predictive skill. This relative gain of predictability due to the elimination of phase decorrelation must be balanced against the reduction of climatological variance due to the projection transformation. The purpose of the present study is to investigate the trade-off between these competing efforts in a simple climate model which generates quasi-geostrophic turbulence.

EXPERIMENTAL DESIGN

As mentioned above, the model used here is the simplest model which allows a realistic generation of quasi-two-dimensional turbulence: the two-layer quasi-geostrophic model. The domain is spherical in order to allow "realistic" spatial scales and topographic effects.

1 Presented at the Workshop on the Predictability of Fluid Motions, La Jolla Institute, Scripps Institute of Oceanography, La Jolla, CA 92038, February 1983.

2 The National Center for Atmospheric Research is sponsored by the National Science Foundation.

The governing equations are then:

$$\xi_t + J(\psi,\xi+f) + J(\tau,q) = D_m + F_m$$

$$q_t + J(\psi,q) + J(\tau,\xi+f) = D_s + F_s$$

where ψ is the barotropic streamfunction ($\psi = (\psi_u+\psi_L)/2$), τ is the baroclinic streamfunction ($\tau = (\psi_u-\psi_L)/2$), $\xi = \nabla^2\psi$, $q = (\nabla^2-r^2)q$, $r^2 = f^2/gH_s$, $H_s = 1$ km, and F_m and F_s (D_m and D_s) are the barotropic and baroclinic forcing (dissipative) terms further explained below. The forcing is taken to be representing two physical processes of radiative heating (through the use of Newtonian relaxation to a zonally symmetric tempeature) and mechanical topographic forcing; thus $F_m = -J(\psi_4, fh/H_o)$ and $F_s = +J(\psi_4, fh/H_o) + r^2(\tau-\tau_E)/T_R$ with $\tau_E = -\alpha_E \sin\phi$. The dissipative mechanisms are also of two types, surface Ekman drag and an eighth order eddy diffusion parameterization so that $D_m = -\nabla^8\xi - \hat{c}_d(\xi - 2\nabla^2\tau)$ and $D_s = -\nu\nabla^8 q - \hat{c}_d(2\nabla^2\tau - \xi)$. The various constants in the above terms are listed in Table 1. Lastly, the equations are solved globally using the spectral transform method truncated at T31.

The model is integrated for 100 days starting from the forced zonally symmetric flow plus a small random perturbation in order to spinup to a statistically steady climatology, as deduced from quasi-steady total energetics. The model is then integrated from 180 days with fixed external parameters for the control experiment (an ensemble of three control integrations of 60 days' duration each). At 60-day intervals, random perturbations of RMS amplitude of 1m/sec and spectrally distributed to linearly increase with two-dimensional wavenumber, are introduced into both the barotropic and baroclinic portions of the flow; the 60-day integrations of these initial states constituted the perturbed ensemble of forecasts. The steady part (180-day average of the control experiment) is removed from both the control ensemble and the perturbed integrations and then both the control and perturbed fields are temporally filtered using the 31 point bandpass filters introduced by Blackmon (1972). These filters use data at twelve hour intervals and have response characteristics shown in Fig. 1. As shown, the temporal variability is broken up into three frequency bands, low frequency (periods $\gtrsim$ 10 days), medium or synoptic frequency (2.5 days $\lesssim$ period $\lesssim$ 5 days), and high frequency (period $\lesssim$ 2 days). The ensemble of control and perturbed forecasts (three realizations each) are averaged and the results are decomposed two-dimensional wavenumber and freauency band.

Table 1 - Model Constraints

$TR = (10 \text{ days})^{-1}$	$C_d = (10 \text{ days})^{-1}$
$\alpha_E = 1.0 \times 10^8 m^2/sec$	$\nu = 5.9 \times 10^{35} m^9/sec$
$H_o = 10$ km	

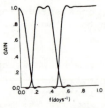

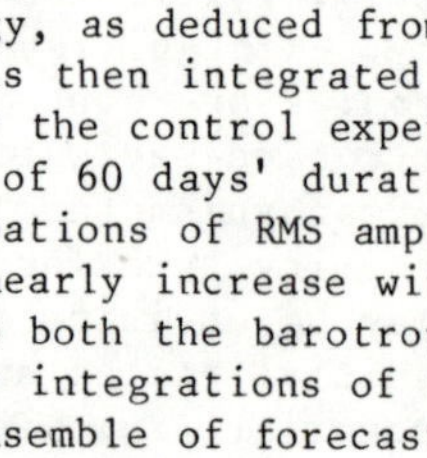

Figure 1 - Gain vs. frequency of Blackmon Filters

RESULTS

The necessary first order of business in presenting the results of the present study is the documentation of the ensemble average climatology of the model both in the frequency domain and in wavenumber decomposition. Figure 2 illustrates the ensemble average of the 60-day average of the total barotropic kinetic energy and the total baroclinic kinetic plus available potential energy as a function of two-dimensional wavenumber. Also shown in the figure are the low, medium and high frequency decomposition of these spectra. By and large these are as might be anticipated with the low and synoptic frequencies predominent in the low wavenumber variance and the synoptic and high frequencies accounting for the majority of the variance in the smaller scales for both the barotropic and baroclinic modes.

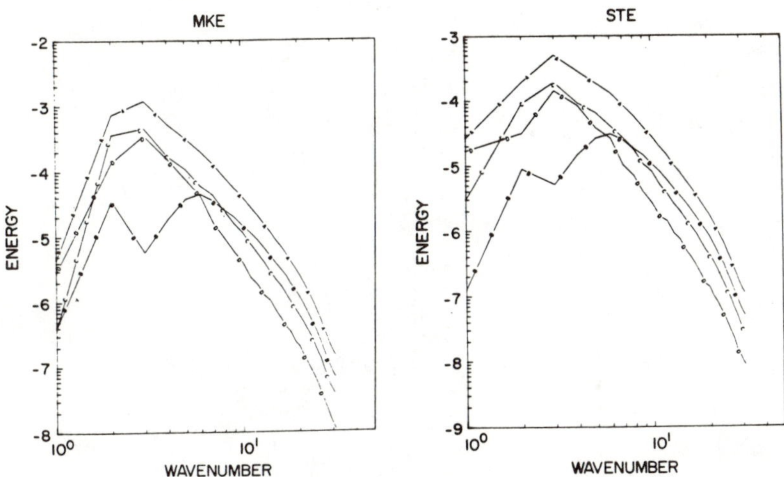

Figure 2: Climatological spectral distribution of barotropic (left) and baroclinic (right) energies, curves intersect the wavenumber 1 axis in the following order from top: Total, low frequency, synotpic frequency, and high frequency energy.

Next, the ensemble average of the predictability experiment is described; the results depicted in Figure 3. In this figure, the ensemble average total energy (barotropic kinetic energy plus the total baroclinic energy) is shown along with one half the total energy of the difference between the perturbed and control integrations, both decomposed with respect to frequency band. As can be seen, all frequencies double their energy at approximately identical rates (~ 2.5 $days^{-1}$) initially. However, by day 20, only the low frequency portion of the perturbation energy has failed to reach its control energy level.

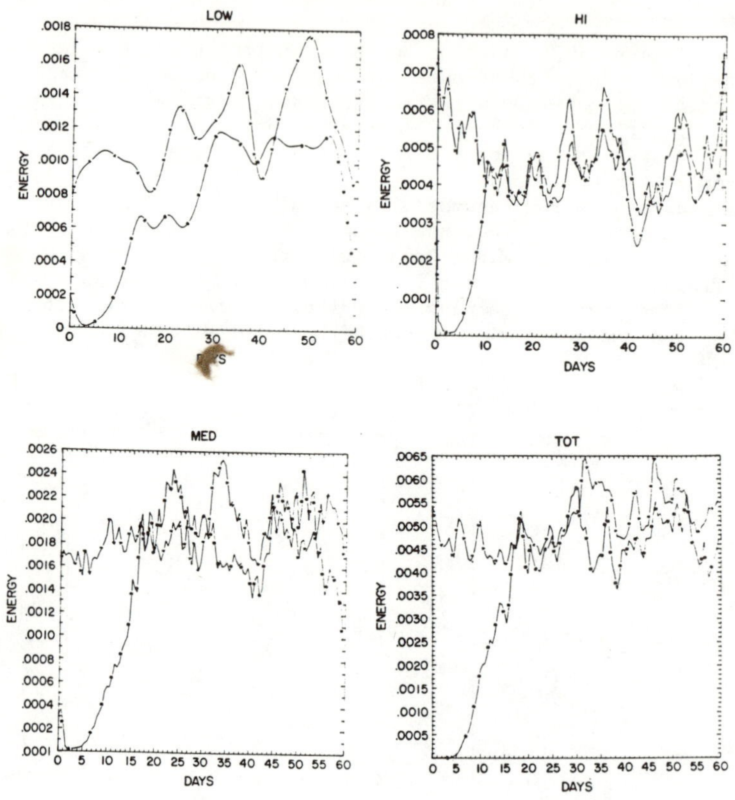

Figure 3: Ensemble average of control energy and perturbation energy as a function of time broken into frequency bands.

The approach to asymptotic climatological energy levels in each frequency band was further broken down by two-dimensional wavenumbers; shown in Figures 4a-4d. From these it is easily seen that approaches to climatological energy levels by the perturbation ensemble occur in essentially identical fashion for both the high and synoptic frequency bands which contain nearly all the variance at all but the gravest horizontal scale. The low frequency approach, however, occurs in a radically different manner and even after 33 days after perturbation excitation the asymptotic energy level in both the barotropic and baroclinic modes is not attained by wavenumbers 1 through 4. Unfortunately, in examining the approach to asymptotic error energy for the sum of all frequency bands it may be noted that the long-wave low-frequency "predictable" component represents less than 10% of the total variance.

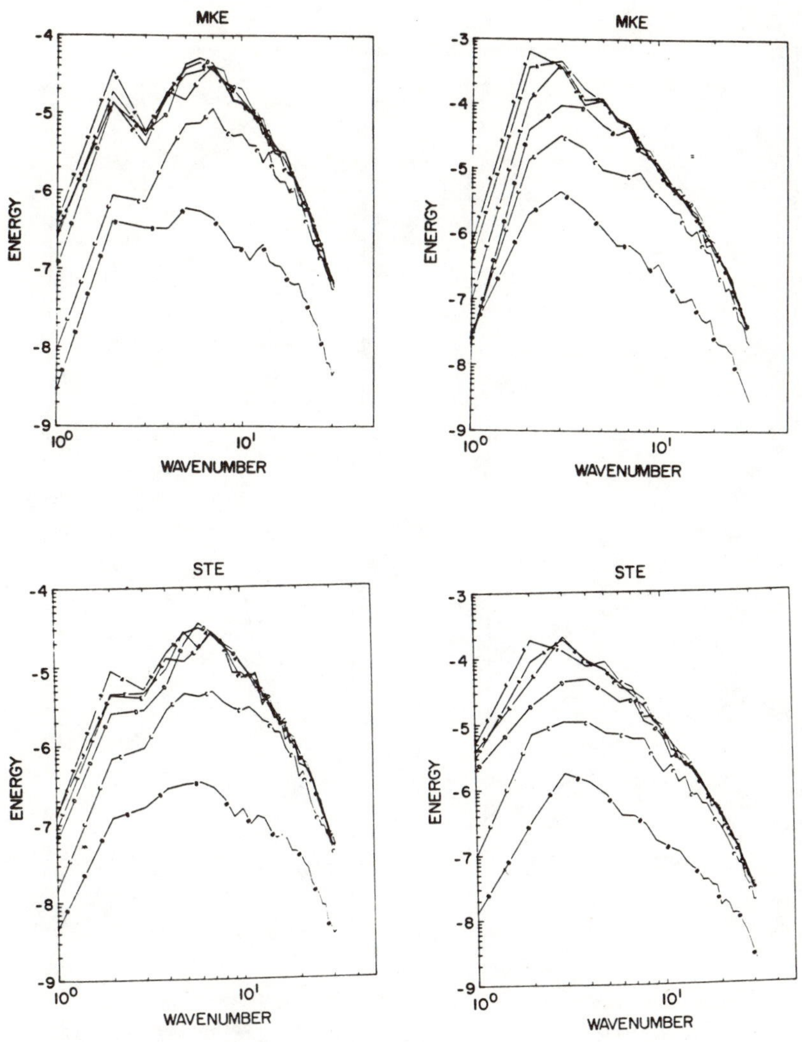

Figures 4a and 4b: Approach to asymptotic variance by perturbation as a function of wavenumber and frequency. Upper spectra are of barotropic kinetic energy and lower spectra are of baroclinic total energy. 4a shows high-pass filtered spectra and 4b shows medium-pass filtered spectra. Curves are at 4-day intervals beginning at day 3.

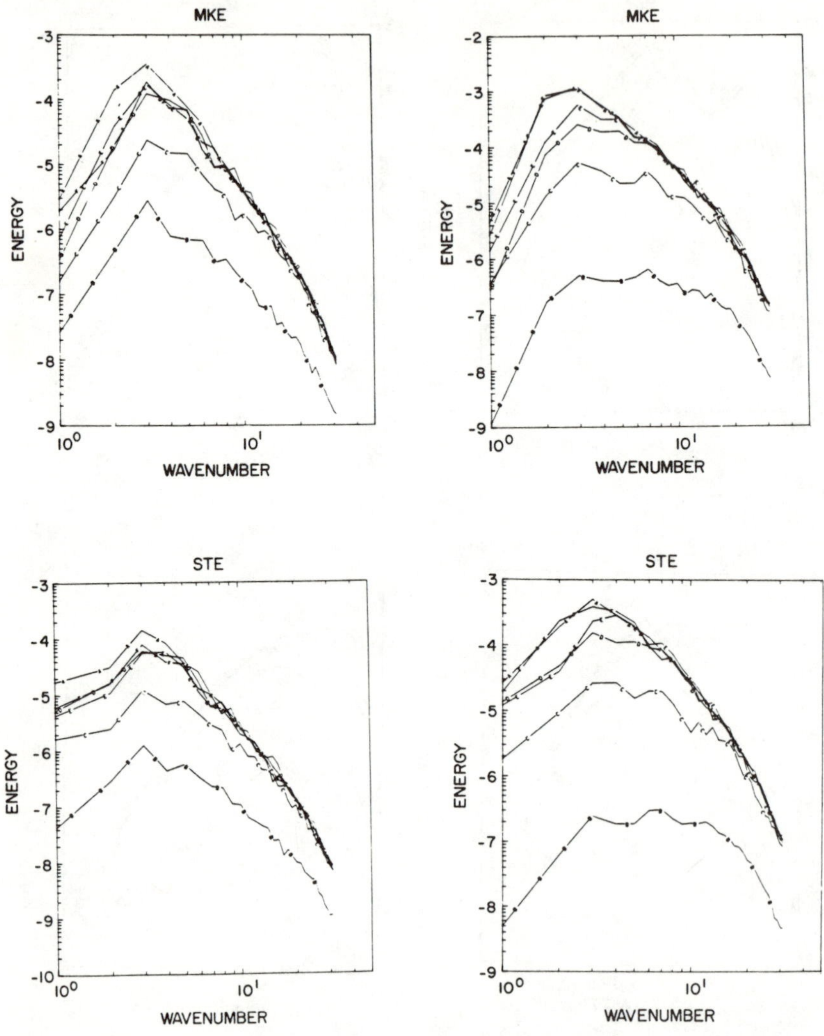

Figures 4c and 4d: As in Figures 4a and 4b, except left spectra are low-pass filtered and right spectra are unfiltered. In 4c spectra are shown at 10-day intervals, while in 4d spectra are depicted at 4-day intervals, both starting at day 3.

Nonetheless, it is of some interest to investigate more closely this small amplitude predictable component. The first question concerning this portion of the flow is whether or not the predictability is associated with phase-locked modes, e.g., is this phenomena associated with amplitude changes of the topographically forced large scale or is the predictable component of the flow dynamically active. To answer this question, the low-frequency long-wave ($n \leq 4$) average of the perturbed and control barotropic streamfunction is examined at days 90 and 120 of the experiment shown in Figure 5. It can be surmised from this figure that the predictable component of the flow is not phase-locked (even in a Lagrangian sense), but seems to be undergoing regime transitions between high and low zonal index flow configurations.

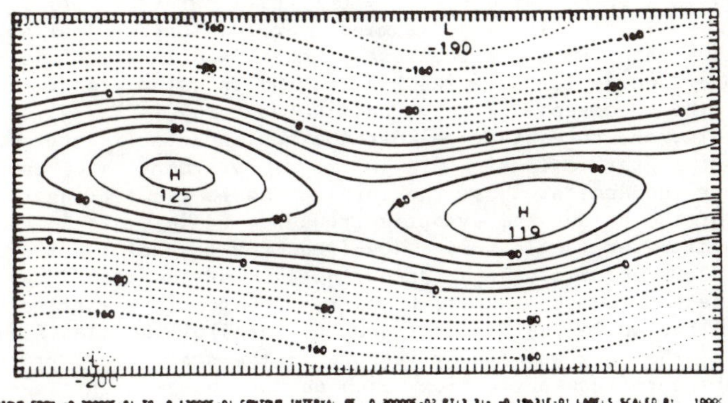

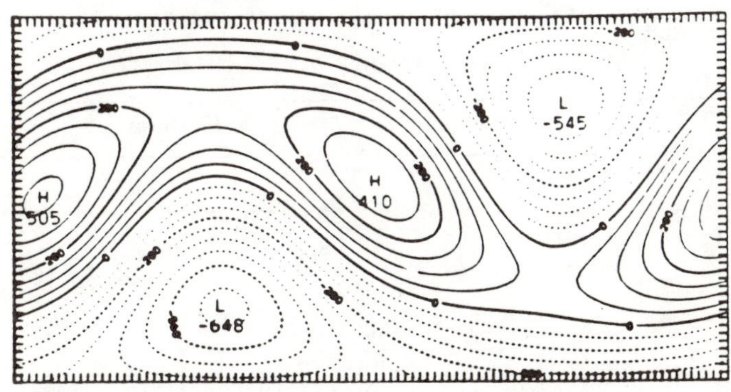

Figure 5: Low-pass filtered long-wave predictable component of the barotropic streamfunction at day 90 (upper) and day 120 (lower).

The dynamical balance associated with these transitions are also examined; two questions are addressed: (1) the importance of the topographic forcing in these transitions, and (2) the extent to which the predictable component is forced by the nonlinear interaction of the medium and high frequency parts of the flows.

These two questions are investigated in a similar manner. First, the time tendencies of each spectral component of the long-wave predictable portion of the barotropic vorticity and baroclinic potential vorticity are obtained from the twelve-hourly time sequence of the control and perturbed integrations, giving $T_o(k,t)$. Second, the temporally and spatially filtered vorticity components are substituted into the prognostic equation of the model to obtain the time tendency of each component due only to the self-interaction of the long-wave low-frequency part of the flow; this is denoted $T_n(k,t)$. Next, the ratio $R(k,t) \equiv (T_o(k,t) - T_n(k,t))/[T_o(k,t) + T_n(k,t)]$ is calculated and summed over the longitudinal wavenumber for each two-dimensional wavenumber, and lastly, this quantity is averaged in time, giving R(n). R(n) is calculated with and without the topographic forcing and is tabulated in Table 2. From this table, one notes that neither the nonlinear self-interaction nor the topographic forcing are important determinants of the low-frequency tendencies; thus, the forcing due to the nonlinear interaction of the height and synoptic frequency eddies must be the primary forcing for the long-wave low-frequency part of the flow. This alone is not surprising; the predictability of the forced components under these circumstances is, however. One possible explanation for this is that while the higher frequency parts of the flow are unpredictable, the long waves only respond to the statistical properties of the higher-frequency eddies which do not change significantly if the averaging time is long enough (i.e., if attention is focused on the low-frequency part of the flow). More diagnosis is needed, however, to say with certainty that this is the case.

Table 2

	Mean			Shear		
n	1	2	3	1	2	3
$R(n)_{TOP}$	.82	.78	.80	.81	.69	.85
$R(n)_{NOTOP}$	.84	.92	1.00	.82	.72	.87

CONCLUSIONS

In this simplest of climate models there exists a small predictable low-frequency component of the flow which is associated with a regime transition from low to high zonal index states. The medium and high frequencies which make up the majority of the power for wavenumbers greater than three are not predictable beyond 15 days, with initial perturbation amplitudes on the order of 1m/sec.

ADDITIONAL COMMENTS ON THE PREDICTABILITY OF FILTERED VARIABLES

During the course of discussion of this work and those of a similar nature, questions were raised as to the tautological nature of the predictability of low-pass filtered (or time-averaged) dynamical variables. The purpose of this section is to attempt to clarify this issue. First, it is important to note that low-pass filtering does not retard even rapid exponential amplification. Thus, if the difference between two forecasts excites an exponentially amplifying, slowly oscillating instability, the two predictions of the low-pass filtered variables will rapidly diverge. This is demonstrated in Figure Al, where the low, medium, and high-pass filters are applied to two time series of the form $f(t) = (\varepsilon + \tanh \beta t) \cos w_{1,2} t$ where $\beta = (2.5 \text{ days})$, $\varepsilon = .01$, $w_1 = 2\pi/(20 \text{ days})$, and $w_2 = 2\pi/(5 \text{ days})$.

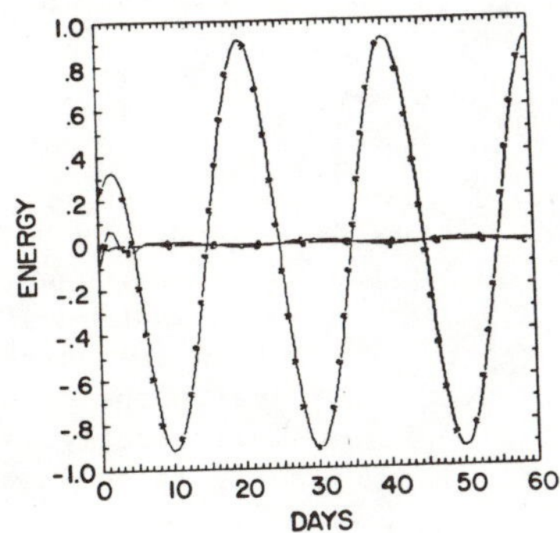

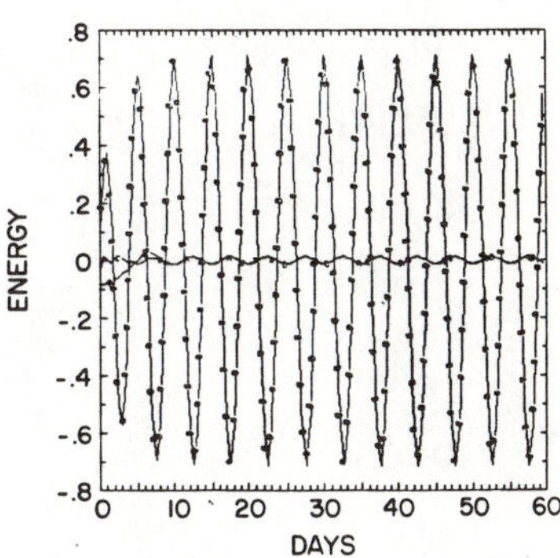

Figure Al: Test of filter response using f(t) as specified in the text. Left $w_1 = 2\pi/(20 \text{ days})$ and right $w_2 = 2\pi/(5 \text{ days})$.

Secondly, it must be noted that even unpredictable time series will show temporal coherence after application of a low-pass filter. As an example of this, consider a random sequence of unbiased coin tosses in which heads is given the value +1 and tails the value -1. The mean of this process is 0 and the standard deviation 1. Filtering this sequence utilizes N consecutive values of the sequence and sums using weights for each element in the sum

i.e., $X_F(K) = \sum_{i=1}^{N} w_i X_{i+(K-N/2)}$. Clearly $\overline{X_F}$ and $\overline{X_F^2} = \sum_{i=1} w_i^2$. Now $V(K) \equiv [X(K+i) - X(i)]^2$ is the variance of the filtered sequence with a K-step lag; its value is given by

$$V(K) = \sum_{i=1}^{N} w_i^2 + \sum_{i=K+1}^{N} (w_{i-K} - w_i)^2 + \sum_{i=N+1-K}^{N} w_i^2$$

which grows to $2\overline{X_F^2}$ at K = N+1. This decorrelation of the unpredictable sequence can be used than as a null hypothesis against which the predictability of filtered sequences can be tested. This is shown in Figure A2 for the low-pass filtered predictability experiments in which the experimental growth of error is compared to that of the null hypothesis. From this one sees that there is predictive skill in the low-pass filtered experiments since the error growth curve is everywhere less than the growth decorrelation growth of the unpredictable sequence.

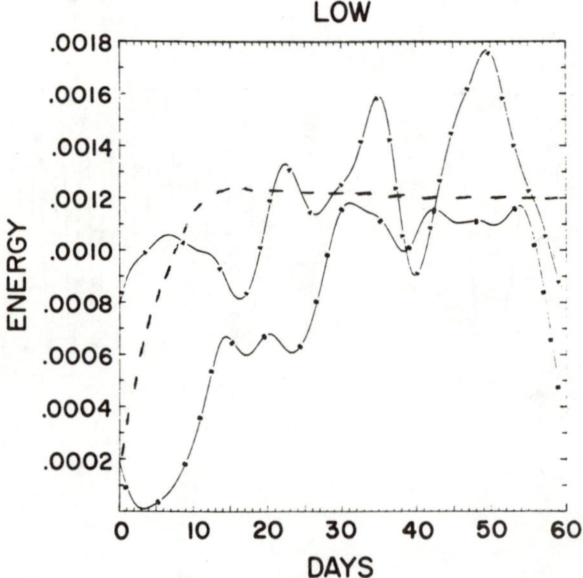

Figure A2: Ensemble average of low frequency perturbation energy growth (as in Figure 3) with null hypothesis of absolutely unpredictable variance growth (---).

REFERENCES

1. C. E. Leith and R. H. Kraichnan, J. Atmos. Sci., 29, 1041-1058 (1972).
2. M. L. Blackmon, J. Atmos. Sci., 36, 2450-2466 (1976).

No.26	High-Energy Physics and Nuclear Structure - 1975 (Santa Fe and Los Alamos)	75-26411	0-88318-125-8
No.27	Topics in Statistical Mechanics and Biophysics: A Memorial to Julius L. Jackson (Wayne State University, 1975)	75-36309	0-88318-126-6
No.28	Physics and Our World: A Symposium in Honor of Victor F. Weisskopf (M.I.T., 1974)	76-7207	0-88318-127-4
No.29	Magnetism and Magnetic Materials - 1975 (21st Annual Conference, Philadelphia)	76-10931	0-88318-128-2
No.30	Particle Searches and Discoveries - 1976 (Vanderbilt Conference)	76-19949	0-88318-129-0
No.31	Structure and Excitations of Amorphous Solids (Williamsburg, VA., 1976)	76-22279	0-88318-130-4
No.32	Materials Technology - 1976 (APS New York Meeting)	76-27967	0-88318-131-2
No.33	Meson-Nuclear Physics - 1976 (Carnegie-Mellon Conference)	76-26811	0-88318-132-0
No.34	Magnetism and Magnetic Materials - 1976 (Joint MMM-Intermag Conference, Pittsburgh)	76-47106	0-88318-133-9
No.35	High Energy Physics with Polarized Beams and Targets (Argonne, 1976)	76-50181	0-88318-134-7
No.36	Momentum Wave Functions - 1976 (Indiana University)	77-82145	0-88318-135-5
No.37	Weak Interaction Physics - 1977 (Indiana University)	77-83344	0-88318-136-3
No.38	Workshop on New Directions in Mossbauer Spectroscopy (Argonne, 1977)	77-90635	0-88318-137-1
No.39	Physics Careers, Employment and Education (Penn State, 1977)	77-94053	0-88318-138-X
No.40	Electrical Transport and Optical Properties of Inhomogeneous Media (Ohio State University, 1977)	78-54319	0-88318-139-8
No.41	Nucleon-Nucleon Interactions - 1977 (Vancouver)	78-54249	0-88318-140-1
No.42	Higher Energy Polarized Proton Beams (Ann Arbor, 1977)	78-55682	0-88318-141-X
No.43	Particles and Fields - 1977 (APS/DPF, Argonne)	78-55683	0-88318-142-8
No.44	Future Trends in Superconductive Electronics (Charlottesville, 1978)	77-9240	0-88318-143-6
No.45	New Results in High Energy Physics - 1978 (Vanderbilt Conference)	78-67196	0-88318-144-4
No.46	Topics in Nonlinear Dynamics (La Jolla Institute)	78-057870	0-88318-145-2
No.47	Clustering Aspects of Nuclear Structure and Nuclear Reactions (Winnepeg, 1978)	78-64942	0-88318-146-0
No.48	Current Trends in the Theory of Fields (Tallahassee, 1978)	78-72948	0-88318-147-9
No.49	Cosmic Rays and Particle Physics - 1978 (Bartol Conference)	79-50489	0-88318-148-7

No.50	Laser-Solid Interactions and Laser Processing - 1978 (Boston)	79-51564	0-88318-149-5
No.51	High Energy Physics with Polarized Beams and Polarized Targets (Argonne, 1978)	79-64565	0-88318-150-9
No.52	Long-Distance Neutrino Detection - 1978 (C.L. Cowan Memorial Symposium)	79-52078	0-88318-151-7
No.53	Modulated Structures - 1979 (Kailua Kona, Hawaii)	79-53846	0-88318-152-5
No.54	Meson-Nuclear Physics - 1979 (Houston)	79-53978	0-88318-153-3
No.55	Quantum Chromodynamics (La Jolla, 1978)	79-54969	0-88318-154-1
No.56	Particle Acceleration Mechanisms in Astrophysics (La Jolla, 1979)	79-55844	0-88318-155-X
No. 57	Nonlinear Dynamics and the Beam-Beam Interaction (Brookhaven, 1979)	79-57341	0-88318-156-8
No. 58	Inhomogeneous Superconductors - 1979 (Berkeley Springs, W.V.)	79-57620	0-88318-157-6
No. 59	Particles and Fields - 1979 (APS/DPF Montreal)	80-66631	0-88318-158-4
No. 60	History of the ZGS (Argonne, 1979)	80-67694	0-88318-159-2
No. 61	Aspects of the Kinetics and Dynamics of Surface Reactions (La Jolla Institute, 1979)	80-68004	0-88318-160-6
No. 62	High Energy e^+e^- Interactions (Vanderbilt , 1980)	80-53377	0-88318-161-4
No. 63	Supernovae Spectra (La Jolla, 1980)	80-70019	0-88318-162-2
No. 64	Laboratory EXAFS Facilities - 1980 (Univ. of Washington)	80-70579	0-88318-163-0
No. 65	Optics in Four Dimensions - 1980 (ICO, Ensenada)	80-70771	0-88318-164-9
No. 66	Physics in the Automotive Industry - 1980 (APS/AAPT Topical Conference)	80-70987	0-88318-165-7
No. 67	Experimental Meson Spectroscopy - 1980 (Sixth International Conference , Brookhaven)	80-71123	0-88318-166-5
No. 68	High Energy Physics - 1980 (XX International Conference, Madison)	81-65032	0-88318-167-3
No. 69	Polarization Phenomena in Nuclear Physics - 1980 (Fifth International Symposium, Santa Fe)	81-65107	0-88318-168-1
No. 70	Chemistry and Physics of Coal Utilization - 1980 (APS, Morgantown)	81-65106	0-88318-169-X
No. 71	Group Theory and its Applications in Physics - 1980 (Latin American School of Physics, Mexico City)	81-66132	0-88318-170-3
No. 72	Weak Interactions as a Probe of Unification (Virginia Polytechnic Institute - 1980)	81-67184	0-88318-171-1
No. 73	Tetrahedrally Bonded Amorphous Semiconductors (Carefree, Arizona, 1981)	81-67419	0-88318-172-X
No.74	Perturbative Quantum Chromodynamics (Tallahassee, 1981)	81-70372	0-88318-173-8

No. 75	Low Energy X-ray Diagnostics-1981 (Monterey)	81-69841	0-88318-174-6
No. 76	Nonlinear Properties of Internal Waves (La Jolla Institute, 1981)	81-71062	0-88318-175-4
No. 77	Gamma Ray Transients and Related Astrophysical Phenomena (La Jolla Institute, 1981)	81-71543	0-88318-176-2
No. 78	Shock Waves in Condensed Matter - 1981 (Menlo Park)	82-70014	0-88318-177-0
No. 79	Pion Production and Absorption in Nuclei - 1981 (Indiana University Cyclotron Facility)	82-70678	0-88318-178-9
No. 80	Polarized Proton Ion Sources (Ann Arbor, 1981)	82-71025	0-88318-179-7
No. 81	Particles and Fields - 1981: Testing the Standard Model (APS/DPF, Santa Cruz)	82-71156	0-88318-180-0
No. 82	Interpretation of Climate and Photochemical Models, Ozone and Temperature Measurements (La Jolla Institute, 1981)	82-071345	0-88318-181-9
No. 83	The Galactic Center (Cal. Inst. of Tech., 1982)	82-071635	0-88318-182-7
No. 84	Physics in the Steel Industry (APS.AISI, Lehigh University, 1981)	82-072033	0-88318-183-5
No. 85	Proton-Antiproton Collider Physics - 1981 (Madison, Wisconsin)	82-072141	0-88318-184-3
No. 86	Momentum Wave Functions - 1982 (Adelaide, Australia)	82-072375	0-88318-185-1
No. 87	Physics of High Energy Particle Accelerators (Fermilab Summer School, 1981)	82-072421	0-88318-186-X
No. 88	Mathematical Methods in Hydrodynamics and Integrability in Dynamical Systems (La Jolla Institute, 1981)	82-072462	0-88318-187-8
No. 89	Neutron Scattering - 1981 (Argonne National Laboratory)	82-073094	0-88318-188-6
No. 90	Laser Techniques for Extreme Ultraviolt Spectroscopy (Boulder, 1982)	82-073205	0-88318-189-4
No. 91	Laser Acceleration of Particles (Los Alamos, 1982)	82-073361	0-88318-190-8
No. 92	The State of Particle Accelerators and High Energy Physics(Fermilab, 1981)	82-073861	0-88318-191-6
No. 93	Novel Results in Particle Physics (Vanderbilt, 1982)	82-73954	0-88318-192-4
No. 94	X-Ray and Atomic Inner-Shell Physics-1982 (International Conference, U. of Oregon)	82-74075	0-88318-193-2
No. 95	High Energy Spin Physics - 1982 (Brookhaven National Laboratory)	83-70154	0-88318-194-0
No. 96	Science Underground (Los Alamos, 1982)	83-70377	0-88318-195-9

No. 97	The Interaction Between Medium Energy Nucleons in Nuclei-1982 (Indiana University)	83-70649	0-88318-196-7
No. 98	Particles and Fields - 1982 (APS/DPF University of Maryland)	83-70807	0-88318-197-5
No. 99	Neutrino Mass and Gauge Structure of Weak Interactions (Telemark, 1982)	83-71072	0-88318-198-3
No. 100	Excimer Lasers - 1983 (OSA, Lake Tahoe, Nevada)	83-71437	0-88318-199-1
No. 101	Positron-Electron Pairs in Astrophysics (Goddard Space Flight Center, 1983)	83-71926	0-88318-200-9
No. 102	Intense Medium Energy Sources of Strangeness (UC-Santa Cruz, 1983)	83-72261	0-88318-201-7
No. 103	Quantum Fluids and Solids - 1983 (Sanibel Island, Florida)	83-72440	0-88318-202-5
No. 104	Physics,Technology and the Nuclear Arms Race (APS Baltimore-1983)	83-72533	0-88318-203-3
No. 105	Physics of High Energy Particle Accelerators (SLAC Summer School, 1982)	83-72986	0-88318-304-8
No. 106	Predictability of Fluid Motions (La Jolla Institute, 1983)	83-73641	0-88318-305-6
No. 107	Physics and Chemistry of Porous Media (Schlumberger-Doll Research, 1983)	83-73640	0-88318-306-4
No. 108	The Time Projection Chamber (TRIUMF, Vancouver, 1983)	83-83445	0-88318-307-2